Modelling, Monitoring, Control and Optimization for Complex Industrial Processes

Modelling, Monitoring, Control and Optimization for Complex Industrial Processes

Editor

Zhiwei Gao

MDPI • Basel • Beijing • Wuhan • Barcelona • Belgrade • Manchester • Tokyo • Cluj • Tianjin

Editor
Zhiwei Gao
Mathematics, Physics and
Electrical Engineering
Northumbria Univresity
Newcastle upon Tyne
United Kingdom

Editorial Office
MDPI
St. Alban-Anlage 66
4052 Basel, Switzerland

This is a reprint of articles from the Special Issue published online in the open access journal *Processes* (ISSN 2227-9717) (available at: www.mdpi.com/journal/processes/special_issues/Complex_Industrial_Processes).

For citation purposes, cite each article independently as indicated on the article page online and as indicated below:

LastName, A.A.; LastName, B.B.; LastName, C.C. Article Title. *Journal Name* **Year**, *Volume Number*, Page Range.

ISBN 978-3-0365-6395-4 (Hbk)
ISBN 978-3-0365-6394-7 (PDF)

Contents

About the Editor . . . . **vii**

Preface to "Modelling, Monitoring, Control and Optimization for Complex Industrial Processes" . . . . **ix**

Zhiwei Gao
Special Issue on "Modelling, Monitoring, Control and Optimization for Complex Industrial Processes"
Reprinted from: *Processes* **2023**, *11*, 207, doi:10.3390/pr11010207 . . . . **1**

Dapeng Zhang, Lifeng Du and Zhiwei Gao
Real-Time Parameter Identification for Forging Machine Using Reinforcement Learning
Reprinted from: *Processes* **2021**, *9*, 1848, doi:10.3390/pr9101848 . . . . **7**

Lijing Ren and Denghui Zhang
A Simple and Effective Modeling Method for 3D Porous Irregular Structures
Reprinted from: *Processes* **2022**, *10*, 464, doi:10.3390/pr10030464 . . . . **27**

Dongtao Liu, Chunshang Qiao, Jun Wan, Yuliang Lu, Jiming Song and Zhenhe Yao et al.
Modelling Method and Application of Anti-Corrosion Pill Particles in Oil and Gas Field Wellbore Casing Annulus Based on the Discrete Element Method
Reprinted from: *Processes* **2022**, *10*, 1164, doi:10.3390/pr10061164 . . . . **39**

Wei Li and Xia Wu
Identification and Analysis of Factors Influencing Green Growth of Manufacturing Enterprises Based on DEMATEL Method—Wooden Flooring Manufacturing Companies as a Case
Reprinted from: *Processes* **2022**, *10*, 2594, doi:10.3390/pr10122594 . . . . **53**

Yanqin Wang, Weijian Ren, Zhuoqun Liu, Jing Li and Duo Zhang
T-S Fuzzy Model-Based Fault Detection for Continuous Stirring Tank Reactor
Reprinted from: *Processes* **2021**, *9*, 2127, doi:10.3390/pr9122127 . . . . **65**

Aihua Zhang, Danlu Yu and Zhiqiang Zhang
TLSCA-SVM Fault Diagnosis Optimization Method Based on Transfer Learning
Reprinted from: *Processes* **2022**, *10*, 362, doi:10.3390/pr10020362 . . . . **81**

Xin Lu, Xiaoxu Liu, Bowen Li and Jie Zhong
Data-Driven State Prediction and Sensor Fault Diagnosis for Multi-Agent Systems with Application to a Twin Rotational Inverted Pendulum
Reprinted from: *Processes* **2021**, *9*, 1505, doi:10.3390/pr9091505 . . . . **113**

Wei Tan, He Wang, Huazhou Hou, Xiaoxu Liu and Meng Zheng
Dynamically Triggering Resilient Control for Networked Nonlinear Systems under Malicious Aperiodic DoS Attacks
Reprinted from: *Processes* **2022**, *10*, 2627, doi:10.3390/pr10122627 . . . . **129**

Feng Liang, Lu Lu, Zhengfeng Li, Fangfang Zhang and Shuaihu Zhang
Tracking Control of a Hyperchaotic Complex System and Its Fractional-Order Generalization
Reprinted from: *Processes* **2022**, *10*, 1244, doi:10.3390/pr10071244 . . . . **145**

Sherif A. Zaid, Hani Albalawi, Hossam AbdelMeguid, Tareq A. Alhmiedat and Abualkasim Bakeer
Performance Improvement of H8 Transformerless Grid-Tied Inverter Using Model Predictive Control Considering a Weak Grid
Reprinted from: *Processes* **2022**, *10*, 1243, doi:10.3390/pr10071243 . . . . **159**

Yang Yuan, Xinyi Tao, Kejin Huang, Haisheng Chen and Xing Qian
An Effective Temperature Control Method for Dividing-Wall Distillation Columns
Reprinted from: *Processes* **2022**, *10*, 1018, doi:10.3390/pr10051018 . . . . **175**

Wenping Jiang, Wenchao Han, Lingyang Wang, Zhouyang Liu and Weidong Du
Linear Golden Section Speed Adaptive Control of Permanent Magnet Synchronous Motor Based on Model Design
Reprinted from: *Processes* **2022**, *10*, 1010, doi:10.3390/pr10051010 . . . . **193**

Ibrahem E. Atawi, Essam Hendawi and Sherif A. Zaid
Analysis and Design of a Standalone Electric Vehicle Charging Station Supplied by Photovoltaic Energy
Reprinted from: *Processes* **2021**, *9*, 1246, doi:10.3390/pr9071246 . . . . **209**

Xiangsong Kong, Changqing Shi, Hang Liu, Pengcheng Geng, Jiabin Liu and Yasen Fan
Performance Optimization of a Steam Generator Level Control System via a Revised Simplex Search-Based Data-Driven Optimization Methodology
Reprinted from: *Processes* **2022**, *10*, 264, doi:10.3390/pr10020264 . . . . **225**

Luobing Chen, Zhiqiang Hu, Fangfang Zhang, Zhongjin Guo, Kun Jiang and Changchun Pan et al.
Remote Wind Farm Path Planning for Patrol Robot Based on the Hybrid Optimization Algorithm
Reprinted from: *Processes* **2022**, *10*, 2101, doi:10.3390/pr10102101 . . . . **247**

Hui Shao, Qimeng Liu and Zhiwei Gao
Material Removal Optimization Strategy of 3D Block Cutting Based on Geometric Computation Method
Reprinted from: *Processes* **2022**, *10*, 695, doi:10.3390/pr10040695 . . . . **269**

Niklas Reinisch, Fridtjof Rudolph, Stefan Günther, David Bailly and Gerhard Hirt
Successful Pass Schedule Design in Open-Die Forging Using Double Deep Q-Learning
Reprinted from: *Processes* **2021**, *9*, 1084, doi:10.3390/pr9071084 . . . . **287**

Yousra El Kihel, Ali El Kihel and Soufiane Embarki
Optimization of the Sustainable Distribution Supply Chain Using the Lean Value Stream Mapping 4.0 Tool: A Case Study of the Automotive Wiring Industry
Reprinted from: *Processes* **2022**, *10*, 1671, doi:10.3390/pr10091671 . . . . **319**

Pedro A. Delou, Leonardo D. Ribeiro, Carlos R. Paiva, Jacques Niederberger, Marcos Vinícius C. Gomes and Argimiro R. Secchi
A Real-Time Optimization Strategy for Small-Scale Facilities and Implementation in a Gas Processing Unit
Reprinted from: *Processes* **2021**, *9*, 1179, doi:10.3390/pr9071179 . . . . **339**

Jinping Chen, Razaullah Khan, Yanmei Cui, Bashir Salah, Yuanpeng Liu and Waqas Saleem
The Effect of Changes in Settings from Multiple Filling Points to a Single Filling Point of an Industry 4.0-Based Yogurt Filling Machine
Reprinted from: *Processes* **2022**, *10*, 1642, doi:10.3390/pr10081642 . . . . **369**

Jing Zheng and Qi Xu
Dynamic Cooperation of the O2O Supply Chain Based on Time Delays and Bidirectional Free-Riding
Reprinted from: *Processes* **2022**, *10*, 2424, doi:10.3390/pr10112424 . . . . **391**

Jun Min, Zhaoqi Liu, Lei Wang, Dongyang Li, Maoqing Zhang and Yantai Huang
Music Generation System for Adversarial Training Based on Deep Learning
Reprinted from: *Processes* **2022**, *10*, 2515, doi:10.3390/pr10122515 . . . . **413**

About the Editor

Zhiwei Gao

Dr. Zhiwei Gao is a recipient of the Alexander von Humboldt Research Fellowship (2004) and Fellow of IEEE (2023). His research interests include augmented observer techniques, fault diagnosis, fault-tolerant control, resilient control, data-driven approaches, intelligent optimization, wind turbine systems, electrical vehicles, power electronics, and bioinformatics.

Preface to "Modelling, Monitoring, Control and Optimization for Complex Industrial Processes"

Industrial automation systems are becoming more complex as well as expensive, leading to higher requirements for operation performance, the quality of products, productiveness, and reliability. Recent developments in modelling, monitoring, control, and optimisation techniques have enhanced the understanding of systems dynamics and boosted the applications of monitoring, control, and optimisation in complex industrial processes. This Special Issue, with papers ranging from theoretical algorithms to the experimental implementation of complex industrial processes, has highlighted advances in this field.

Zhiwei Gao
Editor

processes

Editorial

Special Issue on "Modelling, Monitoring, Control and Optimization for Complex Industrial Processes"

Zhiwei Gao

Faculty of Engineering and Environment, University of Northumbria, Newcastle NE1 8ST, UK; zhiwei.gao@northumbria.ac.uk

Industrial automation systems, such as chemical processes, manufacturing processes, power networks, transportation systems, sustainable energy systems, wireless sensor networks, robotic systems, and biomedical systems, are becoming more complex [1–3], but more expensive, and have higher requirements for operation performance, quality of products, productiveness, and reliability. Stimulated by Industry 4.0, automation industries are keen to improve the reliability and operational performance of complex industrial processes using advanced modelling, monitoring, optimization, and control techniques. Recently, artificial intelligence, data-driven techniques, cyber–physical systems, digital-twin, and cloud computation have further stimulated research and applications of modelling, monitoring, optimization, and control techniques [4–6].

This Special Issue on "Modelling, Monitoring, Control and Optimization for Complex Industrial Processes" (https://www.mdpi.com/journal/processes/special_issues/Complex_Industrial_Processes) aims to provide a forum for researchers and engineers to report their recent results, exchange research ideas, and look over emerging research and application directions in modelling, monitoring, optimization, and advanced control for complex industrial processes. There are 22 papers included in this Special Issue, after a rigorous review process, which are categorised and presented in Table 1.

Table 1. Categories of the paper included in the Special Issue.

Categories	Modelling and Parameter Identification	Monitoring, Diagnosis, and Resilience	Control Applications	Optimisation Applications
Papers	[7–10]	[11–14]	[15–20]	[21–28]

1. Modelling and Parameter Identification for Complex Industrial Processes

It is significant but challenging to identify system parameters of a mechanism system under on-line working conditions as uncertainties exist due to the differences between design requirements and real-time environment. In the paper co-authored by Zhang et al. [7], a reinforcement learning approach was applied to forging machines to attain real-time model parameters, where raw data were used directly, and an online parameter identification algorithm was implemented in a period without the aid of labelled samples as a training database. The addressed parameter identification technique proved to have a powerful capability to adapt a new process without historical data. The effectiveness was validated via a forging machine process.

It is difficult to model porous structures due to their irregular internal morphologies. Conventional CAD modelling approaches fail to represent internal structures and conformations in models, although they can effectively describe the external geometric and topological information of the models. In the work completed by Ren [8], an effective modelling method for 3D irregular porous structures was presented based on a finite element method and thermodynamic analysis, and the key idea was to solve isothermal issues in the modelling of the porosity of porous units. It was shown from experiments

Citation: Gao, Z. Special Issue on "Modelling, Monitoring, Control and Optimization for Complex Industrial Processes". *Processes* **2023**, *11*, 207. https://doi.org/10.3390/pr11010207

Received: 3 January 2023
Accepted: 4 January 2023
Published: 9 January 2023

that the proposed technique can achieve smooth and approximate porous structures from arbitrary irregular 3D surfaces.

The discrete element method has a capability to analyse interactions among particles themselves, and interconnections between particles and mechanical components, to reveal the influencing factors and operating mechanism of the mechanical components. In the work by Liu et al. [9], a discrete element method based modelling technique was used to represent pill particle population which was employed to optimise the anti-corrosion process of oil and gas wellbore casing annuli. A simulation model was built, and the theoretical foundation was established for the further investigation of the pill discharging process and the parameter optimisation of the pill discharging device.

Green growth is defined as a process for a manufacturing enterprise to grow stronger with green strategies and green behaviours to achieve less consumption of resources and energy, less pollution, and more environmentally friendly and healthy products. In the article by Li et al. [10], a conceptual model of the factors influencing the green growth of manufacturing enterprises was established and a method was addressed to further reveal the relevant dynamic mechanisms and essential influencing factors, determined using a decision-making trial and evaluation laboratory strategy. Six key influencing factors were finally verified using a wooden flooring manufacturing company as a case study.

2. Monitoring, Diagnosis, and Resilience for Complex Industrial Processes

Fault diagnosis approaches are categorized into a model-based diagnosis approach, signal-based diagnosis method, and knowledge-based diagnosis approach. A model-based approach is widely used if a model is available to the designer. Continuous stirring tank reactors are widely used in chemical production processes, where there is a nonlinear dynamic process disrupted by time delays and uncertainties. In the article contributed by Wang et al. [11], continuous stirring tank reactors were represented by a T-S fuzzy model with state delays and disturbances. A fuzzy fault detection filter was addressed to detect faults and the design gains were obtained to solve the convex optimization of linear matrix inequalities. The effectiveness of the proposed diagnosis method was demonstrated by simulation studies.

Knowledge-based approaches are based on data driven and machine-learning techniques. Therefore, quantitative knowledge-based approaches are also called data-driven approaches. In the paper co-authored by Zhang et al. [12], a novel fault–diagnosis–classification optimization method was proposed by fusing a sine cosine algorithm, support vector machine, and transfer learning. Intensive simulation studies were carried out, and the proposed algorithm outperformed five existing diagnosis algorithms with higher precision and a faster response time. In addition, the proposed algorithm can run effectively using transfer learning with less failure data.

Multi-agent systems have received attention where multiple agents communicate through a premeditated protocol to operate collectively. A fault in an agent may deteriorate the performance of its neighbouring agents and even the entire network. As a result, it is important to detect an agent fault as early as possible. In the paper co-authored by Lu et al. [13], a fault diagnosis problem was investigated for multi-agent systems, where a neural-network-based state prediction model was built via offline historical data training, and the residuals between the actual outputs and predicted outputs were checked to detect potential faults. The effectiveness of the presented diagnosis algorithm was finally demonstrated via a real experiment on a leader–follower inverted pendulum.

Networked dynamic systems would suffer potential security threats caused by malicious attacks, which would destabilize networked dynamic systems and disturb communications between networked systems. As a result, there is a motivation to discuss the resilience issue in networked systems subjected to cyber-attacks. In the article contributed by Tan et al. [14], a resilient control issue for networked nonlinear dynamic systems with dynamic trigger mechanisms and malicious aperiodic denial-of-service attacks was examined. A resilient dynamically triggering controller was designed to alleviate the effects of cyber-attacks and reduce the usage of communication resources. The proposed approaches were validated by using the well-known nonlinear Chua circuit.

3. Control Applications for Complex Industrial Systems

Chaos is a complex nonlinear phenomenon in nature and chaotic systems have been widely applied to a variety of practical systems, such as secure communications, industrial processes, and ecosystems, etc. Some chaotic behaviours are harmful, which should be supressed. In the article contributed by Liang et al. [15], a tracking controller was designed for hyperchaotic complex systems, and the feasibility of the proposed design was verified from two perspectives, via both mathematical proofs and simulation experiments.

H8 transformer-less inverter can be used to eliminate an earth leakage current, and model predictive control has been a popular control technique in industrial applications. In the paper contributed by Zaid et al. [16], a model predictive control method was used to improve the performance of H8 transformer-less inverters supplied by a photovoltaic energy source. The Hardware-in-the-Loop was implemented using a DSP target LaunchPadXLTMS320F28379D kit to demonstrate the effectiveness of the proposed approach.

Temperature control has been widely used in the control of dividing-wall distillation columns, which has an advantage in dynamic characteristics, but cannot track the steady values well due to its limited accuracy in estimating controlled product purities. Motived by the above, in the paper contributed by Yuan et al. [17], an improved temperature control approach was addressed with the aid of product quality estimation and a genetic algorithm. It was demonstrated by the simulated studies that the proposed control scheme can reduce steady-state deviations in the maintained product purities as well as have better dynamic characteristics, which proved to be a useful tool for the temperature inferential control for dividing-wall distillation columns.

A permanent magnet synchronous motor has a wide industrial application. It is noticed that it is usually challenging to establish an accurate mathematical model for a permanent magnet synchronous motor, and an application of a complex algorithm may pose a challenge for embedded code development. Motived by the above, in the paper co-authored by Jiang et al. [18], a characteristic model for a permanent magnet synchronous motor was built, and a speed control scheme was proposed by integrating a linear golden-section adaptive control and integral compensation. It was shown by the simulation and experimental results that the speed control accuracy using the proposed control algorithm for a permanent magnet synchronous motor was improved by 3.8 times compared with traditional proportional-integral-derivative control algorithms.

Electric vehicles are green modes of transportation, which will replace fossil-fuelled vehicles soon. However, charging stations for electric vehicle batteries may impose a high energy demand on the utility grid. As a result, it is of interest to investigate standalone charging stations for electric vehicles using photovoltaic power sources to support the utility grid. In the paper contributed by Atawi. [19], an isolated electric vehicle charging station model based on a photovoltaic energy source was built, which was composed of a photovoltaic panel, boost converter, energy storage system batteries, DC/DC charging converters, and an electric vehicle battery. The control system was composed of a maximum power tracking controller, electric vehicle charging controller, and storage converter controller, which were, in essence, PI controllers, as well as a single-chip PIC18F4550 microcontroller utilized for control implementation. It was demonstrated by the simulations and experiments that the used controllers can provide good response speeds and satisfactory tracking abilities to their references.

Steam generators are critical devices in nuclear power plants, and their control performances are paramount to maintain normal operations. It is of interest to develop optimal control in a steam generator level process. In the article by Kong et al. [20], a systemic data-driven optimization methodology was proposed, which was used to optimise control system parameters by using control performance measurements directly. The effectiveness of the addressed method was demonstrated via simulations, concluding that the addressed simplex search method was effective in controller parameter optimization to improve control system performance in steam generator level processes.

4. Optimization Applications for Complex Systems

Wind energy plays a leading role in renewable energy industries. To reduce workloads, improve efficiency, and provide better evaluation and judgment, inspection robots have been introduced into wind farms for inspection. It is a prerequisite to produce a path planning for intelligent inspection using robots. In the article by Chen et al. [21], a new path-planning algorithm was proposed based on a chaotic neural network and genetic algorithm. The proposed algorithm was verified via a path planned for patrol robot using the actual locations of 30 wind farms, showing the addressed algorithm can generate a shorter inspection path compared with some existing algorithms.

It is of significance to boost material removal rate and waste reuse rate in a rough processing stage of a three-dimensional stone product with an unusual shape. In the paper contributed by Shao et al. [22], circular saw disc cutting was inspected to cut a convex polyhedron out of a blank box, with reference to a targeted product. It is evident that this problem can be better solved by geometrical methods rather than mathematical methods. An automatic block cutting strategy was proposed by using a series of geometrical optimization approaches. The effectiveness of the proposed method was demonstrated via simulated studies using both MATLAB and the Vericut platform.

It is of importance to have precise process planning to produce an open-die-forged part with a desired final geometry as well as economic production. In the paper contributed by Reinisch et al. [23], a multi-objective optimization-based schedule design was addressed by combining fast process models with a double deep Q-learning algorithm. The produced pass schedules lead to a desired ingot geometry with a minimal number of passes. The addressed methods were validated via a forging experiment, showing the ability of the addressed double deep Q-learning algorithm to achieve an optimal pass schedule in real open-die forging processes.

New opportunities are provided to companies to gain competitiveness with a transformation to Supply Chain 4.0 with the aid of the lean value stream mapping tool. In the work by Kihel et al. [24], a new process design was presented by integrating 4.0 technologies, taking multinational supply chains in Automotive Wiring Equipment Morocco as case study. Using the lean value stream mapping 4.0 tool, all products and information flow in a value chain from suppliers to customers were optimized so that economic, social, and environmental performance were improved.

Real-time optimization is a strategy to maximize a cost function with constraints so that operation can be kept at its optimum point even under conditions subjected to nonlinear behaviours and disturbances. In the work by Delou et al. [25], a small-scale real-time optimisation was investigated for a real industrial case, that is, the Natural Gas Processing Unit. A novel approach was addressed for improving efficiency using a sequential-modular simulator within an optimization framework. It was shown, using the addressed method, that an improvement in stability and an increase in profit were achieved.

Process optimization aims to optimize a set of parameters with constraints to achieve an optimal processing time and production. In the article by Chen et al. [26], the process optimization for an automated yogurt and flavour-filling machine was discussed under two scenarios: multi-filling points (Case I) to filling point (Case II). Mathematical models under different cases were developed by considering optimisation objectives. The models were tested with real data, and it was revealed that Case II was faster than Case I in processing a set of customer orders.

Inter-channel advertising and service cooperation are important research areas in channel convergence, which is an important issue in the online to offline supply chain. In the paper co-authored by Zhang et al. [27], the impacts of time delay and bidirectional free riding on inter-channel service and advertising cooperation strategies were discussed. A differential game model between brands and retailers was established by encompassing delay effect and bidirectional free-riding occurrence. Differential game theory was used to seek the optimal advertising and service decisions of the brand owners and retailers. It was shown that the service strategy, advertising strategy, and brand goodwill of the online to offline supply chain members were optimal under a centralized decision-making system.

Artificial-intelligence-based music generation has attracted much attention. In the work by Min et al. [28], a novel approach was proposed to develop a competitive music generation algorithm by blending a transformer deep-learning model with generative adversarial networks. It is shown that the model based on transformer and generative adversarial networks can reveal the relationship in the notes of long-sequence music samples, and the rules of music composition can be learned well. An optimized transformer and generative adversarial-networks-based model can improve the accuracy of the generated notes.

Funding: This research received no external funding.

Acknowledgments: The guest editor would like to thank all authors who contributed to this Special Issue, and all the volunteering contributions from the reviewers. Special thanks to the *Processes* journal office for their administrative support for this Special Issue.

Conflicts of Interest: The author declare no conflict of interest.

References

1. Gao, Z.; Kong, D.; Gao, C. Modelling and control of complex dynamic systems: Applied mathematical aspects. *J. Appl. Math.* **2012**, *2012*, 869792. [CrossRef]
2. Gao, Z.; Nguang, S.; Kong, D. Advances in modelling, monitoring, and control for complex industrial systems. *Complexity* **2019**, *2019*, 2975083. [CrossRef]
3. Gao, Z.; Chen, M.; Zhang, D. Advances in condition monitoring, optimization and control for complex industrial processes. *Processes* **2021**, *9*, 664. [CrossRef]
4. Gao, Z.; Saxen, H.; Gao, C. Data-driven approaches for complex industrial systems. *IEEE Trans. Ind. Inform.* **2013**, *9*, 2210–2212. [CrossRef]
5. Gao, Z.; Castiglione, A.; Nappi, M. Biometrics in industry 4.0: Open challenges and future perspectives. *IEEE Trans. Ind. Inform.* **2022**, *18*, 9068–9071. [CrossRef]
6. Gao, Z.; Paul, Z.; Wang, X. Digital twinning: Integrating AI-ML and big data analytics for virtual representation. *IEEE Trans. Ind. Inform.* **2022**, *18*, 1355–1358. [CrossRef]
7. Zhang, D.; Du, L.; Gao, Z. Real-time parameter identification for forging machine using reinforcement learning. *Processes* **2021**, *9*, 1848. [CrossRef]
8. Ren, L.; Zhang, D. A simple and effective modelling method for 3D porous irregular structures. *Processes* **2022**, *10*, 464. [CrossRef]
9. Liu, D.; Qiao, C.; Wan, J.; Lu, Y.; Song, J.; Yao, Z.; Wei, X.; Yu, Y. Modelling method and application of anti-corrosion pill particles in oil and gas field wellbore casing annulus based on the discrete element method. *Processes* **2022**, *10*, 1164. [CrossRef]
10. Li, W.; Wu, X. Identification and analysis of factors influencing green growth of manufacturing enterprises based on DEMATEL method—Wooden flooring manufacturing companies as a case. *Processes* **2022**, *10*, 2594. [CrossRef]
11. Wang, Y.; Ren, W.; Liu, Z.; Li, J.; Zhang, D. T-S fuzzy model-based fault detection for continuous stirring tank reactor. *Processes* **2021**, *9*, 2127. [CrossRef]
12. Zhang, A.; Yu, D.; Zhang, Z. TLSCA-SVM fault diagnosis optimization method based on transfer learning. *Processes* **2022**, *10*, 362. [CrossRef]
13. Lu, X.; Liu, X.; Li, B.; Zhong, J. Data-driven state prediction and sensor fault diagnosis for multi-agent systems with application to a twin rotational inverted pendulum. *Processes* **2021**, *9*, 1505. [CrossRef]
14. Tan, W.; Wang, H.; Hou, H.; Liu, X.; Zhang, M. Dynamically triggering resilient control for networked nonlinear systems under malicious aperiodic dos attacks. *Processes* **2022**, *10*, 2627. [CrossRef]
15. Liang, F.; Lu, L.; Li, Z.; Zhang, F.; Zhang, S. Tracking control of a hyperchaotic complex system and its fractional-order generalization. *Processes* **2022**, *10*, 1244. [CrossRef]
16. Zaid, S.; Albalawi, H.; Meguid, H.; Alhmiedat, T.; Bakeer, A. Performance improvement of H8 transformerless grid-tied inverter using model predictive control considering a weak grid. *Processes* **2022**, *10*, 1243. [CrossRef]
17. Yuan, Y.; Tao, X.; Huang, K.; Chen, H.; Qian, X. An effective temperature control method for dividing-wall distillation columns. *Processes* **2022**, *10*, 1018. [CrossRef]
18. Jiang, W.; Han, W.; Wang, L.; Liu, Z.; Du, W. Linear golden section speed adaptive control of permanent magnet synchronous motor based on model design. *Processes* **2022**, *10*, 1010. [CrossRef]
19. Atawi, I.; Hendawi, E.; Zaid, S. Analysis and design of a standalone electric vehicle charging station supplied by photovoltaic energy. *Processes* **2021**, *9*, 1246. [CrossRef]

20. Kong, X.; Shi, C.; Liu, H.; Geng, P.; Liu, J.; Fan, Y. Performance optimization of a steam generator level control system via a revised simplex search-based data-driven optimization methodology. *Processes* **2022**, *10*, 264. [CrossRef]
21. Chen, L.; Hu, Z.; Zhang, F.; Guo, Z.; Jiang, K.; Pan, C.; Ding, W. Remote wind farm path planning for patrol robot based on the hybrid optimization algorithm. *Processes* **2022**, *10*, 2101. [CrossRef]
22. Shao, H.; Liu, Q.; Gao, Z. Material removal optimization strategy of 3d block cutting based on geometric computation method. *Processes* **2022**, *10*, 695. [CrossRef]
23. Reinisch, N.; Rudolph, F.; Günther, S.; Bailly, D.; Hirt, G. Successful pass schedule design in open-die forging using double deep Q-learning. *Processes* **2021**, *9*, 1084. [CrossRef]
24. Kihel, Y.; Kihel, A.; Embarki, S. Optimization of the sustainable distribution supply chain using the lean value stream mapping 4.0 tool: A case study of the automotive wiring industry. *Processes* **2022**, *10*, 1671. [CrossRef]
25. Delou, P.; Riberio, L.; Paiva, C.; Niederberger, J.; Gomes, M.; Secchi, A. A real-time optimization strategy for small-scale facilities and implementation in a gas processing unit. *Processes* **2021**, *9*, 1179. [CrossRef]
26. Chen, J.; Khan, R.; Cui, Y.; Salah, B.; Liu, Y.; Saleem, W. The effect of changes in settings from multiple filling points to a single filling point of an industry 4.0-based yogurt filling machine. *Processes* **2022**, *10*, 1642. [CrossRef]
27. Zhang, J.; Xu, Q. Dynamic cooperation of the O2O supply chain based on time delays and bidirectional free-riding. *Processes* **2022**, *10*, 2424. [CrossRef]
28. Min, J.; Liu, Z.; Wang, L.; Li, D.; Zhang, M.; Huang, Y. Music generation system for adversarial training based on deep learning. *Processes* **2022**, *10*, 2515. [CrossRef]

processes

MDPI

Article

Real-Time Parameter Identification for Forging Machine Using Reinforcement Learning

Dapeng Zhang [1], Lifeng Du [2] and Zhiwei Gao [3,*]

1 School of Electrical and Information Engineering, Tianjin University, Tianjin 300072, China; zdp@tju.edu.cn
2 Tianforging Press Co., Ltd., Tianjin 300142, China; dlf@tjdy.com
3 Faculty of Engineering and Environment, University of Northumbria, Newcastle upon Tyne NE2 8ST, UK
* Correspondence: zhiwei.gao@northumbria.ac.uk

Abstract: It is a challenge to identify the parameters of a mechanism model under real-time operating conditions disrupted by uncertain disturbances due to the deviation between the design requirement and the operational environment. In this paper, a novel approach based on reinforcement learning is proposed for forging machines to achieve the optimal model parameters by applying the raw data directly instead of observation window. This approach is an online parameter identification algorithm in one period without the need of the labelled samples as training database. It has an excellent ability against unknown distributed disturbances in a dynamic process, especially capable of adapting to a new process without historical data. The effectiveness of the algorithm is demonstrated and validated by a simulation of acquiring the parameter values of a forging machine.

Keywords: parameter acquisition; mechanism model; reinforcement learning; forging machine

Citation: Zhang, D.; Du, L.; Gao, Z. Real-Time Parameter Identification for Forging Machine Using Reinforcement Learning. *Processes* **2021**, 9, 1848. https://doi.org/10.3390/pr9101848

Academic Editors: Rodolfo Haber and Jae-Yoon Jung

Received: 6 August 2021
Accepted: 14 October 2021
Published: 18 October 2021

1. Introduction

Complex engineering systems are with a high requirement for system reliability and control and production performance. A variety of technologies are developed to support the monitoring, optimization, and control for complex industrial processes such as chemical processes, manufacturing systems, power, and energy systems [1–3]. The forging process that enhances the mechanical properties by compressing the microstructure of parts [4] is widely applied in the fields of mining equipment, thermal hydro wind power generation equipment, nuclear power equipment, petroleum, and so on. As the key equipment, a forging machine should provide a precise pressing speed with a huge force to achieve the technological requirements of forging pieces. Therefore, the control of the forging machine is the guarantee of high forging quality. The control algorithms have made great progress from conventional PID-based algorithms [5] to advanced model-based control algorithms, including sliding mode control [6,7], back-stepping control [8], and feedback linearization [9], in order to obtain higher performance. However, the effects of these control algorithms strongly depend on the accuracy of the mechanism model. In [10,11], fuzzy-based control was proposed by using fuzzy rules instead of the mechanism model, but it cannot achieve the requirement of high precision. It is worthy to point out that the equivalent models, including regression models [12], neural networks [13], support vector machines [14], and so on [15], are alternatives of the mechanism model. These equivalent models overcome the difficulty of mechanical analysis, but at the cost of the model's extension and physical meanings. Up to now, the mechanism model is still feasible for precision control of the forging machine.

The mechanism knowledge of the forging machine has been mastered based on the related principles such as fluid mechanics, dynamics, and machinery technology. For example, the dynamic behaviors of the forging machine were analyzed according to the mechanism model [16]. A focus of the mechanism model with known structure is to determine the parameters, which is often by the way of offline identification and online correction. Especially for a forging machine, most parameters come from the design handbook of forging machine [17] in which the values of parameters are recorded under the pre-set environment. The others are estimated based on the states of the forging machine by kinds of sensors. A number of offline identification methods such as least square method, maximum likelihood, Bayesian estimation, posteriori estimates, and minimizing maximum entropy were shown in reviews [18–20]. Reference [21] proposed to minimize the entropy of a kernel estimation, constructed from the residuals to deal with the case of not using the maximum likelihood estimation. In reference [22], a system parameter estimation method based on deconvolution of the system output process and explicit Levenberg optimization method was presented. Reference [23] presented a new derivative-free search method for finding models of acceptable data fit in a multidimensional parameter space and made use of the geometrical constructs known as Voronoi cells to derive the search in the parameter space. Reference [24] described a method for estimating the Nakagami distribution parameters by the moment method in which the distribution moments were replaced by their estimates. In order to trace the varying working parameters, the online estimated techniques were developed to improve the accuracy of model. The recursive parameter estimations were introduced to the linear model [25], the bilinear system [26], and the ARMA system [27]. In [28], an estimated noise transfer function was used to filter the input–output data of the Hammerstein system. By combining the key-term separation principle and the filtering theory, a recursive least squares algorithm and a filtering-based recursive least squares algorithm were addressed. Reference [29] proposed a parameter estimation algorithm using the simultaneous perturbation stochastic approximation (SPSA) to modify parameters with only two measurements of an evaluation function regardless of the dimension of the parameter. Reference [30] collected time-series data from an experimental paradigm involving repeated training and investigated the effect of various clustering methods on the parameter estimation. Reference [31] provided a servo press force by employing a novel dual-particle filter-based algorithm, achieving a maximum relative error in the force estimation of 3.6%.

As a foundation, a lot of effective historical data are necessary for parameter identification. Unfortunately, a forging machine is often working on batch processes whose parameters are different in each batch, and are even impossible to be known for new forging pieces. This means the parameters of the mechanism model for a forging machine will need to be determined from as few data as possible. From the perspective of data effectiveness, the classical parameter identification methods, whether offline estimation or online correction, are based on the least squares concept with the assumption of data following a normal distribution. It needs an appropriate window to observe the data because the statistical characteristics hide in the collected data. However, the difference of forging material quality and the variable pressure caused by pipe diameter change and flow rate change will lead to some disturbances that cause the data noise to be in an unknown distribution. So it is a challenge to determine the parameters of a model for a forging machine online to meet the needs of a complex environment.

Reinforcement learning (RL), motivated by psychology, statistics, neuroscience, and computer science, is about learning from interaction how to behave in order to achieve a design goal [32–34]. It will get rid of the limitation of training samples by learning directly from the raw data online. Through the learning process, an optimal action will be achieved to respond to the states. By sensing the current states, the RL does not need the assumption of prior distribution of noise. By episodes training, the action will overcome the overfitting difficulty and become robust due to eliminating the disturbance gradually. If the parameters were taken as the actions, they would be determined by reinforcement learning without thinking about the assumptions and disadvantages of the methods. In the case of a forging machine, it is a feasible approach to find the optimal values of the model parameters in a new condition under disturbances. There are some mature algorithms in the RL family, such as Q-learning [35], actor–critic [36], and deep reinforcement learning [37]. In this study, the Q-learning algorithm is proposed to determine the model parameters under the working condition due to its simplicity. The contributions of this paper can be summarized as follows:

(1) The parameters are identified only based on the information of one period, which is promising for online control.
(2) The values of parameters are determined directly by raw data without any assumptions of noisy characteristics.
(3) The parameters have strong stability through a number of training episodes, which resists the bad influence of disturbance of unknown law.

The rest of this paper is organized as follows. Section 2 gives the model of pressing-down in forging machine that shows the state variables and the parameters. Section 3 describes the RL's procedure and releases the proposed approach. In Section 4, the model parameters are elaborated by the proposed approach and comparisons are made with two classical methods. Finally, conclusions are drawn in Section 5.

2. The Model of the Pressing-Down in Forging Machine

A semisolid metallic confectioning constant-speed isothermal forging is an important forging technique especially for light-weight alloy confectioning in the aerospace industry. The typical structure of the forging machine is illustrated in Figure 1, and the model has been built in our previous work [38]. It is repeated here for integrity.

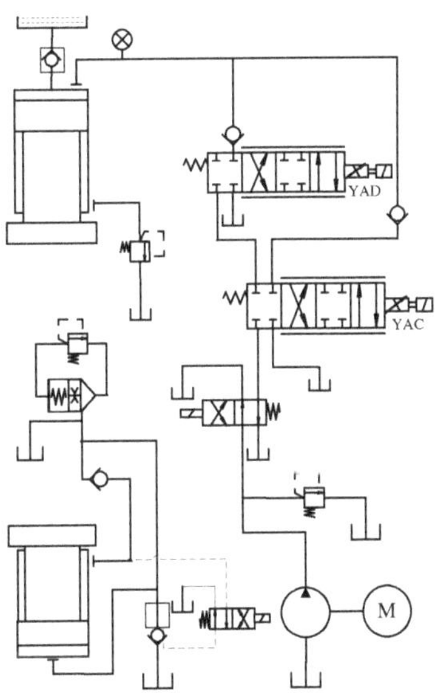

Figure 1. Typical structure of forging machine [38].

The function of the forging machine in pressing-down phase is affected by the oil pipe-line, the proportional servo valve, and the hydraulic cylinder with abandoning the auxiliary attachments.

2.1. The Oil Pipe-Line

The pressing speed in the pressing-down phase is always slow to meet the craft needs, so the oil works in the state of filament flow. Taking a pipe oil column as an object, the pressure balance equation is in the form of Formula (1).

$$\rho S_1 l \frac{d(q_1/S_1)}{dt} = (p_1 - p_s)S_1 - \frac{128\mu l}{\pi d^2} q_1 S_1 \quad (1)$$

Let R $= \frac{32\mu}{\rho}$, so Formula (1) becomes

$$\frac{1}{S_1}\frac{dq_1}{dt} = \frac{p_1 - p_s}{\rho l} + \frac{R}{S_1}q_1 \tag{2}$$

The difference between input volume and output volume is equal to the sum volume of oil compress and pipe swelling. So the oil continuity equation is

$$q_2 - q_1 = \frac{S_1 l}{K}\frac{d(p_1 - p_s)}{dt} \tag{3}$$

where q_1 and q_2 are the oil flow in pipe and the output oil flow of proportional servo valve, p_1 and p_s are the input pressure of proportional servo valve and the pressure of a constant rate pump output, S_1 and l are the sectional area of pipe and the length of oil pipe, and K is the young's modulus of oil equal volume.

2.2. Proportional Servo Valve

The proportional servo valve performs between the servo valve and the proportional valve. It eliminates the dead band by the way of fluid forerunner. The proportional servo valve is widely applied in the ultra-low-speed hydraulic machine to control the oil flow to the hydraulic cylinder. The proportional servo valve is described as

$$\frac{1}{\omega_n^2}\frac{d^2q_2}{dt^2} + \frac{2\xi}{\omega_n}\frac{dq_2}{dt} + q_2 = K_q A \tag{4}$$

where ξ amd ω_n are the damping rate and the inherent frequency of propositional servo valve, respectively, $K_q = K_n\sqrt{\frac{p_1 - p_2}{\Delta p_n}}$ is used to compensate the error between the practical pressure and criterion pressure, and A is the opening of proportional servo valve.

2.3. The Hydraulic Cylinder

The pipe-line between proportional servo valve and the hydraulic cylinder is omitted due to its short distance. The oil continuity equation of hydraulic cylinder is the form of

$$q_2 = S_2 v + \lambda_c p_2 + \frac{V_c}{K}\frac{dp_2}{dt} \tag{5}$$

where S_2 is the plunger's sectional area of exporting cavity of hydraulic cylinder, v is the moving speed of plunger, λ_c is the leak coefficient of hydraulic cylinder, p_2 is the output pressure of proportional servo valve, and V_c is the oil volume of upper cavity of hydraulic cylinder, $V_c = V_0 + vS$.

The dynamic equation of plunger is obtained according to the force analysis with the form of

$$p_2 S_2 + mg = m\frac{dv}{dt} + Bv + F + p_3 S_2 \tag{6}$$

where m is the mass of slider block, g is the acceleration of gravity, B is the viscous damping coefficient, F is the load resistance, and p_3 is the holding pressure of slide block. According to the design of forging machine, the holding power of slide block is equal to the gravity of slide block:

$$p_3 S_2 = mg \tag{7}$$

The Formula (6) is simplified to Formula (8) by substituting Formula (7) for Formula (6):

$$p_2 S_2 = m\frac{dv}{dt} + Bv + F \tag{8}$$

2.4. The Model of the System as a Whole

Let $x_1 = q_1$, $x_2 = p_1 - p_s, x_3 = \frac{dq_2}{dt}, x_4 = q_2, x_5 = p_2$, and $x_6 = v$. By integrating the subsystems together, the global forging machine model can be described in the state–space form

$$\dot{x} = f(x) + g(x)u \tag{9}$$

where $x = [x_1, x_2, x_3, x_4, x_5, x_6]^T, u = A,$

$$f(x) = \begin{bmatrix} \frac{R}{S_1} & \frac{S_1}{\rho l} & 0 & 0 & 0 & 0 \\ -\frac{K}{S_1 l} & 0 & 0 & \frac{K}{S_1 l} & 0 & 0 \\ 0 & 0 & 0 & 1 & 0 & 0 \\ 0 & 0 & -2\xi\omega_n & -\omega_n{}^2 & 0 & 0 \\ 0 & 0 & 0 & \frac{K}{V_c} & -\frac{K\lambda_c}{V_c} & -\frac{KS_2}{V_c} \\ 0 & 0 & 0 & 0 & \frac{S_2}{m} & -\frac{B}{m} \end{bmatrix} \begin{bmatrix} x_1 \\ x_2 \\ x_3 \\ x_4 \\ x_5 \\ x_6 \end{bmatrix},$$

$$g = \left[0, 0, 0, \omega_n{}^2 K_n \sqrt{\frac{x_2 - x_5 + P_s}{\Delta p_n}}, 0, -\frac{F}{B}\right]^T$$

Remark 1. *In the model, most parameters such as the length, the sectional area of oil pipe, the mass of slider block, and the rated flow gain can be valued according to the design. The values of parameters that are influenced by the surrounding or working conditions will result in the inaccuracy of model.*

3. The Proposed Method

3.1. Reinforcement Learning

The basic frame of reinforcement learning is shown in Figure 2. At each time step k, the agent makes observations $x(k) \in X$ and takes action $u(k) \in U$, and receives reward $R(x(k+1), x(k), u(k)) \in \mathbb{R}$.

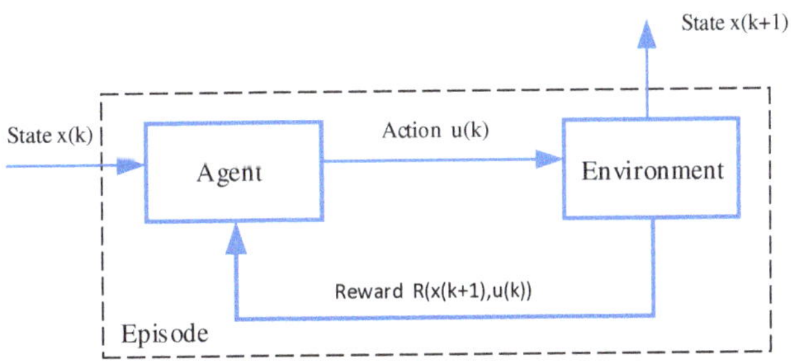

Figure 2. The basic frame of reinforcement learning.

The expected return that is received in the long run is described using the state–action value function $V(x, u)$, under the condition of first taking an arbitrary action $u \in U$ from a certain state $x \in X$ and subsequently acting according to a certain control series π. So the value function $V_\pi(x(k), u(k))$ at time k is defined as

$$V_\pi(x(k), u(k)) = \sum_{t=k}^{\infty} \gamma R(x(k+1), x(t), u(t)) \tag{10}$$

where $\gamma \in [0, 1]$ is the discount factor.

The value function $V_\pi(x(k+1), u(k))$ at time $k+1$ is defined as

$$V_\pi(x(k+1), u(k)) = \sum_{t=k+1}^{\infty} \gamma R(x(t+1), x(t), u(t)) \tag{11}$$

According to the theory of dynamic programming

$$V_{\pi}(x(k),\ u(k)) = R(x(k+1), x(k),\ u(k)) + V_{\pi}(x(k+1),\ u(k)) \tag{12}$$

Unfortunately, the value function $V_{\pi}(x(k),\ u(k))$ and $V_{\pi}(x(k+1),\ u(k))$ is not obtained because no one knows the rewards after time $k+1$. To remove this obstacle, the Q-function is designed with $Q(x(k), u(k))$ and $Q(x(k+1), u(k))$ replacing $V_{\pi}(x(k),\ u(k))$ and $V_{\pi}(x(k+1),\ u(k))$, respectively

Let

$$\delta = R(x(k+1), x(k),\ u(k)) + \gamma Q(x(k+1), u(k)) - Q(x(k), u(k)) \tag{13}$$

The $u(k)$ will be optimized by a process of seeking δ approach to zero.

As an important member of reinforcement learning family, the basic step of Q-algorithm is carried out as Procedure 1 [30].

Procedure 1.

Initialize $Q(x(k), u(k))$ *arbitrarily*
Repeat (for each episode)
Initialize $x(k)$
Repeat (for each step of episode)
Choose $u(k)$ *from* $x(k)$ *using policy derived from Q (e.g.,* $\varepsilon -$ *greedy)*
Take action $u(k)$*, observe* $R(k),\ x(k+1)$
$Q(x(k), u(k)) \leftarrow Q(x(k), u(k)) + \alpha\left[R(k) + \gamma max_{u(k+1)} Q(x(k+1), u(k+1)) - Q(x(k), u(k))\right]$
$x(k) \leftarrow x(k+1)$
until $x(k)$ *is terminal.*

Remark 2. *There is only state information in Procedure 1. One can obtain the optimal action online by using two states,* $x(k)$ *and* $x(k+1)$*, in the process of maximizing the value function. By this way, it makes an online control become possible because this approach gives up the requirement of sliding window length.*

3.2. The Proposed Approach

The scheme of proposed approach is shown in Figure 3.

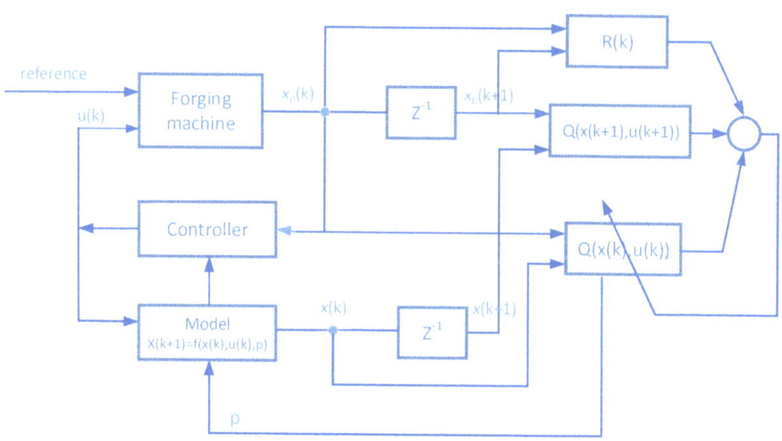

Figure 3. The scheme of proposed approach.

A model that consists of undetermined parameter p ($p \in R^m$) is paralleled to the forging machine under the controller. The state variables of model are recorded as $x(k)$ and $x(k+1)$ at sampling k and $k+1$, which are connected by a delay link z^{-1}. The undetermined parameter p is regarded as the action of Q-algorithm. Therefore, the Q-algorithm following Procedure 1 is applied to determine the parameter p based on $x(k)$ and $x(k+1)$ and finally, the optimal parameter p^* will be obtained when it is convergent.

To explicate Q-algorithm for the acquisition of model parameters, the key concepts of the proposed Q-algorithm are illustrated as follows.

(i). Action space, reward, and value function

The action space is made up of the undetermined parameter p. The values of parameter are usually inconsistent with the working condition, which will disturb with model accuracy. A goal is to determine their values responding to the surroundings.

The forging machine's velocity is designated a constant pressing speed or a given curve of speed during a certain temperature range according to the properties of forging materials, so the reward $\mathrm{R}(k)$ is selected as the reciprocal of change for absolution error between the measured speed and the set speed at adjacent sampling times k and $k+1$

$$\mathrm{R}(k) = \frac{1}{||v(k) - v_{set}(k)| - |v(k+1) - v_{set}(k+1)||} \tag{14}$$

where $v(k)$ and $v_{set}(k)$ are the measured speed and the preset speed at sample k; $v(k+1)$ and $v_{set}(k+1)$ are the measured speed and the preset speed at sample $k+1$. Here, using v instead of x_6 that is the sixth component of state vector x is only to stress the physics meaning.

Let $s = [x;\ u]$ so the value functions $\mathrm{V}(s(k), p(k))$ and $\mathrm{V}(s(k+1), p(k+1))$ from samples k and $k+1$ are defined by Formulas (15) and (16)

$$\mathrm{V}(s(k), p(k)) = \sum_{i=k}^{\infty} \mathrm{R}(i) \tag{15}$$

$$\mathrm{V}(s(k+1), p(k+1)) = \sum_{i=k+1}^{\infty} \mathrm{R}(i) \tag{16}$$

(ii). Q-function

The value functions $\mathrm{V}(s(k), p(k))$ and $\mathrm{V}(s(k+1), p(k+1))$ are replaced by Q-function according to the Q-algorithm because the value functions are not obtained due to the unknown rewards after sample k. The early Q-function that is applied for the discrete space is presented as a look-up table of states row and actions column. When the states or actions are continuous, their discretization will lead to the curse of dimensionality by generating an exponentially increasing complexity of algorithm and insufficient storage. Therefore, the parameterized function is proposed to fit the Q-function with the form

$$\mathrm{Q}(s(k), p(k)) = f(s(k), p(k), \theta) \tag{17}$$

where f and θ are a parameterized mapping and the parameters, respectively. Let $s = [s; p]$, an approximator is used to substitute for the unknown parameterized mapping, and there is

$$\hat{\mathrm{Q}}(s) = \sum_{i=1}^{\mathrm{n}} \phi_i(s)\theta_i \tag{18}$$

where $\phi_i(s, a)$ is usually selected as Gauss radial kernel function due to its simplicity, whose form is

$$\phi_i(s) = e^{-\frac{||s - s_i||^2}{2\sigma_i^2}} \tag{19}$$

in which s_i is the central coordinates of i-th radial kernel function and σ_i is the width of i-th radial kernel function.

(iii). Exploitation and exploration

There are two ways to determine the action in RL. The exploitation is used to get the best action from the Q-function that is based on the reward received. The exploration is used to escape the local optimization of exploitation by randomly giving the action. As a compromise of exploitation and exploration, the $\varepsilon-$ greedy algorithm is proposed to evolve the action. The agent selects the action that maximizes the Q-value function according to the probability ε that is usually a large probability event. In addition, it selects the action randomly according to the probability of $1-\varepsilon$ from the action space, which makes sure the action exploration is within the unknown area. The form of $\varepsilon-$ greedy algorithm is

$$p(k+1) = \begin{cases} \underset{p(k)}{argmax} Q(s(k), p(k), \theta), \ Pr < \varepsilon \\ rand(U), \ Pr \leq 1-\varepsilon \end{cases} \tag{20}$$

where $p(k)$ and $p(k+1)$ are the acquisition parameter at k and $k+1$, respectively, Pr is the probability of select action, and U is the action set.

(iv). The Process of Method

The proposed algorithm is summarized as Procedure 2. In this procedure, the input states are $x(k)$, $u(k)$ and $x(k+1)$, whose physical meanings are shown in Section 2, and the output parameter is p.

Procedure 2.

Step 1: Give a state $x(k)$ and the control $u(k)$ and then construct s according to $s = [x;\ u]$
Step 2: Select parameters $p(k)$ randomly.
Step 3: Observe the next state $x(k+1)$
Step 4: Receive immediate reward $R(k)$ according to Formula (14)
Step 5: select $p(k+1)$ according to Formula (20)
Setp6: Compute $Q(s(k), p(k), \theta)$ and $Q(s(k+1), p(k+1), \theta)$ according to the Formulas (18) based on the model of Formula (9)
Step7: Compute the time series error $\delta(k)$ according to
$$\delta = R(k) + \gamma Q(s(k+1), p(k+1), \theta) - Q(s(k), p(k), \theta)$$
Step 8: Update $Q(s(k), p(k), \theta)$ according to
$$Q(s(k), p(k), \theta) \leftarrow Q(s(k), p(k), \theta) + \alpha\delta$$
Step9: $x(k) \leftarrow x(k+1), u(k) \leftarrow u(k+1)$ and $p(k) \leftarrow p(k+1)$
Step 10: Repeat steps 3 to 9 until it is convergent. The output p is the convergent $p(k)$ in which $p(k) = p(k+1) = p$.

(v). Convergence

The convergence of Q-algorithm can be found in [35,36].

4. Case Studies

The forging machine usually keeps a good state at the early life stage. In this stage, the values of parameters after a fine machine debugging always coincide with the design condition, except for the viscous damping coefficient B because it is prone to be influenced by the temperature and working condition. With time elapsing, the leakage becomes the main uncertainty of the forging machine. A little leakage is permitted for the forging machine if the leakage does not affect the work process. Nevertheless, the forging machine needs to be repaired if there appears much leakage. Therefore, we chose the viscous damping coefficient B and leakage coefficient λ_c as the identification parameters. These two parameters are unmeasurable, which make their values unverifiable in practice. As a result, we conducted a simulation to verify the proposed method.

4.1. Data Source

The state space model of (9) was used to simulate a forging machine. The values of model parameters are shown in Table 1 according to the design condition.

Table 1. The parameters values under the design condition.

ξ	ω_n	R	S_1	ρ	l	m	K	S_2	V_0	K_n	P_s	Δp_n
			m^2	kg/m^3	m	kg		m^2	m^2		Mpa	pa
0.7	70	0.0064	0.0138	870	7	1×10^4	1×10^{10}	0.02463	4.9×10^{-3}	2×10^{-4}	12	3.5×10^6

A controller is necessary for a forging machine to guarantee the quality of pressing process, therefore, a PID controller was used to simulate this situation. We chose a PID controller because here we focus on verifying our proposed method rather than discussing the control method. The PID controller is enough to provide the states and control for the proposed approach. The data series were generated by solving the model (9) with ODE45 that applies the fourth-order Runge Kutta algorithm to provide the candidate solution and the fifth-order Runge Kutta algorithm to control errors. These continuous sequences provided the data source by adding two kinds of noise with uniform distribution or Gaussian distribution as a simulation of real data. The set speed was changed from 0.02 to 0.08 that is consistent with the requirement of a typical pressing process. A typical control process that includes a transition process and a stable process is shown in Figure 4.

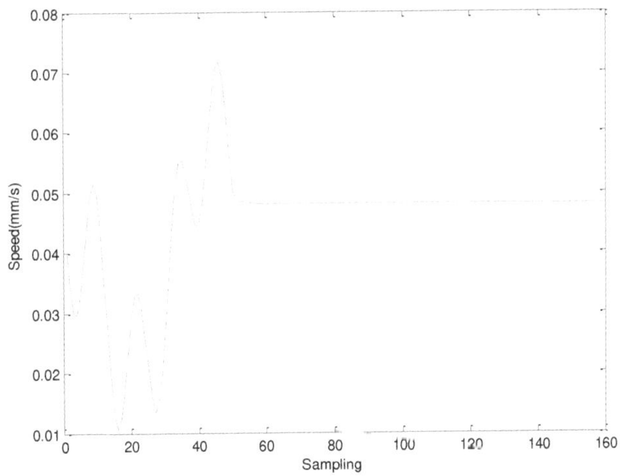

Figure 4. A typical control process (the set speed = 0.05).

The subsequent simulation was carried out at the platform of MatlabR2011b with the computer of Intel® Core™ 2 Duo CPU E7300 @2.66GHz 2.67GHz.

4.2. Acquisition of the Viscous Damping Coefficient

According to experiments, the viscous damping coefficient B is usually during 10–30 for this model. As a result, the value of 15 was chosen as the predetermined value and targeted by the proposed approach according to Procedure 2. The episodes training process is shown in Figure 5, where the subgraph above is with the noises of the uniform distributions and the subgraph below is with the noises of the Gaussian distributions. It is generally believed that the training time is related to the nature of the object and the computer performance. In order to avoid the time difference caused by different computer performance, we used the number of the episodes as an index of training time.

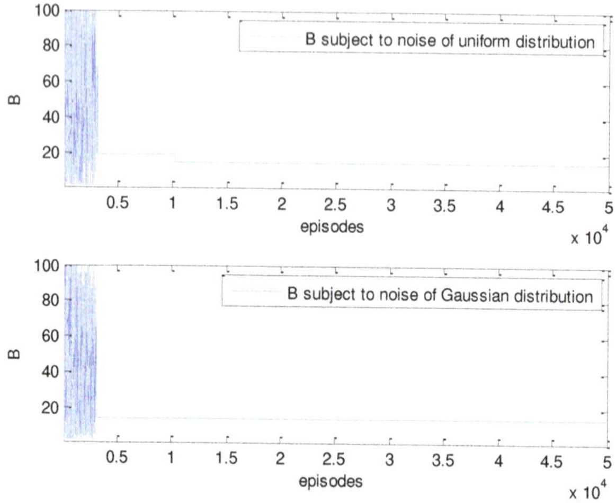

Figure 5. The episode training process of viscous damping coefficient (B was predetermined as 15).

Figure 5 shows there is a trial process at the beginning of training because there is no priori information on B. After a trial of about 3000 episodes, the best historical value of B that indicates 20 for the above subgraph and 15.0626 for the below subgraph appears during the process of seeking the best reward. After about 10,000 episodes, a better value of 14.5000 occurs for the above subgraph. In contrast, a value of 15.0626 for the below subgraph is unchanged until the episodes terminate.

The viscous damping coefficient B was changed from 15 to 20 to test the proposed method. The episodes training process is shown under a uniform distribution (the above subgraph) and under a Gaussian distribution (the below subgraph). Figure 6 shows the training episodes process similar to Figure 5. It is also seen that the trial process of Figure 6 lasts about 3000 episodes.

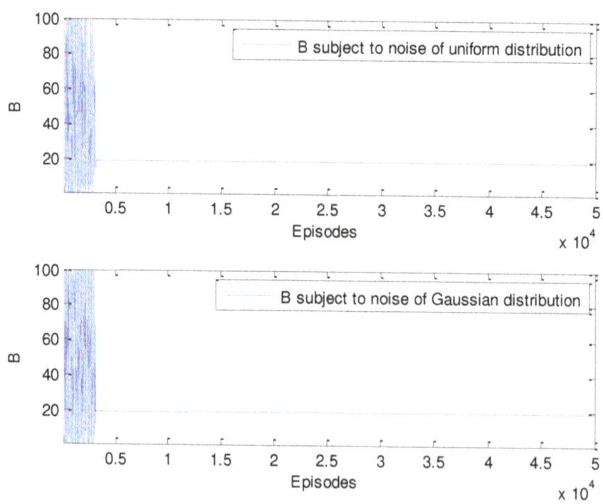

Figure 6. The episode training process of viscous damping coefficient (B was predetermined as 20).

In order to show the accuracy of parameter acquisition, the relative error δ between the estimated value $\hat{B}$ and the predetermined value B_r is defined as a form of

$$\delta = (\hat{B} - B_r)/B_r \tag{21}$$

and the results are shown in Table 2

Table 2. The results of viscous damping coefficient without leakage.

Predetermined Value	Noise Distribution	Acquisition	Relative Error
15	Uniform	15.0626	0.4%
15	Gaussian	14.5000	3.33%
20	Uniform	19.0000	5%
20	Gaussian	19.0000	5%

It is seen from Figures 5 and 6 that the excellent results with relative errors no greater than 5% were obtained in the cases of noises with different distributions.

Further tests under the condition of oil leakage were done to verify the effectiveness of the proposed approach. For a forging machine, the leakage is prone to go into saturation and is limited to a small value, so the leakage coefficients λ_c were assumed as a constant 0.01 and 0.02. The episodes training processes are shown in Figures 7–10. Figures 7 and 8 present the training processes of acquiring the viscous damping coefficient with a goal of 15 and of 20, respectively, under the leakage coefficient of 0.01. Figures 9 and 10 present the training processes of acquiring the viscous damping coefficient with a goal of 15 and of 20, respectively, under the leakage coefficient of 0.02. These figures show the proposed approach will be convergent after episodes training processes, and the final results are listed in Table 3. Table 3 shows the viscous damping coefficient will approach the predetermined value B_r under different coefficients or different noise distributions, showing a maximal relative error less than 2%. For training time, there are some differences for different parameters, such as about 6000 episodes in Figure 7, about 4000 episodes in Figure 9, and about 3000 episodes in Figure 10. Sometimes the different distributions also have an effect on the training speed, which is shown in Figure 8.

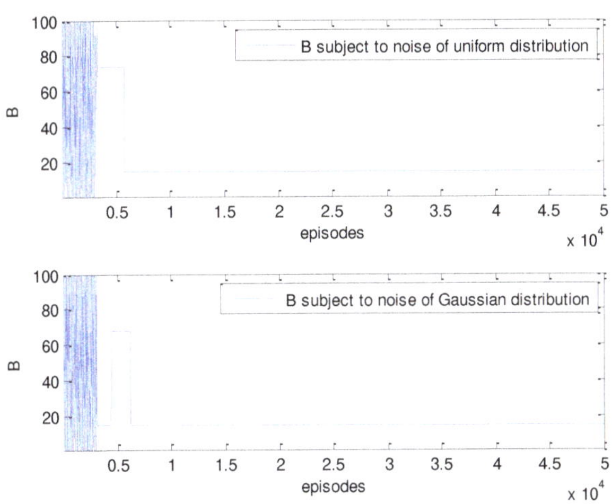

Figure 7. The episode training process with leakage of 0.01 (B_r = 15).

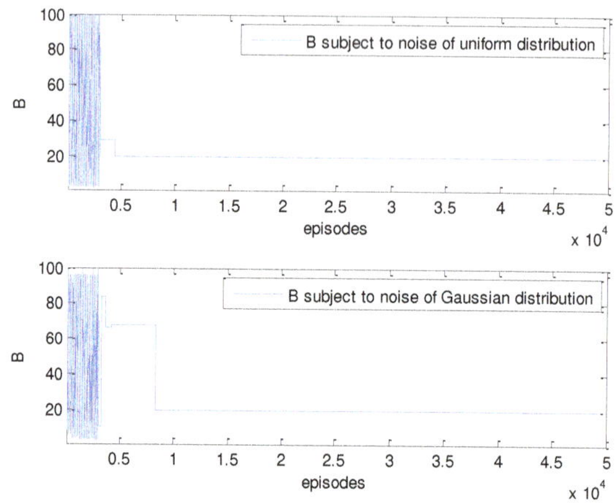

Figure 8. The episode training process with leakage of 0.01 ($B_r = 20$).

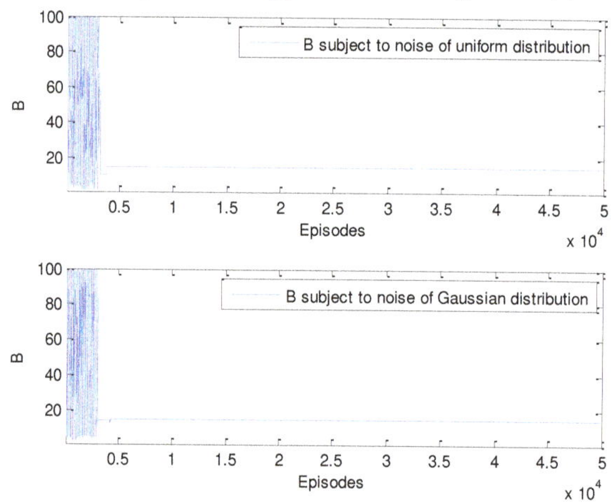

Figure 9. The episode training process with leakage of 0.02 ($B_r = 15$).

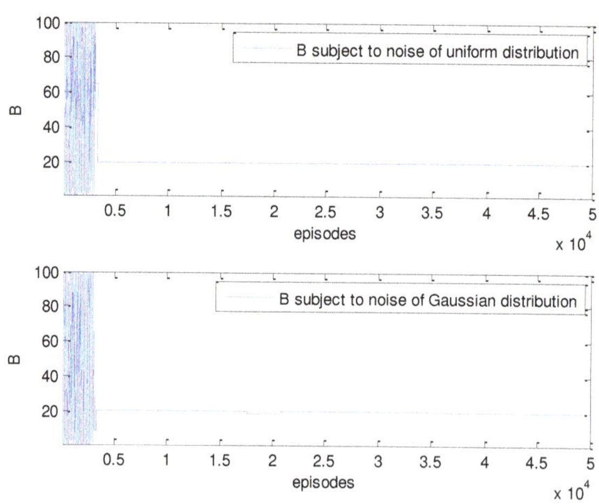

Figure 10. The episode training process with leakage of 0.02 ($B_r = 20$).

Table 3. The results of viscous damping coefficient under leakage.

Leakage Coefficient λ_c	Noise Distribution	Acquisition B	Predetermined Value B	Relative Error
0.01	Uniform	15.0000	15	0%
0.01	Gaussian	15.2500	15	1.67%
0.01	Uniform	20.0000	20	0%
0.01	Gaussian	20.0000	20	0%
0.02	Uniform	15.0000	15	0%
0.02	Gaussian	15.0000	15	0%
0.02	Uniform	20.0000	20	0%
0.02	Gaussian	19.9375	20	0.31%

4.3. Acquisition of the Leakage Coefficient

The leakage that is marked with leakage coefficient λ_c in the model will become the main uncertainty along with the lapsing time of forging machine. The leakage coefficient was predetermined as a constant 0.01 and 0.02. The learning processes with uniform distribution and with Gaussian distribution are shown in Figures 11 and 12, respectively. As for training time, it is affected by different distributions in Figure 11 and about 5000 episodes in Figure 12.

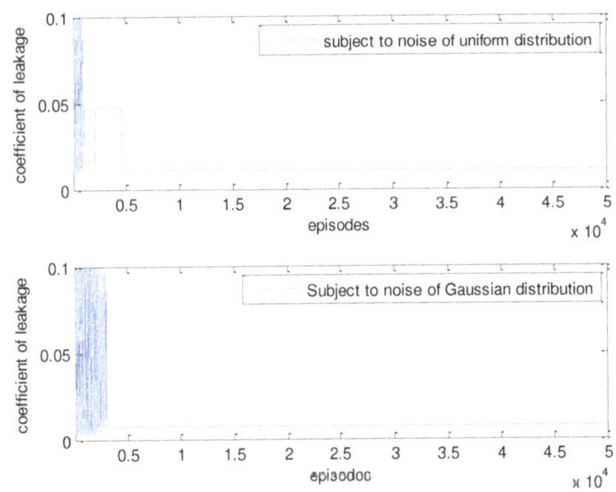

Figure 11. The learning process of leakage coefficient (λ_c was predetermined as 0.01).

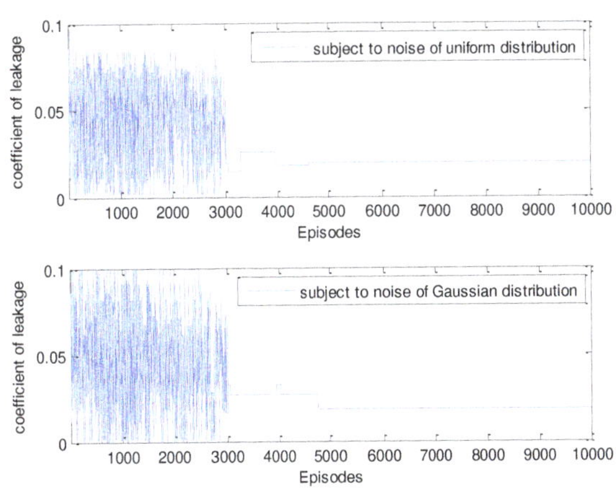

Figure 12. The episode training process of leakage coefficient (λ_c was predetermined as 0.02).

The values of leakage coefficient $\hat{\lambda}_c$ are acquired when the curve becomes stable. Here, the absolute error E with the definition of

$$E = \left|\lambda_c - \hat{\lambda}_c\right| \tag{22}$$

was used to replace the former relative error because the value of leakage coefficient is too small as the denominator of Formula (22), which is prone to an inappropriate relative error. The results are listed in Table 4. Table 4 shows the absolute errors are not more than 0.0015 in the cases of noisy with different distributions.

Table 4. The results of leakage coefficient.

Predetermined Value	Noise Distribution	Acquisition	Absolute Error
0.01	Uniform	0.0114	0.0014
0.01	Gaussian	0.0075	0.0015
0.02	Uniform	0.0200	0
0.02	Gaussian	0.0187	0.0013

4.4. Acquisition of the Viscous Damping Coefficient and the Leakage Coefficient

In order to test higher dimensionality of parameters, an experiment on acquiring concurrently the viscous damping coefficient and the leakage coefficient was done. The parameters of B and λ_c were predetermined as 18 and 0.01, respectively. The learning processes with uniform distribution and with Gaussian distribution are shown in Figures 13 and 14, respectively, and the results are shown in Table 5, which shows both parameters can reach a good estimation concurrently in the cases of noisy conditions. Here, all the training times are less than 5000 episodes.

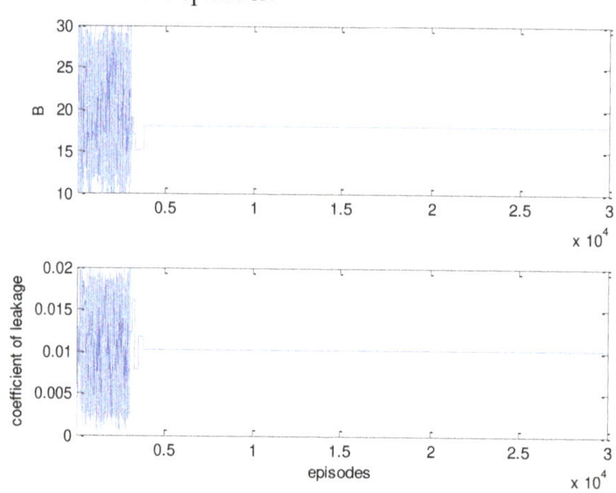

Figure 13. The episode training process of viscous damping coefficient and leakage coefficient subject to noise of uniform distribution.

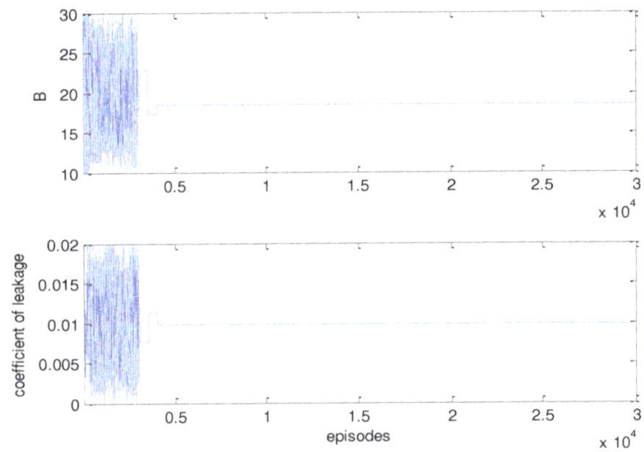

Figure 14. The episode training process of viscous damping coefficient and leakage coefficient subject to noise of Gaussian distribution.

Table 5. The results of the viscous damping coefficient and the leakage coefficient concurrently.

Noise Distribution	Parameters	Predetermined Value	Acquisition
Uniform	Viscous damping coefficient B	18	18.0488
	Leakage coefficient λ_c	0.01	0.0102
Gaussian	Viscous damping coefficient B	18	18.4141
	Leakage coefficient λ_c	0.01	0.0098

4.5. Comparison with Other Methods

A famous BP network approach and the sliding window correlation methods were chosen as a comparison of the proposed approach. The data series with 160 samples that was produced by the model with a controller was considered as the data source to determine the parameters. This data series includes a transient process of 50 and a stable process of 110 based on the viscous damping coefficient B of 15.

As we know, the BP network has a strong nonlinear approximation ability and an excellent estimation of recursion problem, which needs the length of input time series to match the order of the system. Here, we focused on identifying the parameter of viscous damping coefficient B just in one period. After several attempts, the BP network was chosen as a 7-20-1 structure with an input of seven variables (six states and one control in the model of Section 2) and an output of the viscous damping coefficient B. It was trained by the back propagation algorithm based on a train set of 2000 data from different cases in which the set speed was changed from 0.02 to 0.08. The learning rate was 0.001. The well-trained BP network was used to estimate the values of viscous damping coefficient, and the results are shown in Figure 15.

The values of viscous damping coefficient from sampling 1 to sampling 160 that were estimated by the BP network and the proposed approach are shown with the black curve and the red curve. It is seen that the BP network will approach to the viscous damping coefficient in the stable process, but it is bad in the transient process. The proposed approach shows an excellent performance that achieves the 15.0625 approaching to the goal of 15.0000 throughout the whole process.

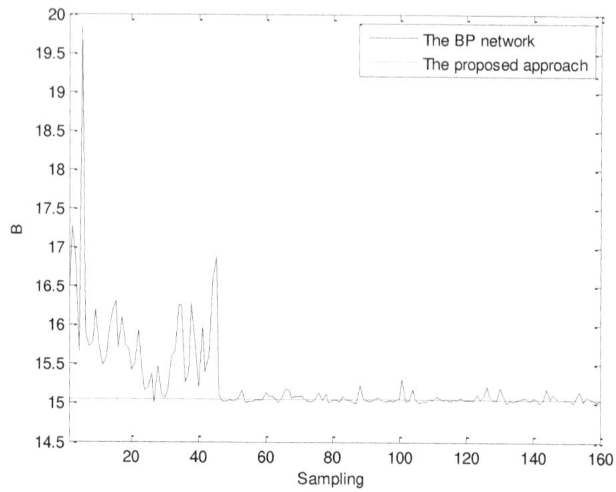

Figure 15. The comparison between BP network and the proposed approach.

The sliding window correlation method, as a kind of conventional parameters identification method for data series, was applied to estimate the values of viscous damping coefficient by an optimization of minimizing the sums of squared errors during each observation window. Considering the sliding window is influenced with the disturbance, it is prone to change the statistical properties of the observation window. The numbers of 2, 5, 10 and 50 were chosen as the length of sliding window, and the results are seen in Figure 16.

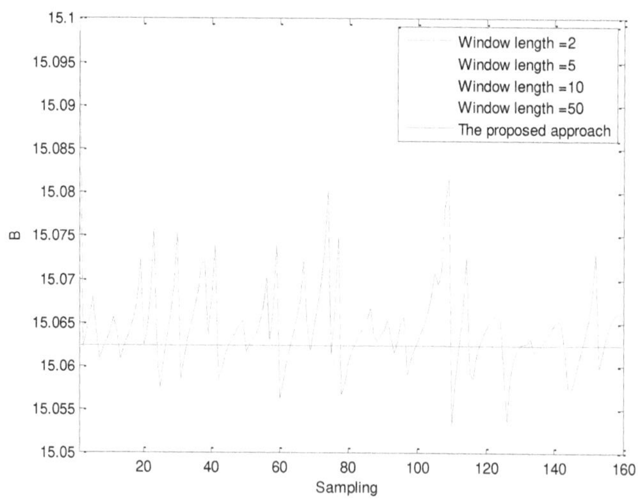

Figure 16. The comparison between the slide window and the proposed approach.

It is seen from Figure 16 that the sliding window correlation method and the proposed approach have a similar accuracy throughout the process from sampling 1 to sampling 160. However, there are some fluctuations for the sliding window correlation method according to different window length. The shorter the length of the slide window, the more sensitive the result, and vice versa. In contrast, the proposed approach shows a fine stability owing to its episodes training.

The advantages and disadvantages of three methods are summarized in Table 6.

Table 6. The comparisons of three methods.

	Advantage	Disadvantage
The BP networks	Learning algorithm, high accuracy in steady state	Worse in transient state
The sliding window correlation method	Optimization algorithm, high accuracy in steady state and transient state	Related to the length of the window and affected by disturbance
The proposed approach	high accuracy in steady state and transient state, only using the data during a period	Long training time

The proposed approach has the ability to obtain a high accuracy of viscous damping coefficient in steady state and transient state during only a period. To our best knowledge, there are no other approaches to implement the identification of model parameters with so little information, which is beneficial to the online control. However, it is limited to a slow process of the forging machine due to a long training time, though some improvements have been made, such as eligibility traces and heuristic search. A hardware implementation of this proposed approach is an attractive request for broader industrial processes.

5. Conclusions

In this paper, reinforcement learning has been addressed to identify optimal parameters values online by directly using raw data in one period. Compared with the BP network approach, the proposed technique has a good accuracy throughout the whole process. Compared with the sliding window correlation method, the proposed method has a similar accuracy but has a better ability to resist the influence of noise. As a result, the proposed approach has been demonstrated to be effective for online parameter identification in a simulation of real-time process of a forging machine.

Author Contributions: Conceptualization and methodology, D.Z. and Z.G.; formal analysis, L.D.; writing—original draft preparation, D.Z.; writing—review and editing, Z.G.; All authors have read and agreed to the published version of the manuscript.

Funding: This research was funded by the Program of Science and Technology Commissioner, and National Nature Science Foundation of China, grant number 61673074.

Institutional Review Board Statement: Not applicable.

Informed Consent Statement: Not applicable.

Data Availability Statement: Not applicable.

Acknowledgments: The authors would like to acknowledge the research support from the School of Electrical Engineering and Automation at Tianjin University, and the E&E faculty at the University of Northumbria at Newcastle.

Conflicts of Interest: The authors declare no conflict of interest.

Nomenclature

Symbol	Meanings
ρ	Density of oil
μ	Dynamic viscosity
λ_c	Leak coefficient of hydraulic cylinder
ξ	Damping rate of propositional servo valve
ω_n	Inherent frequency of propositional servo valve
B	Viscous damping coefficient
d	Diameter of pipe
F_l	Load resistance
K	Young's modulus of oil equal volume
K_n	Rated flow gain

Symbol	Meanings
K_q	Flow gain of propositional servo valve
l	Length of oil pipe
m	Mass of slider block
p_1	Input pressure of proportional servo valve
p_2	Output pressure of proportional servo valve
P_s	Pressure of a constant rate pump output
Δp_n	Valve port pressure drop
q_1	Oil flow in pipe
q_2	Output oil flow of proportional servo valve
R	Intermediate coefficient
$S1$	Sectional area of pipe
$S2$	Plunger's sectional area of exporting cavity of hydraulic cylinder
u	Control voltage of proportional servo valve
v	Moving speed of plunger
V_0	Initial oil volume of upper cavity of hydraulic cylinder
V_c	Current oil volume of upper cavity of hydraulic cylinder

References

1. Gao, Z.; Chen, M.Z.Q.; Zhang, D. Special Issue on "Advances in condition monitoring, optimization and control for complex industrial processes". *Processes* **2021**, *9*, 664. [CrossRef]
2. Gao, Z.; Liu, X. An overview on fault diagnosis, prognosis and resilient control for wind turbine systems. *Processes* **2021**, *9*, 300. [CrossRef]
3. Gao, Z.; Dai, X.; Breikin, T.; Wang, H. Novel parameter identification by using a high-gain observer with application to a gas turbine engine. *IEEE Trans. Ind. Inform.* **2008**, *4*, 271–279. [CrossRef]
4. Available online: https://www.forging.org/producers-and-suppliers/technology/vision-of-the-future#importance (accessed on 1 May 2021).
5. Lu, X.; Huang, M. System-decomposition-based multilevel control for hydraulic press machine. *IEEE Trans. Ind. Electron.* **2012**, *59*, 1980–1987. [CrossRef]
6. Jia, C.; Wu, A.; Du, C.; Zhang, D. Variable structure control with sliding mode for a class of hydraulic nonlinear system. In Proceedings of the World Congress on Intelligent Control and Automation (WCICA), Jinan, China, 7–9 July 2010.
7. Ho, T.; Ahn, K. Speed control of a hydraulic pressure coupling drive using an adaptive fuzzy sliding-mode control. *IEEE-ASME Trans. Mechatron.* **2012**, *17*, 976–986. [CrossRef]
8. Li, C.; Wu, A.; Du, C. Speed control of hydraulic press via adaptive back-stepping. *Appl. Mech. Mater.* **2011**, *40–41*, 46–51. [CrossRef]
9. Zhang, D.; Wu, A.; Zhang, G.; Du, C. Application of the differential geometric feedback linearization to the speed control of forging machine. In Proceedings of the 2010 Chinese Control and Decision Conference, Xuzhou, China, 26–28 May 2010.
10. Lee, Y.; Kopp, R. Application of fuzzy control for a hydraulic forging machine. *Fuzzy Sets Syst.* **2001**, *118*, 99–108. [CrossRef]
11. Duan, X.; Deng, H.; Li, H. A saturation-based tuning method for fuzzy PID controller. *IEEE Trans. Ind. Electron.* **2013**, *60*, 5177–5185. [CrossRef]
12. Azari, A.; Poursina, M.; Poursina, D. Radial forging force prediction through MR, ANN, and ANFIS models. *Neural Comput. Appl.* **2014**, *25*, 849–858. [CrossRef]
13. Bharti, P.S. Process modelling of electric discharge machining by back propagation and radial basis function neural network. *J. Inf. Optim. Sci.* **2019**, *40*, 263–2778. [CrossRef]
14. Fan, B.; Lu, X.; Huang, M. A novel LS-SVM control for unknown nonlinear systems with application to complex forging process. *J. Cent. South Univ.* **2017**, *24*, 2524–2531. [CrossRef]
15. Hong, J.; Yeh, W. Application of response surface methodology to establish friction model of upset forging. *Adv. Mech. Eng.* **2018**, *10*, 1–9. [CrossRef]
16. Pan, Q.; Li, Y.; Huang, M. Estimation of dynamic behaviors of hydraulic forging press machine in slow-motion manufacturing process. *Nonlinear Dyn.* **2019**, *96*, 339–362. [CrossRef]
17. China Society for Technology of Plasticity CMES. *Forging Manual*; China Machine Press: Beijing, China, 2013.
18. Koksal, G.; Batmaz, I.; Murat, C. A review of data mining applications for quality improvement in manufacturing industry. *Expert Syst. Appl.* **2011**, *38*, 13448–13467. [CrossRef]
19. Olivier, A.; Smyth, A.W. Review of nonlinear filtering for SHM with an exploration of novel higher-order Kalman filtering algorithms for uncertainty quantification. *J. Eng. Mech.* **2017**, *143*, 04017128. [CrossRef]
20. Lu, X.; Huang, M. Novel multi-level modeling method for complex forging processes on hydraulic press machines. *Int. J. Adv. Manuf. Technol.* **2015**, *79*, 1869–1880. [CrossRef]

21. Pronzato, L.; Thierry, E. Entropy minimization for parameter estimation problems with unknown distribution of the output noise. In Proceedings of the IEEE International Conference on Acoustics, Speech, and Signal Processing, Saltlake City, UT, USA, 7–11 May 2001.
22. Slivinskas, V.; Simonyte, V. Modeling of a mechanical system using output data of the hammer blow sequence response. *J. Vibroeng.* **2009**, *11*, 120–129.
23. Ratnaweera, A.; Halgamuge, S.K.; Watson, H.C. Self-organizing hierarchical particle swarm optimizer with time-varying acceleration coefficients. *IEEE Trans. Evol. Comput.* **2004**, *8*, 240–255. [CrossRef]
24. Artyushenko, V.M.; Volovach, V.I. Nakagami distribution parameters comparatively estimated by the moment and maximum likelihood methods. *Optoelectron. Instrum. Data Process.* **2019**, *55*, 237–242. [CrossRef]
25. Ram, S.S.; Veeravalli, V.V.; Nedic, A. Distributed and Recursive parameter estimation in parametrized linear state-space models. *IEEE Trans. Autom. Control* **2010**, *55*, 488–492. [CrossRef]
26. Zhang, X.; Ding, F. Recursive parameter estimation and its convergence for bilinear systems. *IET Control Theory Appl.* **2020**, *14*, 677–688. [CrossRef]
27. Chen, F.; Ding, F.; Sheng, J. Maximum likelihood based recursive parameter estimation for controlled autoregressive ARMA systems using the data filtering technique. *J. Frankl. Inst.* **2015**, *352*, 5882–5896. [CrossRef]
28. Wang, Z.; Shen, Y.; Ji, Z. Filtering based recursive least squares algorithm for Hammerstein FIR-MA systems. *Nonlinear Dyn.* **2013**, *73*, 1045–1054. [CrossRef]
29. Hirokami, T.; Maeda, Y.; Tsukada, H. Parameter estimation using simultaneous perturbation stochastic approximation. *Electr. Eng. Jpn.* **2006**, *154*, 30–39. [CrossRef]
30. Ozdemir, M.C.; Eggert, T.; Straube, A. Improving the repeatability of two-rate model parameter estimations by using autoencoder networks. *Prog. Brain Res.* **2019**, *249*, 189–194. [PubMed]
31. Olaizola, J.; Bouganis, C.S.; De, A. Real-time servo press force estimation based on dual particle filter. *IEEE Trans. Ind. Electron.* **2020**, *67*, 4088–4097. [CrossRef]
32. Kaelbling, L.; Littman, M.; Moore, A. Reinforcement learning: A survey. *J. Artif. Intell. Res.* **1996**, *4*, 237–285. [CrossRef]
33. Sutton, R.; Barto, A. *Reinforcement Learning: An Introduction*; The MIT Press: Cambridge, MA, USA; London, UK, 2005.
34. Farias, V.; Moallemi, C.; Van, B.; Weissman, T. Universal Reinforcement Learning. *IEEE Trans. Inf. Theory* **2010**, *56*, 2441–2454. [CrossRef]
35. Watkins, J.; Dayan, P. Q-learning. *Mach. Learn.* **1992**, *8*, 279–292. [CrossRef]
36. Vamvoudakis, K.G.; Lewis, F.L. Online actor-critic algorithm to solve the continuous-time infinite horizon optimal control problem. *Automatica* **2010**, *46*, 878–888. [CrossRef]
37. Schmidhuber, J. Deep learning in neural networks: An overview. *Neural Netw.* **2015**, *61*, 85–117. [CrossRef] [PubMed]
38. Zhang, D.; Gao, Z.; Lin, Z. An online control approach for forging machine using reinforcement learning and taboo search. *IEEE Accesses* **2020**, *8*, 158666–158678. [CrossRef]

processes

Article

A Simple and Effective Modeling Method for 3D Porous Irregular Structures

Lijing Ren [1,2] **and Denghui Zhang** [1,*]

1 Cyberspace Institute of Advanced Technology, Guangzhou University, Guangzhou 510006, China; ren.lijing@foxmail.com
2 School of Traffic and Transportation, Shijiazhuang Tiedao University, Shijiazhuang 050043, China
* Correspondence: denghui.zhang@gzhu.edu.cn

Abstract: Porous structures are kinds of structures with excellent physical properties and mechanical characteristics through components and internal structure. However, the irregular internal morphology of porous structures poses new challenges to product modeling techniques. Traditional computer-aided design (CAD) modeling methods can only represent the external geometric and topological information of models, lacking the description of the internal structure and conformation, which limits the development of new porous products. In this paper, a new simple and effective modeling method for 3D irregular porous structures is proposed, which improves the controllability of pore shape and porosity, thus overcoming the limitations of existing methods in 3D and concave structures. The key idea is to solve isothermal for modeling the porosity of porous units. Experimental results show that the method can easily obtain smooth and approximate porous structures from arbitrary irregular 3D surfaces.

Keywords: porous structure; finite element analysis; CAD; centroidal Voronoi diagram

Citation: Ren, L.; Zhang, D. A Simple and Effective Modeling Method for 3D Porous Irregular Structures. *Processes* **2022**, *10*, 464. https://doi.org/10.3390/pr10030464

Academic Editor: Zhiwei Gao

Received: 17 January 2022
Accepted: 21 February 2022
Published: 25 February 2022

1. Introduction

With the rapid development of aviation, mechatronics, biology, and medical technologies, many fields have put forward requirements for structures with special properties including light weight and strong rigidity, which regular structures cannot satisfy. Mechanical products are evolving from a single structure to products with more complex internal components [1]. Porous objects are structures with special physical properties and mechanical characteristics through components and conformations [2]. Porous structures have stronger scalability in physical and mechanical properties and have solved the problems of light-weighting, explosion-proof, and tribology in engineering [3,4]. By changing the topology of these primitive cells and the proportion of each unit, various composite materials with extreme properties can be designed, such as low thermal expansion coefficient materials with planar fine structure, negative Poisson's ratio materials, high-performance piezoelectric ceramic materials [5]. However, due to the irregularity and intricacy in cells and internal components, porous structures pose a new challenge to traditional CAD and manufacturing techniques [6].

In recent years, the rapid development of additive manufacturing (AM) has solved the problem of manufacturing products with complex internal components such as porous and porous structures [7,8]. AM (also known as 3D printing) is a technology that uses a gradual accumulation of materials to manufacture solid parts. Compared with traditional subtractive and isotactic manufacturing technologies such as cutting, casting, and stamping, AM does not require tooling fixtures and complex machining processes [9]. It can rapidly produce intricate internal structures, enabling the free manufacturing of parts and solving the manufacturing problem of complex structures like porous structures [10]. More and more artificial porous structures such as truss, honeycomb, and foam structures are prepared with AM [11].

The modeling of porous structures is a challenging task due to the highly irregularity and complexity in geometries [12,13]. The 3D model of a product is the basis for manufacturing [14,15], which determines the structure and performance of the product. Using trial-and-error methods to determine the fine structure is often time-consuming and tedious and does not guarantee an optimal design [16,17]. Although porous structures exist in nature and have been prepared by experimental and reconstructed methods, there are currently no effective CAD methods and tools [18,19].

The modeling methods for products can be divided into two categories: forward modeling and reverse modeling [20]. Reverse modeling is known as a technique of obtaining a 3D digital model of an object through scanning or measuring the physical model. This method can effectively shorten the product design and development cycle and reduce the cost and risk of new product development. However, these advantages are based on the existing physical prototype, which does not facilitate the exploration of new porous structures. The forward modeling, including constructive solid geometry (CSG) and boundary representation (Brep), is a process from scratch, which has more design freedom and is more conducive to exploiting the performance advantages of porous structures. However, existing CAD modeling methods are advantageous in representing simple and regular homogeneous structures and lack effectiveness in modeling porous structural entities with irregular structures or varying material distribution. The lack of porous structure design methods has become an urgent problem in the process of new product development.

To deal with above problems, we propose a novel CAD method called Modeling Complex Porous Objects using the Finite Element Method (MPFEM) for modeling 3D irregular porous structures. MPFEM first tessellates an original surface with 3D centroidal Voronoi diagrams, then generates porous structures with varying components and morphologies by extracting isothermal surfaces of each cell. The main idea behind MPFEM is to convert the design of porous structures into the extraction of isothermal surfaces. Compared with existing methods, our method provides higher design freedom and estimates the stress concentration in irregular 3D porous structures.

The main contributions of this paper are:

(i) We propose a novel modeling method for irregular 3D porous structures based on the FEM and thermodynamic analysis.
(ii) We build up 3D porous structures for both intricate convex and concave structures by combining adjacent Voronoi cells.
(iii) We implementat a prototype system which performs kinds of modeling experiments and shows the effectiveness of the proposed method.

The remainder of this paper is organized as follows: Section 2 briefly reviews background work about modeling methods for porous and heterogeneous objects. Then, Section 3 combines the existing results in the Voronoi tessellation and finite element analysis (FEA) to build up 3D irregular porous structures. Section 4 gives case studies to show the effectiveness of the proposed method. At last, Section 5 concludes the paper and discusses future work.

2. Related Work

The research of CAD and objects modeling has evolved from single and homogeneous productions to composite and heterogeneous objects [6,7,21]. Through the combination of basic parameters, the performance of macroscopic objects consisting of fine structures can be fabricated to meet versatile engineering requirements. These parameters can be the physical properties like stiffness, coefficient of thermal expansion of material particles in fine material cells, or the geometric parameters including distribution of materials, shape, volume fraction [22]. To improve the overall performance of the product, it is necessary to combine the mechanical, physical, and biological advantages of components through the interaction of external form and internal components.

Several theories have been developed for the design and optimization of porous structures. Kim et al. [4] obtained super-strength, low-density, and large-ductility composites by adjusting the dispersion and morphology of the face-center cubic matrix to alleviate the harmful effect on the ductility of the alloy, which is even stronger than the known titanium alloys with the highest strength. Gan et al. [23] developed a new data-driven approach to derive an ideal microstructure and mechanical properties, which combines the thermal-fluid dynamics model, physical-based models, and experimental measurements. Homogenization theory is a common approach in the optimal design of porous structures, which discretizes the structure into periodic pore microstructures. After reviewing the application cases of topological optimization methods based on homogenization theory in metamaterial modeling, Sigmund et al. [5] gave a structural optimization method with periodic fine structure using FEA to model metamaterials, and the optimization criterion method to find the lightest microstructure that satisfies the prescribed properties.

For the design of porous objects, Wang et al. [16] used skin and internal rigid frames to represent the printable 3D model. The exterior of the model is a skin with a certain thickness and the interior is an optimized truss structure. To minimize the volume of truss structures, the authors propose an iterative algorithm by optimizing the rod radius, node position, and the number of truss structures. Lu et al. [24] adopted a honeycomb as the internal structure of the product model to ensure the strength of the model while reducing material loss. The generated 3D porous cells are similar to those obtained with our methods, but it has limitations for modeling concave porous cells. Xu et al. [25] combined the traditional numerical analysis method with model stress distribution and allowable stress information of the material to present a topology optimization algorithm for the force transfer path of the product. To accelerate optimization, the authors introduce a multi-resolution technology from the coarse tetrahedral meshes to fine meshes. Inspired by the random colloid aggregation model, Kou et al. [26] used Voronoi tessellation and cell merging to generate irregular convex and concave pores. However, this design method is mainly aimed at 2D porous structures.

For the fabrication of porous objects, Liu et al. [27] designed a cost-effective method for cellular tissue culture by fused deposition molding, demonstrating the ability of 3D printing technology to rapidly fabricate a variety of different tubular scaffolds. Compton et al. [22] proposed a 3D printing technology to fabricate honeycomb composites in nanoclay sheets doped with epoxy resin filler, which can recover their original shape after external forces are released.

3. A Simple and Effective Modeling Method for 3D Irregular Structures

3.1. Problems of Existing Methods

It is a challenging task to establish the rich solid model of a product and give full play to the advantages of physical and mechanical properties [28]. A regular porous structure can be formed by linear or array combination of simple and primitive cells, while it is difficult to model the sizes and distributions of irregular 3D pores. Reverse modeling can only reconstruct the existing model, which is limited by stimuli and constraints and lacks flexibility. Although there are some methods for modeling 3D irregular porous structures, two significant drawbacks remain.

Figure 1 shows the problems of existing modeling methods for 3D porous objects and our improvements. Conventional methods obtain smooth pores from a convex surface including the Catmull-Clark subdivision based on triangular meshes and loop subdivision based on quadrilateral meshes [26,29]. A surface subdivision is a refinement scheme to recursively produce underlying inner mesh from a coarser polygonal mesh. It generates a new set of control vertices including face points, edge points, and vertices from the original control grid, and new points form the new control grid. This iteration will result in a more rounding surface. However, as shown in the red box on the porous structure Figure 1b which is generated from Figure 1a using the recursive subdivision algorithm, the Catmull-Clark subdivision may converge to its limit surface, which can lead to stress concentration. While our method can generate a smoother porous surface as shown in Figure 1c.

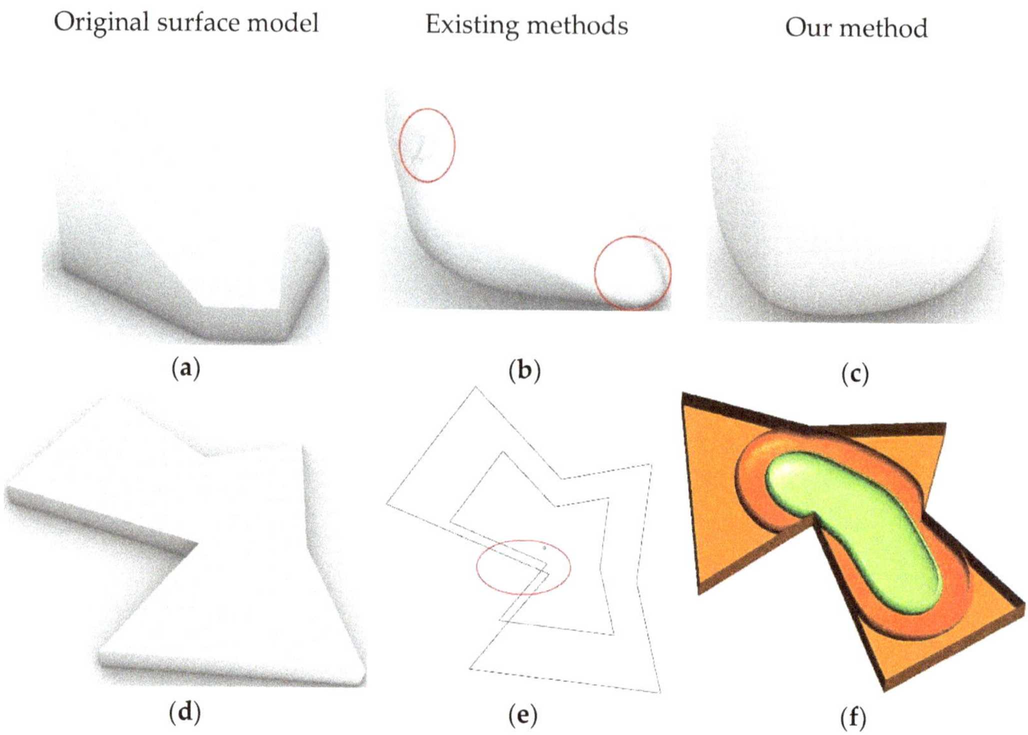

Figure 1. Problems of current 3D porous structure modeling methods and our improvements. (**a**) is a 3D surface model; (**b**) is the generated pore with the subdivision surface algorithm from the convex model (**a**) which leads to stress concentration phenomena; (**c**) is the generated pore from (**a**) with our proposed method; (**d**) is an extrusion surface model; (**e**) is the cross-scaling surface from the concave model (**d**); (**f**) is generated isothermal surfaces from (**d**) with our method. The different colors in (**f**) represent different isothermal surfaces, and the same color means that the geometric units in them have the same temperature.

To model porous structures, we need to build not only convex but also concave pores. We can obtain concave structures by simply merging convex polyhedrons. It needs to scale the original concave polyhedron to obtain concave pores with different porosities. As shown in Figure 1c, the scaled polyhedron may exceed the boundary of the original model. Therefore, the simple scaling approach is not feasible. While as shown in Figure 1f, the 3D polyhedrons generated by our method not only have a smoother surface but are all inside the original model.

Algorithm 1 shows the flow of our proposed porous structure modeling method. The whole flow consists of three procedures: Voronoi tessellation, concave porous structures generation, and isosurface extraction. In the next sections, we will detail our improvements.

Algorithm 1: Porous Structure Modeling Method.

Data: A 3D solid model M
Result: 3D Porous Structures$\{P_1, P_2 \ldots, P_n\}$
Distribute n sites randomly inside S
Tessellate with CVT and generate n Voronoi Units $(V_i, i \leq n)$
for *each* V_i **do**
  for *each* V_j **do**
    if V_i *and* V_j *are a pair* **then**
      $V_i \leftarrow Solid - Union \mathrm{S}_i$ and S_j
      Delete V_j
    end
  end
end
for *each* V_i **do**
  $C \leftarrow thecentroidsof \mathrm{V}_i$
  Extract the surface of $M \rightarrow$S
  $M_e \leftarrow Discretize \mathrm{M} witha3Dfiniteelementmeshgenerator$
  Define the EBC and assign T_h meshes inside S, T_l to meshes inside C, respectively
  Solve the poisson equation to generate the post-processing model P
  $V_{iin} \leftarrow Aiso - surfaceextracted from the predefined temperature from \mathrm{P}$
  $P_i \leftarrow Solid - Diffierence between S and \mathrm{V}_{iin}$
end

3.2. Voronoi Tessellation

The core of CAD for porous structures lies in the geometric representation of pore units (meta-structures). To generate porous structure from a single surface model, it needs to tessellate original models. We can find a variety of subdivision schemes for geometric design and graphics applications, where Voronoi tessellation is most widely used [18]. A Voronoi diagram is a tessellation where each polygon (also called Voronoi cell) represents the set of points closest to a central site. In the engineering field, the Voronoi diagram can simulate the foaming process of porous materials to study the various mechanical properties of porous materials. Voronoi diagrams can partition the space into compartments according to the location of the original points (also called sites). Let $S = \{s_i\}$ be a set of distinct sites in a connected compact region Ω. The Voronoi region/cell of Ω_i of s_i can be defined as:

$$\Omega_i = \{x \in \Omega | \|x - s_i\| \leq \|x - s_j\|, \forall i \neq j\} \quad (1)$$

If the region is restricted to a finite domain like a square, then all of these divisions are closed regions. However, the Voronoi facets obtained are irregular with a large gap between the shape and area of each facet, and the corresponding triangulation is not good.

To optimize the position of sites for getting a Voronoi facet with more consistent and better triangulation, we use the Centroidal Voronoi Tessellation (CVT) [30] and Lloyd's relaxation algorithm to make the shape and volume of the dissected polyhedron converge, that is:

$$s_i = c_i = \frac{\int_{\Omega_i} p(x) x d\sigma}{\int_{\Omega_i} p(x) d\sigma} \quad (2)$$

where $p(x)$ is a density function greater than 0, and $d\sigma$ is the area differential on Ω_i. This process can be simply understood as updating the sites by recalculating the midpoint of each facepiece, which repeatedly finds the centroid of each set in the partition and then re-partitions the input according to which of these centroids is closest. In this setting, the mean operation is an integral over a region of space, and the nearest centroid operation results. Lloyd relaxation redraws these polygons after moving the seed of each cell to the centroid of that cell. The iteration of this process causes the seeds to space out evenly and settle into a stable state. Finally, each site is moved to the centroid of its Voronoi cell.

3.3. Isosurface Extraction

After obtaining the cells, we need to subdivide the polyhedron to obtain smoother surfaces. As discussed in Section 3.1, Voronoi cells differ from natural objects and suffer from stress concentrations at the vertices and edges of surfaces. If we can get the isothermal surface information inside the Voronoi cell, we can use it as a boundary to get a smoother pore without introducing discontinuities along prescribed boundaries.

To extract the isosurface, we adopt the finite element method (FEM) [31]. FEM is a numerical method for solving partial differential equations (PDEs) based on the mathematical theory of the weak solution of PDEs. To solve a system of a PDE and compute a temperature distribution, we express the resulting weak form for Poisson's problem to find the unknown function $T \in V(\Omega)$ for any test function s such that:

$$\begin{cases} \int_{\Omega} c\nabla s \cdot \nabla t = 0, \ \forall s \notin (\partial\Omega \cup C) \\ T(t) = T_h, \ \forall t \in \partial\Omega_{out} \\ T(t) = T_l, \forall t \in \partial\Omega_{in} \end{cases} \tag{3}$$

where $\Omega \in R^n$ is the solution domain with the boundary $\partial\Omega = \partial\Omega_{out} \cup \partial\Omega_{in}$. $\partial\Omega$ are also called essential boundary conditions (EBC), $\partial\Omega_{in}$ is the defined cell's centroids which can be either a single point or a sequence of points. $\partial\Omega_{out}$ is the surface of the model. s should be zero on domains $\partial\Omega$. c is the material constant which does not influence the solution for our simulation. The boundary conditions place restrictions on the finite element formulation and result in a unique solution to the problem. We specify that a temperature of 100 is applied to the T_h and 0 to the T_l.

After defining the boundary conditions, we use the available finite element tools to perform meshing and finite element analysis. Finally, in the post-processing process, we extract all the meshes on the specified temperature values and stitch them together to make the boundary of a pore. since the polyhedron we use is closed, the final generated pore is also closed. We perform finer meshing to obtain smoother pores. Meanwhile, due to the smoothness of the isothermal surface, the constructed porous structure model eliminates the stress concentration phenomenon.

We can obtain a more adaptive isosurface by defining a more complex temperature distribution. When solving a Poisson equation of temperature diffusion, the temperature field is linear. If the isothermal surface is extracted directly from the temperature values, the porosity of the generated pore model is not linear. To address this issue, we extract the isothermal surface using the following transformation:

$$\begin{cases} iso_t = iso_{min} + (p_v)^{1/3} \times (iso_{max} - iso_{min}) \\ iso_{min} = T_l + \alpha \times (T_h - T_l) \\ iso_{max} = T_h \end{cases} \tag{4}$$

where p_v is the porosity prescribed by users and iso_t is the corresponding temperature value. Note that the temperature values obtained do not strictly correspond to p_v. We can also iterate one by one on the isothermal surface to invert the exact temperature value corresponding to the porosity exactly. However, modeling the exact internal geometric and topological structure of pore objects is often not necessary and this method requires iterations for each pore model, which will increase the computational afford. In contrast, the above equation provides a general approximation to extract isosurface from T_l. We truncate the lowest temperature value since the temperature field generated by the FEM solver is concentrated near the centroid sites, which will gradually thin out at the outer boundary because the volume is a cubic function. to remove the small isothermal surface near sites, we require the porosity of porous structures to be greater than 0.5, and we set $\alpha = 0.8$. We can also change the truncation threshold according to different requirements.

3.4. Concave Porous Structures

Voronoi tessellation can only produce convex polyhedrons [30]. Although concave polyhedrons can be obtained by merging multiple Voronoi cells, we still need to solve the problem of centroids of the merge concave pores. The proposed method allows users to specify the pairing relationship of neighboring Voronoi cells. If two or more adjacent Voronoi cells are specified as a pair, we treat them as one whole-cell and merge the paired convex structures by a Boolean-Union operation. In this case, if T_h is simply applied to the merged outer surface and T_l is applied to the separate center of the merged concave cell, the generated isothermal surface will be clustered near the center and deviate from the outer original non-adjacent surface because of the concave shape property, which will also create a stress concentration problem. As shown in Figure 2a, if we use the centroids of original cells directly, although the boundary is concave, the distribution of internal isothermal surfaces still obeys the convex cell, which does not meet our requirement of generating concave isothermal surfaces.

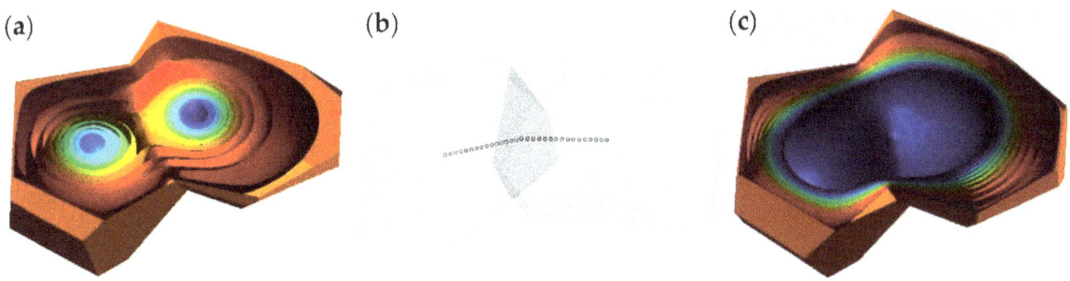

Figure 2. The construction of constraints for concave polyhedron. (**a**) is the concave surface and separate isothermal surfaces; (**b**) is generated centroids of a concave polyhedron from origin centroids and the common face; (**c**) is extracted isothermal surfaces using the EBC from (**b**).

When adjusting the porosity, the scaled concave polygon may exceed the original boundary and overlap with other concave pores. It is necessary to adjust the scaled concave polygon boundary to confine it inside the original surface and ensure that each pore does not overlap. To improve the design freedom of pores and better simulate the porous structure, the outsourced polyhedron of the concave pore structure can be obtained by combining multiple neighboring cells. We extract the boundary of each paired Voronoi diagram and remove the duplicated faces to obtain the concave pore contour after Voronoi tessellation. To make the generated isothermal surface fit better with the outer surface, we calculate the center C_0 of the paired adjacent face, and then generate the line segment C_1C_0 and C_2C_0, where C_1 and C_2 are original centroids of paired Voronoi cells, respectively. As shown in Figure 2b, we sample the two-line segments to generate multiple centers $C_l = tC_iC_0$, $i = \{1,2\}$. In the finite element analysis stage, we apply T_l to the all mesh containing C_l. As shown in Figure 2c, the generated isothermal surface is closer to the outsourcing profile. We can generate more realistic pores by simply resampling sites.

4. Modeling Experiments

In this section, we will perform several kinds of modeling experiments to verify the effectiveness of the proposed method. The proposed method benefits from the mature and widely used methods including Voronoi tessellation and FEM, which makes our method easy to implement with existing software and tools. During the solid model construction and Voronoi tessellation, we utilize the *Grasshopper* [32] tool, which is a graphical algorithm editor plugin for Rhino. For meshing, we use the *Gmsh* [33] tool. *Gmsh* is an open-source 3D finite element mesh generator whose fast and lightweight parametric features facilitate high-quality network delineation. We use the Python finite element library *sfepy* [34] to compute the thermal map. *sfepy* is a flexible Python finite element analysis toolkit for the rapid implementation and testing of finite element models in research. These tools support secondary development based on Python, so the whole modeling process of porous structures can be automated using the Python language, thus improving modeling efficiency. Note that although thermodynamic diffusion is a time-series problem, we do not need to define a time-stepping solver in *sfepy*, because the isothermal problem is stationary and the default solver can solve it once.

As shown in Figure 3a, the isosurfaces generated by FEM are consistent with the scaled sphere. It infers the accuracy of the proposed method. The different color surfaces of Figure 3b,c represent the isosurfaces of polyhedrons, where we sample fewer isosurfaces to see the extracted isothermal surface more clearly. Figure 3d shows the isosurfaces from a concave polyhedron. Figure 3e,f are the resulting pore structures with different porosities. By extracting isothermal surfaces at different temperatures such as 99 and 95, we can obtain pore structures with the required porosity. For convex and concave structures, the closer to the center, the more spherical the isosurface tends to be. At the boundary, the isosurface is the fusion of polyhedron and enveloping sphere.

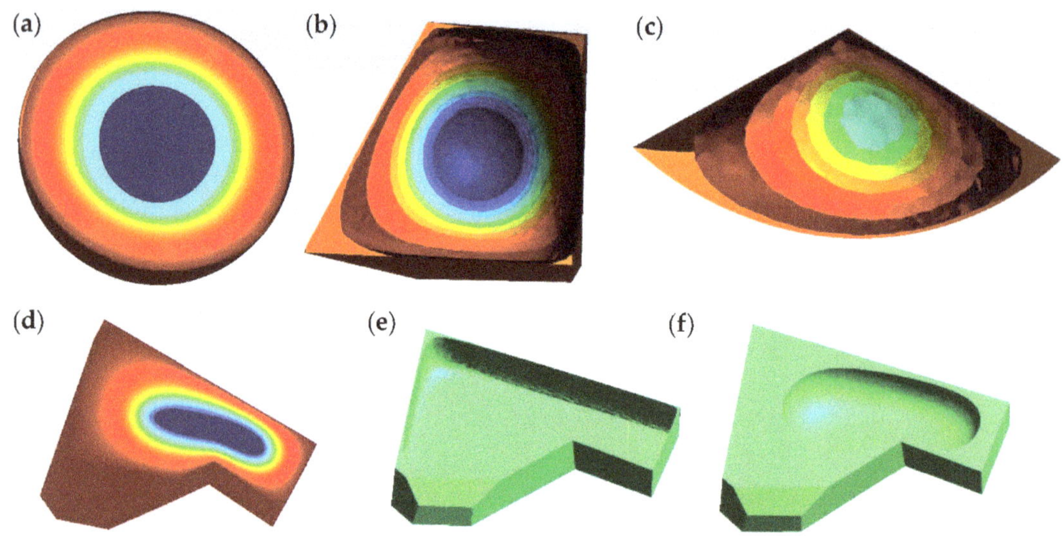

Figure 3. Isosurfaces generated by convex polyhedrons. (**a**) sphere; (**b**) polyhedron; (**c**) section polyhedron; (**d**) concave polyhedron; (**e**) is a pore generated from the 99 isothermal surfaces. (**f**) is a pore generated from the 95 isothermal surfaces. The different color represents different isosurfaces (isothermals).

Figure 4 shows the process of generating porous structures from cuboid and more complex kitten models. The blue, light green, and dark yellow colors in Figure 4g represent the corresponding 99.9, 99, and 98 isothermals, respectively. Although the kitten model has a complex surface, the generated pores are still smooth and natural as shown in Figure 4d,f. From the modeling experiments above, we can conclude that we can obtain both smooth and approximate porous structures from any irregular 3D structures with the method proposed in this paper.

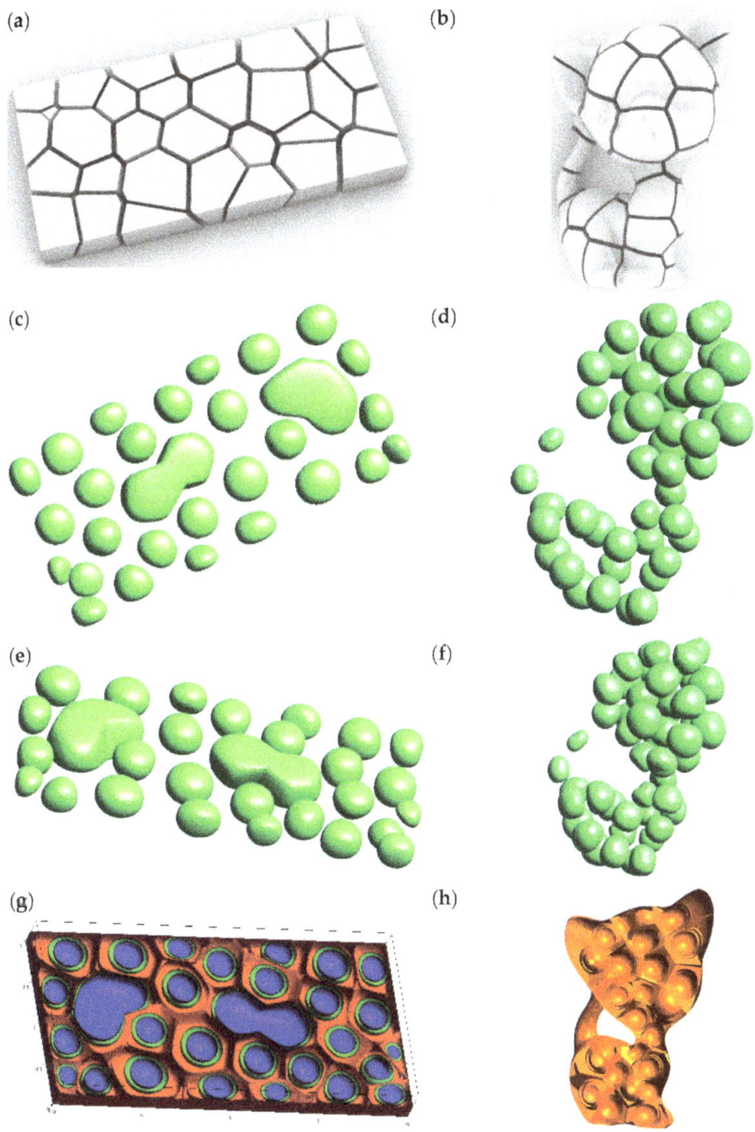

Figure 4. Porous structures generated from cuboid and kitten surfaces. (**a**) is a tessellated cuboid Voronoi cells; (**b**) a tessellated cat Voronoi cells; (**c**) is generated pores from (**a**) with 98 isothermals; (**d**) is generated pores from (**b**) with 98 isothermals; (**e**) is generated pores from (**a**) with 99 isothermals; (**f**) is generated pores from (**b**) with 99 isothermals; (**g**) are resulting porous structures for (**a**); (**h**) are resulting porous structures for (**b**).

5. Conclusions and Future Work

Recently, porous structures have become widely used in the fields of aviation, mechatronics, biology, and mechanics. The ability to design new products relies on the representations of models. However, intricate inner components, which enable excellent physical properties and mechanical characteristics in porous structures, pose new challenges for traditional CAD methods. In this paper, we combine the Voronoi tessellation and FEA methods to propose a novel modeling method called MPFEM to generate irregular and smooth 3D pores from polyhedrons. Different from previous modeling methods aimed at 2D geometric representation of convex structures, our method can process both 3D and complex concave structures. MPFEM can generate convave porous structures which are combined adjacent Voronoi cells and smooth surfaces which estimate a stress concentration. A series of convex, concave, and complex polyhedron modeling experiments show that this proposed method is easy and feasible, and the results show good performance in terms of smoothness and fitness of geometric representation.

As future work, first, although the purpose of this paper is modeling solids of porous structures, we can extend it to other fields such as the current popular functionally graded material (FGM). Since the generated isothermal surfaces are smoother, the problem of mismatching material interface properties can be further eliminated. By distributing different materials on different isothermal surfaces, the proposed method can also provide a new idea for the modeling of FGM. Second, through an evolutionary or neural network (NN) [35] algorithms to obtain the distribution, combination, and porosity of porous structures under specific load and boundary constraints, we can realize function designs like the lightweight. Since the optimization goal becomes site locations and porosities, which eliminates model meshing in traditional FEA, our method will significantly improve design efficiency. Last but not least, we will utilize the AM including direct light processing (DLP) to fabricate generating 3D porous models.

Author Contributions: Conceptualization, investigation, methodology, and writing: D.Z.; validation and data curation: L.R. All authors have read and agreed to the published version of the manuscript.

Funding: This work is supported in part by the National Key Research and Development Program of China (2019YFB1706003), and the Natural Science Foundation of Guangdong Province, China (grant number 2214050004382).

Institutional Review Board Statement: Not applicable.

Informed Consent Statement: Not applicable.

Data Availability Statement: The data presented in this study are available on request from the corresponding author.

Conflicts of Interest: The authors declare that they have no known competing financial interests or personal relationships that could have appeared to influence the work reported in this paper.

References

1. Gao, Z.; Nguang, S.K.; Kong, D.-X. Advances in Modelling, Monitoring, and Control for Complex Industrial Systems. *Complexity* **2019**, *2019*, 2975083. [CrossRef]
2. Liu, W. Progress in additive manufacturing on complex structures and high-performance materials. *J. Mech. Eng.* **2019**, *55*, 128–151. [CrossRef]
3. Zhang, D.; Zhao, Z.; Zhou, Y.; Guo, Y. A novel complex network-based modeling method for heterogeneous product design. *Clust. Comput.* **2019**, *22*, 7861–7872. [CrossRef]
4. Kim, S.-H.; Nack, J.; Hansoo, K. Brittle intermetallic compound makes ultrastrong low-density steel with large ductility. *Nature* **2015**, *518*, 77–79. [CrossRef]
5. Sigmund, O. Systematic design of metamaterials by topology optimization. In Proceedings of the IUTAM Symposium on Modelling Nanomaterials and Nanosystems, Aalborg, Denmark, 19–22 May 2009. [CrossRef]
6. Gupta, V. CAD modeling and rapid manufacturing of heterogeneous objects: A review. *Int. J. Tech. Res. Appl.* **2014**, *3*, 30–35.
7. Li, B.; Fu, J.; Feng, J.; Shang, C.; Lin, Z. Review of heterogeneous material objects modeling in additive manufacturing. *Vis. Comput. Ind. Biomed.* **2020**, *3*, 1–18. [CrossRef]
8. Dichen, L.I. Additive Manufacturing: Integrated Fabrication of Macro/Microstructures. *J. Mech. Eng.* **2013**, *49*, 129. [CrossRef]
9. Denghui, Z.; Guo, Y.; Zhao, Z.; Zhou, Y. Co-authorship Networks in Additive Manufacturing Studies Based on Social Network Analysis. *Br. J. Appl. Sci. Technol.* **2016**, *15*, 1–6. [CrossRef]
10. Haleem, A.; Javaid, M. Additive Manufacturing Applications in Industry 4.0: A Review. *J. Ind. Integr. Manag.* **2019**, *4*, 1930001. [CrossRef]
11. Nagarajan, B.; Hu, Z.; Song, X.; Zhai, W.; Wei, J. Development of Micro Selective Laser Melting: The State of the Art and Future Perspectives. *Engineering* **2019**, *5*, 702–720. [CrossRef]

12. Gao, L.; Shen, W.; Li, X. New Trends in Intelligent Manufacturing. *Engineering* **2019**, *5*, 619–620. [CrossRef]
13. Gao, Z.; Kong, D.; Gao, C. Modeling and Control of Complex Dynamic Systems: Applied Mathematical Aspects. *J. Appl. Math.* **2012**, *2012*, 869792. [CrossRef]
14. Rosen, D.W. Thoughts on Design for Intelligent Manufacturing. *Engineering* **2019**, *5*, 609–614. [CrossRef]
15. Wang, L. From Intelligence Science to Intelligent Manufacturing. *Engineering* **2019**, *5*, 615–618. [CrossRef]
16. Wang, W.; Wang, T.Y.; Yang, Z.; Liu, L.; Tong, X.; Tong, W.; Deng, J.; Chen, F.; Liu, X. Cost-effective printing of 3D objects with skin-frame structures. *ACM Trans. Graph. (TOG)* **2013**, *32*, 1–10. [CrossRef]
17. Koeppe, A.; Padilla, C.A.H.; Voshage, M.; Schleifenbaum, J.H.; Markert, B. Efficient numerical modeling of 3D-printed lattice-cell structures using neural networks. *Manuf. Lett.* **2018**, *15*, 147–150. [CrossRef]
18. Martínez, J.; Dumas, J.; Lefebvre, S. Procedural voronoi foams for additive manufacturing. *ACM Trans. Graph.* **2016**, *35*, 1–12. [CrossRef]
19. Liu, L.G.; Xu, W.P.; Wang, W.M.; Yang, Z.W.; Liu, X.P. Survey on geometric computing in 3D printing. *Chin. J. Comput.* **2015**, *38*, 1243–1267. [CrossRef]
20. Andrews, J.; Jin, H.; Sequin, C. Interactive Inverse 3D Modeling. *Comput.-Aided Des. Appl.* **2013**, *9*, 881–900. [CrossRef]
21. Kou, X.Y.; Tan, S.T. Heterogeneous object modeling: A review. *Comput. Aided Des.* **2007**, *39*, 284–301. [CrossRef]
22. Compton, B.G.; Lewis, J.A. 3D-printing of lightweight cellular composites. *Adv. Mater.* **2014**, *26*, 5930–5935. [CrossRef] [PubMed]
23. Gan, Z.; Li, H.; Wolff, S.J.; Bennett, J.L.; Hyatt, G.; Wagner, G.J.; Cao, J.; Liu, W.K. Data-Driven Microstructure and Microhardness Design in Additive Manufacturing Using a Self-Organizing Map. *Engineering* **2019**, *5*, 730–735. [CrossRef]
24. Lu, H.Z.L.; Wei, A.S.Y.; Fan, Q.; Chen, X.; Savoye, Y.; Tu, C.; Cohen-Or, D.; Chen, B. Build-to-last: Strength to weight 3D printed objects. *ACM Trans. Graph. (Proc. SIGGRAPH)* **2014**, *33*, 1–10. [CrossRef]
25. Xu, W.; Wang, W.; Li, H.; Yang, Z.; Liu, L. Topology Optimization for Minimal Volume in 3D Printing. *J. Comput. Res. Dev.* **2015**, *52*, 38. [CrossRef]
26. Kou, X.Y.; Tan, S.T. A simple and effective geometric representation for irregular porous structure modeling. *Comput.-Aided Des.* **2010**, *42*, 930–941. [CrossRef]
27. Liu, K.; Jiang, L. Bio-inspired design of multiscale structures for function integration. *Nano Today* **2011**, *6*, 155–175. [CrossRef]
28. Zhang, D.; Zhou, Y.; Guo, Y. A modeling and optimization method for heterogeneous objects based on complex networks theory. *Clust. Comput.* **2019**, *22*, 2645–2654. [CrossRef]
29. Ma, W. Subdivision surfaces for CAD—An overview. *Comput.-Aided Des.* **2005**, *37*, 693–709. [CrossRef]
30. Wang, X.; Ying, X.; Liu, Y.-J.; Xin, S.-Q.; Wang, W.; Gu, X.; Mueller-Wittig, W.; He, Y. Intrinsic computation of centroidal Voronoi tessellation (CVT) on meshes. *Comput.-Aided Des.* **2015**, *58*, 51–61. [CrossRef]
31. Tornabene, F.; Fantuzzi, N.; Ubertini, F.; Viola, E. Strong formulation finite element method based on differential quadrature: A survey. *Appl. Mech. Rev.* **2015**, *67*, 020801. [CrossRef]
32. Roudsari, M.S.; Pak, M.; Smith, A. Ladybug: A parametric environmental plugin for grasshopper to help designers create an environmentally-conscious design. In Proceedings of the 13th International IBPSA Conference, Lyon, France, 26–28 August 2013; pp. 3128–3135.
33. Geuzaine, C.; Remacle, J.-F. Gmsh: A 3-D finite element mesh generator with built-in pre- and post-processing facilities. *Int. J. Numer. Methods Eng.* **2009**, *79*, 1309–1331. [CrossRef]
34. Cimrman, R.; Lukeš, V.; Rohan, E. Multiscale finite element calculations in Python using SfePy. *Adv. Comput. Math.* **2019**, *45*, 1897–1921. [CrossRef]
35. Villarrubia, G.; de Paz, J.F.; Chamoso, P.; la Prieta, F.D. Artificial neural networks used in optimization problems. *Neurocomputing* **2018**, *272*, 10–16. [CrossRef]

processes

Article

Modelling Method and Application of Anti-Corrosion Pill Particles in Oil and Gas Field Wellbore Casing Annulus Based on the Discrete Element Method

Dongtao Liu [1,*], Chunshang Qiao [1], Jun Wan [2], Yuliang Lu [1], Jiming Song [1], Zhenhe Yao [1], Xinjie Wei [1] and Yajun Yu [3,*]

1 CNOOC EnerTech-Drilling & Production Co., Shenzhen 518067, China; qiaochsh@cnooc.com.cn (C.Q.); luyl@cnooc.com.cn (Y.L.); songjm2@cnooc.com.cn (J.S.); yaozhh@cnooc.com.cn (Z.Y.); weixj6@cnooc.com.cn (X.W.)
2 Shenzhen Branch, CNOOC (China) Co., Ltd., Shenzhen 518067, China; wanjun2@cnooc.com.cn
3 School of Biological and Agricultural Engineering, Jilin University, Changchun 130022, China
* Correspondence: liudt@cnooc.com.cn (D.L.); yuyajun@jlu.edu.cn (Y.Y.); Tel.: +86-138-2879-7030 (D.L.); +86-151-4317-3701 (Y.Y.)

Abstract: This study uses a self-developed anti-corrosion pill particle as the research object and develops the pill particle population modelling method in order to optimize the anti-corrosion process of oil and gas wellbore casing annuli. The shape of the pill particle is similar to a cylinder, according to the test and analysis of geometrical characteristics, and can be simplified into three types based on height, namely pill particles A (5.4 mm), B (5.8 mm), and C (6.2 mm). The multi-sphere approach is then used to create models of three different types of pill particles with varying degrees of precision. The feasibility and effectiveness of the modelling method for pill particle populations are proven by comparing the simulation results of the bulk density test and the angle of repose test. The results show that the 12-sphere models of pill particles A, B, and C are accurate representations of genuine pill particle morphologies and are adequate for simulating particle mechanics and flow processes. The applicability and practical use of the modelling method are then demonstrated using an example of a self-designed pill particle discharging mechanism. The results show that the modelling method can accurately simulate the pill discharging process and provide an accurate simulation model and theoretical basis for the optimization of the structural parameters, dimension parameters, and operating parameters of the discharging device.

Keywords: discrete element method; particle modelling; multi-sphere method; simulation analysis; anti-corrosion pill particle

Citation: Liu, D.; Qiao, C.; Wan, J.; Lu, Y.; Song, J.; Yao, Z.; Wei, X.; Yu, Y. Modelling Method and Application of Anti-Corrosion Pill Particles in Oil and Gas Field Wellbore Casing Annulus Based on the Discrete Element Method. *Processes* **2022**, *10*, 1164. https://doi.org/10.3390/pr10061164

Academic Editor: Zhiwei Gao

Received: 2 May 2022
Accepted: 8 June 2022
Published: 9 June 2022

Publisher's Note: MDPI stays neutral with regard to jurisdictional claims in published maps and institutional affiliations.

1. Introduction

The anti-corrosion of oil and gas field wellbore casing annuli is a critical link in the integrity management of oil and gas field wellbores. Traditional anti-corrosion fluid perfusion is harmed by the tiny area and buildup of oil sludge, and it has a number of disadvantages, including difficulties in lowering the pipeline and restricted anti-corrosion fluid perfusion. Our organization has created a solid slow-release anti-corrosion pill particle for this purpose. However, due to the complexity of the working conditions, such as a single route for pill discharging and a small inner diameter space for the casing, optimizing the pill discharging device to minimize clogging and fragmentation of the pill discharging process and to increase its stability and uniformity has become a critical technical bottleneck to overcome [1–3]. The discrete element method (DEM) [4], based on particle dynamics, is capable of analyzing the interaction between particles and between particles and mechanical components from a microscopic perspective, obtaining microscopic data such as particle displacement, velocity, acceleration, and so on, and revealing the influencing factors and working mechanism of mechanical components. Thus, DEM has developed into a potent tool for optimizing mechanism design and overcoming technological constraints in industrial and agricultural output [5–10].

Particle modelling, being the central issue in DEM, is a critical aspect in determining the accuracy of simulation results. Currently, the polyhedron method [11], super-quadric equation method [12], and multi-sphere method [13–15] are the most frequently utilized particle modelling methods. The multi-sphere approach is more extensively utilized in industrial and agricultural production due to the simplicity of the contact detection algorithm. For example, Danesh et al. [16] used a multi-sphere method to create ballast particle shapes to investigate the macroscopic and microscopic mechanics shear behavior of railroad ballast and demonstrated that the particle shape has an effect on the formation of shear bands via the microscopic mechanics response. Tekeste et al. [17] modeled the soil particles using a multi-sphere approach and developed a proportionate connection between the soil response force and the bulldozer blade length. Horabik et al. [18] constructed one-, three-, four-, and six-sphere models for wheat kernels using a multi-sphere method and demonstrated that the four-sphere model could better reproduce the transmission of particle–particle forces from the vertical to the lateral directions through loading and unloading cyclic compression tests. Zeng et al. [19] used the multi-sphere approach to construct a 16-sphere model of rice kernels and disclosed the rice crushing mechanism by the simulated study of the rice milling process. Due to the manufacturing process, the size of the pill particle generated by our organization varies. Thus, how to examine the shape and size characteristics of pill particles through testing, taking into account the unpredictability of their scale distribution, and establish a technique for modelling the pill particle population must be thoroughly investigated.

On the other hand, validation of the modelling method is a critical problem that must be addressed. At the moment, the majority of modelling approaches are verified by simulating static and dynamic experimental processes [20–22]. For example, Zhang et al. [23] utilized the multi-sphere technique to create four different kinds of soil particle models and then analyzed the operation of a deep loosening machine working on the soil using a combined DEM–MBD simulation, demonstrating the modelling method's utility. Tao et al. [24] employed a multi-sphere technique to construct three representations of ellipsoidal particles, and the modelling method's practicality was shown by moving bed experiments. Zhou et al. [25] used a multi-sphere method to model four different shapes of maize seed particles, including the horse-tooth shape, the truncated triangular pyramid shape, the ellipsoid cone shape, and the spheroid shape, and validated the proposed modelling method for maize seeds using the bulk density, the angle of repose, and self-flow screening. Sun et al. [26] used the multi-sphere method to construct a double-ellipsoidal 13-, 17-, 25-, and 33-sphere model for wheat seeds, as well as a single-ellipsoidal 5-, 9-, 13-, 17-, and 21-sphere model for wheat seeds. They validated the wheat seed particle model proposed in this paper by comparing experimental data to simulation results using the static angle of repose, self-flow screening, and dynamic angle of repose tests The proposed pill particle population modelling method requires an in-depth study of how to select an appropriate validation method to reveal the effect of the number of filled spheres on the particle population accumulation process and flow behavior, as well as to verify the modelling method's feasibility, validity, and applicability.

In order to address the aforementioned issues, this work uses self-developed anti-corrosion pill particles as the research object, measuring and analyzing their geometric form and dimension characteristics. On this premise, the multi-sphere approach was used to build particle models of three different forms of pill particles. The bulk density and angle of repose simulation results were compared to actual data to determine the feasibility and usefulness of the suggested approach for modelling the pill particle population in this work. Finally, the modelling method's applicability and practical use are shown using the self-designed pill discharge device as an example. The work presented in this article establishes an accurate simulation model and theoretical foundation for the future investigation of the pill discharging process and optimization of the pill discharging device's structural characteristics, dimensional parameters, and operating parameters.

2. Materials and Modelling

2.1. Shape and Size Analysis

The self-developed annular air anti-corrosion pill particle is used as the research object, with 200 pill particles randomly chosen. The geometric forms and size characteristics of pill particles were determined, as well as the mass ratio of various shapes of pill particles. The findings indicated that the pill particle was approximately cylindrical in shape, as shown in Figure 1a, with a constant diameter (D) and a nearly uniform height (H) dispersed according to three scales. In the particle population, the mass ratios of pill particles A, B, and C are 35%, 35%, and 30%, respectively. Three scales of pill particles are modeled in this article using DEM and distributed according to their mass ratio in order to build a technique for modelling pill particle populations.

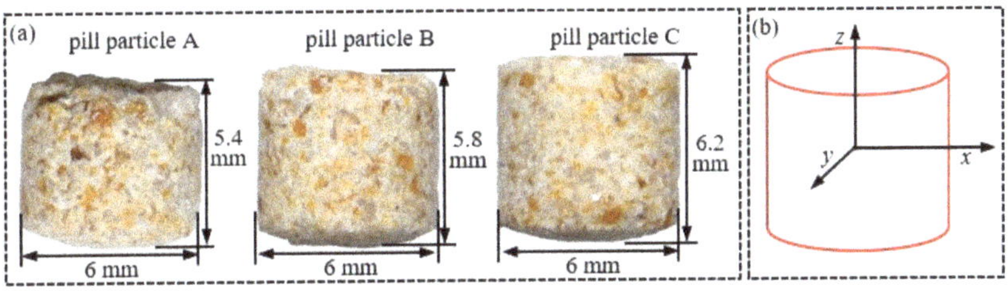

Figure 1. The form and coordinate system of pill particles: (**a**) Three different forms of pill particles and their associated size characteristics; (**b**) The pill particle's coordinate system.

2.2. Modelling Methods of Pill Particles

The pill particles are modeled in this article using a multi-sphere technique. Single-layer, double-layer, and triple-layer particles are used to characterize the pill particle form in order to analyze the influence of the number of filled spheres on the modeling accuracy and flow properties of the pill particles. The coordinate system of the pill particles is shown in Figure 1b, where the coordinate origin is the mass center of the particle, the x- and z-axis are the diameter and height directions of the particle, respectively, and the *y*-axis is defined by the right-hand rule.

2.2.1. Modelling Methods for Pill Particle A and B

Because the particle diameter is smaller than the height for particles A and B, a 4-sphere model is generated using the particle diameter as the filling sphere's diameter. Based on the tangency and overlap of the particles, a 12-sphere double layer model and a 20-sphere triple layer model are created. Each layer of the triple layer model fills 6 spheres, with 2 more filled spheres in the centre of the top and bottom layers. The detailed modelling method is described below.

The point (D/2−H/2, 0) in the xoz coordinate plane is filled by a sphere O_1 of radius H/2 and tangent to the cylindrical surface to the x axis, while spheres O_2 to O_4 are generated by an array of spheres O_1 arranged at a 90° angle around the z axis in the oxyz coordinate system, as shown in Figure 2a.

To minimize inaccuracy, the pill particle model was constructed in the z-axis direction using a double-layer filled sphere. At the point (D/2H/3, H/6) in the xoz coordinate plane, a sphere O_1 with radius H/3 and tangent to the cylindrical surface to the x axis fills the space; spheres O_2 to O_6 are formed by an array of spheres O_1 arranged at a 60° angle around the z axis in the oxyz coordinate system. The spheres O_7 to O_{12} are filled with the mirror image of the spheres O_1 to O_6, forming the 12-sphere model of the pill particle A depicted in Figure 2b.

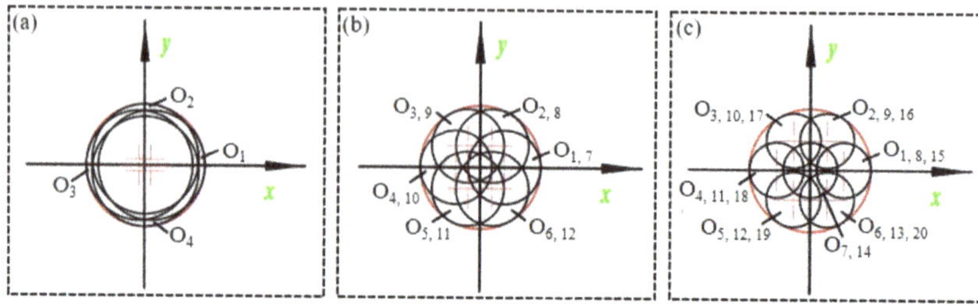

Figure 2. Diagram illustrating the procedure of filling a pill particle A: (**a**) A 4-sphere model; (**b**) A 12-sphere model; and (**c**) A 20-sphere model.

To further decrease inaccuracy, three layers of filled spheres were used to simulate the pill particles in the z-axis direction. At the point (D/2−H/4, H/4) in the xoz coordinate plane, a sphere O_1 with radius H/4 and tangent to the cylindrical surface to the x axis fills the space; spheres O_2 to O_6 are formed by an array of spheres O_1 arranged at a 60° angle around the z axis in the oxyz coordinate system. A sphere O_7 of radius H/4 and tangent to the cylindrical surface to the z axis is filled at the position (0, H/4) in the xoz coordinate plane. The spheres O_8 to O_{14} are filled with the mirror image of the spheres O_1 to O_7 in the xoy coordinate plane. The point (D/2−H/4, 0) in the xoz coordinate plane is filled by a sphere O_{15} with radius H/4 and tangent to the cylindrical surface to the x axis, while spheres O_{16} to O_{20} are generated by an array of spheres O_{15} arranged at a 60° angle around the z axis in the oxyz coordinate system, as shown in Figure 2c.

As seen in Figure 3, the same filling procedure was employed to model the pill particles B.

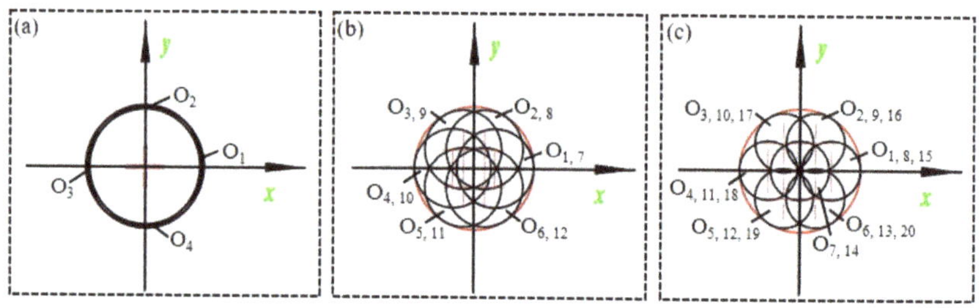

Figure 3. Diagram illustrating the procedure of filling a pill particle B: (**a**) A 4-sphere model; (**b**) A 12-sphere model; and (**c**) A 20-sphere model.

2.2.2. Modelling Methods for Pill Particle C

Because the particle diameter is greater than the height for particle C, a 2-sphere model with the particle height as the filled ball's diameter is developed. The tangency and overlap of the particles are the foundations of the 12-sphere double layer and 18-sphere triple layer models.

At the point (0, H/2−D/2) in the xoz coordinate plane, a sphere O_1 with radius D/2 and tangent to the cylindrical surface parallel to the *z* axis is filled; spheres O_2 are filled with the mirror image of spheres O_1 in the xoy coordinate plane, thereby constructing the two-sphere model of the pill particle C shown in Figure 4a.

At the point (D/2−H/3, H/6) in the xoz coordinate plane, a sphere O_1 with radius H/3 and tangent to the cylindrical surface to the *x* axis is filled; spheres O_2 to O_6 are formed by an array of spheres O_1 arranged around the *z* axis at an angle of 60° in the oxyz coordinate system. As illustrated in Figure 4b, the 12-sphere model of the pill particle C is constructed by filling spheres O_7 to O_{12} with the mirror image of spheres O_1 to O_6 with respect to the xoy coordinate plane.

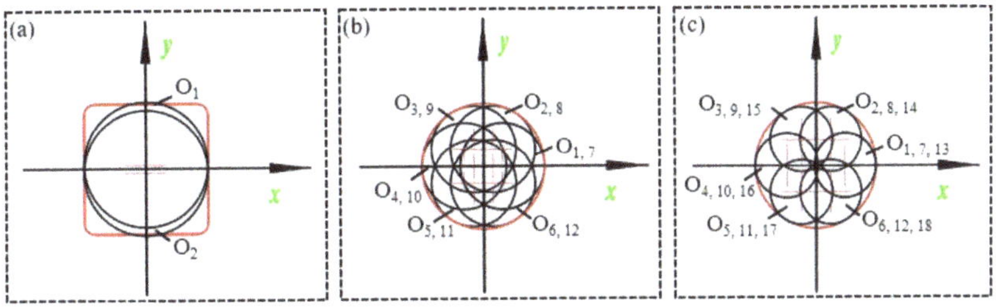

Figure 4. Diagram illustrating the procedure of filling a pill particle c: (**a**) A 2-sphere model; (**b**) A 12-sphere model; and (**c**) A 18-sphere model.

To further decrease inaccuracy, three layers of filled spheres were used to simulate the pill particles in the *z*-axis direction. At the point (D/2−H/4, H/4) in the xoz coordinate plane, a sphere O_1 with radius H/4 and tangent to the cylindrical surface to the x axis fills the space; spheres O_2 to O_6 are formed by an array of spheres O_1 arranged at a 60° angle around the z axis in the oxyz coordinate system. The spheres O_7 to O_{12} are filled with the mirror image of the spheres O_1 to O_6 in the xoy coordinate plane. The point (D/2−H/4, 0) in the xoz coordinate plane is filled by a sphere O_{13} with radius H/4 and tangent to the cylindrical surface to the x axis, and spheres O_{14} to O_{18} are generated by an array of spheres O_{13} arranged at a 60° angle around the z axis in the oxyz coordinate system, thereby generating the 18-sphere model of the pill particle C shown in Figure 4c.

Figure 5 illustrates the 4-, 12-, and 20-sphere particle models of the pill particles A and B, the pill particle C's 2-, 12-, and 18-sphere particle models.

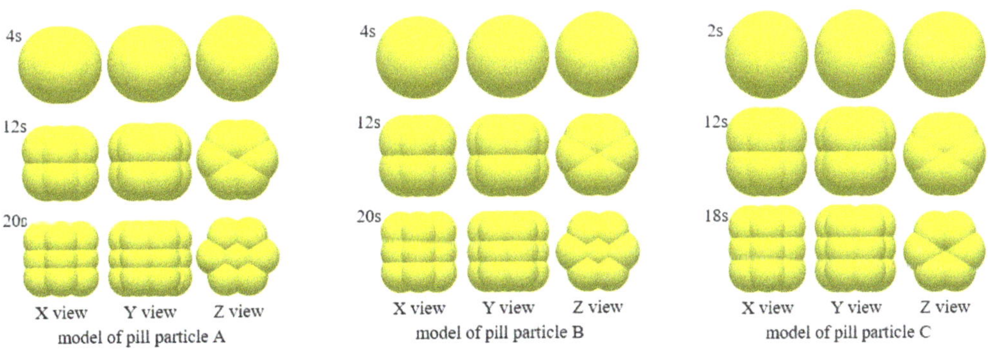

Figure 5. Three pill particle numerical models.

3. Experimental Verification and Simulation Analysis

The feasibility and effectiveness of the pill particle population modelling approach are shown in this study by a comparison of simulation findings for the bulk density and angle of repose tests to experiment data.

3.1. Simulation Model

In this paper, the Hertz–Mindlin contact model is used to simulate the bulk density test and the angle of repose test based on the material and physical properties of the pill particle. This contact model is one of DEM's most fundamental and well-known models, and it is commonly used to study the contact and mechanical behavior of non-viscous particles [27–30]. The motion of a particle is solved by the basic theories adopted by DEM, such as Newton's second law, Euler's equation, and dynamic relaxation method [30–33], given by

$$m_i \frac{d\mathbf{v}_i}{dt} = \sum_j (\mathbf{F}_{ij}^n + \mathbf{F}_{ij}^s) + m_i \mathbf{g} \tag{1}$$

and

$$I_i \frac{d\omega_i}{dt} = \sum_j \left(\mathbf{R}_i \times \mathbf{F}_{ij}^s - \mu_r R_i \left| \mathbf{F}_{ij}^n \right| \hat{\omega}_i \right) \tag{2}$$

where $\mathbf{v}_i$, ω_i and I_i are the translational and angular velocities, and moment of inertia of particle i, respectively. $\mathbf{R}_i$ is a vector running from the centre of the particle to the contact point with its magnitude equal to particle radius R_i. μ_r is the coefficient of rolling friction. $\mathbf{F}_{ij}^n$ and $\mathbf{F}_{ij}^s$ are the normal contact force and tangential contact force between particle i and j, given by

$$\mathbf{F}_{ij}^n = \left(\frac{4}{3} E^* \sqrt{R^*} \delta_n^{\frac{3}{2}} - 2\sqrt{\frac{5}{6}} \beta \sqrt{S_n m^*} (\mathbf{v}_{ij} \cdot \hat{\mathbf{n}}_{ij}) \right) \hat{\mathbf{n}}_{ij} \tag{3}$$

and

$$\mathbf{F}_{ij}^s = \min \left(-S_t \delta_t - 2\sqrt{\frac{5}{6}} \beta \sqrt{S_t m^*} (\mathbf{v}_{ij} \cdot \hat{\mathbf{s}}_{ij}), \mu_s \mathbf{F}_{ij}^n \right) \hat{\mathbf{s}}_{ij} \tag{4}$$

where the equivalent Young's Modulus E^* and the equivalent radius R^* are defined as $E^* = \left[\left(1 - v_i^2\right)/E_i + \left(1 - v_j^2\right)/E_j \right]^{-1}$ and $R^* = \left[1/R_i + 1/R_j \right]^{-1}$ with E_i, v_i, R_i and E_j, v_j, R_j, being the Young's Modulus, Poisson ratio, and Radius of particles i and j, respectively; δ_n is the normal overlap; the damping factor β, the normal stiffness S_n and the equivalent mass m^* are given by $\beta = -\ln e / \sqrt{\ln^2 e + \pi^2}$, $S_n = 2E^* \sqrt{R^* \delta_n}$ and $m^* = \left[1/m_i + 1/m_j \right]^{-1}$ with e, m_i, and m_j being the coefficient of restitution and the mass of each particle in contact; $\mathbf{v}_{ij}$ is the relative velocity between particle i and j; the unit vector $\hat{\mathbf{n}}_{ij}$ is calculated as $\hat{\mathbf{n}}_{ij} = (\mathbf{R}_i - \mathbf{R}_j) / \left| \mathbf{R}_i - \mathbf{R}_j \right|$; the tangential stiffness S_t is given by $S_t = 8G^* \sqrt{R^* \delta_n}$ with G^* being the equivalent shear modulus; δ_t is the tangential overlap; μ_s is the coefficient of static friction; $\hat{\mathbf{s}}_{ij}$ is the unit tangent vector.

3.2. Measurement and Calibration of Physical and Mechanical Parameters

The determination of simulation parameters has a significant impact on the stability and accuracy of the simulation calculation, according to the literature and preliminary research [34,35]. Hence, to verify the correctness of the simulation test findings, it is necessary to correctly test the physical and mechanical characteristics of the pill particles. The ASAE standard specified the Poisson's ratio of the pill particles [36]. The density of the pill particles was determined using the hydrometer technique [37]. The elastic modulus of the pill particles was determined using a compression test on an electronic universal testing equipment [38]. The coefficients of static friction between the pill particles and between the pill particles and the wall surface were determined using the slope technique [39] (ABS plastic, organic glass, and galvanized steel plate). The restitution coefficients between pill particles and between pill particles and the wall surface were determined using a high-speed camera and a drop test and a single pendulum test, respectively [40].

Due to the non-sphere surface solid nature of the pill particles, it is impossible to directly quantify the coefficients of rolling friction between the pill particles and between the pill particle and the wall surface. This research calibrates the simulation parameters using an angle of repose test and simulation in order to get them closer to their actual values. The findings indicate that the angle of repose is unaffected by the coefficient of rolling friction between the pill particle and the wall surface. As a result, the paper's first selection of the coefficient of rolling friction is between pill particle and ABS plastic, organic glass, and galvanized steel plate. The coefficient of rolling friction between the pill particles was then established by comparing the angle of repose test results to those obtained from the simulation. Table 1 contains the simulation parameters.

Table 1. Parameter selection for simulation.

Parameters	Pill Particle	ABS Plastic	Organic Glass	Galvanized Steel
Density ρ, kg/m^3	1380	1050	1800	7865
Poisson's ratio υ	0.350	0.394	0.350	0.300
Elastic modulus E, Pa	1.100×10^8	3.189×10^9	1.300×10^9	7.900×10^{10}
Restitution coefficient e	0.201	0.299	0.279	0.305
Coefficient of static friction μ_s	0.466	0.577	0.533	0.511
Coefficient of rolling friction μ_r	0.080	0.120	0.050	0.070

3.3. Bulk Density Test and Simulation

3.3.1. Bulk Density Test Setup

Bulk density is a critical statistic for evaluating the physical characteristics of granular materials, since it is directly connected to the shape and size of the particles. As a result, this paper employs the bulk density test to confirm the approach of pill particle modelling. The bulk density of the pill particles was determined in this research using a cylindrical measuring cylinder (organic glass) with a volume of 0.26 L. The procedure is as follows: first, the pill particles are released above the measuring cylinder and fall naturally; second, when the pill particles exceed the cylinder's height, the excess particles are scraped off with an organic glass scraper; and finally, the mass of the remaining particles in the cylinder is calculated, and the bulk density is calculated using the mass of the pill particles divided by the volume of the cylinder. As seen in Figure 6a, each set of tests was performed three times.

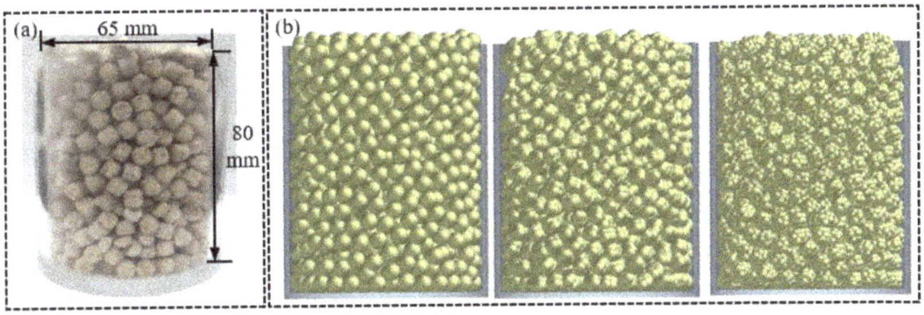

Figure 6. Bulk density test procedure and simulation analysis: (**a**) Bulk density test procedure; (**b**) Bulk density test simulation analysis of three pill particles.

3.3.2. Bulk Density Simulation Setup

The bulk density simulations of pill A (4-, 12-, and 20-sphere), pill B (4-, 12-, and 20-sphere), and pill C (2-, 12-, and 18-sphere) are performed using EDEM 2018 software with the Hertz–Mindlin (no-slip) contact model and the parameters listed in Table 1. The measuring cylinders used in the experiment are identical. When the simulation starts, a cylindrical region with a diameter of 20 mm and a height of 20 mm is placed as the particle factory 5 mm above the top of the measuring cylinder. Second, the pill particles are formed in the particle factory and gradually accumulate in the cylinder until they reach the cylinder's maximum height. When the pill particles have stabilized, the scraper is used to remove any remaining particles. Finally, the mass of the particles left in the measuring cylinder is determined. Each series of tests was done three times; the bulk density test simulation analysis is given in Figure 6b.

3.4. Angle of Repose Test and Simulation

3.4.1. Angle of Repose Test Setup

The angle of repose is also a critical parameter for assessing the physical and mechanical characteristics of granular materials, as it is directly connected to the form, size, and friction coefficient of the particles. As a result, the angle of repose test is utilized to confirm the approach of pill particle modelling. The lifting technique is utilized in this article to examine the pill particle stacking process. The following are the test steps: To begin, a lifting cylinder (ABS plastic) with a diameter of 65.7 mm and a height of 150 mm is inserted into the center of a flat plate (galvanized steel) with a side length of 500 mm, and 0.5 kg of pill particles are poured into the cylinder, maintaining a flat surface on the particles; second, the cylinder is lifted upwards at a speed of 200 mm/min, causing the pill particles to flow out of the cylinder and make contact with the galvanized steel plate; finally, the particles are stabilized, forming a conical pile, and the angle of repose is determined using an image processing method. As seen in Figure 7a, each set of tests was performed three times.

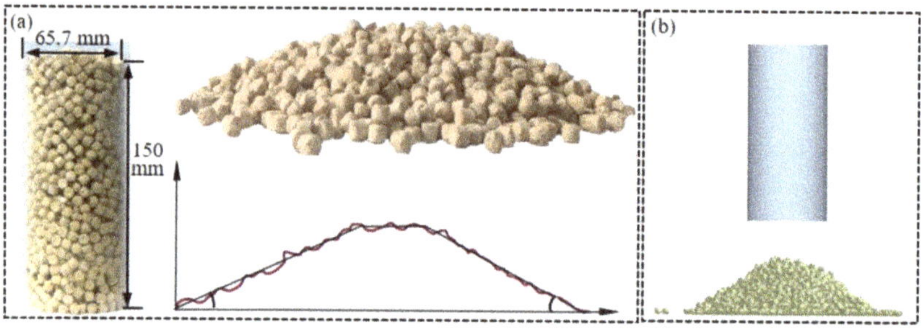

Figure 7. Angle of repose test procedure and simulation analysis: (**a**) Angle of repose test procedure and image processing method; (**b**) Angle of repose test simulation analysis.

3.4.2. Angle of Repose Simulation Setup

Angle of repose simulations of various pill particle models are carried out using EDEM 2018 software with the Hertz–Mindlin (no-slip) contact model and the parameters listed in Table 1. The test devices are identical to those used throughout the experiment. When the simulation starts, a cylindrical region with a diameter of 60 mm and a height of 20 mm is set up as the particle factory, 5 mm above the cylinder's top. Second, 0.5 kg of nearly 3000 pill particles are formed in the particle factory and gradually accumulate in the cylinder until they reach the cylinder's maximum height. Once the pill particles are stable, the cylinder is elevated at a 200 mm/min pace. Finally, after the pill particles have ceased to flow, a conical particle heap is produced, and the angle of repose on both sides is determined using an image processing technique. Each pair of tests is done three times; the angle of repose test simulation analysis is displayed in Figure 7b.

4. Results Analysis and Discussion

4.1. Results Analysis and Discussion of Bulk Density Test

4.1.1. Bulk Density Test Results Analysis

Figure 8 illustrates the fluctuation in the bulk density test simulation results with the number of filled spheres for various pill particle models, and the green region in the figure depicts the bulk density test results and standard deviation.

The change in the bulk density of the pill particles A as a function of the number of filled spheres is seen in Figure 8a. As the number of filled spheres grows from 4 to 20, the particle bulk density simulation results first increase and then decline, which is more consistent with the experiment data. Further research reveals that the relative errors of the simulation and experiment data are 1.11%, 0.06 percent, and 2.93%, respectively, for the number of filled spheres of 4, 12, and 20. As can be seen, the pill particle model's bulk density is affected by the number of filled spheres.

The change in the bulk density of the pill particles B as a function of the number of filled spheres is seen in Figure 8b. As seen in the image, the results of the discrete bulk density simulation grow and then decline as the number of filled spheres increases. The relative errors between simulation and experiment data were 2.10%, 1.28%, and 3.89%, respectively, for the number of filled spheres of 4, 12, and 20.

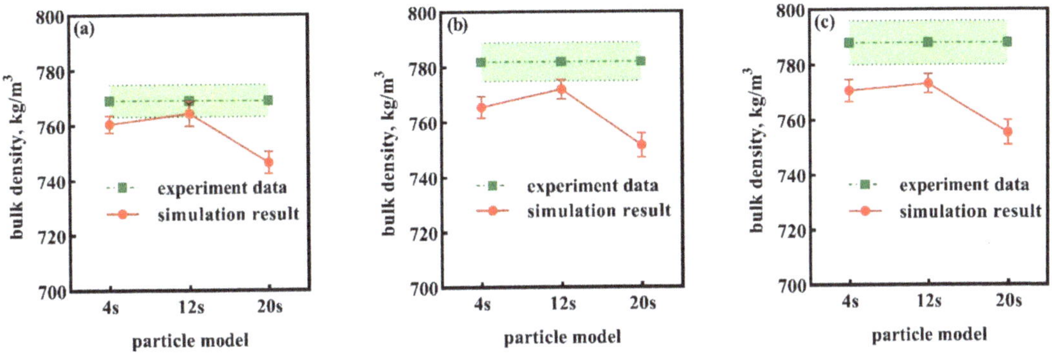

Figure 8. The findings of a simulation of the bulk density test of a pill particle model with varying numbers of filled spheres: (**a**) Results of the simulation for pill particle A; (**b**) Results of the simulation for pill particle B; and (**c**) Results of the simulation for pill particle C.

The fluctuation in the bulk density of the pill particles C as a function of the number of filled spheres is seen in Figure 8c. When shown in the figure, as the number of filled spheres increases, the particle bulk density simulation results initially rise and then decline, eventually approaching the experimental data, which is compatible with the changing trend of pill particles A and B. The relative errors between the simulation and experiment data were 2.20%, 1.88%, and 4.14%, respectively, for the number of filled spheres of 2, 12, and 18.

4.1.2. Bulk Density Test Discussion

According to the bulk density simulation and test results, when the number of filled spheres is 4 or 2, the space around the model is more defective than the actual pill particles, but the simulation results are closer to the test data due to the particles' reduced self-locking and increased sliding and rolling. The particle model becomes more accurate as the number of filled spheres increases, and the simulation results progressively converge to the test findings. When the number of filled spheres is 20 or 18, the filled spheres in the center of the particles increase their frictional impact, which has an influence on their movement and reduces the simulation results. The simulation findings validate the model suggested in this paper's accuracy for varied pill particle populations.

Furthermore, because the bulk density test is relatively simple, the simulation process takes less time, with different models taking less than 1 h to simulate. In conclusion, based on the simulation findings of the pill particle bulk density test, other modelling techniques, with the exception of the pill particle A and B 20-sphere models and the pill particle C 18-sphere model, have a poor correlation with the experiment results.

4.2. Results Analysis and Discussion of Angle of Repose

4.2.1. Angle of Repose Test Results Analysis

Figure 9 illustrates the relationship between the angle of repose test simulation results and the number of filled spheres for various pill particle models, while the green region in the figure depicts the angle of repose experiment findings and standard deviation.

Figure 9a illustrates the relationship between the angle of repose of pill particle A and the number of filled spheres in the simulation. As seen in the image, the particle angle of repose simulation results steadily rise as the number of filled spheres grows from four to twenty. The simulation findings for the four-sphere model are insignificant and do not match the experiment data. The angle of repose simulation findings for the 12- and 20-sphere models are within the standard deviation of the experiment data and near to the mean, however the change in the number of filled spheres in the pill particles model resulted in considerable angle of repose variations.

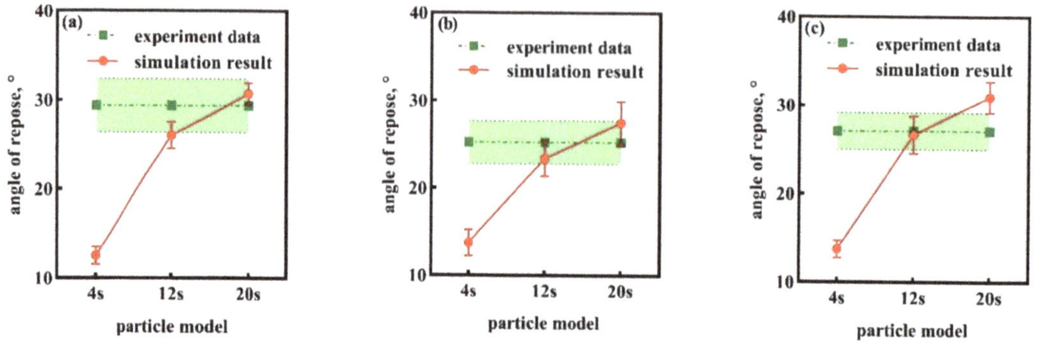

Figure 9. Angle of repose test results for the pill particle model as a function of the number of filled spheres: (**a**) Results of the simulation for pill particle A; (**b**) Results of the simulation for pill particle B; and (**c**) Results of the simulation for pill particle C.

Figure 9b illustrates the relationship between the angle of repose of pill particle B and the number of filled spheres in the simulation. As seen in the image, the particle angle of repose simulation results steadily rise as the number of filled spheres increases. The simulation findings for 12 and 20 filled spheres are within the standard deviation of the experiment data.

The angle of repose of the pill particle C is shown in Figure 9c as a function of the number of filled spheres. As shown in the image, the particle angle of repose simulation results steadily increases as the number of filled spheres increases, which is similar with the pattern seen for pill particles A and B. When the number of filled spheres is 12, the smallest relative error between the simulation and experiment data is 1.59%.

4.2.2. Angle of Repose Test Discussion

The test and simulation findings demonstrate that the intricacy of the pill's particle form results in the particles self-locking throughout the piling process. The number of particle-filled spheres has a significant influence on the geometry of the particle model, which in turn has an effect on the angle of repose simulation results. When the number of particle-filled spheres is small, the sphericity of the particles rises and the roughness and self-locking of the particles decrease, increasing the particles' sliding and rolling. As the number of filled spheres rises, the particle form eventually converges to that of genuine pill particles, increasing both the particles' self-locking behavior and the pile's angle of repose. The simulation findings further demonstrate the correctness of the approach for modelling pill particle populations suggested in this work.

Moreover, since the angle of repose test is slightly more complex than the bulk density test, the simulation time increases as the number of filling spheres increases, but the total simulation time for all models is less than 5 h. To summarize, based on the computational efficiency and the analysis of the simulation results for the angle of repose test of the pill particles, the simulation results for the 12- and 20-sphere models of pill particle A and B, as well as the 12-sphere model of pill particle C, are within the standard deviation of the experiment results. According to a comprehensive analysis of the bulk density and angle of repose test simulation results, the 12-sphere models of pill particles A, B, and C are close to the actual pill particle shapes, and the simulation results have small errors in comparison to the experimental data, making them suitable for simulating the mechanics behavior of particles and their flow processes during the actual pill discharging process.

5. Example Applications

On the basis of the preceding work, the applicability of the suggested modelling technique for pill particle population and its practical use are shown by simulation results of the pill particle discharge process.

5.1. Pill Discharging Process Device

The research is based on a self-designed pill discharge mechanism, which includes the primary functioning components depicted in Figure 10a, including the material box, the pill wheel, and the frame. The material box is stamped from a galvanized steel plate, and its conical top half-angle shape is inspired by the friction properties of pill particles. The pill wheel is made of ABS technical plastic alloy and has eight identical medication feeding grids, each measuring 100 mm in length.

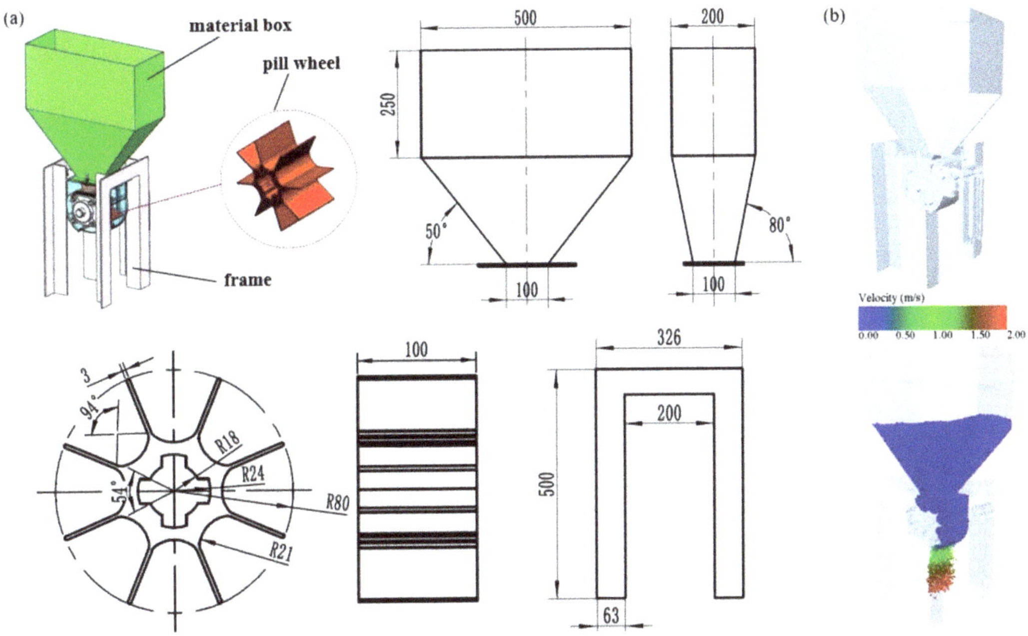

Figure 10. Test device and simulation analysis of the pill discharge process: (**a**) Test device for the pill discharge process; (**b**) DEM model of the pill discharge device and simulation analysis of the pill discharge process.

5.2. Simulation Setup of Pill Discharging Process

The pill particle discharge processes are simulated using EDEM 2018 software with the Hertz–Mindlin (no-slip) contact model and the parameters listed in Table 1. The particle models of 12-sphere pills A, B, and C are utilized in proportions of 35%, 35%, and 30%, respectively. Figure 10b depicts the DEM model of a pill particle discharge device. When the simulation starts, a box area with a length of 400 mm and a width of 150 mm is placed as the particle factory 5 mm above the top of the work bin. Second, 10 kg of nearly 60,000 pill particles are produced in the particle factory and gradually accumulate in the hopper. When the pill particles reached a state of stability, the pill wheel started to spin (i.e., 10, 15, and 30 rpm). Finally, the mass of particles ejected by the pill wheel spinning one grid every 10 s is determined. Thus, the discharge capacity and variation coefficient of pill particles are determined. Each pair of tests is done three times; Figure 10b depicts a simulation study of the pill particle discharge process.

5.3. Results Analysis and Discussion of Discharge Process

The mass of the pills discharged and its variation coefficient at various rotating speeds throughout the pill release process are shown in Figure 11.

Figure 11a illustrates the simulation findings for the time-dependent change in pill particle discharging at various rotation speeds. As seen in the figure, the mass of discharging pill particles grows almost linearly with the simulation duration at various rotating speeds. The pill wheel's speed was raised from 10 to 30 rpm, and the amount of pills discharged steadily rose, while the filling mass of the medication feeding grid was reduced from 0.25 to 0.23 kg.

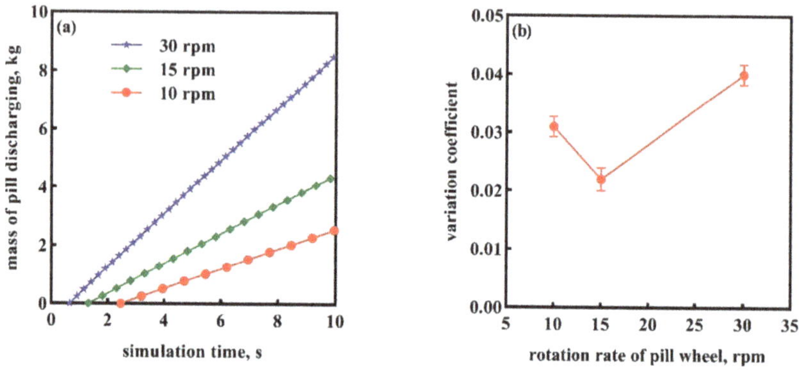

Figure 11. The simulation results for pill particle discharging mass and its variation coefficient at various rotation speeds during the pill discharging process: (**a**) Simulation results for pill particle discharging mass; (**b**) Simulation results for pill particle discharging mass variation coefficient.

Figure 11b illustrates the simulation findings for the change in the discharge efficiency of pill particles as a function of the pill wheel's rotation speed. When seen in the figure, the mass variation coefficient drops from 0.03 to 0.02 as the rotation speed of the pill wheel increases from 10 to 15 revolutions per minute. When the rotation speed is increased to 30 rpm, the variation coefficient of the pill discharge mass rises from 0.02 to 0.04. As can be observed, the variation coefficient of the pill wheel's rotation speed has a substantial influence on the mass of pill particles discharged and their stability.

As a result of the simulation findings, it is clear that the rotation speed of the pill wheel is the most critical operating parameter of the pill discharging device, as it has a significant effect on the mass and stability of the pills discharged. When the rotation speed is low, it has a negligible influence on the filling of the pill discharging grid, while increasing the rotation speed marginally enhances the discharging capability. When the rotation speed exceeds a particular threshold, it has an influence on the filling mass of the pill discharging grid, which in turn has an effect on the discharging stability, with the effect steadily rising as the rotation speed increases. The simulation findings in this work are compatible with those in the literature on fertilizer discharge processes [5,41], demonstrating the applicability of the pill particle population modelling approach presented in this paper and its practical use. For further in-depth investigations of the pill discharging process, it offers an accurate simulation model and theoretical foundation for optimizing the structural parameters, dimensions parameters, and operating parameters of the pill discharging device.

6. Conclusions

This paper uses self-developed anticorrosive pill particles as the research object, proposes a pill particle population modelling method based on DEM by testing and analyzing the pill particles' shape and size parameters, and verifies the modelling method's feasibility, efficacy, and applicability via experimental research and simulation analysis. The following are the study's major conclusions:

(1) The particle shape and size parameters were evaluated and analyzed to approximate the cylindrical shape of the pill particles, and the particle population was classified into pill particles A (5.4 mm), B (5.8 mm), and C (6.2 mm) based on their height, with the mass ratio of particles accounting for 35%, 35%, and 30%, respectively.

(2) This work proposes a population modelling approach for pill particles based on DEM. Multi-sphere particle models were created for pill particle A (4, 12, and 20 spheres), pill particle B (4, 12, and 20 spheres), and pill particle C (2, 12, and 18 spheres). It serves as a guide for modeling cylindrical and irregular particles.
(3) Using the bulk density and angle of repose tests as examples, the pill particle population modelling approach was utilized to deduce the mechanism by which the number of pill particle-filled spheres affects the particle accumulation process and flow behavior. By comparing the simulation findings to the test data, the feasibility and efficacy of the pill particle population modelling approach were established.
(4) Using the independently built pill discharging device as an example, the 12-sphere model of pill particles A, B, and C was utilized to deduce the process by which the wheel's rotation speed affects the pill discharging performance. The method's applicability and practical use were shown by assessing the simulation results of the pill discharging process and establishing the groundwork for future improvements of the pill discharging device.

Author Contributions: Conceptualization, D.L. and Y.Y.; methodology, D.L. and Y.Y.; validation, C.Q. and Y.L.; resources, C.Q.; writing—original draft preparation J.S.; writing—review and editing, D.L.; supervision, Z.Y. and X.W.; project administration, J.W. All authors have read and agreed to the published version of the manuscript.

Funding: The research was financially supported by Scientific Research Project of the CNOOC EnerTech-Drilling & Production Co., grant number GCJSXMHT-T2101.

Institutional Review Board Statement: Not applicable.

Informed Consent Statement: Not applicable.

Data Availability Statement: Not applicable.

Conflicts of Interest: The authors declare no conflict of interest.

References

1. Gao, Z.W.; Nguang, S.K.; Kong, D.X. Advances in modelling, monitoring, and control for complex industrial systems. *Complexity* **2019**, *2019*, 2975083. [CrossRef]
2. Gao, Z.W.; Kong, D.X.; Gao, C.H. Modeling and control of complex dynamic systems: Applied mathematical aspects. *J. Appl. Math.* **2012**, *2012*, 869792. [CrossRef]
3. Gao, Z.W.; Saxen, H.; Gao, C.H. Special section on data-driven approaches for complex industrial systems. *IEEE Trans. Ind. Inform.* **2013**, *9*, 2210–2212. [CrossRef]
4. Cundall, P.A.; Strack, O.D.L. Discrete numerical model for granular assemblies. *Géotechnique* **1979**, *29*, 47–65. [CrossRef]
5. Yu, Y.J.; Fu, H.; Yu, J.Q. DEM-based simulation of the corn threshing process. *Adv. Powder Technol.* **2015**, *26*, 1400–1409. [CrossRef]
6. Yu, Y.J.; Li, L.S.; Zhao, J.L.; Wang, X.G. Discrete element simulation based on elastic-Plastic damping model of corn kernel-cob bonding force for rotation speed optimization of threshing component. *Processes* **2021**, *9*, 1410. [CrossRef]
7. Sun, J.F.; Chen, H.M.; Wang, Z.M.; Ou, Z.W.; Yang, Z.; Liu, Z.; Duan, J.L. Study on plowing performance of EDEM low-resistance animal bionic device based on red soil. *Soil Tillage Res.* **2020**, *196*, 104336. [CrossRef]
8. Li, Y.D.; Xu, H.G.; Jing, C.H.; Jiang, J.H.; Hou, X.Y. A novel heat transfer model of biomass briquettes based on secondary development in EDEM. *Renew. Energy* **2019**, *131*, 1247–1254. [CrossRef]
9. Yu, Y.J.; Li, L.S.; Zhao, J.L.; Wang, X.G.; Fu, J. Optimal design and simulation analysis of spike tooth threshing component based on DEM. *Processes* **2021**, *9*, 1163. [CrossRef]
10. Yang, L.W.; Chen, L.S.; Zhang, J.Y.; Liu, H.J.; Sun, Z.C.; Sun, H.; Li, Y.D.; Zheng, L.H. Fertilizer sowing simulation of a variable-rate fertilizer applicator based on EDEM. *IFAC-Pap.* **2018**, *51*, 418–423. [CrossRef]
11. Neto, A.G.; Hudobivnik, B.; Moherdaui, T.F.; Wriggers, P. Flexible polyhedra modeled by the virtual element method in a discrete element context. *Comput. Methods Appl. Mech. Eng.* **2021**, *387*, 114163. [CrossRef]
12. Yan, B.; Regueiro, R.A.; Sture, S. Three-dimensional ellipsoidal discrete element modeling of granular materials and its coupling with finite element facets. *Eng. Comput.* **2010**, *27*, 519–550. [CrossRef]
13. Kim, K.C.; Jiang, T.; Kim, N.L.; Kwon, C. Effects of ball-to-powder diameter ratio and powder particle shape on EDEM simulation in a planetary ball mill. *J. Indian Chem. Soc.* **2022**, *99*, 100300. [CrossRef]
14. Zhou, H.; Chen, Y.H.; Li, H.S.; Xu, Z.Y.; Dong, H.R.; Wang, W.P. Effect of particles micro characteristics destroyed by ball milling on fly ash electrostatic separation. *Adv. Powder Technol.* **2022**, *33*, 103449. [CrossRef]

15. Tavares, L.M.; Rodriguez, V.A.; Sousani, M.; Padros, C.B.; Ooi, J.Y. An effective sphere-based model for breakage simulation in DEM. *Powder Technol.* **2021**, *392*, 473–488. [CrossRef]
16. Danesh, A.; Mirghasemi, A.A.; Palassi, M. Evaluation of particle shape on direct shear mechanical behavior of ballast assembly using discrete element method (DEM). *Transp. Geotech.* **2020**, *23*, 100357. [CrossRef]
17. Tekeste, M.Z.; Way, T.R.; Syed, Z.; Schafer, R.L. Modeling soil-bulldozer blade interaction using the discrete element method (DEM). *J. Terramechanics* **2020**, *88*, 41–52. [CrossRef]
18. Horabik, J.; Wiącek, J.; Parafiniuk, P.; Bańda, M.; Kobyłka, R.; Stasiak, M.; Molenda, M. Calibration of discrete-element-method model parameters of bulk wheat for storage. *Biosyst. Eng.* **2020**, *200*, 298–314. [CrossRef]
19. Zeng, Y.; Mao, B.Q.; Jia, F.G.; Han, Y.L.; Li, G.R. Modelling of grain breakage of in a vertical rice mill based on DEM simulation combining particle replacement model. *Biosyst. Eng.* **2022**, *215*, 32–48. [CrossRef]
20. Wang, T.Z.; Quan, Q.Q.; Li, M.X.; Huang, J.C.; Zhao, Z.J.; Guo, F.; Meng, L.Z.; Wang, G.Q.; Tang, D.W.; Deng, Z.Q. An asteroid anchoring method based on cross-drilling geometric force closure of ultrasonic drill. *Acta Astronaut.* **2021**, *178*, 813–823. [CrossRef]
21. Han, D.D.; Zhang, D.X.; Jing, H.R.; Yang, L.; Cui, T.; Ding, Y.Q.; Wang, Z.D.; Wang, Y.X.; Zhang, T.L. DEM-CFD coupling simulation and optimization of an inside-filling air-blowing maize precision seed-metering device. *Comput. Electron. Agric.* **2018**, *150*, 426–438. [CrossRef]
22. Stoimenov, N.; Ruzic, J. Analysis of the particle motion during mechanical alloying using EDEM software. *IFAC-Pap.* **2019**, *52*, 462–466. [CrossRef]
23. Zhang, L.; Zhai, Y.B.; Chen, J.N.; Zhang, Z.E.; Huang, S.Z. Optimization design and performance study of a subsoiler underlying the tea garden subsoiling mechanism based on bionics and EDEM. *Soil Tillage Res.* **2022**, *220*, 105375. [CrossRef]
24. Tao, H.; Zhong, W.Q.; Jin, B.S. Comparison of construction method for DEM simulation of ellipsoidal particles. *Chin. J. Chem. Eng.* **2013**, *21*, 800–807. [CrossRef]
25. Zhou, L.; Yu, J.Q.; Wang, Y.; Yan, D.X.; Yu, Y.J. A study on the modelling method of maize-seed particles based on the discrete element method. *Powder Technol.* **2020**, *374*, 353–376. [CrossRef]
26. Sun, K.; Yu, J.Q.; Liang, L.S.; Wang, Y.; Yan, D.X.; Yu, Y.J. A DEM-based general modelling method and experimental verification for wheat seeds. *Powder Technol.* **2022**, *401*, 117353. [CrossRef]
27. Langston, P.A.; Tuzun, U.; Heyes, D.M. Continuous potential discrete particle simulations of stress and velocity-fields in hoppers—transition from fluid to granular flow. *Chem. Eng. Sci.* **1994**, *49*, 1259–1275. [CrossRef]
28. Langston, P.A.; Tuzun, U.; Heyes, D.M. Discrete element simulation of granular flow in 2D and 3D hoppers—dependence of discharge rate and wall stress on particle interactions. *Chem. Eng. Sci.* **1995**, *50*, 967–987. [CrossRef]
29. Langston, P.A.; Tuzun, U.; Heyes, D.M. Discrete element simulation of internal-stress and flow-fields in funnel flow hoppers. *Powder Technol.* **1995**, *85*, 153–169. [CrossRef]
30. Zhu, H.P.; Zhou, Z.Y.; Yang, R.Y.; Yu, A.B. Discrete particle simulation of particulate systems: Theoretical developments. *Chem. Eng. Sci.* **2007**, *62*, 3378–3396. [CrossRef]
31. Iwashita, K.; Oda, M. Rolling resistance at contacts in simulation of shear band development by DEM. *J. Eng. Mech. ASCE* **1998**, *124*, 285–292. [CrossRef]
32. Zhou, Y.C.; Wright, B.D.; Yang, R.Y.; Xu, B.H.; Yu, A.B. Rolling friction in the dynamic simulation of sandpile formation. *Phys. A Stat. Mech. Its Appl.* **1999**, *269*, 536–553. [CrossRef]
33. Zhu, H.P.; Yu, A.B. Averaging method of granular materials. *Phys. Rev. E* **2002**, *66*, 021302. [CrossRef] [PubMed]
34. Coetzee, C.J. Review: Calibration of the discrete element method. *Powder Technol.* **2017**, *310*, 104–142. [CrossRef]
35. Coetzee, C.J. Particle upscaling: Calibration and validation of the discrete element method. *Powder Technol.* **2019**, *344*, 487–503. [CrossRef]
36. American Society of Agricultural Engineers. *Compression Test of Food Materials of Convex Shape*; ASAE Standards: St. Joseph, MI, USA, 2017; pp. 1–15.
37. Zhang, X.H. Computer Numerical Simulation of Pellet Feed in Cooling Process. Master's Thesis, Huazhong Agricultural University, Wuhan, China, 2004.
38. Yan, Y.F.; Meng, D.X.; Song, Z.H.; Liu, L.Q.; Li, F.D. Particle kinetic simulation and experiment for flute-wheel feeding machine. *Trans. Chin. Soc. Agric. Mach.* **2016**, *47*, 249–253.
39. González-Montellano, C.; Fuentes, J.M.; Ayuga-Téllez, E.; Ayuga, F. Determination of the mechanical properties of maize grains and olives required for use in DEM simulations. *J. Food Eng.* **2012**, *111*, 553–562. [CrossRef]
40. LoCurto, G.J.; Zhang, X.; Zakirov, V.; Bucklin, R.A.; Vu-Quoc, L.; Hanes, D.M.; Walton, O.R. Soybean impacts: Experiments and dynamic simulations. *Trans. ASAE* **1999**, *40*, 789–794. [CrossRef]
41. Ding, S.P.; Bai, L.; Yao, Y.X.; Yue, B.; Fu, Z.L.; Zheng, Z.Q.; Huang, Y.X. Discrete element modelling (DEM) of fertilizer dual-banding with adjustable rates. *Comput. Electron. Agric.* **2018**, *152*, 32–39. [CrossRef]

processes

Article

Identification and Analysis of Factors Influencing Green Growth of Manufacturing Enterprises Based on DEMATEL Method—Wooden Flooring Manufacturing Companies as a Case

Wei Li and Xia Wu *

College of Applied Science and Technology, Beijing Union University, Beijing 100012, China
* Correspondence: yykjtwuxia@buu.edu.cn

Abstract: It is significant to scientifically identify what factors influence the green growth of manufacturing enterprises and analyze the relationship among these factors, thus promoting green growth. Firstly, the corresponding conceptual model is designed; then, the DEMATEL method and steps used to identify the influencing factors are introduced; finally, the DEMATEL method is adopted to empirically analyze wooden flooring manufacturing companies so as to identify influencing factors of their green growth. According to the results, there are six reason factors, namely environmental standard constraints, green market demand, market competition, green technology advancement, upstream and downstream synergy of green industrial chain, and policy support, which provide the most important external support to enterprises' green growth and main driving power to wooden flooring manufacturing ones.

Keywords: green growth; influencing factor; factor identification; DEMATEL method; manufacturing enterprises

Citation: Li, W.; Wu, X. Identification and Analysis of Factors Influencing Green Growth of Manufacturing Enterprises Based on DEMATEL Method—Wooden Flooring Manufacturing Companies as a Case. *Processes* **2022**, *10*, 2594. https://doi.org/10.3390/pr10122594

Academic Editors: Hideki Kita and Jochen Strube

Received: 9 October 2022
Accepted: 22 November 2022
Published: 5 December 2022

1. Introduction

Crucial to the industrial economy, manufacturing is a typical and fundamental symbol of comprehensive national strength and international status. Over the past 40 years since the reform and opening-up of China, China's manufacturing industry has made great progress, now with more than 30% of the world's total manufacturing output, a comprehensive industrial system with all sectors, and a complete industrial chain [1], However, despite the huge volume, China is still not a manufacturing powerhouse [2] subject to scale expansion [3]; therefore, transformation is urgent to upgrade China's manufacturing industry.

To pursue a higher quality of industrial development, green growth has become an important symbol [4]. In 2015, the Chinese government officially put forward the "Made in China 2025" Initiative, in which green growth was a guideline for strategic implementation, emphasizing that sustainable economic development should be coordinated with the natural environment towards fully green manufacturing [5]. In particular, manufacturing enterprises are required to shift from a traditional development model that was at the expense of the environment to foster green transformation and upgrading to realize green growth [6,7].

"Green growth" refers to the process that manufacturing enterprises grow stronger through green strategies and green behaviors, fewer pollutant residues, less consumption of resources and energy, and more environmentally friendly, safe, and healthy products, together with ever-increasing green competitiveness. In particular, the leading concept is green development throughout the production and management practices of manufacturing enterprises relying on relevant technological and management innovations, featured by less environmental pollution and higher resource efficiency [8,9].

It is clear that green growth essentially emphasizes sustainable development to realize both economic growth and environmental protection [10]. Enterprise behaviors should follow the basic green premise and corresponding requirements under green constraints, which is necessary for their survival and development.

Such green growth of manufacturing enterprises is a "systematic project" with a wide range, rich contents, and many influencing factors [11]. As a result, it is pragmatic to identify and refine these key factors.

This paper aims to effectively identify the influencing factors of manufacturing enterprises' green growth, then further define and utilize key ones to promote green growth. Major contributions made in this paper include a conceptual model of the factors influencing the green growth of manufacturing enterprises and a method to further explore the relevant dynamic mechanisms and key influencing factors identified through the DEMATEL method and verify the effectiveness of the DEMATEL method. Finally, this paper concludes with six verified key influencing factors. In particular, the introduction of the "industrial chain" factor and the proposed "collaboration within the industrial chain" factor have improved comprehensiveness in analyzing green growth dynamics and provided more perspectives apart from previous individual enterprises.

2. The Research Objectives and Method

2.1. Setting up a Conceptual Model of Influencing Factors

Enterprise management pursues more values and shareholders' maximum benefits. That is, green behaviors cannot sustain without more business values or better financial performance [12]; therefore, based on rational human assumption, manufacturing enterprises driven by their interests (including short-term and long-term gains) will be more willing to adopt green behaviors with various "green motivation" factors that add values. In this sense, effective identification of external factors that can improve such "green behavior willingness" [13] is necessary; thus, a conceptual model of influencing factors is designed as shown in Figure 1.

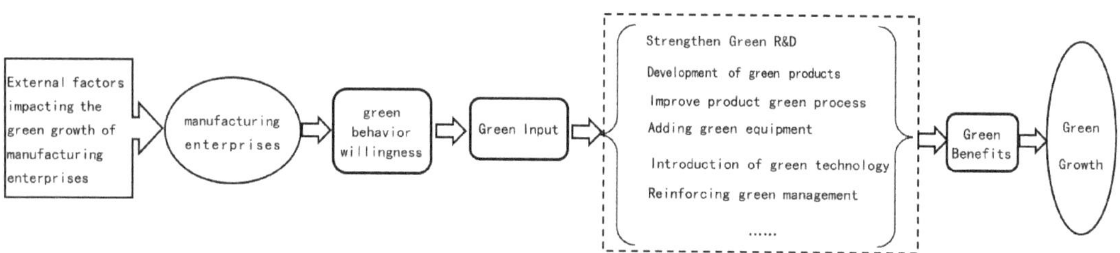

Figure 1. Conceptual model of the main factors influencing green growth.

The model logic is that due to external influencing factors, manufacturing enterprises have a stronger willingness to green behaviors driven by their interests. Then these green behaviors realize green benefits and green growth in different ways. That is, "external influencing factors → manufacturing enterprises → stronger willingness to green behaviors → green behaviors → more revenues → green growth."

Here are definitions of several relevant concepts in Figure 1.

(1) Green behavior willingness, with manufacturing enterprises as the subject. It refers to enterprises' subjective readiness for green behaviors under comprehensive influencing factors, which is crucial to connect external and internal factors.

(2) Green input, mainly including capital input and technical staff input. The former one reflects emphasis and implementation of environmental protection, while the latter is key to ensuring deliveries and green competitive advantages.

(3) Green behaviors, namely a series of environmentally friendly actions. There are mainly four categories: Firstly, green product development, that is, relying on green technologies to develop new green products with less resource or energy consumption, emissions, and pollution. Secondly, green process improvement, including clean production and end-of-pipe management, that is, better new technologies, processes, and equipment are adopted to save energy and control pollution. Next is the research and development of green technologies so as to improve green competitiveness in pollution prevention and control. Finally, stronger green management. Through building up green corporate culture, staff training, and a green management system, employees can have and implement this concept with stronger green awareness and consciously fulfill their environmental responsibilities.

(4) Green benefits, namely, more environmental and economic benefits as a result of green behaviors. Specifically, environmental benefits are obvious in saving energy and reducing consumption, and further enhancing green image indirectly. However, economic benefits refer to revenue and profit increases, along with fewer costs and expenses.

2.2. Methods and Steps to Identify Influencing Factors

This paper adopts the widely recognized DEMATEL method to identify influencing factors of manufacturing enterprises' green growth. Compared to methods such as structural equations, linear regression analysis, and system dynamics, the DEMATEL method can not only analyze the influence relationship between individual factors, but also show corresponding specific influence levels [14]. Simply, this method is powerful in simplifying intricate relationships. Firstly, the direct impact matrix is established by judging the logical relationships between factors in the system with the help of professional expertise and rich experience. The influence level of each factor on other factors and the degree of being influenced are analyzed using this matrix, thus calculating the centrality and the reason degree of each factor [15]. This helps to identify key influencing factors for system optimization decisions.

The major steps are as follows:

Firstly, selecting out influencing factors of green growth. As numerous relevant factors constrain and interact with each other, it is neither practical nor necessary to examine them one by one. Instead, "literature reference+ expert consultation" is more feasible to conduct preliminary screening and form the "alternative sets" $\{f_1, f_2, \cdots, f_n\}$, so as to have more sensible analysis and decisions from theoretical or practical perspectives.

Next is to determine the relationship between these factors through comparisons between each other. A panel of experts was invited to score each group from 0 and 4 according to "influence level".

Corresponding figures are shown in Table 1.

Table 1. Judgment basis of impact degree.

Influence Level	No influence	Little Influence	Moderate Influence	Stronger Influence	Huge Influence
Score	0	1	2	3	4

Thirdly, direct impact matrix. Based on scores, direct impact matrix A of these influencing factors can be set up as $A = [a_{ij}]_{n \times n}$, a_{ij} representing the influence level of factor f_i on factor f_j.

The fourth step is to normalize the direct impact matrix so as to obtain the normalized influence matrix B: $B = [b_{ij}]_{n \times n}$.

$$b_{ij} = a_{ij} \times \frac{1}{\max\limits_{1 \le i \le n} \left(\sum_{j=1}^{n} a_{ij} \right)} \quad (i, j = 1, 2, \ldots, n) \tag{1}$$

Then, the next step is to calculate the comprehensive impact matrix T according to the formula $T = B(I - B)^{-1}$, where I is the unit matrix. That is, $T = B(I - B)^{-1} = [t_{ij}]_{nn}$, t_{ij} indicates the level of direct and indirect influence of factor f_i on factor f_j.

The sixth step is to calculate corresponding levels of influence and being influenced. According to the comprehensive impact matrix T, the relationship between each influencing factor is determined, specifically the influencing level D_i and the level being influenced F_i. The calculation formula is:

$$D_i = \sum_{j=1}^{n} t_{ij}(i = 1,\ 2, \ldots,\ n) \tag{2}$$

$$F_i = \sum_{i=1}^{n} t_{ij}(i = 1,\ 2, \ldots,\ n) \tag{3}$$

D_i is a row-wise sum of the elements in T, which represents the comprehensive value of the influence level of X_i on other factors.

F_i is a column sum of the elements in T, which represents the comprehensive value of how much X_i is influenced by other factors.

Finally, the centrality and reason degree of each factor are calculated. The formulas for the centrality H_i and the reason degree J_i are as follows:

$$H_i = D_i + F_i(i = 1,2,\ldots,n) \tag{4}$$

$$J_i = D_i - F_i(i = 1,2,\ldots,n) \tag{5}$$

The centrality H_i is the total sum of D_i and F_i, which indicates the position of the factor X_i in the system. A larger H_i indicates that X_i has a higher position in the system and X_i plays a larger role [16].

The reason degree J_i is the difference between D_i and F_i. It indicates how the influence is realized among influencing factors, literally whether a factor is to influence or to be influenced. If $J_i > 0$, it is called a reason factor, indicating that factor X_i has a strong influence on other factors; but if $J_i < 0$, it is called a result factor, indicating that factor X_i is strongly influenced by other factors [17].

In summary, the DEMATEL method not only helps to identify key influencing factors, but also provides a preliminary analysis of the interaction mechanism among these factors according to centrality and reason degree, thus providing a reference for exploring the green growth mechanism of manufacturing enterprises.

3. Empirical Analysis of Wooden Flooring Manufacturing Enterprises as a Case

Wooden flooring manufacturing is a typical traditional type. This industry in China grew rapidly from the 1980s, with an average production scale of 400 million m^2 for many years and an annual output value of nearly 100 billion RMB, making China the world's largest producer and consumer of wooden flooring [18]; however, along with rapid expansion, this industry also faces some environmental problems that cannot be ignored or yet to be fundamentally solved, especially low resource utilization, high unit energy consumption, harmful emissions, free formaldehyde residues [19]. The environmental pressure and utilization chain are shown in Figure 2.

Concerning increasing pressure on resources and the environment, the consumer demand for green wooden flooring products is growing. Faced with both challenges and opportunities brought by "greenness", enterprises urgently need to identify and grasp key influencing factors of green growth so as to empower their sustainable growth [20].

3.1. The Main Factors Influencing Green Growth of Wooden Flooring Manufacturing Enterprises

Berry and Rondinelli [21] held that government, customer, employee, and competitor pressure are driving enterprises' shift to proactive environmental management. Bansal and Roth [22] proposed competitiveness, legitimacy, and ecological responsibility as three main elements leading to the ecological responsiveness of enterprises. Zhu [23] identified through factor analysis that corporate awareness of laws and regulations, environmental strategies, supply chain pressure, market demand, and the cost of green activities are the major factors of pressure/motivation and practice of green supply chain management in companies, also pointed out that green supply chain management had become an effective means for companies to improve their competitiveness. Hao et al. [24] used the "entropy decision model" to investigate 30 enterprises and concluded that expected benefits, environmental regulations, ecological environment, cluster network characteristics, and corporate social responsibility are key factors affecting green behavior decisions of enterprises in resource-based industrial clusters. Jiang [25] found through an empirical study that demand pressure, competitive pressure, policy opportunities, demand opportunities, and competitive opportunities all contribute to green performance. Furthermore, Zeng [18] also found through an empirical study that command-and-control policy instruments in environmental regulation, international market pull in demand-pull factors, and ISO14001-certified firms in supply-side factors are all key factors influencing enterprises to engage in green innovations.

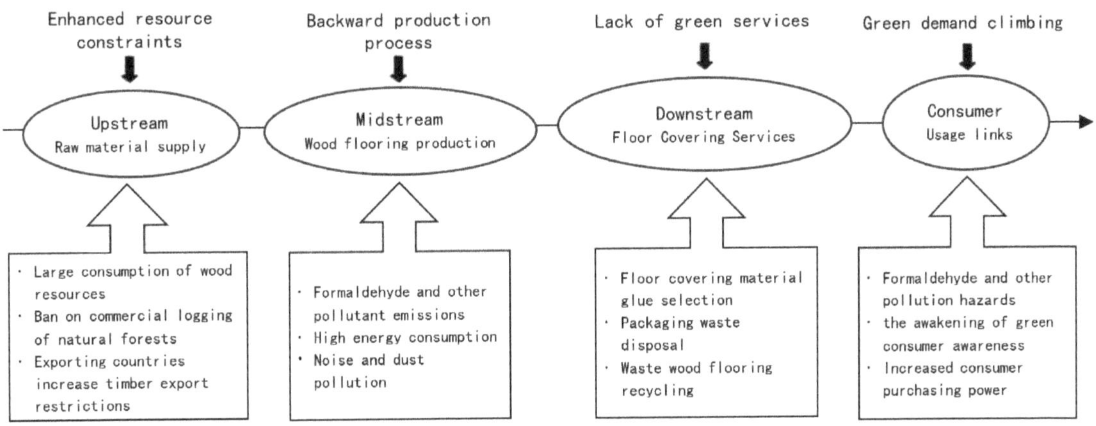

Figure 2. Environmental pressure from manufacturing to consumption.

Based on relevant literature and characteristics, together with many rounds of discussions by the CGE, a prepared set of factors influencing the green growth of wooden flooring manufacturing enterprises was formed as follows.

(1) Policies, basically government policy support and environmental standard constraints. The former mainly promotes enterprises to adopt green behaviors and implement green growth through favorable, subsidy, and incentive policies; the latter sets relevant environmental standards to constrain and regulate enterprises' behaviors [26].

(2) Industry, such as green market demand, market competition, green technology advancement, and local support. Among them, green market demand is a necessary precondition for green growth, which is mainly reflected comprehensively by population, purchasing power, and purchasing desire. Moreover, market competition mainly focuses on the status of competition between various industrial chains, with wooden flooring manufacturing enterprises as the core. Green technology advancement is crucial to support green innovations and a decisive factor for quality "greenness" [27]. In terms of local support, namely industrial support and other "hardware and software" in the area where an enterprise is located, it provides protection to enterprises' green growth. Specifically, such support includes public service facilities, production factors' trading market, logistics' supporting network, local economy, and government–industry–university–research cooperation and innovation, along with industrial information environment.

(3) Industrial chain. This mainly refers to the green synergy of enterprises in the industry chain, which is represented by their cooperation level, the "green requirements" for wooden flooring manufacturing enterprises, and response levels of upstream and downstream enterprises.

(4) Green behavior willingness. It demonstrates how strong enterprises' readiness are to adopt green behaviors, which is the key link to transforming the "external factors" of green growth into "internal ones".

(5) Green input. This shows the quantity and quality of input resources and generates economic and environmental benefits through the process of "green input—green output", which is mainly reflected by the green input intensity (such as upgrading green products, manufacturing processes, production equipment, end-of-pipe pollution control, etc.) and the number of technicians.

(6) Green management level. Led by the sustainable development idea, environmental protection is integrated into the whole process of enterprise production and operation so as to control pollution, save resources, shape the green image, and finally achieve sustainable growth of enterprises embodied in "green" comprehensive management capacity. To define such a level, symbol factors mainly refer to green strategies and their implementation, higher green quality of products, and enterprises' green images.

(7) Green output. This refers to outcomes of enterprises' green governance and green R&D through green inputs; for example, the number of patent applications, especially invention patents that can reflect the comprehensive strength of enterprise scientific research, which can be used as an important indicator reflecting green outputs.

Building a Direct Impact Matrix.Firstly, the major influencing factors of wooden flooring manufacturing enterprises' green growth are named and listed in Table 2.

Table 2. The main factors impacting the green growth of wooden flooring manufacturing companies.

Dimensionality	Impact Factors
Policy	Government Policy Support f_1, Environmental Standard Constraints f_2
Industrial environment	Green Market Demand f_3, Market Competition f_4, Green Technology Advancement f_5, Local Support f_6
Industry Chain	Green Synergy Between Industry Upstream and Downstream f_7
Green Behavior Willingness	Green Behavior Willingness f_8
Green Input	Green Input Intensity f_9, Number of Technical Staff f_{10}
Green Management Level	Green Strategy Formulation and Implementation f_{11}, Product Green Quality f_{12}, Corporate Green Image f_{13}
Green Output	Number of Patent Applications f_{14}

Next, a panel of nine professionals in business growth and green development was established, including four university professors, two researchers from research institutions, two senior consultants from consulting organizations, and one top executive from a wooden flooring manufacturing enterprise. They were invited to evaluate the above 14 influencing factors and score them according to Table 1, thus forming the quantitative relationship values between these factors. Each set of quantified influence relationship values indicates the direct effect of an influencing factor on another.

Finally, the experts' scores are averaged and rounded to form a direct impact matrix A'. Then A' is sent back to experts for confirmation and correction so as to gain direct impact matrix A, as shown in Table 3.

Table 3. The direct impact matrix A.

	f_1	f_2	f_3	f_4	f_5	f_6	f_7	f_8	f_9	f_{10}	f_{11}	f_{12}	f_{13}	f_{14}
Government Policy Support f_1	0	0	1	2	1	2	2	4	3	1	3	3	3	2
Environmental Standard Constraints f_2	0	0	2	3	4	2	2	4	4	2	4	4	3	3
Green Market Demand f_3	2	2	0	4	3	2	3	4	4	2	4	4	4	3
Market Competition f_4	1	3	0	0	3	2	3	4	4	3	4	4	4	3
Green Technology Advancement f_5	0	4	2	2	0	2	1	4	4	3	3	3	2	3
Local Support f_6	1	0	2	1	2	0	0	1	1	1	1	1	0	1
Green Synergy Between Industry Upstream and Downstream f_7	2	3	2	2	1	1	0	4	3	1	4	4	3	2
Green Behavior Willingness f_8	1	1	0	2	1	1	3	0	4	4	4	4	4	3
Green Input Intensity f_9	0	0	0	2	2	0	1	3	0	4	2	3	3	4
Number of Technical Staff f_{10}	0	0	0	1	1	0	0	1	2	0	2	3	2	3
Green Strategy Formulation and Implementation f_{11}	1	1	0	1	1	0	2	3	4	3	0	4	3	2
Product Green Quality f_{12}	0	0	1	2	1	0	2	3	3	2	3	0	4	2
Corporate Green Image f_{13}	0	0	0	1	0	0	1	3	3	2	3	2	0	2
Number of Patent Applications f_{14}	0	0	0	2	3	0	0	2	2	2	2	4	4	0

3.2. *Calculations and Results*

3.2.1. The Comprehensive Influence Matrix

Based on matrix A, the normalized influence matrix B is calculated according to Equation (1). Then, according to $T = B(I - B)^{-1} = [t_{ij}]_{nn}$, the comprehensive influence matrix T is obtained, as shown in Table 4.

Table 4. The comprehensive influence matrix T.

	f_1	f_2	f_3	f_4	f_5	f_6	f_7	f_8	f_9	f_{10}	f_{11}	f_{12}	f_{13}	f_{14}
f_1	0.0215	0.0363	0.0444	0.1165	0.0840	0.0712	0.1085	0.2058	0.1918	0.1264	0.1838	0.1976	0.1926	0.1485
f_2	0.0288	0.0559	0.0782	0.1694	0.1814	0.0832	0.1320	0.2533	0.2655	0.1916	0.2526	0.2730	0.2415	0.2144
f_3	0.0788	0.1068	0.0347	0.2019	0.1670	0.0882	0.1647	0.2719	0.2834	0.2039	0.2707	0.2919	0.2822	0.2282
f_4	0.0499	0.1204	0.0326	0.0976	0.1556	0.0808	0.1513	0.2507	0.2623	0.2107	0.2505	0.2707	0.2608	0.2122
f_5	0.0247	0.1376	0.0746	0.1376	0.0838	0.0793	0.0996	0.2337	0.2455	0.1979	0.2128	0.2322	0.2009	0.1999
f_6	0.0345	0.0212	0.0589	0.0604	0.0794	0.0151	0.0310	0.0810	0.0848	0.0734	0.0800	0.0874	0.0612	0.0748
f_7	0.0724	0.1140	0.0738	0.1361	0.1022	0.0564	0.0789	0.2358	0.2237	0.1498	0.2355	0.2521	0.2235	0.1734
f_8	0.0448	0.0619	0.0233	0.1240	0.0919	0.0477	0.1343	0.1291	0.2285	0.2059	0.2193	0.2374	0.2307	0.1854
f_9	0.0148	0.0309	0.0155	0.1051	0.0983	0.0187	0.0726	0.1642	0.1051	0.1804	0.1443	0.1810	0.1767	0.1805
f_{10}	0.0084	0.0174	0.0092	0.0607	0.0567	0.0103	0.0315	0.0846	0.1121	0.0566	0.1065	0.1386	0.1153	0.1247
f_{11}	0.0395	0.0534	0.0183	0.0875	0.0778	0.0201	0.1002	0.1723	0.2019	0.1628	0.1045	0.2084	0.1828	0.1418
f_{12}	0.0179	0.0323	0.0386	0.1059	0.0747	0.0196	0.0987	0.1675	0.1755	0.1364	0.1683	0.1127	0.1986	0.1361
f_{13}	0.0121	0.0203	0.0098	0.0662	0.0379	0.0120	0.0611	0.1380	0.1452	0.1134	0.1393	0.1288	0.0794	0.1120
f_{14}	0.0124	0.0290	0.0150	0.0988	0.1137	0.0175	0.0468	0.1347	0.1424	0.1274	0.1346	0.1887	0.1865	0.0800

3.2.2. The Levels of Influence, Being Influenced, the Reason Degree, and Centrality

According to Equations (2)–(5), the levels of influence, being influenced, reason degree, and centrality of each factor are calculated, respectively [18], as shown in Table 5.

Table 5. D_i, F_i, H_i, J_i of each index.

Factor	D_i	F_i	J_i	H_i
Government Policy Support f_1	1.7289	0.4607	1.2682	2.1896
Environmental Standard Constraints f_2	2.4207	0.8374	1.5833	3.2581
Green Market Demand f_3	2.6743	0.5270	2.1473	3.2014
Market Competition f_4	2.4061	1.5676	0.8385	3.9736
Green Technology Advancement f_5	2.1600	1.4044	0.7556	3.5644
Local Support f_6	0.8432	0.6202	0.2230	1.4634
Green Synergy Between Industry Upstream and Downstream f_7	2.1277	1.3112	0.8165	3.4389
Green Behavior Willingness f_8	1.9642	2.5225	−0.5584	4.4867
Green Input Intensity f_9	1.4881	2.6678	−1.1797	4.1558
Number of Technical Staff f_{10}	0.9325	2.1365	−1.2040	3.0690
Green Strategy Formulation and Implementation f_{11}	1.5714	2.5027	−0.9313	4.0740
Product Green Quality f_{12}	1.4828	2.8004	−1.3176	4.2832
Corporate Green Image f_{13}	1.0754	2.6327	−1.5573	3.7080
Number of Patent Applications f_{14}	1.3277	2.2120	−0.8843	3.5397

Based on Table 5, factors' positions in the plane coordinate system were marked to form a diagram of their integrated influence relationship. In this figure, centrality is the horizontal coordinate, reason degree is the vertical coordinate, the intersection of the horizontal and vertical coordinates is [k,0], and the distances from k to the maximum and minimum values of centrality are equal—see Figure 3.

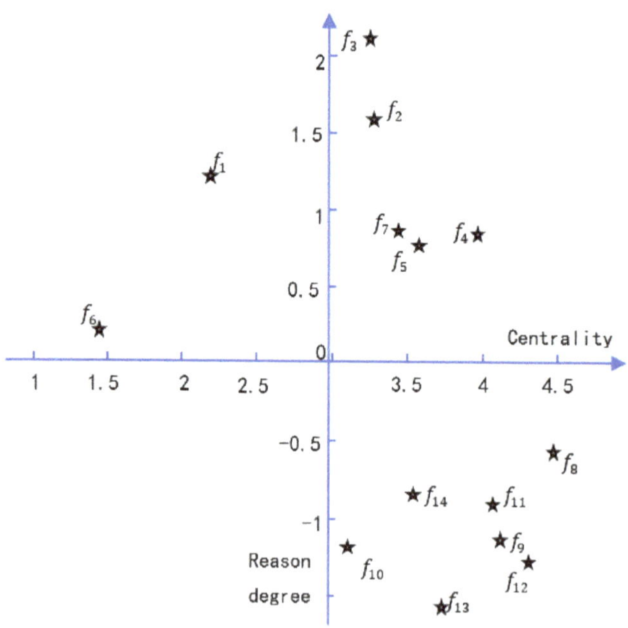

Figure 3. Schematic diagram of comprehensive impacting factors.

3.3. Analysis of Results

By calculating factors in comprehensive influence matrix T, the levels of influence, being influenced, reason degree, and centrality of each factor are derived. Through further analysis of the results, the following conclusions were obtained.

(1) In terms of reason degree, each factor has positive and negative values, which indicates that how each factor influences wooden flooring manufacturing enterprises' green growth is complicated. Among them, $f_1 - f_7$ are positive or reason factors and $f_8 - f_{14}$ are negative or result factors.

Reason factors (reason level greater than 0) are Green Market Demand f_3 > Environmental Standard Constraints f_2 > Government Policy Support f_1, Market Competition f_4 > Upstream and Downstream Green Synergy f_7 > Green Technology Advancement f_5 > Local Support f_6 according to importance.

Result factors (reason level lower than 0) can contribute to green growth through the influence exerted by reason factors.

(2) Concerning centrality, Green Behavior Willingness f_8 demonstrates the largest value. Other influencing factors with a centrality level greater than 4 are Product Green Quality f_{12}, Green Input Intensity f_9, and Green Strategy Formulation and Implementation f_{11}, which should be the focus of corporate management.

(3) According to Figure 3, Environmental Standard Constraints f_2, Green Market Demand f_3, Market Competition f_4, Green Technology Advancement f_5, Upstream and Downstream Green Synergy f_7 in the first quadrant and are Driving Factors, with the greatest influence and most critical role in promoting the green growth of wooden flooring manufacturing enterprises [28–30].

Government Policy Support f_1 and Local Support f_6 in the second quadrant are called Voluntariness, which plays a supportive role in the model. Specifically, Local Support f_6 has the lowest centrality value, indicating that this factor has little influence on wooden flooring manufacturing enterprises' green growth, which is the same as its reason level. As result, this factor can be excluded from the analysis. The reason level of Government Policy Support f_1 is higher with a certain centrality level, which will promote green growth.

Located in the fourth quadrant, Green Behavior Willingness f_8, Green Input Intensity f_9, Number of Technical Staff f_{10}, Green Strategy Formulation and Implementation f_{11}, Product Green Quality f_{12}, Corporate Green Image f_{13}, and Number of Patent Applications f_{14} are called Core Problems. They are key elements vulnerable to other factors' influence, which are involved in different ways in enterprise production and operation to promote green growth. Among them, Green Behavior Willingness f_8 is crucial to connect enterprises' external motivating factors and internal influencing factors by transforming external motivation into internal actions.

To sum up, six factors, namely environmental standard constraints, green market demand, market competition, green technology advancement, upstream and downstream green synergy, and government policy support, work together to enhance enterprises' willingness to conduct green behaviors, then generating green benefits to promote green growth of wooden flooring manufacturing enterprises. Therefore, driven by the ultimate goal of profit maximization, the above six factors provide the most important external support for the green growth of enterprises, especially the key driving force for wooden flooring manufacturing ones.

4. Conclusions

The green growth of manufacturing enterprises is certainly affected by interactions and joint influence of multiple factors, which is very complex and challenging to analyze the corresponding relationship.

This study uses the DEMATEL method to identify factors influencing the green growth of wooden flooring manufacturing enterprises, concluding with six factors, namely environmental standard constraints, green market demand, market competition, green technology advancement, upstream and downstream green synergy, together with government policy support as reason factors. They are the most important external support for enterprises' green growth, particularly major driving factors for wooden flooring manufacturing ones.

The DEMATEL method is relatively easy to operate and can generate clear and straightforward outcomes; however, there are also limitations, such as the subjective part of experts' scores and relatively few samples. In future studies, a larger scope and more samples can be utilized to obtain more reliable data. Apart from the DEMATEL method, fuzzy sets theory and the interpretative structural modeling method (ISM) can also be used to further analyze factors influencing the green growth of manufacturing enterprises [31].

Author Contributions: Conceptualization, W.L. and X.W.; methodology, W.L.; software, W.L.; validation, W.L. and X.W.; formal analysis, X.W.; investigation, W.L.; re-sources, W.L.; data curation, X.W.; writing—original draft preparation, W.L.; writing—review and editing, X.W.; visualization, W.L. and X.W.; supervision, X.W.; project administration, W.L.; funding acquisition, W.L. All authors have read and agreed to the published version of the manuscript.

Funding: This research was funded by the Research Project of Beijing Union University "Research on financial support for green technology innovation in manufacturing enterprises in Beijing, Tianjin and Hebei" (grant number: SK30202102).

Data Availability Statement: The data used to support findings in this study are available from corresponding authors upon request.

Acknowledgments: Appreciation to BUU for funding support. Thanks to the expert panel members for their professional ratings. Thanks to the reviewers for their rigorous review comments. We also thank the editors for their hard work in revising and typesetting.

Conflicts of Interest: The authors declare no conflict of interest.

References

1. Liu, W. Economic Growth and the Goal of Well-Off Society in an All-Round Way under the Impact of Epidemic Situation. *J. Manag. World* **2020**, *36*, 1–8.
2. He, W.B. Countermeasures for Digital Economy to Promote the Upgrading of China's Manufacturing Industry under the Visual of Global-Value-Chain. *Asia Pac. Econ. Rev.* **2020**, 3, 115–130.
3. Kong, D.J.; Yan, J.L.; Yang, X.Y.; Qu, X.M. A Study on the Issues of Conducting China's Manufacturing Power Strategy. *Chin. J. Eng. Sci.* **2017**, 3, 6–13.
4. Shi, D.; Li, P. Quality Evolution and Assessment of China's Industry over the Past Seven Decades. *China Ind. Econ.* **2019**, *9*, 5–23.
5. Gao, Q.S.; Li, T. Progress and Review in 'China Manufacturing 2025'. *J. Ind. Technol. Econ.* **2018**, *10*, 59–66.
6. Lin, H.; Hsu, I.; Lin, T.; Tung, L.; Ling, Y. After the Epidemic, Is the Smart Traffic Management System a Key Factor in Creating a Green Leisure and Tourism Environment in the Move towards Sustainable Urban Development? *Sustainability* **2022**, *14*, 3762. [CrossRef]
7. Shen, L.; Fan, R.; Wang, Y.; Yu, Z.; Tang, R. Impacts of Environmental Regulation on the Green Transformation and Upgrading of Manufacturing Enterprises. *Int. J. Environ. Res. Public Health* **2020**, *17*, 7680. [CrossRef]
8. Liu, X.; Tong, D.; Huang, J.; Zheng, W.; Kong, M.; Zhou, G. What Matters in the E-commerce Era? Modelling and Mapping Shop Rents in Guangzhou, China. *Land Use Policy* **2022**, *123*, 106430. [CrossRef]
9. Qiu, L.; Jie, X.W.; Wang, Y.N.; Zhao, M.J. Green Product Innovation, Green Dynamic Capability, and Competitive Advantage: Evidence from Chinese Manufacturing Enterprises. *Corp. Soc. Responsib. Environ. Manag.* **2020**, *27*, 146–165. [CrossRef]
10. Hong, D.Y. Economic Growth, Environmental Protection and Ecological Modernization: A Perspective from Environmental Sociology. *Soc. Sci. China* **2012**, *9*, 82–99.
11. Qu, C.; Shao, J.; Cheng, Z.H. Can Embedding in Global Value Chain Drive Green Growth in China's Manufacturing Industry? *J. Clean. Prod.* **2020**, *268*, 121962. [CrossRef]
12. Alfaro, M.; Teresa, B.; Diaz, R.A. Impact of Green Initiatives on the Financial Performance of Small and Medium Enterprises: The Case of Manufacturing Firm in Central Luzon. *Work. Pap. Oper. Manag.* **2021**, *12*, 28–41. [CrossRef]
13. Seth, D.; Rehman, M.A.A.; Shrivastava, R.L. Green Manufacturing Drivers and Their Relationships for Small and Medium (SME) and Large Industries. *J. Clean. Prod.* **2018**, *198*, 1381–1405. [CrossRef]
14. Zhu, F.F.; Hu, H.; Xu, F. Risk Assessment Model for International Construction Projects Considering Risk Interdependence Using the DEMATEL Method. *PLoS ONE* **2022**, *17*, e0265972. [CrossRef] [PubMed]
15. Yuan, T.; Wang, G.H.; Zhou, J.L. Evaluation on Growth of Enterprises in Technological-Cluster: From Perspective of Social Capital. *Technol. Econ.* **2014**, *6*, 1–6.
16. Yang, W. *Green Growth Key Influence Factors Study of 3PL Enterprises in China*; Dalian University of Technology: Dalian, China, 2017.
17. Ma, L.R.; Ma, D.C.; Li, X.F. Analysis on Influence Factors of Characteristic Agricultural Product Environmental Logistics Development Based on the DEMATEL in Gansu Province. *Logist. SciTech* **2016**, 3, 102–105.
18. Zeng, Y.Q. International Wood Industry. *Int. Wood Ind.* **2015**, *10*, 4–7.
19. Lu, P.; Wen, Y.; Li, Z.H.; Ding, Y.; Chen, X.Y.; Han, L.L. The Impacts of Environmental Regulation on Regional Green Productivity Growth in China: From the Perspective of Local-neighborhood Effects. *Econ. Res. Ekon. Istraz.* **2022**. [CrossRef]

20. Wang, M.; Yin, S.; Lian, S. Collaborative Elicitation Process for Sustainable Manufacturing: A Novel Evolution Model of Green Technology Innovation Path Selection of Manufacturing Enterprises under Environmental Regulation. *PLoS ONE* **2022**, *17*, e0266169. [CrossRef] [PubMed]
21. Berry, M.A.; Rondinelli, D.A. Proactive Corporate Environmental Management: A New Industrial Revolution. *AMP* **1998**, *2*, 38–50. [CrossRef]
22. Bansal, P.; Roth, K. Why Companies Go Green: A Model of Ecological Responsiveness. *Acad. Manag. J.* **2000**, *4*, 717–736. [CrossRef]
23. Zhu, Q.H. Models and Approaches of Green Supply Chain Based Product Eco-Design. *Chin. J. Manag.* **2008**, *3*, 360–365.
24. Hao, Z.T.; Yan, L.; Xie, X.B.; Duan, X.H. Identification and Analysis on Critical Influential Factors of Green Behavior Decision-making for Enterprises in Resource-Based Industry Cluster. *China Popul. Resour. Environ.* **2014**, *10*, 170–176.
25. Jiang, S.Y. The Impact of External Environmental Pressures and Perceived Opportunities on Corporate Green Performance. *Sci. Technol. Prog. Policy* **2015**, *11*, 72–76.
26. Qian, Y.; Liu, J.; Forrest, J.; Lin, Y. Impact of Financial Agglomeration on Regional Green Economic Growth: Evidence from China. *J. Environ. Plan. Manag.* **2022**, *65*, 1611–1636. [CrossRef]
27. Corrocher, N.; Malerba, F.; Morrison, A. Technological Regimes, Patent Growth, and Catching-up in Green Technologies. *Ind. Corp. Chang.* **2021**, *30*, 1084–1107. [CrossRef]
28. Cuang, H.M.; Lin, C.K.; Chen, D.R.; Chen, Y.S. Evolving MCDM Applications Using Hybrid Expert based ISM and DEMATEL Models: An Example of Sustainable Ecotourism. *Sci. World J.* **2013**, *2013*, 751728.
29. Lee, Y.C.; Li, M.L.; Yen, T.M.; Huang, T.H. Analysis of Adopting an Integrated Decision Making Trial and Evaluation Laboratory on a Technology Acceptance Model. *Expert Syst. Appl.* **2010**, *37*, 1745–1754. [CrossRef]
30. Wu, K.J.; Liao, C.J.; Tseng, M.L.; Chiu, A.S.F. Exploring Decisive Factors in Green Supply Chain Practices under Uncertainty. *Int. J. Prod. Econ.* **2015**, *159*, 147–157. [CrossRef]
31. Jalilibal, Z.; Bozorgi-Amiri, A. A Hybrid Grounded Theory, Fuzzy DEMATEL and ISM Method for Assessment of Sustainability Criteria for Project Portfolio Selection Problems. *Iran. J. Manag. Stud.* **2022**, *15*, 425–442.

processes

MDPI

Article

T-S Fuzzy Model-Based Fault Detection for Continuous Stirring Tank Reactor

Yanqin Wang [1,2,3], Weijian Ren [3,4,*], Zhuoqun Liu [5], Jing Li [1] and Duo Zhang [3,4]

1 School of Mechatronics Engineering, Daqing Normal University, Daqing 163712, China; wyqdqpi@126.com (Y.W.); lj_9009@163.com (J.L.)
2 Artificial Intelligence Energy Research Institute, Northeast Petroleum University, Daqing 163318, China
3 Heilongjiang Provincial Key Laboratory of Networking and Intelligent Control, Northeast Petroleum University, Daqing 163318, China; mail_zhangduo@126.com
4 School of Electrical and Information Engineering, Northeast Petroleum University, Daqing 163318, China
5 School of Electrical Engineering, University of Rostock, 18051 Rostock, Germany; liuzhuoqun8577@gmail.com
* Correspondence: renwj@126.com

Abstract: Continuous stirring tank reactors are widely used in the chemical production process, which is always accompanied by nonlinearity, time delay, and uncertainty. Considering the characteristic of the actual reaction of the continuous stirring tank reactors, the fault detection problem is studied in terms of the T-S fuzzy model. Through a fault detection filter performance analysis, the sufficient condition for the filtering error dynamics is obtained, which meets the exponential stability in the mean square sense and the given performance requirements. The design of the fault detection filter is transformed into one that settles the convex optimization issue of linear matrix inequality. Numerical analysis shows the effectiveness of this scheme.

Keywords: continuous stirring reactors; fault detection; T-S fuzzy model; channel fading

Citation: Wang, Y.; Ren, W.; Liu, Z.; Li, J.; Zhang, D. T-S Fuzzy Model-Based Fault Detection for Continuous Stirring Tank Reactor. *Processes* **2021**, 9, 2127. https://doi.org/10.3390/pr9122127

Received: 19 October 2021
Accepted: 19 November 2021
Published: 25 November 2021

1. Introduction

Continuous stirring tank reactors (CSTR) are the most widely used chemical reactors in chemical production [1]. The CSTR reaction process is an important chemical production process, and the complexity and risk of its operation are determined by the nonlinearity, time delay, and uncertainty of the reaction process. With the development of chemical equipment being geared towards integration and larger scales, the importance of fault detection (FD) for the reaction process has increased and the technology used in its performance is continuously being improved [2]. The nonlinear dynamic equation of CSTR can be established according to the equilibrium formula of reaction materials. However, in the actual production process, most of the systems are uncertain nonlinear systems, and the uncertainty is represented by model error, parameter perturbation, and unknown disturbance, which increases the complexity and difficulty of FD.

As is well known, the task of FD is to check whether there is a fault in the system and to determine the time of the fault occurrence [3]. During the past several decades, the technology for detecting faults has already been widely adopted in industrial processes and has gradually become a significant method of enhancing both system security and reliability [4–11]. For linear systems, the FD issue has been discussed since the 1970s, and several applicable FD methods have been developed [12–16]. Nevertheless, numerous industrial systems exhibit inherent nonlinearity. Nonlinearity is known to be a primary factor that impacts system performance. The existence of nonlinearity raises the system complexity, which simultaneously brings significant challenges to the issue of system analysis and synthesis. Note that these problems can no longer be solved by using the former FD approaches for linear systems. So far, the problem of FD for nonlinear systems has not been discussed enough [17–19].

On the other hand, the fuzzy set theory has been proven to be a powerful method in dealing with nonlinear systems, and a considerable number of reports have been published on it [20,21]. More particularly, a substantial amount of attention has been paid to the Takagi–Sugeno (T-S) fuzzy model for the reason that it can approach any smooth nonlinear system reaching an arbitrarily designated accuracy inside any compact set. This approach has been employed in numerous fields, e.g., electrical controlling, quantitative modeling, signal processing and pattern recognition, intelligent decision-making, and robot investigation [22,23]. Compared with the extensive research on controller and filter design problems with regard to the T-S fuzzy system, the corresponding FD problem has not been investigated thoroughly [24].

The channel fading phenomenon unavoidably occurs in systems linked through wireless and shared connections. As is known, the fading effect is one of the major features of wireless transmission. Diffraction, reflection, and scattering seriously affect signal power, which results in fading or attenuation. Some scholars have paid attention to the problem of channel fading, and some works have emerged. For instance, [25] studied the filtering problem of linear systems subject to channel fading. An event-based state-feedback controller is designed in [26] for interval type-2 fuzzy systems over fading channels. Nevertheless, despite the large number of research findings about filtering and control issues in the case of channel fading [27], the FD problem still has not received enough attention.

Inspired by the aforementioned statements, this paper is devoted to dealing with the FD issue in CSTR with regard to parameter uncertainty and channel fading within a networked environment and in terms of the T-S fuzzy model. We are to realize the FD by carrying out the fuzzy FD filtering, which presents a residual signal in order to obtain the estimate of the fault signal. The primary principle is to decrease the error between the residual and the fault to the minimum. Distinct from other published results in previous papers, the highlights of this paper are as follows: (1) the issue discussed is novel in view of the fact that this paper represents the first of a few endeavors to settle the H_∞ fault detection issue against parameter uncertainties, channel fading, and delays for the CSTR reaction process; (2) the considered system is comprehensive and reflects the reality of the CSTR reaction process, which involves the Takagi-Sugeno fuzzy model, parameter uncertainties, time delay, and channel fading; and (3) a specific fault detection scheme is proposed, which ensures that CSTR fuzzy systems achieve exponential stability in the mean square and H_∞ performance.

The rest of this paper is organized as follows. The T-S fuzzy model of CSTR is established in Section 2. The performance of an FD dynamic system is analyzed in Section 3. A fuzzy FD filter is designed in Section 4. Section 5 presents a numerical example. A conclusion is given in Section 6.

2. Model of CSTR

The material enters CSTR at a certain concentration and temperature for exothermic reaction. The operational goal is to continuously adjust the coolant temperature to make the product concentration and reactor temperature meet the production requirements, as shown in Figure 1. Based on the law of energy conservation and the principle of chemical dynamics, the dimensionless mechanism model of the CSTR system is as follows [1]:

$$\dot{x}_1(t) = D_\alpha[1 - x_1(t)]\exp\left[\frac{x_2(t)}{1 + x_2(t)/\gamma_0}\right] - \frac{1}{\lambda}x_1(t) + \left(\frac{1}{\lambda} - 1\right)x_1(t - d(t))$$

$$\begin{aligned}\dot{x}_2(t) &= HD_\alpha[1 - x_1(t)]\exp\left[\tfrac{x_2(t)}{1+x_2(t)/\gamma_0}\right] - (\tfrac{1}{\lambda} + \beta)x_2(t) + \left(\tfrac{1}{\lambda} - 1\right)x_2(t - d(t)) + \zeta w(t) \\ &+\delta f(t)\end{aligned}$$

where $x_1(t) = \frac{C_0 - C_a(t)}{C_0}$ and $x_2(t) = \frac{\gamma_0(T_a(t) - T_0)}{T_0}$ represent the dimensionless product concentration and reactor temperature, respectively.

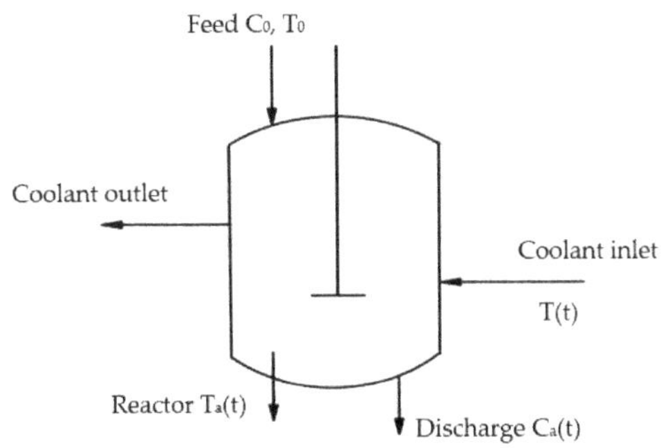

Figure 1. Schematic of CSTR.

The symbols in the formula are explained as follows: λ, D_α, γ_0, H, β, T_0 are dimensionless system parameters, ζ is the disturbance coefficient, $w(t)$ is the external disturbance, $d(t)$ is the term of variable time delay. In this paper, the T-S fuzzy model is adopted in order to approach the mechanism model. The reactor temperature, which is easier to measure online, is chosen as the precursor variable, and the linear processing is carried out near each steady-state equilibrium point. Then, considering the parameter uncertainty, the T-S fuzzy model is obtained, which is expressed as follows:

Plant Rule i: IF $\theta_1(k)$ is M_{i1}, $\theta_2(k)$ is M_{i2}, ... $\theta_p(k)$ is M_{ip}, then

$$\begin{cases} x(k+1) = (A_i + \Delta A_i)x(k) + (A_{di} + \Delta A_{di})x(k-d(k)) + D_{1i}w(k) + G_i f(k) \\ y(k) = C_i x(k) + D_{2i}w(k) \\ x(k) = \psi(k),\ \forall k \in \left[-\bar{d}, 0\right] \end{cases} \tag{1}$$

where r is the IF-THEN rule number; M_{ij} is the fuzzy set; $\theta(k) = [\theta_1(k), \theta_2(k), \cdots, \theta_p(k)]$ is the premise variable vector; $x(k) \in \mathbb{R}^n$ is the state vector; $y(k) \in \mathbb{R}^m$ is the measurement output; $w(k) \in \mathbb{R}^q$ is the disturbance input; $f(k) \in \mathbb{R}^l$ is the fault signal; $w(k)$ and $f(k)$ belong to $l_2[0,\infty)$; $0 \le d(k) \le \bar{d}$ represents time delay; system matrices A_i, C_i, D_{1i}, D_{2i}, and G_i are given real-valued matrices with appropriate dimensions; $\psi(k)$, $k \in \left[-\bar{d}, 0\right]$ is the given initial state and satisfies $\sup_{k\in[-\bar{d},0]} \mathbb{E}\left\{\|\psi(i)\|^2\right\} < \infty$; ΔA_i and ΔA_{di} represent norm-bounded parameter uncertainties, which satisfy the following formula:

$$\begin{bmatrix} \Delta A_i & \Delta A_{di} \end{bmatrix} = H_i F(k) \begin{bmatrix} E_a & E_d \end{bmatrix} \tag{2}$$

where $F(k)$ is the unknown matrix that satisfies $F^T(k)F(k) \le I$, and H_i, E_a, E_d stand for known matrices with appropriate dimensions.

For the T-S fuzzy system (1), the defuzzified output is denoted as follows:

$$\begin{cases} x(k+1) = \sum\limits_{i=1}^{r} h_i(\theta(k))[(A_i + \Delta A_i)x(k) + (A_{di} + \Delta A_{di})x(k-d(k)) \\ \qquad\qquad + D_{1i}w(k) + G_i f(k)] \\ y(k) = \sum\limits_{i=1}^{r} h_i(\theta(k))[C_i x(k) + D_{2i}w(k)] \\ x(k) = \psi(k),\ \forall k \in \mathbb{Z}^- \end{cases} \tag{3}$$

where the fuzzy basis functions are described as

$$h_i(\theta(k)) = \frac{\vartheta_i(\theta(k))}{\sum\limits_{i=1}^{r} \vartheta_i(\theta(k))}$$

with $\vartheta_i(\theta(k)) = \prod_{j=1}^{p} M_{ij}(\theta_j(k))$, $\vartheta_i(\theta(k)) \geq 0, i = 1, 2, \cdots, r$, $\sum_{i=1}^{r} \vartheta_i(\theta(k)) > 0$, $M_{ij}(\theta_j(k))$ denoting the membership of $\theta_j(k)$ in M_{ij}, understandably.

$$h_i(\theta(k)) \geq 0,\ i = 1, 2, \cdots, r,\ \sum_{i=1}^{r} h_i(\theta(k)) = 1$$

For simplicity, we denote $h_i = h_i(\theta(k))$.

Considering that the fading phenomenon occurs in the transmission process of the measurement signal from the sensor to the FD filter, based on the Lth-order rice fading model, the measurement signal obtained by the fault detection filter is expressed in the following form:

$$y_f(k) = \sum_{s=0}^{\ell} \beta_s(k) y(k-s) + E_y \xi(k) \tag{4}$$

where ℓ is a given positive scalar and $\beta_k^s (s = 0, 1, \cdots, \ell)$ represent the channel coefficients, and they are mutually independent. Moreover, β_k^s own the probability density function over the interval [0, 1], which has the expectation $\overline{\beta}_s$ and variance $\widetilde{\beta}_s^*$. $\xi_k \in l_2([0, \infty); \mathbb{R}^m)$ stands for external noise and E_y denotes a given real-valued matrix with a proper dimension.

Remark 1. *In this paper, channel fadings are characterized* via *the improved Lth-order Rice model. Such a model has been extensively utilized in fields of signal processing and remote control due to its capacity to describe both channel fadings and random time-delays at the same time. Differing from the conventional model of channel fadings, in model (4), the channel coefficients are described by random variables obeying an arbitrary probabilistic distribution over the interval [0, 1]. Note that the consideration of channel fadings increases the complexity of acquiring the FD filter.*

Taking into account the physical object described by (1) and (2), an FD filter is constructed with the following expression:

Filter Rule i: IF $\theta_1(k)$ is M_{i1}, $\theta_2(k)$ is M_{i2}, ... $\theta_p(k)$ is M_{ip}, then

$$\begin{cases} \hat{x}(k+1) = A_{fi}\hat{x}(k) + B_{fi} y_f(k) \\ r(k) = C_{fi}\hat{x}(k) + D_{fi} y_f(k) \end{cases} \tag{5}$$

where $\hat{x}(k) \in \mathbb{R}^n$ denotes the state vector of the filter, $r(k) \in \mathbb{R}^l$ represents the residual signal being compatible with the fault signal $f(k)$, A_{fi}, B_{fi}, C_{fi}, and D_{fi} are appropriately dimensioned filter gains to be decided. Therefore, the whole fuzzy fault detection filter is constructed in the following formulation:

$$\begin{cases} \hat{x}(k+1) = \sum_{i=1}^{r} h_i[A_{fi}\hat{x}(k) + B_{fi} y_f(k)] \\ r(k) = \sum_{i=1}^{r} h_i[C_{fi}\hat{x}(k) + D_{fi} y_f(k)]. \end{cases} \tag{6}$$

In what follows, we denote

$$\sum_{a_1, a_2, \cdots, a_s = 1}^{r} h_{a_1} h_{a_2} \cdots h_{a_s} = \sum_{a_1=1}^{r} h_{a_1} \sum_{a_2=1}^{r} h_{a_2} \cdots \sum_{a_s=1}^{r} h_{a_s}, \forall s \geq 1$$

$$\eta(k) = \left[x^T(k)\ \hat{x}^T(k)\right]^T,\ v(k) = \left[w^T(k)\ \xi^T(k)\ f^T(k)\right]^T,\ \overline{r}(k) = r(k) - f(k),$$

$$\eta^*(k) = \left[\eta^T(k-1)\ \eta^T(k-2)\ \cdots\ \eta^T(k-\ell)\right]^T,\ \hat{v}(k) = \left[\ v^T(k)\quad v^{*T}(k)\ \right]^T$$

$$v^*(k) = \left[v^T(k-1)\ v^T(k-2) \cdots v^T(k-\ell)\right]^T.$$

By (3) and (6), the following FD dynamic system can be obtained:

$$\begin{cases} \eta(k+1) = \sum\limits_{i,j=1}^{r} h_i h_j \Big[(\overline{A}_{ij} + \Delta\overline{A}_{ij} + \widetilde{\beta}_0(k)\hat{A}_{ij})\eta(k) + (\overline{A}_{di} + \Delta\overline{A}_{di})\eta(k-d(k)) \\ \qquad + (\overline{\Lambda}_l A^*_{ij} + \widetilde{\Lambda}_l(k) A^*_{ij})\eta^*(k) + (\overline{B}_{ij} + \widetilde{\beta}_0(k)\hat{B}_{ij})v(k) \\ \qquad + (\overline{\Lambda}_l B^*_{ij} + \widetilde{\Lambda}_l(k) B^*_{ij})v^*(k) \Big] \\ \overline{r}(k) = \sum\limits_{i,j=1}^{r} h_i h_j [(\overline{C}_{ij} + \widetilde{\beta}_0(k)\hat{C}_{ij})\eta(k) + (\overline{\Lambda}_l C^*_{ij} + \widetilde{\Lambda}_l(k) C^*_{ij})\eta^*(k) \\ \qquad + (\overline{D}_{ij} + \widetilde{\beta}_0(k)\hat{D}_{ij})v(k) + (\overline{\Lambda}_l D^*_{ij} + \widetilde{\Lambda}_l(k) D^*_{ij})v^*(k)] \end{cases} \tag{7}$$

where

$$\overline{A}_{ij} = \begin{bmatrix} A_i & 0 \\ B_{fj}C_i & A_{fj} \end{bmatrix}, \Delta\overline{A}_i = \begin{bmatrix} \Delta A_i & 0 \\ 0_i & 0 \end{bmatrix}, \hat{A}_{ij} = \begin{bmatrix} 0 & 0 \\ B_{fj}C_i & 0 \end{bmatrix}, \overline{B}_{ij} = \begin{bmatrix} D_{1i} & 0 & G_i \\ \overline{\beta}_0 B_{fj} D_{2i} & B_{fj}E_y & 0 \end{bmatrix},$$

$$\hat{B}_{ij} = \begin{bmatrix} 0 & 0 & 0 \\ B_{fj}D_{2i} & 0 & 0 \end{bmatrix}, \overline{A}_{di} = \begin{bmatrix} A_{di} & 0 \\ 0 & 0 \end{bmatrix}, \Delta\overline{A}_{di} = \begin{bmatrix} \Delta A_{di} & 0 \\ 0 & 0 \end{bmatrix}, \overline{C}_{ij} = \begin{bmatrix} \overline{\beta}_0 D_{fj} C_i & C_{fj} \end{bmatrix},$$

$$\hat{C}_{ij} = \begin{bmatrix} D_{fj}C_i & 0 \end{bmatrix}, \overline{D}_{ij} = \begin{bmatrix} \overline{\beta}_0 D_{fj} D_{2i} & E_y & -I \end{bmatrix}, \hat{D}_{ij} = \begin{bmatrix} D_{fj}D_{2i} & 0 & 0 \end{bmatrix},$$

$$A^*_{ij} = \text{diag}\{\underbrace{\hat{A}_{ij}, \cdots, \hat{A}_{ij}}_{\ell}\}, B^*_{ij} = \text{diag}\{\underbrace{\hat{B}_{ij}, \cdots, \hat{B}_{ij}}_{\ell}\}, \overline{\Lambda}_l = [\overline{\beta}_1 I, \cdots, \overline{\beta}_l I],$$

$$\widetilde{\Lambda}_l(k) = [\widetilde{\beta}_1 I, \cdots, \widetilde{\beta}_l I], C^*_{ij} = \text{diag}\{\underbrace{\hat{C}_{ij}, \cdots, \hat{C}_{ij}}_{\ell}\}, D^*_{ij} = \text{diag}\{\underbrace{\hat{D}_{ij}, \cdots, \hat{D}_{ij}}_{\ell}\},$$

$$\widetilde{\alpha}_m(k) = \alpha_m(k) - \overline{\alpha}_m, \mathbb{E}\{\widetilde{\alpha}_m(k)\} = 0, \mathbb{E}\left\{\widetilde{\alpha}^2_m(k)\right\} = \overline{\alpha}_m(1-\overline{\alpha}_m),$$

$$\mathbb{E}\left\{\widetilde{\beta}^2_s(k)\right\} = \widetilde{\beta}^*_s, \widetilde{\beta}_s(k) = \beta_s(k) - \overline{\beta}_s (s = 0, 1, \ldots, l).$$

Definition 1. *With the FD dynamic system (7) and each initial condition ψ, in the situation of $\hat{v}(k) = 0$, system (7) is said to be exponentially mean-square stable if there are constants $\delta > 0$ and $0 < \kappa < 1$, which achieve the following [28].*

$$\mathbb{E}\left\{\|\eta(k)\|^2\right\} \leq \delta\kappa^k \sup_{l \in Z} \mathbb{E}\left\{\|\psi(i)\|^2\right\}, \forall k \geq 0.$$

Thus, the ideal FD filter is designed via the following steps:

Step (1) Introduce a residual signal. With system (2), a fuzzy FD filter expressed as (5) is designed to produce a residual signal $r(k)$. Then, the filter is devised to guarantee that the whole FD system (6) achieves exponential stability in the mean square and the following H_∞ performance under the zero-initial condition:

$$\sum_{k=0}^{\infty} \mathbb{E}\left\{\|\overline{r}(k)\|^2\right\} \leq \gamma^2 \sum_{k=0}^{\infty} \|\hat{v}(k)\|^2 \tag{8}$$

where $\hat{v}(k) \neq 0$ and $\gamma > 0$ are made as small as possible in the feasibility of (8).

Step (2) Establish a residual evaluation stage containing an evaluation function $J(k)$ and a threshold J_{th} as follows [29]:

$$J(k) = \left\{ \sum_{k=s-L}^{k=s} r^T(k) r(k) \right\}^{\frac{1}{2}}, \; J_{\text{th}} = \sup_{w \in l_2, f=0} \mathbb{E}\{J(k)\} \tag{9}$$

where L is the length of the finite evaluating time horizon. Based on (9), whether a fault occurs is detected according to the rule below:

$$J(k) > J_{\text{th}} \rightarrow \text{ fault occurs and alarm}$$

$$J(k) \leq J_{\text{th}} \rightarrow \text{no fault occurs.}$$

3. Performance Analysis of an FD Dynamic System

In this part, we are concerned with the performance analysis of the FD filter for the T-S fuzzy system, as stated previously. Before proceeding, we present several useful lemmas:

Lemma 1. *(Schur Complement) Given constant matrices* $X = \begin{bmatrix} X_{11} & X_{12} \\ X_{21} & X_{22} \end{bmatrix}$*, where* X_{11} *is* $r \times r$*, the following three conditions are equivalent:*

(i) $X < 0$;

(ii) $X_{11} < 0, X_{22} - X_{12}^T X_{11}^{-1} X_{12} < 0$;

(iii) $X_{22} < 0, X_{11} - X_{12} X_{22}^{-1} X_{12}^T < 0$.

Lemma 2. *(S-procedure) Given matrix* $E = E^T$*,* M *and* N *are real matrices with suitable dimensions, and* F *satisfies* $F^T F \leq I$*, then the sufficient condition for* $E + MFN + N^T F^T M^T < 0$ *is that there is a positive number, so that*

$$E + \mu M M^T + \mu^{-1} N^T N < 0 \text{ or } \Pi = \begin{bmatrix} E & \mu M & N^T \\ \mu M^T & -\mu I & 0 \\ N & 0 & -\mu I \end{bmatrix} < 0.$$

Lemma 3. *For any real matrices* X_{ij}*,* $i, j = 1, 2, \cdots, r$ *and* $\Lambda > 0$ *with proper dimensions, one has [30].*

$$\sum_{i=1}^{r} \sum_{j=1}^{r} \sum_{k=1}^{r} \sum_{l=1}^{r} h_i h_j h_k h_l X_{ij}^T \Lambda X_{kl} \leq \sum_{i=1}^{r} \sum_{j=1}^{r} h_i h_j X_{ij}^T \Lambda X_{ij} \tag{10}$$

The following analysis outcome provides a theoretical basis for the subsequent discussion.

Theorem 1. *For the fuzzy CSTR system (2) with known filter parameters and a specified* H_∞ *performance* $\gamma > 0$*. The fuzzy FD system (6) becomes exponentially stable in the mean square with a disturbance attenuation level* γ *if there are positive definite matrices* $P > 0$ and $Q > 0$ *satisfying*

$$\Pi_{ii}^T \widetilde{P} \Pi_{ii} + \widetilde{\Pi}_{ii}^T \acute{P} \widetilde{\Pi}_{ii} + \overline{P}_{ii} < 0 \tag{11}$$

$$2(\overline{P}_{ii} + \overline{P}_{jj}) + (\Pi_{ij} + \Pi_{ji})^T \widetilde{P} (\Pi_{ij} + \Pi_{ji}) + (\widetilde{\Pi}_{ij} + \widetilde{\Pi}_{ji})^T \acute{P} (\widetilde{\Pi}_{ij} + \widetilde{\Pi}_{ji}) < 0 \tag{12}$$

where

$$\Pi_{ij} = \begin{bmatrix} \overline{A}_{ij} + \Delta\overline{A} & \overline{A}_{di} + \Delta\overline{A}_{di} & \overline{\Lambda}_l A_{ij}^* & \overline{B}_{ij} & \overline{\Lambda}_l B_{ij}^* \\ \overline{C}_{ij} & 0 & \overline{\Lambda}_l C_{ij}^* & \overline{D}_{ij} & \overline{\Lambda}_l D_{ij}^* \end{bmatrix},$$
$$\widetilde{\Pi}_{ij} = \begin{bmatrix} \widetilde{\Pi}_{1ij}^T & \widetilde{\Pi}_{2ij}^T & \widetilde{\Pi}_{3ij}^T & \widetilde{\Pi}_{4ij}^T \end{bmatrix}^T,$$
$$\widetilde{\Pi}_{1ij} = \begin{bmatrix} \check{\beta}\hat{A}_{ij} & 0 & 0 & \check{\beta}\hat{B}_{ij} & 0 \end{bmatrix}, \widetilde{\Pi}_{2ij} = \begin{bmatrix} 0 & 0 & \check{\beta}_l A_{ij}^* & 0 & \check{\beta}_l B_{ij}^* \end{bmatrix},$$
$$\widetilde{\Pi}_{3ij} = \begin{bmatrix} \check{\beta}\hat{C}_{ij} & 0 & 0 & \check{\beta}\hat{D}_{ij} & 0 \end{bmatrix}, \widetilde{\Pi}_{4ij} = \begin{bmatrix} 0 & 0 & \check{\beta}_l C_{ij}^* & 0 & \check{\beta}_l D_{ij}^* \end{bmatrix},$$
$$\acute{P} = \text{diag}\{P, P_\ell, I, I\}, R = I_{\ell+2} \otimes P, \overline{P}_{is} = \text{diag}\{\overline{Q}, -Q, -\gamma^2 I\},$$
$$\overline{Q} = -P + (\overline{d} + 1)Q + \sum_{l=1}^{\ell} R_l,$$
$$\widetilde{P} = \text{diag}\{P, I\}, R_l = \text{diag}\{R_1, \cdots, R_\ell\}, \check{\beta} = \sqrt{\widetilde{\beta}_0^*} I, \check{\beta}_l = \text{diag}\left\{\sqrt{\widetilde{\beta}_1^*} I, \cdots, \sqrt{\widetilde{\beta}_\ell^*} I\right\},$$
$$\widetilde{\Lambda}_l^* = \text{diag}\left\{\widetilde{\beta}_1^*, \cdots, \widetilde{\beta}_l^*\right\}, \widetilde{\beta}_l^* = \text{E}\left\{\widetilde{\beta}_l^2(k)\right\},$$
$$\text{E}\left\{\widetilde{\Lambda}_l^T(k) P \widetilde{\Lambda}_l(k)\right\} = \text{diag}\left\{\widetilde{\beta}_1^* P, \cdots, \widetilde{\beta}_l^* P\right\} \triangleq \widetilde{\Lambda}_l^* \otimes P.$$

Proof. For simplicity, denote $\hat{\eta}(k) = \begin{bmatrix} \eta^T(k) & \eta^T(k-d(k)) & \eta^{*T}(k) & v^T(k) & v^{*T}(k) \end{bmatrix}^T$. With the dynamic system (7), define the following Lyapunov function:

$$V(k) = \sum_{i=1}^{4} V_i(k) \tag{13}$$

where

$$V_1(k) = \eta^T(k)P\eta(k),\ V_2(k) = \sum_{i=k-d(k)}^{k-1} \eta^T(i)Q\eta(i),$$
$$V_3(k) = \sum_{n=-\bar{d}+1}^{0} \sum_{i=k+n}^{k-1} \eta^T(i)Q\eta(i),\ V_4(k) = \sum_{l=1}^{\ell} \sum_{i=k-l}^{k-1} \eta^T(i)R_l\eta(i)$$

where $P > 0$ and $Q > 0$ denote unknown matrices yet to be decided. By (7), one has

$$\begin{aligned}
&\mathbb{E}\{\Delta V_1(k)\} = \mathbb{E}\{\eta^T(k+1)P\eta(k+1) - \eta^T(k)P\eta(k)\} \\
=\ &\mathbb{E}\{ \sum_{i,j,s,t=1}^{r} h_ih_jh_sh_t[\eta^T(k)((\overline{A}_{ij} + \Delta\overline{A}_i)^T P(\overline{A}_{st} + \Delta\overline{A}_s) + \widetilde{\beta}_0^* \hat{A}_{ij}^T P\hat{A}_{st} - P)\eta(k) \\
&+2\eta^T(k)(\overline{A}_{ij} + \Delta\overline{A}_i)^T P\overline{\Lambda}_l A_{st}^*\eta^*(k) + 2\eta^T(k)(\overline{A}_{ij} + \Delta\overline{A}_i)^T P\overline{B}_{st}v(k) \\
&+2\widetilde{\beta}_0^*\eta^T(k)\hat{A}_{ij}^T P\hat{B}_{st}v(k) + 2\eta^T(k)(\overline{A}_{ij} + \Delta\overline{A}_i)^T P\overline{\Lambda}_l B_{st}^* v^*(k) \\
&+2\eta^T(k)(\overline{A}_{ij} + \Delta\overline{A}_i)^T P(\overline{A}_{ds} + \Delta\overline{A}_{ds})\eta(k-d(k)) \\
&+\eta^{*T}(k)A_{ij}^{*T}\overline{\Lambda}_l^T P\overline{\Lambda}_l A_{st}^*\eta^*(k) + \eta^{*T}(k)A_{ij}^{*T}(\widetilde{\Lambda}_l^* \otimes P)A_{st}^*\eta^*(k) \\
&+2\eta^{*T}(k)A_{ij}^{*T}\overline{\Lambda}_l^T P\overline{B}_{st}v(k) + 2\eta^{*T}(k)A_{ij}^{*T}\overline{\Lambda}_l^T P\overline{\Lambda}_l B_{st}^* v^*(k) \\
&+2\eta^{*T}(k)A_{ij}^{*T}(\widetilde{\Lambda}_l^* \otimes P)B_{st}^* v^*(k) + 2\eta^{*T}(k)A_{ij}^{*T}\overline{\Lambda}_l^T P(\overline{A}_{ds} + \Delta\overline{A}_{ds})\eta(k-d(k)) \\
&+v^T(k)\overline{B}_{ij}^T P\overline{B}_{st}v(k) + \widetilde{\beta}_0^* v^T(k)\hat{B}_{ij}^T P\hat{B}_{st}v(k) + 2v^T(k)\overline{B}_{ij}^T P\overline{\Lambda}_l B_{st}^* v^*(k), \\
&+2v^T(k)\overline{B}_{ij}^T P(\overline{A}_{ds} + \Delta\overline{A}_{ds})\eta(k-d(k)) + v^{*T}(k)B_{ij}^{*T}\overline{\Lambda}_l^T P\overline{\Lambda}_l B_{st}^* v^*(k) \\
&+v^{*T}(k)B_{ij}^{*T}(\widetilde{\Lambda}_l^* \otimes P)B_{st}^* v^*(k) + 2v^{*T}(k)B_{ij}^{*T}\overline{\Lambda}_l^T P(\overline{A}_{ds} + \Delta\overline{A}_{ds})\eta(k-d(k)) \\
&+\eta^T(k-d(k)(\overline{A}_{di} + \Delta\overline{A}_{di})P(\overline{A}_{ds} + \Delta\overline{A}_{ds})\eta(k-d(k))],
\end{aligned} \tag{14}$$

$$\begin{aligned}
&\mathbb{E}\{\Delta V_2(k)\} = \mathbb{E}\{V_2(k+1) - V_2(k)\} \\
&\leq [\eta^T(k)Q\eta(k) - \eta^T(k-d(k))Q\eta(k-d(k)) + \sum_{i=k-\bar{d}+1}^{k} \eta^T(i)Q\eta(i)],
\end{aligned} \tag{15}$$

$$\begin{aligned}
&\mathbb{E}\{\Delta V_3(k)\} = \mathbb{E}\{V_3(k+1) - V_3(k)\} \\
&\leq \mathbb{E}\left[\bar{d}\eta^T(k)Q\eta(k) - \sum_{i=k-\bar{d}+1}^{k} \eta^T(i)Q\eta(i)\right],
\end{aligned} \tag{16}$$

$$\begin{aligned}
&\mathbb{E}\{\Delta V_4(k)\} = \mathbb{E}\{V_4(k+1) - V_4(k)\} \\
&= \sum_{l=1}^{\ell}\left\{ \sum_{i=k+1-l}^{k} \eta^T(i)R_l\eta(i) - \sum_{i=k-l}^{k-1} \eta^T(i)R_l\eta(i) \right\} \\
&= \sum_{l=1}^{\ell} \{\eta^T(k)R_l\eta(k) - \eta^T(k-l)R_l\eta(k-l)\}.
\end{aligned} \tag{17}$$

In the next stage, firstly, we are to verify the exponential stability of the FD dynamic system (7) with $\hat{v}(k) = 0$. By (14)–(17) and Lemma 1, we acquire the following:

$$\begin{aligned}
&\mathbb{E}\{\Delta V_1(k)|\hat{v}(k)=0\}\\
\leq\ &\mathbb{E}\sum_{i,j,s,t=1}^{r} h_ih_jh_sh_t[\eta^T(k)((\overline{A}_{ij}+\Delta\overline{A}_i)^T P(\overline{A}_{st}+\Delta\overline{A}_s)+\widetilde{\beta}_0^*\hat{A}_{ij}^T P\hat{A}_{st}-P)\eta(k)\\
&+2\eta^T(k)(\overline{A}_{ij}+\Delta\overline{A}_i)^T P\overline{\Lambda}_l A_{st}^*\eta^*(k)+2\eta^T(k)(\overline{A}_{ij}+\Delta\overline{A}_i)^T P\overline{\Lambda}_l B_{st}^* v^*(k)\\
&+2\eta^T(k)(\overline{A}_{ij}+\Delta\overline{A}_i)^T P(\overline{A}_{ds}+\Delta\overline{A}_{ds})\eta(k-d(k))+\eta^{*T}(k)A_{ij}^{*T}\overline{\Lambda}_l^T P\overline{\Lambda}_l A_{st}^*\eta^*(k)\\
&+\eta^{*T}(k)A_{ij}^{*T}(\widetilde{\Lambda}_l^*\otimes P)A_{st}^*\eta^*(k)+2\eta^{*T}(k)A_{ij}^{*T}\overline{\Lambda}_l^T P(\overline{A}_{ds}+\Delta\overline{A}_{ds})\eta(k-d(k))\\
&+\eta^T(k-d(k)(\overline{A}_{di}+\Delta\overline{A}_{di})P(\overline{A}_{ds}+\Delta\overline{A}_{ds})\eta(k-d(k))].
\end{aligned} \tag{18}$$

Denote $\mathrm{A}_{ij}=[\overline{A}_{ij}+\Delta\overline{A}\overline{A}_{di}+\Delta\overline{A}_{di}\overline{\Lambda}_l A_{ij}^*]$, $\widetilde{\mathrm{A}}_{ij}=\mathrm{diag}\left\{\check{\beta}\hat{A}_{ij},0,\check{\beta}_l A_{ij}^*\right\}$, then

$$\begin{aligned}
&\mathbb{E}\{\Delta V(k)|\hat{v}(k)=0\}\\
\leq\ &\mathbb{E}\{\sum_{i,j,s,t=1}^{r} h_ih_jh_sh_t\hat{\eta}^T(k)(\mathrm{A}_{ij}{}^T\mathrm{PA}_{st}+\widetilde{\mathrm{A}}_{ij}^T\mathcal{P}\widetilde{\mathrm{A}}_{st}+\hat{P}_{is})\hat{\eta}(k)\}\\
\leq\ &\sum_{i,j=1}^{r} h_ih_j\hat{\eta}^T(k)\mathrm{A}_i{}^T\mathrm{PA}_{ij}+\widetilde{\mathrm{A}}_{ij}^T P\widetilde{\mathrm{A}}_{ij}+\hat{P}_{ii})\hat{\eta}(k)\\
\leq\ &\sum_{i=1}^{r} h_{ij}{}^2\hat{\eta}^T(k)\mathrm{A}_{ii}{}^T P\mathrm{A}_{ii}+\widetilde{\mathrm{A}}_{ii}^T P\widetilde{\mathrm{A}}_{ii}+\hat{P}_{ii})\hat{\eta}(k)\\
&+\tfrac{1}{2}\sum_{i,j=1,i<j}^{r} h_ih_j\hat{\eta}^T(k)[(\mathrm{A}_{ij}=\mathrm{A}_{ji})^T P(\mathrm{A}_{ij}=\mathrm{A}_{ji})\\
&+(\widetilde{\mathrm{A}}_{ij}=\widetilde{\mathrm{A}}_{ji})^T R(\widetilde{\mathrm{A}}_{ij}=\widetilde{\mathrm{A}}_{ji})+2(\hat{P}_{ii}+\hat{P}_{jj})]\hat{\eta}(k)
\end{aligned} \tag{19}$$

where $\hat{P}_{is}=\mathrm{diag}\{\overline{Q},-Q,-R_l\}$, $\overline{Q}=-P+(\bar{d}+1)Q+\sum_{l=1}^{\ell}R_l$.

By Theorem 1, we have $\Omega<0$. Furthermore, according to the method used in the proof in reference [31], it is observed that system (7) reaches exponential stability. Next, the H_∞ performance of fuzzy dynamic system (7) is analyzed. Suppose zero initial conditions and construct the exponential function as follows:

$$\begin{aligned}
J(n)&=\mathbb{E}\sum_{k=0}^{n}[\bar{r}^T(k)\bar{r}(k)-\gamma^2\hat{v}^T(k)\hat{v}(k)]\\
&\leq\mathbb{E}\sum_{k=0}^{n}[\bar{r}^T(k)\bar{r}(k)-\gamma^2 v^T(k)v(k)-\gamma^2 v^{*T}(k)v^*(k)+\Delta V(k)].
\end{aligned} \tag{20}$$

It can be deduced from (7) that

$$\begin{aligned}
\bar{r}^T(k)\bar{r}(k)=\ &\sum_{i,j,s,t=1}^{r} h_ih_jh_sh_t[\eta^T(k)(\overline{C}_{ij}^T\overline{C}_{st}+\widetilde{\beta}_0^*\hat{C}_{ij}^T\hat{C}_{st})\eta(k)\\
&+2\eta^T(k)(\overline{C}_{ij}^T\overline{\Lambda}_l C_{st}^*)\eta^*(k)+2\eta^T(k)(\overline{C}_{ij}^T\overline{D}_{st}+\widetilde{\beta}_0^*\hat{C}_{ij}^T\hat{D}_{st})v(k)\\
&+2\eta^T(k)(\overline{C}_{ij}^T\overline{\Lambda}_l D_{st}^*)v^*(k)+\eta^{*T}(k)(C_{ij}^{*T}\overline{\Lambda}_l^T\overline{\Lambda}_l C_{st}^*+\widetilde{\Lambda}_l^* C_{ij}^{*T}C_{st}^*)\eta^*(k)\\
&+2\eta^{*T}(k)((\overline{\Lambda}_l C_{ij}^*)^T\overline{D}_{st})v(k)+2\eta^{*T}(k)(C_{ij}^{*T}\overline{\Lambda}_l^T\overline{\Lambda}_l D_{st}^*+\widetilde{\Lambda}_l^* C_{ij}^{*T}D_{st}^*)v^*(k)\\
&+v^T(k)(\overline{D}_{ij}^T\overline{D}_{st}+\widetilde{\beta}_0^*\hat{D}_{ij}^T\hat{D}_{st})v(k)+2v^T(k)(\overline{D}_{ij}^T\overline{\Lambda}_l D_{st}^*)v^*(k)\\
&+v^{*T}(k)(D_{ij}^{*T}\overline{\Lambda}_l^T\overline{\Lambda}_l D_{st}^*+\widetilde{\Lambda}_l^* D_{ij}^{*T}D_{st}^*)v^*(k)].
\end{aligned} \tag{21}$$

Denote

$$\hat{\eta}(k)=\begin{bmatrix}\eta^T(k) & \eta^T(k-d(k)) & \eta^{*T}(k)\end{bmatrix}^T,\ \widetilde{\eta}(k)=\begin{bmatrix}\hat{\eta}^T(k) & v^T(k) & v^{*T}(k)\end{bmatrix}^T.$$

By (19) and (20) and Lemma 1, we have

$$
\begin{aligned}
J(n) \leq \mathbb{E}\left\{ \sum_{k=0}^{n} \sum_{i,j,s,t=1}^{r} h_i h_j h_s h_t \widetilde{\eta}^T(k)(\Pi_{ij}^T \widetilde{P} \Pi_{st} + \widetilde{\Pi}_{ij}^T \acute{P} \widetilde{\Pi}_{ij} + \overline{P}_{is})\widetilde{\eta}(k) \right\} \\
\leq \sum_{i,j=1}^{r} h_i h_j \widetilde{\eta}^T(k)(\Pi_{ij}^T \widetilde{P} \Pi_{ij} + \widetilde{\Pi}_{ij}^T \acute{P} \widetilde{\Pi}_{ij} + \overline{P}_{ii})\widetilde{\eta}(k) \\
\leq \sum_{i=1}^{r} h_i^2 \widetilde{\eta}^T(k)(\Pi_{ii}^T \widetilde{P} \Pi_{ii} + \widetilde{\Pi}_{ii}^T \acute{P} \widetilde{\Pi}_{ii} + \overline{P}_{ii})\widetilde{\eta}(k) \\
+\frac{1}{2} \sum_{i,j=1,i<j}^{r} h_i h_j \widetilde{\eta}^T(k)[(\Pi_{ij} + \Pi_{ji})^T \widetilde{P}(\Pi_{ij} + \Pi_{ji}) \\
+(\widetilde{\Pi}_{ij} + \widetilde{\Pi}_{ji})^T \acute{P}(\widetilde{\Pi}_{ij} + \widetilde{\Pi}_{ji}) + 2(\overline{P}_{ii} + \overline{P}_{jj})]\widetilde{\eta}(k).
\end{aligned}
\tag{22}
$$

With Theorem 1, $J(n) \leq 0$, then (8) is obtained, and the proof is complete. □

4. Fuzzy FD Filter Design

In this section, on the basis of the previous analysis, the fuzzy FD filter design problem will be settled by the subsequent theorem.

Theorem 2. *Consider the fuzzy dynamic system (7) and make $\gamma > 0$ a known scalar. If there are matrices $P > 0$, $Q > 0$, $X$and K satisfying the following linear matrix inequality (LMI):*

$$
\begin{bmatrix} \overline{\Gamma}_1 & * & * \\ M_i^T & -\varepsilon I & * \\ \varepsilon N & 0 & -\varepsilon I \end{bmatrix} < 0 \tag{23}
$$

$$
\begin{bmatrix} \overline{\Gamma}_2 & * & * \\ M_i^T + M_j^T & -\varepsilon I & * \\ \varepsilon N & 0 & -\varepsilon I \end{bmatrix} < 0 \tag{24}
$$

then the FD filter in the form of (6) exists with the following:

$$
\overline{\Gamma}_1 = \begin{bmatrix} \overline{P}_{ii} & * \\ \overline{Z}_{1ii} & -\overline{P} \end{bmatrix}, \overline{\Gamma}_2 = \begin{bmatrix} 2(\overline{P}_{ii} + \overline{P}_{jj}) & * \\ \overline{Z}_{1ij} + \overline{Z}_{1ji} & -\overline{P} \end{bmatrix},
$$

$$
\overline{Z}_{1ij} = \begin{bmatrix}
P\hat{A}_{0i} + X_j\hat{R}_{1i} & P\overline{A}_{di} & \overline{\Lambda}_l \otimes (X_j\hat{R}_{2i}) & P\hat{B}_{0i} + X_j\hat{R}_{3i} & \Lambda_l \otimes (X_j\hat{R}_{4i}) \\
K_j\hat{R}_{1i} & 0 & \overline{\Lambda}_l \otimes (K_j\hat{R}_{2i}) & \overline{D}_0 + K_j\hat{R}_{3i} & \overline{\Lambda}_l \otimes (K_j\hat{R}_{4i}) \\
\check{\beta}X_j\hat{R}_{2i} & 0 & 0 & \check{\beta}X_j\hat{R}_{4i} & 0 \\
0 & 0 & \check{\beta}_l \otimes (X_j\hat{R}_{2i}) & 0 & \check{\beta}_l \otimes (X_j\hat{R}_{4i}) \\
\check{\beta}K_j\hat{R}_{2i} & 0 & 0 & \check{\beta}K_j\hat{R}_{4i} & 0 \\
0 & 0 & \check{\beta}_l \otimes (K_j\hat{R}_{2i}) & 0 & \check{\beta}_l \otimes (K_j\hat{R}_{4i})
\end{bmatrix},
$$

$$
M_i = \begin{bmatrix} 0 & 0 & 0 & 0 & 0 & \overline{H}_i^T P & 0 & 0 & 0 & 0 & 0 \end{bmatrix}^T,
$$

$$
N = \begin{bmatrix} \overline{E}_a & \overline{E}_d & 0 & 0 & 0 & 0 & 0 & 0 & 0 & 0 & 0 \end{bmatrix},
$$

$$
\hat{E} = \begin{bmatrix} 0 & I \end{bmatrix}^T, \overline{D}_0 = \begin{bmatrix} 0 & -I \end{bmatrix}, \hat{A}_{0i} = \begin{bmatrix} A_i & 0 \\ 0 & 0 \end{bmatrix},
$$

$$
\hat{B}_{0i} = \begin{bmatrix} D_{1i} & G_i \\ 0 & 0 \end{bmatrix}, \hat{R}_{1i} = \begin{bmatrix} 0 & I \\ C_i & 0 \end{bmatrix}, \hat{R}_{2i} = \begin{bmatrix} 0 & 0 \\ D_{2i} & 0 \end{bmatrix}
$$

If P, Q, X_j and K_j are feasible solutions to (23) and (24), then the FD filter gains of (5) are computed via the following formula:

$$
[A_{fj}\, B_{fj}] = (\hat{E}^T P \hat{E})^{-1} \hat{E}^T X_j, \; [C_{fj}\, D_{fj}] = K_j.
$$

Proof. For the purpose of avoiding splitting the matrix P, Q_m, and R_l, the parameters in Theorem 1 are rewritten as follows:

$$\overline{A}_{ij} = \hat{A}_{0i} + \hat{E}L_j\hat{R}_{1i},\ \overline{B}_{ij} = \hat{B}_{0i} + \hat{E}L_j\hat{R}_{2i}, \overline{C}_{ij} = K_j\hat{R}_{1i},\ \overline{D}_{ij} = \overline{D}_0 + K_j\hat{R}_{2i}$$

where $L_j = [A_{fj}\, B_{fj}]$, $K_j = [C_{fj}\, D_{fj}]$.

Then, according to Lemma 1, (11) and (12) are rewritten as follows:

$$\begin{bmatrix} \overline{P}_{ii} & * \\ \widetilde{Z}_{ii} & -\overline{P}^{-1} \end{bmatrix} < 0 \tag{25}$$

$$\begin{bmatrix} 2(\overline{P}_{ii} + \overline{P}_{jj}) & * \\ \widetilde{Z}_{ij} + \widetilde{Z}_{ji} & -\overline{P}^{-1} \end{bmatrix} < 0 \tag{26}$$

where $1 \le i < j \le r\ (i, j \in R)$.

$$\widetilde{Z}_{ij} = \begin{bmatrix} \hat{A}_{0i} + \hat{E}L_j\hat{R}_{1i} + \Delta\overline{A}_i & \overline{A}_{di} + \Delta\overline{A}_{di} & \overline{\Lambda}_l \otimes (X_j\hat{R}_{2i}) & \hat{B}_{0i} + \hat{E}L_j\hat{R}_{3i} & \overline{\Lambda}_l \otimes (\hat{E}L_j\hat{R}_{4i}) \\ K_j\hat{R}_{1i} & 0 & \overline{\Lambda}_l \otimes (K_j\hat{R}_{2i}) & \overline{D}_0 + K_j\hat{R}_{3i} & \overline{\Lambda}_l \otimes (K_j\hat{R}_{4i}) \\ \check{\beta}\hat{E}L_j\hat{R}_{2i} & 0 & 0 & \check{\beta}\hat{E}L_j\hat{R}_{4i} & 0 \\ 0 & 0 & \check{\beta}_l \otimes (\hat{E}L_j\hat{R}_{2i}) & 0 & \check{\beta}_l \otimes (\hat{E}L_j\hat{R}_{4i}) \\ \check{\beta}K_j\hat{R}_{2i} & 0 & 0 & \check{\beta}K_j\hat{R}_{4i} & 0 \\ 0 & 0 & \check{\beta}_l \otimes (K_j\hat{R}_{2i}) & 0 & \check{\beta}_l \otimes (K_j\hat{R}_{4i}) \end{bmatrix}$$

Pre- and post-multiply inequalities (25) and (26) by $\operatorname{diag}\{I, \overline{P}\}$, respectively, and denote $X_j = P\hat{E}L_j$, one acquires the following:

$$\Gamma_1 = \begin{bmatrix} \overline{P}_{ii} & * \\ \overline{Z}_{ii} & -\overline{P} \end{bmatrix} < 0 \tag{27}$$

$$\Gamma_2 = \begin{bmatrix} 2(\overline{P}_{ii} + \overline{P}_{jj}) & * \\ \overline{Z}_{ij} + \overline{Z}_{ji} & -\overline{P} \end{bmatrix} < 0 \tag{28}$$

where

$$\overline{Z}_{ij} = \begin{bmatrix} P\hat{A}_{0i} + X_j\hat{R}_{1i} + P\Delta\overline{A}_i & P\overline{A}_{di} + P\Delta\overline{A}_{di} & \overline{\Lambda}_l \otimes (X_j\hat{R}_{2i}) & P\hat{B}_{0i} + X_j\hat{R}_{3i} & \overline{\Lambda}_l \otimes (X_j\hat{R}_{4i}) \\ K_j\hat{R}_{1i} & 0 & \overline{\Lambda}_l \otimes (K_j\hat{R}_{2i}) & \overline{D}_0 + K_j\hat{R}_{3i} & \overline{\Lambda}_l \otimes (K_j\hat{R}_{4i}) \\ \check{\beta}X_j\hat{R}_{2i} & 0 & 0 & \check{\beta}X_j\hat{R}_{4i} & 0 \\ 0 & 0 & \check{\beta}_l \otimes (X_j\hat{R}_{2i}) & 0 & \check{\beta}_l \otimes (X_j\hat{R}_{4i}) \\ \check{\beta}K_j\hat{R}_{2i} & 0 & 0 & \check{\beta}K_j\hat{R}_{4i} & 0 \\ 0 & 0 & \check{\beta}_l \otimes (K_j\hat{R}_{2i}) & 0 & \check{\beta}_l \otimes (K_j\hat{R}_{4i}) \end{bmatrix}$$

According to the expression of the uncertainty parameters, we have $\Delta\overline{A}_i = \overline{H}_iF(k)\overline{E}_a, \Delta\overline{A}_{di} = \overline{H}_iF(k)\overline{E}_d, \overline{H}_i = \begin{bmatrix} H_i^T & 0 \end{bmatrix}^T, \overline{E}_a = \begin{bmatrix} E_a & 0 \end{bmatrix}, \overline{E}_d = \begin{bmatrix} E_d & 0 \end{bmatrix}$.

Equations (27) and (28) can be rewritten as follows:

$$\overline{\Gamma}_1 + MF(k)N + N^TF^T(k)M_i^T < 0 \tag{29}$$

$$\overline{\Gamma}_2 + (M_i + M_j)F(k)N + N^TF^T(k)(M_i + M_j)^T < 0 \tag{30}$$

where $1 \le i < j \le r\ (i, j \in R)$; the parameters therein are defined in Theorem 2. In accordance with the S-procedure in Lemma 2, (23) and (24) are obtained, and the proof is now complete. □

Remark 2. *Until now, the H_∞ fault detection filter design has been accomplished for the CSTR reaction process subject to parameter uncertainties, channel fadings, and delays. The main results of this paper are thus highlighted as follows. In Section 3, Lemmas 1–3 lay a necessary foundation for later analysis and design, and Theorem 1 realizes the performance analysis (exponential stability in the mean square of the error dynamics of the fault detection filter and the H_∞ disturbance rejection level of the residual filtering error against external disturbances). In Section 4, the fault detection filter design is fulfilled in Theorem 2, the gain expression of the desired fault detection filter is acquired by virtue of the feasible solution to certain LMIs. More specifically, Theorem 2 contains all the system parameters such as delay bound, parameters in the parameter uncertainties, and statistical characteristics of the channel coefficient.*

Remark 3. *The main work of this paper is further emphasized as follows: (1) constructing a fuzzy T-S model to reflect the CSTR reaction process on the basis of the dimensionless mechanism model; (2) the channel fading phenomenon is considered in the transmission process of CSTR measurement signal from the sensor to the FD filter, which is characterized by the improved Lth-order Rice fading model by reflecting the actual situation of signal transmission more accurately; and (3) a reinforced stochastic analysis technique is implemented in order to conform to the H_∞ performance of the fault detection filter concerning the CSTR fuzzy systems, except for the constraint of exponential stability in the mean square.*

5. Numerical Example

The chosen CSTR system parameters are the following: $\gamma_0 = 20$, $H = 8$, $\beta = 1$, $D_\alpha = 0.072$, and $\lambda = 0.8$. Let $\bar{d} = 5$, $D_{11} = D_{12} = D_{13} = \begin{bmatrix} 0 & 1 \end{bmatrix}^T$. In the reaction, the CSTR system has three equilibrium points: $\hat{x}_{01} = \begin{bmatrix} 0.1440 & 0.8862 \end{bmatrix}^T$, $\hat{x}_{02} = \begin{bmatrix} 0.4472 & 2.7520 \end{bmatrix}^T$, $\hat{x}_{03} = \begin{bmatrix} 0.7646 & 4.7052 \end{bmatrix}^T$, the following T-S fuzzy rules are then employed to expand near the three equilibrium points.

Rule 1: If $x_2(k)$ is small ($x_2(k)$ is about 0.8862), then

$$x(k+1) = (A_1 + \Delta A_1)x(k) + (A_{d1} + \Delta A_{d1})x(k - d(k)) + D_{11}w(k) + G_1 f(k);$$

Rule 2: If $x_2(k)$ is medium ($x_2(k)$ is about 2.7520), then

$$x(k+1) = (A_2 + \Delta A_2)x(k) + (A_{d2} + \Delta A_{d2})x(k - d(k)) + D_{12}w(k) + G_2 f(k);$$

Rule 3: If $x_2(k)$ is large ($x_2(k)$ is about 4.7052), then

$$x(k+1) = (A_3 + \Delta A_3)x(k) + (A_{d3} + \Delta A_{d3})x(k - d(k)) + D_{13}w(k) + G_3 f(k).$$

Here, $x(k)$ and $x(k - d(k))$ are the set of differences between the temperature state value and the corresponding equilibrium point temperature value. According to the selected parameters, there are

$$A_1 = \begin{bmatrix} 0.0418 & 0.0132 \\ 0.0346 & -0.0194 \end{bmatrix}, A_2 = \begin{bmatrix} 0.0590 & 0.0346 \\ -0.0472 & 0.0515 \end{bmatrix}, A_3 = \begin{bmatrix} 0.0498 & -0.0167 \\ 0.0983 & 0.0758 \end{bmatrix},$$
$$A_{d1} = A_{d2} = A_{d3} = \text{diag}\{0.25, 0.25\}, F(k) = \sin(0.6k), C_2 = \begin{bmatrix} -0.79 & 0.65 \end{bmatrix},$$
$$H_1 = H_2 = H_3 = \begin{bmatrix} 0.2 \\ 0.01 \end{bmatrix}, E_a = \begin{bmatrix} 0 & 0.15 \end{bmatrix}, E_d = \begin{bmatrix} 0 & 0.2 \end{bmatrix}, G_1 = \begin{bmatrix} 0.21 \\ -0.14 \end{bmatrix}, C_3 = \begin{bmatrix} -0.81 & 0.65 \end{bmatrix},$$
$$G_2 = \begin{bmatrix} 0.20 \\ -0.12 \end{bmatrix}, G_3 = \begin{bmatrix} 0.19 \\ -0.15 \end{bmatrix}, D_{21} = D_{22} = D_{23} = 0.02, C_1 = \begin{bmatrix} -0.8 & 0.65 \end{bmatrix}.$$

The membership functions are shown in Figure 2.

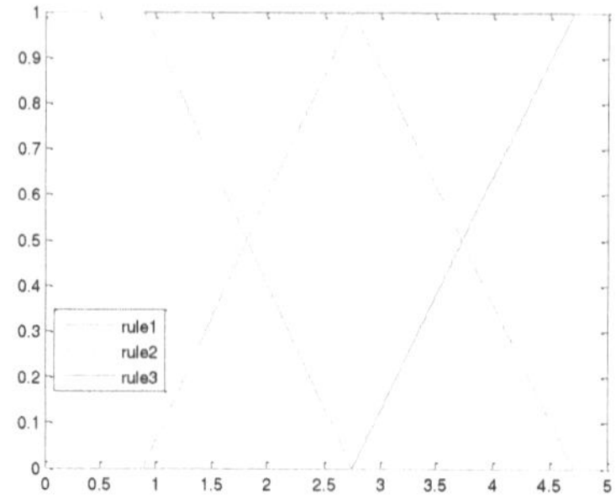

Figure 2. Membership function.

The order of the fading model is $\ell = 2$, the probability quality function of the channel coefficient is as follows:

$$f(\beta_0) = 0.0005(e^{9.89\beta_0} - 1),\ 0 \le \beta_0 \le 1$$
$$f(\beta_1) = \begin{cases} 10\beta_1, & 0 \le \beta_1 \le 0.20 \\ -2.50(\beta_1 - 1), & 0.20 < \beta_1 \le 1 \end{cases}$$
$$f(\beta_2) = 8.5017e^{-8.5\beta_2},\ 0 \le \beta_2 \le 1$$

The mathematical expectations $\overline{\beta}_s (s = 0, 1, 2)$ are acquired as 0.8991, 0.4000, and 0.1174, the variance $(\widetilde{\beta}_s)^2$ are 0.0133, 0.0467, and 0.01364, respectively. In terms of the above parameters and using the LMI toolbox in the Matlab software, the gains of the FD filter can be calculated by solving the feasible solution to matrix inequalities (23) and (24). The obtained gains of the fault detection filter (5) are shown in Table 1.

Table 1. The computed gains of the fault detection filter.

	A_{fi}	B_{fi}	C_{fi}	D_{fi}
$i = 1$	$\begin{bmatrix} -0.8972 & 0.7112 \\ -0.4820 & 0.4996 \end{bmatrix}$	$\begin{bmatrix} 0.0145 \\ 0.0017 \end{bmatrix}$	$\begin{bmatrix} 0.0893 & -0.4515 \end{bmatrix}$	2.4358
$i = 2$	$\begin{bmatrix} 0.4122 & -0.3698 \\ -0.6324 & 0.5753 \end{bmatrix}$	$\begin{bmatrix} -0.0166 \\ 0.0139 \end{bmatrix}$	$\begin{bmatrix} 0.4414 & -0.3681 \end{bmatrix}$	−5.2732
$i = 3$	$\begin{bmatrix} 0.0014 & -0.0012 \\ 0.0002 & -0.0002 \end{bmatrix}$	$\begin{bmatrix} -0.0489 \\ -0.0214 \end{bmatrix}$	$\begin{bmatrix} 0.1356 & -0.0628 \end{bmatrix}$	−7.4592

The initial state is taken as $x(0) = \begin{bmatrix} 0.9 & 0.9 \end{bmatrix}^T$, noise $w(k) = \begin{cases} 0.2\text{rand}(1,1), & 30 \le k \le 130 \\ 0, & \text{else} \end{cases}$, and the fault signal $f(k)$ is chosen as follows:

$$f(k) = \begin{cases} 1, & 50 \le k \le 100 \\ 0, & \text{else} \end{cases}$$

Figure 3 plots measurement curves, in which the dashed line denotes the ideal measurement output, and the solid line represents the signal actually received by the fault detection filter. It can be seen that the amplitude change of the received signal is more intense than that of the ideal measurement, which validates that channel fadings may lead to the signal distortion (signal missing and delays). Additionally, the occurrence and existence of faults cause the abnormal values of the measurement signals. Figures 4 and 5 describe the residual signal curves with and without noise, respectively. We notice that the residual signal curve without noise is smoother than the one with noise, and the influence of both faults and channel fadings on the residual signal is obvious, which is in accordance with Equation (5). In terms of Equation (9), Figures 6 and 7 reflect the evolution of the residual evaluation function curves with and without noise, respectively. It is shown that there are more fluctuations in the residual evaluation function with noise than those without noise. In Figure 6 (or Figure 7), the dashed line and the solid line depict the residual evaluation function with and without faults, respectively. It is also illustrated that the value of the residual evaluation function increases due to the existence of faults, which lay a basis for the fault detection.

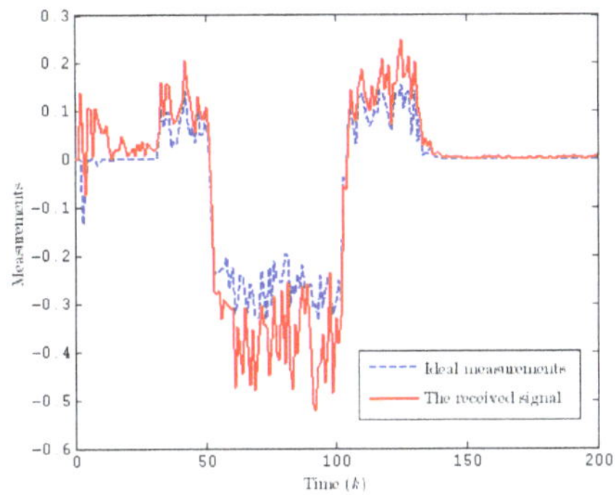

Figure 3. Ideal and practical measurement outputs.

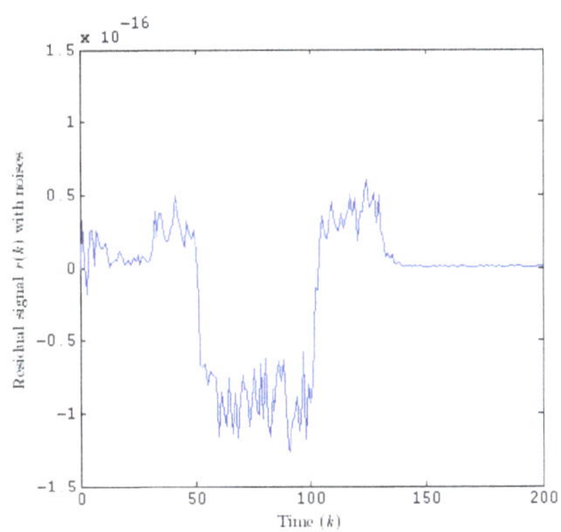

Figure 4. Residual signal with noise.

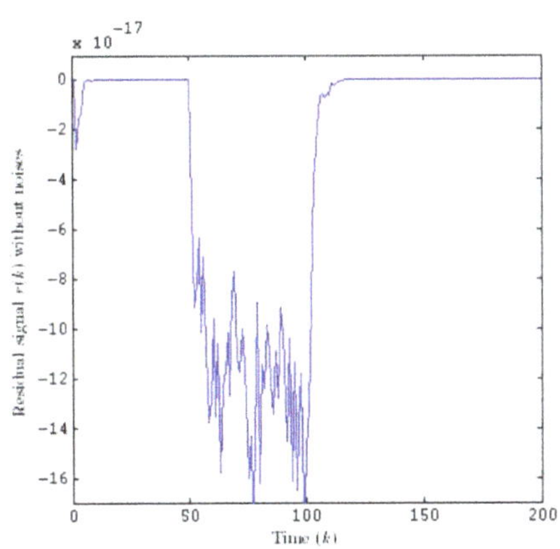

Figure 5. Residual signal without noise.

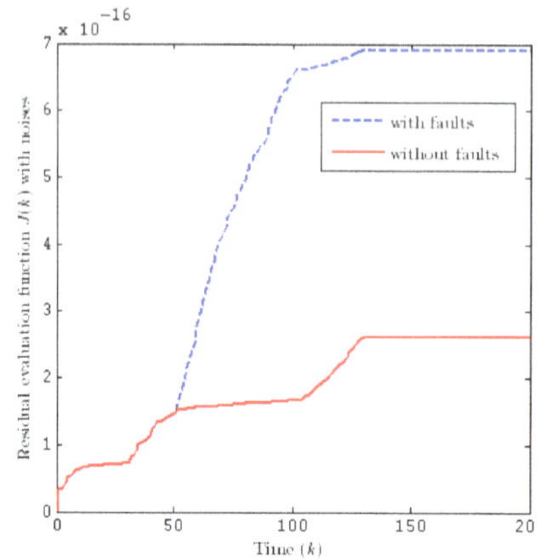

Figure 6. Residual evaluation function with noise.

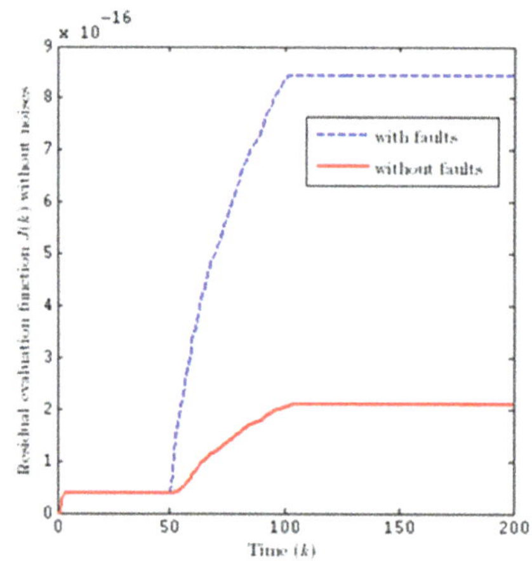

Figure 7. Residual evaluation function without noise.

Assuming the threshold $J_{th} = \sup_{f=0}\sqrt{\sum_{h=0}^{200} r^T(h)r(h)}$, after 200 fault-free simulation runs, the average threshold is then $J_{th} = 0.2622$. It can be recognized from Figure 6 that $2.519 = J(59) < J_{th} < J(60) = 2.772$, i.e., the fault is detected in step 10, after it occurs. It can be concluded that the residual can not only reflect the fault in time, but also detect the fault in the case of disturbance.

6. Conclusions

In this paper, the FD issue for CSTR with respect to time delay, uncertainty parameters, and channel fadings was investigated in terms of the T-S fuzzy model. Norm-bounded uncertainties were adopted to describe parameter imprecision caused by modelling errors. The phenomenon of channel fadings was considered while the measurement output signal was transmitted from the sensor to the FD filter, which was then reflected with an improved L-th Rice fadings model. The performance constraints to be met by the constructed fault detection filter were both the exponential stability in the mean square of the filtering error system and the H_∞ disturbance rejection level of the residual filtering error in resistance to external disturbances. With the help of the Lyapunov stability theory and reinforced stochastic analysis techniques, the analysis of the performance and the design of the fault detection filter were carried out for the CSTR. As a result, a sufficient condition was put forward, ensuring the existence of a satisfactory FD filter. Simultaneously, a direct expression was acquired from the FD filter in accordance with the feasible solution to a specified LMI, which is solved conveniently via the standard Matlab software. Lastly, a simulation example demonstrated that faults can be reflected and detected in time under circumstances of disturbance by choosing the thresholds appropriately, which validates the effectiveness and the correctness of the developed FD strategy for CSTR in this paper. For subsequent research topics, we would like to deal with fault estimation, fault prognosis, and related issues therein [32].

Author Contributions: Model construction, system performance analysis, filter design, and writing—original draft preparation, Y.W.; investigation, D.Z.; writing—review and editing, Z.L.; supervision and project administration, W.R.; numerical simulation, J.L. All authors have read and agreed to the published version of the manuscript.

Funding: This research was funded by the Starting Fund for Doctoral Research of Daqing Normal University, grant number 19ZR09, and the Natural Science Foundation of Heilongjiang Province of China, grant number LH2021F006.

Institutional Review Board Statement: Not applicable.

Informed Consent Statement: Not applicable.

Data Availability Statement: Not applicable.

Conflicts of Interest: Authors declare no conflict of interest.

References

1. Li, Y.; Feng, Y. T-S model based L2-L∞ control for continuous stirring tank reactor. *J. Jilin Univer. (Inform. Sci. Ed.)* **2014**, *329*, 247–253.
2. Chen, Z.; Cao, Y.; Ding, S.X.; Zhang, T. A Distributed canonical correlation analysis-based fault detection method for plant-wide process monitoring. *IEEE Trans. Ind. Inform.* **2019**, *15*, 2710–2720. [CrossRef]
3. Gao, Z.; Cecati, C.; Ding, S.X. A survey of fault diagnosis and fault-tolerant techniques—Part I: Fault diagnosis with model-based and signal-based approaches. *IEEE Trans. Ind. Electron.* **2015**, *62*, 3757–3767. [CrossRef]
4. Gao, Z.; Cecati, C.; Ding, S.X. A survey of fault diagnosis and fault-tolerant techniques—Part II: Fault diagnosis with knowledge-based and hybrid/active approaches. *IEEE Trans. Ind. Electron.* **2015**, *62*, 3768–3774. [CrossRef]
5. Gao, Z.; Sheng, S. Real-time monitoring, prognosis, and resilient control for wind turbine systems. *Renew. Energy* **2017**, *116*, 1–4. [CrossRef]
6. Hou, N.; Wang, Z.; Ho, D.W.; Dong, H. Robust partial-nodes-based state estimation for complex networks under deception attacks. *IEEE Trans. Cybern.* **2020**, *50*, 2793–2802. [CrossRef]
7. Chen, H.; Jiang, B.; Chen, W.; Yi, H. Data-driven detection and diagnosis of incipient faults in electrical drives of high-speed trains. *IEEE Trans. Ind. Electron.* **2019**, *66*, 4716–4725. [CrossRef]
8. Gao, Z.; Liu, X. An Overview on Fault Diagnosis, Prognosis and resilient control for wind turbine systems. *Processes* **2021**, *9*, 300. [CrossRef]
9. Ali, N.; Hong, J. Failure detection and prevention for cyber-physical systems using ontology-based knowledge base. *Computers* **2018**, *7*, 68. [CrossRef]
10. Dong, H.; Hou, N.; Wang, Z. Fault estimation for complex networks with randomly varying topologies and stochastic inner couplings. *Automatica* **2020**, *112*, 108734. [CrossRef]
11. Zhang, Z.; Yang, G. Distributed fault detection and isolation for multiagent systems: An interval observer approach. *IEEE Trans. Syst. Man Cybern. Syst.* **2020**, *50*, 2220–2230. [CrossRef]
12. Ye, D.; Chen, M.; Yang, H. Distributed adaptive event-triggered fault-tolerant consensus of multiagent systems with general linear dynamics. *IEEE Trans. Cybern.* **2019**, *49*, 757–767. [CrossRef]
13. Zhu, X.; Liu, Y.; Fang, J.; Zhong, M. Fault detection for a class of linear systems with integral measurements. *Sci. China Inform. Sci.* **2021**, *64*, 1–10. [CrossRef]
14. Taoufik, A.; Defoort, M.; Djemai, M.; Busawon, K.; Sánchez-Torres, J.D. Distributed global actuator fault-detection scheme for a class of linear multi-agent systems with disturbances. *IFAC-PapersOnLine* **2020**, *53*, 4202–4207. [CrossRef]
15. Alikhani, H.; Meskin, N. Event-triggered robust fault diagnosis and control of linear Roesser systems: A unified framework. *Automatica* **2021**, *128*, 109575. [CrossRef]
16. Zhong, M.; Ding, S.X.; Zhou, D.; He, X. An Hi/ H∞ optimization approach to event-triggered fault detection for linear discrete time systems. *IEEE Trans. Autom. Control* **2020**, *65*, 4464–4471. [CrossRef]

17. Ren, W.; Gao, M.; Kang, C. Non-fragile h-infinity fault detection for nonlinear systems with stochastic communication protocol and channel fadings. *Int. J. Control Autom. Syst.* **2021**, *19*, 2150–2162. [CrossRef]
18. Ma, H.; Xu, L. Cooperative fault diagnosis for uncertain nonlinear multiagent systems based on adaptive distributed fuzzy estimators. *IEEE Trans. Cybern.* **2020**, *50*, 1739–1751. [CrossRef]
19. Gao, Z. Estimation and compensation for Lipschitz nonlinear discrete-time systems subjected to unknown measurement delays. IEEE. *Trans. Indust. Electron.* **2015**, *62*, 5950–5961. [CrossRef]
20. Liu, X.; Gao, Z.; Zhang, A. Observer-based fault estimation and tolerant control for stochastic Takagi–Sugeno fuzzy systems with Brownian parameter perturbations. *Automatica* **2019**, *102*, 137–149. [CrossRef]
21. Ji, W.; Wang, T.; Qiu, J.; Fu, S. Distributed fuzzy H∞ filtering for nonlinear multi-rate networked double-layer industrial processes. *IEEE Trans. Indust. Electron.* **2017**, *64*, 5203–5211.
22. Su, X.; Shi, P.; Wu, L.; Basin, M.V. Reliable filtering with strict dissipativity for T–S fuzzy time-delay systems. *IEEE Trans. Cybern.* **2017**, *44*, 2470–2483.
23. Ali, M.S.; Gunasekaran, N.; Zhu, Q. State estimation of T–S fuzzy delayed neural networks with Markovian jumping parameters using sampled-data control. *Fuzzy Sets Syst.* **2017**, *306*, 87–104.
24. Chibani, A.; Chadli, M.; Ding, S.X.; Braiek, N.B. Design of robust fuzzy fault detection filter for polynomial fuzzy systems with new finite frequency specifications. *Automatica* **2018**, *93*, 42–54. [CrossRef]
25. Shen, Y.; Wang, Z.; Shen, B.; Fuad, E. Alsaadi. Event-based recursive filtering for a class of nonlinear stochastic parameter systems over fading channels. *Int. J. Gen. Syst.* **2018**, *47*, 401–415. [CrossRef]
26. Zhang, Z.; Su, S.; Niu, Y. Dynamic event-triggered control for interval type-2 fuzzy systems under fading channel. *IEEE Trans. Cybern.* **2020**, *99*, 1–10. [CrossRef]
27. Wang, Y.; Ren, W.; Lu, Y. Fault detection for complex systems with channel fadings, randomly occurring multiple delays and infinitely distributed delays. *J. Syst. Sci. Complex.* **2018**, *31*, 419–435. [CrossRef]
28. Dong, H.; Wang, Z.; Lam, J.; Gao, H. Fuzzy-model-based robust fault detection with stochastic mixed time delays and successive packet dropouts. *IEEE Trans. Syst. Man Cybern. Syst.* **2012**, *42*, 365–376. [CrossRef]
29. Li, M.; Yu, D.; Chen, Z.; XiaHou, K.; Ji, T.; Wu, Q.H. Data-driven residual-based method for fault diagnosis and isolation in wind turbines. *IEEE Trans. Sustain. Energy* **2019**, *10*, 895–904. [CrossRef]
30. Guan, X.P.; Chen, C.L. Delay-dependent guaranteed cost control for T-S fuzzy systems with time delays. *IEEE Trans. Fuzzy Syst.* **2004**, *12*, 236–249. [CrossRef]
31. Wang, Z.; Ho, D.; Liu, Y.; Liu, X. Robust H∞ control for a class of nonlinear discrete time-delay stochastic systems with missing measurements. *Automatica* **2009**, *45*, 684–691. [CrossRef]
32. Djeziri, M.; Nguyen, T.-B.-L.; Benmoussa, S.; M'Sirdi, N. Fault prognosis based on physical and stochastic models. In Proceedings of the 2016 European Control Conference, Aalborg, Denmark, 29 June–1 July 2016; pp. 2269–2274.

processes

Article

TLSCA-SVM Fault Diagnosis Optimization Method Based on Transfer Learning

Aihua Zhang [1,*], Danlu Yu [2] and Zhiqiang Zhang [1]

[1] College of Physical Science and Technology, Bohai University, Jinzhou 121013, China; zhangzhiqiang@qymail.bhu.edu.cn
[2] College of Control Science and Engineering, Bohai University, Jinzhou 121013, China; yudanlu15178672625@163.com
* Correspondence: jsxinxi_zah@163.com; Tel.: +86-138-4063-8593

Abstract: In fault-diagnosis classification, a pressing issue is the lack of target-fault samples. Obtaining fault data requires a great amount of time, energy and financial resources. These factors affect the accuracy of diagnosis. To address this problem, a novel fault-diagnosis-classification optimization method, namely TLSCA-SVM, which combines the sine cosine algorithm and support vector machine (SCA-SVM) with transfer learning, is proposed here. Considering the availability of fault data, this thesis uses the data generated by analog circuits from different faults for analysis. Firstly, the data signal is collected from different faults of the analog circuit, and then the characteristic data are extracted from the data signals by the wavelet packets. Secondly, to employ the principal component analysis (PCA) reduces the feature-value dimension. Lastly, as an auxiliary condition, the error-penalty item is added to the objective function of the SCA-SVM classifier to construct an innovative fault-diagnosis model namely TLSCA-SVM. Among them, the Sallen–Key bandpass filter circuit and the CSTV filter circuit are used to provide the data for horizontal- and vertical-contrast classification results. Comparing the SCA with the five optimization algorithms, it is concluded that the performance of SCA optimization parameters has certain advantages in the classification accuracy and speed. Additionally, to prove the superiority of the SCA-SVM classification algorithm, the five classification algorithms are compared with the SCA-SVM algorithm. Simulation results showed that the SCA-SVM classification has higher precision and a faster response time compared to the others. After adding the error penalty term to SCA-SVM, TLSCA-SVM requires fewer fault samples to process fault diagnosis. Ultimately, the method which is proposed could not only perform fault diagnosis effectively and quickly, but also could run effectively to achieve the effect of transfer learning in the case of less failure data.

Keywords: TLSCA-SVM; optimization method; fault diagnosis; transfer learning

Citation: Zhang, A.; Yu, D.; Zhang, Z. TLSCA-SVM Fault Diagnosis Optimization Method Based on Transfer Learning. *Processes* **2022**, *10*, 362. https://doi.org/10.3390/pr10020362

Academic Editor: Jie Zhang

Received: 10 January 2022
Accepted: 8 February 2022
Published: 14 February 2022

1. Introduction

In terms of practical application, fault diagnosis is primarily used in industrial failure. Among such incidents, 80% of industrial failures come from analog circuits. Therefore, analog circuit fault diagnosis is the research focus of industrial fault diagnosis. Compared with digital circuits, analog circuits are more complicated and have more stringent requirements for fault diagnosis. Training using analog circuits for diagnostic models in fault diagnosis is more valuable. With the continuous refinement of various data-analysis methods, the troubleshooting technology for analog circuits is also improving; common troubleshooting techniques, including PCA [1], Search Grid [2], particle swarm optimization (PSO) [3], ant colony algorithm (ACA) [4], simulated annealing (SA) [5], genetic algorithm (GA) [6], Back Propagation Neural Network (BP) [6], Self-organizing Maps (SOM) [7], Extreme Learning Machine (ELM) [8], decision tree [9], random forest [10] and SVM [11] all have good classification results to some extent [12].

For the fault diagnosis of analog circuits [13], the following operations are required. Firstly, an analog circuit model is built, the fault is set in the analog circuit and the failure data is collected. Secondly, the fault data is processed, using methods such as feature-value extraction, dimensionality reduction, linearization, etc. Finally, the processed data is used in the modelling of fault diagnosis to judge the performance of the model [14]. In this article, the data of Sallen–Key circuit is used for model comparison and the more complex CSTV circuit is used for validation to determine the versatility of the model. Here, the data of both analog circuits are used for horizontal and vertical comparison experiments of the algorithm in order to obtain more practical conclusions. When there are fewer fault samples, the auxiliary condition, namely the error penalty item, is added to the objective function of the SCA-SVM [15] classifier to construct a new fault diagnosis model, which is TLSCA-SVM. The model not only has the advantages of the SCA-SVM fault diagnosis model, i.e., the fault classification speed is fast, the accuracy is higher, but also the transfer-learning [16] ability is better.

As a new concept of machine learning [17], transfer learning brings new research directions to machine learning. This paper uses the data from different faults in the analog circuit for fault diagnosis [18], performs a series of data processing, combines the SCA optimization method with SVM to obtain the classifier SCA-SVM and adds an error penalty item to build a new fault diagnosis model, namely the TLSCA- SVM model. This method imports the data processed by the wavelet packet and PCA into the improved SCA-SVM classifier for training and prediction, thereby improving the speed and accuracy of diagnosis [19].

The controversy about the SCA algorithm [20] concerns the local search of the optimization algorithm. It is not difficult to understand that the optimal solution can be guaranteed using global search [21]. However, it is inevitable that the local optimal solution may appear in the local search, performance at a certain point is best, and the farther away from the point, the worse the performance. In response to this debate, this paper effectively avoids the occurrence of the problem. In this paper, the parameters of SVM are optimized by using the same characteristics of local search and global search probability of the SCA algorithm in optimizing parameters. The SCA-SVM classifier and a good classification effect are obtained.

As machine learning continues to innovate, the concept of transfer learning has been introduced through research. Traditional machine learning uses the continuous autonomous acquisition of knowledge from the data. This mechanism requires a lot of data and iterations to achieve data-driven effects. In machine learning, when a trained model is applied to a new field, the effect is often poor. Transfer learning, as an important branch of machine learning, focuses on applying knowledge that has already been learned to new problems, so that the knowledge transfer can allow trained models to be applied to new areas. Transfer learning includes zero-shot learning [22], one-shot learning [23] and few-shot learning [24]. Since the zero-shot learning is too ideal to exist in actual data processing, the limit of few-shot learning is zero-shot learning. Therefore, in this paper about analog circuits fault diagnosis, few-shot learning is used for fault diagnosis [25], as well as transfer learning, in order to achieve the effect of fault prediction [26].

Transfer learning refers to the transfer of knowledge from one field to another, and it is common to apply the experience gained from learning to new learning, which is the influence of one learning state on another. The earliest application of transfer learning was in educational psychology, which put forward that the implementation of transfer learning must combine old with new knowledge. In terms of machine learning, transfer-learning development has a better development prospect. Since 1995, NIPS launched a "Learning to Learn" professional seminar in the field of machine learning [27], and since then transfer-learning research has entered a period of rapid development. Transfer learning is to apply the knowledge extracted from the source task [28] to the target task. Compared with multitasking, the most important thing in transfer learning is the target task [29], rather than extracting knowledge of all source tasks at the same time. Of course, in educational psychology, the implementation object of transfer learning is human, while the implementation object in machine learning is the corresponding algorithm model. In this regard, humans have excellent transfer-learning capabilities. For example, in human childhood, when young children are provided with oral descriptions by their parents, the children can find the corresponding objects. Even in the learning process, as long as the solution is mastered, similar problems can be solved. However, the application of transfer learning in machine learning is not particularly easy. This is because, unlike humans, machines cannot achieve the so-called independent thinking and can only achieve the desired results through continuous data optimization. It is worth mentioning that in the face of massive data analysis, the introduction of migration learning can reduce a lot of the unnecessary workload, which has certain research significance.

At present, there are many scholars focused on transfer learning research and most people research the algorithm related to transfer learning. They use different technologies to achieve the requirement of transfer learning. Many transfer-learning ideas are applied to image recognition [30], such as image features being extracted to a training set, the knowledge of the source task is transferred to the target task, and finally the image recognition in the test set can be realized.

This article uses transfer learning for the purpose of fault diagnosis. Obtaining fault data in industrial fault diagnosis requires a lot of manpower and material resources. The fault of actual industrial equipment is a kind of damage. Therefore, the application of fault-diagnosis methods in transfer learning is typical. The use of normal data and a some fault data can effectively diagnose faults and achieve predictive effects. This method can maintain equipment safety and ensure the normal operation of actual industrial equipment. Considering the availability of fault data, this paper uses the data generated by analog circuits from different faults for analysis. An error penalty is added to the SCA-SVM classifier to build an innovative fault-diagnosis model, which is TLSCA-SVM. By comparing the experiments, it is concluded that the model can diagnose the fault with less fault data, achieve the effect of transfer learning, and finally realize the prediction of failure.

2. SCA-SVM Algorithm

The SCA optimization parameter algorithm is a novel random optimization method based on population. The SCA was proposed by Australian scholar Mirjalili in 2016 [31], the essence of which is to optimize the parameter by using random probability. SCA creates randomly generating multiple initial candidate solutions, and uses sine and cosine functions to make these initial candidate solutions have the same probability of moving either in the direction of the optimal solution or reverse. This method not only guarantees the accuracy of global search optimization parameters, but also ensures the speed of local search optimization parameters, so that the optimal parameters can be found quickly and accurately in the model. Because the essence of the SCA optimization algorithm is the population optimization algorithm, it also meets the general law of the two stages of the population optimization algorithm, which is exploration and utilization. When the algorithm is in the exploration stage, the algorithm randomly searches with a large probability gradient to ensure sufficient search space. When the algorithm is in the utilization stage, the random probability gradient decreases gradually, ensuring that the optimal solution can be found accurately. In SCA [32], the optimization process is divided into global search and local search. These two parts promote and restrain each other, as global search is used to quickly locate the optimal solution of the range, and local search is used to find the optimal solution. These two parts reach a dynamic balance can find the global optimal solution. If there is only a global search, the optimization speed is slow and the equipment configuration is high. If there is only a local search, it is easy to obtain a local optimal solution. The SCA optimization parameter algorithm is determined by the different value situation of the sine function global search and local search.

The two important parameters of SVM are kernel function [33] and the penalty factor [34]. These two parameters directly affect the performance of SVM [35], so finding the optimal parameters of SVM becomes the core problem of constructing a classification model. How to find the optimal parameters quickly and effectively becomes the key to the optimization algorithm. This article introduces SCA to optimize the SVM parameters. SCA randomly creates multiple initial candidate solutions and uses sine and cosine functions [36] to determine the search method. This method ensures that the initial candidate solution moves toward or away from the optimal solution with the same probability. This way of finding the optimal solution not only ensures the accuracy of the global search optimization parameters, but also ensures the speed of the local search optimization parameters, so that the optimal parameters can be found quickly and accurately in the model [37].

2.1. Principles of SCA Optimization Parameters

In this essay, the SCA optimization parameter algorithm can be used to optimize two important parameters in SVM. The essence of the SCA algorithm is to use the unique characteristics of the trigonometric function to make the probability of each optimization the same, so that the optimization effect meets the accuracy of the global search and the speed of the local search. The specific operation is that the value of the sine and cosine function is used to determine whether to perform a global search. For example, when the distance between the value of the sine and cosine function and the abscissa exceeds 0.5, that is the function value is distributed between 0.5 and 1 or −0.5 and −1, the optimization of the global search method is achieved. When the distance from the sine and cosine function value to the horizontal coordinates is less than 0.5, which is the function value is distributed between −0.5 and 0.5, the SCA algorithm conducts a local search. SCA randomly creates multiple initial-candidate solutions and uses sine and cosine functions to move these initial candidates in the same probability of moving or reversing in the direction of the best solution, which not only ensures the accuracy of the global-search optimization parameters, but also ensures the speed of the local search [11].

In the SCA optimization algorithm, the trigonometric function determines the position of the next iteration point and is iterate according to the following formula:

$$X_i^{t+1} = X_i^t + \theta_1 + \sin\theta_2 + \left|\theta_3 P_i^t - X_i^t\right| \tag{1}$$

$$X_i^{t+1} = X_i^t + \theta_1 + \cos\theta_2 + \left|\theta_3 P_i^t - X_i^t\right| \tag{2}$$

In the equation, X_i^t is the position of the current iteration, $\theta_1\theta_2$ and θ_3 are random numbers of random values in iteration that decide the trigonometric function, P_i^t is the best selected scenario. In the actual optimization process, the above positions are selected appropriately and the equation can be updated with the sine cosine function:

$$\begin{cases} X_i^{t+1}=X_i^t+\theta_1+\sin\theta_2+|\theta_3 P_i^t-X_i^t|,\ \theta_4<0.5 \\ X_i^{t+1}=X_i^t+\theta_1+\cos\theta_2+|\theta_3 P_i^t-X_i^t|,\ \theta_4\geq 0.5 \end{cases} \tag{3}$$

In the equation, θ_4 is a number randomly selected each time in the range of 0 and 1, ensuring that the probability is the same when optimizing.

In the above formula, according to parameter θ_1, the algorithm can determine the direction of the next iteration. The operation occurs when the distance from the parameter value to the abscissa exceeds 0.5, that is, the value is distributed between 0.5 and 1 or between −0.5 and −1, in the outer ring area shown in Figure 1. When the distance between the parameter value and the abscissa is less than 0.5, that is, the function value is distributed between −0.5 and 0.5, the search is performed in the circle area shown in Figure 1. The parameter is a number randomly selected between 0 and π each time. Parameter θ_3 is used to determine the random allocation of the enhanced or weakened relationship, and a size comparison with 1 has been used to determine whether the relationship is enhanced or weakened. As can be seen from the Formula (3), the parameter value of θ_4 can select the type of the sine and cosine function, when $\theta_4 < 0.5$, the equation containing only the sine function is selected for the next iteration. When $\theta_4 \geq 0.5$, the equation containing only the cosine function is selected for the next iteration. Through the above iteration, it can be guaranteed that the algorithm has the same probability in global search and local search, so that the algorithm itself can combine the benefits of both search modes itself to achieve better results.

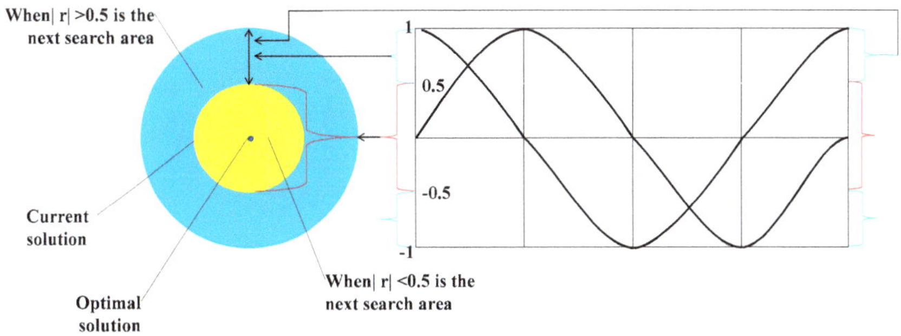

Figure 1. SCA optimization algorithm iterative schematic.

In order to satisfy the global search and the local search with the same probability conditions, and make the optimization solution converge, the SCA optimization algorithm self-adjusts using Formula (4):

$$\theta_1 = a - t\frac{a}{T} \tag{4}$$

2.2. The Classification Principle of SVM

SVM was originally used to deal with dichotomous problems. In order to meet more classification problems, slack variables and maximum separation hyperplane theory were continuously introduced and applied to classify nonlinear problems. Due to the special theoretical concept, the SVM classifier effect can be applied to transfer learning, so this paper used the SVM for fault diagnosis based on transfer learning.

The most important problem of SVM is the mapping of kernel functions. It is a nonlinear indistinguishable problem. It is classified by hyperplane and mapped to a simple linear problem, where the largest hyperplane satisfies the following conditions:

$$\begin{cases} \min\phi(\omega) = \frac{1}{2}\|\omega\|^2 \\ s.t.y_i(\omega^T x + b) \geq 1, i = 1,2,3,\ldots,n \end{cases} \tag{5}$$

ω is the general vector of the maximum hyperplane, and b is the offset of the maximum hyperplane, parameter ξ ($\xi > 0$) determines the boundaries.

$$y_i[\omega \cdot \phi(x_i) + b] + \xi_i \geq 1, i = 1, 2, \ldots, n \tag{6}$$

The penalty factor C is applied to alter the proportion of relaxation variable ξ_i, and Formula (5) is altered to:

$$\begin{cases} \min(\omega) = \frac{1}{2}\|\omega\|^2 + C\sum_{i=1}^{n}\xi_i \\ y_i[\omega \cdot \phi(x_i) + b] + \xi_i \geq 1, i = 1,2,\ldots,n \end{cases} \tag{7}$$

By introducing Lagrange multiplier α, the maximum hyperplane problem as a pair problem is reflected as follows:

$$\begin{cases} \max L(\alpha) = \sum_{i=1}^{n}\alpha_i + \frac{1}{2}\sum_{i=1}^{n}\sum_{j=1}^{n}\alpha_i\alpha_j y_i y_j K(x_i \cdot x_j) \\ s.t. \sum_{i=1}^{n}\alpha_i y_i = 0, 0 \leq \alpha_i \leq C, i = 1,2,\ldots,n \end{cases} \tag{8}$$

Solving the upper class, the decision function of the SVM is shown below:

$$f(\mathrm{x}) = \mathrm{sgn}\left[\sum_{i=1}^{n} \alpha_i \mathrm{y}_i K(\mathrm{x}_i \cdot \mathrm{x}_j) + b\right] \quad (9)$$

In this paper, the important parameters penalty factor C and kernel parameter in SVM were perfected by SCA optimization parameters, and a fault classifier with excellent classification performance is obtained.

3. Fault Diagnosis Model of TLSCA-SVM Algorithm

3.1. Principles of Transfer Learning

Transfer learning is the novel branch of machine learning, which has a critical influence on the data processing of artificial intelligence. In 1995, the NIPS professional seminar "Learning to Learn" was launched in the field of machine learning 26. Since then, transfer-learning research has entered a period of rapid development. The main principle of transfer learning is to apply the knowledge of the source task to the target task. Compared with multi-task learning, the most important thing in transfer learning is the target task, rather than learning the source task that applies to the target task.

Transfer learning mainly includes the following two concepts: domain D and task H. The two parts of D are composed of two parts: the feature space E and the edge probability distribution $P(e)$, where $\{e_1, \ldots, e_n\} \in E$, for a given domain, H also consists of two parts.

D_s is given a labeled source domain data, H_s is the corresponding source task. D_t is a target domain with very few tags, H_t is the corresponding learning task. The purpose is to transfer the knowledge of D_s and H_s to D_t, so as to perfect the performance of the target function ($D_s \neq D_t, H_s \neq H_t$).

3.2. TLSCA-SVM Algorithm

The SCA optimization algorithm uses the equality constraints to optimize the parameters of the SVM classifier. This method transforms the solution of the optimization problem into the solution of a set of linear equations. For a set of input samples $\{(d_i, g_i)\}, i = 1, \ldots, n, d_i \in R^n$. SVM maps the nonlinear inseparable problem to a high-dimensional space into a linearly separable problem through kernel function mapping, where the function discriminant is:

$$f(d) = u^T \varphi(d) + q \quad (10)$$

To propose the migration algorithm of the auxiliary data set, the auxiliary data sets are similar to the target set to enhance the accuracy of the classification model. Given that $\{(d_i, g_i)\}_{i=1}^{n}$ is the target sample data, $\{(d_i, g_i)\}_{i=n+1}^{n}$ is similar sample data. On the basis of the original SCA-SVM optimization problem, the TLSCA-SVM method adds auxiliary data that is an error penalty term, to the target formula to realize the transfer of knowledge. The improved objective function can be expressed as:

$$\begin{cases} \min \mathrm{J}(u,\lambda) = \frac{1}{2}u^T u + \frac{\beta_t}{2}\sum_{i=1}^{n}\lambda^2 + \frac{\beta_s}{2}\sum_{i=n+1}^{n+m}\lambda_i^2 \\ s.t. g_i(u^T\varphi(d_i)+q)+1-\lambda_i, i=1,2,\ldots,n,\ldots,n+m \end{cases} \quad (11)$$

In the formula, β_t and β_s are the penalty parameters of the target domain and source domain data sets, respectively, λ_i is the prediction error.

When the Lagrange multiplier is introduced, the dual equation of Equation (10) is derived, multiply each equation constraint with the Lagrange multiplier α_i, Then, the Lagrange function is established by adding the objective function:

$$L(u, q, \lambda_i, \alpha_i) = \mathrm{J}(u, \lambda) - \sum_{i=1}^{n+1} \alpha_i \left[g_i(u^T\varphi(d_i) + q) - 1 + \lambda_i\right] \quad (12)$$

The necessary condition for taking the extreme value is to find the partial derivative of each variable and set it to zero to obtain the Formula (13).

$$\begin{cases} \frac{\partial L}{\partial u} = 0 \Rightarrow u = \sum\limits_{i=1}^{n+m} \alpha_i g_i \varphi(d_i) \\ \frac{\partial L}{\partial \lambda_i} = 0 \Rightarrow \alpha_i = \begin{cases} \beta_t \lambda_i, i = 1, 2, \ldots, n \\ \beta_s \lambda_i, i = n+1, \ldots, n+m \end{cases} \\ \frac{\partial L}{\partial q} = 0 \Rightarrow \sum\limits_{i=1}^{n} \alpha_i g_i = 0 \\ \frac{\partial L}{\partial \alpha_i} = 0 \Rightarrow g_i(u^T \varphi(d_i) + q) - 1 + \lambda_i = 0 \end{cases} \tag{13}$$

Eliminate variables u and λ_i to get the matrix equation:

$$\begin{bmatrix} \Omega + \frac{1}{\beta} & G \\ G^T & 0 \end{bmatrix} \begin{bmatrix} \alpha \\ q \end{bmatrix} = \begin{bmatrix} e \\ 0 \end{bmatrix} \tag{14}$$

In the formula, $\Omega = g_i g_j k(d_i, d_j)$, $k(d_i, d_j) = \varphi(d_i)^T \varphi(d_j)$ is the kernel function. $G = [g_1, \ldots, g_{n+m}]^T$ and $e = [1, \ldots, 1]^T$ are $n+m$ dimensional column vectors, $\beta = diag[\beta_t, \ldots, \beta_t, \beta_s, \ldots, \beta_s]$.

The expression of TLSCA-SVM decision function is obtained by Formula (14).

$$G(d) = \text{sgn}\left[\sum_{i=1}^{l} \alpha_i g_i \varphi(d_i)^T \varphi(d_i) + q\right] \tag{15}$$

The Gaussian kernel function has an excellent anti-interference ability to the noise of the data. In this paper, the Gaussian kernel is used as the kernel function of TLSCA-SVM. The expression is:

$$k(d, d_i) = \exp\left(\frac{-\|d - d_i\|^2}{2\delta^2}\right) \tag{16}$$

where δ is the parameter of the Gaussian kernel width.

Finally, combined with the previous SCA-SVM algorithm, the decision function of the TLSCA-SVM classifier is obtained by the improved algorithm.

3.3. Fault Diagnosis Process of TLSCA-SVM Algorithm

In this paper, the TLSCA-SVM algorithm is applied to an analog circuit fault analysis, which is accomplished via the following main four steps (shown from step 1–step 4): building analog circuit model, setting and collecting fault data, using wavelets for data feature extraction, using PCA dimension reduction, using TLSCA-SVM classifier for troubleshooting, and comparing the actual fault data with the fault diagnosis results to arrive at the fault diagnosis rate.

The programming in this article is based on the mutual control of multiple modules. The data processing uses a five-layer wavelet packet and PCA technology. The classifier part uses SCA to perfect the parameters in the SVM to achieve fast and accurate classification results. When the classification algorithm adds auxiliary conditions, the algorithm has the ability of transfer learning. The specific fault handling process is shown in Figure 2.

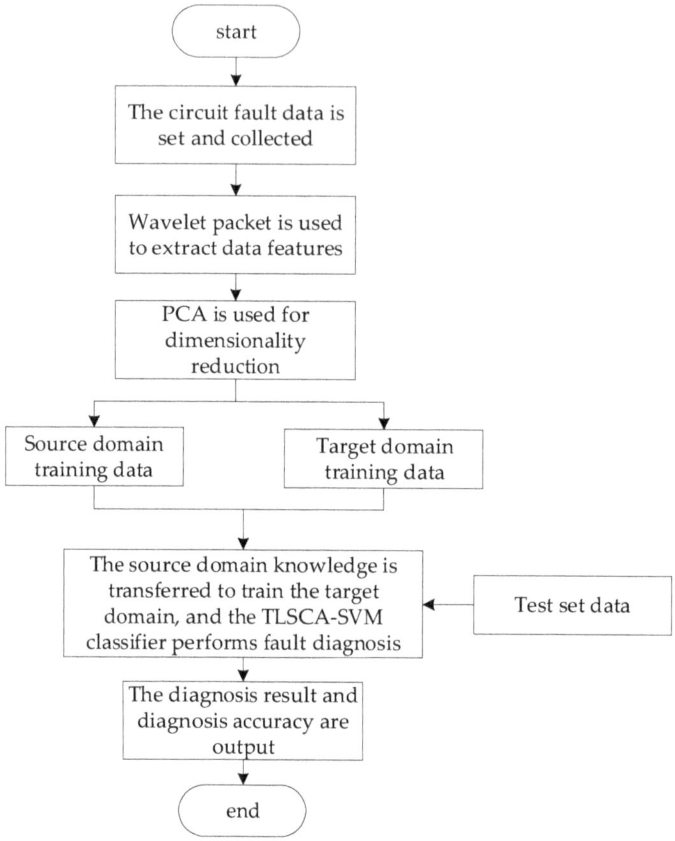

Figure 2. TLSCA-SVM fault diagnosis specific operation flowchart.

In this paper, the steps of fault diagnosis are shown below. The entire diagnostic process is not changed when comparing algorithms, but the comparison is made with the classifier and the optimized parameters of the classifier. This ensures that when comparing algorithms, the external conditions are consistent, and only the diagnostic effect of the classifier itself is considered. In order to ensure the consistency of the external conditions of the comparison algorithm, the same failure training set data is the same when using different classifiers for troubleshooting.

Step 1. The circuit model is established to collect and process data from different faults in the analog circuit. Input variables are obtained and the input data undergoes a series of data processing steps such as wavelet packet extraction, feature-value processing and PCA dimensionality reduction.

Step 2. After the above data processing is completed, the data is put into the pre-built classification model to achieve the effect of classifying the fault data. In this paper, the function of SVM is realized by LIBSVM. The penalty factor and kernel parameter in SVM are perfected by the SCA optimization method. In the SCA optimization method, multiple initial candidate solutions are randomly created. These initial candidates are used with the same probability of moving or reversing in the direction of the best solution. When looking for a relatively good solution, this method chooses to keep and proceed to the next iteration.

Step 3. An auxiliary condition is added to the objective function of the already debugged SCA-SVM classifier that is an error penalty term. An innovative fault diagnosis model, namely TLSCA-SVM, is constructed. The knowledge of the source domain is transferred to train the target domain, and the TLSCA-SVM classifier performs fault diagnosis. Fewer fault data is used as a training set and put into the classifier for fault diagnosis

Step 4. The trained TLSCA-SVM classifier model is used to classify faults on the test set, output the diagnosis results and judge the accuracy of the fault diagnosis results.

In this article, it is worth mentioning that the problem of a single-fault diagnosis in analog circuits is considered. Single fault refers to changing the parameters of only one component in the circuit while the parameters of other components in the circuit remain unchanged. The situation in which multiple component parameters are changed at the same time is referred to as a multiple-fault diagnosis problem. The data processing method of multi-fault diagnosis is similar to that of single-fault diagnosis. Multiple faults are changing the parameters of only two or more components while the parameters of other circuit components remain unchanged. Single fault is the basis of fault diagnosis in analog circuit fault diagnoses. In the actual analog circuit fault, the occurrence probability of a single fault is more than 80%, and the occurrence probability of multiple faults is relatively low. Additionally, a multi-fault diagnosis can be regarded as multiple single-fault problems occurring at the same time. That is, a multi-fault problem can be decomposed into multiple single-fault problems. Single-fault diagnosis and multi-fault diagnosis have many similarities in analog circuit fault diagnosis and multi-fault problems with less probability can be decomposed into single-fault for processing. Considering all of these aspects, this paper does not analyze the multi-fault diagnosis.

4. Acquisition and Process Fault Samples of Analog Circuits

This article takes the Sallen–Key circuit [38] and CSTV circuit [39] as diagnostic examples. Sallen–Key circuits and CSTV circuits are typical circuits that are often used to analog-circuit fault diagnosis. In the simulation experiment, considering the selectivity of the simulated circuit data, for more rigorous consideration, the faults of the Sallen–Key circuit and the CSTV circuit were set according to the literature [40–42].

At the same time, the Sallen–Key circuit is used because it is relatively simple as a second-order circuit and can use a public dataset. The Sallen–Key circuit is widely used in the fault diagnosis of analog circuits, so the Sallen–Key circuit was used to verify the validity of the algorithm. The CSTV circuit is a fourth-order circuit, which is more complex than the Sallen–Key circuit. The application of this circuit can show that the algorithm itself has universal applicability

4.1. Data Processing of Sallen-Key Band-Pass Filter Circuit with Injected Fault

In this essay, the data were studied by extracting the data in different modes of the Sallen–Key circuit. Circuit structure, individual component types and values are shown in Figure 3. Here, the parameter value is set to deviate from the original value by 50%, the capacity is set to deviate from the original value by 10%, and the resistance is set to deviate from the original value by 5%.

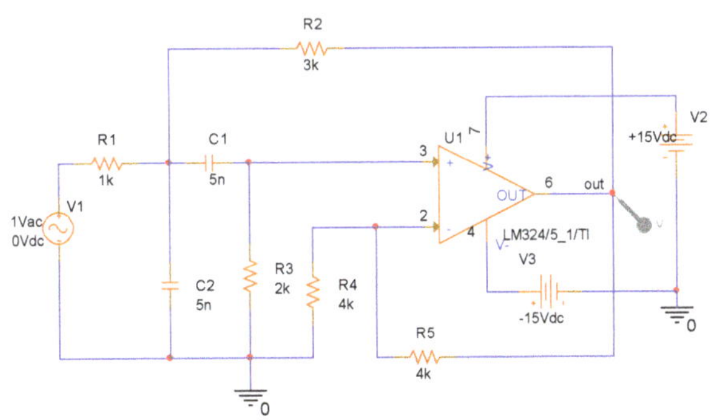

Figure 3. Sallen-Key band-pass filter circuit.

The excitation-signal parameters of the analog circuit are shown in Figure 3. After sensitivity analysis, R2, R3, C1, and C2 are selected as faulty components. The fault type and parameter list are shown in Table 1. The data of Sallen−Key circuit contains eight failure modes, which includes R2+, R2−, R3+, R3−, C1+, C1−, C2+ and C2−. The symbols − and + denote additional or little data and normal state (no failure, NF).

Table 1. Fault type and encoding of Sallen-Key band-pass filter circuit.

Malfunction Coding	Failure Mode	Nominal Value	Fault Value
F0	NF	-	-
F1	R2+	3 KΩ	4.5 KΩ
F2	R2−	3 KΩ	1.5 KΩ
F3	R3+	2 KΩ	3 KΩ
F4	R3−	2 KΩ	1 KΩ
F5	C1+	5 nF	7.5 nF
F6	C1−	5 nF	2.5 nF
F7	C2+	5 nF	7.5 nF
F8	C2−	5 nF	2.5 nF

In this paper, the data of the analog circuit output signal waveform is collected, and the nine modes are set, which are the signal data of the normal state and the signal data of the eight fault states. Each mode collects 100 sample points, so there are a total of 900 state sample points, which are randomly selected. A total of 600 sample points were used as the training set, and the remaining 300 sample points were used as the test set.

4.2. Data Processing of the CSTV Filter Circuit Injected into the Fault

To reflect the generality of the SCA-SVM classifier to fault diagnosis, a relatively complex circuit, CSTV filter, was selected as the verification analog circuit. The circuit element name and value are shown in Figure 4, and the status type is shown in Table 2. A total of 1800 samples, 900 of which were collected as the training set, and the other 900 samples were used as the test set.

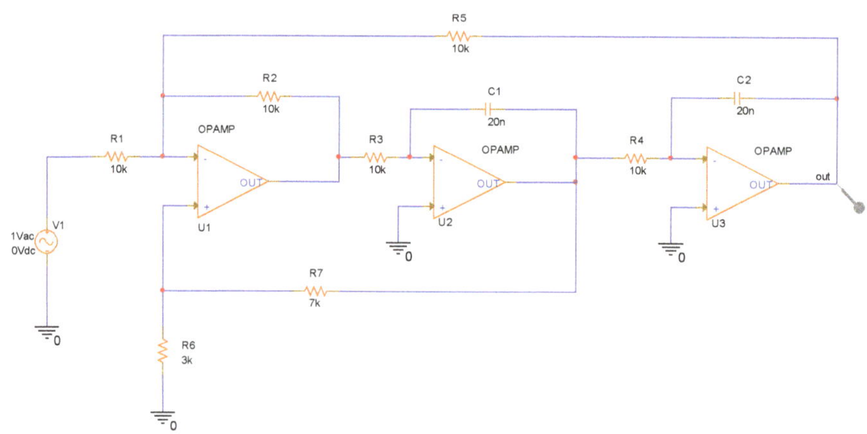

Figure 4. CSTV filter circuit.

The specific operation is to compare the classification algorithm and optimize the parameter method. The common second-order filter Sallen-Key circuit is applied for comparison experiments. To verify the versatility of the comparison algorithm conclusions, the multi-stage filter, which is a CSTV filter, is used. As verification, a more general result can be obtained.

Because the experimental simulation environment and experimental data are different, the experimental results will also be different. Therefore, when simulating the soft fault of the circuit, a more general experimental environment and data set were selected. According to the above parameter settings, each of the nine modes are analyzed using 100 MC, the start frequency is set to 100 Hz, and the cutoff frequency is set to 10k Hz. A voltage probe is set at the output node of the circuit to collect the output voltage. This ensures that the experimental results studied in this paper are more extensive and reliable.

Table 2. Fault type and encoding of CSTV filter circuit.

Malfunction Coding	Failure Mode	Nominal Value	Fault Value
F0	NF	-	-
F1	R1+	10 KΩ	15 KΩ
F2	R1−	10 KΩ	5 KΩ
F3	R2+	10 KΩ	15 KΩ
F4	R2−	10 KΩ	5 KΩ
F5	C1+	20 nF	30 nF
F6	C1−	20 nF	10 nF
F7	C2+	20 nF	30 nF
F8	C2−	20 nF	10 nF

4.3. Feature Processing of Analog Circuit Fault Data

In this paper, the processing of analog circuit fault data is divided into the three steps, which includes the extraction of fault signals, wavelet packet extraction of data features and PCA dimensionality reduction.

4.3.1. Extract Fault Signal

Multiple fault control groups were established, the original data set was simulated, a voltage probe was set on the output node of the circuit, the output was processed by Monte Carlo and the output voltage was collected. The output voltage of each mode was used as the original data set. To visually compare the characteristics of circuit failures in different modes, the simulation results of the above two circuits were selected, and the signal data of the eight failure modes and the normal mode were displayed. Figures 5 and 6 present the results of circuit simulation of different fault modes in the Sallen–Key circuit and CSTV circuit. As can be seen from the figure, there are differences in the output signal of the circuit in different fault modes.

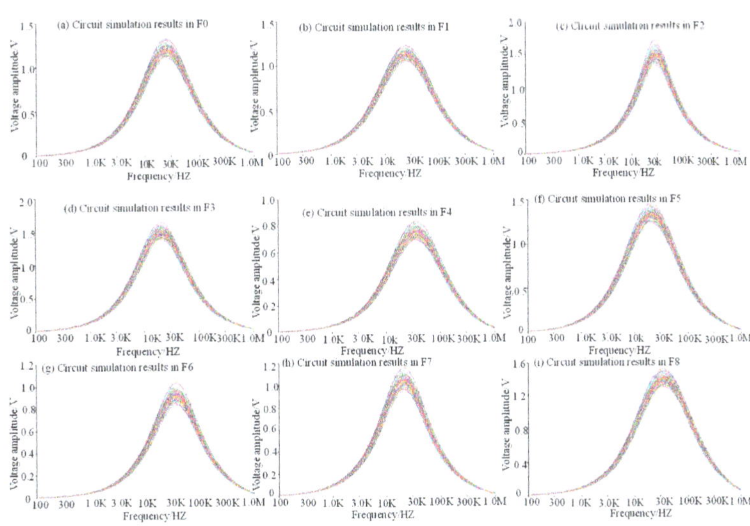

Figure 5. The output signal of Sallen-Key band-pass filter circuit.

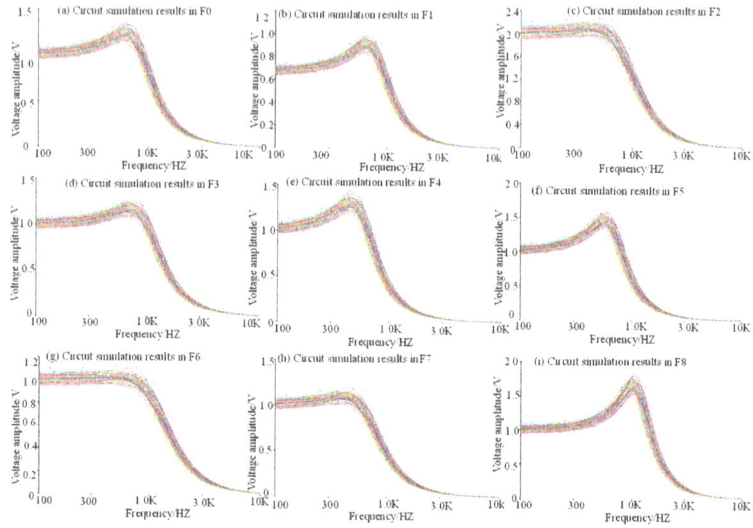

Figure 6. The output signal of CSTV filter circuit.

Figures 5 and 6 reflect the states of different faulty circuit signals. Since the analog circuit fault data is set by changing the values of different elements of the circuit. Therefore, the circuit signals corresponding to each fault have a certain similarity. In each subplot in Figures 5 and 6, each image is similar, but each subplot is slightly different due to the different values of the components.

4.3.2. Feature Extraction and Dimensionality Reduction of Fault Signals

In this paper, the wavelet packet was used for feature extraction. The feature extraction is representative data extracted from a large amount of data, so the original signal is similar but not identical to the signal after feature extraction. In Figure 7, it can be seen that Figure 7a is similar but not the same as Figure 7d. The main function in the step of extracting features is to obtain highly representative data in the overall data, that is to remove redundant data. In Figure 7, it can be seen that Figure 7c contrasts Figure 7b with the redundant signal data removed.

After the analog circuit is built, the relevant parameters are set. All failure modes are analyzed by 100 MC, the start frequency is set to 100 Hz, and the cutoff frequency is set to 10 kHz. At the output node of the circuit, a voltage probe is used to collect the output voltage. The output voltage is used as the original data set, the wavelet packet is used to extract the features, and the PCA is used to reduce the dimensionality. Finally, the feature-processed analog circuit fault data is obtained, which can be used in the fault diagnosis classifier.

When the fault signal is extracted, the output voltage in each mode is obtained as the original data set, and the features are extracted by wavelet packet, and then the PCA dimension reduction is performed. Finally, the fault data of the analog circuit after feature processing is obtained, which is used in the fault-diagnosis classifier. Among them, the wavelet packet is mainly used to extract the eigenvalues, and the wavelet packet comes from the wavelet, which is a more perfect decomposition method based on the wavelet decomposition. Compared with the wavelet transformation, the wavelet packet transformation can analyze the signal in all directions, can further decompose the wavelet transform without subdividing, and evenly distribute the corresponding two frequency bands according to the same spacing in the high frequency part in the range. This article uses a five-layer wavelet packet, which can effectively decompose the input signal to extract the characteristic value. Figure 7 shows the changes before and after the signal when the wavelet packet is used in this article to extract the characteristic value.

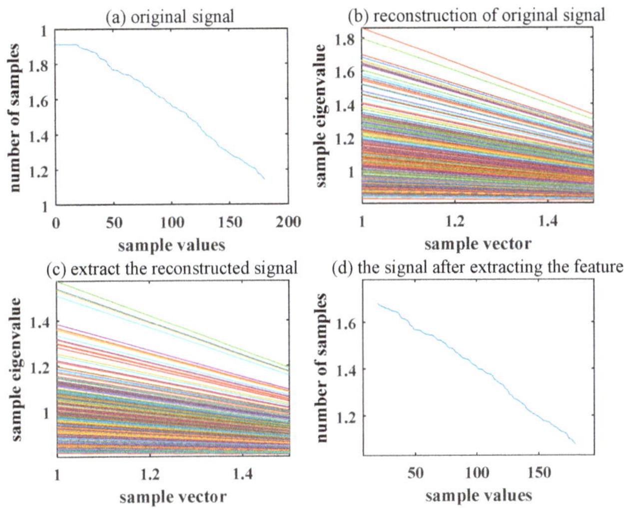

Figure 7. Comparison before and after extraction of eigenvalue signal by wavelet packet, where (**a**) original signal; (**b**) reconstruction of original signal; (**c**) extract the reconstructed signal; (**d**) the signal after extracting the feature.

Regarding the PCA technique, the main purpose of this essay is to use PCA to perform dimensionality reduction operations. The number of selected Principal components is five, as shown in Figure 8, which reflects the changes in the data before and after the PCA analysis. When using PCA to reduce the dimensionality, the threshold is set to 85%. When the cumulative contribution rate reaches 85%, the pivot is no longer selected. After all the original variables are processed by dimensionality reduction, the pivot can be obtained. The pivot is the linearity of the original variable combination.

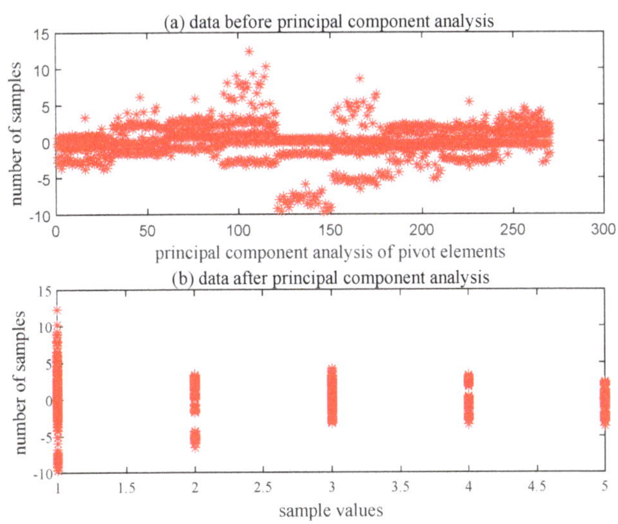

Figure 8. Data comparison before and after principal component analysis.

5. Algorithm Horizontal and Vertical Comparison Experiment Results

5.1. SCA Optimization Parameter Comparison

5.1.1. Comparison of Optimized Parameters under Sallen-Key Band-Pass Filter Circuit

The fault-diagnosis method in this article is based on SVM. There are two important parameters in SVM, namely the penalty factor C and the parameter coefficient g of the kernel function. The two parameters determine the classification performance of the SVM, and a way to quickly and effectively obtain the optimal parameter value becomes the key consideration of the optimization algorithm. This paper compares optimization parameter algorithms such as Grid Search, GA, PSO, ACA, SA and SCA, by comparing the results, and the SCA method is considered to be the best way to optimize parameters. The main operation is to compare Grid Search-SVM, GA-SVM, PSO-SVM, ACA-SVM, SA-SVM and SCA-SVM. The classifiers of different optimization algorithms are used in the fault diagnosis of the Sallen-Key circuit, and there are certain advantages in determining the SCA-optimization parameters. The experimental results are shown in Table 3, which compares the results of fault diagnosis with different optimization algorithms. It can be concluded that the SCA-optimization parameter algorithm can enhance the optimization speed and reduce the optimization time under the premise of meeting the parameter optimization requirements, so that the performance of the entire model not only has the rapidity of classification, but also has the accuracy of classification.

Table 3. Sallen-Key bandpass filter circuit optimization parameter algorithm comparison.

Optimization Parameter Algorithm	Accuracy Rating/%	Elapsed Time/s
Grid Search	100	62.37
GA	87.04	31.35
PSO	99.67	19.87
ACA	98.13	30.52
SA	89.65	17.54
SCA	100	10.85

Fitness curves are introduced to reflect the performance of different optimization algorithms. In Figures 9–14, the fitness curves of different optimization algorithms applied to the SVM parameter optimization of the Sallen-Key circuit fault data are detailed.

The optimization parameter comparison experiment in this article aimed to perfect the parameters of the classification algorithm under the premise of SVM, in which no changes are made to the various modules of the fault classification, and only the optimization-parameter module is compared to ensure that the external conditions are consistent. The experimental performance is compared below, and the comparison results are shown in Table 3.

The experimental results can be deduced by comparing optimization-parameter algorithms such as Grid Search, GA, PSO, ACA and SA. The performance of SCA-optimization parameters is superior than other algorithms in terms of accuracy and classification speed.

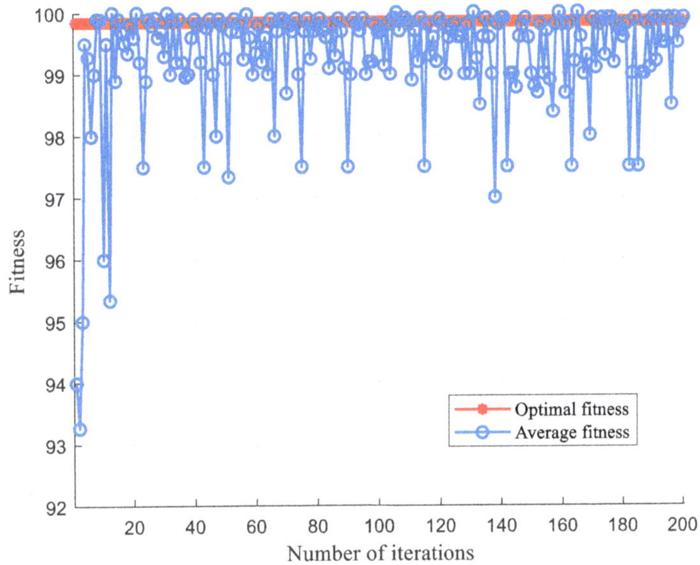

Figure 9. Grid Search-SVM fitness curve from Sallen-Key band-pass filter.

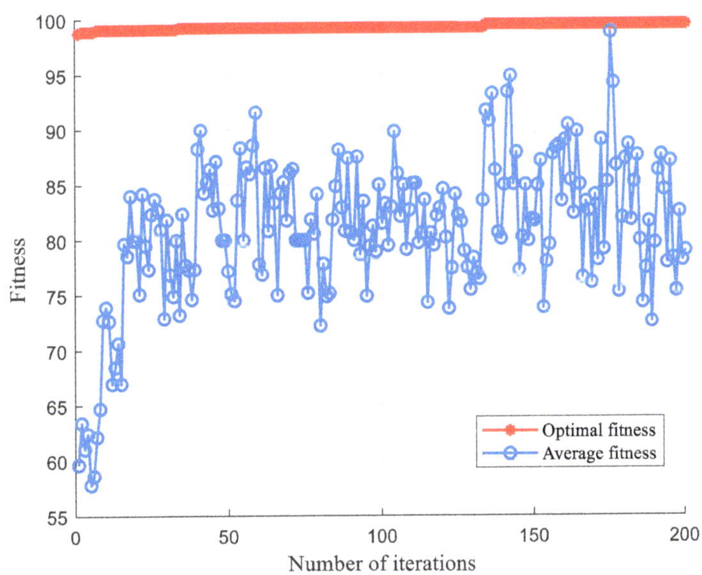

Figure 10. GA-SVM fitness curve from Sallen-Key band-pass filter.

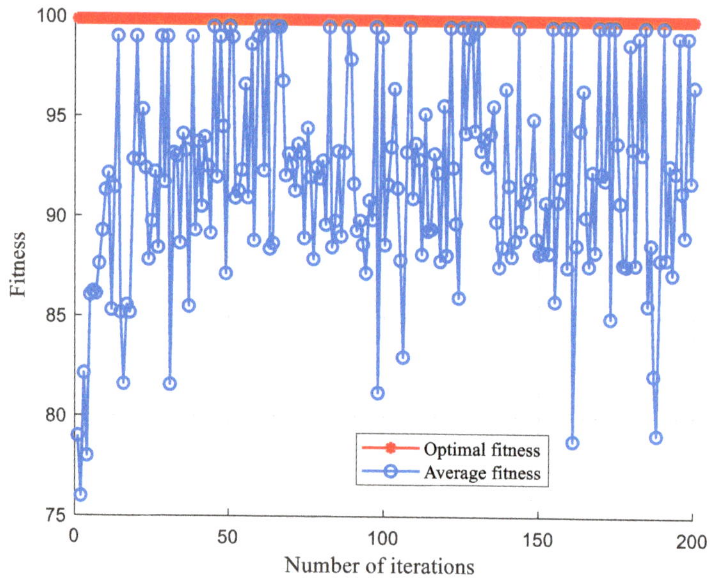

Figure 11. PSO-SVM fitness curve from Sallen-Key band-pass filter.

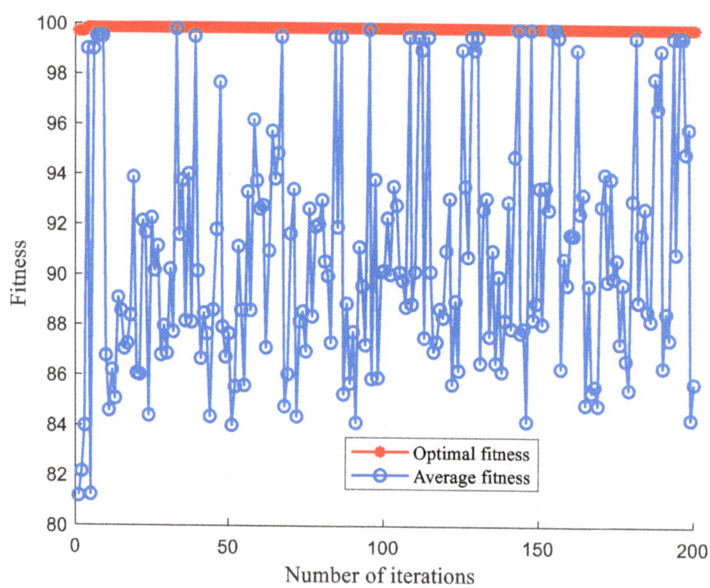

Figure 12. ACA-SVM fitness curve from Sallen-Key band-pass filter.

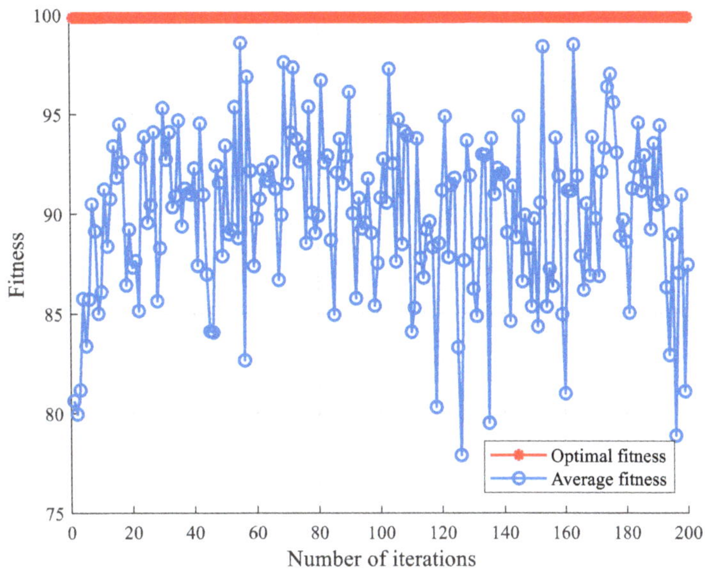

Figure 13. SA-SVM fitness curve from Sallen-Key band-pass filter.

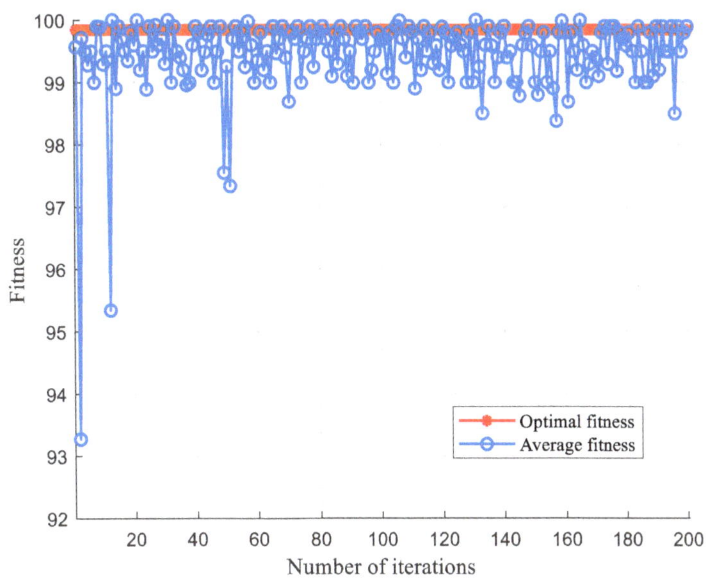

Figure 14. SCA-SVM fitness curve from Sallen-Key band-pass filter.

5.1.2. Comparison of Optimized Parameters under CSTV Filter Circuit

In the method of comparing and optimizing parameters, the common Sallen-Key filter is used for comparison experiments. To verify the generality of the comparison method conclusions, the multi-stage filter, which is a CSTV circuit, is used for verification. Grid Search-SVM, GA-SVM, PSO-SVM, ACA-SVM, SA-SVM and SCA-SVM are compared, and the fault classifier is used with the CSTV filter-circuit data to ensure that the fault-diagnosis classifier has universal applicability. In Table 4, comparison results of optimized parameter algorithms by the CSTV filter circuit are presented.

Table 4. CSTV filter circuit optimization parameter algorithm comparison.

Optimization Parameter Algorithm	Accuracy Rating/%	Elapsed Time/s
Grid Search	99.85	73.07
GA	81.54	34.08
PSO	97.08	27.15
ACA	95.38	39.87
SA	83.66	26.31
SCA	99.89	18.49

Fitness curves are introduced to reflect the performance of different optimization algorithms. As shown in Figures 15–20, the fitness curves of different optimization algorithms applied to the SVM parameter optimization of the Sallen–Key circuit fault data are shown in detail. When comparing the algorithms for optimizing parameters, the number of iteration steps is uniformly set to 200.

Finally, in order to draw a more general conclusion, the CSTV filter circuit is used as verification. It can be concluded that the performance of the SCA-optimization parameters has certain advantages, when considering the classification accuracy and classification speed. SCA not only meets the requirements of parameter optimization, but also improves the optimization speed and reduces the optimization time, so that the performance of the entire algorithm has both the rapidity of classification and the accuracy of classification.

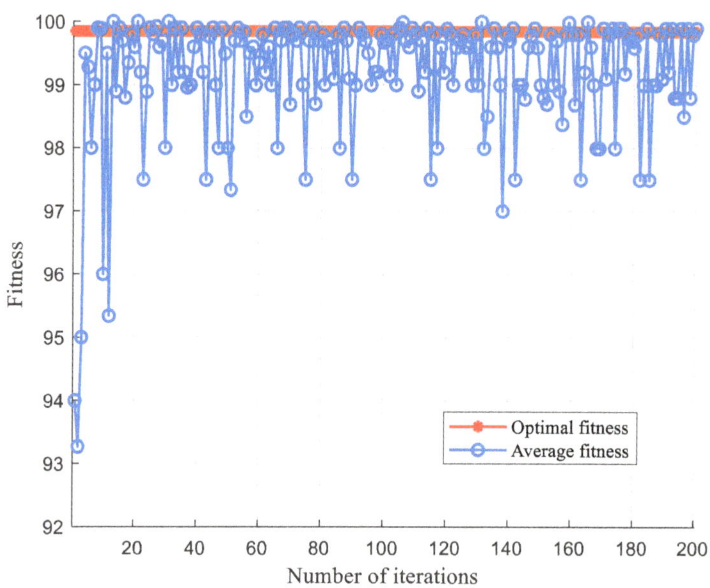

Figure 15. Grid Search-SVM fitness curve from CSTV filter.

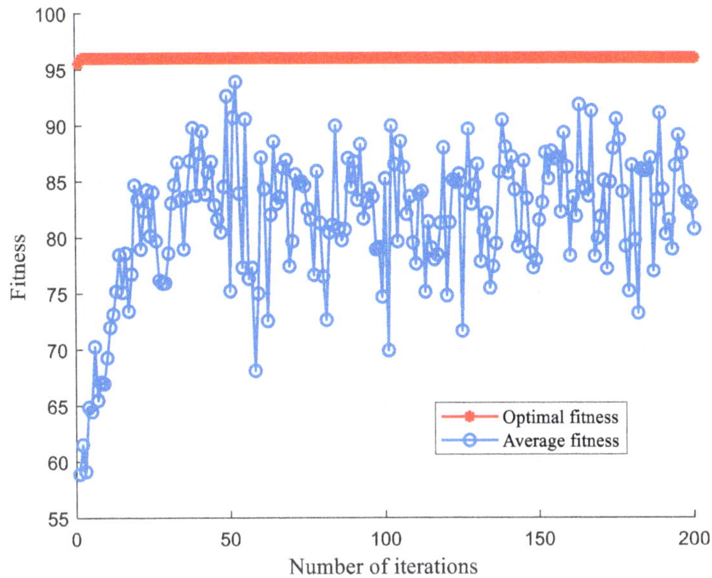

Figure 16. GA-SVM fitness curve from CSTV filter.

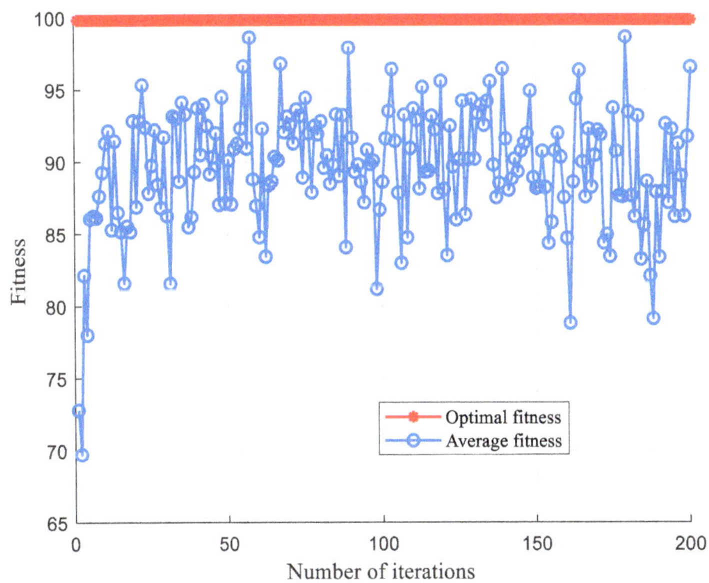

Figure 17. PSO-SVM fitness curve from CSTV filter.

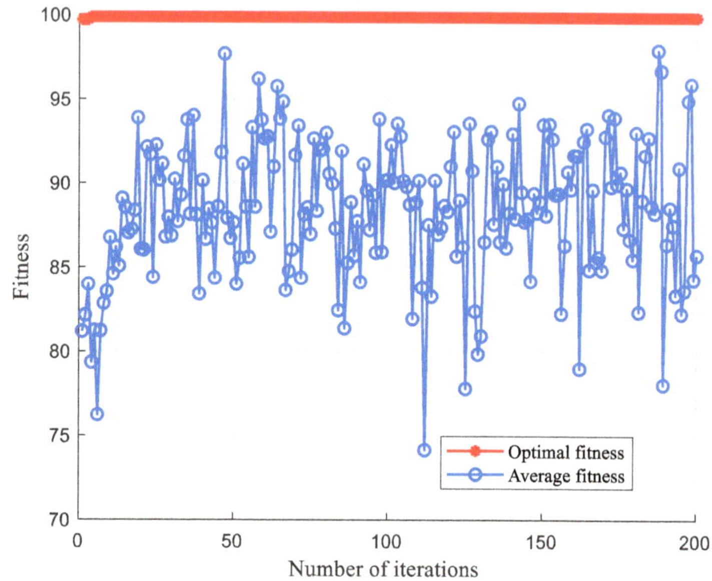

Figure 18. ACA-SVM fitness curve from CSTV filter.

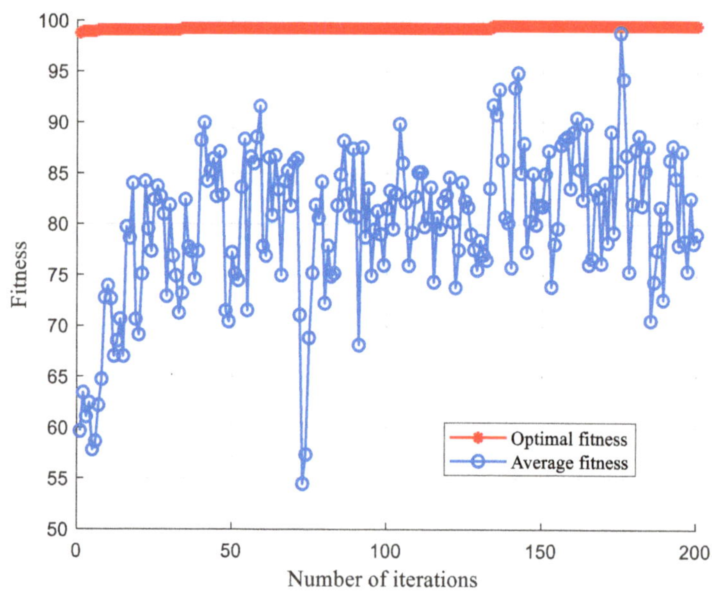

Figure 19. SA-SVM fitness curve from CSTV filter.

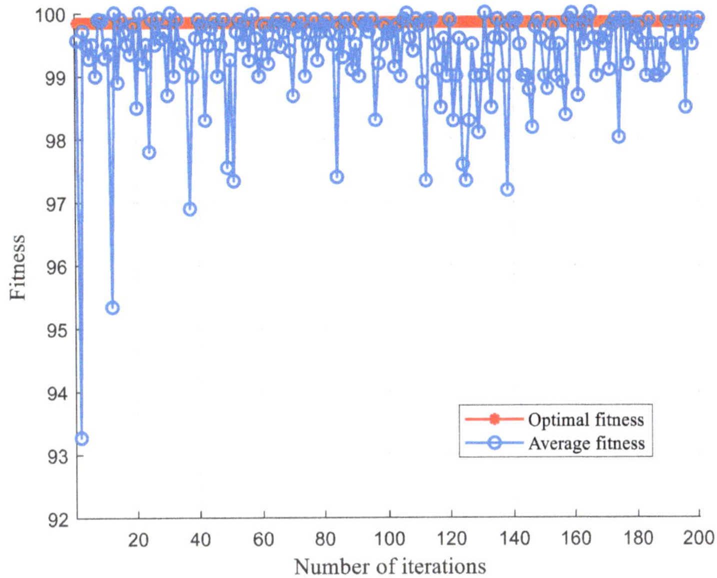

Figure 20. SCA-SVM fitness curve from CSTV filter.

5.2. Comparison of SCA-SVM Classification Algorithms

5.2.1. Comparison of Classification Algorithms under Sallen-Key Bandpass Filter Circuit

According to the fault data of the Sallen-Key circuit, different classification algorithms are used, such as BP, SOM, ELM, decision tree, random forest and SCA-SVM. Through comparative experiments, it is concluded that the SCA-SVM classification algorithm is superior to other classification algorithms in fault diagnosis. As shown in Figures 21–26, the classification effect of the Sallen-Key circuit fault data by different classification algorithms is shown in detail.

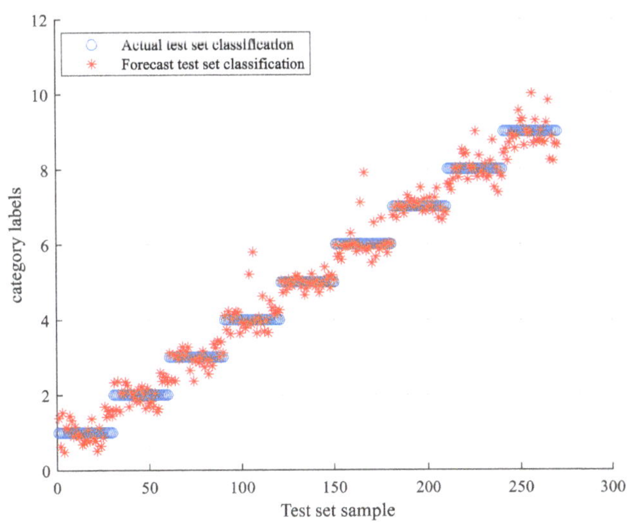

Figure 21. Classification effect of BP algorithm from Sallen-Key bandpass filter.

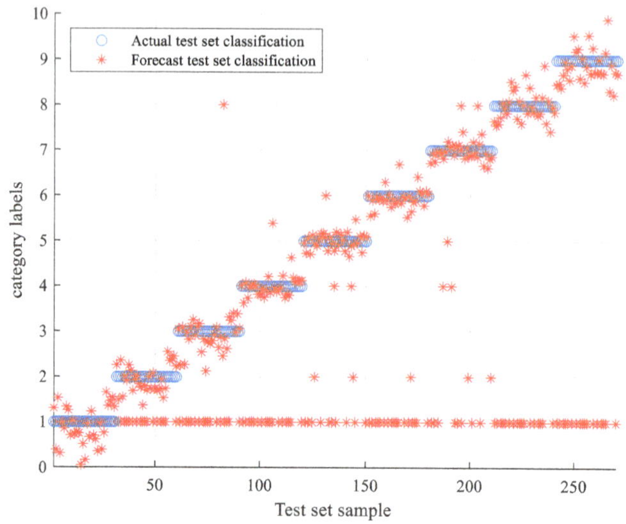

Figure 22. Classification effect of SOM algorithm from Sallen-Key bandpass filter.

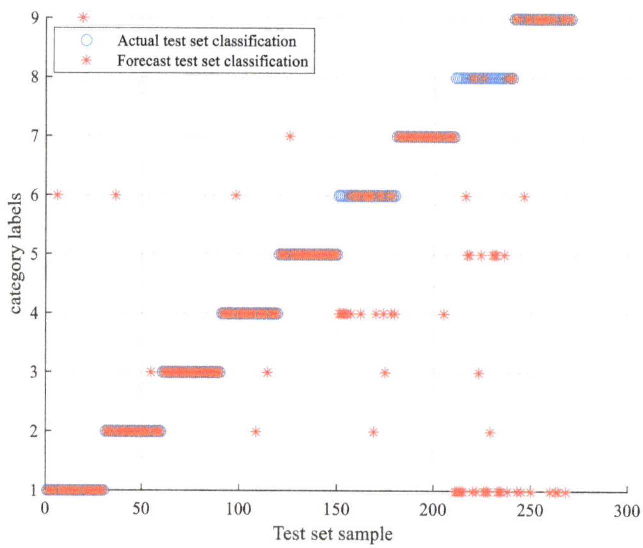

Figure 23. Classification effect of ELM algorithm from Sallen-Key bandpass filter.

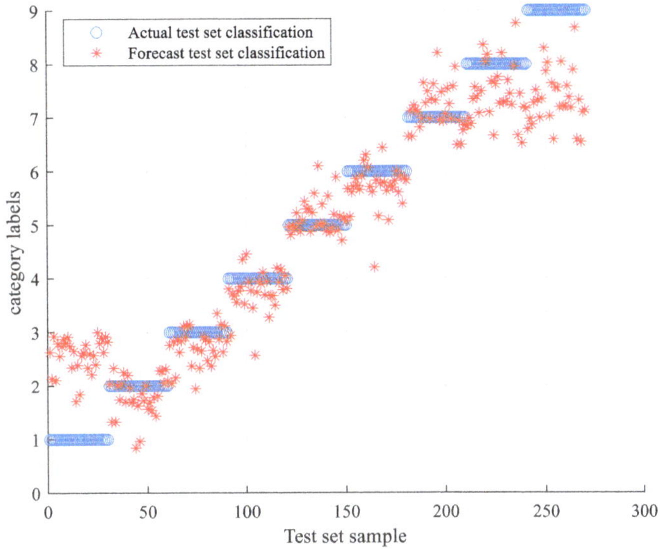

Figure 24. Classification effect of decision tree algorithm from Sallen-Key bandpass filter.

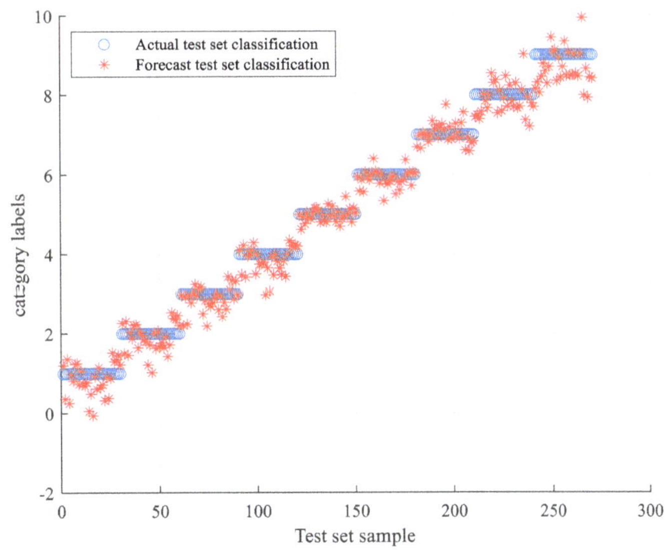

Figure 25. Classification effect of random forest algorithm from Sallen-Key bandpass filter.

It is worth mentioning that when comparing the classification algorithms, the neural networks [43] used for comparison include BP, RBF, GRNN, PNN, competitive neural network, and SOM. The characteristics of the experimental data are combined. The BP and the SOM have the best comprehensive classification performance. After comprehensive consideration, it was decided to use the BP neural network and the SOM neural network as representative of the neural network-classification algorithm for comparison.

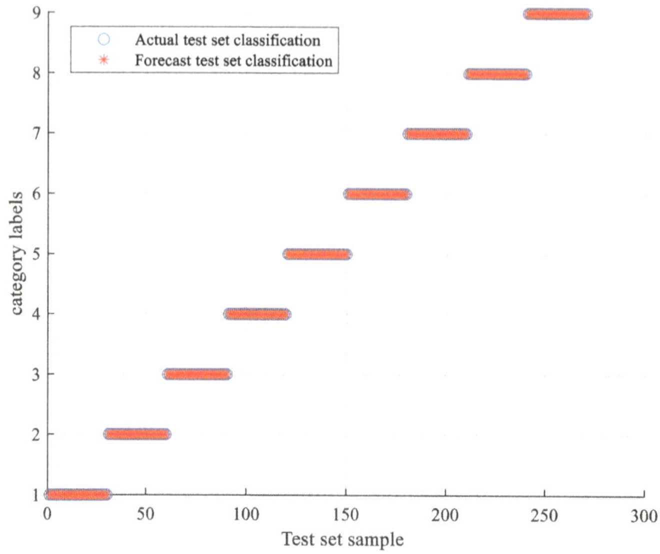

Figure 26. Classification effect of SCA-SVM algorithm from Sallen-Key bandpass filter.

In terms of the classification effect, the accuracy of the SCA-SVM classification algorithm is superior to other algorithms. Table 5 is obtained by digitizing the experimental statements. As presented in Table 5, by comparing other classification algorithms such as BP, SOM, ELM, decision trees, random forests, etc., the conclusion is obtained. It draws a conclusion that when considering classification accuracy, the performance of SCA-SVM is better than that other algorithms. SCA is very effective for optimizing SVM parameters fo4 fault diagnosis, and the classification effect of the Sallen–Key circuit fault data can reach 100%.

Table 5. Comparison of Sallen-Key bandpass filter circuit classification algorithms.

Classification Algorithm	Accuracy Rating/%	Elapsed Time/s
BP	99.25	31.57
SOM	82.76	7.20
ELM	94.70	1.43
Decision Tree	93.07	4.26
Random Forest	97.88	9.75
SCA-SVM	100	10.85

5.2.2. Comparison of Classification Algorithms under CSTV Filter Circuit

In the comparison classification algorithm, the common second-order filter, which is the Sallen–Key circuit, is used for comparison experiments. To verify the versatility of the conclusions of the comparison method, a multi-stage filter, which is a CSTV filter circuit, is used for verification. Since the CSTV filter circuit is more complex and there are more aspects to consider in the fault diagnosis, the data of the CSTV filter circuit is added for a better analysis. So, for the CSTV filter circuit, a total of 1800 samples were collected, of which 900 samples constitute the training data set, and the remaining 900 samples constitute the test data set. In Figures 27–32, the classification effect of CSTV filter circuit fault data with different classification algorithms is shown in detail.

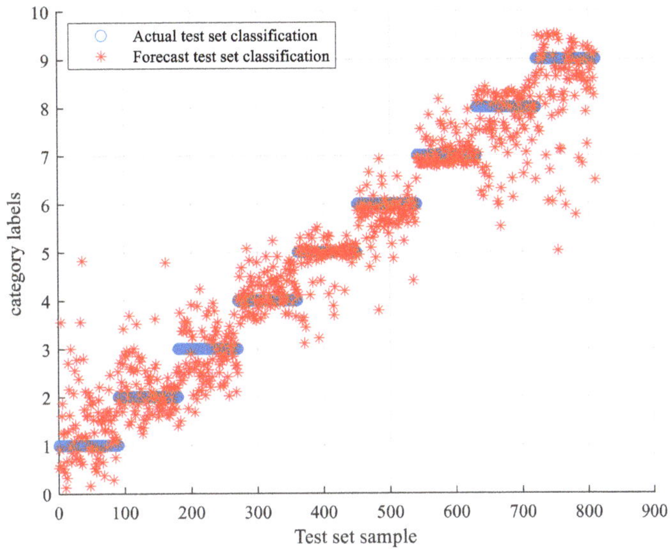

Figure 27. Classification effect of BP algorithm from CSTV filter.

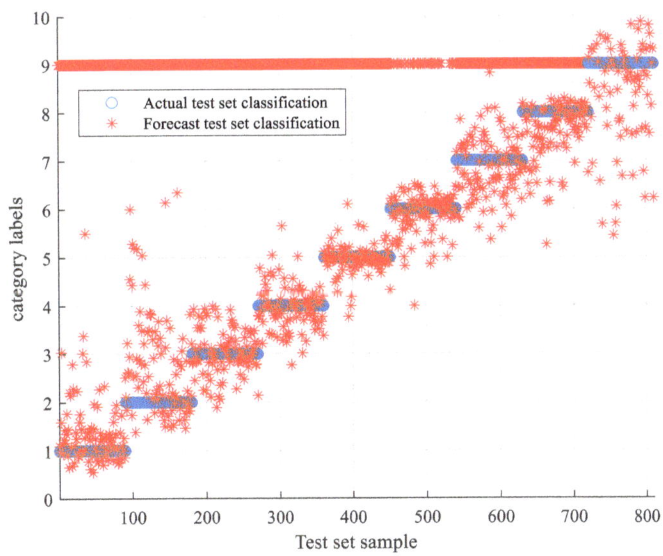

Figure 28. Classification effect of SOM algorithm from CSTV filter.

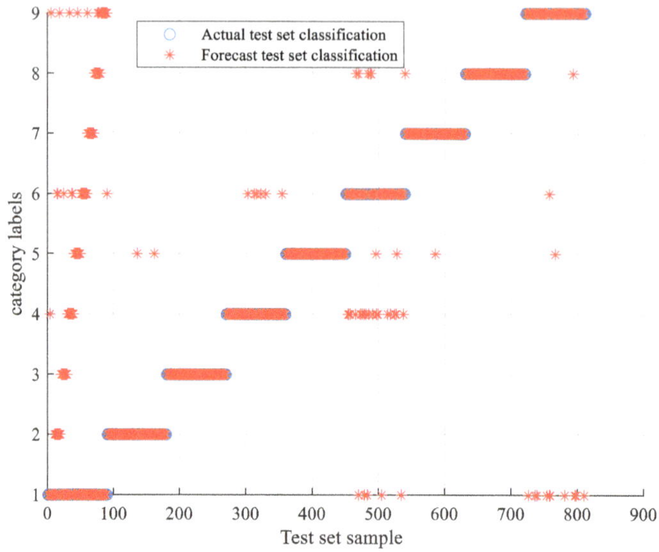

Figure 29. Classification effect of ELM algorithm from CSTV filter.

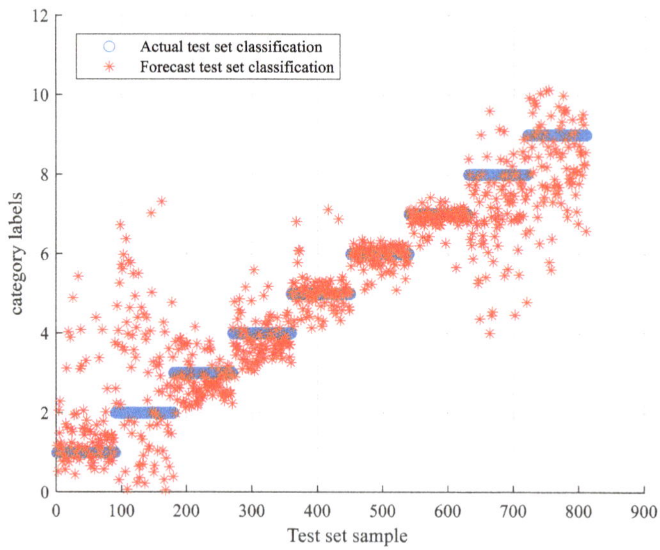

Figure 30. Classification effect of decision tree algorithm from CSTV filter.

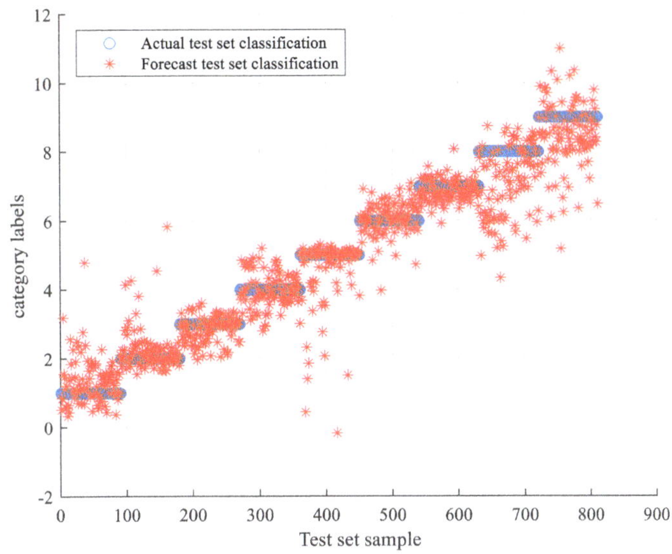

Figure 31. Classification effect of random forest algorithm from CSTV filter.

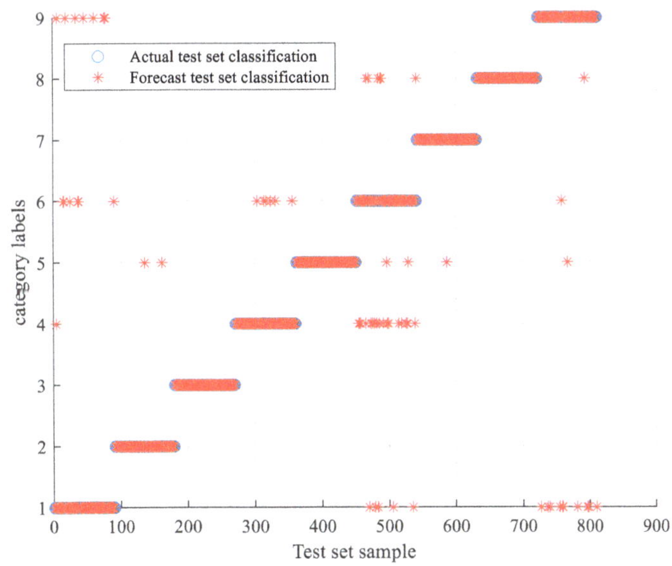

Figure 32. Classification effect of SCA-SVM algorithm from CSTV filter.

It can be seen from Table 6 that by comparing other classification algorithms such as BP, SOM, ELM, decision trees, random forests, etc., when considering accuracy, the performance of SCA-SVM is better than other algorithms. SCA is very effective for optimizing SVM parameters for fault diagnosis, and the classification effect of the CSTV filter circuit fault data can reach 99.89%.

Table 6. Comparison of CSTV filter circuit classification algorithms.

Classification Algorithm	Accuracy Rating/%	Elapsed Time/s
BP	96.89	40.69
SOM	74.15	17.49
ELM	91.85	9.12
Decision Tree	89.45	19.99
Random Forest	95.12	10.90
SCA-SVM	99.89	18.49

In order to draw a more general comparison algorithm conclusion, the multi-stage filter, that is, the CSTV filter circuit, is used as a verification object and a more general conclusion can be drawn. When the target is analog circuit-fault diagnosis, SCA-SVM is considered to have certain advantages at the level of classification accuracy.

5.3. TLSCA-SVM Comparative Test Results

The aforementioned SCA-SVM fault classifier can be effectively applied to fault diagnosis, but when there are too few fault samples or only normal samples, there will be a problem of inaccurate fault diagnosis. Therefore, this paper introduces the TLSCA-SVM algorithm. An auxiliary condition, that is, an error penalty term, is added to the objective function of the SCA-SVM classifier to construct a new fault-diagnosis model so that the fault diagnosis satisfies the ability to effectively classify faults when the fault samples are not complete. It combines the advantages of the SCA-SVM classifier in fault diagnosis with high accuracy, fast diagnosis speed and good stability.

According to the type of training samples, transfer learning can be divided into zero-shot learning and few-shot learning. This paper changes the proportion of faulty data in the training set by changing the database to reduce the proportion of faulty data in the training set. Faulty data and normal data are kept in the test set to perform transfer learning. It is worth mentioning here that the limit of few-shot learning in transfer learning is zero-shot learning. Zero-shot learning can be achieved under an extremely idealized model, but zero-shot learning is unrealizable in real data processing, so the transfer learning of this algorithm is embodied in the fault classification of few-shot learning. This paper compares the SCA-SVM classifier used in traditional machine learning with the TLSCA-SVM classifier, based on the transfer-learning theory by constantly changing the proportion of the failure samples in the training set. Figure 33 is obtained. There is a point to note here. Since the transfer learning ability of the comparison algorithm requires the use of a lot of data and a relatively simple circuit form, in the comparison experiment, a simple Sallen-Key band-pass filter circuit with more circuit data was used.

The conclusion can be drawn from Figure 33 that when the training sample is relatively small, the traditional SCA-SVM classification algorithm cannot effectively perform fault diagnosis. The classifier trained by the non-transfer learning model is not effective in the fault diagnosis of few-shot learning. Classifiers with transfer ability have certain advantages in few-shot learning. As the proportion of fault samples increases, the effect of transfer learning becomes weaker. When the proportion of fault samples reaches 50%, the effect of transfer learning and non-transfer-learning fault diagnosis is the same. When the proportion of faulty samples does not reach 50%, the classifier with transfer learning ability shows better classification performance in fault diagnosis. This article classifies the performance of the TLSCA-SVM classification algorithm in fault diagnosis as better than SCA-SVM fault samples when the proportion of faults does not reach 50%.

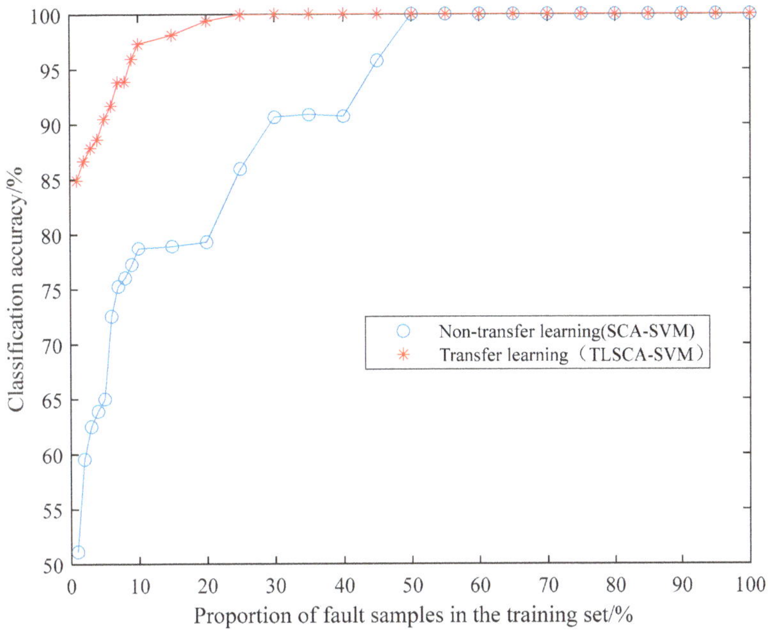

Figure 33. Comparison of transfer learning and non-transfer learning.

6. Conclusions

This article uses SCA to optimize SVM parameters, and proposes a classification-diagnosis method using SCA-SVM. Finally, it is proposed that the TLSCA-SVM classification algorithm can be used in the case of fewer fault samples in the training set. The data of the Sallen–Key circuit is used as experimental data, the data of the CSTV circuit is used to verify the versatility of the method, and a more general conclusion is drawn as follows:

(a) By comparing optimization parameter algorithms such as Grid Search, GA, PSO, ACA, SA, etc., the SCA is proposed. The SCA optimization-parameter algorithm can improve the optimization speed and shorten the optimization time on the premise of meeting the parameter-optimization requirements. The entire algorithm has both high classification accuracy and fast classification performance.

(b) The SCA-SVM fault-classification method has superior performance in the fault diagnosis. A complex CSTV circuit is used to verify the versatility of the method. Several comparison experiments show that the method is not only superior, but also universal in performance to other algorithms. Different classification algorithms are used, such as BP, SOM, ELM, decision tree, random forest and SCA-SVM to compare. It can be concluded that the accuracy of the SCA-SVM classification algorithm is superior to other comparison algorithms in terms of the classification effect.

(c) With regard to the problems of most optimization algorithms, this paper reasonably avoids them. The classification algorithm of this paper is analyzed. This paper uses the SVM classifier as the main body for fault diagnosis. The SVM classifier itself has a good classification effect, and the difference between the important parameter's penalty factor C and kernel parameter will affect the classification effect of the SVM. The objective of the SCA algorithm is to obtain appropriate parameters. The search method is randomly determined each time an optimal solution is found. That is to say, in the next search, both the local search and the global search are random, i.e., the probability is the same. Each time the optimal solution is approximated, the approximation method is randomly determined. Such an optimization method can avoid local optimal solutions and shorten the optimization time. The classifier formed after finding suitable parameters has a good classification effect in fault classification, and the classification efficiency improved.

(d) Various optimization algorithms were compared, such as Gray Wolf Optimization (GWO), Gravitational Search Algorithm (GSA), competitive swarm optimizer (CSO), etc. The advantages and disadvantages of different optimization algorithms were discovered. Most of the shortcomings focus on non-global search. When the optimization algorithm performs a non-global search, local optimal solutions may appear. After continuous exploration, some algorithms were optimized by combining the characteristics of multiple algorithms. For example, the GWO was combined with SCA to optimize parameters. These issues deserve to be studied in future.

(e) When the training data is deficient, the TLSCA-SVM classification algorithm can effectively diagnose the fault. Because the TL-SCASV algorithm adds an auxiliary condition to the objective function of the SCA-SVM classifier, that is, an error penalty term to construct a new fault diagnosis model, the fault diagnosis is satisfactory. When the fault samples are not complete, it can still effectively classify the faults. It combines the advantages of the SCA-SVM classifier with high accuracy, fast diagnosis speed and good stability in fault diagnosis. The algorithm not only achieves high fault-diagnosis accuracy, but can also operate effectively in the case of a lack of fault samples, and can effectively perform fault classification in multiple circuits.

To sum up, the TLSCA-SVM classifier was constructed. When there are more fault samples, the fault classification effect of the classifier was found to be better by a cross-sectional comparison. The classifier was also effective in diagnosis when there were fewer fault samples. It has certain versatility in fault-data diagnosis and broad application prospects.

Author Contributions: Conceptualization, A.Z. and D.Y.; methodology, A.Z. and D.Y.; software, D.Y.; validation, A.Z., D.Y. and Z.Z.; formal analysis, D.Y.; investigation, D.Y.; resources, A.Z.; data curation, D.Y.; writing—original draft preparation, D.Y.; writing—review and editing, A.Z. and D.Y.; visualization, D.Y.; supervision, A.Z. and Z.Z.; project administration, A.Z.; funding acquisition, A.Z. All authors have read and agreed to the published version of the manuscript.

Funding: This works is partly supported by the Natural Science Foundation of Liaoning, China under Grant 2019MS008, Education Committee Project of Liaoning, China under Grant LJKZ1011 and LJ2019003.

Data Availability Statement: Not applicable.

Conflicts of Interest: The authors declare no conflict of interest.

References

1. Wen, X.; Xu, Z. Wind turbine fault diagnosis based on ReliefF-PCA and DNN. *Expert Syst. Appl.* **2021**, *178*, 115016. [CrossRef]
2. Li, Y.; Miller, M.S. Seismic Evidence for Thermal and Chemical Heterogeneities in Region Beneath Central America From Grid Search Modeling. *Geophys. Res. Lett.* **2021**, *48*, e2021GL092493. [CrossRef]
3. Ebrahim, M.A.; Ayoub, B.A.A.; Nashed, M.N.F.; Osman, F.A.M. A Novel Hybrid-HHOPSO Algorithm Based Optimal Compensators of Four-Layer Cascaded Control for a New Structurally Modified AC Microgrid. *IEEE Access* **2020**, *9*, 4008–4037. [CrossRef]
4. Ding, Y.; Meng, R.; Yin, H.; Hou, Z.; Sun, C.; Liu, W.; Hao, S.; Pan, Y.; Wang, B. Keratin-A6ACA NPs for gastric ulcer diagnosis and repair. *J. Mater. Sci. Mater. Med.* **2021**, *32*, 66. [CrossRef] [PubMed]
5. Han, H.; Xu, L.; Cui, X.; Fan, Y. Novel chiller fault diagnosis using deep neural network (DNN) with simulated annealing (SA). *Int. J. Refrig.* **2021**, *121*, 269–278. [CrossRef]
6. Tobi, M.A.; Bevan, G.; Wallace, P.; Harrison, D.; Okedu, K.E. Using MLP-GABP and SVM with wavelet packet transform-based feature extraction for fault diagnosis of a centrifugal pump. *Energy Sci. Eng.* **2021**, 1–14. [CrossRef]
7. Fu, J.; Che, G. Fusion Fault Diagnosis Model for Six-Rotor UAVs Based on Conformal Fourier Transform and Improved Self-Organizing Feature Map. *IEEE Access* **2021**, *9*, 14422–14436. [CrossRef]
8. Jain, M.; William, A.; Mark, S. CNN vs ELM for Image-Based Malware Classification. *arXiv* **2021**, preprint. arXiv:2103.13820.
9. Li, Z.; Wang, L.; Huang, L.; Zhang, M.; Cai, X.; Xu, F.; Wu, F.; Li, H.; Huang, W.; Zhou, Q.; et al. Efficient management strategy of COVID-19 patients based on cluster analysis and clinical decision tree classification. *Sci. Rep.* **2021**, *11*, 9626. [CrossRef]
10. Wab, Y.; Yang, Y.; Xu, M.; Huang, W. Intelligent fault diagnosis of planetary gearbox based on refined composite hierarchical fuzzy entropy and random forest. *ISA Trans.* **2021**, *109*, 340–351.

11. Yu, D.; Zhang, A.; Wei, M. SCA-SVM Fault Diagnosis of Analog Circuits Based on Transfer Learning. In Proceedings of the 2021 IEEE 10th Data Driven Control and Learning Systems Conference (DDCLS), Suzhou, China, 14–16 May 2021; IEEE: Piscataway, NJ, USA, 2021; pp. 818–823.
12. Yin, S.; Ding, S.X.; Xie, X.; Luo, H. A review on basic data-driven approaches for industrial process monitoring. *IEEE Trans. Ind. Electron.* **2014**, *61*, 6418–6428. [CrossRef]
13. Zhong, T.; Qu, J.; Fang, X.; Li, H.; Wang, Z. The intermittent fault diagnosis of analog circuits based on EEMD-DBN. *Neurocomputing* **2021**, *436*, 74–91. [CrossRef]
14. Su, X.; Cao, C.; Zeng, X.; Feng, Z.; Shen, J.; Yan, X.; Wu, Z. Application of DBN and GWO-SVM in analog circuit fault diagnosis. *Sci. Rep.* **2021**, *11*, 7969. [CrossRef] [PubMed]
15. Jing, Z.; Liang, Y. Electronic circuit fault diagnosis based on SCA-SVM. In Proceedings of the 2018 10th International Conference on Communications, Circuits and Systems (ICCCAS), Chengdu, China, 22–24 December 2018; IEEE: Piscataway, NJ, USA, 2018; pp. 44–49.
16. Gianvito, P.; Mignone, P.; Magazzù, G.; Zampieri, G.; Ceci, M.; Angione, C. Integrating genome-scale metabolic modelling and transfer learning for human gene regulatory network reconstruction. *Bioinformatics* **2022**, *38*, 487–493.
17. Lago, V.; Alegre, A.; Lago, V.; Carot-Sierra, J.M.; Bme, A.T.; Montoliu, G.; Domingo, S.; Alberich-Bayarri, Á.; Martí-Bonmatí, L. Machine Learning ased Integration of Prognostic Magnetic Resonance Imaging Biomarkers for Myometrial Invasion Stratification in Endometrial Cancer. *J. Magn. Reson. Imaging* **2021**, *54*, 987–995.
18. Gao, Z.; Liu, L.X. An overview on fault diagnosis, prognosis and resilient control for wind turbine systems. *Processes* **2021**, *9*, 300. [CrossRef]
19. Li, L.; Ding, S.X.; Luo, H.; Peng, K.; Yang, Y. Performance-Based Fault-Tolerant Control Approaches for Industrial Processes with Multiplicative Faults. *IEEE Trans. Ind. Inform.* **2020**, *16*, 4759–4768. [CrossRef]
20. Niu, P.; Niu, S.; Liu, N.; Chang, L. The defect of the Grey Wolf optimization algorithm and its verification method. *Knowl.-Based Syst.* **2019**, *171*, 37–43. [CrossRef]
21. Singh, N.; Singh, S.B. A novel hybrid GWO-SCA approach for optimization problems. *Eng. Sci. Technol. Int. J.* **2017**, *20*, 1586–1601. [CrossRef]
22. Xie, C.; Xiang, H.; Zeng, T.; Yang, Y.; Yu, B.; Liu, Q. Cross knowledge-based generative zero-shot learning approach with taxonomy regularization. *Neural Netw.* **2021**, *139*, 168–178. [CrossRef]
23. Ntalampiras, S. One-shot learning for acoustic diagnosis of industrial machines. *Expert Syst. Appl.* **2021**, *178*, 114984. [CrossRef]
24. Liu, T.; Yang, Y.; Fan, W.; Wu, C. Few-shot learning for cardiac arrhythmia detection based on electrocardiogram data from wearable devices. *Digit. Signal Process.* **2021**, *116*, 103094. [CrossRef]
25. Gao, Z.; Cecati, C.; Ding, S.X. A survey of fault diagnosis and fault-tolerant techniques—Part I: Fault diagnosis with model-based and signal-based approaches. *IEEE Trans. Ind. Electron.* **2015**, *62*, 3757–3767. [CrossRef]
26. Luo, H.; Li, K.; Kaynak, O.; Yin, S.; Huo, M.; Zhao, H. A Robust Data-Driven Fault Detection Approach for Rolling Mills with Unknown Roll Eccentricity. *IEEE Trans. Control. Syst. Technol.* **2019**, *28*, 2641–2648. [CrossRef]
27. Chaehan, S. Exploring Meta Learning: Parameterizing the Learning-to-learn Process for Image Classification. In Proceedings of the 2021 International Conference on Artificial Intelligence in Information and Communication (ICAIIC), Jeju Island, Korea, 13–16 April 2021; IEEE: Piscataway, NJ, USA, 2021; pp. 199–202.
28. Blau, F.; Kahn, L.; Comey, M.; Eng, A.; Meyerhofer, P.; Willén, A. Culture and gender allocation of tasks: Source country characteristics and the division of non-market work among US immigrants. *Rev. Econ. Househ.* **2020**, *18*, 907–958. [CrossRef]
29. Tao, D.; Diao, X.; Wang, T.; Guo, J.; Qu, X. Freehand interaction with large displays: Effects of body posture, interaction distance and target size on task performance, perceived usability and workload. *Appl. Ergon.* **2021**, *93*, 103370. [CrossRef]
30. Du, X. Complex environment image recognition algorithm based on GANs and transfer learning. *Neural Comput. Appl.* **2020**, *32*, 16401–16412. [CrossRef]
31. Mirjalili, S. SCA: A sine cosine algorithm for solving optimization problems. *Knowl.-Based Syst.* **2016**, *96*, 120–133. [CrossRef]
32. Khawaja, A.W.; Kamari, N.; Zainuri, M. Design of a Damping Controller Using the SCA Optimization Technique for the Improvement of Small Signal Stability of a Single Machine Connected to an Infinite Bus System. *Energies* **2021**, *14*, 2996. [CrossRef]
33. Wang, B.; Zhang, X.; Xing, S.; Sun, C.; Chen, X. Sparse representation theory for support vector machine kernel function selection and its application in high-speed bearing fault diagnosis. *ISA Trans.* **2021**, *118*, 207–218. [CrossRef]
34. Yong, S.B.; Lee, S.; Filatov, M.; Choi, C.H. Optimization of Three State Conical Intersections by Adaptive Penalty Function Algorithm in Connection with the Mixed-Reference Spin-Flip Time-Dependent Density Functional Theory Method (MRSF-TDDFT). *J. Phys. Chem. A* **2021**, *125*, 1994–2006.
35. Zhou, J.; Huang, S.; Wang, M.; Qiu, Y. Performance evaluation of hybrid GA–SVM and GWO–SVM models to predict earthquake-induced liquefaction potential of soil: A multi-dataset investigation. *Eng. Comput.* **2021**, *2*, 1–19. [CrossRef]
36. Li, Y.; Zhao, Y.; Liu, J. Dynamic sine cosine algorithm for large-scale global optimization problems. *Expert Syst. Appl.* **2021**, *177*, 114950. [CrossRef]
37. Huang, J.; Kumar, G.S.; Ren, J.; Zhang, J.; Sun, Y. Accurately predicting dynamic modulus of asphalt mixtures in low-temperature regions using hybrid artificial intelligence model. *Constr. Build. Mater.* **2021**, *297*, 123655. [CrossRef]
38. Zhang, C.; He, Y.; Yuan, L.; He, W.; Xiang, S.; Li, Z. A Novel Approach for Diagnosis of Analog Circuit Fault by Using GMKL-SVM and PSO. *J. Electron. Test.* **2016**, *32*, 531–540. [CrossRef]

39. Zhang, B.; Zhang, L.; Zhang, B.; Yang, B.; Zhao, Y. A fault prediction model of adaptive fuzzy neural network for optimal membership function. *IEEE Access* **2020**, *8*, 101061–101067. [CrossRef]
40. Soltan, A.; Ahmed, G.; Soliman, M. Fractional order Sallen-Key and KHN filters: Stability and poles allocation. *Circuits Syst. Signal Process.* **2015**, *34*, 1461–1480. [CrossRef]
41. Bao, H.; Wu, P.; Bao, B.-C.; Chen, M.; Wu, H. Sallen–Key low-pass filter-based inductor-free simplified Chua's circuit. *J. Eng.* **2017**, *2017*, 653–655. [CrossRef]
42. Song, G.; Li, Q.; Luo, G.; Jiang, S.; Wang, H. Analog circuit fault diagnosis using wavelet feature optimization approach. In Proceedings of the 2015 12th IEEE International Conference on Electronic Measurement & Instruments (ICEMI), Qingdao, China, 16–18 July 2015; IEEE: Piscataway, NJ, USA, 2015; Volume 1.
43. Rahimilarki, R.; Gao, Z.; Jin, N.; Zhang, A. Convolutional neural network fault classification based on time-series analysis for benchmark wind turbine machine. *Renew. Energy* **2022**, *185*, 916–931. [CrossRef]

processes

Article

Data-Driven State Prediction and Sensor Fault Diagnosis for Multi-Agent Systems with Application to a Twin Rotational Inverted Pendulum

Xin Lu [1], Xiaoxu Liu [1,*], Bowen Li [2] and Jie Zhong [3]

[1] Sino-German College of Intelligent Manufacturing, Shenzhen Technology University, Shenzhen 518118, China; luwenkai67109@outlook.com
[2] School of Computer Science, Nanjing University of Posts and Telecommunications, Nanjing 210023, China; bowenli@njupt.edu.cn
[3] College of Mathematics and Computer Science, Zhejiang Normal University, Jinhua 321004, China; jiezhong0615math@zjnu.edu.cn
* Correspondence: liuxiaoxu@sztu.edu.cn

Abstract: When a multi-agent system is subjected to faults, it is necessary to detect and classify the faults in time. This paper is motivated to propose a data-driven state prediction and sensor fault classification technique. Firstly, neural network-based state prediction model is trained through historical input and output data of the system. Then, the trained model is implemented to the real-time system to predict the system state and output in absence of fault. By comparing the predicted healthy output and the measured output, which can be abnormal in case of sensor faults, a residual signal can be generated. When a sensor fault occurs, the residual signal exceeds the threshold, a fault classification technique is triggered to distinguish fault types. Finally, the designed data-driven state prediction and fault classification algorithms are verified through a twin rotational inverted pendulum system with leader-follower mechanism.

Keywords: data-driven; state prediction; fault classification; multi-agent system

Citation: Lu, X.; Liu, X.; Li, B.; Zhong, J. Data-Driven State Prediction and Sensor Fault Diagnosis for Multi-Agent Systems with Application to a Twin Rotational Inverted Pendulum. *Processes* **2021**, *9*, 1505. https://doi.org/10.3390/pr9091505

Academic Editor: Jie Zhang

Received: 22 July 2021
Accepted: 23 August 2021
Published: 26 August 2021

1. Introduction

Monitoring the condition of complex systems in real-time can save valuable time and cost to maintain the system. Fault diagnosis can detect process anomalies and classify the types of anomalies, and has hence drawn enormous attention (e.g., [1–3]). In survey papers [4,5], the methods of fault diagnosis are divided into model-based, signal-based, knowledge-based, and hybrid/active methods. Knowledge-based method is also named data-driven method, where a fault diagnosis model is built through historical data rather than precise mathematical model. Therefore, a data-driven method is suitable for complex systems that are difficult to obtain an accurate model or whose signal is unknown. Data-driven fault diagnosis has been applied to real systems such as wind turbine system [6], high-speed trains [7], and induction motor drive system [8], etc.

On the other hand, many modern engineering systems are modeled as multi-agent systems (MASs), where two or more agents are communicated through a designed protocol to work cooperatively [9,10]. Due to the communication, a fault in one agent can degrade performance of its neighbors, and even the whole network. Therefore, an effective fault diagnosis technique is crucial for MAS. Furthermore, a fault alarm from one agent can be induced by its neighboring agents, hence, fault diagnosis for multi-agent system is more challenging compared with single agent system. A variety of fault diagnosis approaches have been developed for MAS recently [11,12]. Most existing work of MAS is based on a precise state-space model of each agent as well as their communication, e.g., [13–15]. However, the communication between agents can be unknown. Thus, it is difficult to establish an accurate mathematical model. As a result, data-driven fault diagnosis plays an important role in complex MAS.

Among various data driven fault diagnosis methods [16–20], the neural network can convert fault diagnosis into a multi-label classification problem, and automatically learn the features of the original data. However, storing and leaning a large amount of data in real-time is challenging for the computation and communication device/software. In order to deal with the limited capability of the device/software, event-triggered mechanize [21,22] and distributed methods [23] have been hot topics in recent years. Specifically, event-triggered fault diagnosis methods have been developed in [21,22], where the mathematical model of the system is assumed to be known. Nevertheless, when model and communication of MAS are not available, the above event-triggered methodologies are not applicable. Therefore, it is motivated to develop event-triggered data driven fault diagnosis for MAS with unknown mathematical model and unknown communication.

In this paper, a residual-triggered fault diagnosis technique is proposed for MAS. Specifically, a neural network-based state prediction model is established through training historical data offline. Then, online comparison of real state/output and the predicted state/output can generate a residual signal, which indicates whether there is a fault. If the residual exceeds the threshold, it triggers a fault classification training process to identify and locate the fault. This residual-triggered fault diagnosis method does not depend on a mathematical model and communication information. Moreover, online identification of a fault is implemented only in case of fault, hence the data transmission and calculation are reduced. A real experiment on leader-follower inverted pendulum demonstrates the effectiveness of the developed algorithm. The contribution includes: 1. Residual-triggered data-driven fault diagnosis for MAS is a novel topic, where data calculation can be reduced; 2. The designed fault classifiers are distributed, where a fault in one agent can be identified by fault classifier of its neighbor; 3. The communication among agents are internal in the agents but unknown (not available) in state prediction and fault classification, which implies that the designed state prediction and fault diagnosis techniques are fully distributed. It should be mentioned that many existing estimation/prediction models of MAS rely on communication information among agents, such as the adjacency matrix [13–15], nevertheless, the adjacency matrix consists of the overall communication information, which makes the developed methods centralized rather than distributed. In this article, only input and output data is required in the developed state prediction and fault classification method, and communication is not used.

The organization of the paper is as follows. After the introduction section, the data-driven state prediction algorithm is introduced in Section 2. Based on the prediction model, a residual-triggered fault classification technique is proposed in Section 3. Section 4 presents the experimental results in a twin rotational inverted pendulum system with leader-follower mechanism. The paper is ended by Section 5 with the conclusion and future researches.

2. Data-Driven State Prediction for Multi-Agent System

In this section, we introduce the establishment of a neural network model to predict the state of a multi-agent system with unknown communication. To be precise, the controller of each agent and communication protocol among the agents are pre-designed to guarantee the performance of a multi-agent system (i.e., consensus and robustness) in a fault-free case, and the design of the controller and communication is not of concern in this paper. The physical models of the agents are unknown or highly nonlinear. Moreover, the communication protocol is internal to the system, but not available for the prediction model.

The diagram of the prediction model for the multi-agent system is shown in Figure 1.

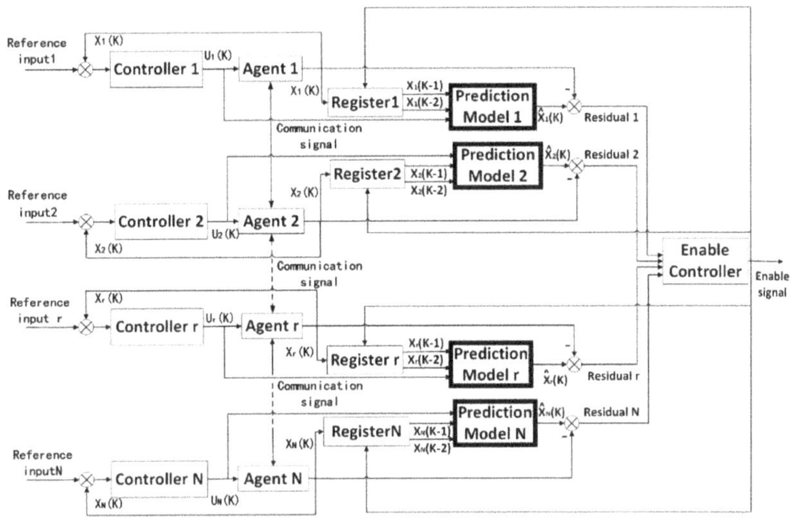

Figure 1. Prediction model for the multi-agent system.

In Figure 1, X_r and U_r represent state and control input of agent r, $r = 1,2,\ldots N$, and N is the number of agents; K represents the time of KT, where T is the sampling time; $K-1$ and $K-2$ represent the time of $(K-1)T$ and $(K-2)T$, respectively, $\hat{X}_r(K)$ is the prediction of $X_r(K)$. Firstly, the state of each Agent r is recorded in the corresponding Register r at the past two sampling times, namely $X_r(K-1)$ and $X_r(K-2)$ are obtained. Then, $X_r(K-1)$, $X_r(K-2)$ and control input of Agent r at current time $U_r(K)$ are used to train the Prediction Model r. The output of the prediction model is the predicted state at the current time $\hat{X}_r(K)$. By comparing the real state $X_r(K)$ and the predicted state $\hat{X}_r(K)$ Residual $r = \hat{X}_r(K) - X_r(K)$ can be generated. The residual values are sent into Enable Controller, which is responsible for deciding whether the residual exceeds the threshold. To be precise, when it exceeds the threshold, it is recognized that there is a fault in the system. At this time, the enable signal stops the prediction model and triggers fault diagnosis algorithms, which will be presented in Section 3.

The enable control algorithm is described as follows:

$$\text{if Residual 1} > \beta_1 \text{ or Residual 2} > \beta_2 \text{ or} \ldots \text{Residual N} > \beta_N enable = 1$$
$$\text{else}$$
$$enable = 0$$

where, β_r represents the residual threshold of Agent r, $enable$ is the output of Enable Controller.

Remark 1. It should be mentioned that communication among agents is not used in the prediction model. The "unknown communication" in this paper means the communication is internal to the MAS, but cannot be used in the prediction/fault diagnosis. Moreover, the controllers are predesigned for the MAS, which is not under concern in this paper.

The network structure used to build the prediction model is the back propagation (BP) neural network, which is known as a multilayer feedforward neural network trained by error back propagation algorithm. It can learn and store a large number of input–output pattern mapping relations without concrete mathematical functions. A neural network is composed of a number of neurons, and the BP neural network of a single neuron for predicting the concerned model is shown in Figure 2.

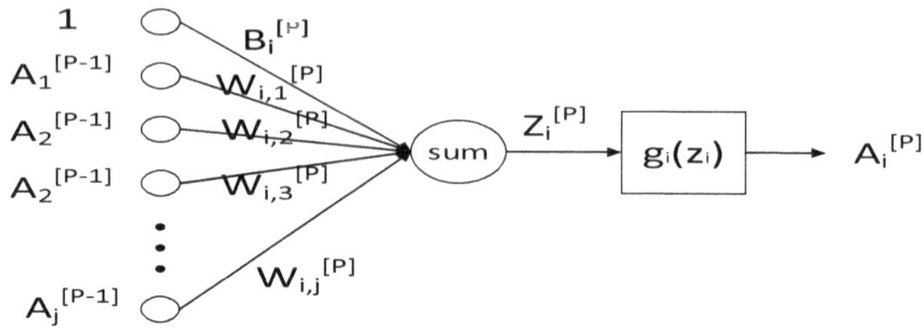

Figure 2. Schematic diagram of single neuron.

In the diagram, $W_{ij}^{[P]}$ and $B_i^{[P]}$ represent the weight parameter and bias parameter between hidden layers, respectively; P represents the number of current layers; i and j represent the number of current nodes in the current layer and the number of current nodes in the upper layer, respectively. Z represents the input of the neuron and the output of the weighted multiplication summation. A represent the input or output of the neuron. Where:

$$Z^{[P]} = W^{[P]}.A^{[P-1]} + B^{[P]}. \tag{1}$$

The hidden layer takes the Tansig function as the excitation function $g_1(x)$, where:

$$g_1(x) = \frac{2}{1+e^{-2x}} - 1. \tag{2}$$

The reason for using the Tansig function is that the training data changes periodically in $[-1, 1]$. Using Tansig can accelerate the decline of training gradient.

The output of the neural network is the predicted value of system state $\hat{X}_r(K)$ in a fault-free scenario. Therefore, the output layer uses the Purelin function as the activation function, which is defined as $g_2(x)$, and

$$g_2(x) = x. \tag{3}$$

The predicted state $\hat{X}_r(K)$ is compared with the actual system state $X_r\ (K)$ and the network topology structures and training parameter should be designed to make $\hat{X}_r(K)$ closed to $X_r\ (K)$.

In the healthy state, the residual between $\hat{X}_r(K)$ and $X_r\ (K)$ is convergent. However, when the system is in the fault state, the residual will exceed the threshold. At this time, it is deemed to be in the fault state and start fault diagnosis.

Root mean square error (RMSE) between the predicted value and the actual value is used as the evaluation standard of the prediction accuracy. In BP neural network, the gradient descent is used to update the $W_{ij}^{[P]}$ and $B_i^{[P]}$ until the RMSE between $\hat{X}_r(K)$ and $X_r\ (K)$ is locally minimum. As a result, the optimal weight and bias parameters of the neural network are calculated.

There are a variety of network structures and learning rates. In order to obtain optimized performance of the state prediction, RMSEs of different hierarchical structures under the same training parameters and the same training time are generated and compared. Generally speaking, smaller the RMSE value indicates better training performance, however, the generalization capability should also be considered to avoid over fitting. Accordingly, the network structure can be determined. Subsequently, learning rates are determined by comparing their accuracy with the selected network structure.

Then, the developed state prediction model can be implemented to a real-time system to predict the state in absence of fault. By comparing real state and the predicted health state, a residual signal can be generated. This residual signal can indicate whether a fault occurs, and if the residual signal excesses a threshold, it triggers a fault classification mechanism, which is designed in Section 3.

3. Sensor Fault Classification

The fault of one sensor may lead to the fault of the whole system [23]. Therefore, it is very important to diagnose the fault of the sensor.

In this section, a data-driven sensor fault detection and classification technique is presented. Three typical sensor faults are under consideration: zero-output fault, drift fault, and deviation fault. Figures 3–5 are schematic diagrams of the three types of sensor faults. Moreover, the three types of faults can exist in different sensors and different agents. The objective of this section is to use a neural network classifier to identify and locate different types of faults.

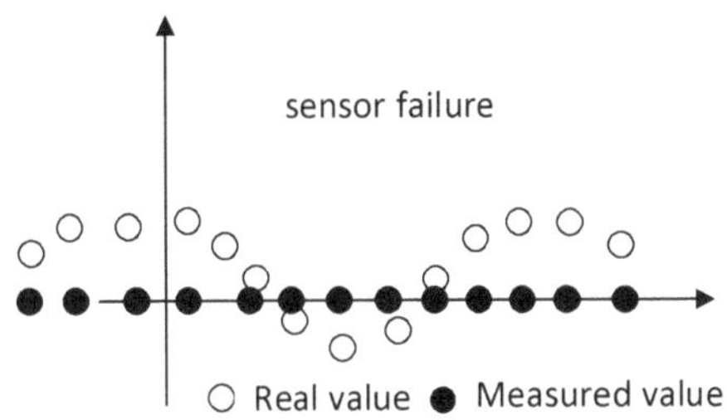

Figure 3. Zero-output sensor fault diagram.

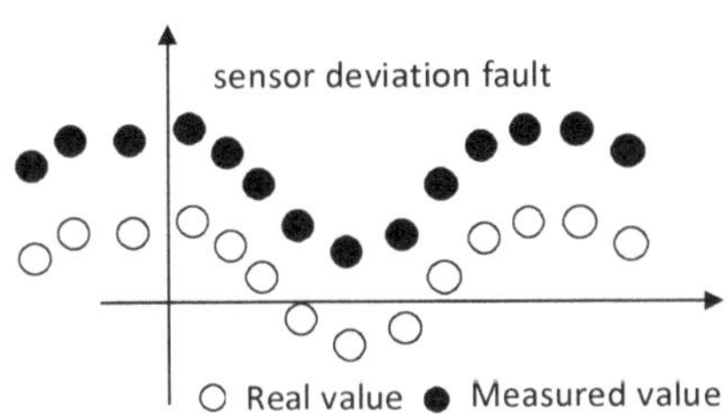

Figure 4. Sensor deviation fault diagram.

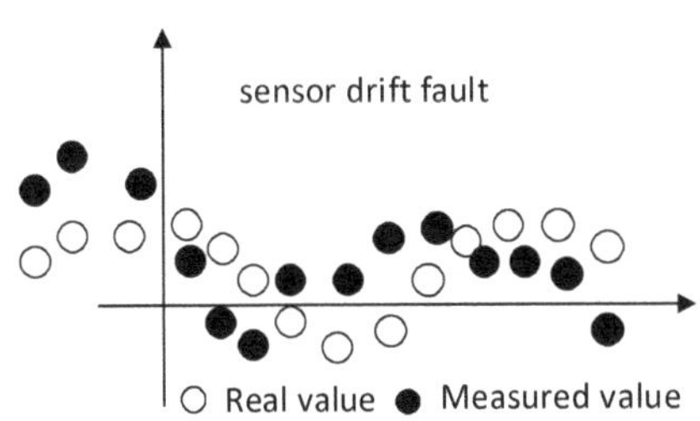

Figure 5. Sensor drift fault diagram.

Specifically, the zero-output sensor fault [24] is molded as:

$$f_s(t) = \begin{cases} 0, \; t < t_0 \\ -y(t), \; t \geq t_0 \end{cases} \tag{4}$$

where $f_s(t)$ represents sensor fault, t_0 denotes the time that a sensor fault occurs, $y(t)$ is the real system output. In engineering, it is easy to occur when the signal is open circuited. A deviation fault is molded as:

$$f_{de}(t) = \begin{cases} 0, \ t < t_0 \\ d, \ t \geq t_0 \end{cases} \tag{5}$$

where $f_{de}(t)$ represents deviation fault and d is a bounded constant. The deviation fault is easy to appear in the current or voltage sensor [25]. A drift fault is molded as:

$$f_{dr}(t) = \begin{cases} 0, \ t < t_0 \\ n(t), \ t \geq t_0 \end{cases} \tag{6}$$

where $f_{dr}(t)$ represents drift fault and $n(t)$ is an irregular bounded disturbance signal, which is a sensor noise (due to the influence of external environment and internal factors of the sensor) [26].

The data used to train the classifier is $X_r(K)$. The procedure to select an appropriate network structure and learning rate is the same with state prediction. The output of the classifier is the probability of each fault category, therefore, the last output layer activation function is replaced by the Softmax function. Through non-maximum suppression, the original network output is fuzzed, and the fault type and location with the highest probability can be determined. The network structure diagram of a fault classification model can be found in Figure 6.

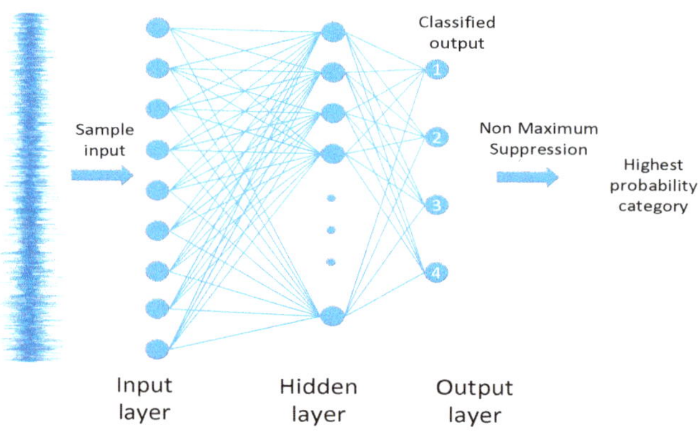

Figure 6. Network structure diagram of fault classification model.

In the fault classification model, the amount of network input data can be large. Identification of such an amount of data in real-time brings a challenge to the computation ability. As a result, a triggering mechanism is designed to active the identification. Specifically, the prediction model introduced in Section 2 is implemented in the system to predict the system state and output in absence of fault. By comparing the predicted healthy output and the measured output, which can be abnormal in the case of sensor faults, a residual signal can be generated. When a sensor fault occurs, the residual signal exceeds the threshold, and the fault diagnosis model of the neural network is triggered to identify and locate the fault types. The state prediction triggered fault classification mechanism is illustrated in Figure 7.

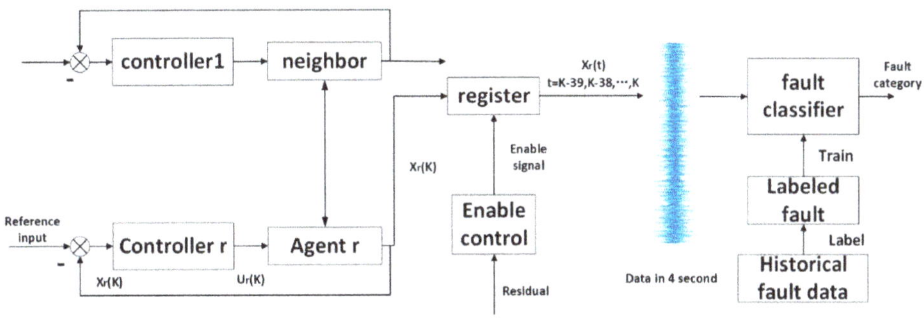

Figure 7. State prediction triggered fault classification.

When the residual in Figure 1 is greater than the set threshold, Enable Controller sends an enable signal to the register of fault classifier in Figure 7, and the register starts to record the abnormal state data of the agent for 4 s. The stored data is then sent to the fault diagnosis network. The fault diagnosis network is obtained by labeling historical fault data and off-line supervised learning. The diagnosis model can classify the faults in agent r and its neighbor through the output of agent r. Moreover, communication is not utilized in the fault classifier.

4. Experimental Results

4.1. System and Fault Description

In this section, the designed data-driven state prediction and the sensor fault classification techniques are implemented to the collaborative system to verify the effectiveness. We use two Quanser Servo 2 rotating inverted pendulum hardwares to build a multi-agent system with internal communication. The communication protocol is a leader-follower mechanism. The inverted pendulums transfer sensor data to Matlab Simulink in real-time through USB, and the control protocol is pre-designed in Simulink. The specific hardware-in-the-loop control diagram is shown in Figure 8. There are four states of each agent, which are introduced in Table 1.

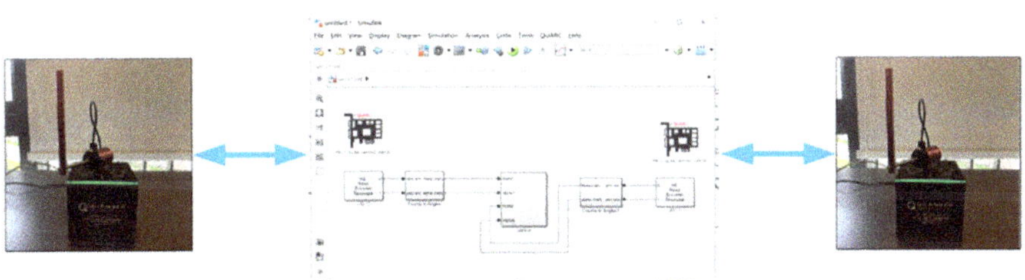

Figure 8. Leader-follower system control diagram.

Table 1. Parament and meaning.

Parameter	Representative Meaning
θ	Horizontal displacement of inverted pendulum
α	Vertical displacement of inverted pendulum
$\dot{\theta}$	Horizontal velocity of inverted pendulum
$\dot{\alpha}$	Vertical velocity of inverted pendulum

It is assumed that the faults occur in the horizontal displacement sensor, and seven scenarios of faults are under investigation: fault-free, leader's zero-output sensor fault, leader's sensor deviation fault, leader's sensor drift fault, follower's zero-output sensor fault, follower's sensor deviation fault, and follower's sensor drift fault.

Remark 2. The equipment is working in a real laboratory environment. Thus, the data collected is subjected to noises/disturbances due to equipment noises, environment noises, data conversion uncertainties, etc. On the other hand, drift fault can also be regarded as disturbances with relatively big amplitude. In order to avoid alarm by acceptable noises in the data, we select the threshold parameters for the enable control as $\beta_1 = \beta_5 = 0.5$; $\beta_2 = \beta_6 = 0.006$; $\beta_3 = \beta_7 = 0.3$; $\beta_4 = \beta_8 = 0.25$.

4.2. Data Acquisition and Data Expansion

The data acquisition of the system is carried out through Simulink, then a hardware-in-the-loop experiment can be implemented. The data sampling is carried out according to the sampling time of 0.005 s. Due to the limited storage capacity of MATLAB, 29 s of effective data can be collected in each experiment.

In order to further improve the generalization ability of the model, a large number of data is needed to train the neural network. Nevertheless, it is often impossible to collect sufficient data in reality. Therefore, this paper is motivated to employ sliding window data sampling to complete data amplification. As shown in Figure 9, if the length of the sampling window is $\boldsymbol{f}$, the moving step of the sampling window is $\boldsymbol{S}$, and the total length of the data is $\boldsymbol{L}$, the number of data $\boldsymbol{n}$ can be obtained as:

$$n = [\frac{L - f}{S}] \quad (7)$$

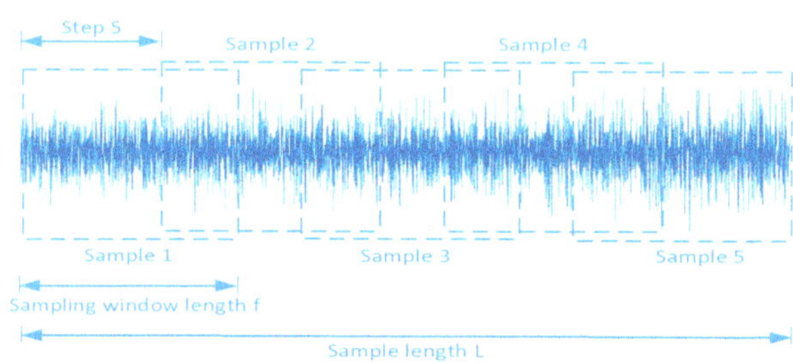

Figure 9. Data expansion diagram.

The original data is collected from each fault during 29 s, and the sampling time is 0.005 s. The total length of the signal is 5800 sampling points ($L = 5800$). By selecting 800 sampling points ($f = 800$) with the length of the sampling window and one sampling point in step ($S = 1$), 5000 groups of data ($n = 5000$) in each fault state can be obtained, and a total of 35,000 groups of 7 kinds of fault scenarios can be obtained. Compared with the original method with 40 sampling window length, the amount of data is increased by 114.28 times.

4.3. Experimental Results and Analysis

4.3.1. Neural Network-Based State Prediction

The historical healthy and stable operation data are selected as the network training input of state estimation. The training process is offline. The process of recognition is to connect the offline trained model into the system to complete online prediction.

Neural network models with different hidden layer nodes, learning rate and momentum factor, and the training effect of the final network are compared in Table 2, where the performance of state prediction is evaluated by measuring RMSE.

Table 2. Performance comparison of neural networks with different parameter structures.

Number of Hidden Layer Nodes	Learning Rate	Momentum Factor	RMSE
8/4	0.001	0.99	0.0623
	0.001	0.95	0.0652
	0.010	0.99	0.0641
	0.010	0.95	0.0627
15/4	0.001	0.99	0.0598
	0.001	0.95	0.0592
	0.010	0.99	0.0604
	0.010	0.95	0.0597
15/8/4	0.001	0.99	0.0579
	0.001	0.95	0.0553
	0.010	0.99	0.0517
	0.010	0.95	0.0571

The basic structure of the BP shallow neural network for predicting the concerned model is shown in the Figure 10.

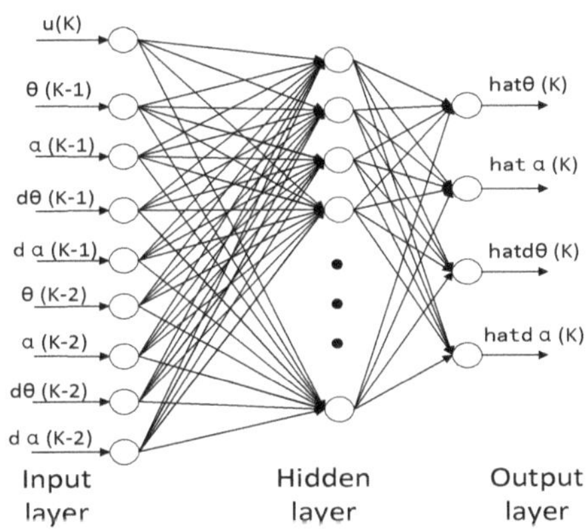

Figure 10. Neural network structure diagram.

From Table 2, we can notice that the most accurate state prediction model is the three-layer neural network with RMSE equal to 0.0517. The structure of the network is 15/8/4 from input to output in turn. However, the neural network will appear over the fitting phenomenon when the model is too accurate, which can cause the divergence of the system when processing the data that does not appear in the training set. To be precise, the data that does not appear in the training set refers to the data that appear in normal operation but that is not in the training set. Identifying these data requires the network to have a certain generalization ability. As a result, this paper selects a two-layer neural network with the middle accuracy. Its parameters are: a learning rate of 0.001, momentum factor of 0.95, and layer series from input to output of 15 and 4.

Figures 11–14 compare actual states and predicted states. As shown in the results, the neural network can accurately predict the full states of an inverted pendulum, which can be used as a healthy signal and compared with the actual output to monitor whether the system is under fault-free case or not. In case of sensor faults, the residual signal can be generated immediately to trigger the fault identification and classification process.

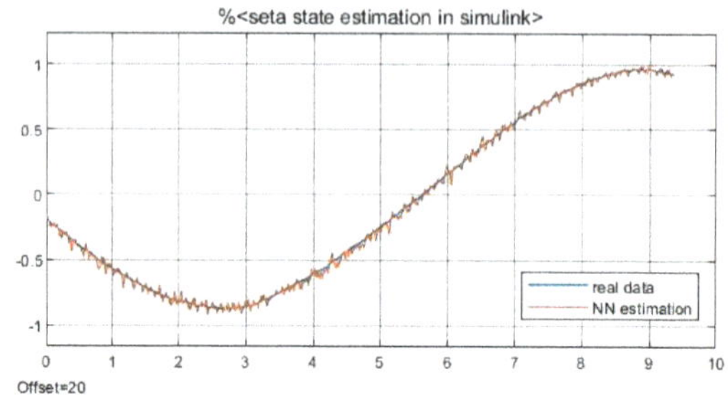

Figure 11. Comparison of actual and predicted values of θ.

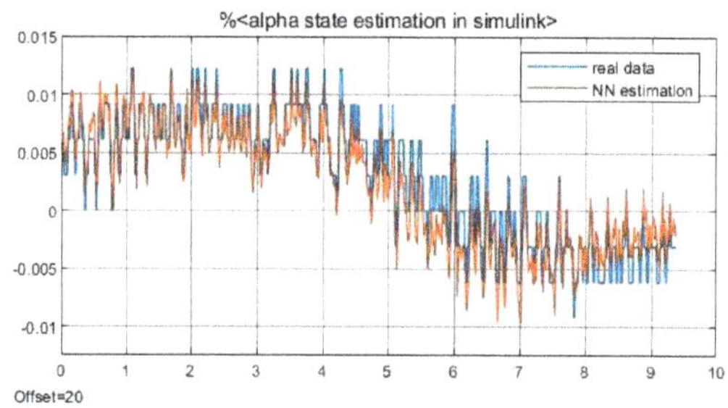

Figure 12. Comparison of actual and predicted values of α.

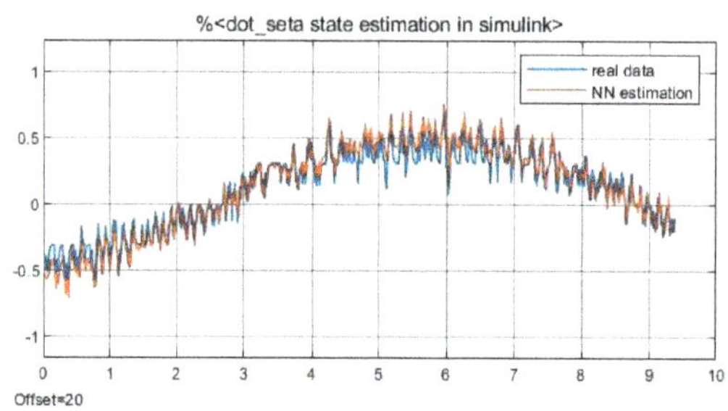

Figure 13. Comparison of actual and predicted values of $\dot{\theta}$.

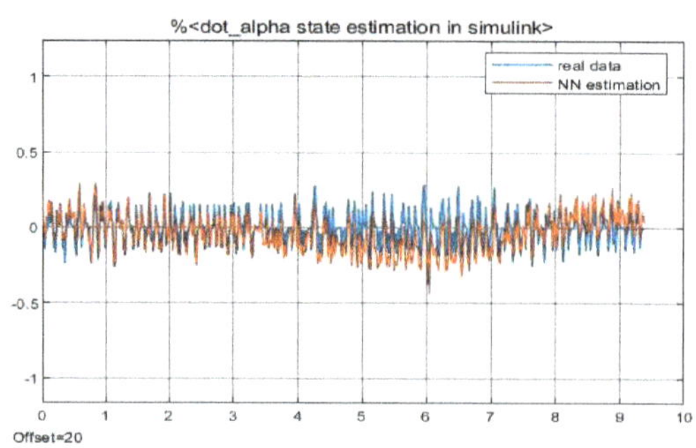

Figure 14. Comparison of actual and predicted values of $\dot{\alpha}$.

4.3.2. Fault Classification

Through the method introduced in Section 3, we can build the neural network for fault classification. The training data is divided into two parts: 70% and 30%. Seventy percent of the data is used to train the network, update the model weight parameters, and the remaining 30% is used to evaluate the model performance. According to the fault detection of the horizontal displacement sensor of the leader-follower system, the faults can be divided into seven types.

We stipulate that all collection time of data is 29 s and the sampling time is 0.005 s. Thus, 5800 sampling points can be collected within 29 s. The cycle time of inverted pendulum motion is 7 s, and 1400 sampling points need to be collected when we use a sampling time of 0.005 s. If there are m sensors in the system, there are $m \times 1400$ neural network inputs, which require a lot of operation for training. However, the calculation ability of software is limited. In order to reduce data calculation, we expend the sampling time of the sliding window after data expansion to 0.1 s. The length of the sliding window is 4 s (40 sampling points), which is more than half a cycle of the system. According to Formula (7), the number of total data is 5000. Because the data acquisition is carried out just when the fault occurs, the data of the first 40 minimum sampling points (0.2 s) are filtered as the signal delay. All subsequent data segments contain the fault characteristic information, except that the fault characteristics of some faults only last for a few seconds. In this scenario, the whole data acquisition time cannot be filled, and the edges of the data need to be filtered to retain the parts with fault characteristics. For the fault requiring edge screening, several groups of fault data shall be collected to supplement 4960 groups of data. The parameter is provided in Table 3:

Table 3. The type of faults and category label.

Sample Type	Sample Length	Number of Sample	Category Label
leader and follower work normally	40	4960	1
leader's zero-output sensor fault	40	4960	2
leader's sensor deviation fault	40	4960	3
leader's sensor drift fault	40	4960	4
follower's zero-output sensor fault	40	4960	5
follower's sensor deviation fault	40	4960	6
follower's sensor drift fault	40	4960	7

In order to enhance the result, we did experiments with different number of nodes in different hidden layers, and the fault classification performances are compared in Tables 4 and 5. To be precise, Table 4 records the average accuracy and standard deviation of the training set of the network model under the same learning rate but with different random initialization conditions and different number of nodes. Accordingly, the average value accuracy and standard deviation of test set are shown in Table 5. Through the above experiments, we can find a network structure with the highest accuracy, which is achieved when the number of the hidden layers is 80-25, As a result, we chose the 80-25 hidden layer structure.

Table 4. The train set result of different bp model (%).

Run No.	BP-25-12	BP-50-25	BP-50-12	BP-80-50	BP-80-25	BP-80-12
1	98.50	98.15	98.65	97.32	98.71	99.49
2	97.96	98.85	98.37	98.69	97.96	97.73
3	96.49	97.32	96.75	99.09	99.01	98.00
4	99.35	97.98	98.25	99.06	99.06	98.35
5	98.53	96.50	95.43	97.32	98.42	99.11
Mean	98.16	97.76	97.49	98.30	98.63	98.54
Std	1.0603	0.8902	1.3683	0.9048	0.4552	0.7435

Table 5. The test set result of different bp model (%).

Run No.	BP-25-12	BP-50-25	BP-50-12	BP-80-50	BP-80-25	BP-80-12
1	84.72	81.40	85.58	79.91	82.51	71.92
2	77.13	84.12	81.31	79.11	88.78	79.54
3	79.56	80.97	83.83	84.72	87.29	70.26
4	76.50	84.12	91.53	83.17	87.46	81.14
5	89.44	75.33	81.74	79.91	90.87	80.37
Mean	81.47	81.19	84.80	81.36	87.38	76.65
Std	5.5061	3.5917	4.1348	2.4410	3.0770	5.1370

After the network structure is determined, the accuracy of the model can be further enhanced by selecting the appropriate learning rate. Figure 15 records the number of iterations and loss function values corresponding to different learning rates, and the accuracy is compared in Table 6. From Figure 15 and Table 6, the gradient decreases the fastest when the learning rate is 0.001. However, the corresponding test accuracy is only 88.38%. This is due to the overfitting phenomenon in deep learning. From overall consideration, the learning rate is determined as 0.0001, where the gradient descent speed is the second fastest, and the test accuracy is the highest.

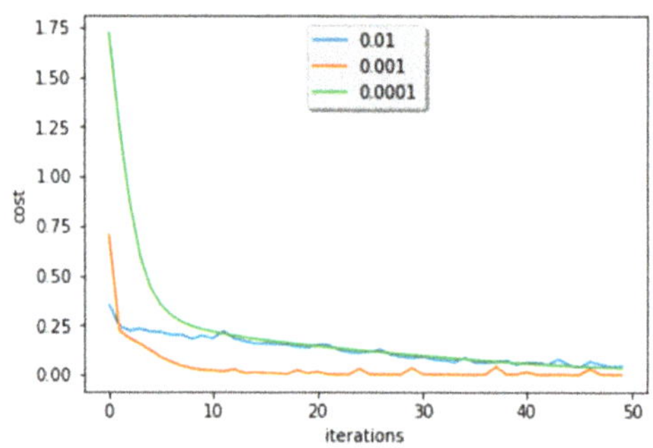

Figure 15. The train cost of different learning rate.

Table 6. The result of different learning rate (%).

Learning Rate α	0.01	0.001	0.0001
Train accuracy	98.42	100	99.56
Test accuracy	90.87	88.38	91.81

Until now, the network structure and learning parameters are determined. Then, the test set of different fault scenarios is input to the determined neuro-network-based fault classifier, and the results are illustrated in Table 7. It can be seen that the classifier can achieve 100% recognition rate for types 2 and 5, and more than 90% recognition rate for types 1, 3, 4, and 6. The recognition rate of type 7 is only 58.72%, which is not ideal. In order to show the performance of BP neural network algorithm on sensor fault diagnosis of leader-follower fault system, the fault misclassification matrix is drawn in Figure 16.

Table 7. The result of different fault type (%).

Label	1	2	3	4	5	6	7
Accuracy	97.2	100	96.8	98.4	100	91.6	58.72

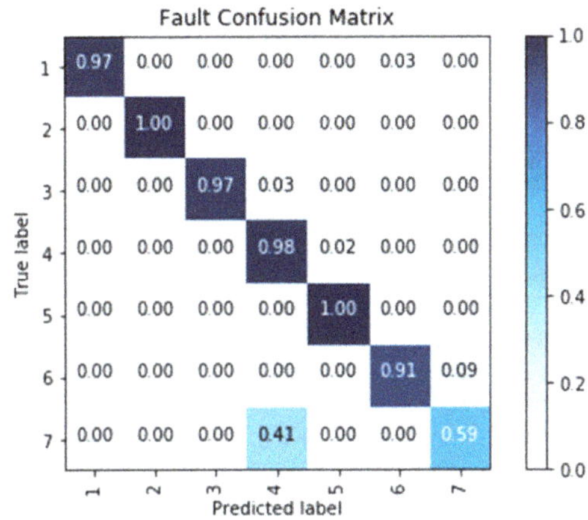

Figure 16. Fault misclassification matrix.

In Figure 16, the coordinate values from 1 to 7 are the label numbers in Table 3, representing different fault types of the leader-follower system. The number in the shadow is the number of actual sample tags that match the predicted sample tags. It shows that the probability of misclassifying most types of faults is not big. However, the error rate of type 7 is significant, and it cannot be distinguished from type 4. The occurrence of misclassification is due to the similar characteristics between the corresponding types. For example, types 4 and 7 have no significant difference in amplitude characteristics, but their frequency characteristics are different. Moreover, the amplitude is small, namely, drift fault is like disturbance, which is challenging for classification.

4.4. Discussions

4.4.1. Delay of Fault Diagnosis

The developed state prediction is implemented in real-time, and there is nearly no delay. When the state varies fast, tracking errors exist, and this phenomenon is general in many estimation/prediction problems. The tracking errors in the experiments is small and acceptable. When we label fault types, the faults occur for a period of time, hence, a complete fault feature is recorded in data sequence during this period. When the residual triggers the fault classifier, there is a period of delay such that complete data of the fault can be stored in the register. It generally takes 2–3 s for complete fault features to appear. The fault diagnosis module can identify the corresponding fault only after a complete fault feature is recorded in the register. Therefore, the delay is also acceptable.

4.4.2. A Limitation of Performance and Further Research

Through the above, we can find that the BP network model is more accurate for amplitude type feature recognition, but not ideal for frequency type feature recognition. Because there are different amplitude characteristics and frequency characteristics in the seven types of faults. Under limited calculation ability of the software, amplitude features can be effectively preserved, however, the frequency characteristics will be partially lost with the increase of the sampling interval. Therefore, faults with similar amplitude but different frequencies, namely drift faults, are difficult to be identified. This leads to a decrease in recognition accuracy. In future research, an alternative network will be investigated to classify faults with the same and small amplitude but a different frequency.

It can be noticed that the developed state prediction and fault classification techniques are distributed, namely the techniques are potential to be generalized in many MASs where the number of agents can be large. In addition, the mathematical model is not required, and only input and output data is utilized in the methods. Therefore, the methods are extendable for many other MASs where the type of agents can be diverse, such as cooperative manipulators (4–6 freedoms), cooperative unmanned aerial vehicles, etc.

5. Conclusions

This research presents a data-driven state prediction and fault classification method by the BP neural network model. The main contribution is to establish a state prediction model for a multi-agent system with unknown communication, and a residual-triggered fault classifier for sensor faults. The developed techniques are implemented in a real physical system. Specifically, for the leader-follower system with communication coupling, the fault diagnosis of the leader can be achieved by observing the follower. RMSE can reach 0.0592 for the state estimation of a leader-follower system. In terms of fault diagnosis, observing the follower to realize the fault diagnosis of the leader is an innovation. Investigation of data-driven state prediction and residual-triggered fault classification of multi-agent systems with unknown communication is a new topic; identification of fault in one agent only through data of its neighbors is a contribution to the distributed fault problem. In the future, more fault types will be considered, such as actuator faults or communication faults. Moreover, improving the fault recognition rate is also in our further research.

Author Contributions: Conceptualization, X.L. (Xiaoxu Liu) and X.L. (Xin Lu); methodology, X.L. (Xin Lu); software, X.L. (Xin Lu); validation, X.L. (Xiaoxu Liu), X.L. (Xin Lu), and B.L.; formal analysis, X.L. (Xiaoxu Liu) and X.L. (Xin Lu); investigation, X.L. (Xiaoxu Liu) and X.L. (Xin Lu); resources, X.L. (Xiaoxu Liu) and X.L. (Xin Lu); data curation, X.L. (Xin Lu) and J.Z.; writing—original draft preparation, X.L. (Xin Lu); writing—review and editing, X.L. (Xiaoxu Liu); visualization, X.L. (Xiaoxu Liu), B.L. and J.Z.; supervision, X.L. (Xiaoxu Liu); project administration, X.L. (Xiaoxu Liu); funding acquisition, X.L. (Xiaoxu Liu). All authors have read and agreed to the published version of the manuscript.

Funding: This work was supported in part by National Nature Science Foundation of China (Grant No. 62003218); Guangdong Basic and Applied Basic Research Foundation (Grant No. 2019A1515110234); Shenzhen Science and Technology Program (Grant No. RCBS20200714114921371); Natural Science Foundation of Top Talent of SZTU (Grant No. 2020106).

Institutional Review Board Statement: Not applicable.

Informed Consent Statement: Not applicable.

Data Availability Statement: Not applicable.

Conflicts of Interest: The authors declare no conflict of interest.

References

1. Liu, R.; Yang, B.; Enrico, Z.; Chen, X. Artificial intelligence for fault diagnosis of rotating machinery: A review. *Mech. Syst. Signal Process.* **2018**, *108*, 33–47. [CrossRef]
2. Liu, X.; Gao, Z.; Zhang, A. Observer-based fault estimation and tolerant control for stochastic Takagi–Sugeno fuzzy systems with Brownian parameter perturbations. *Automatica* **2019**, *102*, 137–149. [CrossRef]
3. Ji, C.; Ma, F.; Wang, J.; Wang, J.; Sun, W. Real-time industrial process fault diagnosis based on time delayed mutual information analysis. *Processes* **2021**, *9*, 1027. [CrossRef]
4. Gao, Z.; Cecati, C.; Ding, S.X. A survey of fault diagnosis and fault-tolerant techniques—Part II: Fault diagnosis with knowledge-based and hybrid/active approaches. *IEEE Trans. Ind. Electron.* **2015**, *62*, 3768–3774. [CrossRef]
5. Gao, Z.; Liu, X. An overview on fault diagnosis, prognosis and resilient control for wind turbine systems. *Processes* **2021**, *9*, 300. [CrossRef]
6. Li, M.; Yu, D.; Chen, Z.; XiaHou, K.; Ji, T.; Wu, Q.H. Data-driven residual-based method for fault diagnosis and isolation in wind turbines. *IEEE Trans. Sustain. Energy* **2019**, *10*, 895–904. [CrossRef]
7. Chen, H.; Jiang, B.; Chen, W.; Yi, H. Data-driven detection and diagnosis of incipient faults in electrical drives of high-speed trains. *IEEE Trans. Ind. Electron.* **2019**, *66*, 4716–4725. [CrossRef]
8. Gou, B.; Xu, Y.; Xia, Y.; Wilson, G.; Liu, S. An intelligent time-adaptive data-driven method for sensor fault diagnosis in induction motor drive system. *IEEE Trans. Ind. Electron.* **2018**, *66*, 9817–9827. [CrossRef]
9. Xu, W.; Wang, Z.; Ho, W.C. Finite-horizon H-infinity consensus for multi-agent systems with redundant channels via an observer-type event-triggered scheme. *IEEE Trans. Cybern.* **2018**, *48*, 1567–1576. [CrossRef] [PubMed]

10. Du, H.; Wen, G.; Wu, D.; Cheng, Y.; Lü, J. Distributed fixed-time consensus for nonlinear heterogeneous multi-agent systems. *Automatica* **2020**, *113*, 108797. [CrossRef]
11. Ma, H.; Xu, L. Cooperative fault diagnosis for uncertain nonlinear multiagent systems based on adaptive distributed fuzzy estimators. *IEEE Trans. Cybern.* **2020**, *50*, 1739–1751. [CrossRef]
12. Zhang, Z.; Yang, G. Distributed fault detection and isolation for multiagent systems: An interval observer approach. *IEEE Trans. Syst. Man Cybern. Syst.* **2020**, *50*, 2220–2230. [CrossRef]
13. Ye, D.; Chen, M.; Yang, H. Distributed adaptive event-triggered fault-tolerant consensus of multiagent systems with general linear dynamics. *IEEE Trans. Cybern.* **2019**, *49*, 757–767. [CrossRef] [PubMed]
14. Zhang, K.; Jiang, B.; Shi, P. Adjustable parameter-based distributed fault estimation observer design for multiagent systems with directed graphs. *IEEE Trans. Cybern.* **2017**, *47*, 306–314. [CrossRef]
15. Khalili, M.; Zhang, X.; Polycarpou, M.; Parisini, T.; Cao, Y. Distributed adaptive fault-tolerant control of uncertain multi-agent systems. *Automatica* **2018**, *87*, 142–151. [CrossRef]
16. Lu, R.; Hong, S.; Yu, M. Demand Response for Home Energy Management Using Reinforcement Learning and Artificial Neural Network. *IEEE Trans. Smart Grid* **2019**, *10*, 6629–6639. [CrossRef]
17. Wen, L.; Li, X.; Gao, L.; Zhang, Y. A New Convolutional Neural Network-Based Data-Driven Fault Diagnosis Method. *IEEE Trans. Ind. Electron.* **2017**, *65*, 5990–5998. [CrossRef]
18. Wang, C.; Chen, X.; Cao, J. Neural network-based distributed adaptive pre-assigned finite-time consensus of multiple TCP/AQM networks. *IEEE Trans. Circ. Syst. I Regul. Pap.* **2021**, *68*, 387–395. [CrossRef]
19. Chen, Z.; Cao, Y.; Ding, S.X.; Zhang, T. A Distributed canonical correlation analysis-based fault detection method for plant-wide process monitoring. *IEEE Trans. Ind. Inform.* **2019**, *15*, 2710–2720. [CrossRef]
20. Ali, N.; Hong, J. Failure Detection and Prevention for Cyber-Physical Systems Using Ontology-Based Knowledge Base. *Computers* **2018**, *7*, 68. [CrossRef]
21. Alikhani, H.; Meskin, N. Event-triggered robust fault diagnosis and control of linear Roesser systems: A unified framework. *Automatica* **2021**, *128*, 109575. [CrossRef]
22. Zhong, M.; Ding, S.X.; Zhou, D.; He, X. An H-i/H-infinity optimization approach to event-triggered fault detection for linear discrete time systems. *IEEE Trans. Autom. Control* **2020**, *65*, 4464–4471. [CrossRef]
23. Ding, S.X. Data-Driven Fault Detection in Large-Scale and Distributed Systems. In *Advanced Methods for Fault Diagnosis and Fault-Tolerant Control*; Springer: Berlin/Heidelberg, Germany, 2020; pp. 315–424.
24. Diez-Gonzalez, J.; Alvarez, R.; Prieto-Fernandez, N.; Perez, H. Local Wireless Sensor Networks Positioning Reliability Under Sensor Failure. *Sensors* **2020**, *20*, 1426. [CrossRef] [PubMed]
25. Xia, J.; Guo, Y.; Dai, B.; Zhang, X. Sensor Fault Diagnosis and System Reconfiguration Approach for an Electric Traction PWM Rectifier Based on Sliding Mode Observer. *IEEE Trans. Ind. Appl.* **2017**, *53*, 4768–4778. [CrossRef]
26. Han, X.; Jiang, J.; Xu, A.; Bari, A.; Pei, C.; Sun, Y. Sensor Drift Detection Based on Discrete Wavelet Transform and Grey Models. *IEEE Access* **2020**, *8*, 204389–204399. [CrossRef]

processes

Article

Dynamically Triggering Resilient Control for Networked Nonlinear Systems under Malicious Aperiodic DoS Attacks

Wei Tan [1,*], He Wang [1], Huazhou Hou [1], Xiaoxu Liu [2] and Meng Zheng [3]

[1] School of Mathematics, Southeast University, Nanjing 211189, China
[2] Sino-German College of Intelligent Manufacturing, Shenzhen Technology University, Shenzhen 518118, China;
[3] Shenyang Institute of Automation, Chinese Academy of Sciences, Shenyang 110016, China
* Correspondence: tanweilina@163.com

Abstract: Networked nonlinear systems (NNSs) have great potential security threats because of malicious attacks. These attacks will destabilize the networked systems and disrupt the communication to the networked systems, which will affect the stability and performance of the networked control systems. Therefore, this paper aims to deal with the resilient control problem for NNSs with dynamically triggering mechanisms (DTMs) and malicious aperiodic denial-of-service (DoS) attacks. To mitigate the impact from DoS attacks and economize communication resources, a resilient dynamically triggering controller (RDTC) is designed with DTMs evolving an adaptive adjustment auxiliary variable. Thus, the resulting closed-loop system is exponentially stable by employing the piecewise Lyapunov function technique. In addition, according to the minimum inter-event time, the Zeno behavior can be excluded. Finally, the merits of the proposed controllers and theory are corroborated using the well-known nonlinear Chua circuit.

Keywords: resilient dynamically triggering controller (DTRC); dynamically triggering mechanisms (DTMs); denial-of-service (DoS); networked nonlinear systems (NNSs); resilient dynamically event-triggering (RDET)

Citation: Tan, W.; Wang, H.; Hou, H.; Liu, X.; Zheng, M. Dynamically Triggering Resilient Control for Networked Nonlinear Systems under Malicious Aperiodic DoS Attacks. *Processes* **2022**, *10*, 2627. https://doi.org/10.3390/pr10122627

Academic Editor: Xiong Luo

Received: 30 October 2022
Accepted: 1 December 2022
Published: 7 December 2022

1. Introduction

Recently, due to the irreplaceable position of communication in the network, many scholars have devoted their attention to the study of networked control systems. More specifically, they are focusing on the data transmission of networked control systems because of the advantages of the information interaction based on the interconnection of the different systems. Naturally, communication-based issues of control and optimization are emerging and developing rapidly [1–3]. Although the advantages of communication technology have brought seismic shocks to academia and industry, there exists a problem. The limitations of the periodic sampling technique subject to guaranteeing desired system performance by reducing the sampling period create a large amount of redundant sampled data, resulting in network congestion and executing control tasks periodically after the system is stabilized, which results in wasting network bandwidth and computation resources. To tackle these limitations, an event-triggering technique has emerged at this historical moment and developed rapidly based on supervising the controller's update [4,5]. Herein, it is worth noting that the event triggering techniques not only ensure the desired performance from control tasks but also decrease the update frequency of the controller, resulting in energy saving in system communication. In the past decades, different types of the event triggering techniques have been proposed, such as the static event-triggering technique [6–8], dynamic event-triggering technique [9–11], stochastic event-triggering technique [12–14], and switched event-triggering technique [15,16].

In addition, unreliable communication channels cause much concern in the discussion of the stability and performance maintenance for networked control systems. In this regard, there exist some innovative works [17–19]. In particular, security-based networked control systems resisting malicious attacks have been given attention in the past years [20,21]. Herein, it is worth noting that on the premise of ensuring system stability and desired performance, the so-called security is the elasticity of resisting malicious cyber attacks. Although there exist a few innovative and groundbreaking results for networked control systems with co-design of the event triggering techniques and DoS attacks [22–24], the co-design with dynamically triggering techniques and DoS attacks for NNSs is still a challenging problem.

Up to now, although many innovative and groundbreaking results have sprung up for the control and optimization in the framework of co-design of dynamically triggering techniques and DoS attacks [25–27], they primarily focus on linear systems. Moreover, some research results of nonlinear systems with DTMs and DoS attacks appear sporadically [28–31], the existing DTMs have shown certain limitations in theorem research and industrial practice. Inspired by the aforementioned discussion, this paper deals with the resilient control problem for NNSs with DTMs and DoS attacks. More specifically, a DTRC is designed with DTMs with an adaptive adjustment auxiliary variable, which can result in the closed-loop system being exponentially stable by employing the piecewise Lyapunov function technique. Meanwhile, a minimal inter-event time is obtained to ensure it is Zeno-free under aperiodic DoS attacks. In addition, the innovations of this paper are as follows:

- Different from the static trigger strategy in [22–24], a novel dynamically triggering strategy is proposed for NNSs with aperiodic DoS attacks. Because of the longer trigger intervals compared with the static trigger intervals, this strategy further reduces the sampling data transmission rate and improves the usage of network resources.
- Compared with the trigger strategy in [23], the dynamically triggering resilient control strategy is introduced into nonlinear systems to obtund the influence of aperiodic DoS attacks. In addition, the sampled data cannot be transmitted even if condition (7) is satisfied since aperiodic DoS attacks will result in the loss of control input during the DoS attacks range.
- Compared with [23], a new piecewise Lyapunov function is designed to ensure the exponential stability of the networked control system under DoS attacks. In particular, the minimum inter-event time excludes Zeno behavior in the resilient controller. Moreover, the proposed method not only releases Assumption 4 in [23] but also reduces the conservative of the system.

The structure of this article is as follows. First, NNSs and problem statements are presented in Section 2. Then, the conditions for the stability of NNSs under DoS attacks are driven in Sections 3. Furthermore, the satisfactory and better performance of the RDET controller designed than the existing ones is provided in Section 4. Finally, the conclusion is shown in Section 5.

Notations: $\mathbb{R}^+$ and $\mathbb{Z}^+$ represent the set of the positive real numbers and the set of the positive integer numbers, respectively. $\mathbb{R}^n$ and $\mathbb{R}^{n\times m}$ indicate the space of real n-vectors and $n \times m$ matrices. x^{-1} is the inverse of x (function or matrix). $\|\cdot\|$ means the 2-norm.

2. Problem Formulation

Figure 1 shows wireless NNSs under aperiodic DoS attacks. First, aperiodic DoS attack scenarios are typically depicted by the sleeping intervals and DoS attack intervals in Figure 2. Then, the dynamically triggering resilient control strategy and switching controller are designed for NNSs, respectively. Next, based on these descriptions, we give out the problem statement.

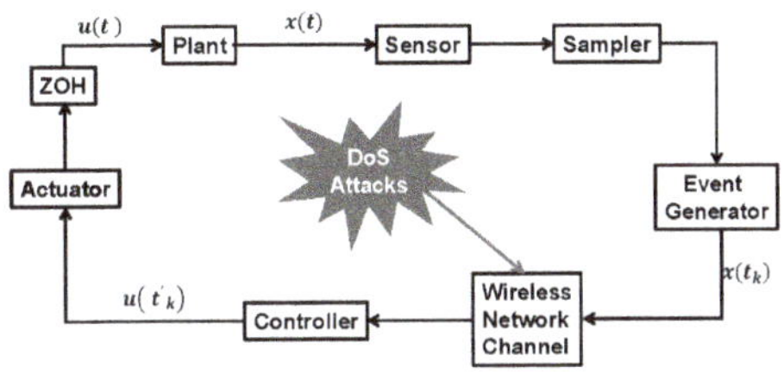

Figure 1. Schematic representation of dynamic event-triggered control for nonlinear systems under DoS attacks.

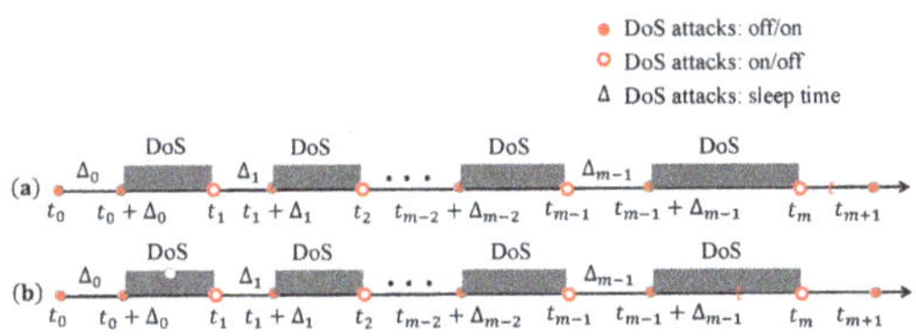

Figure 2. Situation of aperiodic DoS attacks. (**a**) presents the current instant in the sleeping interval, and (**b**) presents the current instant in the DoS attack interval.

2.1. Networked Nonlinear Systems

Consider the following class of NNSs

$$\begin{cases} \dot{x}(t) = f(x(t), u(t)) \\ x(0) = x_0 \end{cases} \tag{1}$$

where $x(t) \in \mathbb{R}^n$ is the system state; $u(t) \in \mathbb{R}^m$ represents the control input; x_0 denotes the initial condition; the Lipschitz continuous function $f : \mathbb{R}^n \times \mathbb{R}^m \to \mathbb{R}^n$ satisfies $f(0,0) = 0$ for all $t \in \mathbb{R}^+$.

For the sake of later analysis, we provide the following definitions and assumptions.

Definition 1 ([32]).

1. *A function $\alpha : [0,\infty) \to [0,\infty)$ is called a class of $\mathcal{K}$ if it is continuous, strictly increasing and $\alpha(0) = 0$. If $\alpha \in \mathcal{K}$ and also $\alpha(s) \to [0,\infty)$ as $s \to [0,\infty)$, then it is said to be of class $\mathcal{K}_\infty$.*
2. *A function $\beta : [0,\infty) \times [0,\infty) \to [0,\infty)$ is called a class of $\mathcal{KL}$ if the function $\beta(\cdot,t) \in \mathcal{K}$ for all fixed $t > 0$ and the function $\beta(s,\cdot)$ is decreasing and $\beta(s,t) \to 0$ as $t \to \infty$ for all fixed $s \in \mathbb{R}^+$.*

Definition 2 ([32]). *System (1) is said to be globally weakly exponentially stable (GWES) if there exist functions $\alpha_1, \alpha_2 \in \mathcal{K}_\infty$ and constants $\delta > 0, M \geq 1$ such that for any initial value $x(0)$, the solution $x(t)$ satisfies $\alpha_1(\|x(t)\|) \leq Me^{-\delta t}\alpha_2(\|x(0)\|), \forall t \geq 0$. In particular, when $\alpha_1(\|x\|) = \alpha_2(\|x\|) = \|x\|^m, m \in \mathbb{Z}^+$, it is said to be globally exponentially stable (GES).*

Definition 3 ([32]). *A function $V : \mathbb{R}^n \to \mathbb{R}_{\geq 0}$ is an ISS-Lyapunov function if there exist some $\mathcal{K}_\infty$ functions α_1, α_2, and γ that satisfy*

$$\alpha_1(\|x\|) \leq V(x) \leq \alpha_2(\|x\|)$$
$$\nabla V(x) f(x,u) \leq -cV(x) + \gamma(\|u\|)$$

where $c \in \mathbb{R}^+$.

Definition 4 ([32]). *Given a local Lipschitz function $V : \mathbb{R}^n \rightarrow \mathbb{R}^+$, the upper left-hand Dini derivative of V along system* (1) *is defined by*

$$D^- V[f] = \lim_{h \rightarrow 0^-} \sup \frac{1}{h}\{V(x+hf) - V(x)\}.$$

2.2. Aperiodic DoS Attacks

In this section, for the convenience of analysis and design, we assume that the DoS attacks only occur on the measurement channel and that no packet loss or no delay occurs during the sleeping intervals. As shown in Figure 2, $\{t_m\}(t_0 \geq 0)$ and $\{\Delta_m\}_{m\in\mathbb{N}_0}(\Delta_m \geq 0)$ represent the sequence of DoS on/off transitions times and the duration of the mth sleep status, respectively. Correspondingly, for simplicity, let $\mathcal{H}_m$ and $\mathcal{D}_m$ present the sleeping interval and the attack interval, respectively.

$$\begin{cases} \mathcal{H}_m = [t_m, t_m + \Delta_m) \\ \mathcal{D}_m = [t_m + \Delta_m, t_{m+1}) \end{cases} \tag{2}$$

In addition, let $\Xi_s(0,t)$ and $\Xi_a(0,t)$ represent, respectively, all single sleeping ranges and all single attack ranges

$$\begin{cases} \Xi_s(0,t) = \{t_m\} \bigcup \mathcal{H}_m \bigcap [0,t) \\ \Xi_a(0,t) = \bigcup \mathcal{D}_m \bigcap [0,t). \end{cases} \tag{3}$$

To characterize the constraints of limited energy on DoS attacks, it is necessary to give the following two assumptions for the frequency and the duration of DoS attacks, respectively.

Assumption 1 (*DoS Duration* [22]). *There exist $T_0 \in \mathbb{R}_{\geq 0}$ and $T \in \mathbb{R}_{>1}$ for all $t \in \mathbb{R}_{\geq 0}$, which makes the following inequality hold*

$$|\Xi_a(0,t)| \leq T_0 + \frac{t}{T}. \tag{4}$$

Assumption 2 (*DoS Frequency* [22]). *There exist $m(t)$, $T_{D_0} \in \mathbb{R}_{\geq 0}$ and $T_D \in \mathbb{R}_{>0}$ for all $t \in \mathbb{R}_{\geq 0}$, which makes the following inequality hold*

$$m(0,t) \leq T_{D_0} + \frac{t}{T_D}. \tag{5}$$

Remark 1. *The intent of the DoS attack is not generally sporadic and periodic, but aperiodic (stochastic). Therefore, the periodic DoS attack is not realistic in theoretical research and practical industrial production. To tackle this issue, this paper focuses on the more realistic aperiodic DoS attacks in the following part. In addition, Assumption 1 excludes the situation of continuous DOS attacks, which makes the considered system open-loop and uncontrollable.*

In this subsection, we will illustrate the proposed procedures for a resilient dynamically triggering strategy in favor of the aperiodic DoS attacks. To achieve this goal, denote e the difference between the last successfully transmission state $x(t_i)$ and the current state $x(t)$

$$e = x(t_i) - x(t), \forall t \in [t_i, t_{i+1}), i \in \mathbb{Z}, t_0 = 0 \tag{6}$$

where t_i is determined by the upcoming DTM (7).

$$t_{i+1} = \inf\{t > t_i, t \in R | g(e) \leq 0\}. \tag{7}$$

Herein, similar to [9], using the trigger function $g(e) = \eta + \theta(\rho(1-c_1)\alpha(x) - \gamma(\|e\|)), \theta \in \mathbb{R}^+$. Additionally, the internal dynamic variable η is to be defined before a new dynamic triggering strategy sprung up, which is

$$\dot{\eta} = -\lambda\eta + \rho(1-c_1)\alpha(\|x\|) - \gamma(\|e(t)\|),\ \eta(0) = 0 \tag{8}$$

where η is a locally Lipschitz continuous $\mathcal{K}_\infty$ function. Intuitively, η may be regarded as a filtered value of $\alpha(1-c_1)\alpha(\|x\|) - \gamma(\|e(t)\|)$ (refers to [5]). In particular, the filter (8) is possibly nonlinear if the η is nonlinear. The dynamic event-triggered strategy (7) reduces to event-triggered strategy in [29] when θ goes to $+\infty$ (detailed analysis refers to [9]).

Based on the discussion before, the following lemma is needed to guarantee $\eta \geq 0$.

Lemma 1. *The variable η defined in* (8) *is always non-negative.*

Proof. According to (8), which corresponds to the following inequality:

$$\eta + \theta(\rho(1-c_1)\alpha(\|x\|) - \gamma(\|e(t)\|)) \geq 0. \tag{9}$$

First, if $\theta = 0$, then $\eta \geq 0$ is true.

Second, if $\theta \neq 0$, by combing (8) and (9), one has

$$\dot{\eta} + \lambda\eta = \rho(1-c_1)\alpha(\|x\|) - \gamma(\|e(t)\|) \geq -\frac{\eta}{\theta},\ \eta(0) \geq 0. \tag{10}$$

Then, solve (10) for $t \in [0, +\infty)$, one has

$$\eta(t) \geq \eta(0)e^{-(\lambda+\frac{1}{\theta})t}$$

which means that η is lower bound by a positive exponential signal, so one can obtain $\eta \geq 0$. □

In addition, taking DoS attacks into consideration, as we all know, the measurement data will be lost even if the condition (7) is satisfied. To alleviate the effects of DoS attacks, a resilient strategy will be presented in the following. In particular, combining with (7), one defines a novel RDET communication strategy as follows:

$$t_{i+1} = \{t_{i+1} \text{ satisfies } g(e) \leq 0 | t_{i+1} \in \mathcal{H}_n\} \cup \{t_m\}. \tag{11}$$

Remark 2. *With the opening of network control system communication, the system is more vulnerable to all kinds of malicious attacks. In order to eliminate the adverse effects of the attack and ensure better performance of the system, the elastic control technology based on dynamic event triggering plays an important role. This is especially true in many industrial controls, such as power systems [27], Chua circuits [30], and vehicle systems [31].*

Next, we use the following DoS attacks as follows:

$$\mathcal{W}(t) = \begin{cases} 0, & t \in [t_m, t_m + \Delta_m) \\ 1, & t \in [t_m + \Delta_m, t_{m+1}). \end{cases} \tag{12}$$

In this paper, the state-dependent control input $u(t) = k(x(t))$ under DoS attacks can be represented as

$$u(t) = (1 - \mathcal{W}(t))k(x(t)). \tag{13}$$

Based on the above analysis, in what follows, combining (1), (12), and (13), the switched version of system (1) can be represented as

$$\dot{x}(t) = f(x(t), (1 - \mathcal{W}(t))k(x(t))), \tag{14}$$

In the following sections, the conditions for the exponential stability of system (14) with DoS attacks will be provided.

3. Main Result

This section aims to develop a piecewise Lyapunov function for NNS under DoS attacks. The resilient analysis of nonlinear switched system (14) is discussed, and the related parameters are obtained. Theorem 1 is presented to guarantee the RDET control for NNS under DoS attacks. In addition, it is worth noting that the Zeno behavior is excluded in Theorem 2.

Theorem 1. *Consider the NNSs* (1) *under DoS attacks satisfying Assumptions 1 and 2, under the switched controller* (13) *with dynamic event-triggered condition* (7). *If some $\mathcal{K}_\infty$ functions α_1, α_2 and γ hold, then the switched Lyapunov function $V_j, j = 1,2$ of* (14) *satisfies the following inequality*

$$\alpha_1(\|x\|) \leq V_j(x) \leq \alpha_2(\|x\|) \tag{15}$$

$$\nabla V_1(x) \cdot f(x, k(x+e)) \leq -\omega_1 V(x) + \gamma(\|e\|), \quad t \in \Xi_s(0,t) \tag{16}$$

$$\nabla V_2(x) \cdot f(x, 0) \leq \omega_2 V_2(x), t \in \Xi_a(0,t) \tag{17}$$

$$V_1(x) \leq \mu_1 V_2(x^-), V_2(x) \leq \mu_2 V_1(x^-) \tag{18}$$

where $\mu_1, \mu_2 \geq 1$, $\omega_1 > 0, \omega_2 > 0$ and the parameters T_D in (6) *and T in* (4) *satisfy*

$$\frac{\ln(\mu_1 \mu_2)}{T_D} + \frac{(\omega_1 + \omega_2)}{T} \leq \omega_1. \tag{19}$$

Then, system (14) *is GWES. In particular*

$$\|x(t)\| \leq \alpha_1^{-1}\left(Me^{-\beta_1 t}\alpha_2(\|x(0)\|)\right). \tag{20}$$

Proof. To show the complete theoretical analysis of the above theorem, we will deal with it in two steps.

Case 1: Assume there are no DoS attacks.

For $t \in \mathcal{H}_m$, according to (16) and DTM (11), the derivative of $V_1(x(t))$ is subject to

$$\begin{aligned}
\dot{V}_1(x(t)) &\leq -c_1 V_1(x(t)) + \gamma(\|e\|) + \dot{\eta} \\
&\leq -\rho c_1 V_1(x(t)) - c_1(1-\rho)V_1(x(t)) + \gamma(\|e\|) \\
&\quad - \lambda\eta + c_1(1-\rho)V_1(x(t)) - \gamma(\|e\|) \\
&\leq -\rho c_1 V_1(x(t)) - \lambda\eta \\
&\leq -\omega_1 V_1(x(t))
\end{aligned} \tag{21}$$

where $\rho \in (0,1)$ and $\omega_1 = c_1\rho$.

Case 2: Assume there are DoS attacks.

For $t \in \mathcal{D}_m$, based on (17), it is easy to obtain the derivative of $V_2(x(t))$ as follows

$$\dot{V}_2(x(t)) \leq \omega_2 V_2(x(t)). \tag{22}$$

Hence, combing (21) with (22) gives a piecewise Lyapunov functional, $t \in \mathcal{H}_m$ and $t \in \mathcal{D}_m$, respectively, can be found as below

$$V(x(t)) = \begin{cases} e^{-\omega_1(t-t_m)}V(x(t_m)), t \in \mathcal{H}_m \\ e^{\omega_2(t-t_{m-1}-\Delta_{m-1})}V(x(t_{m-1}+\Delta_{m-1})), \\ \qquad t \in \mathcal{D}_m. \end{cases}$$

First, assume $t \in \Xi_s(0,t)$, according to definition 4, one has

$$\begin{aligned} V_1(x(t)) \leq & e^{-\omega_1(t-t_m)}V_1(x(t_m)) \\ \leq & \mu_1 e^{-\omega_1(t-t_m)}V_2(x^-(t_m)) \\ \leq & \mu_1 e^{-\omega_1(t-t_m)}e^{\omega_2(t_m-t_{m-1}-\Delta_{m-1})} \\ & V_2(x(t_{m-1}+\Delta_{m-1})) \\ \leq & \mu_1\mu_2 e^{-\omega_1(t-t_m)}e^{\omega_2(t_m-t_{m-1}-\Delta_{m-1})} \\ & V_1(x^-(t_{m-1}+\Delta_{m-1})) \\ & \vdots \\ \leq & (\mu_1\mu_2)^{m(0,t)}e^{-\omega_1(t-t_m+\Delta_{m-1}+\cdots+\Delta_0)} \\ & e^{\omega_2(t_m-\Delta_{m-1}-\cdots-\Delta_0)}V(x(0)) \\ \leq & (\mu_1\mu_2)^{T_{D0}+\frac{t}{T_D}}e^{(\omega_1+\omega_2)T_0} \\ & e^{(\omega_1+\omega_2)\frac{t}{T}-\omega_1 t}V(x(0)) \\ = & M_1 e^{[\frac{\ln(\mu_1\mu_2)}{T_D}+\frac{(\omega_1+\omega_2)}{T}-\omega_1]t}V(x(0)) \end{aligned} \tag{23}$$

where $M_1 = (\mu_1\mu_2)^{T_{D0}}e^{(\omega_1+\omega_2)T_0}$.

Then, assume $t \in \Xi_a(0,t)$, according to definition 4 again, one has

$$\begin{aligned} V_2(x(t)) \leq & e^{\omega_2(t-t_{m-1}-\Delta_{m-1})}V_2(x(t_{m-1}+\Delta_{m-1})) \\ \leq & \mu_2 e^{\omega_2(t-t_{m-1}-\Delta_{m-1})}V_1(x^-(t_{m-1}+\Delta_{m-1})) \\ \leq & \mu_2 e^{\omega_2(t-t_{m-1}-\Delta_{m-1})}e^{\ \omega_1\Delta_{m-1}}V_1(x(t_{m-1})) \\ \leq & \mu_1\mu_2 e^{\omega_2(t-t_{m-1}-\Delta_{m-1})}e^{-\omega_1\Delta_{m-1}}V_2(x^-(t_{m-1})) \\ & \vdots \\ \leq & \mu_1^{m(0,t)-1}\mu_2^{m(0,t)}e^{-\omega_1(\Delta_{m-1}+\cdots+\Delta_0)} \\ & e^{-\omega_2(t-\Delta_{m-1}-\cdots-\Delta_0)}V(x(0)) \\ \leq & \mu_1^{m(0,t)-1}\mu_2^{m(0,t)}e^{-\omega_1(t-\Xi_a)}e^{-\omega_2\Xi_a}V(x(0)) \\ = & M_2 e^{[\frac{\ln(\mu_1\mu_2)}{T_D}+\frac{(\omega_1+\omega_2)}{T}-\omega_1]t}V(x(0)) \end{aligned} \tag{24}$$

where $M_2 = \mu_1^{-1}(\mu_1\mu_2)^{T_{D0}}e^{(\omega_1+\omega_2)T_0}$.

Finally, combining (23) and (24), one has

$$V(x(t)) \leq Me^{-[\omega_1-\frac{\ln(\mu_1\mu_2)}{T_D}-\frac{(\omega_1+\omega_2)}{T}]t}V(x(0)) \tag{25}$$

where $M = \max\{M_1, M_2\}$.

In what follows, using (15), the above inequality (25) can be modified as

$$\begin{aligned}\alpha_1(\|x(t)\|) \le & V(x(t)) \le Me^{-\beta_1 t}V(x(0)) \\ \le & Me^{-\beta_1 t}\alpha_2(\|x(0)\|) \end{aligned} \tag{26}$$

where $\beta_1 = \omega_1 - \frac{\ln(\mu_1\mu_2)}{T_D} - \frac{(\omega_1+\omega_2)}{T}$. Based on (26), if DoS attacks satisfy $\frac{\ln(\mu_1\mu_2)}{T_D} + \frac{(\omega_1+\omega_2)}{T} \le \omega_1$, then, system (1) with control input (11) under DoS attacks is GWES. Further, one has

$$\|x(t)\| \le \alpha_1^{-1}(Me^{-\beta_1 t}\alpha_2(\|x(0)\|)).$$

The proof is completed. □

Remark 3. *This theorem characterizes the system's resilience issue. Moreover,* $\frac{\ln(\mu_1\mu_2)}{T_D} + \frac{(\omega_1+\omega_2)}{T} \le \omega_1$ *shows that the stability of NNSs can be guaranteed in the event they suffer from more DoS attacks that satisfy some constraints of attack interval and attack frequency.*

Remark 4. *It is worth pointing out that the inequality constraint on α_1 and γ in Assumption 4 in [23] is unnecessary in our work. More specifically, in this article, with the aid of introducing a piecewise Lyapunov function, Assumption 4 in [23] is removed, which reduces the conservatism of the system.*

Next, we will give the conditions that void the Zeno behavior. Before continuing the discussion, we impose an assumption.

Assumption 3 ([5]). *Because of the Lipschitz continuity of the function $f(x,u)$, there exists a constant L_1 which satisfies the following inequality*

$$\begin{aligned}\|f(x,u\| = & \|f(x+k(x+e_i)\| \\ \le & L_1(\|x\| + \|e_i\|) \end{aligned} \tag{27}$$

where $u(t) = k(x(t_i)), t \in [t_i, t_{i+1})$.

Theorem 2. *For the NNSs* (1) *under the event-triggered strategy* (11) *and the controller* (13), *there exists a minimal inter-event time τ ensures that Zeno behavior does not exist, where τ is given by*

$$\tau \ge \frac{1}{2L_1}\ln\left(\frac{(2\gamma^{-1}(\eta/\theta + \rho(1-c_1)\alpha(x))}{\|x(t_i)\|} + 1\right).$$

Proof. First, we define the inter-execution time $\tau = t_{i+1} - t_i$. According to the ZOH scheme, there is $\dot{e}_i = 0$ when $t = t_i$. Meanwhile, with the inequality (27), one has

$$\begin{aligned}\|\dot{x}\| = & \|f(x, u(x+e_i)\| \\ \le & L_1(\|x\| + \|e_i\|). \end{aligned} \tag{28}$$

Furthermore, for $\forall\, t \in [t_i, t_{i+1})$, it is easy to get the following equation

$$\dot{x} = -\dot{e}_i. \tag{29}$$

Next, combining (6), (28), and (29), one has

$$\|\dot{e}_i\| \le L_1(\|x(t_i) - e_i\| + \|e_i\|).$$

Herein, using the comparison lemma, we have

$$\|e_i\| \le \frac{\|x(t_i)\|(e^{2L_1(t-t_i)}-1)}{2}. \tag{30}$$

With (11) associated with

$$\eta + \theta(\rho(1-c_1)\alpha(x) - \gamma(\|e_i\|)) \le 0,$$

from this inequality (31), we get

$$\|e_i\| \ge \gamma^{-1}(\eta/\theta + \rho(1-c_1)\alpha(x)). \tag{31}$$

Finally, combining with (30), (31), and $t - t_i \le \tau, t \in [t_i,\ t_{i+1})$, one has

$$\tau \ge \frac{1}{2L_1}\ln\Big(\frac{(2\gamma^{-1}(\eta/\theta + \rho(1-c_1)\alpha(x))}{\|x(t_i)\|} + 1\Big),$$

where $\eta > 0,\ \theta > 0$. The proof is completed. □

4. Simulation

In this section, the practical merits of the proposed controllers and theory are corroborated using the well-known nonlinear Chua circuit, as shown in Figure 3. Considering the control input $u = (u_1; u_2; u_3)$, its dynamics are generated as

$$\begin{cases} \dot{v}_1 = \frac{v_2 - v_1}{RC_1} - \frac{f(v_1)}{C_1} + u_1, \\ \dot{v}_2 = \frac{v_1 - v_2}{RC_2} + \frac{v_3}{C_2} + u_2, \\ \dot{i}_3 = -\frac{v_2}{L} - \frac{R_0 v_3}{L} + u_3 \end{cases} \tag{32}$$

where v_1 and v_2 are voltages across C_1 and C_2, respectively. i_3 is current through the inductance. $f(v_1) = g_1 v_1 + g_2 v_1^3$ is characteristic of the nonlinear resistance R_N.

Next, let $x_1 = v_1, x_2 = v_2, x_3 = i_3$, and $x = (x_1; x_2; x_3)$. Then, (32) can be transferred into the following as

$$\begin{cases} \dot{x}_1 = \frac{x_2 - x_1}{RC_1} - \frac{g_1 x_1 + g_2 x_1^3}{C_1} + u_1, \\ \dot{x}_2 = \frac{x_1 - x_2}{RC_2} + \frac{x_3}{C_2} + u_2, \\ \dot{x}_3 = -\frac{x_2}{L} - \frac{R_0 x_3}{L} + u_3 \end{cases} \tag{33}$$

According to [33], there exists a chaotic attractor (see Figure 4) when some parameters are fixed at $C_1 = 0.7; C_2 = 7.8; L = 1.897; R_0 = 0.01499; g_1 = -0.59; g_2 = 0.02; R = 2.1$; $u = 0$. Meanwhile, system (33) is rewritten as

$$\begin{cases} \dot{x}_1 = 0.1626x_1 + 0.6803x_2 - 0.0286x_1^3 + u_1, \\ \dot{x}_2 = 0.0611x_1 - 0.0611x_2 + 0.1282x_3 + u_2, \\ \dot{x}_3 = -0.5271x_2 - 0.079x_3 + u_3. \end{cases} \tag{34}$$

Set the initial state $x_0 = (-0.6061; -0.3483; 0.6013)$. System (34) is unstable without the control input u. Then, set $\theta = 0.1; \eta = 0.1; \lambda = 0.5; \rho = 0.01$; simulation time $[0, 200s]$ with sampling period $h = 0.05s$.

Case 1: According to Theorem 1, one can design a controller $u = [-0.01x_1; 0; -0.1x_3]$ under the RDET strategy (11) to stabilize system (34) without aperiodic DoS attacks. The state response of system (34) is shown in Figure 5. Moreover, the number of triggered packets to be transmitted is 82 times. The event interval time of the event generator is depicted in Figure 6. In particular, one can design another controller $u = [-0.1x_1; 0; -0.1x_3]$ under event-triggered communication scheme (7) to stabilize system (34) without DoS attacks. Meanwhile, a stable periodic solution will be presented in Figure 7. The event intervals of the event generator are depicted in Figure 8.

Case 2: In the sequence, under the same circumstances, once system (34) suffers from malicious aperiodic attacks, the system is unstable in Figure 9, where the gray areas represent the DoS attack time intervals. In addition, release time intervals are depicted in Figure 10 with 756 triggered packets to be successfully transmitted to the controller under the event-triggered communication scheme (7).

Case 3: The system is not GAS with $u = 0$. The stabilizing control law is $u = [-2x_1; -2x_2; -x_3]$. We select $V_1(x) = \frac{1}{2}\|x\|^2$ as a Lyapunov function when $t \in \mathcal{H}_m$, so that the $V_1(x) > 0$ holds true if $\|x\| \neq 0$. Notice that

$$\begin{aligned}\nabla V_1(x)f(x,u) =& x_1 \cdot \dot{x}_1 + x_2 \cdot \dot{x}_2 + x_3 \cdot \dot{x}_3 \\ =& -1.8374x_1^2 + 0.7414x_1x_2 - 2.0611x_2^2 \\ & -1.079x_3^2 - 0.3989x_2x_3 - 0.0286x_1^4 \\ & -2x_1e_1 - 2x_2e_2 - x_3e_3 \\ \leq & -0.4667x_1^2 - 0.4909x_2^2 - 0.8795x_3^2 \\ & + e_1^2 + e_2^2 + e_3^2 \\ \leq & -0.403V_1(x) + \|e\|^2. \end{aligned} \quad (35)$$

Next, We select $V_2(x) = \frac{1}{2}(x_1^2 + x_2^2 + 2x_3^2)$ as Lyapunov function when $t \in \mathcal{D}_m$, so that the $V_2(x) > 0$ holds true if $\|x\| \neq 0$. Notice that

$$\begin{aligned}\nabla V_2(x)f(x,0) =& x_1 \cdot \dot{x}_1 + x_2 \cdot \dot{x}_2 + 2x_3 \cdot \dot{x}_3 \\ \leq & 0.5333x_1^2 + 0.7726x_2^2 + 0.3050x_3^2 \\ \leq & 0.0314V_2(x). \end{aligned} \quad (36)$$

According to the aforementioned analysis, there exist $\alpha_1(\cdot) = x_1^2 + x_2^2 + x_3^2$ and $\alpha_2(\cdot) = 2x_1^2 + 2x_2^2 + 2x_3^2$, which satisfy (15) in Theorem 1. Meanwhile, (35) and (36) allow $\omega_1 = 0.403$, $\omega_2 = 0.0314$ to satisfy (16) and (17) in Theorem 1, respectively. In addition, set $\beta_1 = 0.1165$, $\mu_1 = 1$ and $\mu_2 = 2$, which satisfy (18) in Theorem 1. Let $T_0 = 0.1$, $T = 2$, $T_{D_0} = 0.1$, and $T_D = 10$, one has $\frac{\ln 2}{10} + \frac{(0.403+0.0314)}{2} = 0.2865 < 0.403$ and $\|x(t)\| \leq \alpha_1^{-1}(Me^{-\beta_1 t}\alpha_2(\|x(0)\|))$ satisfy (19) and (20), respectively. Moreover, $m \leq 0.1 + \frac{200}{10} = 20.1$ and $\|x(t)\| \leq 0.1204e^{-0.233t}$.

Herein, Figure 11 depicts the state responses of system (34) under DoS attacks, and which shows that system (34) is stable. The release instants are depicted in Figure 12, and there are 716 sampled packets transmitted successfully Under the supervision of the DTM (11). Figure 13 presents the aperiodic DoS attack sequence. Next, Table 1 presents a comparison of the different triggering strategies.

First, before analyzing Table 1, we give out design formulas on a triggering rate, which is expressed as $\frac{\text{event}_{\text{number}}}{\text{sample}_{\text{number}}}$. Second, from Table 1, it is obvious that once the system suffers from aperiodic DoS attacks, the number of triggering events and the triggering rate will add the same parameters designed before to the framework. On contrary, the average interval will become small to add the number of triggering events and compensate for lost packets due to DoS attacks.

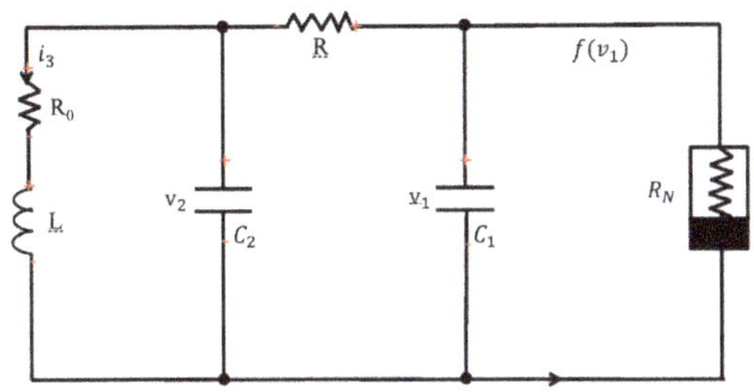

Figure 3. Diagram of a nonlinear Chua circuit.

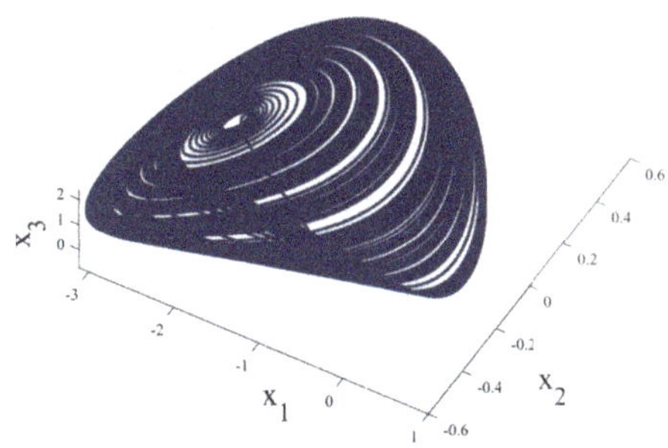

Figure 4. The chaotic attractor for the Chua circuit.

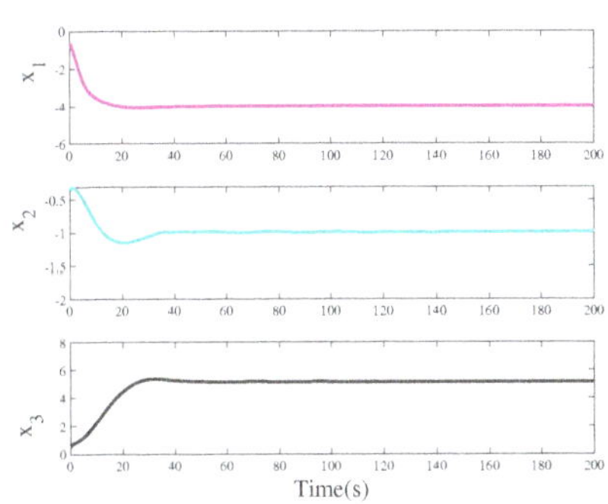

Figure 5. State responses of triggering control systems without aperiodic DoS attacks.

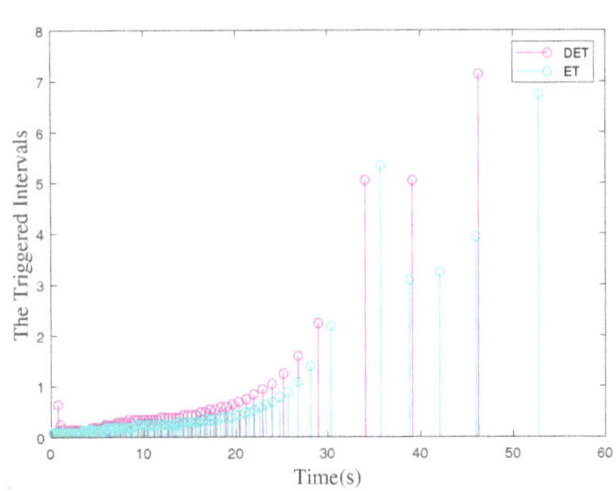

Figure 6. The triggered intervals of control systems without aperiodic DoS attacks.

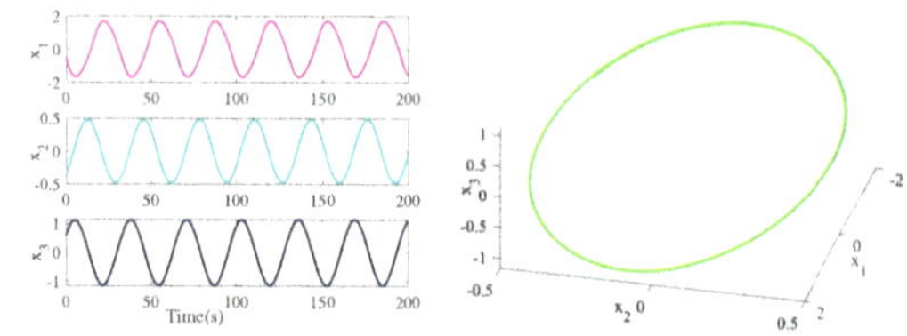

Figure 7. Stable periodic solution of triggering control systems without aperiodic DoS attacks.

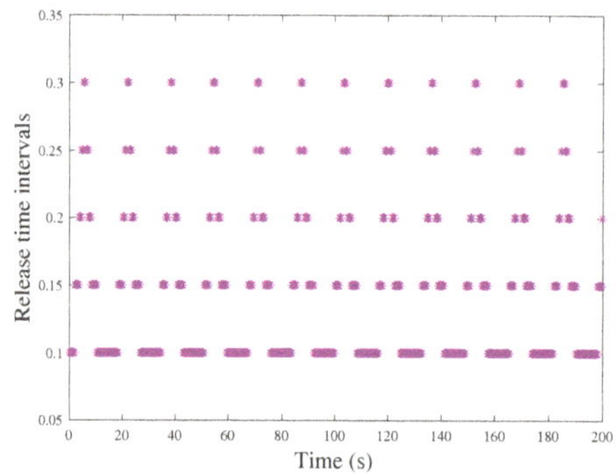

Figure 8. Release time intervals corresponding to a stable periodic solution.

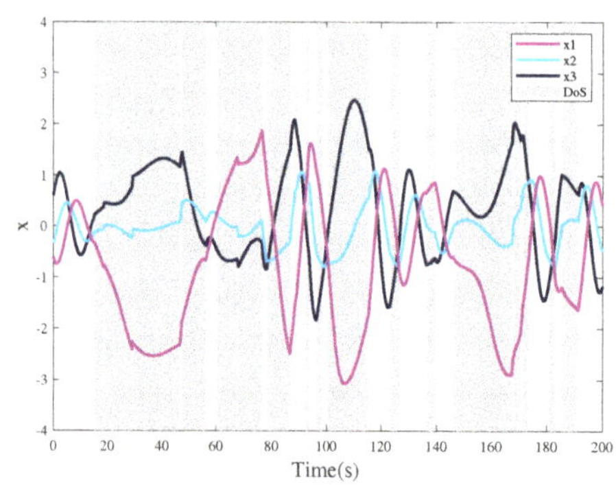

Figure 9. The state trajectories of unstable triggering control systems under aperiodic DoS attacks.

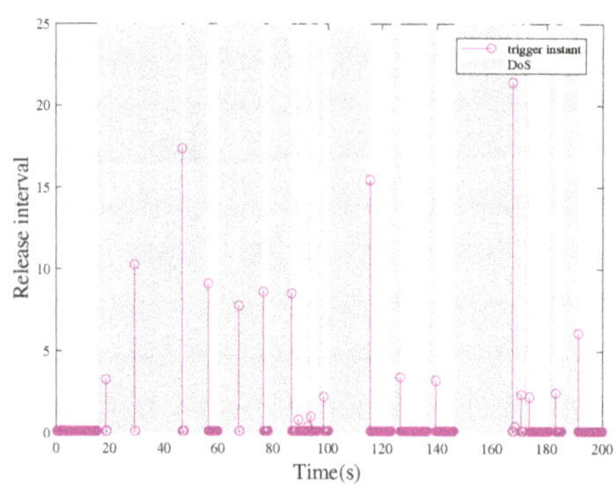

Figure 10. Period of triggering control systems under aperiodic DoS attacks.

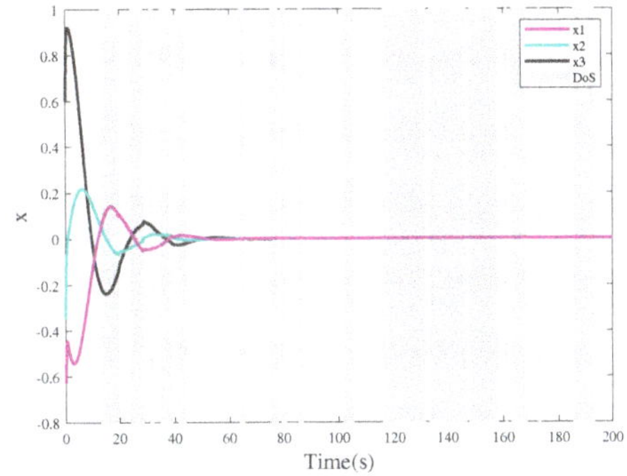

Figure 11. The stable state trajectories under aperiodic DoS attacks.

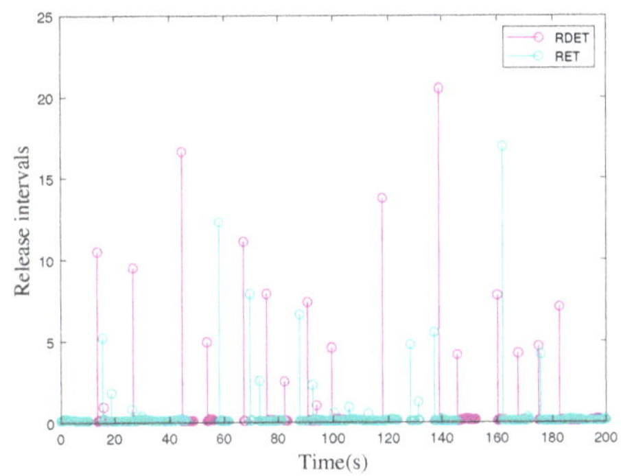

Figure 12. Triggered instants and release intervals under aperiodic DoS attacks.

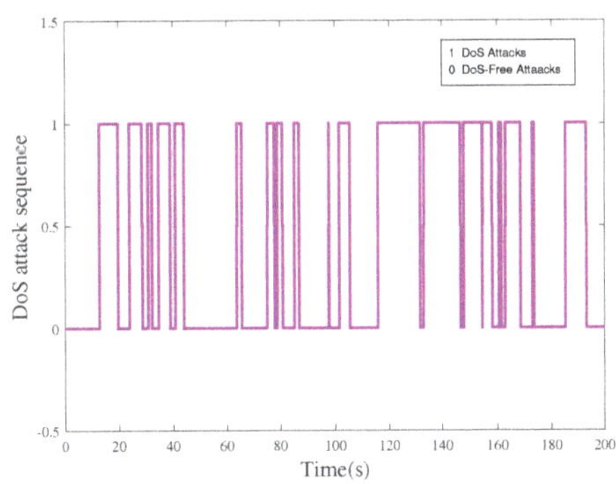

Figure 13. Aperiodic DoS attacks.

Remark 5. *In the simulation, the modeling and generation of DoS attacks is similar to [22]. Since the attacks are affected by energy constraints, they are intermittent non-periodic attacks, and only focus on the single communication channels (sensor-to-controller).*

Remark 6. *Figures 6 and 12 display the triggered intervals of the four strategies. Figure 6 indicates RDET strategy can generate a bigger average interval than the ET strategy without aperiodic DoS attacks. Moreover, a similar result is presented in Figure 12, namely, the RDET strategy can generate a bigger average interval than the RET strategy suffering aperiodic DoS attacks. This result is consistent with Table 1.*

Table 1. Comparing different control schemes.

Strategies	Trigger	Average Interval	Triggering Rate
ET in [5]	120	1.67	3%
DET in [9]	82	2.44	2.05%
RET in [22]	960	0.21	24%
DTRC in this work	716	0.28	17.9%

5. Conclusions

In this paper, we have designed a DRTC to stabilize NNSs under malicious aperiodic DoS attacks. Furthermore, the stability criterion is obtained under malicious aperiodic DoS attacks based on Lyapunov theory. In addition, the minimal inter-event time τ excludes Zeno behavior for the controller (13) with dynamically triggering strategy (11). Finally, the merits of the proposed controllers and theory are corroborated using the well-known nonlinear Chua circuit. Based on our current work, in the future, we will consider security-based event-triggered learning control for NNSs subject to stochastic attacks.

Author Contributions: Conceptualization, W.T.; methodology, W.T.; validation, W.T., H.W. and H.H.; formal analysis, H.W., H.H. and X.L.; investigation, W.T. and H.H.; writing—original draft preparation, W.T.; writing—review and editing, W.T., H.W. and H.H.; visualization, W.T., H.W., X.L. and M.Z.; supervision, H.W., H.H. and X.L.; project administration, H.W. and H.H.; funding acquisition, H.W., H.H., X.L. and M.Z. All authors have read and agreed to the published version of the manuscript.

Funding: This work was supported by the National Natural Science Foundation of China under Grant Nos. 61673107, 62073076, 62203109, 62003218, and 62022088, International Partnership Program of Chinese Academy of Sciences under Grant 173321KYSB20200002, the Jiangsu Provincial Key Laboratory of Networked Collective Intelligence under Grant No. BM2017002, Guangdong Basic, Applied Basic Research Foundation under Grants 2019A1515110234 and 2020A1515110148, and Natural Science Foundation of Jiangsu Province under grant numbers BK20210216 and BK20220812.

Institutional Review Board Statement: Not applicable.

Informed Consent Statement: Not applicable.

Data Availability Statement: Not applicable.

Conflicts of Interest: The authors declare no conflict of interest.

References

1. Chen, C.; Nagananda, K.G.; Xiong, G.; Kishore, S.; Snyder, L.V. A communication-based appliance scheduling scheme for consumer-premise energy management systems. *IEEE Trans. Smart Grid* **2013**, *4*, 56–65. [CrossRef]
2. Saleh, M.; Esa, Y.; Mohamed, A.A. Communication-based control for DC microgrids. *IEEE Trans. Smart Grid* **2018**, *10*, 2180–2195. [CrossRef]
3. Song, H.; Gao, S.; Li, Y.; Liu, L.; Dong, H. Train-centric communication based autonomous train control system. *IEEE Trans. Intell. Vehicles* **2022** . [CrossRef]
4. Heemels, W.P.M.H.; Johansson, K.H.; Tabuada, P. An introduction to event-triggered and self-triggered control. In Proceedings of the 2012 IEEE 51st IEEE Conference on Decision and Control (CDC), Maui, HI, USA, 10–13 December 2012; pp. 3270–3285.
5. Tabuada, P. Event-triggered real-time scheduling of stabilizing control tasks. *IEEE Trans. Autom. Control* **2007**, *52*, 1680–1685. [CrossRef]
6. Heemels, W.P.M.H.; Donkers, M.C.F.; Teel, A.R. Periodic event-triggered control for linear systems. *IEEE Trans. Autom. Control* **2012**, *58*, 847–861. [CrossRef]
7. Wang, W.; Postoyan, R.; Nešić, D.; Heemels, W.P. Periodic event-triggered control for nonlinear networked control systems. *IEEE Trans. Autom. Control* **2019**, *65*, 620–635. [CrossRef]
8. Sun, J.; Yang, J.; Zeng, Z. Predictor-based periodic event-triggered control for nonlinear uncertain systems with input delay. *Automatica* **2022**, *136*, 110055. [CrossRef]
9. Girard, A. Dynamic Triggering Mechanisms for Event-Triggered Control. *IEEE Trans. Autom. Control* **2015**, 60, 1992–1997. [CrossRef]

10. Luo, S.; Deng, F.; Chen, W.H. Dynamic event-triggered control for linear stochastic systems with sporadic measurements and communication delay. *Automatica* **2019**, *107*, 86–94. [CrossRef]
11. Dolk, V.S.; Borgers, D.P.; Heemels, W.P.M.H. Output-based decentralized dynamic event-triggered control with guaranteed $\mathcal{L}_p$-gain performance and zeno-freeness. *IEEE Trans. Autom. Control* **2017**, *62*, 34–49. [CrossRef]
12. Demirel, B.; Leong, A.S.; Gupta, V.; Quevedo, D.E. Tradeoffs in stochastic event-triggered control. *IEEE Trans. Autom. Control* **2018**, *64*, 2567–2574. [CrossRef]
13. Peng, C.; Wu, J.; Tian, E. Stochastic event-triggered H_∞ control for networked systems under denial of service attacks. *IEEE Trans. Syst. Man Cybern. Syst.* **2021**, *52*, 4200–4210. [CrossRef]
14. Han, D.; Mo, Y.; Wu, J.; Weerakkody, S.; Sinopoli, B.; Shi, L. Stochastic event-triggered sensor schedule for remote state estimation. *IEEE Trans. Autom. Control* **2015**, *560*, 2661–2675. [CrossRef]
15. Huang, Y.; Liu, Y. Switching event-triggered control for a class of uncertain nonlinear systems. *Automatica* **2019**, *108*, 108471. [CrossRef]
16. Fan, Y.; Huang, X.; Shen, H. Switching event-triggered control for global stabilization of delayed memristive neural networks: An exponential attenuation scheme. *Neural Netw.* **2019**, *117*, 216–224. [CrossRef]
17. Li, H.; Chen, Z.; Wu, L.; Lam, H.K. Event-triggered control for nonlinear systems under unreliable communication links. *IEEE Trans. Fuzzy Syst.* **2016**, *25*, 813–824. [CrossRef]
18. Li, Z.; Hu, B.; Li, M.; Luo, G. String stability analysis for vehicle platooning under unreliable communication links with event-triggered strategy. *IEEE Trans. Veh. Technol.* **2019**, *68*, 2152–2164. [CrossRef]
19. Wang, J.; Ma, F.; Yang, Y.; Nie, J.; Bilin, A.G.; Guvenc, L. Adaptive event-triggered platoon control under unreliable communication links. *Trans. Intell. Transp. Syst.* **2020**, *23*, 1924–1935. [CrossRef]
20. Xiao, S.; Han, Q.L.; Ge, X.; Zhang, Y. Secure distributed finite-time filtering for positive systems over sensor networks under deception attacks. *IEEE Trans. Cybern.* **2020**, *50*, 1220–1229. [CrossRef]
21. Liu, J.L.; Yin, T.T.; Yue, D.; Karimi, H.R.; Cao, J.D. Event-based secure leader-following consensus control for multi-agent systems with multiple cyber-attacks. *IEEE Trans. Cybern.* **2020**, *51*, 162–173. [CrossRef]
22. Persis, C.D.; Tesi, P. Input-to-state stabilizing control under denial-of-service. *IEEE Trans. Autom. Control* **2015**, *60*, 2930–2944. [CrossRef]
23. Persis, C.D.; Tesi, P. Networked control of nonlinear systems under Denial-of-Service. *Syst. Control Lett.* **2016**, *96*, 124–131. [CrossRef]
24. Hu, S.; Yue, D.; Han, Q.L.; Xie, X.; Chen, X.; Dou, C. Observer-based event-triggered control for networked linear systems subject to denial-of-service attacks. *IEEE Trans. Cybern.* **2020**, *50*, 1952–1964. [CrossRef] [PubMed]
25. Zhao, N.; Shi, P.; Xing, W. Dynamic event-triggered approach for networked control systems under denial of service attacks. *Int. J. Robust Nonlinear Control* **2021**, *31*, 1774–1795. [CrossRef]
26. Amini, A.; Asif, A.; Mohammadi, A. A Unified Optimization for Resilient Dynamic Event-Triggering Consensus Under Denial of Service. *IEEE Trans. Cybern.* **2020**, *52*, 2872–2884. [CrossRef] [PubMed]
27. Wang, J.; Wang, D.; Su, L.; Park, J.H.; Shen, H. Dynamic Event-Triggered H_∞ Load Frequency Control for Multi-Area Power Systems Subject to Hybrid Cyber Attacks. *IEEE Trans. Syst. Man Cybern. Syst.* **2022**. [CrossRef]
28. Dolk, V.S.; Tesi, P.; Persis, C.D.; Heemels, W.P.M.H. Event-Triggered Control Systems under Denial-of-Service Attacks. *IEEE Trans. Control Netw. Syst.* **2017**, *4*, 93–105. [CrossRef]
29. Liu, F.; Zhao, C.; Wang, C.; Chen, T. Dynamic Event-triggered Control for Nonlinear NCSs Subject to DoS Attacks. *J. Frankl. Inst.* **2022**, *359*, 5385–5411. [CrossRef]
30. Ai, Z.; Peng, L.; Zong, G.; Shi, K. Impulsive Control for Nonlinear Systems Under DoS Attacks: A Dynamic Event-Triggered Method. *IEEE Trans. Circuits Syst. II Exp. Briefs* **2022**, *69*, 3839–3843. [CrossRef]
31. Chen, X.; Chang, X. Resilient filter of nonlinear network systems with dynamic event-triggered mechanism and hybrid cyber attack. *Appl. Math. Comput.* **2022**, *434*, 127419. [CrossRef]
32. Khalil, H.K. *Nonlinear Systems*, 3rd ed.; Prentice-Hall: Englewood Cliffs, NJ, USA, 2003.
33. Galias, Z. Continuation-Based Method to Find Periodic Windows in Bifurcation Diagrams With Applications to The Chua's Circuit With A Cubic Nonlinearity. *IEEE Trans. Circuits Syst. I Reg. Papers* **2021**, *68*, 3784–3793. [CrossRef]

processes

Article

Tracking Control of a Hyperchaotic Complex System and Its Fractional-Order Generalization

Feng Liang [1,2], Lu Lu [2,†], Zhengfeng Li [3], Fangfang Zhang [3,*] and Shuaihu Zhang [3]

1 Department of Computer and Software Engineering, Shandong College of Electronic Technology, Jinan 250200, China; lf1839@163.com
2 College of Artificial Intelligence and Big Data for Medical Sciences, Shandong First Medical University, Shandong Academy of Medical Sciences, Jinan 250117, China; squarelu@hotmail.com
3 School of Information and Automation Engineering, Qilu University of Technology (Shandong Academy of Sciences), Jinan 250353, China; zhengf_li@163.com (Z.L.); 18437902858@163.com (S.Z.)
* Correspondence: zhff4u@qlu.edu.cn; Tel.: +86-151-6916-3922
† These authors contributed equally to this work.

Abstract: Hyperchaotic complex behaviors often occur in nature. Some chaotic behaviors are harmful, while others are beneficial. As for harmful behaviors, we hope to transform them into expected behaviors. For beneficial behaviors, we want to enhance their chaotic characteristics. Aiming at the harmful hyperchaotic complex system, a tracking controller was designed to produce the hyperchaotic complex system track common expectation system. We selected sine function, constant, and complex Lorenz chaotic system as target systems and verified the effectiveness by mathematical proof and simulation experiments. Aiming at the beneficial hyperchaotic complex phenomenon, this paper extended the hyperchaotic complex system to the fractional order because the fractional order has more complex dynamic characteristics. The influences order change and parameter change on the evolution process of the system were analyzed and observed by MATLAB simulation.

Keywords: hyperchaotic complex system; tracking control; fractional order

Citation: Liang, F.; Lu, L.; Li, Z.; Zhang, F.; Zhang, S. Tracking Control of a Hyperchaotic Complex System and Its Fractional-Order Generalization. *Processes* **2022**, *10*, 1244. https://doi.org/10.3390/pr10071244

Academic Editor: Zhiwei Gao

Received: 25 May 2022
Accepted: 16 June 2022
Published: 22 June 2022

1. Introduction

Chaos is a complex nonlinear phenomenon in nature. The chaotic system has extensive applications in the fields of secure communication [1], industrial process [2], ecosystem [3], and so on. With the application of chaos theory in increasingly more fields, there are increasingly more requirements for chaotic systems. For example, people expect higher dimensional and more complex chaotic systems to describe industrial processes. Therefore, following this, scholars put forward the hyperchaotic system, complex chaotic system, and hyperchaotic complex system [4–6]. The hyperchaotic complex system in particular has higher dimensions and more controllable parameters [7,8] that can more accurately describe some chaotic phenomena in the industrial process. The chaotic system will experience four states: stable point, period, chaos, and divergence. When the system is in the periodic or period doubling state, the phase plane will form a closed trajectory, that is, the limit cycle. Generally speaking, the chaotic system will evolve from a periodic state to a chaotic state, and the corresponding phase diagram will evolve from limit cycle to chaotic attractor [9].

However, some chaotic behaviors are harmful, while some are beneficial. As for harmful behaviors, we hope to transform them into expected behaviors. For example, permanent magnet motor under some parameters can produce chaotic behaviors and disturb the normal operation of the motor. Therefore, it is necessary to add a controller to make it track the desired motion trajectory [10,11]. This means the tracking control for chaos, which can be used to obtain the desired output and improve the performance of the system [12,13]. The research on the tracking control of hyperchaotic complex systems is of great value.

At present, most of the research focuses on the tracking control of real chaotic systems [14–17], and few scholars have studied the tracking control of hyperchaotic complex systems. Gao proposed a novel tracking control method for Lorenz systems by using single-state feedback [18]. Loria addressed the problem of controlled synchronization of a class of uncertain chaotic systems [19]. Zhang presented the tracking control method and the parameter identification procedure, aiming at CVCSs with complex parameters [20]. Chaudhary et al. investigated a hybrid projective combination–combination synchronization scheme (HPCCSS) in four different hyperchaotic (HC) systems via active control technique (ACT) [21]. Abbasi proposed a robust resilient design methodology for stabilization and tracking control for a class of chaotic dynamical systems [22]. Zhao realized tracking control and synchronization of the fractional hyper chaotic Lorenz system [23]. Nagy et al. investigated the combination synchronization phenomena of various fractional-order systems using the scaling matrix [24]. In this paper, the tracking control of complex hyperchaotic system was realized.

As for beneficial chaotic behavior, we hope to enhance its chaotic characteristics. For example, more complex chaotic behaviors can make the stirring more sufficient in industrial process. Moreover, in secure communication, complex chaotic signal can increase the confidentiality of transmitted signal.

In order to obtain more complicated chaotic behaviors, we can extend the integer-order hyperchaotic complex systems to fractional order, because fractional order can increase the degree of freedom of the chaotic system and make its dynamic behaviors more complicated. As an extension of the integer-order complex hyperchaos system, the fractional- order complex hyperchaos system also has higher accuracy in describing processes in many fields and can more accurately describe various irregular physical phenomena. Therefore, many scholars have carried out a large amount of research on the fractional order. Ma investigated a new 4D incommensurate fractional-order chaotic system [25]. Jahanshahi investigated a multi-stable fractional-order chaotic system [26]. Rahman presented a new three-dimensional fractional-order complex chaotic system [27].

However, the above references fractional-order extension of real chaotic systems, but few scholars have studied the fractional-order hyperchaotic complex systems. In this paper, the complex hyperchaotic system was extended to fractional order.

On the basis of the above discussion, the main innovations of this paper are as follows:

(1) The tracking controller for the hyperchaotic complex system is designed. Three state variables of the hyperchaotic complex system track the sine function, constant, and complex Lorenz chaotic system individually, which realize the control of harmful chaotic behaviors. The stability and feasibility of the controller were verified by mathematical proof and simulation experiments.
(2) The hyperchaotic complex system is extended to fractional order, which enhances the beneficial chaotic behaviors. The effects of initial value, order, and parameters on the fractional hyperchaotic complex system are discussed.

The rest of this paper is structured as follows: In Section 2, we introduce the model of hyperchaotic complex system. In Section 3, we designed a controller to realize the tracking control of the hyperchaotic complex system and verified its feasibility from two aspects of a mathematical proof and simulation experiment. In Section 4, we extended the complex hyperchaotic system to fractional order and analyzed its initial value sensitivity and the system evolution process of the fractional complex hyperchaotic system with order and parameters. In the last section, we conclude the paper.

2. The Model of the Hyperchaotic Complex System

In 2021, Li et al. constructed a new hyperchaotic complex system by adding feedback term and introducing complex variables [7]. The mathematical model is as follows:

$$\begin{cases} \dot{x}_1 = a(x_2 - x_1) + x_2x_3 + x_4 \\ \dot{x}_2 = cx_1 - x_2 - x_1x_3 + x_4 \\ \dot{x}_3 = 1/2(\overline{x}_1x_2 + x_1\overline{x}_2) - bx_3 \\ \dot{x}_4 = 1/2(\overline{x}_1x_2 + x_1\overline{x}_2) - dx_4 \end{cases} \tag{1}$$

where $\dot{x}_1 = u_1 + ju_2, \dot{x}_2 = u_3 + ju_4$ is a complex variable and $\dot{x}_3 = u_5, \dot{x}_4 = u_6$ is a real variable. We separated the real part and imaginary part of the variable to obtain the equivalent mathematical model, as shown in system (2):

$$\begin{cases} \dot{u}_1 = a(u_3 - u_1) + u_3u_5 + u_6 \\ \dot{u}_2 = a(u_4 - u_2) + u_4u_5 \\ \dot{u}_3 = cu_2 - u_3 - u_1u_5 + u_6 \\ \dot{u}_4 = cu_2 - u_4 - u_2u_5 \\ \dot{u}_5 = u_1u_3 + u_2u_4 - bu_5 \\ \dot{u}_6 = u_1u_3 + u_2u_4 - du_6 \end{cases} \tag{2}$$

When $a = 10, b = 8/3, c = 20, d = 15$ and the initial value is (1, 2, 3, 4, 5, 6), the system shows obvious chaotic characteristics. The Lyapunov exponents of system (2) is as follows: $LE_1 = 1.7555$, $LE_2 = 0.1175$, $LE_3 = 0$, $LE_4 = -11.6914$, $LE_5 = -14.8827$, $LE_6 = -22.3471$, which is (+, +, 0, −, −, −), so the system is in a hyperchaotic state. The attractor phase diagram of system (2) is shown in Figure 1.

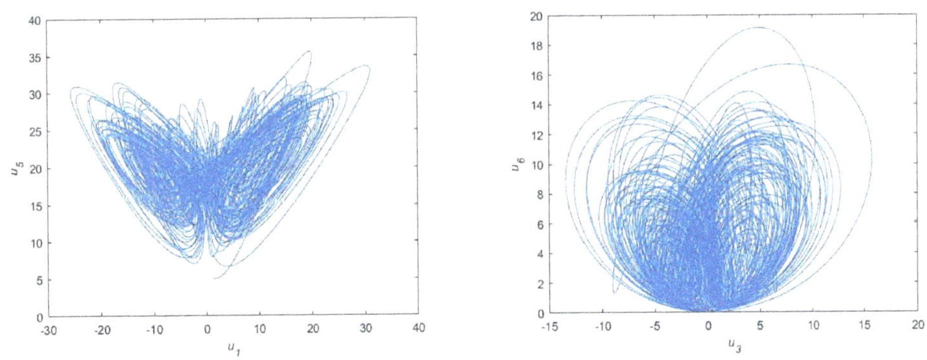

(a) (b)

Figure 1. Attractor phase diagram of system (2). (**a**) $u_1 - u_5$ phase diagram of system (2); (**b**) $u_3 - u_6$ phase diagram of system (2).

3. Tracking Control

In this section, the controller is designed so that the three state variables of the system (2) can track the sine function, constant 5, and the fourth dimension of the complex Lorenz chaotic system individually.

The system model of complex Lorenz chaotic system is as follows:

$$\begin{cases} \dot{x}_1 = a_1(x_2 - x_1) \\ \dot{x}_2 = a_2x_1 - x_1x_3 - a_3x_2 \\ \dot{x}_3 = -a_4x_3 + 1/2(\overline{x}_1x_2 - x_1\overline{x}_2) \end{cases} \tag{3}$$

Separating the real part and imaginary part of the complex Lorenz chaotic system, the following equivalent mathematical model can be obtained:

$$\begin{cases} \dot{x}_1^r = a_1(x_2^r - x_1^r) \\ \dot{x}_1^i = a_1(x_2^i - x_1^i) \\ \dot{x}_2^r = a_2x_1^r - x_1^r x_3 - a_3x_2^r \\ \dot{x}_2^i = a_2x_1^i - x_1^i x_3 - a_3x_2^i \\ \dot{x}_3 = -a_4x_3 + (x_1^r x_2^r + x_1^i x_2^i) \end{cases} \tag{4}$$

We add the controller to system (2) and obtain system (5)

$$\begin{cases} \dot{u}_1 = a(u_3 - u_1) + u_3u_5 + u_6 + v_1 \\ \dot{u}_2 = a(u_4 - u_2) + u_4u_5 + v_2 \\ \dot{u}_3 = cu_2 - u_3 - u_1u_5 + u_6 + v_3 \\ \dot{u}_4 = cu_2 - u_4 - u_2u_5 + v_4 \\ \dot{u}_5 = u_1u_3 + u_2u_4 - bu_5 + v_5 \\ \dot{u}_6 = u_1u_3 + u_2u_4 - du_6 + v_6 \end{cases} \tag{5}$$

where $V = (v_1, v_2, v_3, v_4, v_5, v_6)^T$ is the designed controller vector. We obtain the following Theorem 1.

Theorem 1. *As for system (5), if we design the controller as system (6),*

$$\begin{cases} v_1 = \dot{r}_1 + r_1 + (a-1)u_1 - au_3 - u_3u_5 - u_6 \\ v_2 = \dot{r}_1 + r_1 + (a-1)u_2 - au_4 - u_4u_5 \\ v_3 = \dot{r}_2 + r_2 - cu_2 + u_1u_5 - u_6 \\ v_4 = \dot{r}_3 + r_3 - cu_2 + u_2u_5 \\ v_5 = \dot{r}_2 + r_2 - u_1u_3 - u_2u_4 + (b-1)u_5 \\ v_6 = \dot{r}_2 + r_2 - u_1u_3 - u_2u_4 + (d-1)u_6 \end{cases} \tag{6}$$

then the state variable u_1 can track sine function, the state variable u_2 can track constant 5, and the state variable can track x_2^i of the complex Lorenz chaotic system.

Proof: Set the error as e_1, e_2, e_3, then

$$e_1 = u_1 - r_1, e_2 = u_3 - r_2, e_3 = u_4 - r_3 \tag{7}$$

The expected goals are expressed as follows,

$$\begin{cases} r_1 = \sin t \\ r_2 = 5 \\ r_3 = x_2^i \end{cases} \Rightarrow \begin{cases} \dot{r}_1 = \cos t \\ \dot{r}_2 = 0 \\ \dot{r}_3 = a_2x_1^i - a_3x_2^i - x_1^i x_3 \end{cases} \tag{8}$$

Select Lyapunov function $V = \frac{1}{2}(e_1 + e_2 + e_3)^2 > 0$. Substituting (5)–(8) into Lyapunov function V, we can obtain

$$\begin{aligned} \dot{V} &= e_1\dot{e}_1 + e_2\dot{e}_2 + e_3\dot{e}_3 \\ &= (u_1 - r_1)(u_1 - r_1)' + (u_3 - r_2)(u_3 - r_2)' + (u_4 - r_3)(u_4 - r_3)' \\ &= (u_1 - \sin t)(\sin t - u_1) + (u_3 - r_2)(5 - u_3) \\ &\quad + (u_4 - x_2^i)\left(cu_2 - u_4 - u_2u_5 + v_4 - a_2x_1^i + a_3x_2^i + x_1^i x_3\right) \\ &= (u_1 - \sin t)(\sin t - u_1) + (u_3 - 5)(5 - u_3) + (u_4 - x_2^i)(x_2^i - u_4) \\ &= -(u_1 - \sin t)^2 - (u_3 - 5)^2 - (u_4 - x_2^i)^2 < 0 \end{aligned}$$

According to the Lyapunov stability theorem, the error of tracking control approaches 0, and the proof is completed. □

The tracking results are shown in Figures 2–4.

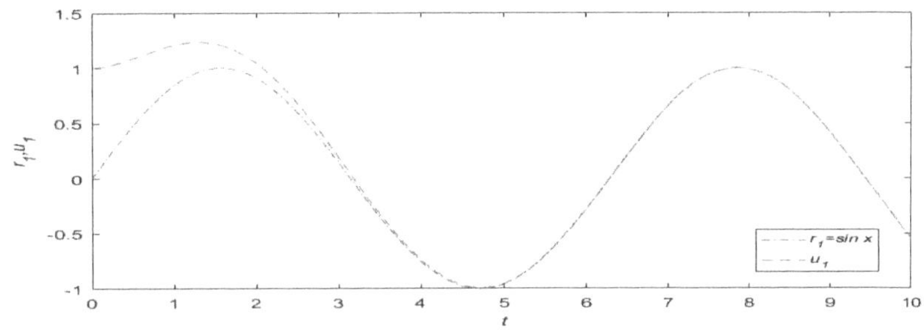

Figure 2. Tracking sine function.

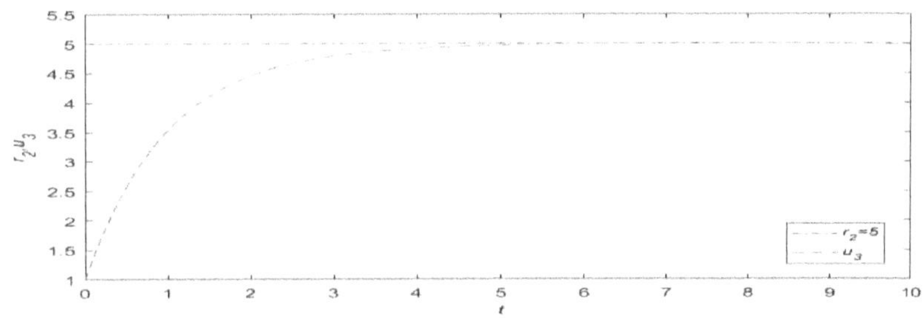

Figure 3. Tracking constant 5.

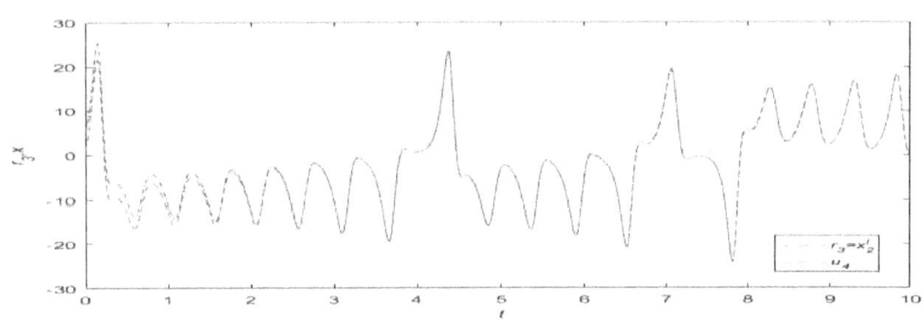

Figure 4. Tracking the fourth dimension of the complex Lorenz chaotic system.

The simulation results are consistent with the theoretical analysis, and the effectiveness of the controller is verified from two aspects of simulation and mathematical analysis.

4. Fractional-Order Generalization

In order to make the beneficial chaotic behavior in industrial process more complicated, this paper extended the hyperchaotic complex system to fractional order, which is a useful and simple method to enhance the beneficial chaotic behaviors.

4.1. Mathematical Background

Fractional order refers to any order of calculus. In a sense, fractional calculus is a generalized form of integer calculus. For fractional calculus, there are three main definitions: Grunwald Letnikov definition, Riemanu Liouville definition, and Caputo definition. Since Caputo definition includes initial conditions and initial values, Caputo calculus is considered in engineering calculations. In this paper, we chose the Caputo definition.

Definition 1. *Caputo fractional differential form is as follows:*

$$ {}_0^C D_t^{\alpha} f(t) = \frac{d^m}{dt^m} J_{m-\alpha} = \begin{cases} \frac{1}{\Gamma(m-\alpha)} \int_0^t \frac{f(m)(\tau)}{(t-\tau)^{\alpha-m+1}} d\tau, & m-1 < \alpha < m \\ \frac{d^m}{dt^m} f(t), & \alpha = m \end{cases} \tag{9} $$

where $\alpha = [m]+1$, $[m]$ *is the integer part of* m, $\Gamma(*)$ *is the gamma function, and* D_t^{α} *is the gamma function* α *order differential operator. In this paper,* D_*^{α} *is used to represent* ${}_0^C D_*^{\alpha}$ *, and we mainly consider the case of* $0 < \alpha < 1$.

4.2. Fractional-Order System Model

In this section, the hyperchaotic complex system is extended to fractional order, and the following fractional-order new hyperchaotic complex system is constructed.

$$ \begin{cases} D_*^{\alpha_1} x_1 = a(x_2 - x_1) + x_2 x_3 + x_4 \\ D_*^{\alpha_2} x_2 = c x_1 - x_2 - x_1 x_3 + x_4 \\ D_*^{\alpha_3} x_3 = 1/2(\overline{x}_1 x_2 + x_1 \overline{x}_2) - b x_3 \\ D_*^{\alpha_4} x_4 = 1/2(\overline{x}_1 x_2 + x_1 \overline{x}_2) - d x_4 \end{cases} \tag{10} $$

where $D_*^{\alpha_l}$ is the Caputo operation of order α_l, and α_l is the order of relevant variables of $x_l (l = 1,2,3,4)$. According to fractional linear operation, we can obtain $D_*^{\alpha_1} x_1 = D_*^{\alpha_1}(u_1 + ju_2) = D_*^{\alpha_1} u_1 + jD_*^{\alpha_1} u_2, D_*^{\alpha_2} x_2 = D_*^{\alpha_2}(u_3 + ju_4) = D_*^{\alpha_2} u_3 + jD_*^{\alpha_2} u_4$ $D_*^{\alpha_3} x_3 = D_*^{\alpha_3} u_5, D_*^{\alpha_4} x_4 = D_*^{\alpha_4} u_6$.

The above system can be transformed into the following forms:

$$ \begin{cases} D_*^{\alpha_1} u_1 = a(u_3 - u_1) + u_3 u_5 + u_6 \\ D_*^{\alpha_1} u_2 = a(u_4 - u_2) + u_4 u_5 \\ D_*^{\alpha_2} u_3 = c u_1 - u_3 - u_1 u_5 + u_6 \\ D_*^{\alpha_2} u_4 = c u_2 - u_4 - u_2 u_5 \\ D_*^{\alpha_3} u_5 = u_1 u_3 + u_2 u_4 - b u_5 \\ D_*^{\alpha_4} u_6 = u_1 u_3 + u_2 u_4 - d u_6 \end{cases} \tag{11} $$

4.3. Fractional-Order Attractor

We select the initial value of system (11) as $(1,1,1,1,1,1)$, $a = 10, b = 8/3, c = 30, d = 12$, and fractional-order $\alpha_l = 0.95 (l = 1,2,3,4)$. On the basis of the definition of Caputo, the system (11) is simulated by MATLAB, and the attractor phase diagram of system (11) is obtained as shown in Figure 5. It can be seen that the attractor of the system presents obvious chaotic characteristics. Comparing Figure 5 with Figure 1, the interval and shape of the attractor are found to be different.

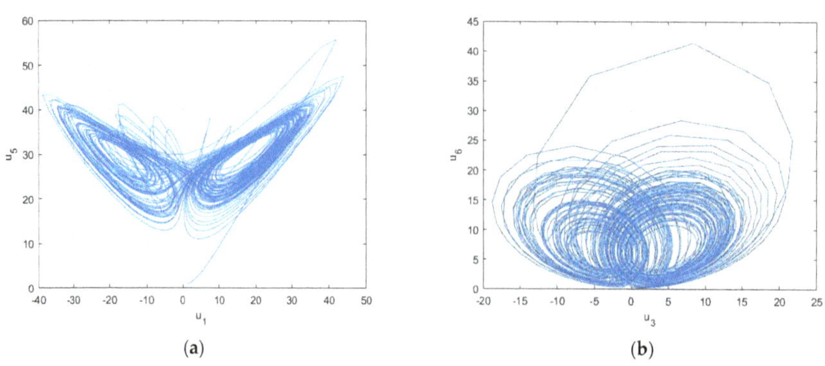

Figure 5. Attractor phase diagram of system (11). (**a**) $u_1 - u_5$ phase diagram of system (11); (**b**) $u_3 - u_6$ phase diagram of system (11).

4.4. 0-1 Test

Gottwald and Melbourne proposed a reliable and effective binary test method to test whether the system is chaotic, which is called the "0-1 test" [28]. The basic idea is to establish a stochastic dynamic process for data and then study how the scale of the stochastic process changes with time. Next, we used this method to test and analyze the chaotic characteristics of system (11), as shown in Figure 6.

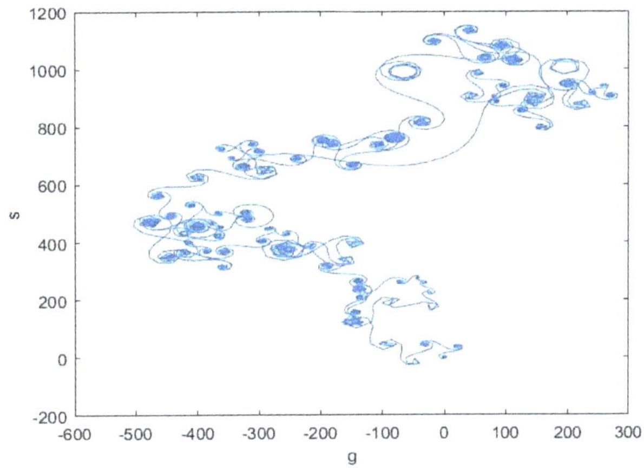

Figure 6. The "0-1"test.

It can be seen from Figure 6 that the new hyperchaotic complex system shows obvious unbounded motion, similar to Brownian motion. Therefore, it is chaotic.

4.5. Order α Impact on System Status

In this section, we all took $u_1 - u_5$ to observe the evolution of the attractor of the system (11).

When $\alpha < 0.82$, the system is in a divergent state.

When $\alpha \in (0.82, 0.94)$, the system converges to a stable point.

When $\alpha = 0.95$, the system changes from stable point to chaotic state, and at this time, it presents obvious butterfly attractor shape.

When $\alpha = 0.99$, the system is still in chaos, but under the same number of cycles, the shape of the attractor is fuller.

When $\alpha = 1.01$, the order of the system is a fractional order greater than 1. At this time, the system is still in a chaotic state, but the attractor forms are different and sparse.

When $\alpha = 1.03$, the system is in a chaotic state, and the attractor shape becomes fuller with the increase in order.

When $\alpha = 1.05$, the system is in period doubling limit cycle state.

When $\alpha > 1.07$, system divergence occurs.

The detailed evolution process is shown in Figure 7. Through observation, it is found that the fractional complex hyperchaotic system shows a more complex system evolution process with the change of order, such as the position and shape of the attractor having changed to some extent.

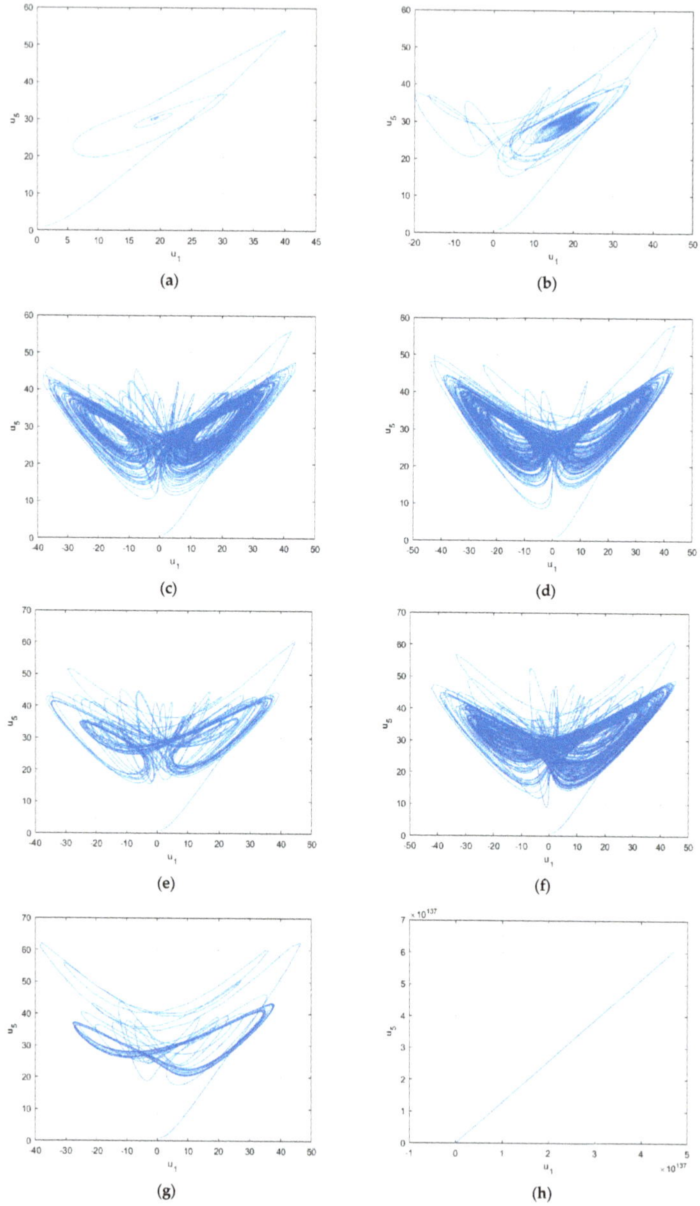

Figure 7. The evolution process of system (11) with order. (**a**) $\alpha_{1(l=1,2,3,4)} = 0.82$; (**b**) $\alpha_{1(l=1,2,3,4)} = 0.94$; (**c**) $\alpha_{1(l=1,2,3,4)} = 0.95$; (**d**) $\alpha_{1(l=1,2,3,4)} = 0.99$; (**e**) $\alpha_{1(l=1,2,3,4)} = 1.01$; (**f**) $\alpha_{1(l=1,2,3,4)} = 1.03$; (**g**) $\alpha_{1(l=1,2,3,4)} = 1.05$; (**h**) $\alpha_{1(l=1,2,3,4)} = 1.07$.

4.6. Influence of Parameter Change on System Attractor

It can be seen from Section 4.4 that when order $\alpha = 0.95$, the system attractor presents an obvious chaotic attractor form. In this section, we selected order $\alpha = 0.95$ and changed the values of system parameters a, b, c, and d individually to observe the influence of parameters on the system state and system attractor. In this section, we all took the phase diagram of $u_1 - u_5$ to observe the evolution of the attractor of the system.

4.6.1. Parameter *a* Change

We kept parameters *b*, *c*, and *d* unchanged; changed the value of parameter *a*; and selected the same initial value as the integer order to observe the evolution process of the system with parameter *a*, as shown in Figure 8. It can be seen that the system entered the chaotic state from the limit cycle state and continued to evolve in the chaotic state. The attractor changed from sparse to full and then to sparse, and then returned to the limit cycle state and finally diverged.

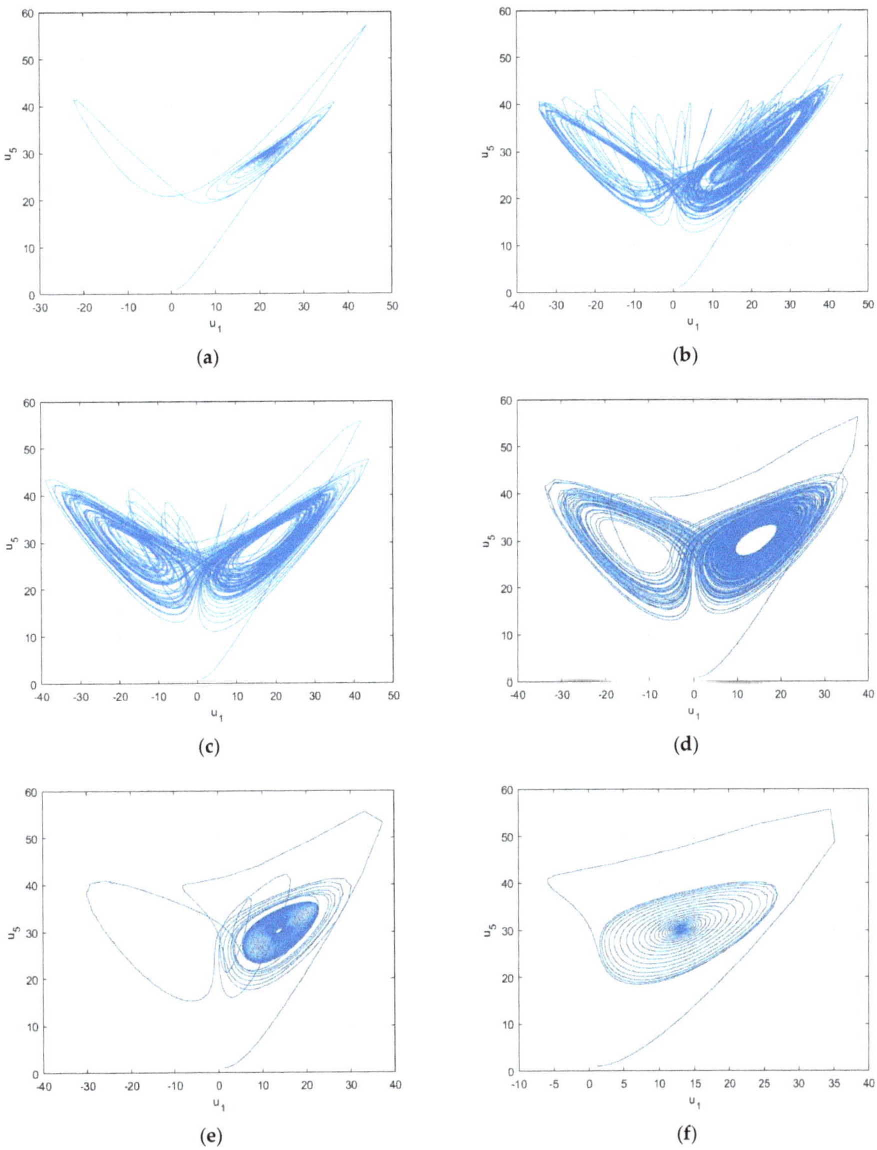

Figure 8. The evolution process of system (11) with parameter *a*. (**a**) $a = 6$, limit cycle; (**b**) $a = 8$, chaos; (**c**) $a = 10$, chaos; (**d**) $a = 20$, chaos; (**e**) $a = 23$, chaos; (**f**) $a = 28$, limit cycle.

4.6.2. Parameter *b* Change

We kept the parameters *a*, *c*, and *d* unchanged; changed the value of parameter *b*; and selected the same initial value as the integer order to observe the evolution process of the system with the change of parameter *b*, as shown in Figure 9. It can be seen that the system entered the chaotic state from the limit cycle state and continued to evolve in the chaotic state. The attractor changed from sparse to full, then returned to the limit cycle state and finally diverged.

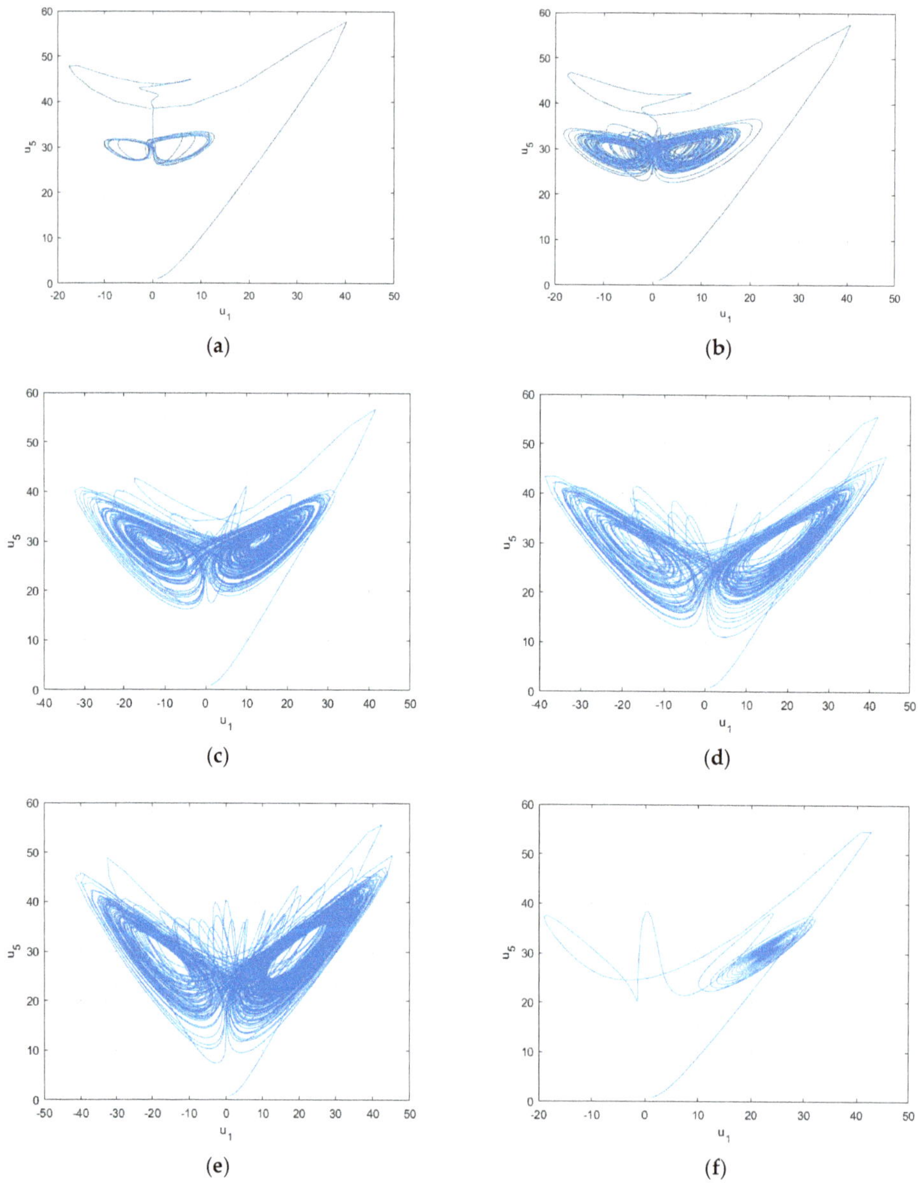

Figure 9. The evolution process of system (11) with parameter *b*. (**a**) $b = 0.2$, limit cycle; (**b**) $b = 0.5$, chaos; (**c**) $b = 1.5$, chaos; (**d**) $b = 8/3$, chaos; (**e**) $b = 3$, chaos; (**f**) $b = 4$, limit cycle.

4.6.3. Parameter c Change

We kept parameters a, b, and d unchanged; changed the value of parameter c; and selected the same initial value as the integer order to observe the evolution process of the system with the change of parameter c, as shown in Figure 10. It can be seen that the system entered the chaotic state from the limit cycle state, evolved continuously in the chaotic state, and finally returned to the limit cycle state and diverged.

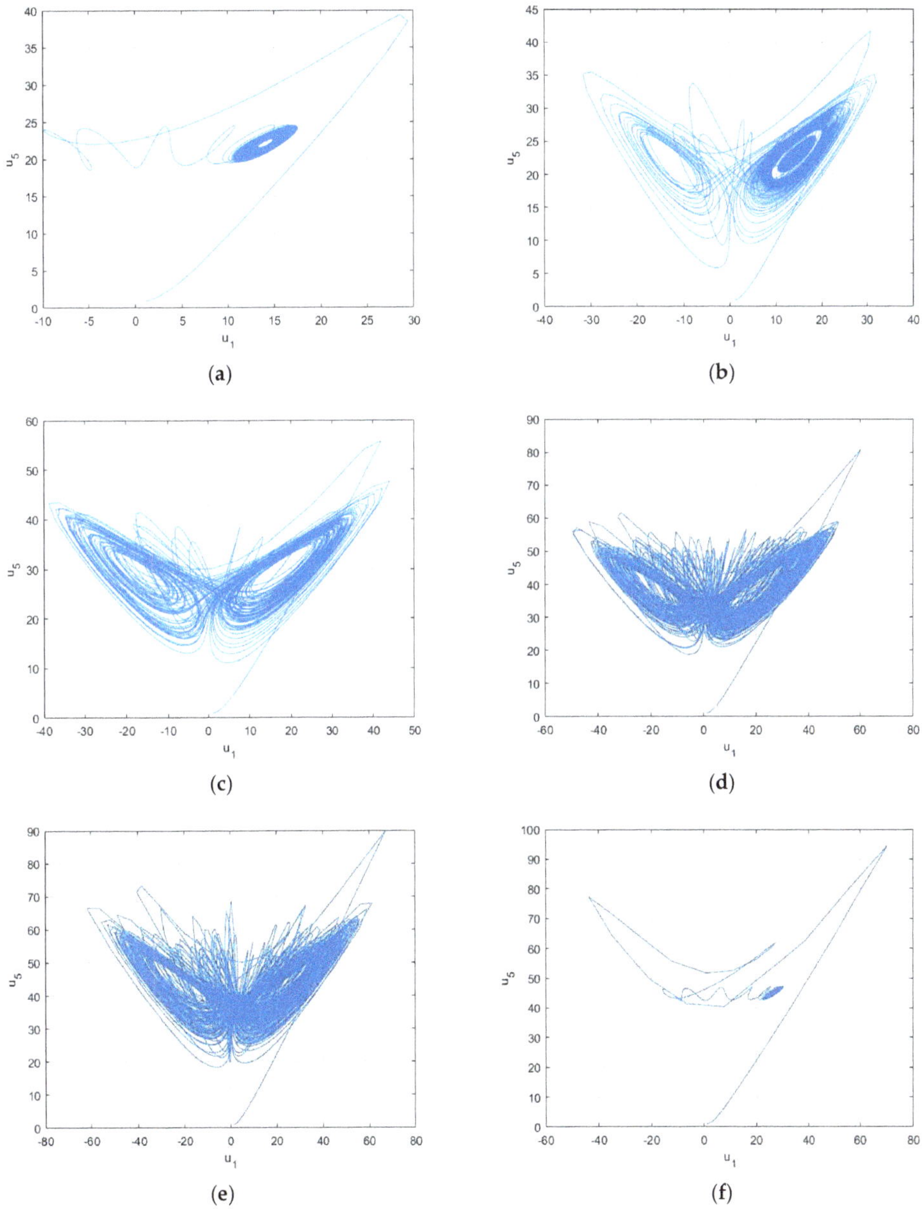

Figure 10. The evolution process of system (11) with parameter c. (**a**) $c = 22$, limit cycle; (**b**) $c = 23$, chaos; (**c**) $c = 30$, chaos; (**d**) $c = 40$, chaos; (**e**) $c = 44$, chaos; (**f**) $c = 44.8$, limit cycle.

4.6.4. Parameter d Change

We kept parameters a, b, and c unchanged; changed the value of parameter d; and selected the same initial value as the integer order to observe the evolution process of the system with parameter d, as shown in Figure 11. It can be seen that the system entered the chaotic state from the limit cycle state and evolved continuously in the chaotic state. It can be seen that when the system was in the chaotic state, compared with parameters a, b, c, parameter d had the largest value range, and finally the system returned to the limit cycle state and finally diverged.

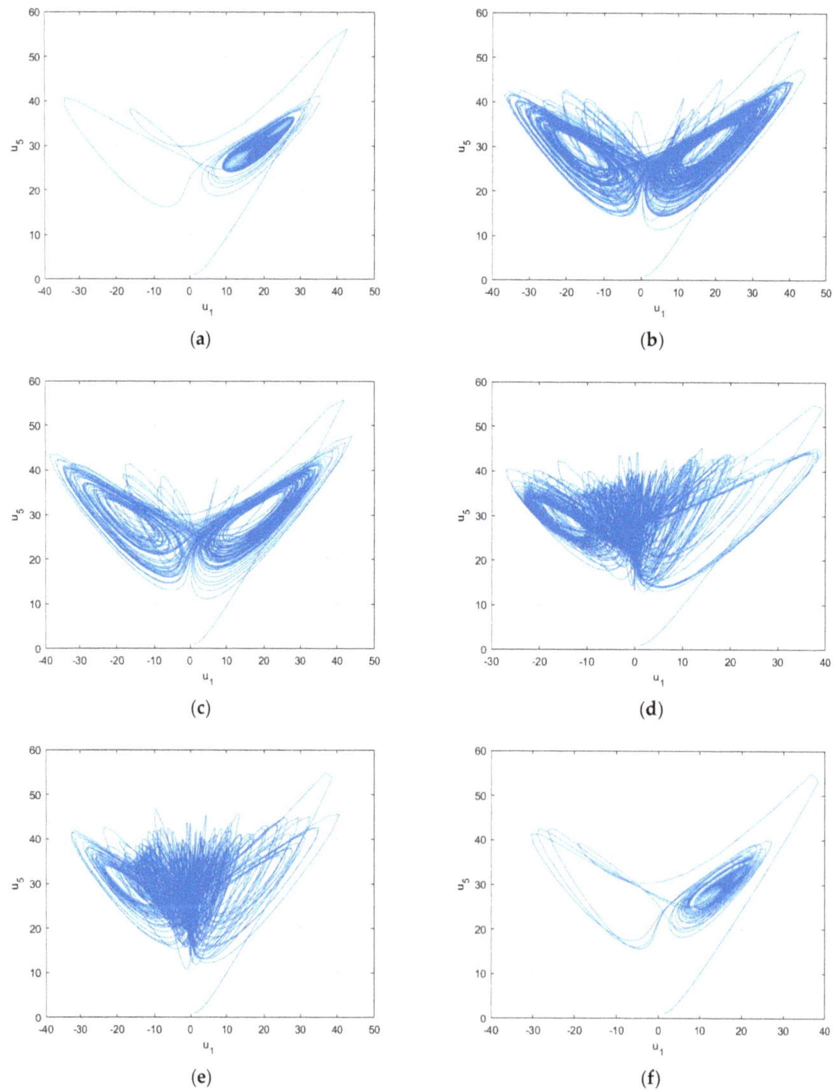

Figure 11. The evolution process of system (11) with parameter d. (**a**) $d = 6$, limit cycle; (**b**) $d = 9$, chaos; (**c**) $d = 10$, chaos; (**d**) $d = 60$, chaos; (**e**) $d = 100$, chaos; (**f**) $d = 151.6$, limit cycle.

It was found that the fractional hyperchaotic complex system showed a more complex system evolution process with the change of parameters, such as the position and shape of the attractor having changed to some extent.

5. Conclusions

In this paper, aiming at the harmful hyperchaotic complex behavior in industrial process, a tracking controller was designed for the hyperchaotic complex system so that the three state variables of the hyperchaotic complex system can track the controller of sine function, constant 5, and complex Lorenz chaotic system individually, and its stability was proven to realize the tracking control of the system. In order to make the intentional chaotic behavior in the industrial process more complex, we extended the hyperchaotic complex system to fractional order. The effects of initial value, order, and parameter changes on the fractional hyperchaotic complex system were discussed and studied. When the order and parameters changed, the detailed evolution process of the system state was given. It was found that there were no coexistence attractors and parameter attractors in the system.

In this paper, the application of the hyperchaotic complex system was studied. For the harmful chaotic system, the controller was designed to convert it into the desired system; for the beneficial chaotic system, this paper extended it to fractional order, which made its chaotic behavior more complex.

There are several prospects for the study of chaos theory: (1) research on the physical background of the chaotic system, or it can show more abundant dynamic behavior; (2) chaos theory can be applied to some complex systems, such as weather forecasting and industrial processes; (3) on the basis of chaotic systems, new chaotic cryptographic algorithms or chaotic neural networks can be formed and applied in various fields.

Author Contributions: Conceptualization, F.L., Z.L. and F.Z.; methodology, Z.L. and L.L.; software, F.L., Z.L. and S.Z.; validation, F.L., L.L. and Z.L.; formal analysis, F.L. and L.L.; investigation, F.L. and F.Z.; resources, F.L. and F.Z.; data curation, F.L. and Z.L.; writing—original draft preparation, F.L., L.L. and Z.L.; writing—review and editing, F.L., Z.L. and F.Z.; visualization, F.L., L.L., Z.L. and S.Z.; supervision, F.Z.; project administration, F.Z.; funding acquisition, F.L., Z.L. and F.Z. All authors have read and agreed to the published version of the manuscript.

Funding: This work is in part supported by the Shandong Provincial Natural Science Fund (2022HWY Q 081), the National Science Foundation of China Key Project (U21A20488), Academic Promotion Project of Shandong First Medical University, the International Collaborative Research Project of Qilu University of Technology (QLUTGJHZ2018020), and major scientific and technological innovation projects of Shandong Province under grant number (2019JZZY010731 and 2020CXGC010901).

Institutional Review Board Statement: Not applicable.

Informed Consent Statement: Not applicable.

Data Availability Statement: Not applicable.

Conflicts of Interest: The authors declare no conflict of interest.

References

1. Zhang, F.; Gao, R.; Huang, Z.; Jiang, C.; Chen, Y.; Zhang, H. Complex modified projective difference function synchronization of coupled complex Chaotic systems for secure communication in WSNs. *Mathematics* **2022**, *10*, 1202. [CrossRef]
2. Gao, C.; Qian, J. Evidence of chaotic behavior in noise from industrial process. *IEEE Trans. Signal Process.* **2007**, *55*, 2877–2884. [CrossRef]
3. Hassan, M.H.; Kamel, S.; Salih, S.Q.; Khurshaid, T.; Ebeed, M. Developing chaotic artificial ecosystem-based optimization algorithm for combined economic emission dispatch. *IEEE Access* **2021**, *9*, 51146–51165. [CrossRef]
4. Mahmoud, G.M.; Mahmoud, E.E.; Ahmed, M.E. On the hyperchaotic complex Lü system. *Nonlinear Dyn.* **2009**, *58*, 725–738. [CrossRef]
5. Mahmoud, E.E. Dynamics and synchronization of new hyperchaotic complex Lorenz system. *Math. Comput. Model.* **2012**, *55*, 1951–1962. [CrossRef]
6. Jiang, C.; Liu, S. Generalized combination complex synchronization of new hyperchaotic complex Lü-like systems. *Adv. Differ. Equ.* **2015**, *2015*, 214. [CrossRef]

7. Li, Z.; Zhang, F.; Zhang, X.; Zhao, Y. A new hyperchaotic complex system and its synchronization realization. *Phys. Scr.* **2021**, *96*, 45208. [CrossRef]
8. Zhang, F.; Li, Z.; Sun, K.; Zhang, X.; Ji, P. A new hyperchaotic complex system with parametric attractors. *Fractals* **2021**, *29*, 2150230. [CrossRef]
9. Azimi, S.; Panchal, C.; Mizera, A.; Petre, I. Multi-Stability, Limit Cycles, and Period-Doubling Bifurcation with Reaction Systems. *Int. J. Found. Comput. Sci.* **2018**, *28*, 1007–1020. [CrossRef]
10. Gao, Z.; Kong, D.; Gao, C. Modeling and control of complex dynamic systems: Applied mathematical aspects. *J. Appl. Math.* **2012**, *2012*, 869792. [CrossRef]
11. Gao, Z.; Nguang, S.K.; Kong, D.X. Advances in Modelling, Monitoring, and Control for Complex Industrial Systems. *Complexity* **2019**, *2019*, 2975083. [CrossRef]
12. Chen, S.H.; Xie, J.; Lu, J.A.; Liu, J. Tracking control and synchronization of the Rossler's chaotic system. *Acta Phys. Sin.* **2002**, *51*, 749–752.
13. Chen, L.; Wang, D.S. Adaptive tracking control of the Chen system. *Acta Phys. Sin.* **2007**, *56*, 5661–5664. [CrossRef]
14. Jiang, Y.; Jin, X.; Wang, H.; Fu, Y.; Ge, W.; Yang, B.; Yu, T. Optimal nonlinear adaptive control for voltage source converters via memetic salp swarm algorithm: Design and hardware implementation. *Processes* **2019**, *7*, 490. [CrossRef]
15. Ji, H.; Liu, S. Position deviation control of drilling machine using a nonlinear adaptive backstepping controller based on a disturbance observer. *Processes* **2021**, *9*, 237. [CrossRef]
16. Yang, S.K.; Chen, C.L.; Yau, H.T. Control of chaos in Lorenz system. *Chaos Solitons Fractals* **2002**, *13*, 767–780. [CrossRef]
17. Fu, Y.; Gao, Z.; Liu, Y.; Zhang, A.; Yin, X. Actuator and sensor fault classification for wind turbine systems based on fast Fourier transform and uncorrelated multi-linear principal component analysis techniques. *Processes* **2020**, *8*, 1066. [CrossRef]
18. Gao, R. A novel track control for Lorenz system with single state feedback. *Chaos Solitons Fractals* **2019**, *122*, 236–244. [CrossRef]
19. Loria, A.; Zavala-Rio, A. Adaptive tracking control of chaotic systems with applications to synchronization. *IEEE Trans. Circuits Syst. I Regul. Pap.* **2007**, *54*, 2019–2029. [CrossRef]
20. Zhang, F.; Sun, K.; Chen, Y.; Zhang, H.; Jiang, C. Parameters identification and adaptive tracking control of uncertain complex-variable chaotic systems with complex parameters. *Nonlinear Dyn.* **2019**, *95*, 3161–3176. [CrossRef]
21. Chaudhary, H.; Sajid, M. Controlling hyperchaos in non-identical systems using active controlled hybrid projective combination-combination synchronization technique. *J. Math. Comput. Sci.* **2021**, *12*, 30.
22. Abbasi, W.; Liu, Y.C. Robust and resilient stabilization and tracking control for chaotic dynamical systems with uncertainties. *Int. J. Dyn. Control.* **2021**, *9*, 1746–1756. [CrossRef]
23. Zhao, L.D.; Hu, J.B.; Liu, X.H. Adaptive tracking control and synchronization of fractional hyper-chaotic Lorenz system with unknown parameters. *Acta Phys. Sin.* **2010**, *59*, 2305–2309. [CrossRef]
24. Nagy, A.M.; Makhlouf, A.B.; Alsenafi, A.; Alazemi, F. Combination Synchronization of Fractional Systems Involving the Caputo–Hadamard Derivative. *Mathematics* **2021**, *9*, 2781. [CrossRef]
25. Ma, C.; Mou, J.; Liu, J.; Yang, F.; Yan, H.; Zhao, X. Coexistence of multiple attractors for an incommensurate fractional-order chaotic system. *Eur. Phys. J. Plus* **2020**, *135*, 95. [CrossRef]
26. Xiong, P.Y.; Jahanshahi, H.; Alcaraz, R.; Chu, Y.M.; Gómez-Aguilar, J.F.; Alsaadi, F.E. Spectral Entropy Analysis and Synchronization of a Multi-Stable Fractional-Order Chaotic System using a Novel Neural Network-Based Chattering-Free Sliding Mode Technique. *Chaos Solitons Fractals* **2021**, *144*, 110576. [CrossRef]
27. Rahman, Z.A.S.; Jassim, B.H.; Al-Yasir, Y.I. New Fractional-Order Chaotic System: Analysis, Synchronization, and it's Application. *Iraqi J. Electr. Electron. Eng.* **2021**, *17*, 116–123. [CrossRef]
28. Gottwald, G.A.; Melbourne, I. A new test for chaos in deterministic systems. *Proc. R. Soc. London. Ser. A Math. Phys. Eng. Sci.* **2004**, *460*, 603–611. [CrossRef]

processes

Article

Performance Improvement of H8 Transformerless Grid-Tied Inverter Using Model Predictive Control Considering a Weak Grid

Sherif A. Zaid [1,2,*], Hani Albalawi [1,2], Hossam AbdelMeguid [3], Tareq A. Alhmiedat [4] and Abualkasim Bakeer [5]

1 Electrical Engineering Department, Faculty of Engineering, University of Tabuk, Tabuk 47913, Saudi Arabia; halbala@ut.edu.sa
2 Renewable Energy & Energy Efficiency Centre (REEEC), University of Tabuk, Tabuk 47913, Saudi Arabia
3 Mechanical Engineering Department, Faculty of Engineering, University of Tabuk, Tabuk 47913, Saudi Arabia; habdelmeguid@ut.edu.sa
4 Computer Science Department, Faculty of Computer & Information Technology, Industrial Innovation and Robotics Center, University of Tabuk, Tabuk 47913, Saudi Arabia; t.alhmiedat@ut.edu.sa
5 Department of Electrical Engineering, Faculty of Engineering, Aswan University, Aswan 81542, Egypt; abualkasim.bakeer@aswu.edu.eg
* Correspondence: shfaraj@ut.edu.sa

Abstract: There is increasing utilization of photovoltaic (PV) grid-connected systems in modern power networks. Currently, PV grid-connected systems utilize transformerless inverters that have the advantages of being low cost, low weight, a small size, and highly efficient. Unfortunately, these inverters have an earth leakage current problem due to the absence of galvanic isolation. This phenomenon represents safety and electrical problems for those systems. Recently, the H8 transformerless inverter was introduced to eliminate the earth leakage current. The present study proposes improving the performance of an H8 transformerless inverter using model predictive control (MPC). The inverter was supplied by PV energy and attached to the grid through an LCL filter. During system modeling, the grid weakness was identified. The discrete model of the overall system, including the PV panel, the boost converter, the H8 transformerless inverter, and the controllers, was derived. Then, the introduced H8 transformerless inverter system was simulated and analyzed by the Matlab/Simulink program. The proposed system response using MPC was tested under step disturbances in the PV insolation level. Moreover, the effect of the weak and strong grid operations was considered. The simulation results indicate that the MPC controller has better performance and high-quality injected power. Despite the excellent performance of the strong grid, the nearly weak grid performance is acceptable. Moreover, the Hardware-in-the-Loop (HIL) of the proposed system was implemented using the DSP target LaunchPadXLTMS320F28379D kit to validate the simulation results. Finally, the system performance under the parameter variations showed good robustness.

Keywords: photovoltaic; leakage current; common-mode voltage; model predictive control; transformerless inverter; H8; weak grid

Citation: Zaid, S.A.; Albalawi, H.; AbdelMeguid, H.; Alhmiedat, T.A.; Bakeer, A. Performance Improvement of H8 Transformerless Grid-Tied Inverter Using Model Predictive Control Considering a Weak Grid. *Processes* **2022**, *10*, 1243. https://doi.org/10.3390/pr10071243

Academic Editor: Zhiwei Gao

Received: 18 May 2022
Accepted: 18 June 2022
Published: 22 June 2022

1. Introduction

Photovoltaic (PV) energy has become one of the most important energy resources in the world. For most countries, solar energy is available in large amounts without being exhausted. Moreover, it has many merits, such as being noise-free, reliable, a long-life, maintenance-free, and it is clean energy [1]. In recent years, the cost of PV systems has decreased to a level that enables their spread worldwide. In addition, rapid advancements in the power electronics field have aided in the development of grid-tied PV systems, which have the advantages of no storage units, good efficiency, and better cost [2].

Grid-tied PV systems have two categories: transformer or transformerless systems. Nevertheless, transformerless grid-connected PV systems are preferred due to their small size, low cost, and high efficiency [3–5]. Usually, transformerless grid-tied PV systems have an inverter type called a transformerless inverter. These inverters have different topologies and modulation techniques [6]. However, this type of inverter has some problems, such as lack of galvanic isolation with the grid and the earth leakage current not matching the limitation recommended by the standards [7]. The presence of the leakage capacitance of the PV array and the absence of the galvanic isolation with the grid are the general causes of the earth leakage current. For any transformerless inverter topology, the earth leakage current origin is the inverter's Common-Mode Voltage (CMV) variations [8]. The problem-solving direction is to reduce the CMV variations or ideally make them constant [9]. This target can be achieved in two ways: either by modifying the inverter topology or by introducing a modulation scheme that limits the CMV variations.

Many topologies and modulation techniques have been introduced for single-phase transformerless inverters [10]. On the other hand, the three-phase transformerless inverters have a higher leakage current and power rating. Hence, fewer attempts have been made to modulate three-phase transformerless inverters [11]. The first attempts were made by references [12,13], where they introduced many modulation schemes for the conventional three-phase topology. However, they concluded that the modulation techniques were not sufficient to reduce the earth leakage current. Hence, introducing new topologies is an important issue in order to eliminate the earth leakage current.

In the literature review, many topologies of the transformerless inverter have been proposed to reduce the earth leakage current [14–28]. A simple topology that incorporates 4-arms instead of 3-arms was introduced and tested with some modulation schemes [14,15]. Nevertheless, the supplied power quality and the current total harmonic distortion (THD) were low. Moreover, the control system was complicated. The topologies of the multilevel inverters have been adapted to work in transformerless mode with some modifications and modulation techniques [16,17]. However, the number of switches is high, and the overall system efficiency is low. According to references [18–20], the H7 topology is the three-phase version of the single-phase H5 transformerless inverter. There were many modulation schemes suitable for the H7 topology. However, the reduction in the earth's leakage current was limited. A more recent topology named H8 was introduced [21–26]. It consists of the traditional 3-arm transformerless inverter connected to the DC bus via two series power transistors. It may be regarded as the three-phase image of the famous 1-φ H6 transformerless inverter topology. The first proposed H8 inverter topology was created to reduce the Common Mode Voltage (CMV) for electrical drives [21,22]. Then, the idea was used for transformerless PV grid-connected systems. In reference [23], the proposed topology combined the merits of both the AC bypass circuit and DC bypass structures. The results showed a low leakage current and a small THD of the grid current. However, the modulation used was the traditional scheme, and the controller utilized had a low response. The performance of the conventional B6-type voltage source inverter was compared to the H8 topology in [24]. Nevertheless, the efficiency was low, and the system had a poor dynamic response. Reference [25] modified the H8 to obtain zero CMV variations. The technique depended on entering and leaving the zero-voltage vector using a modified configuration and the control scheme. The results of [26] indicated that the H8 topology had better performance than the H7 due to the 50% reduction in CMV amplitude. Hence, the leakage current attenuation improved with the H8 topology. Another version of the H8 inverter called oH8 was proposed [26]. It has been shown that for a given CMV, the parasitic capacitor voltage of oH8, the clamped DC bus version of H8, is higher than that of H8. Hence, the leakage current of oH8 would be higher than for the H8 topology.

Another common problem for grid-connected inverters is the restriction of the grid's weakness. It is recommended in the standards of the distributed generation system that the grid-connected inverters should be tied to power systems that have Short-Circuit Ratio (SCR) >20, which corresponds to a grid impedance of 5% [27,28]. The grid-tied inverters act stably with the power system when the SCR is kept within the previous range. However, the spread of distributed renewable energy resources with long transmission lines causes the utility grid to possess weak grid performance. The power system is considered to be weak if SCR < 3 [29]. Moreover, the voltage at the Point of Common Coupling (PCC) should be kept at a low harmonic content to avoid voltage distortion problems at the PCC. Hence, the grid-tied inverter should be designed to supply current with a high harmonic rejection and satisfy the standard limitations on the injected grid current [30].

Recently, a major focus has been given to the promising control technique, Model Predictive Control (MPC). It has been adapted for the control of the grid tide inverters [31]. Perfect response and simple implementation are the great merits of the MPC controller. It has been applied to control 3-level transformerless inverters to minimize the earth leakage current [32]. However, the proposed system is complex. The MPC control technique has been adapted for neutral point clamped inverters [33]. Though the earth leakage current is reduced, the proposed system efficiency is lower.

In this research paper, a PV energized H8 transformerless inverter that utilizes the MPC scheme as a controller is introduced. A boost converter was attached to the PV terminals to ensure the Maximum Power Point Tracking (MPPT) operation. The output of the boost converter was the DC bus of the power system. It must be controlled to be constant at a certain designed value to achieve whole system stability and ensure power balance. The H8 transformerless inverter was supplied by the DC bus and tied to the grid using an LCL filter to minimize the injected harmonics. As the MPC algorithm relies mainly on the system discrete model, the whole system model was derived. The LCL filter and the grid weakness characteristics were represented by the grid internal impedance determined by the model. Hence, the MPC control scheme was applied to the H8 transformerless inverter. Then, the introduced system was simulated and analyzed using the Matlab/Simulink platform. The proposed system response using the MPC was tested under step disturbances in the PV insolation level. Moreover, the effects of the nearly weak and strong grid operations were considered. The simulation results indicate that the MPC controller has better performance and high-quality injected power. Despite the excellent performance of the strong grid, the nearly weak grid performance is also acceptable. The objectives of this research are to:

- Apply the MPC algorithm to the H8 transformerless inverter.
- Investigate the system discrete model including the LCL filter.
- Discuss the effects of the grid weakness on the system response.
- Study the system response, under the disturbances in the insolation level.
- Test the robustness of the system performance against the parameter variations.
- Implement the proposed system using the HIL validation technique.

The arrangement of the paper is as follows: Section 2 explains the topology and operation of the H8 Transformerless Inverter. Section 3 demonstrates the MPC of the H8 Transformerless Inverter. The system controllers are described in Section 4. Section 5 introduces the weak grid operation of the H8 transformerless inverter. The simulation results are discussed in Section 6, while the paper's conclusions are provided in Section 7.

2. H8 Transformerless Inverter Structure and Operation

Figure 1 presents the power circuit of the introduced system. It was a PV-powered grid-connected H8 transformerless inverter. The PV array represented the power generator for the system. The output of the array was attached to a boost converter. It was used to implement the MPPT conditions for the PV array. The output of the boost converter was the system DC bus that feeds the H8 transformerless inverter. The H8 transformerless inverter may be considered as the 3-φ version of the 1-φ H6 transformerless inverter. Its power circuit had eight switches, as shown in Figure 1. It was introduced to reduce the earth leakage current in the grid-connected systems. The topology of the H8 inverter was formed by adding two additional switches, (Q_7 and Q_8). Their function was to decouple the inverter from the PV during the freewheeling periods of the inverter. Hence, there was no path for an earth leakage current. The H8 transformerless inverter acted as a 3-φ two-level inverter. Therefore, it had eight switching states or voltage vectors ($\overline{V}_0$, $\overline{V}_1$, $\overline{V}_2$,

$\overline{V}_3$, $\overline{V}_4$, $\overline{V}_5$, $\overline{V}_6$, and $\overline{V}_7$). The inverter operating voltage vector depended on the controller objectives. The decoupling switches (Q_7 and Q_8) were on together for all active voltage vectors. Nevertheless, during the null voltage vectors, the decoupling switches (Q_7 and Q_8) were modulated according to the following switching functions [25]:

$$Q_7 = Q_8 = A\overline{B} + B\overline{C} + C\overline{A}, \tag{1}$$

where (A, B, and C) were the logic states of the inverter upper switches.

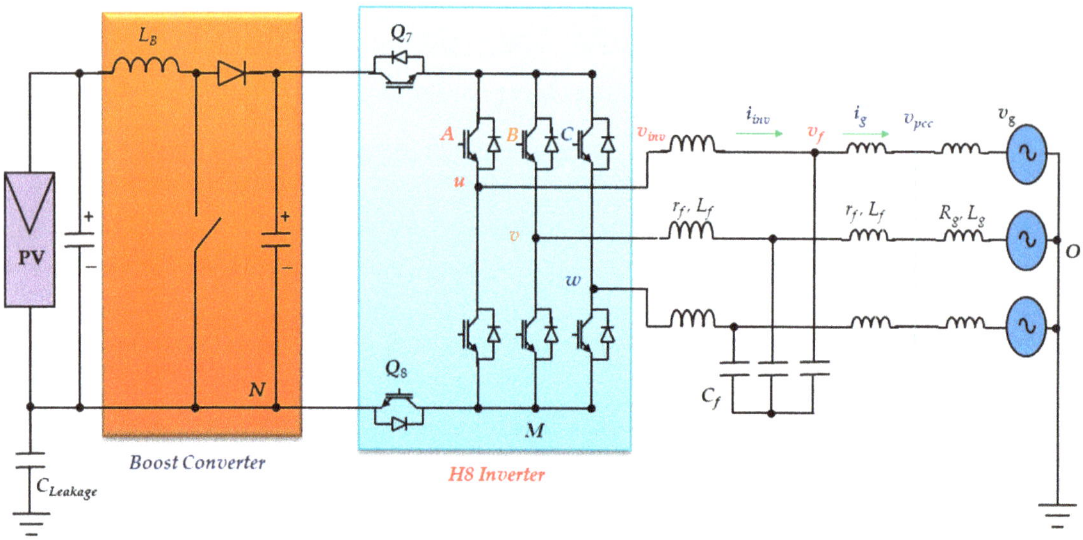

Figure 1. The power circuit of the proposed system.

The configurations of the H8 transformerless inverter for different switching states of the inverter are shown in Figure 2.

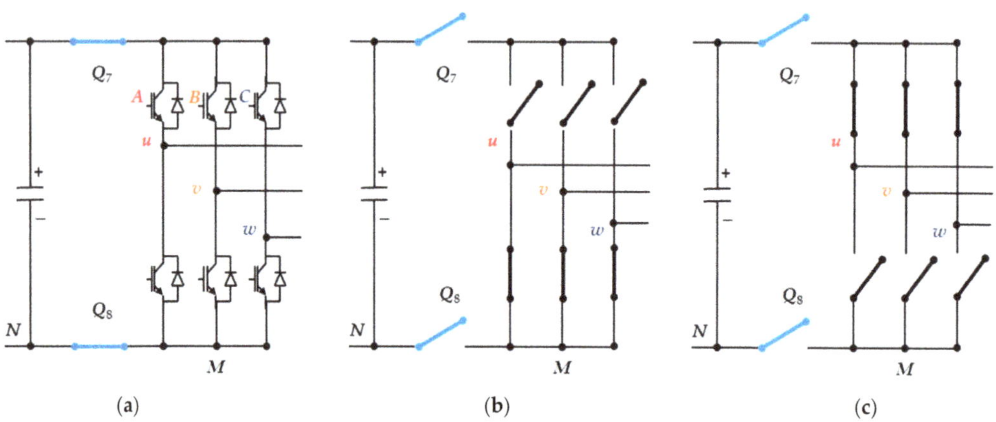

Figure 2. The configurations of the H8 inverter for (**a**) the active states and (**b**,**c**) the zero states.

2.1. CMV Model of the H8 Transformerless Inverter

In this section, the CMV of the H8 inverter is analyzed. The CMV variations of the grid-connected inverters must be minimized or ideally constant. The H8 inverter CMV depends on its terminal voltages (V_{uN}, V_{vN}, and V_{wN},); they can be calculated as follows [20]:

$$V_{CMV} = \frac{V_{uN} + V_{vN} + V_{wN}}{3} \rightarrow V_{CMV} = \frac{V_{uM} + V_{vM} + V_{wM}}{3} + V_{MN}. \quad (2)$$

However, the H8 inverter terminal voltages and the voltage V_{MN} depend on the operating voltage vector or switching state. Based on the switching state, the corresponding CMV value can be determined as shown in Table 1. This can be summarized as follows:

Table 1. The H8 transformerless inverter CMVs.

	$\overline{V}_1$	$\overline{V}_2$	$\overline{V}_3$	$\overline{V}_4$	$\overline{V}_5$	$\overline{V}_6$	$\overline{V}_0$	$\overline{V}_7$
Switching state	(1001)	(0101)	(0011)	(1101)	(0111)	(1011)	(0000)	(1110)
V_{uM}/V_{dc}	1	1	0	0	0	1	0	1
V_{vM}/V_{dc}	0	1	1	1	0	0	0	1
V_{wM}/V_{dc}	0	0	0	1	1	1	0	1
V_{CMV}/V_{dc}	1/3	2/3	1/3	2/3	1/3	2/3	1/3	2/3

- For the active voltage vectors ($\overline{V}_1$, $\overline{V}_3$, and $\overline{V}_5$), the switches Q_7 and Q_8 are on. Hence, the voltage V_{MN} is zero. Hence, the CMV is $V_{dc}/3$.
- For the active voltage vectors ($\overline{V}_2$, $\overline{V}_4$, and $\overline{V}_6$), the switches Q_7 and Q_8 are on. Hence, the voltage V_{MN} is zero. Hence, the CMV is $2V_{dc}/3$.
- For the zero voltage vectors ($\overline{V}_0$, and $\overline{V}_7$), the switches Q_7 and Q_8 are off. Hence, the voltage V_{MN} is not zero and can be determined for each case. Finally, the CMV equals $2V_{dc}/3$ for $\overline{V}_7$ and it equals $V_{dc}/3$ for $\overline{V}_0$.

Therefore, the levels of the CMV of the H8 inverter were limited to ($\frac{1}{3}V_{dc}$ and $\frac{2}{3}V_{dc}$). However, the CMV was limited to ($\frac{1}{3}V_{dc}$, $\frac{2}{3}V_{dc}$, and V_{dc}) for the recent inverter H7 [20]. Consequently, the peak variations of the CMV were reduced, which decreased the leakage current. The idea behind this improvement is the disconnection of the PV from the grid during the freewheeling periods using the switches Q_7 and Q_8.

2.2. Earth Leakage Current Path of the H8 Transformerless Inverter

Figure 3 shows the earth leakage current path the of H8 transformerless inverter with the PV panel [20]. The circuit indicates that the CMV source should be alternating for the earth leakage current to flow. Hence, if the CMV variations remain limited, the serious leakage current will be greatly reduced. The instantaneous CMV is conditional on the switching modulation of the H8 transformerless inverter. Consequently, introducing a new modulation strategy that can reduce the CMV will aid the leakage current reduction. Moreover, the H8 transformerless inverter control technique affects the inverter modulation, which will reduce the leakage current reduction.

The H8 transformerless inverter controller may be either a one-shot controller or a modulator-based controller. In the one-shot controller, the inverter switching states are produced directly from the control algorithm, such as the MPC. However, the modulator-based controller must have a modulator to generate the inverter switching states. Hence, the modulation technique must be selected from the commonly known (SPWM, SVPWM, etc.).

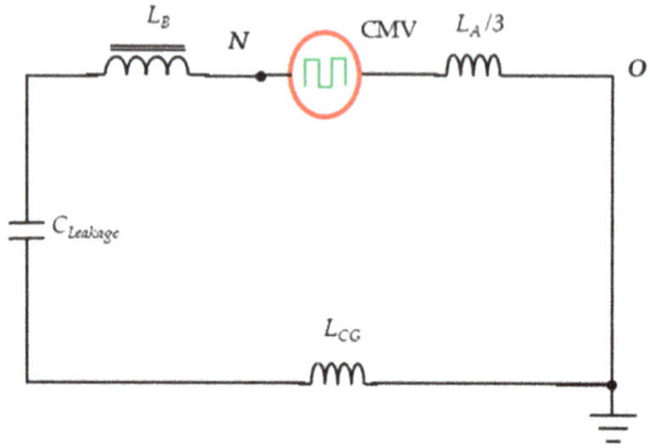

Figure 3. The CMV model of the H8 transformerless inverter.

3. Model Predictive Control of the H8 Transformerless Inverter

The basic structure of the MPC controller for a certain system is based on the system's discrete-model. Hence, the starting point in planning the MPC controller is the investigation of the discrete model of the system. Applying Kirchhoff lows to the inverter output filter shown in Figure 1, we have:

$$\begin{bmatrix} L_{gf}(\frac{d\overline{I}_g}{dt}) \\ L_f(\frac{d\overline{I}_{inv}}{dt}) \\ C_f(\frac{d\overline{V}_C}{dt}) \end{bmatrix} = \begin{bmatrix} \overline{V}_c - \overline{V}_g \\ \overline{V}_{inv} - \overline{V}_C \\ \overline{I}_f - \overline{I}_g \end{bmatrix}, \; L_{gf} = L_g + L_f, \tag{3}$$

where ($\overline{I}_g$, $\overline{V}_g$) are the grid current and voltage space vectors, ($\overline{V}_f$) is the filter capacitor space voltage vector, ($\overline{I}_{inv}$) is the inverter current space vector, ($\overline{V}_{inv}$) is the inverter space vector voltage, (R_g, L_g) is the grid impedance, (r_f, L_f) are the filter inductor parameters, and (C_f) is the filter capacitance.

The state-space form of Equation (3) is:

$$\begin{bmatrix} \frac{d\overline{I}_g}{dt} \\ \frac{d\overline{I}_{inv}}{dt} \\ \frac{d\overline{V}_f}{dt} \end{bmatrix} = \begin{bmatrix} \frac{-(R_g+r_f)}{L_{gf}} & 0 & \frac{1}{L_{gf}} \\ 0 & \frac{-r_f}{L_f} & \frac{-1}{L_f} \\ \frac{-1}{C_f} & \frac{1}{C_f} & 0 \end{bmatrix} \begin{bmatrix} \overline{I}_g \\ \overline{I}_{inv} \\ \overline{V}_f \end{bmatrix} + \begin{bmatrix} 0 \\ \frac{1}{L_f} \\ 0 \end{bmatrix} \overline{V}_{inv} + \begin{bmatrix} \frac{-1}{L_g} \\ 0 \\ 0 \end{bmatrix} \overline{V}_g. \tag{4}$$

This may be simplified as:

$$\frac{d\overline{Y}}{dt} = \alpha\overline{Y} + \beta\overline{V}_{inv} + \gamma\overline{V}_g, \tag{5}$$

where the state vector ($\overline{Y}$) and the constant matrices (α, β, and γ) are defined as:

$$\overline{Y} = \begin{bmatrix} \overline{I}_g \\ \overline{I}_{inv} \\ \overline{V}_f \end{bmatrix}, \; \alpha = \begin{bmatrix} \frac{-(R_g+r_f)}{L_{gf}} & 0 & \frac{1}{L_{gf}} \\ 0 & \frac{-r_f}{L_f} & \frac{-1}{L_f} \\ \frac{-1}{C_f} & \frac{1}{C_f} & 0 \end{bmatrix}, \beta = \begin{bmatrix} 0 \\ \frac{1}{L_f} \\ 0 \end{bmatrix}, \gamma = \begin{bmatrix} \frac{-1}{L_g} \\ 0 \\ 0 \end{bmatrix}. \tag{6}$$

We assumed that the sampling period was (T). The model input was the inverter space voltage vector, and the DC link voltage was considered the disturbance. Hence, the continuous state-space model of Equation (5) can be transformed into the discrete-time state-space model using the backward Euler technique [34]. Although the forward Euler method requires an extra computation at each iteration, the backward Euler method has great stability properties, and its local truncation error is much smaller than using the forward Euler method [35]. Therefore:

$$\overline{Y}(k+1) = e^{\alpha T}\overline{Y}(k) + \int_0^T e^{\alpha\tau}\beta d\tau \overline{V}_{inv} + \int_0^T e^{\alpha\tau}\gamma d\tau \overline{V}_g. \tag{7}$$

Hence, the discrete model of the H8 transformerless inverter was obtained using Equation (7). The algorithm of the MPC utilizes the discrete model to predict the controlled quantities at the next sample. For all possible switching states, a cost function was built and calculated to select the voltage vector that gives the minimum error. The introduced cost function (g) is:

$$g = \left(i_{g\alpha} - i^*_{g\alpha}\right)^2 + \left(i_{g\beta} - i^*_{g\beta}\right)^2, \tag{8}$$

where ($i_{g\alpha}$, $i_{g\beta}$) are the α-β compositions of the grid current, and ($i^*_{g\alpha}$, $i^*_{g\beta}$) are the reference α-β compositions of the grid current.

4. System Controllers

A single-line diagram of the proposed system incorporating the system controllers is shown in Figure 4. It had two main controllers namely the power and the H8 transformerless inverter controllers. The power controller included two controllers called the MPPT and the DC-link voltage controllers. As the set point of the DC link controller was constant, the PID controller would be the best choice due to its high stability and good performance for constant set point systems [36]. However, the set point of the H8 transformerless inverter controller was sinusoidal. Hence, the PI was not suitable, and the MPC controller was adapted. To ensure stability, the MPC response, the controller of the inner loop, should be faster than the DC link controller (PID). These controllers are discussed in detail in the next paragraphs. The whole control system is demonstrated in the following subsections.

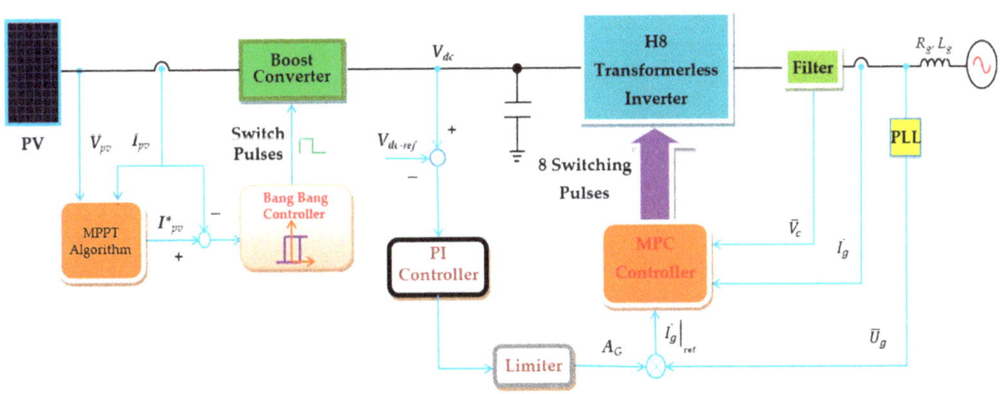

Figure 4. The single line diagram of the proposed system incorporates the system controllers.

4.1. Power Controller

The system incorporated two controllers, the MPPT controller and the DC-link voltage controller. The MPPT controller was utilized to take out the peak power from the PV panel. Therefore, it improved the system utilization. The idea was to control the boost converter input current by regulating the converter duty cycle. The controller type was bang-bang with a ±5% band. The reference current was generated by the MPPT algorithm unit, the incremental conductance [16]. That unit measured the instantaneous PV current and voltage and output the reference current for the MPPT controller.

The DC-link voltage controller forced the V_{dc} to track its reference. This controller regulated the power transfer from the PV to the grid. Its reference voltage value was recommended by the system design values, 650 V for the proposed system [20]. The controller output the reference current for the H8 transformerless inverter. A simple Proportion Integral Derivative (PID) controller was employed. Its output (A_G) and action are represented for discrete operation using:

$$A_G(z) = \left[k_P + k_I \frac{Tz}{z-1} + k_D \frac{z-1}{Tz}\right] e(z), \tag{9}$$

where $e(z)$ is the error signal, and (k_P, k_I, and k_D) are the PID controller gains. Those gains are tuned using the Ziegler-Nichols technique. This tuning algorithm may be implemented using the next sequence [37]:

- Let the integral part be zero and decrease the proportional part to a very small value.
- Increase the proportional part until the output oscillates.
- Measure the period of oscillation (T_s) and the corresponding proportional gain (K_{cp}). Hence, the PI controller gains are calculated using:

$$K_P = 0.35\ K_{cp} \quad \text{and} \quad K_I = 0.8\ K_P/T_s \tag{10}$$

4.2. H8 Transformerless Inverter Controller

The H8 transformerless inverter is a current-controlled voltage source inverter. It has a controller that regulates its output current I_g to be AC, which is compatible with the grid voltage and frequency. Moreover, the output current should be supplied at the unity power factor. This controller utilized the MPC scheme to achieve its goals. The MPC controller output was the optimum switching state for the H8 transformerless inverter. Finally, it produced the H8 inverter switches' pulses. The algorithm of the MPC technique was essentially a discrete control. At startup, the system variables were measured. Then, with the help of the system model, the system variables for the next samples were predicted to optimize the system response. The optimization process for the H8 transformerless inverter was to select the switching state that forced the errors in the controlled variables as close as possible to zero. To accomplish the switching state optimization, an optimization function, sometimes called the cost function, was calculated for each switching state. The switching state that provided the minimum value of the cost function was considered the optimal state and was sent to the converter switches in the following sample. The cost function was adapted to minimize the error in the grid current, which was a 3-φ of zero phase shift with the grid voltage reference generated by a Phase Locked Loop (PLL). The most commonly used PLL for weak grid systems is the Synchronous Reference Frame (SRF) PLL. A block diagram of the SRF PLL is shown in Figure 5a. The PLL measures the grid voltage at the PCC, which may be distorted due to the grid's internal impedance [30]. It generates 3-φ signals ($\overline{U}_g$) synchronized to the ideal grid voltages with unity amplitude. The reference currents ($\overline{I}_g|_{ref}$) for the MPC controller were generated by multiplying the PLL output signal with the reference amplitude (A_G) produced by the DC-link controller output.

$$\overline{I}_g|_{ref}(k) = A_G \overline{U}_g(k), \tag{11}$$

where (k) is the order of the sample. A flowchart for the MPC algorithm is presented in Figure 5b.

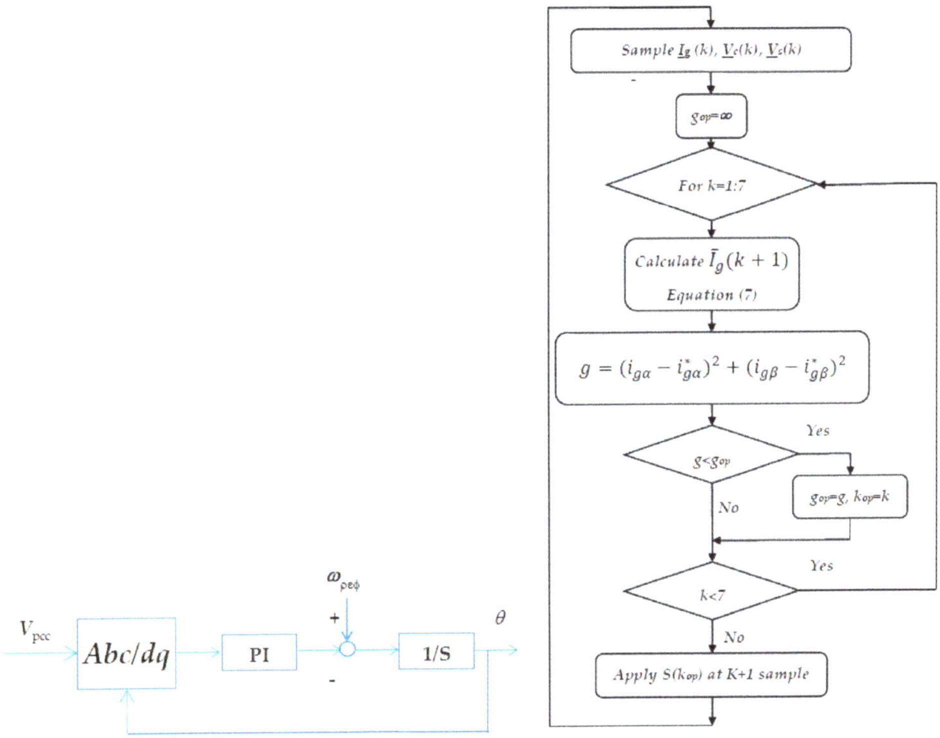

(a) (b)

Figure 5. (**a**) The SRF PLL Block diagram. (**b**) The flowchart of the MPC algorithm for the proposed system.

5. Weak Grid Operation of the H8 Transformerless Inverter

Grid-connected inverters are greatly affected by grid characteristics. The main property that decides its characteristics is the SCR of the grid. Consequently, electrical grids may be classified as strong and weak grids, according to the value of their SCR [30]. Weak grids are characterized by SCR < 3. Grid-connected inverters are recommended to operate stably for grids that have SCR > 20. Many problems have been recorded for the inverters tied to grids that have low SCR values. Instability, the PLL operation disturbed, low power quality, and high distortion at the PCC are some of those problems. To improve the performance of the H8 transformerless inverter when connected to a low SCR grid, a modified PLL was utilized for the proposed system. It has a band-pass filter to assess harmonic attenuation. Moreover, the MPC controller supported the improvement.

6. Simulation Results

Matlab simulations of the proposed grid-connected H8 transformerless inverter supplied by PV were implemented. The system rating and simulation parameters are listed in Table 2. The system photovoltaic array was formed of 6 × 960 cells. Figure 6 shows the response of the proposed system to step variations in the insolation level for the strong grid. These step variations were (100%, 75%, 50%, 30%, and 75%) at the times (0, 0.2 s, 0.4 s, 0.6 s, and 0.8 s). Figure 6a shows the response of V_{pcc}, I_g, V_{dc}, and I_L. V_{pcc} was sinusoidal with constant amplitude and equaled the grid rated values irrespective of the grid current value or the insolation level. This issue is essential for a strong grid. The grid current I_g is also shown. It was also sinusoidal with a unity power factor. However, its amplitude varied according to the MPPT power extracted from the PV. Moreover, the figure showed the V_{dc} response. It had a good response as the maximum overshoot was limited to 1.5%. Finally, the earth leakage current (I_L) response is presented in the figure. It was limited to small values except for some spikes at the disturbance times. However, its maximum RMS value was 100 mA, which is much lower than the standard value [7].

Table 2. Simulation Parameters.

Parameter	Value	Parameter	Value
(I_{SC},V_{OC}) of the PV	24.53 A, 633 V	V_{dc}	650 V
Power	11 KW	DC-link capacitor	2000 μF
$C_{Leakage}$	400 nF	Utility voltage	230 V
C_f	2 μF	Utility frequency	50 Hz
L_{fmmm}	3 mH	Sampling period	33 μs
Weak grid (R_g, L_g)	74.8 mΩ, 238 μH	SCR	5

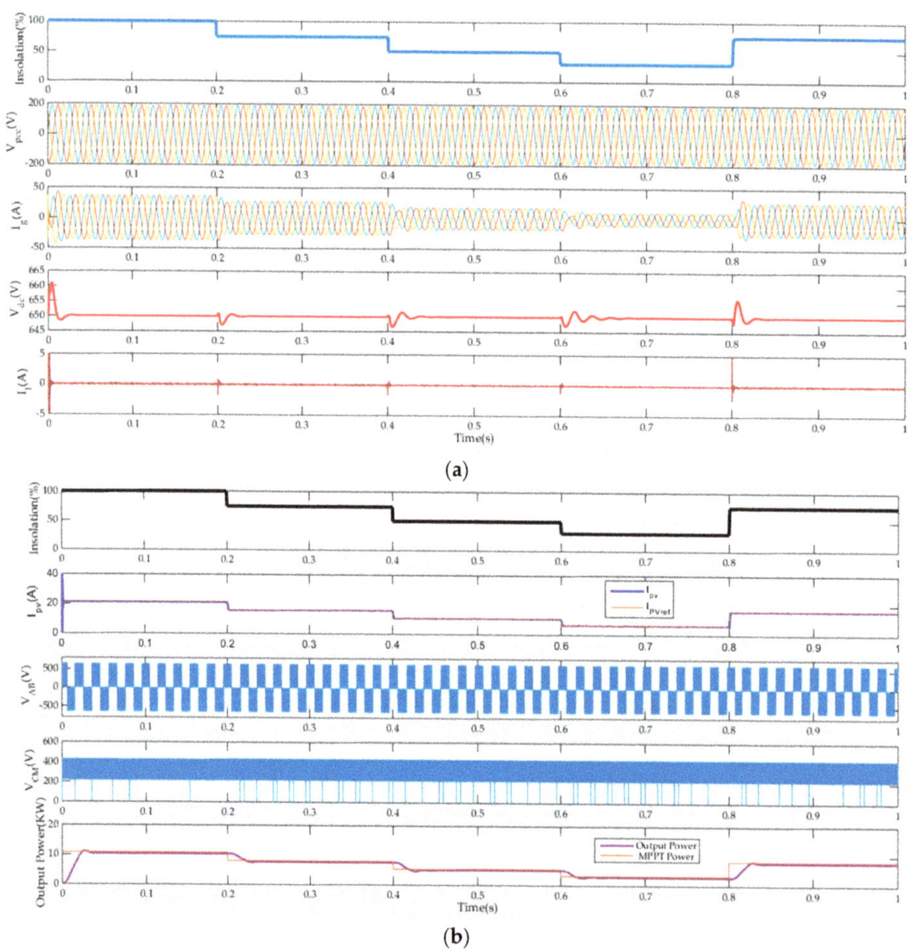

Figure 6. The response of the proposed system to step variations in the isolation level, @ strong-grid, (**a**) V_{pcc}, I_g, V_{dc}, and I_L, (**b**) I_{pv}, V_{AB}, V_{CM}, and the output power.

Figure 6b shows the response of I_{pv}, V_{AB}, V_{CM}, and the output power. The I_{pv} tracked well the reference recommended by the MPPT controller. The H8 inverter output voltage V_{AB} is also demonstrated in the figure. As the conventional 3-φ inverter, its instantaneous value was limited to $\pm V_{dc}$. Further, the levels of the CMV of the H8 inverter were limited to ($\frac{1}{3}V_{dc}$ and $\frac{2}{3}V_{dc}$). However, the output power of the H8 transformerless inverter tracked the MPPT power of the PV with some electrical losses.

Figure 7 shows the response of the proposed system when supplying a weak grid. It was disturbed by the same step variations in the insolation level as that of the strong grid. Figure 7a shows the response of V_{pcc}, I_g, V_{dc}, and I_L. V_{pcc} was sinusoidal with a constant amplitude, slightly higher than the grid rated voltage value to allow the power flow to the grid. The grid current, I_g, is also shown. It was also sinusoidal with a unity power factor. However, its amplitude varied according to the MPPT power extracted from the PV. The V_{dc} response had a better response than the strong grid counterpart. Here, the maximum overshoot was limited to (0.75%). The response of I_L is also presented in the figure. It had a slightly higher instantaneous value than that of the strong. However, its maximum RMS value was 150 mA, which was also within the standard limits.

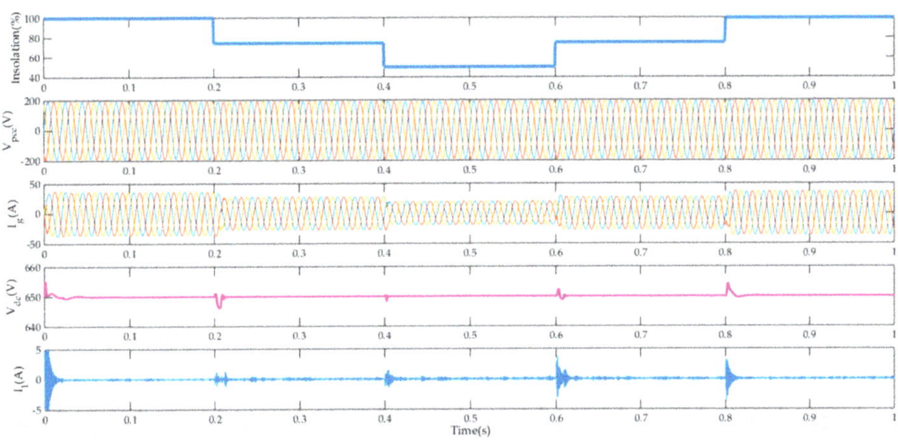

(**a**)

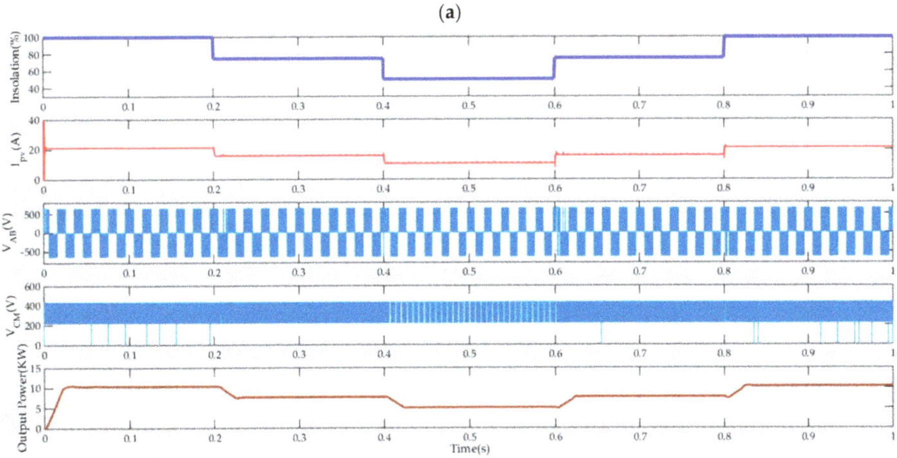

(**b**)

Figure 7. The response of the proposed system to step variations in the isolation level, @ weak-grid, (**a**) V_{pcc}, I_g, V_{dc}, and I_L, (**b**) I_{pv}, V_{AB}, V_{CM}, and the output power.

Figure 7b shows the response of I_{pv}, V_{AB}, V_{CM}, and the output power. The I_{pv} tracked well with the reference recommended by the MPPT controller. The responses of the V_{AB}, CMV, and grid power had a similar figure to that for the strong grid.

The fast Fourier transform analysis of the supplied grid current for the two cases, strong-grid and weak-grid, is presented in Figure 8a,b, respectively. Nevertheless, the harmonics in the case of the weak-grid were higher than in the strong-grid case. The lower-order harmonics were presented with a weak grid. Moreover, the fundamental component of the grid current was slightly higher in the case of the strong grid. The THD of the grid current was 1.36% for the strong grid case, which was better than the corresponding weak grid value, which was 2.06%. Figure 8c shows the Bode plot of the LCL filter based on the simulation parameters.

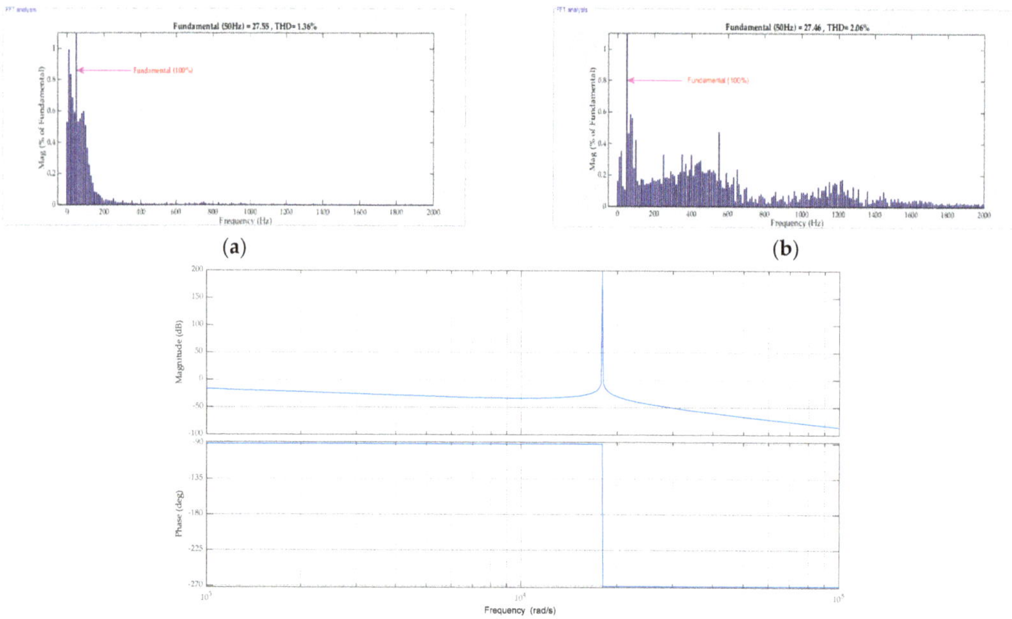

(**a**) (**b**) (**c**)

Figure 8. The spectrum of I_g for (**a**) a strong-grid, (**b**) a weak-grid (@100%insolation), and (**c**) the LCL filter frequency response.

The effect of varying insolation levels and grid SCR on the RMS of the leakage current is shown in Figure 9a. The leakage current decreased with the insolation level. However, it was roughly reduced with the increase in the SCR of the grid. Moreover, the effect of varying insolation levels and grid SCR on the THD of the I_g is shown in Figure 9b. The THD value improved with the increase in both the SCR and the insolation level.

The Hardware-in-the-Loop (HIL) simulator was used to validate the proposed controller of the MPC for the H8 inverter as shown in Figure 10. The power circuit was modelled as a MATLAB model in the host personal computer (Host-PC), while the DSP target executed the proposed MPC after receiving the required signals from the Host-PC. The interface between the Host-PC and the controller board was installed using the virtual serial COM port. The data exchange was triggered at each sampling interval. The results of the HIL simulation in the case of the weak grid are provided in Figure 11. It is clear that the grid current was corresponding to the insolation levels, and the DC-link voltage was maintained constant at around 650 V.

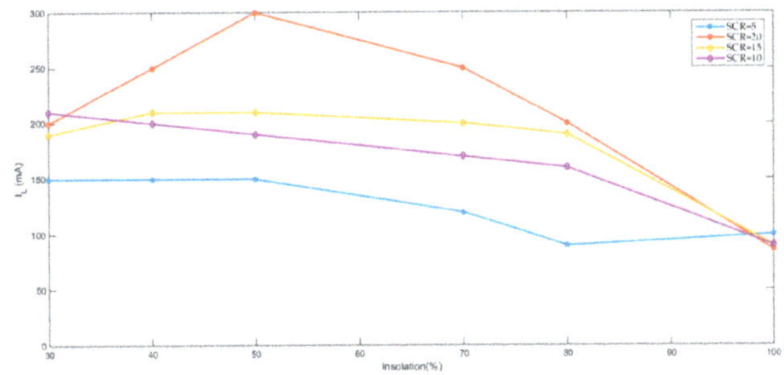

(a)

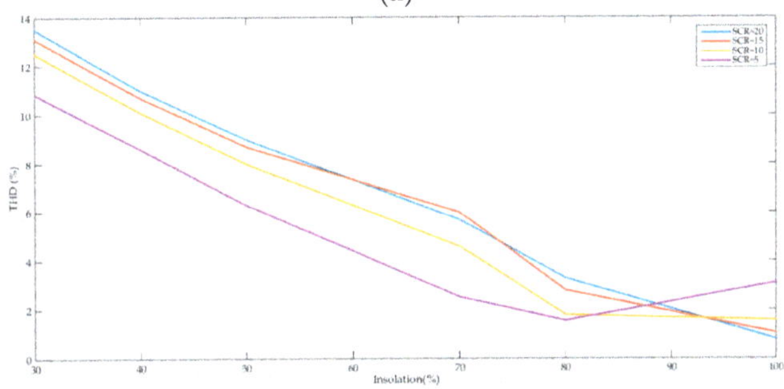

(b)

Figure 9. The effect of varying insolation levels and grid SCR on: (**a**) the RMS of the leakage current and (**b**) the THD of I_g.

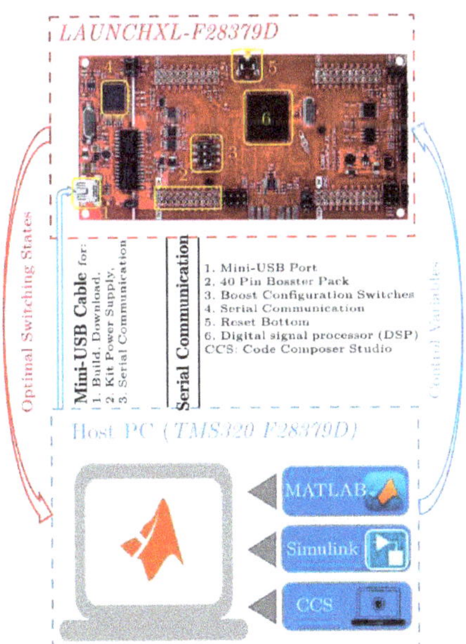

Figure 10. Schematic diagram of the HIL simulation based on the DSP target LaunchPadXLTMS320F28379D kit.

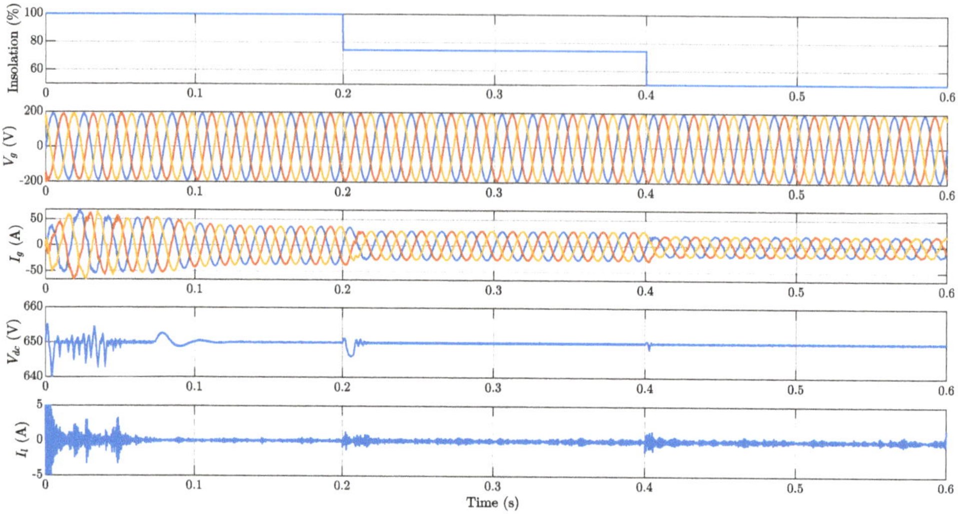

Figure 11. The response of the proposed system to step variations in the isolation level with the HIL simulation, @ weak grid.

In order to test the robustness of the proposed system against system parameter variations, some of the system parameter were forced to change, increasing the PV temperature and series resistance by 10%. Moreover, the grid impedance was decreased by 10%. The simulation results shown in Figure 12 indicate that the grid current and the DC link voltage, for example, had a stable and robust response against parameter variations.

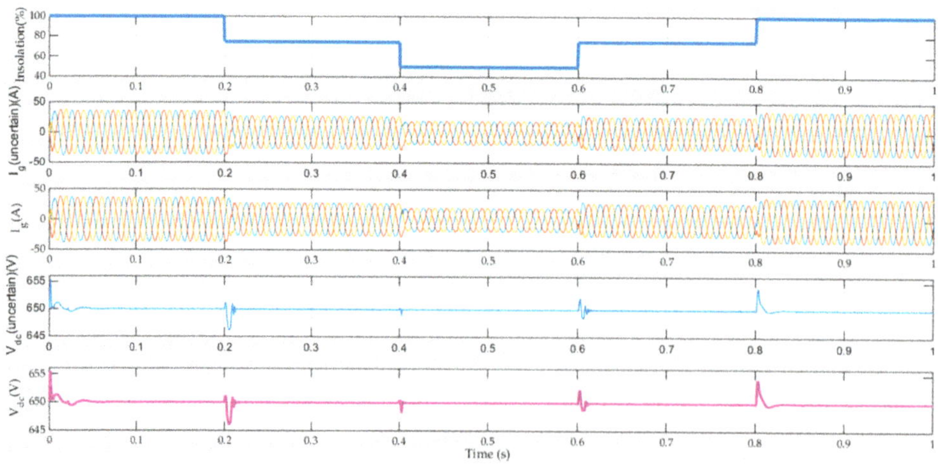

Figure 12. The response of the proposed system under parameters uncertainty, @ weak grid.

7. Conclusions

This research introduced the application of the MPC control algorithm to an H8 transformerless inverter. The inverter was grid-tied and supplied by a PV panel. The MPPT condition of the PV was achieved using a boost converter, and an LCL filter was used to connect the H8 inverter with the grid. The dynamic discrete models of the H8 transformerless inverter, including the filter and the grid weakness, were derived. Hence, the MPC control algorithm was developed for the system. The results of the simulation, using Matlab/Simulink platform, indicated a large reduction in the earth leakage current of the proposed system with the MPC controller. The maximum measured RMS value of the earth leakage currents were 100 mA and 150 mA for the strong and weak grid cases, respectively. These values were much lower than the limiting standard values. At the same time, the injected grid currents were perfect AC with pure active power. These currents had high quality with a THD less than 5% with insolation levels greater than 50%, for the strong and weak grid cases. The effects of varying the solar insolation level and grid SCR on the RMS of the leakage current and the injected current THD were studied. However, the performance was better for high insolation levels than for lower levels. The proposed system stability against model parameters uncertainties was tested by increasing the PV temperature and series resistance by ten percent, and the grid impedance was decreased by ten percent. Despite the modeling errors, the system performance was stable with high accuracy. Moreover, the results from the hardware HIL implementation of the system validated the simulation results.

Author Contributions: S.A.Z. conceived and designed the system, derived the model, and analyzed the results. H.A. (Hossam AbdelMeguid) and T.A.A. helped with collecting the funding, resources validation, and visualization, and H.A. (Hani Albalawi) supported the manuscript proofreading and writing. A.B. helps in the validation and visualization phase. All authors have read and agreed to the published version of the manuscript.

Funding: This research was funded by the University of Tabuk, Grant Number S-1441-0172 at https://www.ut.edu.sa/web/deanship-of-scientific-research/home (accessed on 18 August 2020).

Institutional Review Board Statement: Not applicable.

Informed Consent Statement: Informed consent was obtained from all individual participants included in the study.

Data Availability Statement: Data is available from the authors upon reasonable request.

Conflicts of Interest: The authors declare no conflict of interest.

References

1. Ronanki, D.; Sang, P.H.; Sood, V.; Williamson, S.S. Comparative assessment of three-phase transformerless grid-connected solar inverters. In Proceedings of the 2017 IEEE International Conference on Industrial Technology (ICIT), Toronto, ON, Canada, 22–25 March 2017; pp. 66–71. [CrossRef]
2. Atawi, I.E.; Hendawi, E.; Zaid, S.A. Analysis and Design of a Standalone Electric Vehicle Charging Station Supplied by Photovoltaic Energy. *Processes* **2021**, *9*, 1246. [CrossRef]
3. Zaid, S.A.; Kassem, A.M. Review, analysis and improving the utilization factor of a PV-grid connected system via HERIC transformerless approach. *Renew. Sustain. Energy Rev.* **2017**, *73*, 1061–1069. [CrossRef]
4. Petrone, G.; Spagnuolo, G.; Teodorescu, R.; Veerachary, M.; Vitelli, M. Reliability issues in photovoltaic power processing systems. *IEEE Trans. Ind. Electron.* **2008**, *55*, 2569–2580. [CrossRef]
5. Xiao, H.; Xie, S.; Chen, Y.; Huang, R. An Optimized Transformerless Photovoltaic Grid-Connected Inverter. *IEEE Trans. Ind. Electron.* **2011**, *58*, 1887–1895. [CrossRef]
6. Meneses, D.; Blaabjerg, F.; García, Ó.; Cobos, J.A. Review and Comparison of Step-Up Transformerless Topologies for Photovoltaic AC-Module Application. *IEEE Trans. Power Electron.* **2013**, *28*, 2649–2663. [CrossRef]
7. *VDE 0126-1-1-2006*; Automatic Disconnection Device between a Generator and the Public Low-Voltage Grid. DIN_VDE Normo: Berlin, Germany, 2011.
8. Gonzalez, R.; Lopez, J.; Sanchis, P.; Marroyo, L. Transformerless inverter for single-phase photovoltaic systems. *IEEE Trans. Power Electron.* **2007**, *22*, 693–697. [CrossRef]
9. Kerekes, T.; Teodorescu, R.; Rodríguez, P.; Vázquez, G.; Aldabas, E. A new high-efficiency single-phase transformerless PV inverter topology. *IEEE Trans. Ind. Electron.* **2011**, *58*, 184–191. [CrossRef]
10. Freddy, T.K.S.; Rahim, N.A.; Hew, W.; Che, H.S. Comparison and analysis of single-phase transformerless grid-connected PV inverters. *IEEE Trans. Power Electron.* **2014**, *29*, 5358–5369. [CrossRef]
11. Elbalawi, H.; Zaid, S.A. H5 transformerless inverter for grid-connected pv system with improved utilization factor and simple maximum power point algorithm. *Energies* **2018**, *11*, 2912. [CrossRef]
12. Hou, C.; Shih, C.; Cheng, P.; Hava, A.M. Common-Mode voltage reduction pulsewidth modulation techniques for three-phase grid-connected converters. *IEEE Trans. Power Electron.* **2013**, *28*, 1971–1979. [CrossRef]

13. Kerekes, T.; Teodorescu, R.; Liserre, M.; Klumpner, C.; Sumner, M. Evaluation of three-phase transformerless photovoltaic inverter topologies. *IEEE Trans. Power Electron.* **2009**, *24*, 2202–2211. [CrossRef]
14. Guo, X.; He, R.; Jian, J.; Lu, Z.; Sun, X.; Guerrero, J.M. leakage current elimination of four-leg inverter for transformerless three-phase pv systems. *IEEE Trans. Power Electron.* **2016**, *31*, 1841–1846. [CrossRef]
15. Hasanzad, F.; Rastegar, H.; Pichan, M. Performance evaluation of space vector modulation techniques for reducing leakage current of a three-phase four-leg PV inverter. In Proceedings of the 17th Iranian Conference on Electrical Engineering (ICEE), Tehran, Iran, 12–14 May 2009; pp. 1026–1031.
16. Albalawi, H.; Zaid, S.A. Performance Improvement of a Grid-Tied Neutral-Point-Clamped 3-φ Transformerless Inverter Using Model Predictive Control. *Processes* **2019**, *7*, 856. [CrossRef]
17. Morris, C.T.; Han, D.; Sarlioglu, B. Reduction of common mode voltage and conducted EMI through three-phase inverter topology. *IEEE Trans. Power Electron.* **2017**, *32*, 1720–1724. [CrossRef]
18. Lorenzani, E.; Migliazza, G.; Immovilli, F.; Gerada, C.; Zhang, H.; Buticchi, G. Internal current return path for ground leakage current mitigation in current source inverters. *IEEE Access* **2019**, *7*, 96540–96548. [CrossRef]
19. Wang, W.; Gao, F.; Yang, Y.; Blaabjerg, F. Operation and modulation of h7 current-source inverter with hybrid sic and si semiconductor switches. *IEEE J. Emerg. Sel. Top. Power Electron.* **2018**, *6*, 387–399. [CrossRef]
20. Atawi, I.; Zaid, S. Model Predictive Control of H7 Transformerless Inverter Powered by PV. *Intell. Autom. Soft Comput.* **2022**, *31*, 449–469. [CrossRef]
21. Concari, L.; Barater, D.; Buticchi, G.; Concari, C.; Liserre, M. H8 inverter for common-mode voltage reduction in electric drives. *IEEE Trans. Ind. Appl.* **2016**, *52*, 4010–4019. [CrossRef]
22. Jeong, W.-S.; Kim, S.-H.; Yi, J.; Won, C.-Y. Finite Control Set–Model Predictive Control of H8 Inverter Considering Dead-Time Effect for PMSM Drive Systems with Reduced Conducted Common-Mode EMI and Current Distortions. *IEEE Trans. Power Electron.* **2022**, *37*, 5342–5356. [CrossRef]
23. Gupta, A.K.; Agrawal, H.; Agarwal, V. A novel three-phase transformerless H-8 topology with reduced leakage current for grid-tied solar pv applications. *IEEE Trans. Ind. Appl.* **2019**, *55*, 1765–1774. [CrossRef]
24. Rahimi, R.; Farhangi, S.; Farhangi, B.; Moradi, G.R.; Afshari, E.; Blaabjerg, F. H8 Inverter to Reduce Leakage Current in Transformerless Three-Phase Grid-Connected Photovoltaic systems. *IEEE J. Emerg. Sel. Top. Power Electron.* **2018**, *6*, 910–918. [CrossRef]
25. Guo, X.; Wang, N.; Wang, B.; Lu, Z.; Blaabjerg, F. Evaluation of Three-Phase Transformerless DC-Bypass PV Inverters for Leakage Current Reduction. *IEEE Trans. Power Electron.* **2020**, *35*, 5918–5927. [CrossRef]
26. Xiang, Y.; Pei, X.; Wang, M.; Shi, P.; Kang, Y. An Improved H8 Topology for Common-Mode Voltage Reduction. *IEEE Trans. Power Electron.* **2019**, *34*, 5352–5361. [CrossRef]
27. Li, M.; Zhang, X.; Guo, Z.; Wang, J.; Wang, Y.; Li, F.; Zhao, W. The Control Strategy for the Grid-Connected Inverter Through Impedance Reshaping in q-Axis and its Stability Analysis Under a Weak Grid. *IEEE J. Emerg. Sel. Top. Power Electron.* **2021**, *9*, 3229–3242. [CrossRef]
28. Lasseter, R.H.; Chen, Z.; Pattabiraman, D. Grid-forming inverters: A critical asset for the power grid. *IEEE J. Emerg. Sel. Top. Power Electron.* **2019**, *8*, 925–935. [CrossRef]
29. IEEE Standards Board. *IEEE Standard for Interconnecting Distributed Resources with Electric Power Systems:1547-2003*; IEEE: Piscataway, NJ, USA, 2003; pp. 1–28.
30. Yang, D.; Ruan, X.; Wu, H. Impedance Shaping of the Grid-Connected Inverter with LCL Filter to Improve Its Adaptability to the Weak Grid Condition. *IEEE Trans. Power Electron.* **2014**, *29*, 5795–5805. [CrossRef]
31. Mohamed, I.S.; Zaid, S.A.; Abu-Elyazeed, M.F.; Elsayed, H.M. Improved model predictive control for three-phase inverter with output LC filter. *Int. J. Model. Identif. Control.* **2015**, *23*, 371–379. [CrossRef]
32. Mohamed, I.S.; Zaid, S.A.; Elsayed, H.M.; Abu-Elyazeed, M.F. Implementation of model predictive control for a three-phase inverter with output LC filter on eZdsp F28335 Kit using HIL simulation. *Int. J. Model. Identif. Control* **2016**, *25*, 301–312.
33. Zaid, S.A.; Mohamed, I.S.; Bakeer, A.; Liu, L.; Albalawi, H.; Tawfiq, M.E.; Kassem, A.M. From MPC Based to End-to-End (E2E) Learning-Based Control Policy for Grid-Tied 3L-NPC Transformerless Inverter. *IEEE Access* **2022**, *10*, 57309–57326. [CrossRef]
34. Zaid, S.A.; Albalawi, H. Application of model predictive control to ultra-sparse matrix rectifier. *Int. Rev. Electr. Eng.* **2018**, *13*, 357–364. [CrossRef]
35. Huang, Z.; Li, H.; Li, W.; Liu, J.; Huang, C.; Yang, Z.; Fang, W. A New Trajectory Tracking Algorithm for Autonomous Vehicles Based on Model Predictive Control. *Sensors* **2021**, *21*, 7165. [CrossRef]
36. Tan, N.; Kaya, I.; Yeroglu, C.; Atherton, D.P. Computation of stabilizing PI and PID controllers using the stability boundary locus. *Energy Convers. Manag.* **2006**, *47*, 3045–3058. [CrossRef]
37. Gao, R.; Gao, Z. Pitch control for wind turbine systems using optimization, estimation and compensation. *Renew. Energy* **2016**, *91*, 501–515.

processes

Article

An Effective Temperature Control Method for Dividing-Wall Distillation Columns

Yang Yuan, Xinyi Tao, Kejin Huang, Haisheng Chen * and Xing Qian *

College of Information Science and Technology, Beijing University of Chemical Technology, Beijing 100029, China; yuanyang@mail.buct.edu.cn (Y.Y.); 2021200784@buct.edu.cn (X.T.); huangkj@mail.buct.edu.cn (K.H.)
* Correspondence: chenhs@mail.buct.edu.cn (H.C.); qianx@mail.buct.edu.cn (X.Q.)

Abstract: Temperature control is widely perceived to be superior to direct composition control for the control of dividing-wall distillation columns (DWDCs) due to its advantages in dynamic characteristics. However, because of the limited estimation accuracy to the controlled product purities, the former cannot eliminate the steady-state errors in the maintained product purities as completely as the latter. In order to reduce the steady-state deviations in the maintained product purities, an effective temperature control method is proposed in the current article by means of a kind of simple but effective product quality estimator (PQE). For the proposed PQE, temperatures of three stages located in the controlled column section (T_{I1}, T_{I2}, and T_{I3}) are employed as inputs, and a linear sum of these three inputted stage temperatures ($\alpha \times T_{I1} + \beta \times T_{I2} + \gamma \times T_{I3}$) is given as output. A genetic algorithm with an elitist preservation strategy is used to optimize the locations of the three stage temperatures and the values of α, β, and γ to ensure the estimation accuracy of the PQE. Concerning the controls of two DWDCs, i.e., one Petlyuk DWDC separating an ethanol/propanol/butanol ternary mixture and one Kaibel DWDC separating a methanol/ethanol/propanol/butanol quaternary mixture, the effectiveness of the PQE is assessed through comparing the performance of the temperature inferential control scheme using the PQE and the double temperature difference control scheme. According to the dynamic simulation results obtained, the former control scheme displays not only smaller steady-state deviations in the maintained product purities, but also better dynamic characteristics as compared with the latter control scheme. This result fully demonstrates that the proposed PQE can be a useful tool for the temperature inferential control of the DWDC.

Keywords: dividing-wall distillation column; genetic algorithm; temperature inferential control; temperature difference; quality estimator

Citation: Yuan, Y.; Tao, X.; Huang, K.; Chen, H.; Qian, X. An Effective Temperature Control Method for Dividing-Wall Distillation Columns. *Processes* **2022**, *10*, 1018. https://doi.org/10.3390/pr10051018

Academic Editor: Zhiwei Gao

Received: 29 April 2022
Accepted: 17 May 2022
Published: 20 May 2022

1. Introduction

As a typical complex industrial process [1,2], the dividing-wall distillation column (DWDC) is famous for its excellent energy-saving and investment-reducing capacities [3–5]. However, it is very hard to achieve its stable operation and strict product quality control due to its characteristics of high nonlinearity and large time delay. This control problem constrains greatly the application of the DWDC in chemical industries [6] and reminds us of the importance of studying the control of the DWDC.

Focusing on the control of the DWDC, a wide range of control methods have been studied, including but not limited to direct composition control [7–10], composition-temperature cascade control [11,12], and temperature inferential control [13–16]. Among them, temperature inferential control has the largest application potential in the chemical industry because of its obvious superiorities in dynamic characteristics and equipment investment. The only pity is that temperature inferential control schemes may give rise to large steady-state deviations in the product purities to be maintained due to unpredictable pressure variations within the DWDC. For the purpose of reducing the steady-state deviations in the product purities to be maintained substantially, many studies have been conducted in recent years to give various effective temperature inferential control schemes. For instance, Ling and Luyben gave a temperature difference control (TDC) scheme that involves four temperature difference control loops for controlling the Petlyuk DWDC [17]. Because the effects of pressure variations on the two stage temperatures that form the temperature difference are similar and can offset each other to a certain degree, the estimation accuracy of the temperature difference is less influenced by pressure variations. Therefore, the TDC scheme could achieve stricter product quality control than the conventional temperature control scheme. Wu et al. proposed a double temperature difference control (DTDC) scheme for the control of the Petlyuk DWDC by employing a double temperature difference to estimate the maintained product purity in each control loop [18]. Because the effects of pressure variations on the two temperature differences that form the double temperature difference are similar and can be offset each other, the performance of the DTDC scheme was further improved as compared with the TDC scheme. Not only were the steady-state errors in the maintained product purities obviously reduced, but also the capability of rejecting feed disturbances was improved. Subsequently, Yuan et al. found that it is possible to further increase the estimation accuracy of the double temperature difference to the maintained product purity by carefully coordinating the relationship between the two temperature differences that form the double temperature difference, and developed a new derivation method of the DTDC scheme [19]. According to their dynamic simulation results, the DTDC scheme derived by the new method could result in smaller steady-state errors in the product purities to be maintained as compared with the DTDC scheme derived by the conventional method. Furthermore, Pan et al. developed a pressure-compensated temperature control loop to enhance the control of the lower sidestream product for a Kaibel DWDC [20]. For this pressure-compensated temperature control loop, pressure and temperature of a stage, including primarily binary mixture, were first measured, and the composition of this stage was then inferred to be used as the controlled variable, relying on an equation describing the relationship among the temperature, the pressure, and composition. Although this pressure-compensated temperature control loop is effective, its principle is too complicated to be popular.

Based on the above analysis, we can clearly find that the key to improving the steady-state performance of the temperature inferential control scheme lies in employing appropriate pressure compensation methods. In the current work, we will propose a kind of novel product quality estimator (PQE). It can suppress the adverse effect of pressure variations and provide a more accurate estimation of the product purity to be maintained according to the stage temperatures measured. In the remainder of this article, the structure of the proposed PQE and its design method are firstly elaborated in Section 2. In terms of two DWDCs, including one Petlyuk DWDC separating an ethanol (E)/propanol (P)/butanol (B) ternary mixture (EPB Petlyuk DWDC) and one Kaibel DWDC separating methanol (M)/E/P/B quaternary mixture (MEPB Kaibel DWDC), comparisons between the DTDC scheme derived by the new method (its steady-state and dynamic performances are more excellent than those of the other temperature difference control schemes, and thus it is employed here as a comparison) and the temperature inferential control scheme using the PQE (TC-PQE) are conducted to assess the effectiveness of the PQE in Sections 3 and 4. An objective evaluation of the proposed PQE is provided in Section 5. Some conclusions are briefly given in Section 6.

2. Structure and Design Method of the Proposed PQE

2.1. Structure of the PQE

Figure 1 gives the schematic of the PQE proposed for the control of the DWDC. Its inputs are temperatures of three stages located in the controlled column section (T_{I1}, T_{I2}, and T_{I3}), and its output is equal to a linear sum of these three inputted stage temperatures ($\alpha \times T_{I1} + \beta \times T_{I2} + \gamma \times T_{I3}$). The temperature difference [16,17] and the double temperature difference [18,19] have demonstrated that detecting multiple stage temperatures is an effective method to overcome the adverse effect of pressure variations. This is the main reason why three stage temperatures are measured and inputted in the PQE. Further, someone might ask why the number of the inputted stage temperatures is determined to be three, not two, or a number greater than three. There are mainly the following two reasons: one is that at least three stage temperatures are required if someone wants to clearly and rapidly detect the temperature variations of one column section; the other is that using more temperatures cannot obtain an obvious improvement in detection quality but increases the complexity and equipment cost of the PQE.

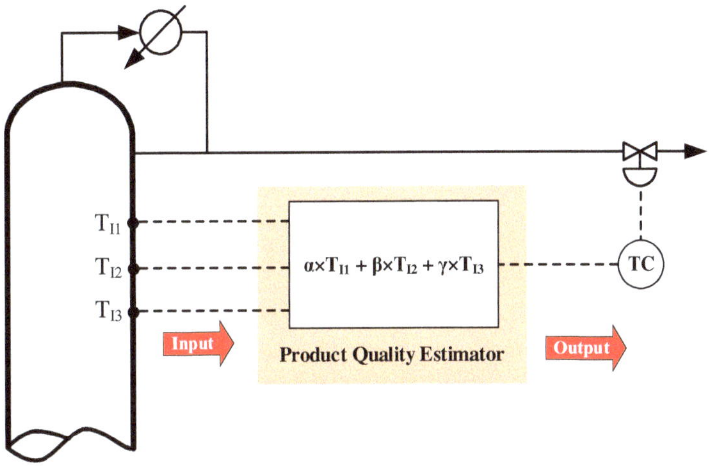

Figure 1. Schematic of the PQE proposed for controlling the DWDC.

2.2. Design Method of the PQE

To ensure the PQE can not only give an accurate inference to the maintained product purity but also has satisfying dynamic characteristics, its six design variables, i.e., the locations of the three inputted stage temperatures and the values of α, β, and γ, must be specially optimized according to the characteristics of the controlled column section before it is used in a certain control loop. To quantitatively evaluate the dynamic characteristics and the estimation accuracy of the PQE, the static gain between the output of the PQE and the manipulated variable and the AAVM of the output of the PQE are employed, respectively. The static gain between the output of the PQE and the manipulated variable can be calculated according to the following procedure. Firstly, a very small variation (0.5%) is given to the manipulated variable; secondly, the steady-state model is run and a new steady state is obtained; finally, the absolute variation in the output of the PQE divided by the absolute variation in the manipulated variable gives the static gain. The bigger static gain means that the PQE has better dynamic characteristics. The AAVM of the output of the PQE indicates the averaged absolute variation magnitude of the output of the PQE for all kinds of disturbances in feed component composition, with the prerequisite of strictly maintaining all controlled product purities to their specifications. The detailed calculation method of the AAVM can be found in our previous article [19] and is no longer given here due to limited space. The smaller AAVM means that the PQE has higher estimation accuracy for the maintained product purity.

A genetic algorithm with an elitist preservation strategy (as shown in Figure 2) is employed to optimize the six design variables of the PQE. Real coding is adopted, and the values of the six design variables of the PQE are determined as the genes of the individual. Thus, every individual can represent a kind of PQE design. To make sure that the PQE obtained finally has satisfied dynamic characteristics, a constraint condition is given to the PQE. Namely, the static gain between the output of the PQE and the manipulated variable must be greater than a pre-set lower limit. If an individual represents a PQE that does not satisfy this constraint condition, it is illegal and should be abandoned during the optimization process. The reciprocal of the AAVM of the output of the PQE is selected as a fitness function to increase the estimation accuracy of the PQE to the greatest degree. In addition, the genetic algorithm employed here differs from the conventional genetic algorithm in that it involves only three steps of selection, crossover, and mutation. An additional elitist preservation strategy is used to improve global searching ability and efficiency. The elitist is defined as the individual that has the biggest fitness in the population, and the fitness of the elitist of the Nth-generation population can be marked Fit_N. After the Nth-evolution round, if the fitness of the elitist of the current generation population (Fit_N) is smaller than that of the last generation population (Fit_{N-1}), the bottom 10% of individuals according to the fitness of the current generation population should be replaced by the elitist of the last generation population.

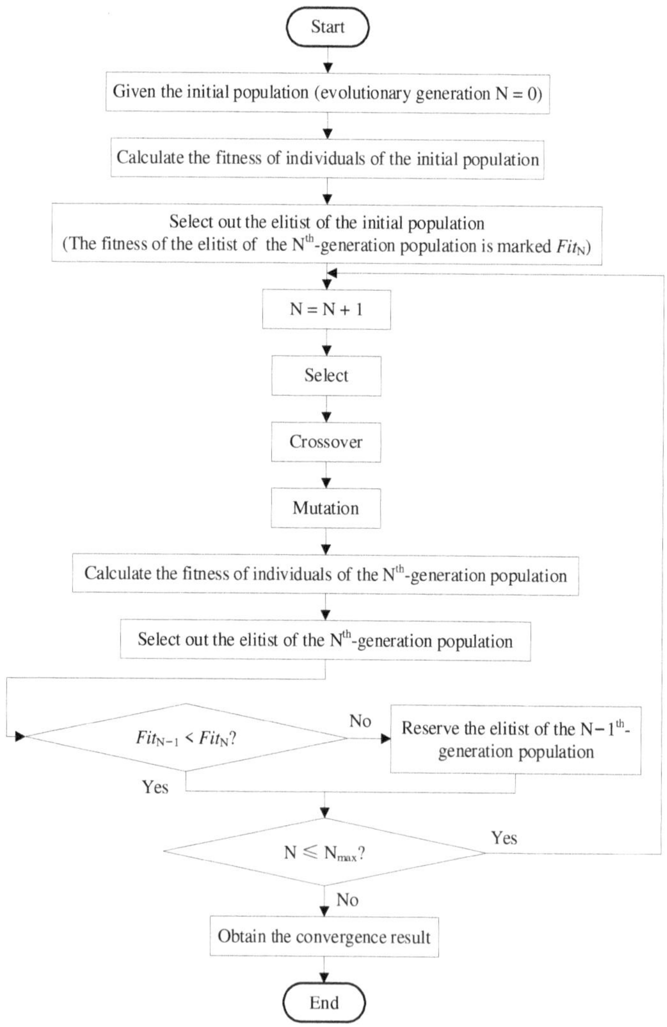

Figure 2. Overview of the genetic algorithm with elitist preservation strategy.

3. Example I: Operation of an EPB Petlyuk DWDC

3.1. Process Design of the EPB Petlyuk DWDC

Figure 3a,b display, respectively, the steady-state design of the EPB Petlyuk DWDC and its temperature profile. The mole ratio between the feed components of E, P, and B is 1:1:1. The specifications for the top, sidestream, and bottom products are all 99 mol%. Throughout the work, Aspen Plus and Aspen Dynamics are respectively employed to build steady-state and dynamic models, and the characteristics of the feed are described with the UNIFAC thermodynamic model. Because there is not an available Petlyuk or Kaibel DWDC module in Aspen Plus, we have to use a combination of four Redfrac modules to simulate them. Stage efficiency is set to the default value of Aspen Plus, i.e., 100%. The diameters of the stages are determined by the "Tray Sizing" function of Aspen Plus, and the sizes of the reflux drum and the column bases are calculated to allow for 5 min of liquid holdup when the vessel is 50% full.

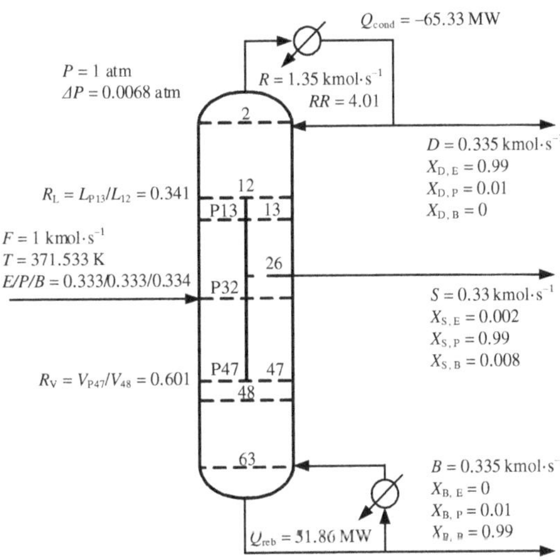

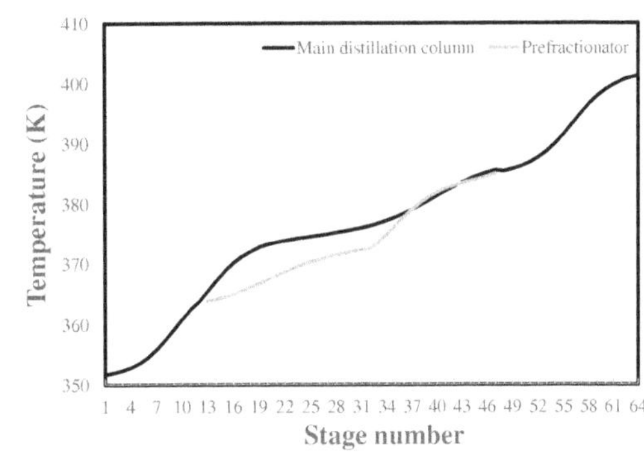

Figure 3. Example of the EPB Petlyuk DWDC: (**a**) steady-state design; (**b**) temperature profile.

3.2. Derivation of the DTDC Scheme

For the DTDC scheme, four double temperature difference control loops with distillation flow rate D, sidestream flow rate S, reboiler heat duty Q_{reb}, and liquid split ratio R_L as manipulated variables are employed to maintain the purities of the top, sidestream, bottom, and prefractionator's top products (liquid composition of component B in the top of the prefractionator), respectively. Figure 4 gives the results of sensitivity analysis and AAVM analysis for the four double temperature difference control loops. With reference to these results, the locations of the sensitive and reference stages of each double temperature difference control loop can be determined, and the resultant DTDC scheme is shown in Figure 5a. As for the detailed method of how to derive the DTDC scheme in terms of the results of sensitivity analysis and AAVM analysis, it can be found in our previous article about the effective derivation of the DTDC scheme [19].

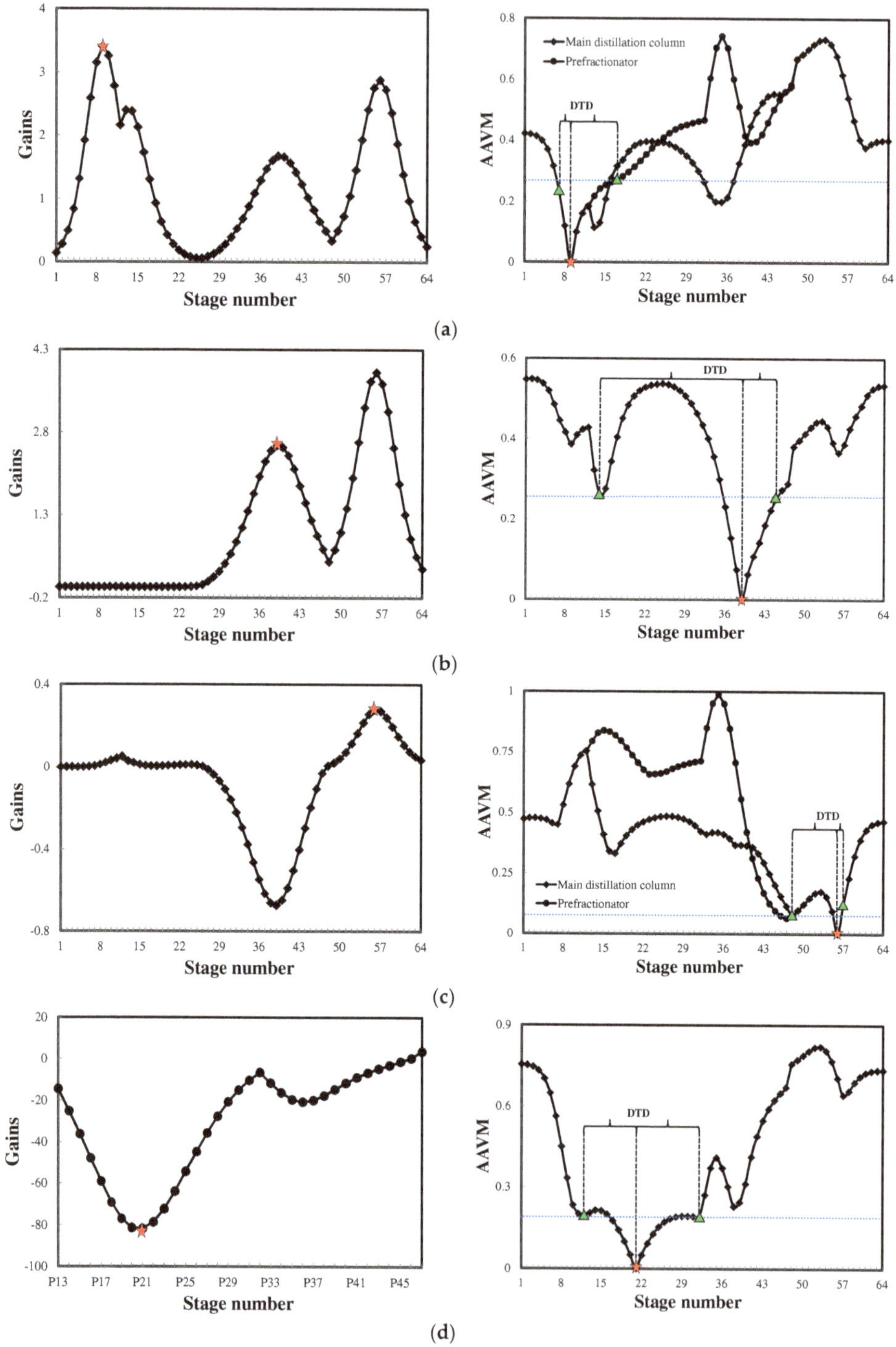

Figure 4. Results of the sensitivity analysis and AAVM analysis (Example I): (**a**) top control loop; (**b**) sidestream control loop; (**c**) bottom control loop; (**d**) prefractionator control loop. (Red pentagram: sensitive stage; green triangle: reference stage).

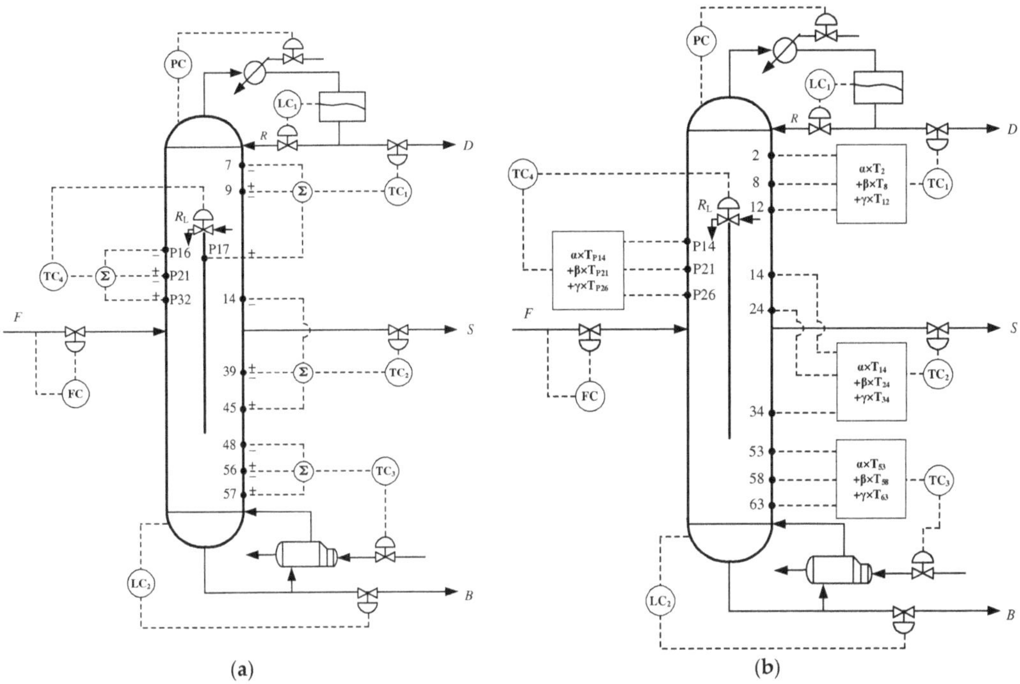

Figure 5. Two temperature inferential control schemes studied (Example I): (**a**) DTDC scheme; (**b**) TC-PQE scheme.

3.3. Derivation of the TC-PQE Scheme

For the sake of fairness, the number of control loops used in the TC-PQE scheme should be the same as that in the DTDC scheme. Therefore, four control loops with D, S, Q_{reb}, and R_L as manipulated variables are employed in the TC-PQE scheme, and four specially designed PQE are used in these four control loops to infer the qualities of the top, sidestream, bottom, and prefractionator's top products, respectively. The design variables of the four PQE are optimized according to the genetic algorithm introduced in Section 2. Here, population size is equal to 2000; evolution generation is equal to 50; crossover and mutation probabilities are respectively equal to between 0.7 and 0.3. For the optimization of the PQE employed in the top, sidestream, bottom, and prefractionator control loops, the locations of the inputted stage temperatures are respectively allowed to be selected from the public rectifying section, the right section of the dividing wall, the public stripping section, and the prefractionator, and the lower limits of the static gains are respectively equal to 1.7027, 1.283, 0.521, and 0.4235 (here, for each control loop, the lower limit is set to be two times the static gain between the sensitive stage temperature and the manipulated variable). Figure 6 displays the optimization processes of the PQE employed in the four control loops using a genetic algorithm. According to the optimization results, four optimal PQE can be obtained, and their detailed parameters are listed in Table 1. Figure 5b displays the resultant TC-PQE scheme.

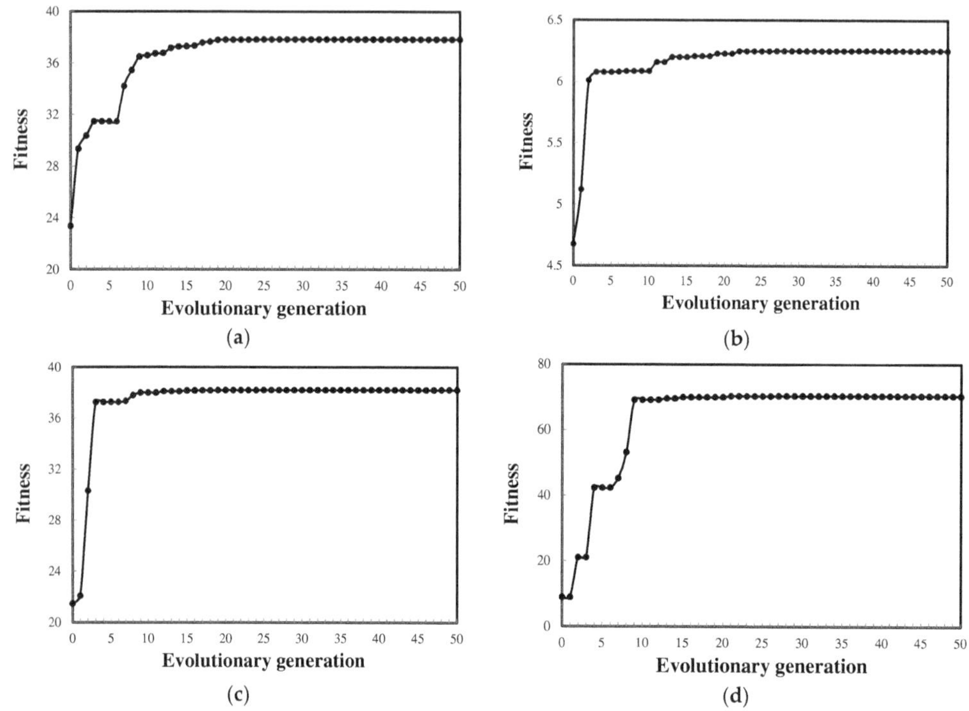

Figure 6. Optimization of the PQE employed in the four control loops of the TC-PQE scheme (Example I): (**a**) top control loop; (**b**) sidestream control loop; (**c**) bottom control loop; (**d**) prefractionator control loop.

Table 1. Optimization results of the PQE employed in the four control loops of the TC-PQE Scheme (Examples I and II).

System	Control Loop	T_{I1}	T_{I2}	T_{I3}	α	β	γ
Example I	Top	T_2	T_8	T_{12}	2.5436	−3.5962	1.7618
	Sidestream	T_{14}	T_{24}	T_{34}	2.1465	4.1458	−3.7670
	Bottom	T_{53}	T_{58}	T_{63}	−1.3321	3.5541	−1.5087
	Prefractionator	T_{P14}	T_{P21}	T_{P26}	−1.5082	3.5383	−1.9755
Example II	Top	T_8	T_{13}	T_{18}	−2.9537	−1.2432	2.2666
	Upper sidestream	T_{21}	T_{22}	T_{40}	−4.9737	−4.9668	2.0480
	Lower sidestream	T_{49}	T_{61}	T_{64}	−0.6192	−2.2745	3.2625
	Prefractionator	T_{P33}	T_{P40}	T_{P45}	−4.0334	3.7216	−2.9252

3.4. Comparison between the DTDC and TC-PQE Schemes

Table 2 tabulates the AAVM of the controlled variable and the static gain between the controlled variable and manipulated variable for each control loop of the DTDC and TC-PQE schemes. It can be clearly found that the control loop of the TC-PQE scheme shows smaller AAVM and bigger static gain than the control loop of the DTDC scheme that maintains the same product purity. This implies that the control loops of the TC-PQE scheme can give not only more accurate estimations of the controlled product purities but also more sensitive responses to the manipulated variables as compared with those of the DTDC scheme.

Table 2. AAVM and static gains for the four control loops of the DTDC and TC-PQE schemes (Examples I and II).

System	Scheme	Control Loop	AAVM	Static Gain
Example I	TC-PQE	Top	0.0264	1.7027
		Sidestream	0.1600	1.2830
		Bottom	0.0262	0.5210
		Prefractionator	0.0142	0.4235
	DTDC	Top	0.0412	1.0155
		Sidestream	0.2633	0.9806
		Bottom	0.0387	0.2620
		Prefractionator	0.0757	0.2832
Example II	TC-PQE	Top	0.0014	1.8430
		Upper sidestream	0.3435	9.5064
		Lower sidestream	0.0119	2.3020
		Prefractionator	0.8165	7.8350
	DTDC	Top	0.0218	0.6165
		Upper sidestream	0.7426	8.7710
		Lower sidestream	0.0529	0.7140
		Prefractionator	1.5454	7.4735

For the DTDC and TC-PQE schemes, all temperature measurements are assumed to have a one-minute deadtime, and PI controllers are used in all temperature inferential control loops. The Tyreus-Luyben tuning rule embedded in Aspen Dynamics is used to tune these PI controllers [21,22]. The bottom control loop is tuned first because it affects all other control loops. Then, with the bottom control loop on automatic, the top control loop is tuned. Next, with the bottom and top control loop on automatic, the sidestream control loop is tuned. Finally, with the bottom, top, and sidestream control loops on automatic, the prefractionator control loop is tuned. Three rounds of tuning are carried out at least to ensure the quality of the parameters obtained. The resultant controller parameters for the DTDC and TC-PQE schemes are listed in Table 3. Figure 7 shows the variations of the product purities as the EPB Petlyuk DWDC controlled under the DTDC and TC-PQE schemes for ±20% variations in feed compositions. It can be seen that, as compared with the DTDC scheme, the TC-PQE scheme has smaller overshoots in the controlled product purities for all six kinds of feed composition disturbances and shorter setting times for the feed composition disturbances in components P and B. Table 4 tabulates the steady-state deviations in the controlled product purities from the initial 99 mol% product specifications. In most scenarios, the TC-PQE scheme shows smaller steady-state deviations than the DTDC scheme. As for the maximum deviation of the controlled product purities (highlighted using bold type and underlining in Table 4), the TC-PQE scheme is also smaller than the DTDC scheme.

Table 3. Controller parameters of the DTDC and TC-PQE schemes (Examples I and II).

System	Scheme	Controller	K_C	T_I (min)
Example I	TC-PQE	TC1	5.30	38.28
		TC2	14.88	27.72
		TC3	2.78	9.24
		TC4	1.28	18.48
	DTDC	TC1	0.91	36.96
		TC2	0.75	23.76
		TC3	0.36	10.56
		TC4	0.31	30.36
Example II	TC-PQE	TC1	3.27	15.84
		TC2	31.31	13.20
		TC3	8.35	25.08
		TC4	7.92	22.44
	DTDC	TC1	0.27	15.84
		TC2	0.30	13.20
		TC3	2.45	30.36
		TC4	0.34	18.48

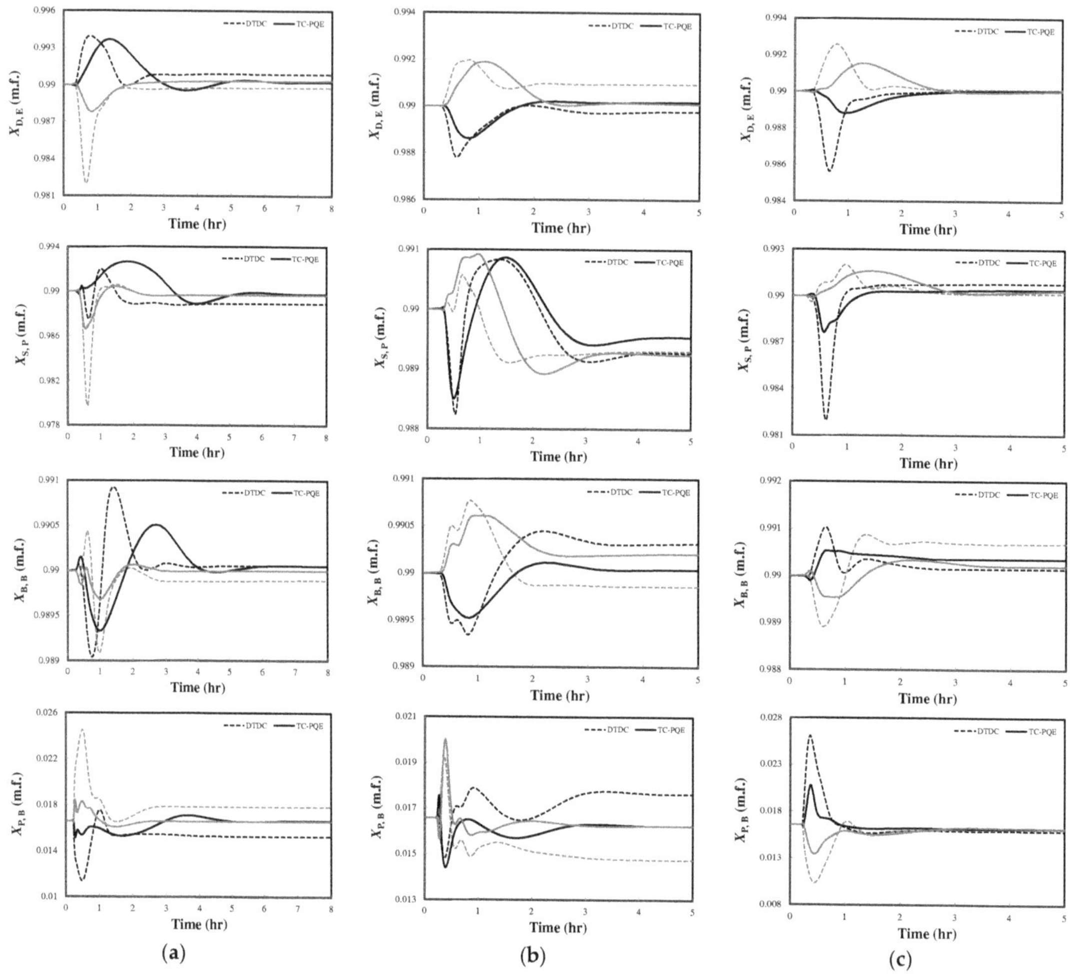

Figure 7. Variations of the product purities as maintained with the DTDC and TC-PQE schemes for ±20% variations in feed compositions (Example I): (**a**) E; (**b**) P; (**c**) B. (Black line: positive disturbance; grey line: negative disturbance).

Furthermore, robustness tests are carried out on the DTDC and TC-PQE schemes to make a deeper performance assessment of the proposed PQE. The magnitude of the concerned feed composition disturbance is gradually increased with a step of 2% until the control system fails. In the current article, the following situations will be identified as control system failures: (1). The controlled product purities do not converge; (2). The steady-state deviations in the controlled product purities are greater than 0.5 mol%; (3). Any controller fails. Table 5 tabulates the maximum feed composition disturbances handled by the DTDC and TC-PQE schemes for the EPB Petlyuk DWDC. It can be easily found that the TC-PQE scheme has better robustness as compared to the DTDC scheme since the dynamic characteristic of the PQE is well guaranteed during the optimization process.

Table 4. Steady-state deviations for ±20% variations in feed compositions of E, P, and B (Example I).

Scenario	Product Purity	Steady-State Deviation (mol%)	
		TC-PQE	DTDC
+20% Z_E	$X_{D, E}$	2.32×10^{-4}	8.76×10^{-4}
	$X_{S, P}$	-2.26×10^{-4}	-10.89×10^{-4}
	$X_{B, B}$	0.44×10^{-4}	0.57×10^{-4}
	$X_{P, B}$	0.14×10^{-4}	-13.32×10^{-4}
−20% Z_E	$X_{D, E}$	3.08×10^{-4}	-2.54×10^{-4}
	$X_{S, P}$	-3.35×10^{-4}	-3.51×10^{-4}
	$X_{B, B}$	0.03×10^{-4}	-1.05×10^{-4}
	$X_{P, B}$	-0.39×10^{-4}	12.35×10^{-4}
+20% Z_P	$X_{D, E}$	1.80×10^{-4}	-2.29×10^{-4}
	$X_{S, P}$	-4.75×10^{-4}	-7.46×10^{-4}
	$X_{B, B}$	0.45×10^{-4}	3.24×10^{-4}
	$X_{P, B}$	-3.29×10^{-4}	10.80×10^{-4}
−20% Z_P	$X_{D, E}$	0.99×10^{-4}	9.35×10^{-4}
	$X_{S, P}$	-7.56×10^{-4}	-6.85×10^{-4}
	$X_{B, B}$	2.10×10^{-4}	-1.36×10^{-4}
	$X_{P, B}$	-3.36×10^{-4}	-18.21×10^{-4}
+20% Z_B	$X_{D, E}$	0.41×10^{-4}	0.91×10^{-4}
	$X_{S, P}$	3.72×10^{-4}	7.88×10^{-4}
	$X_{B, B}$	3.50×10^{-4}	1.41×10^{-4}
	$X_{P, B}$	-4.08×10^{-4}	-6.87×10^{-4}
−20% Z_B	$X_{D, E}$	0.29×10^{-4}	1.27×10^{-4}
	$X_{S, P}$	3.02×10^{-4}	1.41×10^{-4}
	$X_{B, B}$	2.09×10^{-4}	6.79×10^{-4}
	$X_{P, B}$	-3.93×10^{-4}	-5.02×10^{-4}

Table 5. Maximum feed composition disturbances handled by the DTDC and TC-PQE schemes (Examples I and II).

	Example I		Example II	
	TC-PQE	DTDC	TC-PQE	DTDC
$+Z_M$	–	–	40%	38%
$-Z_M$	–	–	−40%	−40%
$+Z_E$	22%	20%	28%	28%
$-Z_E$	−44%	−32%	−58%	−30%
$+Z_P$	30%	28%	26%	20%
$-Z_P$	−28%	−26%	−50%	−40%
$+Z_B$	36%	32%	42%	40%
$-Z_B$	−40%	−34%	−30%	−42%

4. Example II: Operation of a MEPB Kaibel DWDC

4.1. Process Design of the MEPB Kaibel DWDC

Figure 8a,b give, respectively, the steady-state design of the MEPB Kaibel DWDC and its temperature profile. The mole ratio between feed components M, E, P, and B is 1:1:1:1. The specifications for the top, upper sidestream, lower sidestream, and bottom products are all 99 mol%.

4.2. Derivation of the DTDC Scheme

For the DTDC scheme, four double temperature difference control loops with D, upper sidestream flow rate S1, lower sidestream flow rate S2, and R_L as manipulated variables are employed to maintain the purities of the top, upper sidestream, lower sidestream, and prefractionator's top products (liquid composition of component P in the top of the prefractionator), respectively. Figure 9 gives the results of sensitivity analysis and AAVM analysis for the four double temperature difference control loops. According to these results, the locations of the sensitive and reference stages of the four double temperature difference control loops can be determined, and the resultant DTDC scheme is shown in Figure 10a.

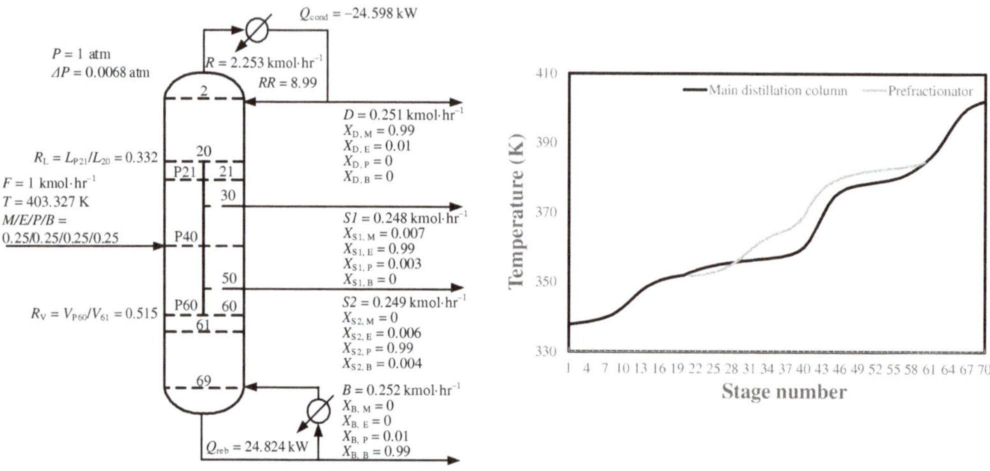

(a) (b)

Figure 8. Example of the MEPB Kaibel DWDC: (a) steady-state design; (b) temperature profile.

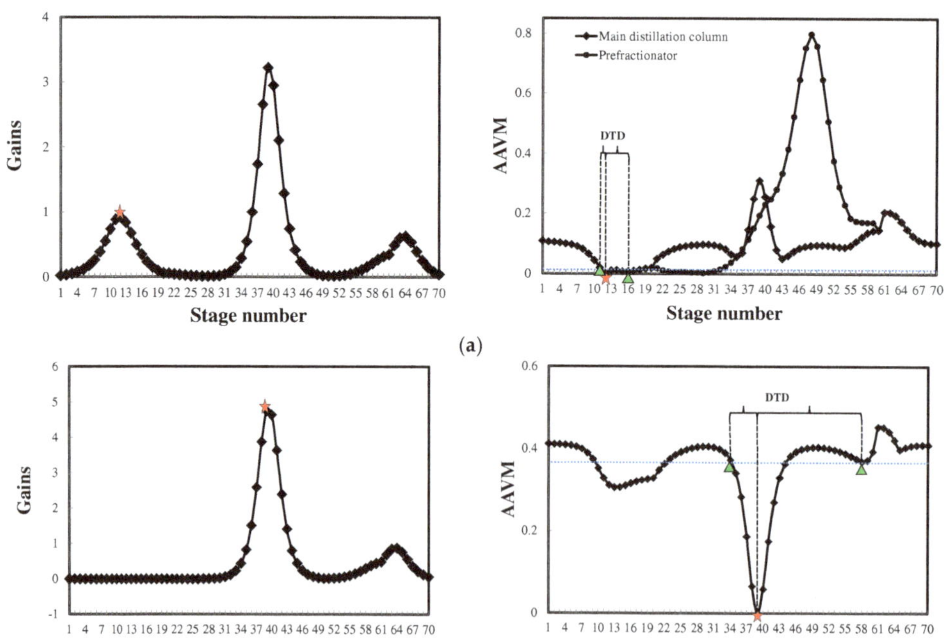

(a)

(b)

Figure 9. *Cont.*

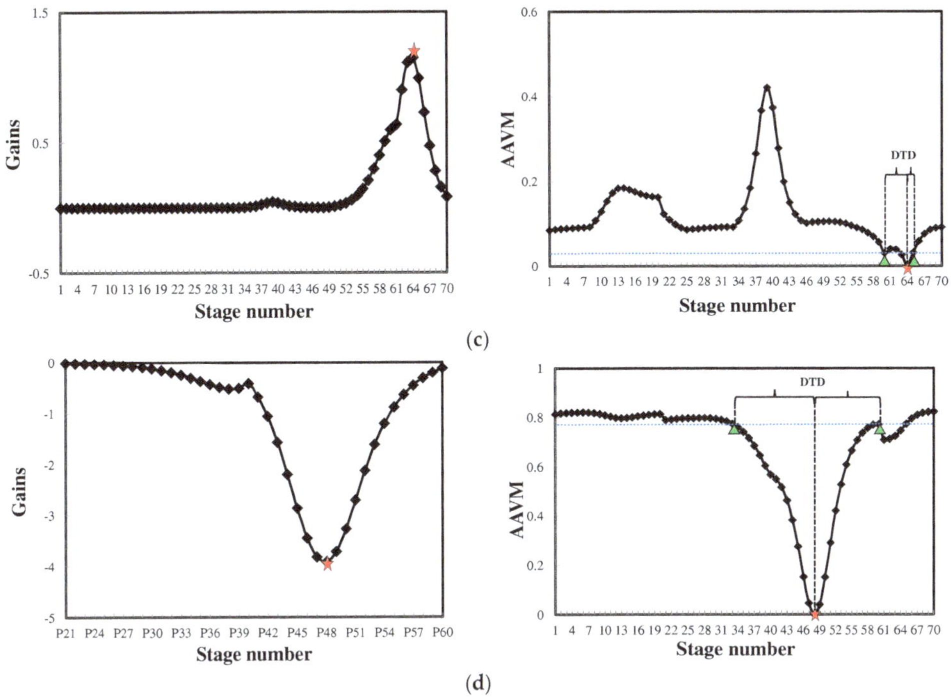

Figure 9. Results of the sensitivity analysis and AAVM analysis (Example II): (**a**) top control loop; (**b**) sidestream control loop; (**c**) bottom control loop; (**d**) prefractionator control loop. (Red pentagram: sensitive stage; green triangle: reference stage).

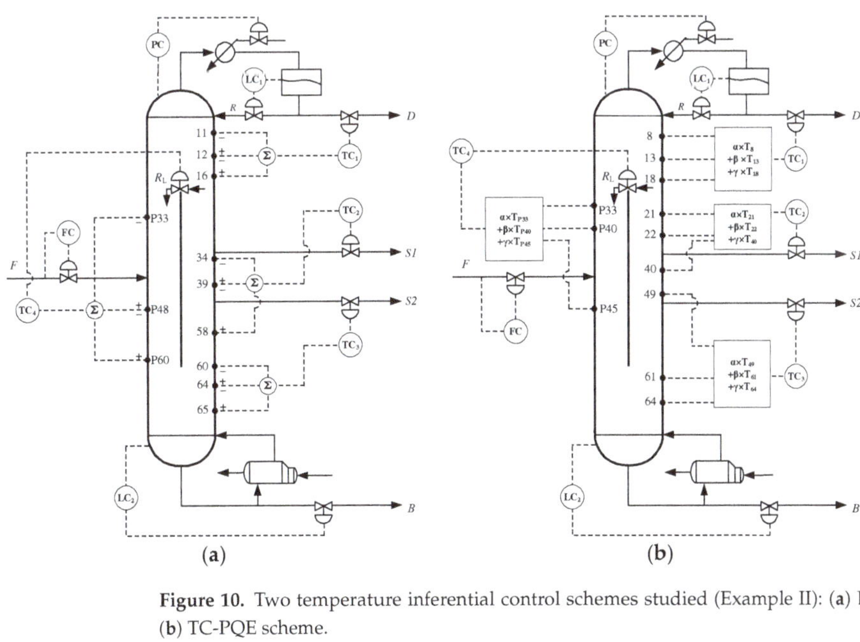

Figure 10. Two temperature inferential control schemes studied (Example II): (**a**) DTDC scheme; (**b**) TC-PQE scheme.

4.3. Derivation of the TC-PQE Scheme

The TC-PQE scheme includes four control loops. Four specially designed PQE are respectively employed in the four control loops to infer the purities of the top, upper sidestream, lower sidestream, and prefractionator's top products. The settings of population size, evolution generation, crossover probability, and mutation probability are the same as those in Example I. For the optimization of the PQE employed in the top, upper sidestream, lower sidestream, and prefractionator control loops, the locations of the inputted stage temperatures are respectively allowed to be selected from the public rectifying section, the section from stage 21 to stage 40, the section from stage 41 to stage 69, and the prefractionator, and the lower limits of the static gains are respectively equal to 1.843, 9.505, 2.302, and 7.835. Figure 11 displays the optimization processes of the PQE designed for the four control loops using a genetic algorithm. According to the optimization results, four optimal PQE can be obtained, and their detailed parameters are listed in Table 1. Figure 10b displays the resultant TC-PQE scheme.

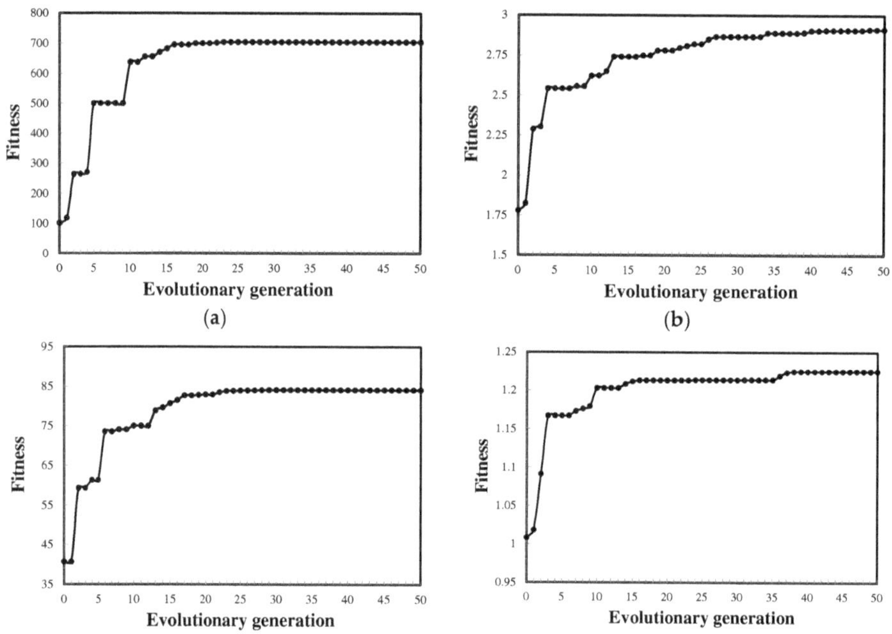

Figure 11. Optimization of the PQE employed in the four control loops of the TC-PQE scheme (Example II): (**a**) top control loop; (**b**) upper sidestream control loop; (**c**) lower sidestream control loop; (**d**) prefractionator control loop.

4.4. Comparison between the DTDC and TC-PQE Schemes

As shown in Table 2, the AAVM of the controlled variable and the static gain between the controlled variable and manipulated variable for each control loop of the DTDC and TC-PQE schemes are given. It is similar to the comparison results obtained in Example I in that the control loop of the TC-PQE scheme shows smaller AAVM and bigger static gain than the control loop of the DTDC scheme that maintains the same product purity.

The four control loops of the TC-PQE and DTDC schemes are tuned in the following order: the top control loop, the upper sidestream control loop, the lower sidestream control loop, and the prefractionator control loop. The controller parameters for the DTDC and TC-PQE schemes are tabulated in Table 3. Figure 12 shows the variations of the product purities as the MEPB Kaibel DWDC controlled under the DTDC and TC-PQE schemes for

±20% variations in feed compositions. For the top, upper sidestream, and lower sidestream products, the dynamic performance of the TC-PQE scheme is obviously better than that of the DTDC scheme, with smaller overshoots and shorter settling times. For the bottom and prefractionator's products, the dynamic performance of the TC-PQE scheme is comparable with that of the DTDC scheme. Table 6 tabulates the steady-state deviations in the controlled product purities from the initial 99 mol% product specifications. As compared with the DTDC scheme, the TC-PQE scheme not only shows smaller steady-state deviations in most scenarios but also has a smaller maximum deviation. Table 5 tabulates the maximum feed composition disturbances handled by the DTDC and TC-PQE schemes for the MEPB Kaibel DWDC. It is the same as the results obtained in Example I that the TC-PQE scheme has better robustness as compared with the DTDC scheme.

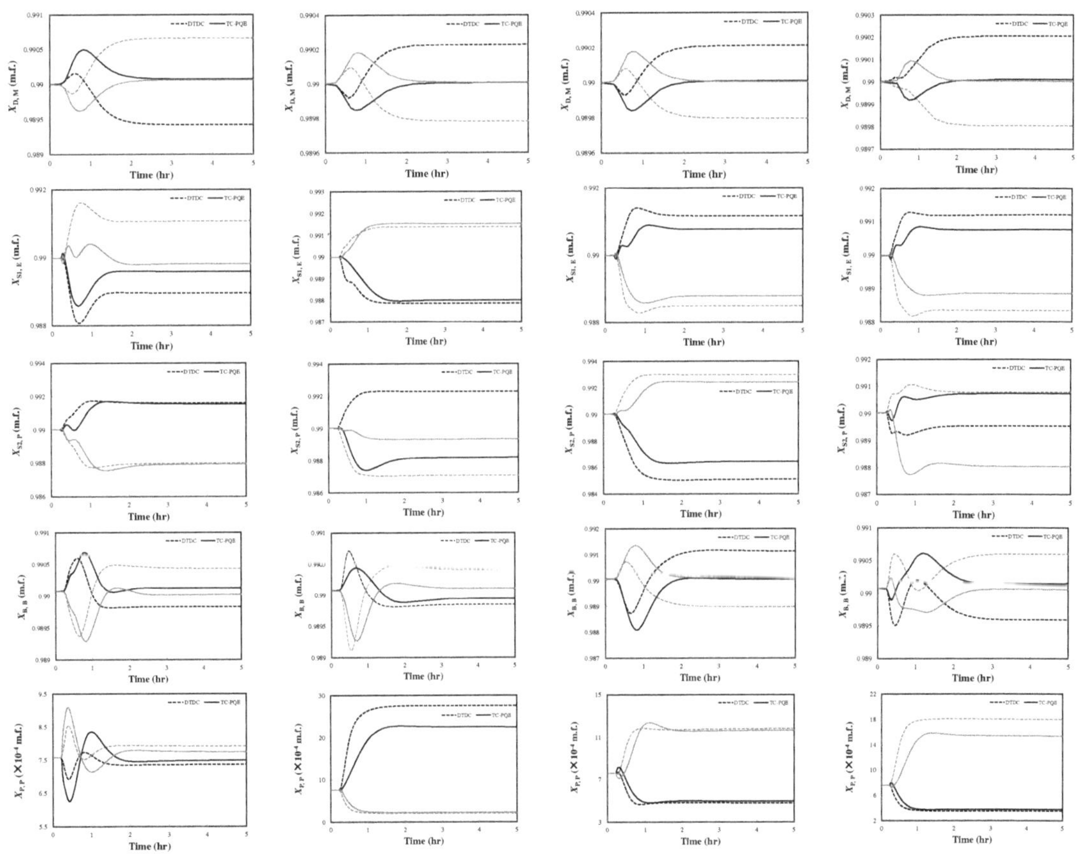

(a) (b) (c) (d)

Figure 12. Variations of the product purities as maintained with the DTDC and TC-PQE schemes for ±20% variations in feed compositions (Example II): (**a**) M; (**b**) E; (**c**) P; (**d**) B. (Black line: positive disturbance; grey line: negative disturbance).

Table 6. Steady-state deviations for ±20% variations in feed compositions of M, E, P, and B (Example II).

Scenario	Product Purity	Steady-State Deviation (mol%)		Scenario	Product Purity	Steady-State Deviation (mol%)	
		TC-PQE	DTDC			TC-PQE	DTDC
+20% Z_M	$X_{D, M}$	0.74×10^{-4}	-5.81×10^{-4}	+20% Z_P	$X_{D, M}$	0.10×10^{-4}	2.18×10^{-4}
	$X_{S1, E}$	-3.99×10^{-4}	-10.29×10^{-4}		$X_{S1, E}$	7.80×10^{-4}	11.71×10^{-4}
	$X_{S2, P}$	15.21×10^{-4}	15.76×10^{-4}		$X_{S2, P}$	-35.73×10^{-4}	-49.46×10^{-4}
	$X_{B, B}$	0.51×10^{-4}	-2.42×10^{-4}		$X_{B, B}$	-0.18×10^{-4}	10.58×10^{-4}
	$X_{P, B}$	-0.08×10^{-4}	-0.21×10^{-4}		$X_{P, B}$	-2.63×10^{-4}	-2.80×10^{-4}
−20% Z_M	$X_{D, M}$	0.63×10^{-4}	6.66×10^{-4}	−20% Z_P	$X_{D, M}$	0.04×10^{-4}	-2.04×10^{-4}
	$X_{S1, E}$	-1.66×10^{-4}	10.87×10^{-4}		$X_{S1, E}$	-12.08×10^{-4}	-15.06×10^{-4}
	$X_{S2, P}$	-21.02×10^{-4}	-20.42×10^{-4}		$X_{S2, P}$	23.90×10^{-4}	29.45×10^{-4}
	$X_{B, B}$	-0.53×10^{-4}	3.60×10^{-4}		$X_{B, B}$	0.44×10^{-4}	-10.87×10^{-4}
	$X_{P, B}$	0.17×10^{-4}	0.34×10^{-4}		$X_{P, B}$	4.08×10^{-4}	4.25×10^{-4}
+20% Z_E	$X_{D, M}$	0.06×10^{-4}	2.36×10^{-4}	+20% Z_B	$X_{D, M}$	0.14×10^{-4}	2.05×10^{-4}
	$X_{S1, E}$	-19.76×10^{-4}	-21.36×10^{-4}		$X_{S1, E}$	7.74×10^{-4}	12.14×10^{-4}
	$X_{S2, P}$	-18.27×10^{-4}	22.55×10^{-4}		$X_{S2, P}$	7.14×10^{-4}	-4.99×10^{-4}
	$X_{B, B}$	-1.31×10^{-4}	-2.25×10^{-4}		$X_{B, B}$	0.74×10^{-4}	-4.80×10^{-4}
	$X_{P, B}$	14.94×10^{-4}	20.05×10^{-4}		$X_{P, B}$	-3.81×10^{-4}	-4.04×10^{-4}
−20% Z_E	$X_{D, M}$	0.06×10^{-4}	-2.20×10^{-4}	−20% Z_B	$X_{D, M}$	0.02×10^{-4}	-1.96×10^{-4}
	$X_{S1, E}$	15.34×10^{-4}	13.96×10^{-4}		$X_{S1, E}$	-11.39×10^{-4}	-16.51×10^{-4}
	$X_{S2, P}$	-6.99×10^{-4}	-29.64×10^{-4}		$X_{S2, P}$	-19.80×10^{-4}	7.50×10^{-4}
	$X_{B, B}$	0.40×10^{-4}	3.30×10^{-4}		$X_{B, B}$	-0.28×10^{-4}	5.16×10^{-4}
	$X_{P, B}$	-5.29×10^{-4}	-5.55×10^{-4}		$X_{P, B}$	7.78×10^{-4}	10.40×10^{-4}

5. Discussion

Based on the two DWDC systems studied, the TC-PQE scheme has been demonstrated to be superior to the DTDC scheme. Although they employ the same number of temperature measurements, the former can lead to not only greatly reduced steady-state deviations in the controlled product purities but also substantially improved dynamic responses and robustness capability. The superiority of the TC-PQE schemes can undoubtedly stem from the fact that the PQE employed in its control loops not only provides very accurate estimations of the product purities to be maintained but also has sensitive responses to the manipulated variables. Admittedly, more calculating efforts are needed to optimize the PQE than to derive the double temperature difference. However, because the performance of the temperature inferential control system can be significantly improved by only paying more computation efforts without increasing additional investment costs, the proposed PQE still has a high performance-price ratio.

In addition, due to the fact that the structure of the PQE is general and the principle of optimizing the PQE is simple, the possible application of the proposed PQE is obviously not only limited to the operation of the Petlyuk and Kaibel DWDCs studied in the current work, but also to the operation of the other types of distillation columns, such as the conventional binary distillation column, the reactive distillation column, and the extractive distillation column et al. Namely, the proposed PQE can be a general tool for the control of various distillation columns. All of the above-mentioned advantages lead to the proposed PQE to be a useful tool in the field of process control.

6. Conclusion

In the current article, a kind of simple but effective PQE is proposed to estimate the controlled product purities for the control of the DWDC. It takes three stage temperatures (T_{I1}, T_{I2}, and T_{I3}) as inputs and uses a linear sum of these three inputted stage temperatures ($\alpha \times T_{I1} + \beta \times T_{I2} + \gamma \times T_{I3}$) as output. By means of a genetic algorithm using an elitist preservation strategy, the locations of the three inputted stage temperatures and the values of α, β, and γ are carefully determined so that the resultant PQE can have not only high estimation accuracy to the maintained product purity but also satisfy dynamic characteristics. With reference to two DWDCs, i.e., one EPB Petlyuk DWDC and one MEPB Kaibel DWDC, the effectiveness of the proposed PQE is assessed by comparing the performance of the TC-PQE and DTDC scheme. According to the dynamic simulation results obtained, the former is better than the latter because of not only the improved dynamic performances but also the greatly reduced steady-state deviations in the controlled product purities. This fully demonstrates the effectiveness of the proposed PQE.

Author Contributions: Conceptualization, Y.Y.; Funding acquisition, K.H., H.C. and X.Q.; Investigation, Y.Y.; Project administration, K.H., H.C. and X.Q.; Visualization, X.T.; Writing—original draft, Y.Y.; Writing—review and editing, X.T., H.C. and X.Q. All authors have read and agreed to the published version of the manuscript.

Funding: This research was funded by National Natural Science Foundation of China, grant number 21878011 and 21808007.

Conflicts of Interest: The authors declare no conflict of interest.

References

1. Gao, Z.; Nguang, S.; Kong, D. Advances in modeling, monitoring, and control for complex industrial systems. *Complexity* **2019**, *2019*, 2975083.
2. Gao, Z.; Chen, M.Z.Q.; Zhang, D. Special issue on "Advances in condition monitoring, optimization and control for complex industrial processes". *Processes* **2021**, *9*, 664. [CrossRef]
3. Dejanović, I.; Matijašević, L.; Olujić, Ž. Dividing wall column—A breakthrough towards sustainable distilling. *Chem. Eng. Process.* **2010**, *49*, 559–580. [CrossRef]
4. Dejanović, I.; Matijašević, L.; Halvorsen, I.J.; Skogestad, S.; Jansen, H.; Kaibel, B.; Olujić, Ž. Designing four-product dividing wall columns for separation of a multicomponent aromatics mixture. *Chem. Eng. Res. Des.* **2011**, *89*, 1155–1167. [CrossRef]
5. Qian, X.; Jia, S.; Huang, K.; Chen, H.; Yuan, Y.; Zhang, L. Optimal design of kaibel dividing wall columns based on improved particle swarm optimization methods. *J. Clean. Prod.* **2020**, *273*, 123041. [CrossRef]
6. Kiss, A.A.; Bildea, C.S. A control perspective on process intensification in dividing-wall columns. *Chem. Eng. Process.* **2011**, *50*, 281–292. [CrossRef]
7. Dwivedi, D.; Halvorsen, I.J.; Skogestad, S. Control structure selection for three-product petlyuk (dividing-wall) column. *Chem. Eng. Process.* **2013**, *64*, 57–67. [CrossRef]
8. Dwivedi, D.; Halvorsen, I.J.; Skogestad, S. Control structure selection for four-product petlyuk column. *Chem. Eng. Process.* **2013**, *67*, 49–59. [CrossRef]
9. Kiss, A.A.; Rewagad, R.R. Energy efficient control of a BTX dividing-wall column. *Comput. Chem. Eng.* **2011**, *35*, 2896–2904. [CrossRef]
10. Yuan, Y.; Huang, K.; Chen, H.; Qian, X.; Zang, L.; Zhang, L.; Wang, S. Comparing composition control structures for Kaibel distillation columns. *Processes* **2020**, *8*, 218. [CrossRef]
11. Wang, S.J.; Wong, D.S.H. Controllability and energy efficiency of a high-purity divided wall column. *Chem. Eng. Sci.* **2007**, *62*, 1010–1025. [CrossRef]
12. Jia, S.; Qian, X.; Yuan, X.; Skogestad, S. Control structure comparison for three-product petlyuk column. *Chin. J. Chem. Eng.* **2018**, *26*, 1621–1630. [CrossRef]
13. Kim, K.; Lee, M.; Park, S. Two-point temperature control structure selection for dividing-wall distillation columns. *Ind. Eng. Chem. Res.* **2012**, *51*, 15683–15695. [CrossRef]
14. Fan, G.; Jiang, W.; Qian, X. Comparison of stabilizing control structures for four-product kaibel column. *Chem. Eng. Res. Des.* **2016**, *109*, 675–685. [CrossRef]
15. Villegas-Uribe, C.A.; Alcántara-Avila, J.R.; Medina-Herrera, N.; Gómez-González, R.; Tututi-Avila, S. Temperature control of a kaibel, agrawal and sargent dividing-wall distillation columns. *Chem. Eng. Process.* **2021**, *159*, 108248. [CrossRef]
16. Qian, X.; Liu, R.; Huang, K.; Chen, H.; Yuan, Y.; Zhang, L.; Wang, S. Comparison of temperature control and temperature difference control for a Kaibel dividing wall column. *Processes* **2019**, *7*, 773. [CrossRef]
17. Ling, H.; Luyben, W.L. Temperature control of the BTX divided-wall column. *Ind. Eng. Chem. Res.* **2010**, *49*, 189–203. [CrossRef]
18. Wu, N.; Huang, K.; Luan, S. Operation of dividing-wall distillation columns. 2. A double temperature difference control scheme. *Ind. Eng. Chem. Res.* **2013**, *52*, 5365–5383. [CrossRef]
19. Yuan, Y.; Huang, K.; Chen, H.; Zhang, L.; Wang, S. Configuring effectively double temperature difference control schemes for distillation columns. *Ind. Eng. Chem. Res.* **2017**, *56*, 9143–9155. [CrossRef]
20. Pan, H.; Wu, X.; Qiu, J.; He, G.; Ling, H. Pressure compensated temperature control of kaibel divided-wall column. *Chem. Eng. Sci.* **2019**, *203*, 321–332. [CrossRef]

21. Luyben, W.L. *Distillation Design and Control Using Aspen Simulation*; John Wiley & Sons: New York, NY, USA, 2006.
22. Gao, R.; Gao, Z. Pitch control for wind turbine systems using optimization, estimation and compensation. *Renew. Energy* **2016**, *91*, 501–515. [CrossRef]

processes

MDPI

Article

Linear Golden Section Speed Adaptive Control of Permanent Magnet Synchronous Motor Based on Model Design

Wenping Jiang [1,*], Wenchao Han [1], Lingyang Wang [1], Zhouyang Liu [2] and Weidong Du [1]

[1] School of Electrical and Electronic Engineering, Shanghai Institute of Technology, Shanghai 201418, China; wenchaohan2022@163.com (W.H.); azureleo7@163.com (L.W.); dd1207570822@163.com (W.D.)
[2] Shanghai Academy of Spaceflight Technology, Shanghai 201109, China; liuzhouyang717@126.com
* Correspondence: jiangwenping@sit.edu.cn

Abstract: Permanent magnet synchronous motor (PMSM) is a multi-variable, strongly coupled, nonlinear complex system. It is usually difficult to establish an accurate mathematical model, and the introduction of new complex algorithms will increase the difficulty of embedded code development. In order to solve this problem, we establish the characteristic model of permanent magnet synchronous motor in this paper, and the speed control scheme of the linear golden-section adaptive control and integral compensation, which is adopted. Finally, using the model-based design (MBD) method, how to build the simulink embedded code automatic generation model is introduced in detail, and then we complete the PMSM speed control physical verification experiment. Simulation and experimental results show that compared with traditional proportional-integral-derivative (PID) control, the speed control accuracy of PMSM is improved about 3.8 times. Meanwhile, the development method based on the model design can increase the PMSM control system physical verification, and then improve the development efficiency.

Keywords: permanent magnet synchronous motor; characteristic model; linear golden-section adaptive control; model-based design

Citation: Jiang, W.; Han, W.; Wang, L.; Liu, Z.; Du, W. Linear Golden Section Speed Adaptive Control of Permanent Magnet Synchronous Motor Based on Model Design. *Processes* **2022**, *10*, 1010. https://doi.org/10.3390/pr10051010

Academic Editor: Jose Carlos Pinto

Received: 10 April 2022
Accepted: 15 May 2022
Published: 19 May 2022

1. Introduction

Permanent magnet synchronous motor (PMSM) has been widely used in automotive, aerospace, and other fields in recent years due to its small size, light weight, and high power density [1]. Because PMSM is a nonlinear, multi-variable, strongly coupled system, coupled with factors such as the parameter changes during operation, it is difficult to establish an accurate mathematical model. Therefore, in recent years the control of PMSM has also become a hot research topic.

Direct torque control (DTC) and field-oriented control (FOC) are the most common basic methods at present for PMSM. The vector control system adopts traditional proportional-integral (PI) control, which has a simple model and is easy to implement. Currently, it is widely used in PMSM. However, traditional PI control cannot solve the contradiction between overshoot and rapidity, and it is easily affected by parameter changes [2,3]. In response to this problem, based on the vector control, there are a series of modern control methods of PMSM. For example, the sliding-model variable structure control using the sliding mode control is insensitive to parameters and has a fast response speed. These two advantages improve the dynamic characteristics of the PMSM [4,5] but the control accuracy is not too high due to the existence of chattering interval. Compared with the traditional PI control, the proportional resonance control eliminates the coupling between the d and q axis components, and the system implementation is more simple [6], but it still cannot solve the contradiction between overshoot and rapidity. Fuzzy control does not require an accurate mathematical model and has strong robustness [7,8], but it relies more on control experience and expert knowledge. Although model predictive control and active disturbance rejection control can improve the speed control accuracy and dynamic performance of the system [9–12], the algorithms are complex and not easy to implement quickly. In [13], a backstepping sliding mode control based on a recurrent radial basis function network (RBFN) for a PMSM is presented, with a novel combination of the backstepping method and sliding mode control, eliminating the chattering effectively without losing the precision. In [14], an online PID parameter adjustment control, combining model predictive control and on-line fuzzy rule adjustment, is proposed. DTC is another common control method for PMSM, but it has its drawbacks with the problem of large torque pulsation particularly evident. In [15], a new DB-DTFC algorithm to solve the stator reference voltage in a stator-flux-oriented coordinate system is proposed. In [16], an analytical motor model taking the spatial harmonics and magnetic saturation characteristics of PMSM into account by reconstructing the numerical solution of magnetic co-energy (MCE) from finite element analysis (FEA) is proposed. In [17], a hybrid decision control strategy based on DNN and DTC (direct torque control) is proposed. In [18], an Ant Colony Optimization (ACO) algorithm was proposed to adjust the PID controller gains of the DTC control.

In the application of the characteristic model for PMSM, Literature [19] used the golden-section of permanent magnet synchronous motor and maintained tracking control of the integrated control method, but this paper did not consider the characteristic model and the sensitivity of the golden-section control of the step signal problems, in the initial stage. The result is that the initial stage has a larger amount of overshoot, and it is not conducive to the stability of the system. Literature [20] used the nonlinear golden-section method, and the introduction of the initial phase transition process gives the system obvious improvement, but it does not spell out the coefficient values of the transition process and the specific influence on the system, and using the nonlinear golden control method also increases the complexity of the system. It is not conducive to the realization of the ascension of the response speed and the actual system.

Based on these studies, this paper starts from the perspective of control accuracy and engineering application. Firstly, characteristic modeling of the PMSM speed control system is carried out, and then the parameters of the characteristic model are identified online by the gradient method. Only linear golden-section controller and integral controller are used to achieve stable tracking. The first-order inertial filter is introduced in the linear golden-section controller as the starting method, and the influence of the filter coefficient on the system is discussed. Finally, the paper introduces the model-based design method in the controller, gives a detailed modeling method, and uses the DSP controller and servo driver to carry out a physical verification test. Experimental results show that the method used in this paper has a high control accuracy, and based on the model design method, can greatly shorten the system verification development cycle, and can be applied to industrial control.

The main contributions to the paper are as follows:

- Based on the characteristic model theory, the rotational speed control characteristic model of PMSM is built.
- Based on the characteristic model theory, a linear golden section and integral compensation PMSM rotational speed adaptive model is built.
- Based on the model design method, the specific modeling method and the verification process of the controller proposed in this paper are proposed.
- The simulation and experiment results show that the proposed method has good performance and is suitable for practical applications.

This paper is organized as follows. In Section 1, the introduction of the control method of PMSM is given. In Section 2, the establishment process of the mathematical model and characteristic model of the PMSM is described. In Section 3, the design of the speed adaptive controller is described. In Section 4, the method of model-based design is described. In Section 5, the simulation and experimental results are given. In Section 6, the detailed conclusions and suggestions for further work are given.

2. Modeling of PMSM

2.1. Mathematical Model of PMSM

Considering the complex relationship between different variables and the complex motion law of permanent magnet synchronous motor, its dynamic mathematical model is nonlinear and multivariate. Therefore, when considering the mathematical model of the three-phase permanent magnet synchronous motor, the following conditions are assumed.

1. Ignore magnetic saturation and hysteresis loss, etc.;
2. Assuming the rotor adopts no damping winding, the external conditions change, and the physical properties of the stator do not change;
3. The conductivity of the rotor permanent magnets and the internal rotor permeability are assumed to be 0;
4. It is assumed that the induced potential in the three-phase winding is a standard sine wave during steady-state operation;
5. Ignore all spatial harmonics in the magnetic field;
6. The windings are distributed symmetrically, the windings turn's number is the same, and the displacement electrical angles between the axes are the same.

Based on the above conditions, the stator flux Equation of permanent magnet synchronous motor in the two-phase rotating coordinate system is molded as:

$$\begin{cases} \psi_{\mathrm{d}} = L_d i_d + \psi_f \\ \psi_q = L_q i_q \end{cases} \quad (1)$$

The voltage Equation is molded as:

$$\begin{cases} u_{\mathrm{d}} = R_s i_{\mathrm{d}} + L_{\mathrm{d}} \frac{di_{\mathrm{d}}}{dt} - \omega \psi_q \\ u_q = R_s i_q + L_q \frac{di_q}{dt} + \omega \psi_d \end{cases} \quad (2)$$

where, ψ, i and u represent flux linkage, current, and voltage respectively, ψ_{f} is permanent magnet flux, R_s is stator phase resistance, L_{d} and L_{q} are synchronous inductors of axis d and q respectively, and ω is angular velocity.

When the motor is running stably, assuming that the steady-state voltage and current of d-axis and q-axis are respectively, and ignore the resistance voltage's drop, the torque Equation in the dq-reference frame is:

$$\tau_{\mathrm{em}} = 1.5 p_n \left[\psi_f i_q + (L_d - L_q) i_d i_q \right] \quad (3)$$

The Equation of motion is:

$$\tau_{\mathrm{em}} = \tau_L + \frac{1}{p_n} B\omega + \frac{1}{p_n} J \frac{d\omega}{dt} \quad (4)$$

where, J is the moment of inertia, ω is the angular velocity, τ_{em} is the electromagnetic torque, τ_{L} is the load torque, p_n is the polar logarithm, B and is the viscous friction coefficient.

In the vector control system, the more application is $i_d = 0$, at this point, the stator current vector on direct axis component to 0, all the current is used for torque control. The rotor magnetic field space vector is perpendicular to the magnetomotive stator force space vector, the electromagnetic torque and the current stator form a first-order linear function relationship, the counter electromotive force, and the same direction, the motor is to achieve the highest efficiency. The size of torque can be controlled through control. In this paper, surface mount PMSM is taken as the control object. For surface mount PMSM, define $L_q = L_d = L$ the mathematical model under $i_d = 0$ control mode is approximately as follows:

$$\begin{cases} \frac{di_q}{dt} = -\frac{R_s}{L} i_q - p_n \omega i_d - \frac{p_n \psi_f}{J} \omega - \frac{1}{L} u_q \\ \frac{d\omega}{dt} = \frac{3 p_n \psi_f}{2J} i_q - \frac{B\omega}{J} - \frac{\tau_L}{J} \end{cases} \quad (5)$$

According to Equation (5), the speed loop of the surface mount PMSM vector control system has a linear relationship between the output speed and the q-axis current, the motor speed can be controlled by control i_q. Figure 1 shows a schematic diagram of the FOC vector control system.

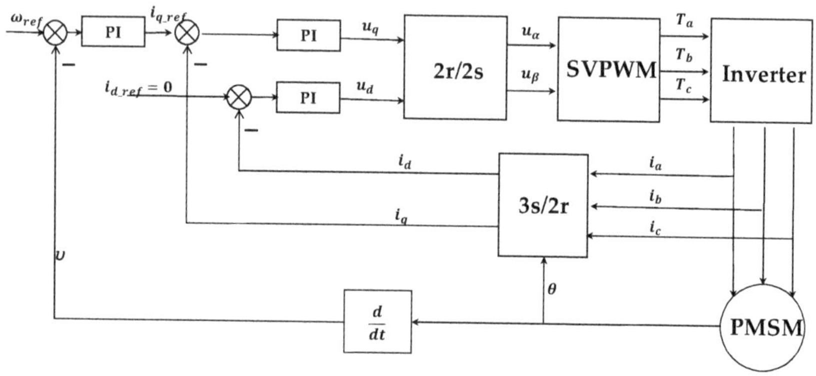

Figure 1. Diagram of FOC vector control system.

2.2. Characteristic Model of PMSM

According to Equation (5), PMSM can be seen as a time-varying linear time-invariant system, let

$$\frac{d\omega}{dt} \approx \frac{\omega(k+1) - \omega(k)}{T} \tag{6}$$

This is,

$$\omega(k+1)+(\frac{B}{J} - 1)\omega(k) = \frac{3Tp_n\psi_f}{2J} i_q(k) - \frac{T\tau_L}{J} + \Delta k \tag{7}$$

In Equation (7), T is the sampling period, Δk is the discretization error. According to the characteristic model theory, for a linear time-invariant system, under a certain sampling period, the characteristic model can be described by a second-order time-varying difference Equation for position holding or tracking control [21] as Equation (8):

$$y(k+1) = f_1(k)y(k) + f_2(k)y(k-1) + g_0(k)u(k) + g_1(k)u(k-1) \tag{8}$$

where, $f_1(k)$, $f_2(k)$, $g_0(k)$ and $g_1(k)$ are the system parameters to be identified, $y(k+n)$ is the output of the system, and $u(k+n)$ is the input of the system. For the convenience of control, the $g_1(k)u(k-1)$ term is discarded within the allowable error range, we can get the characteristic model of PMSM by Equation (8) as Equation (9):

$$\omega(k) = f_1(k)\omega(k-1) + f_2(k)\omega(k-2) + g_0(k)i_q(k-1) \tag{9}$$

In Equation (9), $f_1(k)$, $f_2(k)$ and $g_0(k)$ are the system parameters to be identified, $\omega(k)$ is the speed output value of the system at the moment of k, $\omega(k-n)$ are the speed output value of the system at the moment of $k-n$ and $i_q(k-n)$ are the shaft current feedback value of the system at the moment of $k-n$.

When the sampling time is small enough, the range of characteristic parameters is $f_1(k) \in (1,2]$, $f_2(k) \in [-1,0) g_0(k) \ll 1$ for Equation (9), online identification of characteristic parameters can be carried out according to the input and output values. In this paper, the gradient method is adopted for identification.

$$\begin{aligned} \phi(k) &= [\omega(k)\omega(k-1)\omega(k-2)]^{\mathrm{T}} \\ \theta(k) &= [f_1(k) f_2(k) g_0(k)]^{\mathrm{T}} \end{aligned} \tag{10}$$

Then the identification Equation is [22]:

$$\theta(k) = \theta(k-1) + \frac{\lambda_1 \phi(k)}{\phi^T(k)\phi(k) + \lambda_2} \times \left[\omega(k) - \phi^T(k)\theta(k-1)\right] \tag{11}$$

where, λ_1 and λ_2 is determined by the amount of interference and the speed of convergence. In general, $0 < \lambda_1 < 1$ $0 < \lambda_2 < 4$.

3. Speed Adaptive Control Scheme

Based on the characteristic model, the first by the gradient method parameter online identification characteristics, and then on the basis of the theory of the all-coefficient adaptive into linear golden-section controller, to guarantee the stability of the output can track the reference signal, the integral compensation controller, among them, the reference signal using first-order low-pass filter processing. Meanwhile, the total output is limited to prevent output saturation [23]. Figure 2 shows the overall speed adaptive control system structure [24].

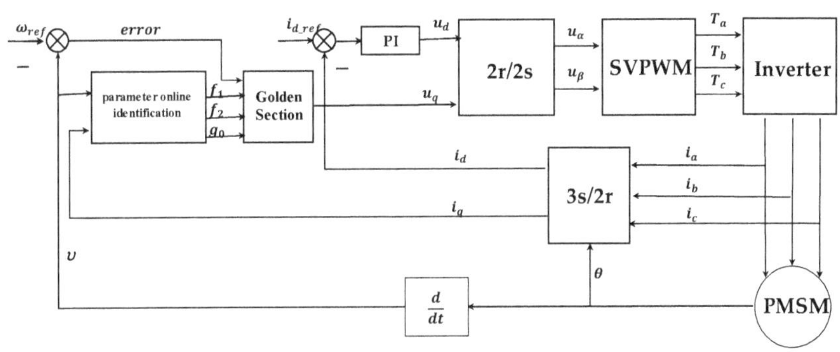

Figure 2. Golden-section speed adaptive system structure diagram.

3.1. Input Signal Processing

For step-type speed signal, due to the large initial error, the output adjustment will be too large, which will cause system oscillation and overshoot, and will also lead to motor torque oscillation, which is not conducive to the stable operation of the motor and will cause adverse effects on the life of the motor. In order to reduce overshoot and oscillation, a first-order low-pass filter is introduced to smooth the step signal of the reference input. The discretized first-order low-pass filtering Equation is:

$$\omega_{ro}(k) = \alpha\omega_{ri}(k) + (1-\alpha)\omega_{ro}(k-1) \tag{12}$$

where, α is the filtering coefficient, $\omega_{ro}(k)$ is the filtered output value at the moment of k, $\omega_{ro}(k-1)$ is the filtered output value at the moment of $k-1$, and $\omega_{ri}(k)$ is the input sampling value at the moment of k.

The first-order low-pass filtering method uses this sampling value and the last filtering output value to be weighted to obtain an effective filtering value so that the output has a feedback effect on the input. Among them, the smaller the filtering coefficient, the smoother the filtering result, but the sensitivity will be reduced; the larger the filtering coefficient, the higher the sensitivity, but the filtering result will be unstable.

3.2. Linear Golden-Section Speed Adaptive Controller

Introducing the golden-section ratio into the control system constitutes the golden-section control. In the characteristic model adaptive control theory, the identification parameters are combined with the golden ratio, and the model parameters of the system are identified online by observing the input and output data. According to Formula (9), a linear golden-section adaptive controller of the PMSM speed control system can be designed, and the formula is:

$$U_L(k) = \frac{-l_1\hat{f}_1(k)e(k) - l_2\hat{f}_2(k)e(k-1)}{\hat{g}_0(k) + k_L} \tag{13}$$

In Equation (13), $l_1 = 0.382$, $l_2 = 0.618$ is the golden-section coefficient, and $\hat{f}_1(k)$, $\hat{f}_2(k)$, $\hat{g}_0(k)$ are the characteristic model coefficients identified online, $e(k)$ is the speed error at time k, $e(k-1)$ is the speed error at time $k-1$, k_L is the adjustable parameter, which determines the stability and immunity of the system, and $0 \le k_L < 1$.

Based on Equation (13), the voltage control quantity of an axis q can be obtained, and a relatively stable speed control system can be achieved through adjustment k_L. However, the system cannot reach the expected tracking value at this time. There is a certain steady-state error, therefore, a compensator needs to be introduced.

3.3. Integral Compensator

The characteristic model and golden-section adaptive control system have a simple structure and are easy to realize. Integral compensation also has the same characteristics and is widely used. Therefore, this paper uses integral compensator as the voltage control quantity of the second q-axis [25,26], denoted as the compensation Equation:

$$U_I(k) = U_I(k-1) + k_I e(k) \tag{14}$$

In Equation (14), $U_I(k)$ is integral compensation output value the time of k, $U_I(k-1)$ is integral compensation output value the time of $k-1$, k_I is the integral coefficient, and, the total axis voltage control quantity is:

$$U_q(k) = U_L(k) + U_I(k) \tag{15}$$

The whole control system has only two adjustable parameters, and the adjustment range is determined, the overall structure is simple, and easy to achieve engineering.

4. PMSM Control System Based on Model Design

4.1. General Process of Model-Based Design

The traditional design is divided into four stages: requirement, design, implementation, and testing [27,28]. It has the disadvantages of low efficiency, high difficulty, and high requirements, which is not conducive to verifying new algorithms. Therefore, model-based design is introduced in this paper.

Compared with traditional design, model-based design connects the four stages and synchronously promotes modeling and verification testing on a unified test platform, which can reduce the migration process and enable engineers to focus on the research of algorithms, greatly shorten the development cycle and reduce the development cost. The general flow of model-based design is shown in Figure 3:

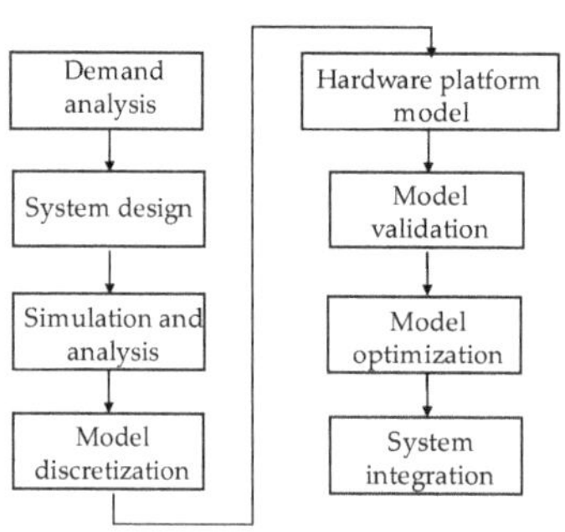

Figure 3. The general process of model-based design.

4.2. DSP Peripheral Configuration Based on Model Design

MATLAB provides many hardware support packages, including TI C2000 processor series, for the basic FOC system, which generally need ePWM module, ADC module, eQEP module and interrupt system [29].

The ePWM module is used to generate three complementary PWM waves. The configurable parameters include period, duty cycle, dead time, input polarity and external trigger events, etc. Set the duty cycle as an external input, and calculate the duty cycle, and then it can be linked to the ePWM input pin of the hardware. The external trigger event sets to start ADC conversion, which means that when the bridge arm turns on, the current value at this moment is collected.

The ADC module is used to configure the ADC value represented by the acquisition current, select the corresponding channel, and set it to ePWMxA to trigger the conversion. The data output type is set to uint16. Finally, the conversion completion interrupt needs to be enabled, and the core program of the algorithm needs to be executed in the interrupt.

The eQEP module is used to configure and collect the encoder signal, and then obtain the rotor position, direction, and speed information. For the quadrature photoelectric encoder, the counting mode can be directly set to quadrature counting, which is equivalent to 4 times the counting frequency of QEP, and then improves speed calculation accuracy. In addition, the module also provides a flag pulse QEP_index, the flag bit will generate a pulse for speed calculation every time the motor rotates one revolution.

4.3. Build PMSM Golden-Section Adaptive Code Generation Model

In this paper, the model-based design method is used to model and verify the golden-section speed adaptive system. First, the characteristic model algorithm is simulated and analyzed in simulink, and then the model is discretized and converted into an embedded code model for modular testing. Finally, the system integration is carried out, and the generated board file is directly downloaded to the corresponding DSP hardware to verify the correctness and efficiency of the model algorithm.

In order to increase the program processing speed, it is necessary to perform fixed-point processing on the core model, mainly processing the data collected by the ADC and eQEP modules and some mathematical calculation modules. In order to accelerate the processing speed of the program, fixed-point processing is needed for the core model, mainly for the data collected by ADC, eQEP module, and part of the mathematical calculation module.

For the acquisition of phase current, it is only necessary to collect AB two-phase current. Since the ADC of C2000 series is unipolar, it cannot collect negative voltage signals. Bipolar signals need to be biased, and the bias voltage is set to 1.65 V. The current signal can be converted into a 0~3.3 V voltage signal that can be processed by the embedded hardware, and then the collected data is shifted to the left by 6 bits to form a normalized Q17 format, and the output data is set to fixdt (1,32,17). The current acquisition and data conversion model is shown in Figure 4:

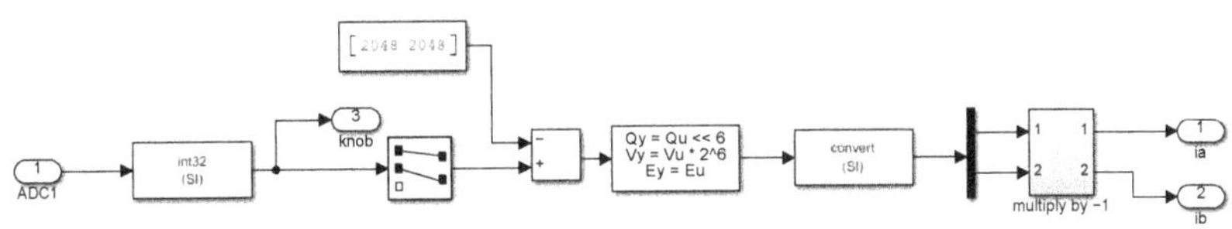

Figure 4. Current acquisition model.

For obtaining rotor position and speed, it is divided into electrical Angle calculation and speed calculation modules. The calculation formula of electrical Angle is:

$$\begin{cases} N_{\mathrm{PU}} = N_{\mathrm{QEP}} - (fix(\frac{N_{QEP}}{N_{\mathrm{PP}} \times R}) \times p_n \times R) \\ ETheta = \frac{N_{\mathrm{PU}} \times p_n}{R} \end{cases} \tag{16}$$

In Equation (16), N_{PU} is the number of counting pulses, N_{QEP} is the number of encoder pulses, p_n is the number of motor poles, R is the number of encoder lines, $ETheta$ is the electrical Angle. Since the encoder has a correction angle when the motor is delivered, which is the deviation between the center line of the rotor magnetic field and the zero point of the encoder, the software correction is also required. The electrical Angle model after adding correction and processing Q17 data format is shown in Figure 5:

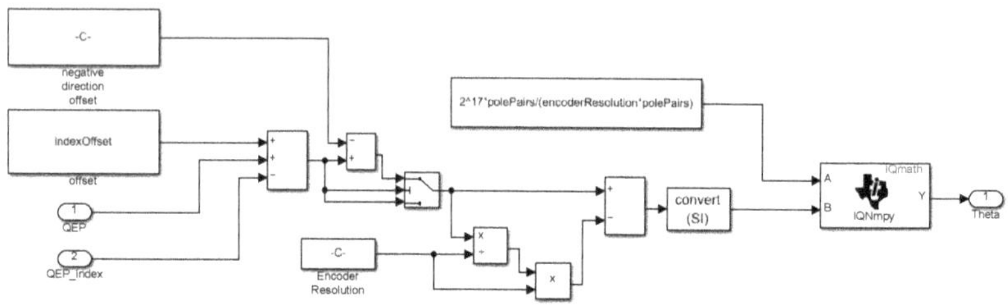

Figure 5. Electrical Angle calculation model.

The rotor speed is calculated by M-method, and the expression is:

$$\omega_e(k) = \frac{\theta_e(k) - \theta_e(k-1)}{f_b \times T} \tag{17}$$

In Equation (17), θ_e is the electrical Angle of the rotor position, f_b is the reference frequency, and T is the sampling period. The speed calculation model is shown in Figure 6:

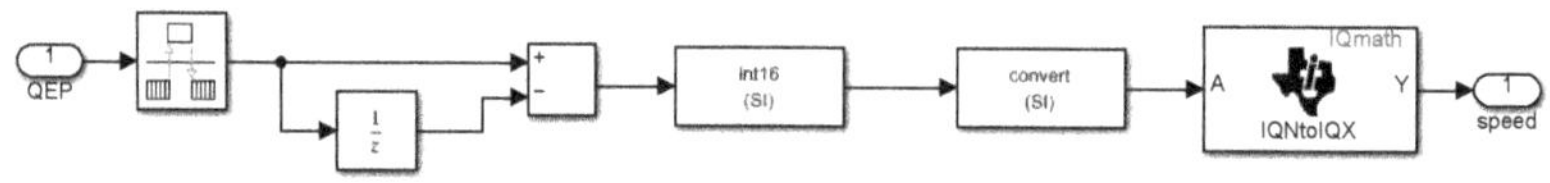

Figure 6. Rotational speed calculation model.

Combining the phase current acquisition and rotor information acquisition models, the FOC model of embedded automatic code generation is built as shown in Figure 7. On this basis, the PI controller of the q-axis is replaced by the discrete model of the golden-section adaptive method described in the paper, which constitutes the automatic code generation model of the golden-section-based rotational speed adaptive control system as shown in Figure 8.

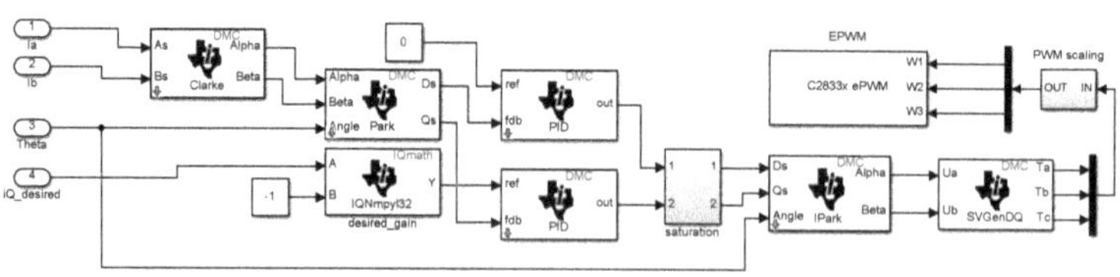

Figure 7. DSP automatic code generation model for FOC system.

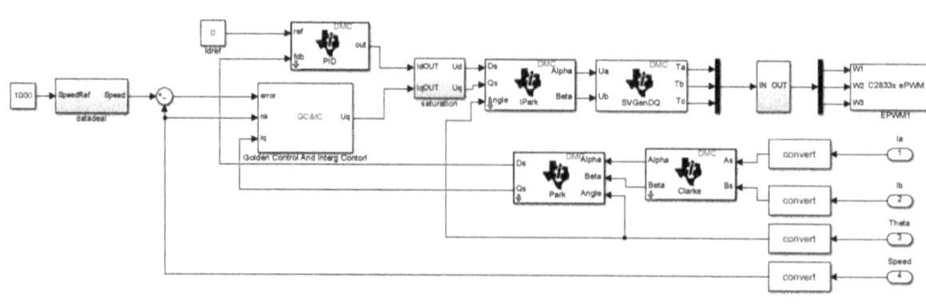

Figure 8. Golden-section speed adaptive code generation model.

5. Simulation and Experimental Research

The control chip used in this paper is TMS320F28335, with a floating point processing unit, and the motor is a surface mount low-voltage permanent magnet synchronous motor, whose specific parameters are shown in Table 1. The actual hardware of the overall system is shown in Figure 9. the speed sensor is an incremental photoelectric encoder, its model is H40-6-0500VL, 500 pulses per circle, and maximum support speed is 8000 rpm. This platform is used as the verification system to verify the PMSM's linear golden-section rotational speed adaptive control system.

Table 1. PMSM parameters.

Parameter	Value
Rated voltage (V)	36
Rated current (A)	4.6
Pole pairs	4
Phase resistance (Ω)	0.38
Phase inductance (mH)	1
Coefficient of viscous friction (N·m·s)	0.0001
Moment of inertia (kg·cm^2)	0.0588
Magnetic flux amplitude (Wb)	0.11867
Rated torque (N)	0.318

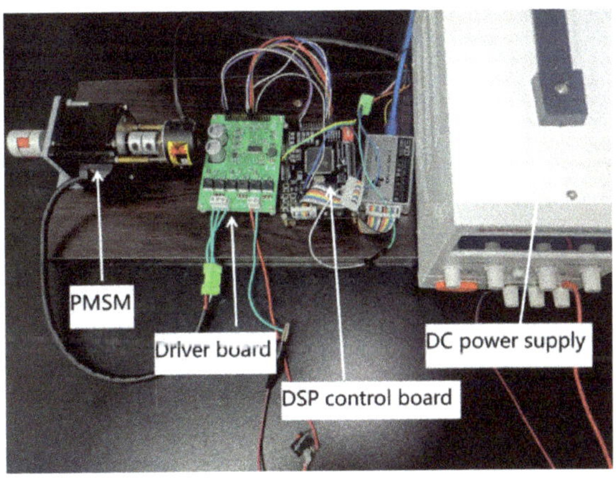

Figure 9. PMSM hardware diagram.

5.1. Simulation Analysis

5.1.1. Influence of Filter Coefficient on the System

The golden-section speed adaptive control scheme is sensitive to step signals, so first-order low-pass filtering is used to process the input reference signals. The filtering coefficient has different influences on the system results. The reference speed is set at 1000 RPM and the simulation time is 0.1 s and Figures 10–13 show the speed response curves under different values.

As can be seen from Figures 10–13, the larger α is, the shorter its adjustment time. Meanwhile, the initial error is large, and the system oscillates. The smaller α is, the more stable its speed tracking, and the smaller the error, but it will increase the system adjustment time and reduce the adjustment sensitivity. Considering the stability and rapidity of the system comprehensively, filter coefficient $\alpha = 0.002$ is selected, and the speed error curve is shown in Figure 14.

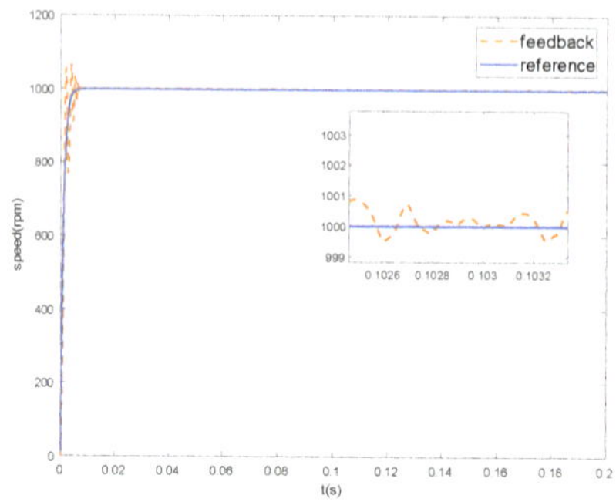

Figure 10. Speed response curve when $\alpha = 0.01$.

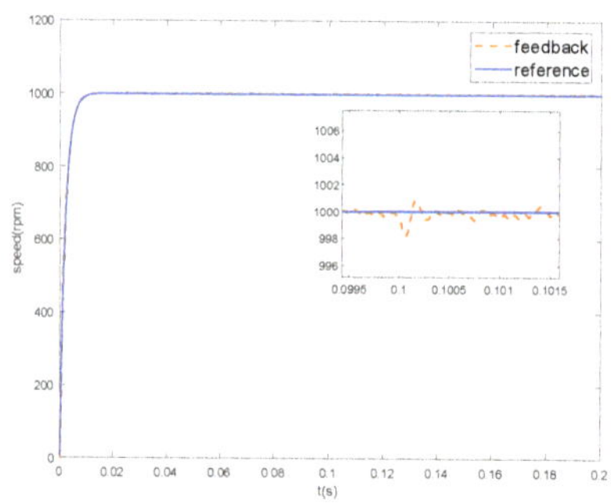

Figure 11. Speed response curve when $\alpha = 0.005$.

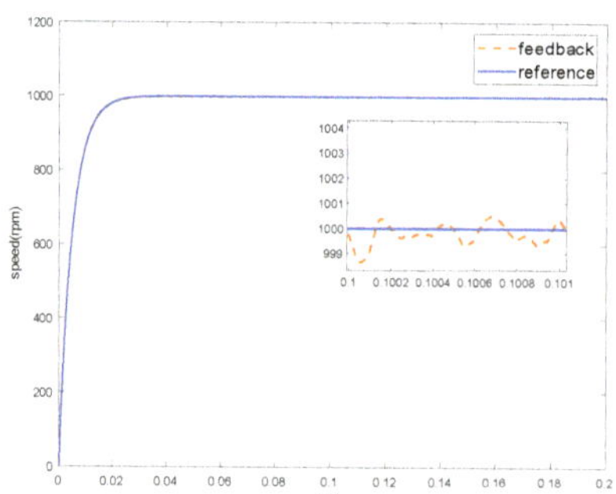

Figure 12. Speed response curve when $\alpha = 0.002$.

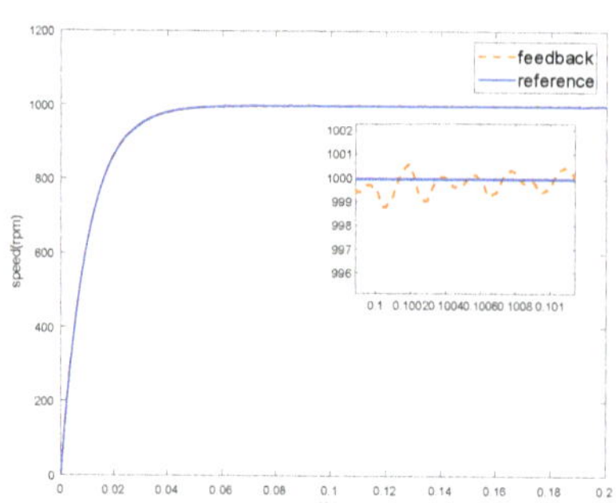

Figure 13. Speed response curve when $\alpha = 0.001$.

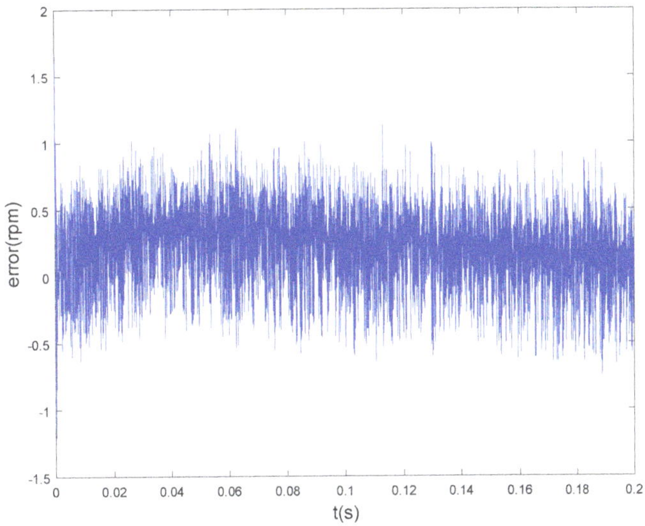

Figure 14. Speed error curve when $\alpha = 0.002$.

5.1.2. Parameter Identification Results and Analysis

When the sampling time is small enough, the range of characteristic parameters $f_1(k) \in (1,2] f_2(k) \in [-1,0) g_0(k) \ll 1$, generally, takes the initial value of parameter identification $f_1(k) = 2 f_2(k) = -1 g_0(k) = 0.001$. Set the simulation time to 0.1 s, the reference speed to 1000 rpm, and the filter coefficient $\alpha = 0.002$. Figures 15–17 show the online parameter identification results of the second-order characteristic model. It can be seen from the figure that the identification parameters have little change after steady-state operation, which further indicates that the speed adaptive system has strong stability.

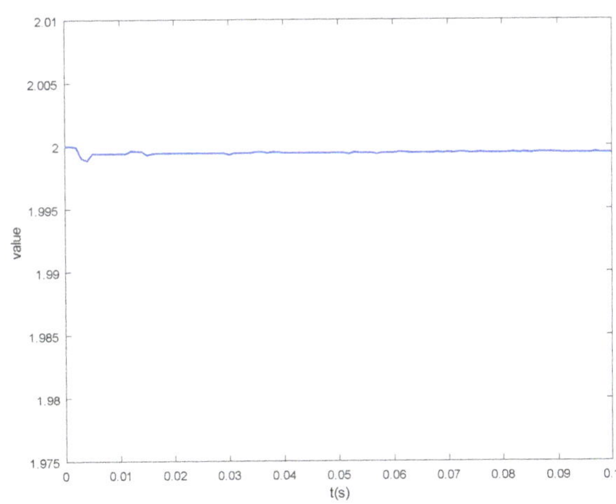

Figure 15. Characteristic parameter identification curve of $f_1(k)$.

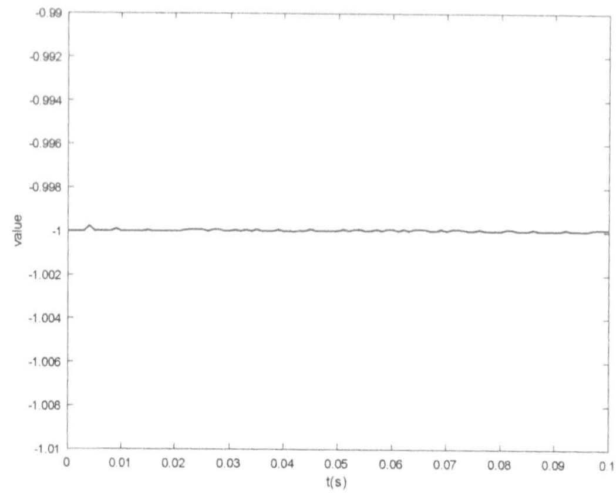

Figure 16. Characteristic parameter identification curve of $f_2(k)$.

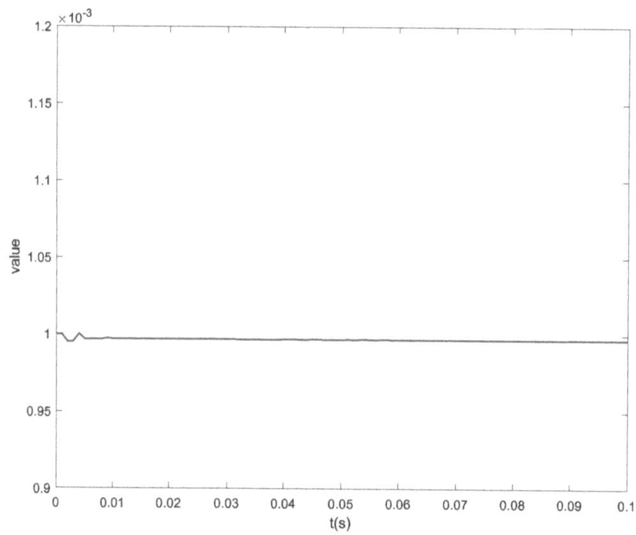

Figure 17. Characteristic parameter identification curve of $g_0(k)$.

5.1.3. Load Performance Analysis

In order to verify the anti-disturbance of the proposed method, a loading simulation experiment was carried out on the system. The rated torque of the motor used was 0.318 N·m, the sudden load was set at 0.1 s, and the loading value was 0.1 N·m. Figure 18 shows the speed response curve, and Figure 19 shows the curve of torque. It can be seen from the figure that in the process of steady-state operation, the impact of sudden load on the system speed can be ignored, indicating that the proposed method has a high anti-disturbance ability.

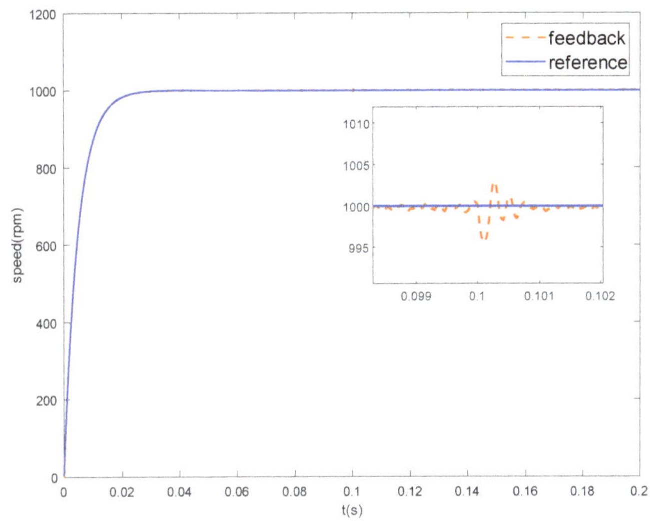

Figure 18. Curve of speed after sudden loading.

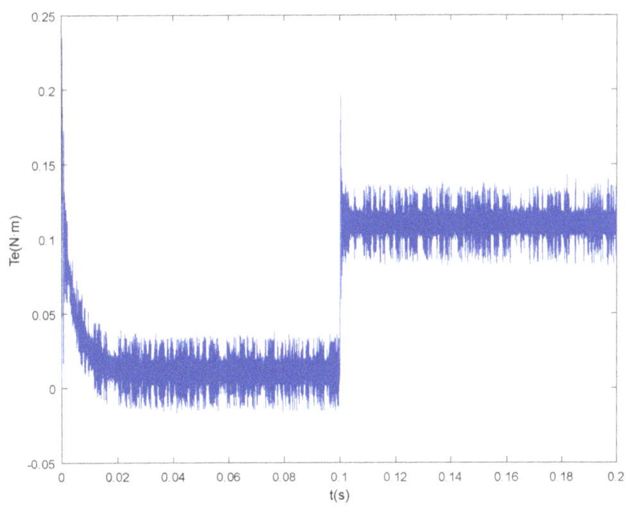

Figure 19. Curve of torque for sudden loading.

5.2. Actual Product to Verify

Physical verification is the key link of model-based design, which determines the feasibility of the proposed method. After the correct simulation verification, the core algorithm needs to be discretized, and then the system motor in the simulation is replaced by the actual motor feedback value which is combined with the hardware.

In this paper, the recursive parameter identification model, the golden-section model, and the integral model are firstly discretized, and then the data is converted into per-unit values. The reference value selects the rated value of the motor. The generated FOC framework interface is connected, and finally, the model is compiled and downloaded to the hardware core unit and optimized to complete the physical verification process. In order to illustrate the performance superiority of the method proposed in this paper, the golden section controller with integral compensation based on the characteristic model is compared with the traditional PID, and the PMSM in Table 1 and the hardware system in Figure 9 are used for verification, and the speed reference value is set to 1000 rpm. Figure 20 is the actual operation effect of the traditional PID and the method proposed in this paper.

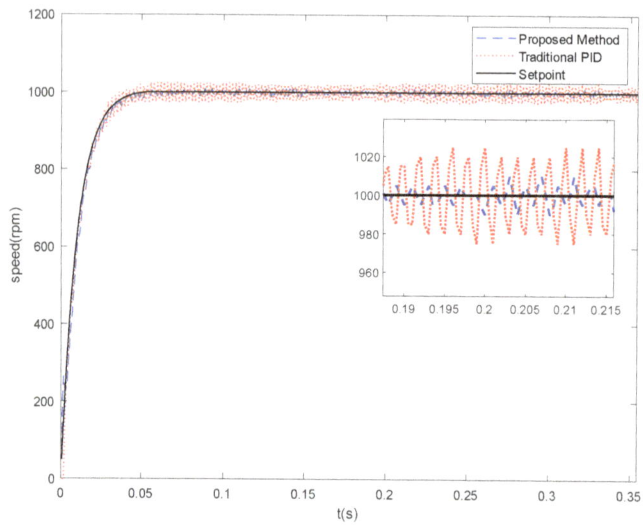

Figure 20. Actual operation effect of traditional PID and proposed method.

Figure 20 shows that the performance of the proposed method is significantly better than that of traditional PID control. However, it is limited by hardware, such as encoder accuracy, current sampling accuracy, etc. The error of actual operation is larger than that of simulation, which can be improved by improving hardware performance.

In order to highlight the actual performance of the proposed method, settling time and average volatility are introduced, where, average volatility is defined as the ratio of the absolute value of the steady-state error to the reference value. Table 2 shows the running results of the proposed method about linear golden-section control (LGSC), nonlinear golden-section control (NGSC), and PID control. They were all carried out on the same experimental platform with reference speed set to 1000 rpm and the frequency of the control system was set to 20 KHz.

Table 2. Performance analysis of different control methods.

Method	Settling Time/ms	Average Volatility/%
LGSC	14	0.23
NGSC	35	0.43
PID	260	1.10

As can be seen from Table 2, our method has better performance than the results of nonlinear golden-section control and traditional PID control. The proposed method does not contain high-order terms, so it has fast regulation speed, which makes it superior to NGSC and PID in rapidity. In terms of average volatility, for the same experimental platform, after the motor enters steady-state operation, the average speed error of the proposed method is 2.3 rpm, the average speed error of NGSC is 4.3 rpm, while the traditional PID is as high as 11 rpm, which shows that the characteristic model and golden-section control have advantages in control accuracy. In conclusion, the proposed method has advantages in rapidity and accuracy.

6. Conclusions

In view of the difficulty of establishing an accurate mathematical model for PMSM, and the introduction of new algorithms increasing the difficulty of embedded code development, this paper adopts the model-based design method on the basis of the characteristic model, and introduces the linear golden-section and integral compensation controller. The speed adaptive control of PMSM is carried out, and the simulation and physical verification tests are carried out. In addition, the first-order low-pass filter is used to process the reference signal, and the influence of different filter coefficients on the result is discussed. Finally, a more suitable coefficient value is selected, which gives the system rapidity but also solves the overshoot problem. Meanwhile, the model-based design method is used to verify the proposed control scheme, and the performance comparison with traditional PID control and nonlinear golden-section control is given. The experiment shows that compared with traditional PID control, the speed control accuracy of PMSM is improved about 3.8 times. In the future, we plan to reduce the order of the characteristic model and use the first-order characteristic model to reduce the identification parameters and further improve the system response speed under the premise of ensuring accuracy.

Author Contributions: Conceptualization and methodology, W.J. and W.H.; software, validation, investigation and writing—original draft preparation, W.H.; formal analysis, L.W.; resources, Z.L.; project administration, W.D. All authors have read and agreed to the published version of the manuscript.

Funding: This research was funded by Shanghai Institute of Technology, grant number XTCX2021-10.

Institutional Review Board Statement: Not applicable.

Informed Consent Statement: Not applicable.

Data Availability Statement: Not applicable.

Conflicts of Interest: The authors declare no conflict of interest.

References

1. Wang, G.; Valla, M.; Solsona, J. Position Sensorless Permanent Magnet Synchronous Machine Drives—A Review. *IEEE Trans. Ind. Electron.* **2019**, *67*, 5830–5842. [CrossRef]
2. Lin, R.; Chen, S.; Ding, X. Modeling and Simulation of Neuron Integral-separated PID Control for Permanent Magnet Synchronous Motor. *J. Fuzhou Univ.* **2009**, *37*, 849–852.
3. Ding, W.; Gao, L. Modeling and Simulation of Permanent Magnet Synchronous Motor Vector Control System. *Micromotors* **2010**, *43*, 66–71. [CrossRef]
4. Liu, Z.G.; Wang, J.Z. Design of Neural Network Adaptive Sliding Mode Controller for Permanent Magnet Synchronous Motor. *Electr. Mach. Control.* **2009**, *13*, 290–295. [CrossRef]
5. Ma, H.C. Design of a New Sliding Mode Speed Controller for Permanent Magnet Low Speed Synchronous. *Motor. Control. Eng. China* **2016**, *23*, 1757–1762. [CrossRef]
6. Chen, Z.; Zhang, Z.H. Current Decoupling and Harmonic Suppression Strategy of Permanent Magnet Synchronous Motor Based on Proportional Resonant Active Disturbance Rejection Control. *Proc. CSEE* **2021**, *58*, 1–10. [CrossRef]
7. Sha, L.; Song, Y.; Li, Z. Permanent Magnet Synchronous Motor Control Method Based on Fuzzy PI and Sliding Mode Observer. *J. Phys. Conf. Ser. IOP Publ.* **2022**, *2218*, 012053. [CrossRef]
8. Ding, X.; Li, R.; Cheng, Y.; Liu, Q.; Liu, J. Design of and Research into a Multiple-Fuzzy PID Suspension Control System Based on Road Recognition. *Processes* **2021**, *9*, 2190. [CrossRef]
9. Wang, A.P.; Huang, X.Z. Variable Weight Coefficient Multi-step Model Predictive Current Control Method for Permanent Magnet Linear Synchronous Motor. *Proc. CSEE* **2022**, *59*, 1–11. [CrossRef]
10. Yang, P.M.; Liu, Y.C. Voltage Stability Control Strategy Based on Model Predictive Current Compensation. *Micromotors* **2022**, *55*, 56–64. [CrossRef]
11. Bai, C.G.; Wei, X.J. Research on Nonlinear Active Disturbance Rejection Compound Control Strategy of Permanent Magnet Synchronous Motor. *Small Spec. Electr. Mach.* **2021**, *49*, 46–50. [CrossRef]
12. Liu, C.; Hu, J.H. Torque Ripple Suppression Strategy of Open-winding Permanent Magnet Synchronous Motor with Common DC Bus Based on Improved Active Disturbance Rejection Control. *Proc. CSEE* **2022**, *59*, 1–12. [CrossRef]
13. Tavoosi, J. PMSM Speed Control Based on Intelligent Sliding Mode Technique. *COMPEL-Int. J. Comput. Math. Electr. Electron. Eng.* **2020**, *39*, 1315–1328. [CrossRef]
14. Tavoosi, J.; Shirkhani, M.; Abdali, A.; Mohammadzadeh, A.; Nazari, M.; Mobayen, S.; Asad, J.H.; Bartoszewicz, A. A New General Type-2 Fuzzy Predictive Scheme for PID Tuning. *Appl. Sci.* **2021**, *11*, 10392. [CrossRef]

15. Chen, J.; Wang, J.; Yan, B. Simulation Research on Deadbeat Direct Torque and Flux Control of Permanent Magnet Synchronous Motor. *Energies* **2022**, *15*, 3009. [CrossRef]
16. Zhong, Z.; You, J.; Zhou, S. Torque Ripple Reduction of DTC Based on An Analytical Model of PMSM. *World Electr. Veh. J.* **2020**, *11*, 28. [CrossRef]
17. Li, Y.-H.; Wu, T.-X.; Zhai, D.-W.; Zhao, C.-H.; Zhou, Y.-F.; Qin, Y.-G.; Su, J.-S.; Qin, H. Hybrid Decision Based on DNN and DTC for Model Predictive Torque Control of PMSM. *Symmetry* **2022**, *14*, 693. [CrossRef]
18. Mahfoud, S.; Derouich, A.; El Ouanjli, N.; Quynh, N.V.; Mossa, M.A. A New Hybrid Ant Colony Optimization Based PID of the Direct Torque Control for a Doubly Fed Induction Motor. *World Electr. Veh. J.* **2022**, *13*, 78. [CrossRef]
19. Dou, X.H.; Wang, Y. Golden Section Adaptive Control of Permanent Magnet Synchronous Motor. *Electron. Technol.* **2015**, *44*, 18–22. [CrossRef]
20. Dou, X.H.; Wang, Y. Nonlinear Golden-section Adaptive Control of Permanent Magnet Synchronous Motor. *J. Syst. Sci. Math. Sci.* **2015**, *35*, 860. [CrossRef]
21. Wu, H.; HU, J. *Intelligent Adaptive Control Based on Characteristic Model*; China Science and Technology Press: Beijing, China, 2009; pp. 55–67.
22. Han, Z.G. A New Method of Dynamic System Forecasting. *Acta Autom. Sin.* **1983**, *21*, 161–168. [CrossRef]
23. Cao, Y.; Guo, J. Adaptive Anti-saturation Control of High Speed Motor Based on Characteristic Model. *Electr. Mach. Control.* **2021**, *25*, 86–93. [CrossRef]
24. Wang, Y.; Yu, H.; Che, Z.; Wang, Y.; Liu, Y. The Direct Speed Control of Pmsm Based on Terminal Sliding Mode and Finite Time Observer. *Processes* **2019**, *7*, 624. [CrossRef]
25. Shi, J.Z.; You, D.M. Golden Section Adaptive Speed Control of Ultrasonic Motors. *Trans. China Electrotech. Soc.* **2013**, *28*, 59–65. [CrossRef]
26. Wu, J.R.; Guo, Y. Adaptive Control of Dual Motor Servo System Based on Characteristic Model. *J. Huazhong Univ. Sci. Technol.* **2013**, *41*, 436–439. [CrossRef]
27. Wang, S.W.; Wang, S.D. Research on Automatic Code Generation Technology of SVPWM Algorithm Based on Model Design. *Electr. Eng.* **2018**, *24*, 134–136. [CrossRef]
28. Wu, Z.J.; Li, J.G. Research on Automatic Code Generation for Permanent Magnet Synchronous Motor Control. *Mach. Des. Manuf.* **2021**, *10*, 182–185. [CrossRef]
29. Yan, F.C.; Zhao, Y.X. Vector Control of Permanent Magnet Synchronous Motor Based on Matlab and CCS. *Micromotors* **2017**, *50*, 60–64. [CrossRef]

processes

Article

Analysis and Design of a Standalone Electric Vehicle Charging Station Supplied by Photovoltaic Energy

Ibrahem E. Atawi [1], Essam Hendawi [2] and Sherif A. Zaid [3,4,*]

[1] Electrical Engineering Department, Faculty of Engineering, University of Tabuk, Tabuk 47913, Saudi Arabia; ieatawi@ut.edu.sa
[2] Electrical Engineering Department, College of Engineering, Taif University, Taif 21944, Saudi Arabia; essam@tu.edu.sa
[3] Electrical Engineering Department, Faculty of Engineering, Cairo University, Cairo 12613, Egypt
[4] Renewable Energy & Energy Efficiency Centre (REEEC), University of Tabuk, Tabuk 47913, Saudi Arabia
* Correspondence: shfaraj@ut.edu.sa or sherifzaid3@yahoo.com

Abstract: Nowadays, there is a great development in electric vehicle production and utilization. It has no pollution, high efficiency, low noise, and low maintenance. However, the charging stations, required to charge the electric vehicle batteries, impose high energy demand on the utility grid. One way to overcome the stress on the grid is the utilization of renewable energy sources such as photovoltaic energy. The utilization of standalone charging stations represents good support to the utility grid. Nevertheless, the electrical design of these systems has different techniques and is sometimes complex. This paper introduces a new simple analysis and design of a standalone charging station powered by photovoltaic energy. Simple closed-form design equations are derived, for all the system components. Case-study design calculations are presented for the proposed charging station. Then, the system is modeled and simulated using Matlab/Simulink platform. Furthermore, an experimental setup is built to verify the system physically. The experimental and simulation results of the proposed system are matched with the design calculations. The results show that the charging process of the electric vehicle battery is precisely steady for all the PV insolation disturbances. In addition, the charging/discharging of the energy storage battery responds perfectly to store and compensate for PV energy variations.

Keywords: electric vehicle; charging station; photovoltaic; maximum power point tracking

Citation: Atawi, I.E.; Hendawi, E.; Zaid, S.A. Analysis and Design of a Standalone Electric Vehicle Charging Station Supplied by Photovoltaic Energy. *Processes* **2021**, *9*, 1246. https://doi.org/10.3390/pr9071246

Academic Editor: Zhiwei Gao

Received: 4 July 2021
Accepted: 16 July 2021
Published: 19 July 2021

1. Introduction

Nowadays, classical internal combustion engine (ICE) vehicles are being replaced by electric vehicles (EVs) [1,2]. In fact, the ICEs have many drawbacks that can be mitigated by EVs. The EVs have negligible pollution, higher energy efficiency, lower noise, and lower maintenance compared to the ICE. However, the charging process of the EV battery still has many obstacles such as the charging time, the charging station's infrastructure, and the effect of these stations on the present electrical power system. The charging time can be greatly reduced to minutes by utilizing fast charging techniques [3–5]. These techniques represent large electrical loads on the utility grid and affect it adversely. Especially, when there is a high number of charging stations connected simultaneously to the utility grid, many problems would be generated such as excessive overload, voltage instability, and voltage variation [6–8]. One of the solutions to these problems is the upgrading of the power system but this will lead to high costs. Another better solution is the use of an energy storage system (ESS) that can act as a buffer between the EV charging station (EVCS) and the utility [9–12]. Nevertheless, the use of ESS will reduce slightly the stress on the utility grid but the expected large number of EVCSs in the future is still a challenge.

The idea that the EV is an environmentally friendly vehicle could not be accepted if the charging stations rely mainly on the grid power that is usually generated from fossil fuels.

Hence, renewable energy sources must be utilized in EV charging stations to emphasize the environmental impacts of the EV. Usually, renewable energy sources are not continuous, it is intermittent. Hence, there is another use for the ESSs in resolving the discontinuity in these sources.

The common renewable energy sources used for the EV charging stations are photovoltaic (PV), biogas, and wind systems [13,14]. The PV energy systems are simple and have higher efficiency than the wind energy systems. Hence, PV energy is more attractive for EV charging stations. Several research papers have been introduced for the PV-based charging stations [15,16]. Ref. [17] has been proposed a high-power EV charging station utilizing PV powered bidirectional charger. Nevertheless, the proposed system could not permit AC charging. Ref. [18], has suggested an integrated PV panel using a multiport converter for the EV charging station. However, the grid current is greatly distorted. Ref. [19] has introduced a PV-powered grid-tied EV charger via a z-source converter. Though the charger gave better performance, it is not suitable for standalone operation. Therefore, it cannot provide EV charging in absence of a grid. A hybrid optimized management for the ESS of a PV-powered EV station is discussed in [20]. Ref. [21] provides an optimization study on the physical scheduling of the EV charging stations. Ref. [22] has presented a PV-based charging station with an ESS to support the system during peak load. In Ref. [23], a new energy management procedure was introduced for use with small EVs in urban environments. Another strategy, in [24], for the EV charging station power management was proposed to reduce the power consumption from utility grid. Furthermore, the strategy serves to store PV energy when the EV is not online with the utility grid. In Ref. [25], a 20 kW charging station for the EVs was designed and introduced using biogas.

In this paper, a new simple analysis and design of a standalone charging station powered by photovoltaic energy. Based on the assumptions, new closed-form equations are derived for the design purpose. The idea of the analysis and the assumptions are new. Modeling, simulation, and experimental verification are carried out to justify the analysis and the design procedure. Simple energy management is tested practically and simulated. The proposed system includes PV panels, a lithium-ion battery representing the electric vehicle EV, and a lead-acid battery representing the energy storage system. A bidirectional converter is employed for charging/discharging the lead-acid battery and a unidirectional converter is utilized for charging the electric vehicle. A new and simple analysis technique is proposed. In addition, simple design equations are generated from the analysis. The proposed system has been implemented physically and simulated using Matlab/Simulink platform. A single-chip PIC18F4550 microcontroller is utilized to realize the operation of the system. The system is tested successfully for 120 W level and it can be extended for a higher power. Besides, it can be applied in urban and remote areas as well. The objectives of this research are:

- Introducing a novel and simple analysis of the energy and power relations of the charging station powered by photovoltaic energy. The analyses produce closed-form equations for the power and energy of the different system components. In addition, relations for the battery state of charge (SOC) and maximum stored energy of the ESS are derived.
- Designing the charging station components with the help of the derived equations in the analysis part. The design includes the power electronic converters used. Further, case study design calculations are provided.
- Simulating and practically implementing the proposed system. Hence, the system performances are studied indicating the effect of solar energy variations on the system response.

The paper structure is as follows: Section 2 presents the proposed system architecture. Analysis and design of the charging station have been described in Section 3. Section 4 presents the charging converters modelling and design. The proposed control system is shown in Section 5. Section 6 presents the proposed system design. The simulation and experimental results have been discussed in Section 7. Finally, Section 8 comes with the conclusions.

2. The Proposed Charging Station Description

The architecture of the proposed EVCS system is shown in Figure 1. The EVCS system is an off-grid type that is powered by solar energy. It is collected by a PV array that generates electrical energy to the EVCS. The PV panel represents the main source of energy for the charging station. However, the generated energy is not steady. It varies according to the solar insolation level and other environmental issues. Hence, ESS batteries are usually used to compensate for the problem of energy intermittence. The output terminals of the PV are connected to a boost converter. Its function is to match the PV voltage level to that of the DC bus and helps in utilizing the maximum power point tracking (MPPT) condition of the PV panel. Two charging converters are connected to the DC bus namely the EV charger and the energy storage converter. These converters are generally DC/DC converters. However, the EV charger is a one quadrant buck converter that is used to charge the EV battery (lithium-ion). It serves in regulating the charging process of the EV battery. However, the energy storage converter is a two-quadrant DC/DC converter. It is used to control the charge/discharge process of the storage battery (lead-acid). In addition, it participates in regulating the DC bus voltage against the variations of the EV load and insolation level. The modelling and theory of operation of these converters will be explained in the next paragraphs.

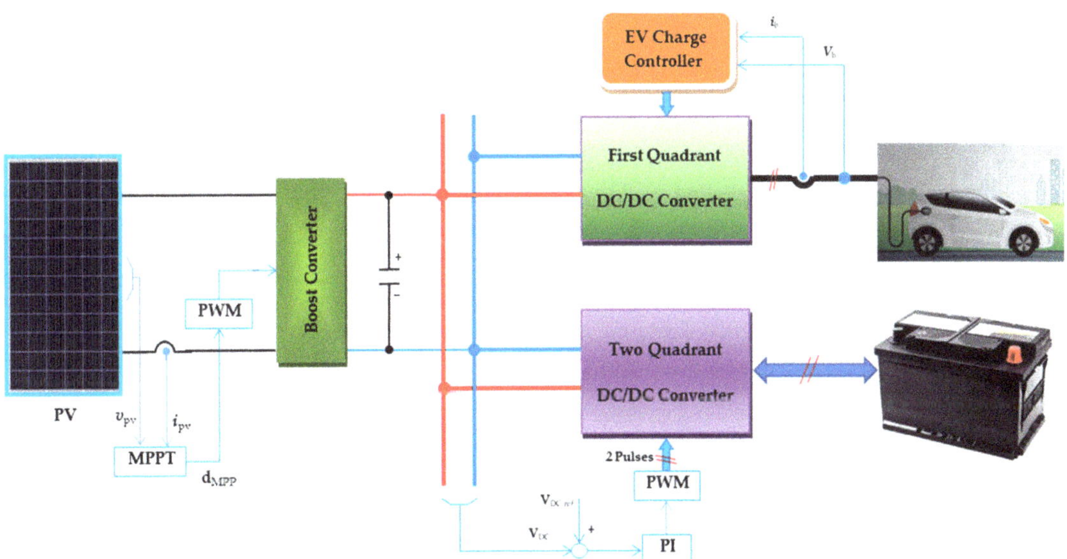

Figure 1. The proposed PV-powered EVCS with the battery storage system.

3. Analysis and Design of the Charging Station

The electrical design of the EVSC means designing all electrical parts of the station such as the PV array, the system converters, the DC bus voltage level, all passive components, and the storage battery. Many aspects must be considered in designing the EVSC:

- The available solar energy on the site and the maximum permissible area for the PV panels.

- The station power is related to the number of EVs to be charged simultaneously. The charging rate is an important issue that affects the charging time. For fast charging, the station should be capable of charging rates of 3 C to 5 C [26].
- The size of the ESS depends on the available PV power and the demand load of the EV.

As a first step for the design process, the relation among the input solar energy, the output charging power to the EV, and the stored energy in the ESS must be studied. The system power and energy relations will be studied in the next subsection.

3.1. Power and Energy Analysis of the System

In this section, the energy and power relations of the system are derived to help in the design process. The assumptions used in this regard are:

- The storage battery's initial energy is "E_i" in Wh.
- The output power of the EV station "P_{EV}" is constant.
- The solar power "P_{pv}" of the PV panel changes with time according to:

$$P_{pv} = \frac{P_{max}}{36}\left(36 - t^2\right) \quad (1)$$

where; "P_{max}" is the PV maximum power and "t" is the time in hours. The relation is sketched in Figure 2. The time axis origin is set at noon where the insolation level is maximum. The period of solar energy is assumed to be 12 H starting at 6:00 a.m.

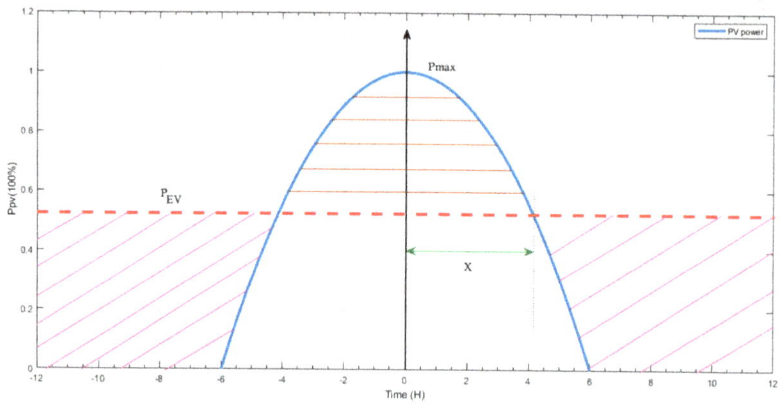

Figure 2. The PV array power distribution for one day.

From the power balance principle and refer to the system power flow diagram shown in Figure 3.

$$P_{EV} = P_{pv} - P_{bat} \quad (2)$$

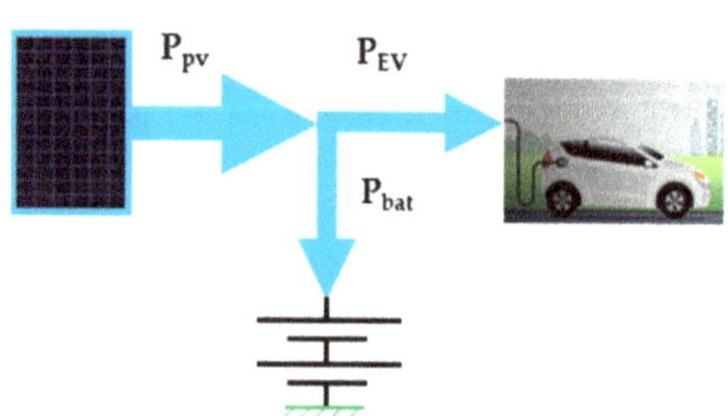

Figure 3. The system power flow diagram.

Hence,

$$P_{bat} = \begin{cases} \frac{P_{max}}{36}(36 - t^2) - P_{EV} & -6 \leq t \leq 6 \\ -P_{EV} & 6 \leq t \leq 18 \end{cases} \tag{3}$$

where; "P_{bat}" is the instantaneous storage battery power. The average power of the storage battery is assumed to be zero along one day period (T = 24 h).

Hence,

$$\int_0^T P_{bat} dt = 0 \tag{4}$$

Hence,

$$\int_{-6}^{6} [\frac{P_{max}}{36}\left(36 - t^2\right) - P_{EV}]dt - \int_{6}^{18} P_{EV} dt = 0 \tag{5}$$

$$P_{max} = 3P_{EV} \tag{6}$$

Battery state of charge and maximum stored energy

The battery energy-stored can be calculated through:

$$E_{bat} = C + \int P_{bat} dt \tag{7}$$

where; "C" is a constant.

The region ($-6 \leq t \leq 6$)

Substitute Equation (3) into Equation (7) and integrate:

$$E_{bat} = C + P_{EV}\left(2t - \frac{t^3}{36}\right) \tag{8}$$

To get the constant "C", we have that at $t = -6 \rightarrow E_{bat} = E_i$. Using Equation (8):

$$C = E_i + 6P_{EV} \tag{9}$$

The instantaneous battery energy is given by:

$$E_{bat} = E_i + P_{EV}\left[6 + 2t - \frac{t^3}{36}\right] \tag{10}$$

The final value of the energy at this region (t = 6):

$$E_{bat}|_{t=6} = E_i + 12P_{EV} \tag{11}$$

This value is the initial value for the next region.

The region ($6 \leq t \leq 18$)

From Equations (7) and (3):

$$E_{bat} = C_1 - P_{EV} t \tag{12}$$

where; "C_1" is constant. From Equations (11) and (12):

$$C_1 = E_i + 18P_{EV} \tag{13}$$

Hence,

$$E_{bat} = E_i + 18P_{EV} - P_{EV} t \tag{14}$$

The maximum stored energy occurs at $t = x$ as shown in Figure 3. It can be shown that the maximum stored energy is given by:

$$E_{bat}|_{max} = E_i + 2P_{EV}[3 + \frac{2\sqrt{24}}{3}] \tag{15}$$

From this analysis, if the EV power P_{EV} is known, then using Equations (1), (6) and (15) the PV power rating and the ESS size can be determined. Hence, the PV array area can be calculated.

4. Charging Converters Modelling and Design

The proposed system has two charging converters. The EV charging converter is a simple buck converter (first-quadrant converter). However, the energy storage battery is a buck-boost converter (two-quadrant converter). The circuit diagrams of the converters are shown in Figure 4. The operation, modeling, and analysis of the two-quadrant converter include that of the first-quadrant converter. The continuous mode is assumed for the converter operation. The converter comprised two IGBT's (S_1, S_2), two antiparallel diodes, and an LC filter. The converter terminals are connected to the DC bus and the storage battery. The battery's internal voltage and resistance are represented by (E_b, r_b). The filter inductance is assumed to be large enough to preserve sufficient energy to charge or discharge the battery. Hence, the discontinuous conduction operation mode is neglected. The converter has two modes namely the buck mode and boost mode. The bidirectional converter serves in the boost mode, when the switch S_2 is active and the switch S_1 acts as a diode, for discharging the battery. However, it serves in the buck mode, when switch S_1 is active and the switch S_2 acts as a diode, resulting in the battery charge mode. The converter dynamic model in the state-space form is described as:

During buck operation:

$$x = \begin{bmatrix} i_l \\ v_c \end{bmatrix} \tag{16}$$

$$\begin{bmatrix} \dot{x}_1 \\ \dot{x}_2 \end{bmatrix} = \begin{bmatrix} 0 & \frac{-1}{L} \\ \frac{1}{C} & \frac{-1}{Cr_b} \end{bmatrix} \begin{bmatrix} x_1 \\ x_2 \end{bmatrix} + \begin{bmatrix} \frac{V_{dc}}{L} \\ 0 \end{bmatrix} u_1 + \begin{bmatrix} 0 \\ \frac{E_b}{Cr_b} \end{bmatrix} \tag{17}$$

where; (i_l, v_c) are the inductor current and the capacitor voltage, u_1 represents the switch S_1 action with PWM taking values from the set of {0:1}, (L, C) are inductance and the capacitance of the filter, V_{dc} is the DC bus voltage.

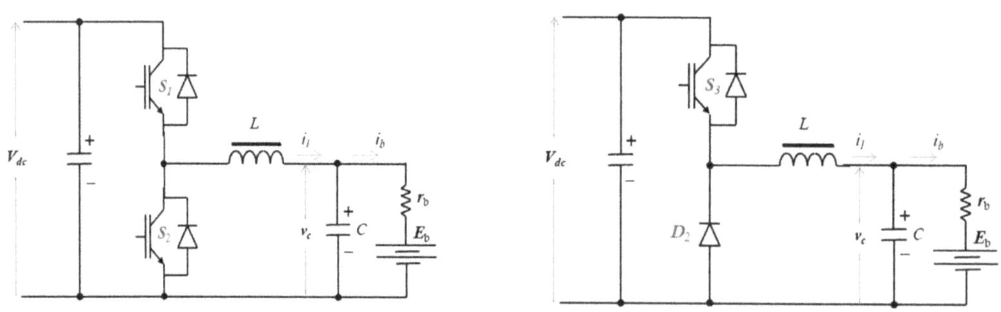

Figure 4. The system converters circuit diagram: (**a**) bidirectional converter, (**b**) buck converter.

During boost operation:

$$\begin{bmatrix} \dot{x}_1 \\ \dot{x}_2 \end{bmatrix} = \begin{bmatrix} 0 & \frac{-1}{L} \\ \frac{1}{C} & \frac{-1}{Cr_b} \end{bmatrix} \begin{bmatrix} x_1 \\ x_2 \end{bmatrix} + \begin{bmatrix} \frac{-V_{dc}}{L} \\ 0 \end{bmatrix} u_2 + \begin{bmatrix} \frac{V_{dc}}{L} \\ \frac{E_b}{Cr_b} \end{bmatrix} \tag{18}$$

where; u_2 represents the switch S_2 action with PWM taking values from the set of {0:1}.

The filter inductor and capacitor design equations are [27]:

$$\frac{(1-d_{min})R_{li}}{2f} \geq \mathrm{L} \leq \frac{E_b}{f\Delta I_{min}} \times d_{max} \quad (19)$$

$$I_{rms} = \sqrt{I_b|_{max} + \frac{(\Delta I_{\max})^2}{3}} \quad (20)$$

$$\mathrm{C} \geq \frac{(1-d_{min})}{8\, r_{\min}\, \mathrm{L}\, f^2} \quad (21)$$

where; (d_{min}, d_{max}) is the minimum and maximum duty cycle, R_{li} is the approximate equivalent resistance of Li-ion battery, f is the switching frequency, $(\Delta I_{min}, \Delta I_{max})$ is the minimum and maximum ripple currents, r is the minimum voltage ripple factor, and $I_b|_{max}$ is the maximum battery current.

The DC bus voltage level is selected according to the charging state of the system as the system converters operate in the buck mode. In this case, the DC bus voltage must be greater than the converter output average voltage. On the other hand, the minimum duty cycle and battery voltage variations affect the value of V_{dc}. These constraints can be written as:

$$\frac{V_{b-max}}{d_{\min}} \geq V_{dc} \geq \frac{V_{b-min}}{d_{\mathrm{n}}} \quad (22)$$

where; (V_{b-min}, V_{b-max}) is the battery minimum and maximum terminal voltage, and d_n is the nominal modulation index. The value of the DC capacitance affects greatly the DC bus stability. Basically, the DC bus capacitor must support a certain voltage and power ripple factor. The DC bus capacitance C_{dc} can be calculated from [28]:

$$C_{dc} = \frac{2\mathrm{P}_{rated}\cdot\Delta x}{\Delta r\cdot V_{dc}^2\cdot f} \quad (23)$$

where; $(\Delta x, \Delta r)$ is the percentage ripple power and voltage; and P_{rated} is the rated power.

5. Control System

The controllers of the proposed system, shown in Figure 5, are the MPPT, the EV charger, and the storage converter controller. The function of the MPPT controller is to keep absorbing the peak power from the PV panel. It outputs the suitable duty ratio to the boost converter that maintains the MPPT. However, the EV converter controller regulates the charging of the EV. Finally, the battery storage converter controller is used to control the DC bus voltage and ESS charge/discharge process.

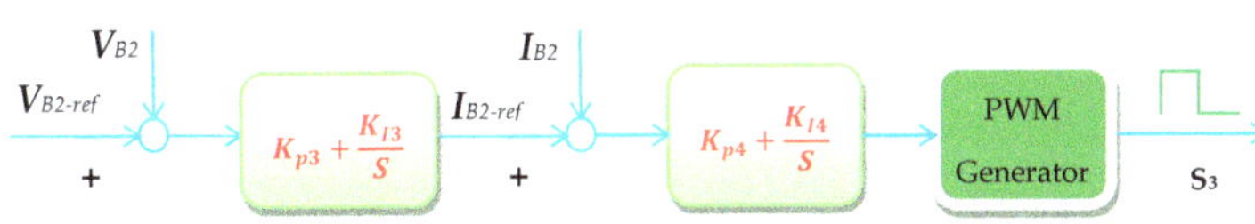

(a)

Figure 5. *Cont.*

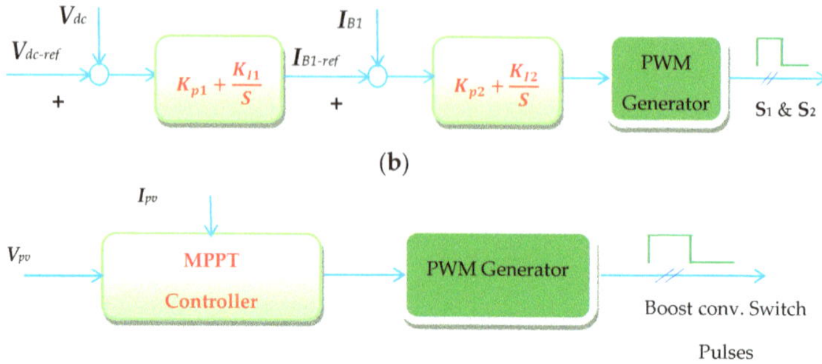

Figure 5. The proposed system controllers: (**a**) MPPT controller, (**b**) EV charge controller, and (**c**) storage battery converter controller.

5.1. MPPT Controller

The MPPT controller is necessary to improve the PV system utilization. The algorithm of the MPPT has been implemented using many approaches [29–32]. For simplicity, the common technique called "perturb and observe" is utilized here. In this technique, a perturb of the boost converter duty ratio is made and the PV voltage and current are measured. After then, the power change is calculated then checked according to the flow chart of Figure 6 [33]. The controller is adapted to generate the necessary duty ratio for the boost converter switch to achieve MPPT conditions.

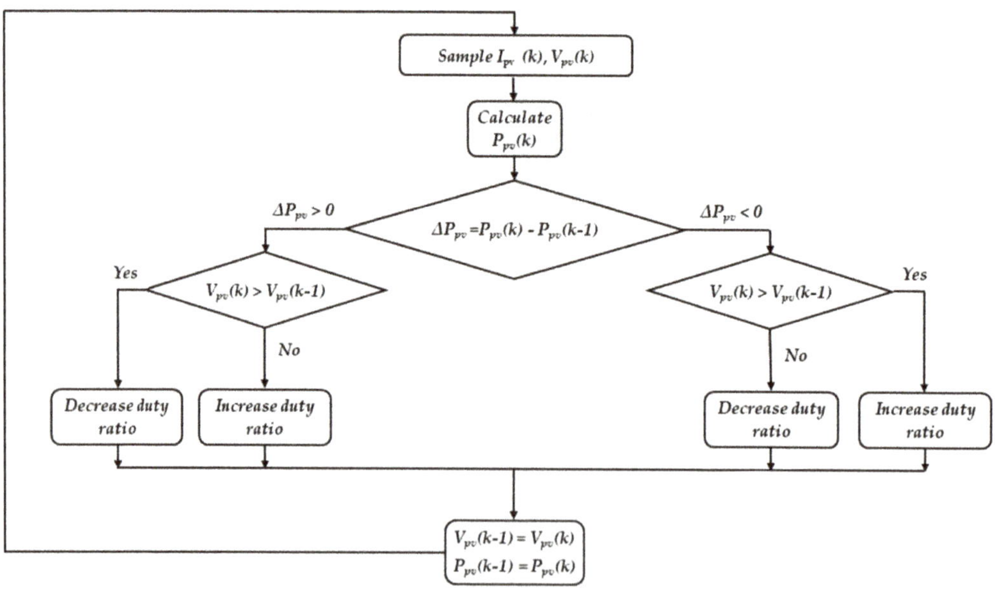

Figure 6. Perturb and Observe MPPT algorithm flowchart.

5.2. EV Converter Controller

The target of this controller is to regulate the charging process of the EV. There are various control techniques such as pulse, constant voltage (CV), constant current (CC), and constant current/constant voltage (CC/CV) techniques. Commonly, the (CC and CC/CV) techniques are preferred for their low charging time. In our system, the (CC/CV) technique is utilized. A simple PI controller is adapted for this technique. The proportion and integral gains of the PI controller are tuned using the Ziegler–Nichols method.

The experimental Ziegler–Nichols tuning algorithm can simply be summarized as the following. First, the PI controller is reduced to only the proportional part with very low proportional gain. Hence, remove the integral actions from the controller by setting it to zero. The proportional gain is then increased until continuous oscillations in the output signal are observed. Note that, when the oscillations occur, the controller should not be hitting limits. After each gain increase, you may need to make a disturbance b changing the setpoint to see if the loop oscillates. The oscillations occur at a critical proportional gain (K_{pc}) with an oscillation period (T_{osc}). By measuring the parameters K_{pc} and T_{osc}, the proportional and integral gains (K_p, K_i) for the PI controller are adjusted as:

$$K_p = 0.35\ K_{pc} \tag{24}$$

$$K_\mathrm{i} = 0.8\ K_p/T_{osc} \tag{25}$$

5.3. Battery Storage Converter Controller

The objective of this controller is to regulate the DC bus voltage at a constant reference value. Consequently, charging and discharging the storage battery are adapted to serve this function. In addition, the PI controller is adapted for this converter. In case that the batteries are fully charged, the converter stops charging except for a small trickle current.

6. The Proposed System Design Calculations

According to the previous analysis, the system parameters can be calculated. Assume that the proposed system data and parameters are listed in Table 1. The design calculations are carried out as follows:

- Design calculations for ESS

As the available PV panel is 120 W, the EV power can be determined from Equation (6). Hence, using Equation (15) the ESS energy can be calculated. The charge capacity of the ESS is 39 Ah. However, the nearest available standard is 65 Ah. Normally, the range of the SOC of the ESS is from 20% to 95% [34].

- Design calculations for EV converter filter

Let that $\Delta I_{min} = 0.3$ A, $V_{dc} = 25\ V$, and $r_{\min} = 0.02$. Hence, substituting in Equation (19), gives the range of inductance as 325 μH and 9375 μH. The standard inductance value of 560 μH is chosen. Then, substituting in Equation (21), gives C $\geq$ 363 μF. The standard inductance value of 1000 μF is chosen.

- Design calculations for Storage battery converter filter

Let that $\Delta I_{min} = 1.3$ A and applying the same procedure to the converter of storage battery giving the range of inductance as 74 μH and 1329 μH. The standard inductance value of 560 μH is chosen. Then, substituting in Equation (21), gives C $\geq$ 446 μF. The standard inductance value of 1000 μF is chosen.

Table 1. System Parameters.

Item	Parameter	Value
PV	SC current I_{sc}	8.82 A
	OC voltage V_{oc}	19.2 V
	MPPT power P_{MPPT}	7.5 × 16 W
ESS Battery	Type	Lead Acid
	Rating	65 Ah, 12 V
	Max. charging current	13 A

Table 1. *Cont.*

Item	Parameter	Value
EV Battery	Type	Lithium-ion
	Rating	6.5 Ah, 3.7 V
	Max. charging current	3 A
	Max. discharging current	5 A
	L	560 μH
	C	1000 μF
	C_{dc}	2200 μF
	PWM carrier frequency	4 kHz

7. Simulation and Experimental Results

The proposed system, shown in Figure 1, is simulated and implemented physically to verify the paper idea. The parameters in Table 1 are used for both the simulation and the experimental setup. The PV panel is the Copex module of model P120. It is a polycrystalline type and has the specifications listed below. The EV battery is formed by four Li-ion batteries connected in series. Its nominal voltage is 14.8 V, the fully charged voltage is 17 V and its cut-off voltage is 12 V. The results are discussed in the following paragraphs.

7.1. Simulation Results

The proposed system is simulated using Matlab/Simulink platform. The system performances according to step changes in the solar irradiation of the PV panel are shown in Figures 7 and 8. Figure 7a shows the irradiation level of solar energy. The DC bus voltage tracks well the reference voltage (25 V), as shown in Figure 7b. The state of charge of the lead-acid battery bank B_2 is shown in Figure 7c. At the first four seconds, the insolation is $\geq$60%. Therefore, the generated PV power is enough to charge the EV battery and store the reserve in the B_2. However, the insolation at the remaining period is $\leq$60% which is not enough to supply the energy to the EV. Hence, the storage battery discharges to compensate for the drop in solar energy. The current of the battery tracks its reference value generated by the DC bus voltage controller, as shown in Figure 7d. In addition, the charging and discharging processes track and compensate for the irradiation level. Figure 7e shows the voltage of the B_2 battery. Its voltage increases with charging and decreases with discharging. The PV current is shown in Figure 7f. The current level is the same as the MPPT conditions.

Figure 8a shows the irradiation level of solar energy. The EV current tracks well the reference current generated by the charge controller, as shown in Figure 8b. Figure 8c shows the voltage of the B_1 battery. It is continuously charging. The state of charge of the EV battery is shown in Figure 8d. Figure 8e,f show the PV power, the power of the B_2 battery, and the EV battery power. At the first four seconds, the irradiation is $\geq$60%. Therefore, the generated PV power is enough to charge the EV battery and store the reserve in the B_2. However, the insolation at the remaining period is $\leq$60% which is not enough to supply the energy to the EV. Hence, the storage battery discharges to compensate for the drop in solar energy. In addition, the charging and discharging processes track and compensate for the irradiation level. Its voltage increases with charging and decreases with discharging. The EV power is constant at all conditions, shown in Figure 8f.

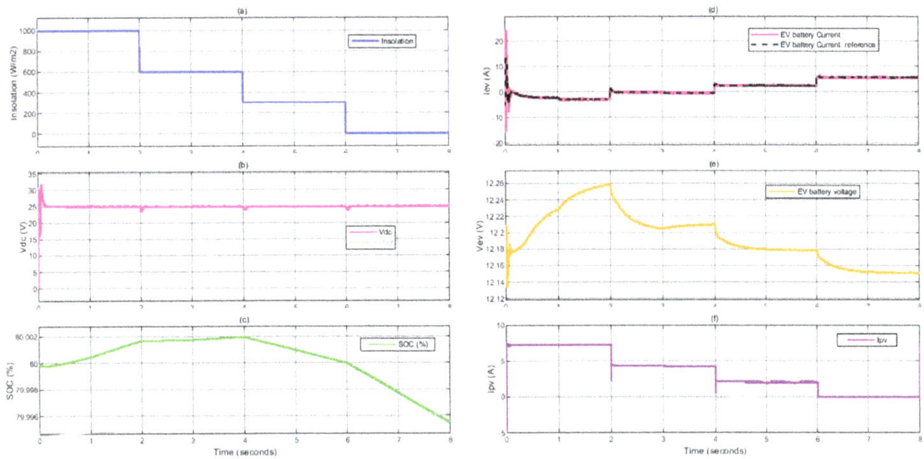

Figure 7. (**a**) PV irradiation level, (**b**) DC bus voltage, (**c**) state of charge of the lead-acid battery bank, (**d**) the lead-acid battery current and reference current, (**e**) the lead-acid battery voltage (**f**) the PV current.

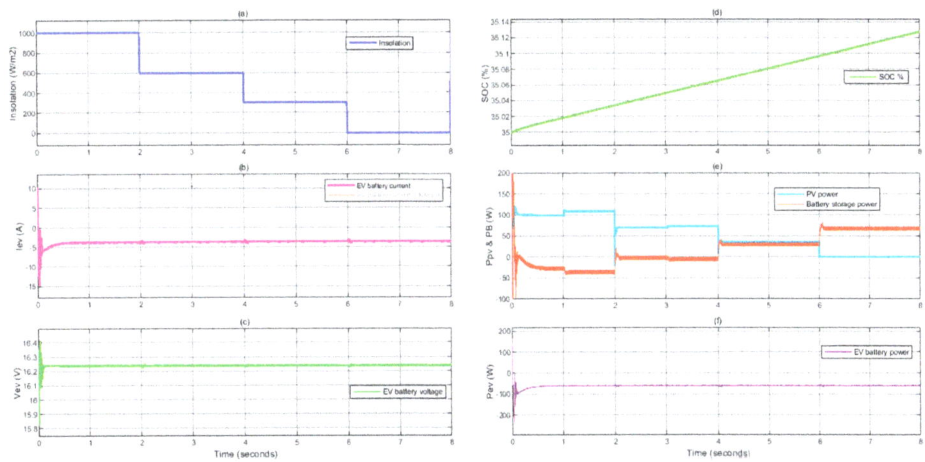

Figure 8. (**a**) PV irradiation level, (**b**) the EV battery bank current and reference current, (**c**) the EV battery voltage, (**d**) state of charge of the EV battery, (**e**) the PV and lead-acid battery power (**f**) the EV charging power.

7.2. Experimental Results

An experimental prototype is built, as shown in Figure 9, to verify the simulation results of the proposed system. The prototype has the same parameters listed in Table 1. A single-chip PIC18F4550 microcontroller, with a 16 MHz oscillator, is utilized to achieve simultaneously five controllers. The five controllers utilize three timers of the microcontroller (T0, T1, and T2) and the PWM module of the microcontroller. The microcontroller generates the control signals for the switches S_1, S_2, S_3, and S_4. The switching frequency is suitable for the converters and gives very good results. All the controllers are included in a single chip microcontroller and all control signals are generated from this microcontroller. Therefore, the execution time of different sections in the codes forces the switching frequency to be not so high. A drive circuit for each control signal is used to raise its level voltage to a suitable level. However, the drive circuits invert the generated control signals, and this inversion is considered in the written code of all controllers. The five controllers are described as follows:

- <u>The boost converter controller</u>

A perturb and observe maximum power point tracker is applied to achieve maximum power extraction of the PV panel. The microcontroller receives the PV voltage and current, performs the perturb-observe algorithm, then generates the control signal to the boost converter switch. This controller uses T2 and CCP modules are operating in the PWM mode.

- <u>The DC bus voltage controller</u>

The voltage of the DC bus is maintained constant during charging both batteries in the sunshine day hours and during night hours (discharging lead-acid battery). The controller gives the required lead-acid current in both modes of operation of the lead-acid battery considering the maximum charging current and maximum discharging current of the lead-acid battery.

- <u>The lead-acid battery voltage controller</u>

During the constant voltage stage of charging the lead-acid battery, its voltage is maintained at a slight value below the gassing voltage of the battery (14.4 V). During discharging mode of the battery, its voltage starts to decrease. The battery voltage is monitored so that the cut-off voltage should not be reached. This controller utilizes the reference battery voltage which is generated by the previous controller and generates the control signals to S_1 and S_2. This controller employs T1 and the interrupt module to achieve its target.

- <u>The lithium-ion voltage and current controllers</u>

The lithium-ion battery pack is charged using the constant current—constant voltage charging method. The maximum allowable charging voltage and current of the lithium-ion battery pack are controlled using the two controllers. The outer loop (voltage controller) controls the battery voltage and generates the suitable battery current taking into consideration the maximum charging current (6.5 A). The inner loop (current controller) generates the control signal to S_3. This controller employs T0 and the interrupt module to realize its target.

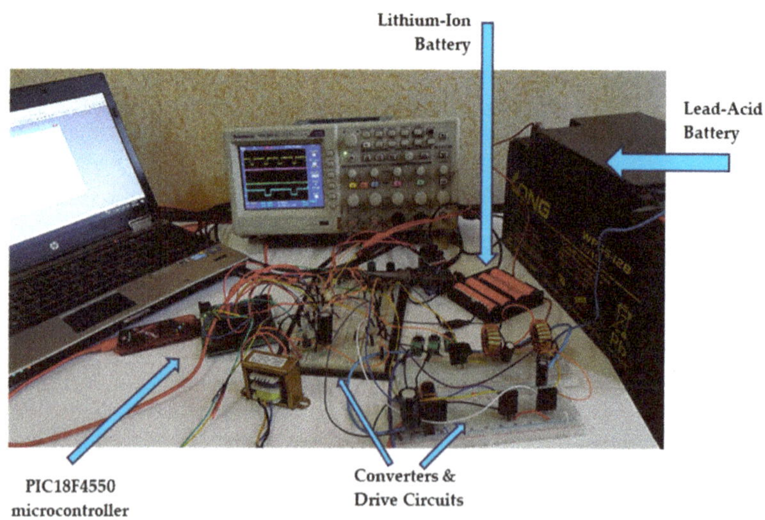

Figure 9. The experimental setup.

The step response of the charge controller of the lithium-ion battery is illustrated in Figure 10. The reference current is stepped at values of 3 A, 2 A, and 1 A. The voltage controller of the battery at the same time keeps the battery voltage at the suitable level corresponding to each current level. The corresponding control signal to S_3 is shown. It is noticed that the duty cycle of S_3 decreases as the reference current decreases. On the other side, the DC bus voltage is maintained constant.

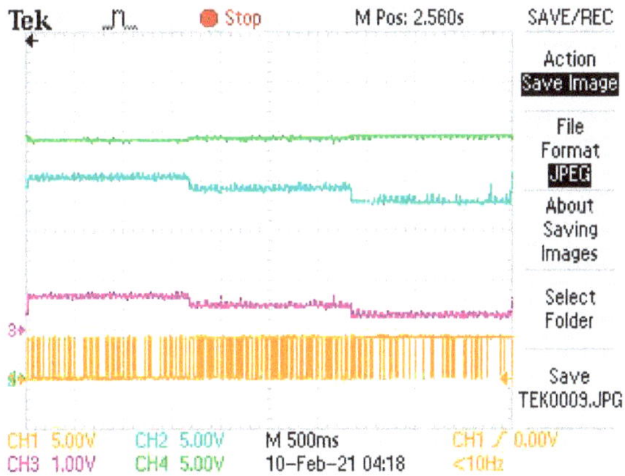

Figure 10. The experimental response of the EV controller: Ch1: Control signal of S_3 (5 V/div) Ch2: Li-ion battery voltage (5 V/div.) Ch3: Li-ion battery current (4 A/div), Ch4: DC bus voltage (5 V/div).

Figure 11 shows the effects of disconnecting the PV panel. The figure presents DC bus voltage, lead-acid battery voltage, lithium-ion battery voltage, and current. The lead-acid battery current changes its direction at the instant of disconnection. The controller keeps the DC bus voltage constant after the instant of disconnecting the PV panel where the lead-acid battery operates in the boost mode in which it starts discharging and its voltage slightly decreases. The charge controller of the lithium-ion battery keeps its level voltage and current at the same values before disconnecting the PV panels.

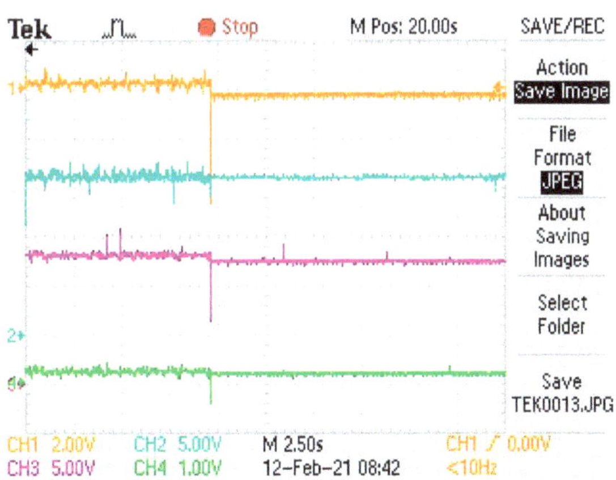

Figure 11. The system response upon disconnecting the PV: Ch1: lead acid battery current (6 A/div) Ch2: Li-ion battery voltage (5 V/div) Ch3: lead acid battery voltage (5 V/div), Ch4: Li-ion battery current (4 A/div).

Disconnecting and reconnecting the PV panel is illustrated in Figure 12a,b. The figures give the DC bus voltage, lead-acid battery, and lithium-ion battery voltage. In Figure 12a, the PV panel is disconnected and the lead-acid battery lonely charges the lithium-ion battery. The DC bus voltage and lithium-ion battery voltage are kept at the original levels while the slightly decreases. Figure 12b shows the reconnection of the PV panel in which the two batteries are being charged and the lead-acid battery voltage slightly increases.

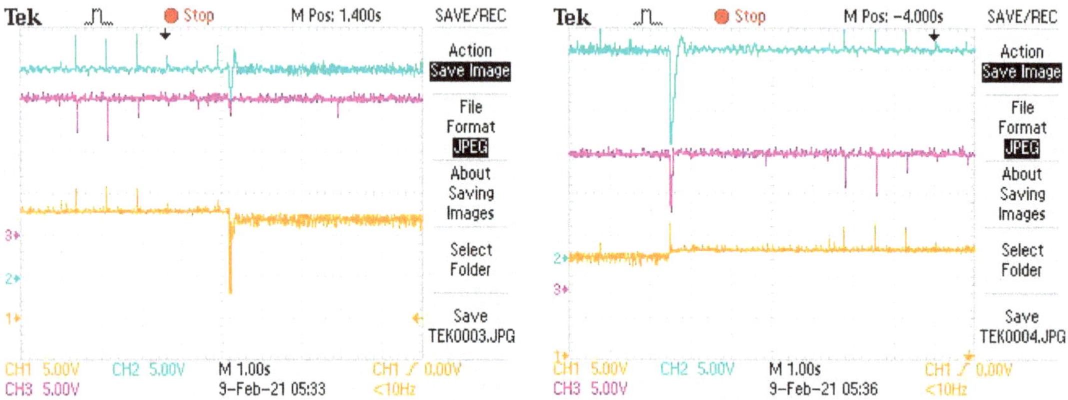

(**a**) (**b**)

Figure 12. The system response upon: (**a**) connecting and (**b**) disconnecting the PV. Ch1: lead–acid battery voltage (5 V/div) Ch2: DC bus voltage (5 V/div) Ch3: Li-ion battery voltage (5 V/div).

Figure 13a,b present the control signals of the switches S_1 through S_4 during connecting and disconnecting the PV, respectively. The experimental results are summarized in Table 2.

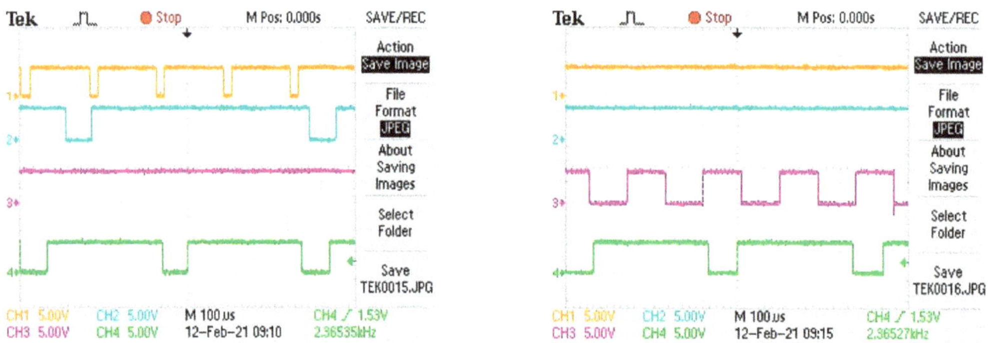

(**a**) (**b**)

Figure 13. Control signals from the microcontroller for the switches: (**a**) PV panel is connected, and the two batteries are being charged, (**b**) PV panel is disconnected. S4 (CH1)—S1 (CH2)—S2 (CH3)—S3 (CH4).

Table 2. The experimental results summary.

Variable	Parameter	Connecting the PV	Disconnecting the PV
DC bus Voltage	Overshoot (%)	20	40
	Setling time (s)	0.2	0.5
ESS Battery Voltage	Overshoot (%)	67	0
	Setling time (s)	0.2	0
EV Battery Voltage	Overshoot (%)	0	40
	Setling time (s)	0	0.1

8. Conclusions

An isolated EV charging station based on a PV energy source is proposed. The system consists of PV panel, boost converter, ESS batteries, two DC/DC charging converters, and an EV battery. The control system consists of three controllers named the MPPT, the EV charger, and the storage converter controller. PI voltage and current controllers are adapted to control charging/discharging of the ESS system and the EV charger as well. The system is simulated and implemented physically. A single-chip PIC18F4550 microcontroller is utilized to realize the system controllers. New simple energy and power analyses procedure has been introduced. Hence, closed-form equations have been derived to help in the design phase. Complete design of the system, including the ESS size, the PV rating, and the filter components, has been proposed. Simulation and experimental results are very close and verify the effectiveness of the proposed system. At different insolation levels, the results show that the charging process of the EV battery is steady without any disturbance. However, the charging/discharging of the ESS battery responds perfectly to store and compensate for PV energy variations. The current and voltage controllers of the converters give good responses and track their references well. In addition, the MPPT controller tracks the peak conditions of the PV precisely.

Author Contributions: I.E.A. collected the funding and resources, E.H. helped in validation and visualization, and S.A.Z. conceived, designed the system model, and analyzed the results. All authors have read and agreed to the published version of the manuscript.

Funding: This research was funded by the University of Tabuk, Grant Number S-1441-0172 at https://www.ut.edu.sa/web/deanship-of-scientific-research/home (accessed on 16 July 2021).

Conflicts of Interest: The authors declare no conflict of interest.

References

1. Irle, R. Global EV Sales for the 1st Half of 2019. EV Volumes. 2019. Available online: http://www.ev-volumes.com/country/total-world-plug-in-vehicle-volumes/ (accessed on 20 November 2019).
2. Sun, X.; Li, Z.; Wang, X.; Li, C. Technology Development of Electric Vehicles: A Review. *Energies* **2020**, *13*, 90. [CrossRef]
3. Luc, Vehicles & Charging Tips. Fastned. 2019. Available online: https://support.fastned.nl/hc/en-gb/sections/115000180588-Cars-charging-tips- (accessed on 30 March 2019).
4. Richard, L.; Petit, M. Fast charging station with a battery storage system for EV: Optimal integration into the grid. In Proceedings of the 2018 IEEE Power & Energy Society General Meeting (PESGM), Portland, OR, USA, 5–10 August 2018; pp. 1–5.
5. Chakraborty, S.; Vu, H.-N.; Hasan, M.M.; Tran, D.-D.; Baghdadi, M.E.; Hegazy, O. DC-DC Converter Topologies for Electric Vehicles, Plug-in Hybrid Electric Vehicles and Fast Charging Stations: State of the Art and Future Trends. *Energies* **2019**, *12*, 1569. [CrossRef]
6. Liu, Y.; Dong, H.; Wang, S.; Lan, M.; Zeng, M.; Zhang, S.; Yang, M.; Yin, S. An Optimization Approach Considering User Utility for the PV-Storage Charging Station Planning Process. *Processes* **2020**, *8*, 83. [CrossRef]
7. Ahmadi, M.; Mithulananthan, N.; Sharma, R. A review on topologies for fast-charging stations for electric vehicles. In Proceedings of the 2016 IEEE International Conference on Power System Technology (POWERCON), Wollongong, Australia, 28 September–1 October 2016; pp. 1–6. [CrossRef]
8. Minh, P.V.; Le Quang, S.; Pham, M.-H. Technical Economic Analysis of Photovoltaic-Powered Electric Vehicle Charging Stations under Different Solar Irradiation Conditions in Vietnam. *Sustainability* **2021**, *13*, 3528. [CrossRef]
9. Rafi, M.A.H.; Bauman, J. A Comprehensive Review of DC Fast Charging Stations with Energy Storage: Architectures, Power Converters, and Analysis. *IEEE Trans. Transp. Electrif.* **2021**, *7*, 345–368. [CrossRef]

10. Yilmaz, M.; Krein, P.T. Review of battery charger topologies, charging power levels, and infrastructure for plug-in electric and hybrid vehicles. *IEEE Trans. Power Electron.* **2013**, *28*, 2151–2169. [CrossRef]
11. Francfort, J.; Salisbury, S.; Smart, J.; Garetson, T.; Karner, D. *Considerations for Corridor and Community DC Fast Charging Complex System Design*; Idaho National Lab. (INL): Idaho Falls, ID, USA, 2017.
12. Nicholas, M.; Hall, D. *Lessons Learned on Early Fast Electric Vehicle Charging Systems;* The National Academies of Sciences, Engineering, and Medicine: Washington, DC, USA, 2018.
13. Gao, Z.; Liu, X. An Overview on Fault Diagnosis, Prognosis and Resilient Control for Wind Turbine Systems. *Processes* **2021**, *9*, 300. [CrossRef]
14. Polimeni, S.; Nespoli, A.; Leva, S.; Valenti, G.; Manzolini, G. Implementation of Different PV Forecast Approaches in a MultiGood MicroGrid: Modeling and Experimental Results. *Processes* **2021**, *9*, 323. [CrossRef]
15. Bhatti, A.R.; Salam, Z. Charging of electric vehi-cle with constant price using photovoltaic based grid-connected system. In Proceedings of the 2016 IEEE International Conference on Power and Energy (PECon), Melaka, Malaysia, 28–29 November 2016; pp. 268–273. [CrossRef]
16. Kumar, V.; Teja, V.R.; Singh, M.; Mishra, S. PV Based Off-Grid Charging Station for Electric Vehicle. *IFAC Pap.* **2019**, 276–281. [CrossRef]
17. Mouli, C.G.R.; Schijffelen, J.; Heuvel, M.; Kardolus, M.; Bauer, P. A 10 kW Solar-Powered Bidirectional EV Charger Compatible With Chademo and COMBO. *IEEE Trans. Power Electron.* **2019**, *34*, 1082–1098. [CrossRef]
18. Monteiro, V.; Pinto, J.G.; Afonso, J.L. Experimental Validation of a Three-Port Integrated Topology to Interface Electric Vehicles and Renewables With the Electrical Grid. *IEEE Trans. Ind. Inform.* **2018**, *14*, 2364–2374. [CrossRef]
19. Singh, S.A.; Carli, G.; Azeez, N.A.; Williamson, S.S. Modeling, Design, Control, and Implementation of a Modified Z-Source Integrated PV/Grid/EV DC Charger/Inverter. *IEEE Trans. Ind. Electron.* **2018**, *65*, 5213–5220. [CrossRef]
20. Chaudhari, K.; Ukil, A.; Kumar, K.N.; Manandhar, U.; Kollimalla, S.K. Hybrid Optimization for Economic Deployment of ESS in PV-Integrated EV Charging Stations. *IEEE Trans. Ind. Inform.* **2018**, *14*, 106–116. [CrossRef]
21. Zhang, Y.P.; You and Cai, L. Optimal Charging Scheduling by Pricing for EV Charging Station With Dual Charging Modes. *IEEE Trans. Intell. Transp. Syst.* **2019**, *20*, 3386–3396. [CrossRef]
22. Hernandez, J.C.; Sutil, F.S. Electric vehicle charg-ing stations feeded by renewable: Pv and train regen-erative braking. *IEEE Lat. Am. Trans.* **2016**, *14*, 3262–3269. [CrossRef]
23. Ibrahim, H.; Sayed, K.; Kassem, A.; Mostafa, R. A new power management strategy for battery electric vehicles. *IET Electr. Syst. Transp.* **2018**, *9*, 65–74.
24. Savio, D.A.; Juliet, V.A.; Chokkalingam, B.; Padmanaban, S.; Holm-Nielsen, J.B.; Blaabjerg, F. Photovoltaic Integrated Hybrid Microgrid Structured Electric Vehicle Charging Station and Its Energy Management Approach. *Energies* **2019**, *12*, 168. [CrossRef]
25. Ashish Kumar Karmaker, M.; Alamgir, H.; Nallapaneni, M.K.; Vishnupriyan, J.; Arunkumar, J.; Biplob, R. Analysis of Using Biogas Resources for Electric Vehicle Charging in Bangladesh: A Techno-Economic-Environmental Perspective. *Sustainability* **2020**, *12*, 2579. [CrossRef]
26. Bai, S.; Yu, D.; Lukic, S. Optimum design of an EV/PHEV charging station with DC bus and storage system. In Proceedings of the 2010 IEEE Energy Conversion Congress and Exposition, Atlanta, GA, USA, 12–16 September 2010; pp. 1178–1184. [CrossRef]
27. Rashid, M. *Power Electronics Handbook*, 2nd ed.; Elsevier Press: Amsterdam, The Netherlands, 2011.
28. Arancibia, A.; Strunz, K. Modeling of an electric vehicle charging station for fast DC charging. In Proceedings of the 2012 IEEE International Electric Vehicle Conference, Greenville, SC, USA, 4–8 March 2012; pp. 1–6. [CrossRef]
29. Elbalawi, H.; Zaid, S.A. H5 Transformerless Inverter for Grid Connected PV System with Improved Utilization Factor and Simple Maximum Power Point Algorithm. *Energies* **2018**, *11*, 2912. [CrossRef]
30. Husain, M.A.; Tariq, A.; Hameed, S.; Arif, M.S.B.; Jain, A. Comparative assessment of maximum power point tracking procedures for photovoltaic systems. *Green Energy Environ.* **2017**, *2*, 5–17. [CrossRef]
31. Zaid, S.A.; Kassem, A.M. Review, analysis and improving the utilization factor of a PV-grid connected system via HERIC transformerless approach. *Renew. Sustain. Energy Rev.* **2017**, *73*, 1061–1069. [CrossRef]
32. Albalawi, H.; Zaid, S.A. Performance Improvement of a Grid-Tied Neutral-Point-Clamped 3-φ Transformerless Inverter Using Model Predictive Control. *Processes* **2019**, *7*, 856. [CrossRef]
33. Sera, D.; Mathe, L.; Kerekes, T.; Spataru, S.V.; Teodorescu, R. On the Perturb-and-Observe and Incremental Conductance MPPT Methods for PV Systems. *IEEE J. Photovolt.* **2013**, *3*, 1070–1078. [CrossRef]
34. Biya, T.S.; Sindhu, M.R. Design and Power Management of Solar Powered Electric Vehicle Charging Station with Energy Storage System. In Proceedings of the 2019 3rd International Conference on Electronics, Communication and Aerospace Technology (ICECA), Coimbatore, India, 12–14 June 2019; pp. 815–820. [CrossRef]

processes

Article

Performance Optimization of a Steam Generator Level Control System via a Revised Simplex Search-Based Data-Driven Optimization Methodology

Xiangsong Kong [1,2,*], Changqing Shi [1], Hang Liu [1], Pengcheng Geng [1,2], Jiabin Liu [1] and Yasen Fan [2]

1 State Key Laboratory of Nuclear Power Safety Monitoring Technology and Equipment, China Nuclear Power Engineering Co., Ltd., Shenzhen 518172, China; shichangqing@cgnpc.com.cn (C.S.); liuhang@cgnpc.com.cn (H.L.); gengpengcheng2020@163.com (P.G.); liujiabin@cgnpc.com.cn (J.L.)
2 School of Electrical Engineering and Automation, Xiamen University of Technology, Xiamen 361021, China; 15259832896@163.com
* Correspondence: xskong@xmut.edu.cn

Abstract: A Steam generator is a crucial device of a nuclear power plant. Control performance of the steam generator level control system is key to its normal operation. To improve its performance, the control system parameters should be optimized by utilizing a proper optimization method. Furthermore, the method's efficiency is critical for its operability in the actual plant. However, the steam generator level process is a complex process, with high nonlinearity and time-varying properties. Traditional parameters tuning methods are experience-based, cumbersome, and time-consuming. To address the challenge, a systemic data-driven optimization methodology based on the model-free optimization with a revised simplex search method was proposed. Rather than the traditional controller parameter tuning method, this method optimizes the control system directly by using control performance measurements. To strengthen its efficiency, two critical modifications were incorporated into the traditional simplex search method to form a knowledge-informed simplex search based on historical gradient approximations. Firstly, with the help of the historical gradient approximations, the revised method could sense the optimization direction more accurately and accomplish the iteration step size tuning adaptively, significantly reducing the optimization cost. Secondly, a revised iteration termination control strategy was developed and integrated to monitor the optimization progress, which can promptly terminate the progress to avoid unnecessary iteration costs. The effectiveness and the efficiency of the revised method were demonstrated through simulation experiments.

Keywords: performance optimization; steam generator level control; knowledge-informed simplex search method; historical iteration information; iteration termination control

Citation: Kong, X.; Shi, C.; Liu, H.; Geng, P.; Liu, J.; Fan, Y. Performance Optimization of a Steam Generator Level Control System via a Revised Simplex Search-Based Data-Driven Optimization Methodology. *Processes* **2022**, *10*, 264. https://doi.org/10.3390/pr10020264

Academic Editor: Zhiwei Gao

Received: 2 December 2021
Accepted: 21 January 2022
Published: 28 January 2022

1. Introduction

The steam generator (SG) is a critical component of the nuclear steam supply system in the nuclear island, which transfers heat from the primary loop to the secondary loop to produce steam [1]. Keeping the water level around preset programmed setpoints during the plant operation is of great importance. The water level requirement violation may jeopardize the plant's availability and safety. It was reported that about 25% of emergency shutdowns in the nuclear power plants were caused by poor control of the SG water level [2]. Hence, the performance of the steam generator level control is vital for plant safety, stability, and economical operation. Meanwhile, achieving the expected performance consists of a series of commissioning tests. The cost for a single test is high; thus, the efficiency of the performance optimization is of great concern. The number of the commissioning tests in optimization should be minimized. However, SG is a highly complex, nonlinear, time-varying system with inverse dynamics [3,4]. One of the dynamic features of the SG is the "shrink and swell" phenomenon. With this phenomenon, the water level may react temporarily in a reverse manner to water mass inventory changes. This complication may be accentuated during the plant start-up or low turbine load [5]. As a result, the performance optimization of the SG level control is challenging to accomplish.

For a nuclear power plant (NPP), the performance of the steam generator level control was greatly determined by several factors in different levels, including the process characteristics, the control system structure, and the controller parameters. To a particular steam generator, the process characteristics were relatively fixed. Once the structure of the control system is determined, the controller parameters are the only vital factors left. In this situation, the control system's performance usually has a close but underlying relationship with them. It is necessary to improve the performance by selecting the optimal settings efficiently. However, as their relationship is difficult to be obtained, searching for the optimal settings is usually experience-based, cumbersome, and time-consuming. There have been increasing demands to develop advanced methods to enhance the optimization efficiency.

Different kinds of methodologies used to search for the optimal controller settings can be summarized as below:

(1) Experience-based method

This type of method relies heavily on the engineers' experience. It does not rely on the controller performance model that it tries to search for the optimal settings directly in a model-free way. It can be called "Primitive model-free methods". These methods may include trial and error, the design of experiments (DOE), and the expert system-based control et al. They have the advantage of easy implementation. However, their disadvantages are apparent: they rely heavily on the experience of operators or experts, and they are usually cumbersome and time-consuming; moreover, they are challenging to obtain the real optimal settings, only suboptimal settings can be achieved.

(2) Model-based optimization

As a mainstream way for system optimization, model-based optimization relies on a performance model that relates the controller parameters and the control system performance. However, for the control parameters optimization, the relationship was too difficult to obtain, such that no accurate model could be built [6,7]. Nevertheless, the Proportional-Integral-Derivative (PID) parameters tuning methods, such as the Ziegler–Nichols tuning method, could be treated as an approximate model-based method in essence [8–10]. For this method, the engineers should first obtain an accurate process model. Then a certain tuning formula can be developed based on the model. Finally, the parameters can be chosen according to the model coefficients and the formula. Hence, it can only provide relatively poor settings. The reason lies in the following: firstly, building an accurate model is often challenging; secondly, the formula is usually experience-dependent; thirdly, the parameter settings obtained are usually suboptimal [11].

(3) Intelligent optimization without model

With the development of artificial intelligence technology, intelligent optimization has gradually attracted more attention. For example, Wu et al. proposed to use a particle swarm optimization algorithm for the controller parameters optimization of an Advanced Passive pressurized water reactor-1000 (AP1000) plant [1]. The experimental results based on the AP1000 SG model indicated that the algorithm could achieve better control performance [1,12]. However, this algorithm needs a lot of batch experiments, which makes it impossible for the online optimization at the actual NPP plant.

(4) Data-driven control methods

As the industrial processes become more and more complex, data-driven control has emerged. The main approaches include iterative feedback tuning (IFT) [13,14], virtual reference feedback tuning (VRFT) [15], and model-free adaptive control (MFAC) [16–18]. However, these methods mainly focus on the control strategy itself, and only covered limited parameters. As a complex process, in the actual NPP plant commissioning, there are various different kinds of parameters to be tuned, such as control logic parameters and time constants of the filters. Therefore, more general optimization methods in the upper level for system optimization that could cover more types of parameters are still needed.

All the above methods have their disadvantages or limitations, which lead to the inefficiency of the control performance optimization. A question emerged: is there a different way to promote the efficiency of the control performance optimization, considering that, by integrating the advantages of the above methods and avoiding their disadvantages, the efficiency may be promoted? Data-driven methods are powerful techniques for complex industrial automation processes [19]. Meanwhile, data-driven techniques have further stimulated research and applications of monitoring, optimization, and control. The data-driven method, integrating efficient optimization algorithms, becomes an attractive way [20–22]. Hence, an analogous philosophy, integrating feasible data-based optimization algorithms without a prior model to formulate a data-driven optimization framework, is proposed to address the optimization of complex systems.

Kong had proposed a systematic and efficient simultaneous perturbation stochastic approximation (SPSA)-based optimization method for the steam generator level control and the injection molding quality control [23–26]. However, as a model-free method, the efficiency of the SPSA-based method still needs to be improved. Another selection was the simplex search-based model-free optimization (MFO), which usually behaves better than the SPSA at the low dimension circumstances. To improve the efficiency of the simplex search method (SS) on the medium-voltage insulator quality control, a revised simplex search method, knowledge-informed simplex search based on historical gradient approximations (GK-SS), has been proposed [27]. Similar to the idea of knowledge-based fault detection and diagnosis [28] and utilization of prior information [29], historical data generated during the data-driven optimization will be mined and utilized. In order to adapt to the characteristics of the control performance optimization problem and beef up the optimization efficiency, two critical modifications were formulated to the SS. Firstly, the historical-gradient approximations were utilized for the optimization search direction compensation. Secondly, an improved iteration termination control strategy, which uses the historical iteration information, was integrated to monitor the progress of the optimization. The two modifications are integrated to form a revised simplex search-based data-driven optimization methodology.

The remainder of this paper is organized as follows. Section 2 illustrates the performance optimization problem for the SG level control system. Section 3 presents an overview of the knowledge-informed optimization strategy, followed by the elementary procedure of the revised GK-SS method. The modifications on the search mechanism and the revised iteration termination control strategy are detailed. In Section 4, the simulation platform of the steam generator is demonstrated. In Section 5, the revised GK-SS method is applied to the performance optimization problem of the SG level control system. Based on the experimental simulation results, a comprehensive discussion was carried out. Finally, the conclusions are summarized.

2. Performance Optimization of the Steam Generator Level Control System

The general structure of the SG level control system in a pressurized water reactor (PWR) can be summarized as shown in Figure 1. According to the PWR commissioning practice, the main goal of the steam generator level control is to maintain the steam generator water level around a preset setpoint with the least power plant commissioning cost. To achieve the goal, the tuning process of the control system should be optimized.

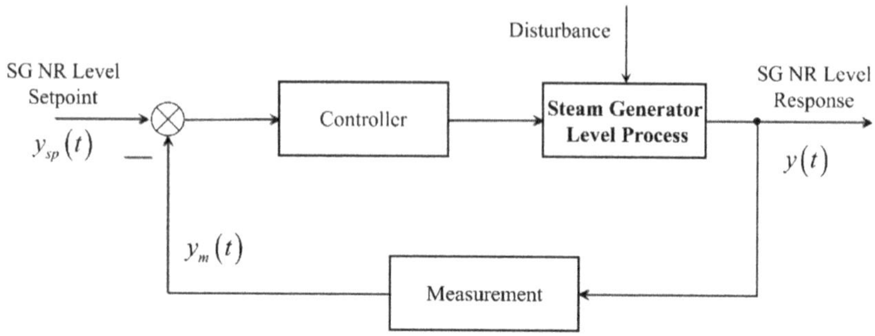

Figure 1. General structure of the steam generator (SG) level control system.

For a PWR plant, once the manufacturing of the steam generator and the control structure of the level controller have been fixed, the control performance of the SG level control system will mainly depend on the controller parameters. There is a close relationship between the control performance and the controller parameters. Without loss of generality, the mathematical relationship between the level control performance and the corresponding controller parameters could be expressed abstractly as below:

$$Perf = f(X) + e, \ \ X = [x_1, \cdots x_n]^T \tag{1}$$

where *Perf* represents the performance index to be optimized, X is an n-dimensional vector representing the SG controller parameters, and x_i $(i = 1, \cdots, n)$ is the ith element of X, n is the dimension of the parameters, e represents the overall process performance uncertainty, $f(x)$ is the theoretical function between the performance and the controller parameters. However, due to the high complexity and nonlinearity of the SG level process, the above function is unable to obtain theoretically.

The performance optimization of the steam generator level control is, therefore, converted to tune the controller parameters and search for the optimal settings. The mathematical formula of the optimization problem can be expressed as below:

$$\begin{aligned} &\max_X Perf \\ &s.t. \ \ X \in \Omega \end{aligned} \tag{2}$$

where Ω is the feasible region of the controller parameters. The schematic of the closed-loop optimization process based on data-driven optimization can be summarized in Figure 2.

To construct the closed-loop optimization scheme, it is necessary to find a suitable way to evaluate the control performance of the steam generator level control system. For a control system, the regulation performance is often expressed by the control error obtained under certain disturbances. Typical step load change can be chosen as the test scenario. Typical control evaluation index can be expressed as below:

$$I = \int_0^{T_{ts}} t^n |e(t)|^m dt \tag{3}$$

where the error is defined as $e = y_{sp} - y(t)$, y_{sp} is the target and $y(t)$ is the actual level response. T_{ts} is a preset time zone for the evaluation of transient performance of the system. n and m are the index parameters, respectively. In this study, the Integral of Time multiply by Absolute Error (ITAE) index was chosen. The ITAE is an index widely used in control performance evaluation with $n = 1$, $m = 1$ in Equation (3). As the magnitude of the ITAE is huge, the logarithm of the ITAE was used to represent the control performance. According to the characteristics of the ITAE, the smaller the ITAE index, the better the control performance.

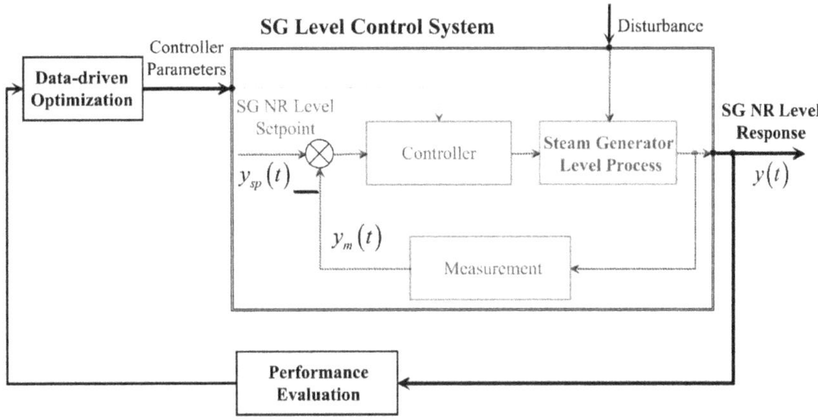

Figure 2. Schematics of the data-driven optimization for the SG level control.

With the above evaluation index, the controller performance optimization problem can be expressed as below:

$$\begin{array}{l} \min\limits_{X} \; -Perf \approx I = \int_0^{T_{ts}} t|e(t)|dt \\ s.t. \; X \in \Omega \end{array} \tag{4}$$

The performance is approximately represented by the evaluation index.

Furthermore, in the PWR commissioning practice, the efficiency of the above optimization is a critical concern. The tuning process should be done as quickly as possible to reduce the commissioning cost and reduce its adverse effects on the plant. A new efficiency index, Optimization Efficiency index (OE), is defined. As a comprehensive efficiency index, it may integrate many aspects, such as experimental cost, time cost, labor cost and performance control deviation et al. Based on the OE index, the performance optimization may be reconstructed as a bi-level optimization problem. The mathematical formulation is as below:

$$\begin{array}{ll} \max\limits_{OptMethods} & OE \\ s.t. & \left\{ \begin{array}{l} \min\limits_{X} \; -Perf \approx I = \int_0^{T_{ts}} t|e(t)|dt \\ s.t. \; X \in \Omega \end{array} \right\}_{OptMethods} \end{array} \tag{5}$$

where *OptMethods* is the available data-driven methods set, which satisfies the optimization framework in Figure 2. The key in this study is to search for a suitable method with high efficiency. Considering the plant commissioning practice, only the experimental cost was included in the OE. A revised simplex search-based model-free optimization method was thus proposed.

3. Revised Simplex Search-Based Data-Driven Optimization

3.1. Performance Optimization via Data-Driven Optimization

According to the characteristics of the optimization problem in Equation (2), the model-free optimization methods could be incorporated for the optimization. The schematics of the model-free optimization are shown in Figure 3. These methods usually achieve their goal by pure historical iterative data. In the framework of the model-free optimization, the data-driven optimization mechanism will iteratively adjust the parameters of the control system to change the performance of the SG liquid level control system dynamically. The optimization iteration process is a batch-by-batch process. The efficiency of the tuning process will mainly depend on the optimization methods. This study adopts the simplex search method as the base method for its high efficiency in nonlinear system optimization. It is a direct search method without explicit gradient information.

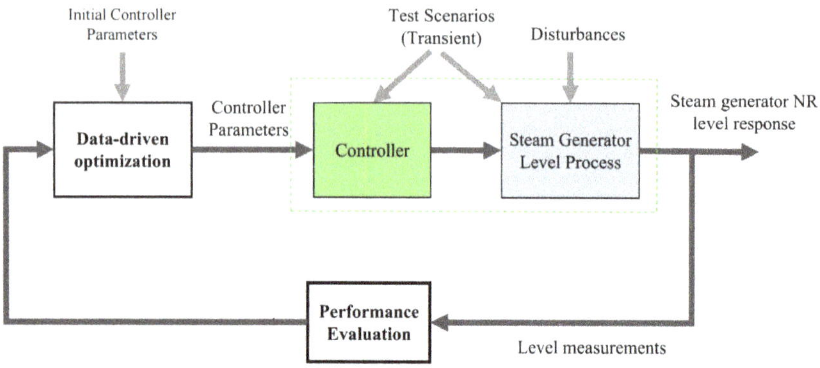

Figure 3. Framework of the model-free optimization (MFO).

However, the optimization cost is critical for a PWR plant's commissioning, and the simplex search method's efficiency should be further improved. Hence, the feasible directions for the performance efficiency promotion were analyzed as below. Figure 4 demonstrates a typical optimization trajectory under the traditional simplex search method. To promote the performance efficiency of the optimization method, the possibilities for the promotion were decomposed into two dimensions. For the first dimension, the optimization could be improved to achieve a rapid descent in the first stage of the optimization. If the optimization algorithms could be revised to accelerate its descent rate, the efficiency would be improved. For the second dimension, as a data-driven optimization method, the method usually cannot appropriately stop the optimization progress immediately. Thus, the optimization often costs more on iterations with only a minor achievement. The minor achievement is intolerable for the vast cost due to the additional iterations. Hence, if the method could terminate timely and intelligently when the optimization progress traps into the stagnation stage, the costs on the optimization could be reduced, and the efficiency thus could be promoted.

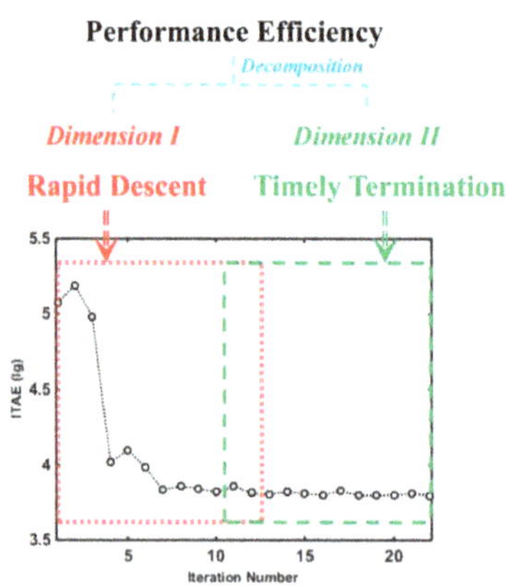

Figure 4. Decomposition of the directions for the performance efficiency promotion.

This study investigated both research dimensions for the simplex search-based model-free optimization. Generally, as shown in Figure 5, the knowledge-guided mechanisms consist of two dimensions and three aspects. In Dimension I, considering that in the iterative process of performance optimization, the optimization process will generate a series of historical iterative process information containing specific process knowledge, in the traditional MFO methods, this kind of information is not fully utilized. In this study, the revised simplex search method may adopt the historical iteration points generated during optimization to adjust the search descent rate. Two aspects were concerned. The first aspect is the search direction of the optimization, with the second aspect being the step size of the search operation. The detailed mechanisms will be demonstrated in the following subsection. In Dimension II, the information of historical iteration points will be incorporated to form a historical sequence, and the optimization progress will be monitored and evaluated based on the sequence. Hence, the optimization progress could be interrupted timely when the method detects itself falling into the stagnation stage.

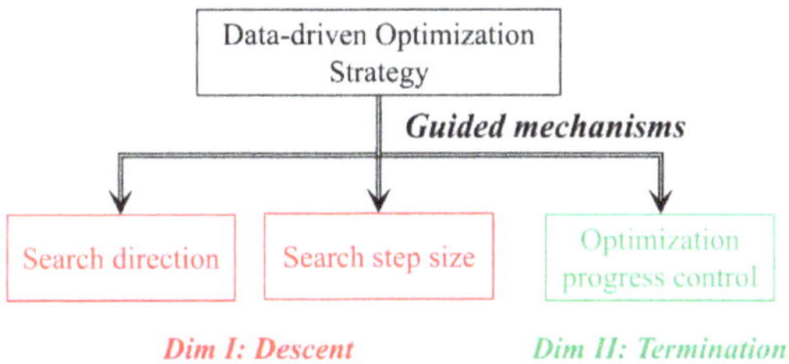

Figure 5. Historical information-guided mechanisms for the data-driven optimization.

From the idea of knowledge-informed optimization fused by iterative knowledge during the optimization, a revised simplex search-based model-free optimization based on the historical iteration information was formulated, and a feasible implementation scheme was put forward.

3.2. Revised Simplex Search Strategy

The simplex search method is a gradient-free method that relies only on the function evaluations [30]. Our team firstly proposed the initial revised simplex search method (GK-SS) to promote the efficiency of the quality control of the medium voltage insulator [27]. In this study, this method was transferred and revised further to adapt to the new optimization occasion of the controller parameters optimization. Furthermore, the GK-SS was reorganized based on the above ideas. The modifications for the traditional simplex search were reorganized into two parts, representing different mechanisms on efficiency promotion.

3.2.1. Procedure of the Traditional Simplex Search Method

The revised GK-SS was built based on the traditional Nelder–Mead method. The procedure of the method is as follows (More detailed procedure may refer to [27]):

Step 1: Methodology initialization.

The coefficients of the methodology, $\{\alpha, \beta, \gamma, \delta\}$, and the initial conditions X_1, are determined in this stage. The initial conditions will be scaled into a normalized range and represented as $\overline{X}_1$.

Step 2: Initial simplex construction.

The initial simplex is constructed obeying the following rule:

$$\begin{cases} \overline{X}_{k+1} = \overline{X}_1 + \tau e_k, & \overline{X}_{1,\,k} \leq 50 \\ \overline{X}_{k+1} = \overline{X}_1 - \tau e_k, & \overline{X}_{1,\,k} > 50 \end{cases}, \ V^s_{k+1} = \overline{X}_{k+1},\ k = 1, \cdots, n \tag{6}$$

where e_k is a special column vector with 1 in the kth element and zeros in the other elements, $\overline{X}_{k+1}$ is the (k + 1)th iteration point, and τ represents the perturbation coefficient. With the construction, n successive iteration points are generated based on the initial point, and the initial simplex is formed as $V^s = \{V_1^s, V_2^s, \cdots, V_{n+1}^s\}$ based on the n + 1 points. s is the simplex iteration number with its initial value set to 0. The corresponding performance response ITAE vector $F^s = \{F_1^s, F_2^s, \cdots F_{n+1}^s\}$ is garnered by a series of experiments.

Step 3: Simplex Sorting.

The current simplex will be sorted according to their corresponding performance index F^s. The sorted simplex will be denoted as V^{s*} with the following relation:

$$F_1^{s*} \leq F_2^{s*} \leq \cdots F_n^{s*} \leq F_{n+1}^{s*} \tag{7}$$

Thus, V_1^{s*} is the vertex with the best response (with the smallest ITAE quality), V_n^{s*} is the vertex with the next-to-the-worst response, V_{n+1}^{s*} is the vertex with the worst response. The simplex iteration number will be updated by $s = s + 1$.

Step 4: Reflection.

The reflection point V_{ref}^s will be generated as below:

$$V_{ref}^s = (1+\alpha)V_c^s - \alpha V_{n+1}^{s*} \tag{8}$$

The centroid of all the vertices except V_{n+1}^{s*} is expressed as:

$$V_c^s = \left(\sum_{i=1}^{n} V_i^{s*}\right)/n \tag{9}$$

Conduct the experiments to evaluate the ITAE performance index F_{ref}^s at the reflection point. If $F_{ref}^s < F_1^{s*}$, the expansion operation should be executed; thus, the procedure goes to Step 5. If $F_1^{s*} \leq F_{ref}^s \leq F_n^{s*}$, the procedure goes to Step 8. If $F_{ref}^s > F_n^{s*}$, the contraction operation should be conducted, the procedure goes to Step 6.

Step 5: Expansion.

The expansion point $V_{\exp}^s$ is generated with the following expansion rule:

$$V_{\exp}^s = (1-\gamma)V_c^s + \gamma V_{ref}^s \tag{10}$$

The experiment will be conducted and the ITAE performance index $F_{\exp}^s$ would be evaluated. If $F_{\exp}^s \leq F_{ref}^s$, the expansion is successful, V_{n+1}^{s*} would be replaced by $V_{\exp}^s$. Otherwise, the worst vertex with the highest ITAE would be substituted by the reflection point. At last, the procedure goes to Step 8.

Step 6: Contraction.

The contraction point V_{ct}^s could be calculated as below:

$$V_{ct}^s = (1-\beta)V_c^s + \beta V_{\max/ref}^s \tag{11}$$

where $V_{\max/ref}^s$ is the reference point for the contraction. The contraction operation consists of two modes: (i) inside contraction, (ii) outside contraction. The choice of the reference $V_{\max/ref}^s$ depends on the contraction mode.

Case 1: Inside contraction.

When $F_{ref}^s \geq F_{n+1}^{k*}$, let $V_{\max/ref}^s = V_{n+1}^{s*}$, $F_{\max/ref}^s = F_{n+1}^{s*}$; conduct the experiments to evaluate the ITAE index F_{ct}^s. If $F_{ct}^s \leq F_{n+1}^{s*}$, the contraction is accepted.

Case 2: Outside contraction.

When $F_n^{k*} \leq Y_{ref} \leq F_{n+1}^{k*}$, let $V_{\max/ref}^s = V_{ref}^s$, $F_{\max/ref}^s = F_{ref}^s$; conduct the experiments to evaluate the ITAE index F_{ct}^s. If $F_{ct}^s \leq F_{ref}^s$, the contraction is accepted.

If the contraction is accepted, replace V_{n+1}^{s*} with V_{ct}^s and go to Step 8. Otherwise, if the contraction is refused, go to Step 7 for a shrink operation.

Step 7: Shrink.

The shrink operation is generated as below:

$$V_i^{s*} = (1-\delta)V_1^{s*} + \delta V_i^{s*}, \ i = 2, \cdots, n+1 \tag{12}$$

The best vertex will be retained. All the vertices, except the best one, are updated according to Equation (12). The ITAE performance indices of the updated vertices will be measured and updated.

Step 8: Termination.

The termination module will judge the termination rule based on the revised iteration termination control strategy.

The traditional simplex-search method was revised according to the knowledge-informed idea in this study. A data-driven method, the GK-SS, was reformulated. The GK-SS deals with the act of optimization as an entirety. The two key modifications are detailed in the following subsections, respectively.

3.2.2. Search Mechanism Modification

The search mechanism was redesigned from the idea of Dimension I of the efficiency promotion. The revised mechanism was conducive to the rapid descent of the optimization process. To build a more efficient knowledge-informed mechanism, the traditional SS was reviewed. The essence of the SS's principle was further revealed and strengthened. The SS is reshaped to construct a new quantity-quasi-gradient estimation from the above perspective. With this new quantity, the quasi-historical gradient estimation information can be recorded and utilized to improve the accuracy of the quasi-gradient estimations for the simplex search method. Once the accuracy of the gradient estimations improved, the simplex-search method's efficiency would be enhanced accordingly. Figure 6. demonstrates the schematic diagram of the quasi-gradient estimation compensation mechanism of the GK-SS. It can be seen from the figure that the compensated search direction will approach the theoretical steepest search direction with the help of the historical quasi-gradient estimations.

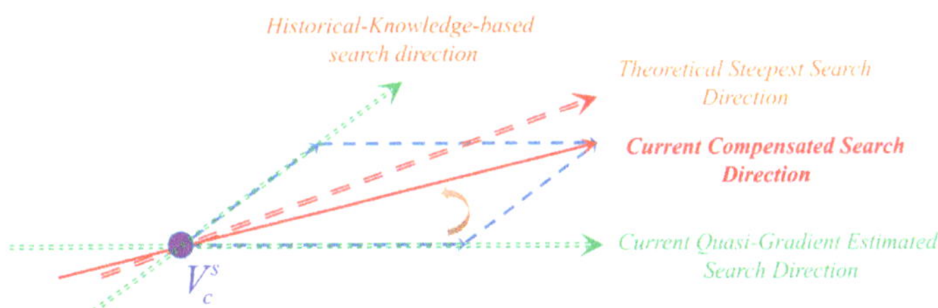

Figure 6. Sketch of the compensation mechanism [27].

Hence, a specific simplex search mechanism was proposed, which incorporates the historical quasi gradient estimations to facilitate the optimization. The quasi-gradient estimation is expressed as follows:

$$\widetilde{G}_c^s = V_{n+1}^{s*} - V_c^s \tag{13}$$

where $\widetilde{G}_c^s$ is the estimated quasi-gradient for current simplex (EGCS) at the sth simplex.

The iteration strategy of the simplex search method is represented as below:

$$V_{vertex\_new}^s = V_c^s - \xi \widetilde{G}_c^s \tag{14}$$

where ξ is the estimated step size of the method, which is determined by the status of the current simplex.

To integrate the historical knowledge to compensate the gradient estimation accuracy at the current simplex, the compensated composite gradient for the current simplex (CCG) is defined as follows:

$$\hat{G}_c^s = \rho_s \hat{G}_c^{s-1} + (1-\rho_s)\widetilde{G}_c^s,\ \hat{G}_c^1 = \widetilde{G}_c^1 \tag{15}$$

where $\hat{G}_c^s$ is the CCG at the sth simplex, $\hat{G}_c^{s-1}$ is the CCG at the $(s-1)$th simplex, $\hat{G}_c^1$ is the CCG at the *1*st simplex. ρ_s is the gradient compensation coefficient at the sth simplex.

The gradient compensation coefficient updating rule is as below:

$$\rho_s = \rho_F - \frac{\Delta\rho_{init}}{s^\tau} \tag{16}$$

where ρ_F is an upper limit that a CCG approached, which is set to 0.5 by default, $\Delta\rho_{init}$ is the initial deviation between ρ_F and its lower limit, which is set to 0.2 by default, and τ is an exponential coefficient.

The original reflection operation, Equation (8), should be substituted as follows:

$$\overline{V}^s_{ref} = V^s_c - \alpha\hat{G}^s_c \tag{17}$$

The expansion operation would be revised as below:

$$\overline{V}^s_{\exp} = (1-\gamma)V^s_c + \gamma\overline{V}^s_{ref} = V^s_c - \alpha\gamma\hat{G}^s_c \tag{18}$$

The outside contraction is as follows:

$$\overline{V}^s_{oct} = (1-\beta)V^s_c + \beta\overline{V}^s_{ref} = V^s_c - \alpha\beta\hat{G}^s_c \tag{19}$$

The inside contraction is as below:

$$\overline{V}^s_{ict} = (1-\beta)V^s_c + \beta\overline{V}^s_{n+1} = V^s_c + \beta\hat{G}^s_c \tag{20}$$

The quasi-gradient information is incorporated with the above modification, and the revised mechanism is formulated. The different search mechanisms of the traditional simplex-search and the revised method are illustrated and compared in Figure 7. The revised GK-SS method is a typical knowledge-informed optimization strategy that utilizes the historical quasi-gradient estimations to compensate for the search direction at each simplex iteration. In a statistical sense, this strategy could significantly accelerate the search rate of the optimization.

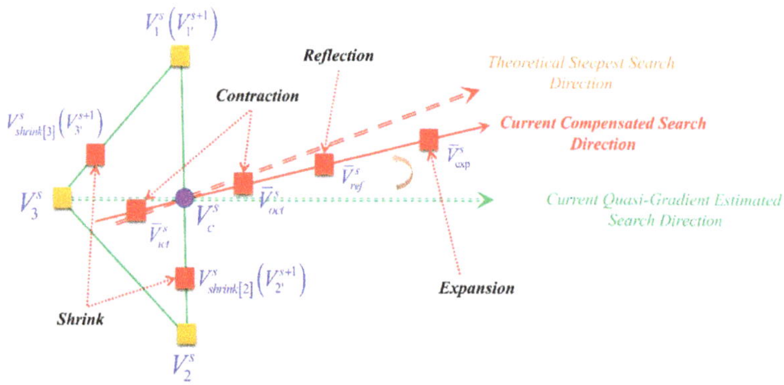

Figure 7. Optimization mechanisms of the knowledge-informed simplex search based on historical gradient approximations (GK-SS) illustrated in a two-dimensional occasion [27].

3.2.3. Iteration Termination Control Modification

Data-driven optimization relies on commissioning experiments. Hence, from Dimension II, the efficiency may be promoted through timely termination of the optimization progress. Usually, when the optimization progress goes into the stagnation zone, the improvement in the optimization becomes significantly slight. Too many experiments may only lead to trivial promotion. To reduce the commissioning costs, proper termination criteria should be developed. If the optimization progress could be monitored, the optimization method may construct a proper termination rule to timely terminate the optimization. In this study, iteration termination control strategy was designed to take advantage of the achievement estimation of the optimization process to discover the proper time to terminate the optimization progress [23].

Figure 8 shows the framework of the revised iteration termination strategy. The procedure can be divided into seven steps as follows:

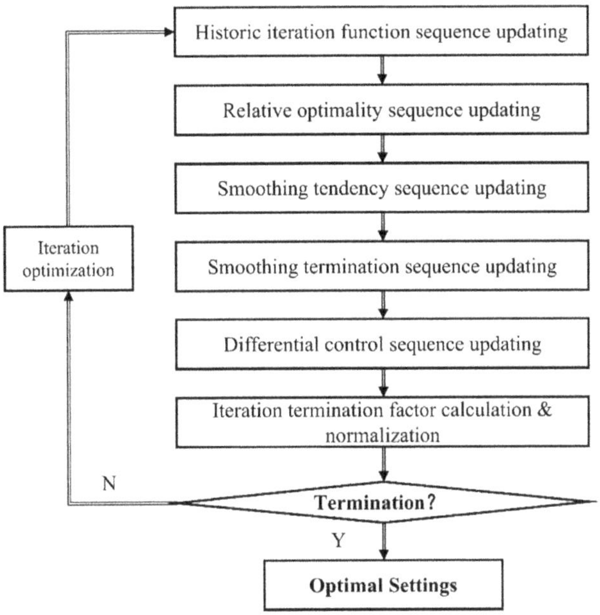

Figure 8. Framework of the Iteration Termination Control.

Step 1: Historic iteration function sequence updating.

Historic iteration function sequence, S_H, is a self-increasing sequence storing the function evaluations of all the historic iteration points. When a new iteration point is generated, and the corresponding experiment is conducted, the evaluation of the ITAE index will be recorded. Then, the Sequence S_H will grow with the new evaluation.

Step 2. Relative optimality sequence updating.

In this stage, the historic iteration function sequence will be sorted according to the magnitude of the control performance index. The best index is gained and recorded iteratively, and a relative optimality sequence, S_{RO}, is updated sequentially with the current best index.

$$S_{RO}(i) = \min(S_H) \tag{21}$$

where i is the current count number of the iterations.

Step 3: Smoothing tendency sequence updating.

The relative optimality sequence is further smoothed by a moving average method. Then, the tendency sequence $\boldsymbol{S_{ST}}$ is obtained as below:

$$S_{ST}(i) = \begin{cases} \frac{1}{i}\sum\limits_{k=1}^{i} S_{RO}(k), & i \in [1,\ \lambda(n+1)); \\ \frac{1}{\lambda(n+1)}\sum\limits_{k=i-\lambda(n+1)-1}^{i} S_{RO}(k), & i \in [\lambda(n+1),\ +\infty). \end{cases} \tag{22}$$

where n is the parameters dimension, λ is the smoothing coefficient.

Step 4: Smoothing termination sequence updating.

To obtain a more meaningful sequence, another smooth action is conducted further. The termination sequence S_{TM}, to be utilized for the termination control, is generated. It is a monotonically decreasing sequence. Its formulation rule is defined as below:

$$S_{TM}(i) = \begin{cases} \frac{1}{i}\sum\limits_{k=1}^{i} S_{ST}(k), & i \in [1,\ \eta(n+1)); \\ \frac{1}{\eta(n+1)} \sum\limits_{k=i}^{i-\eta(n+1)+1} S_{ST}(k)\ , & i \in [\eta(n+1),\ +\infty). \end{cases} \tag{23}$$

Step 5: Differential control sequence updating.

The differential control sequence ΔS_{TM} is formulated as below:

$$\Delta S_{TM}(i) = \begin{cases} 1, i \in [1, \eta(n+1)] \\ S_{TM}(i) - S_{TM}(i - \eta(n+1) + 1), i \in [\eta(n+1), +\infty] \end{cases} \tag{24}$$

Step 6: Iteration termination factor calculation & normalization.

To evaluate the relative progress of the optimization, the termination factor $\xi(i)$ is defined:

$$\xi(i) = \begin{cases} 0, \\ \frac{\Delta S_{TM}(i)}{S_{TM}(i)} \end{cases} . \tag{25}$$

It indicates the ratio of the control performance improvement at the current point.

However, the numerical range of this factor is too broad, which is highly related to the specific problems and the optimization process at different times, so it may appear at different scales when used to judge the status of the optimization process. In fact, in the actual operation of the SG level control performance optimization, the termination factor may vary quite sharply. It may deteriorate the function of the iteration termination control. Hence, a further normalization mechanism was invented to cope with this challenge.

Considering the consistency of the judgment factor, the factors should be reformulated. Each calculated factor is added to form a sequence; thus, a termination factor sequence S_ξ is generated.

A new normalized sequence is defined as below:

$$\varsigma(k) = \frac{\xi(k) - \min(S_\xi)}{\max(S_\xi) - \min(S_\xi)}, k = 1, \cdots, i. \tag{26}$$

At each iteration, every element of the entire sequence will be recalculated, and the sequence is scrolled.

Step 7: Termination criteria judgment.

When the current judgment factor is small enough, i.e., $\varsigma(i) < \varsigma_T$ (ς_T is the tolerance), the progress at the current iteration is so slight that the termination criteria may be satisfied. However, to avoid the prematurity of optimization, further verification should be conducted. The iteration terminate rule is defined as:

$$(\varsigma(i) < \varsigma_T) \cap (\kappa = \kappa_F) \tag{27}$$

where κ_F is the repeating coefficient set by the engineers, κ is a counter representing the number of successive iterations that satisfies the former tolerance.

When Equation (27) is satisfied, the controller parameters optimization process will be terminated. The optimal settings will be set on the controller for the SG level control.

The anatomy of the GK-SS, including the two modifications, is illustrated in Figure 9. The primary mechanism of the GK-SS uses knowledge of historical quasi-gradient information generated during the iteration of the simplices to predict more accurate search directions and uses the historical iteration function evaluations to monitor the optimization progress. The revised method leverages synergies across different mechanisms within the data-driven optimization strategy. With these two modifications, the efficiency of the optimization would be enhanced.

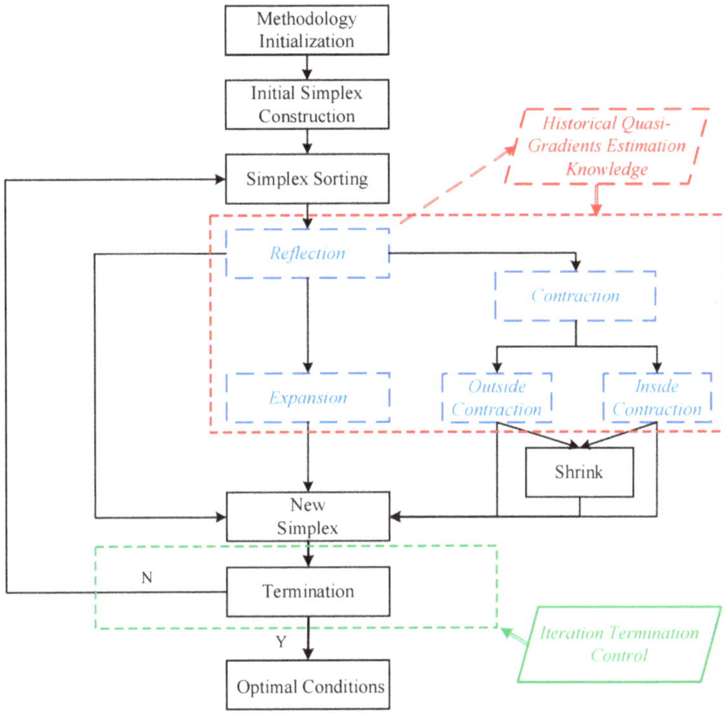

Figure 9. Anatomy of the revised GK-SS.

4. Simulation Experimental Setup

Considering the feasibility of verifying the proposed methodology's effectiveness and efficiency, a simulation platform for the SG level control system was established. The platform consists of the process model of the steam generator, the control scheme, and the controller parameters.

The general structure of a steam generator is illustrated in Figure 10. A highly complex and nonlinear system should adopt a suitable simplified model for the controller design. A widely used simplified steam generator level model was proposed by Irving et al. [2] and was adopted as the process model. The model captures the essential dynamics of the steam generator and is widely used for the simulation and control of the steam generator level process. The transfer function of the model is as follows:

$$Y(s) = \frac{G_1}{s}(Q_e(s) - Q_v(s)) - \frac{G_2}{1+\tau_2 s}(Q_e(s) - Q_v(s)) + \frac{G_3(s)}{\tau_1^{-2} + 4\pi T^{-2} + 2\tau_1^{-1}s + s^2}Q_e(s) \tag{28}$$

where $Y(s)$ is the output of the model, which represents the narrow range water level of the steam generator. $Q_e(s)$ and $Q_v(s)$ are the inputs to the steam generator, where $Q_e(s)$ represents the feed water rate and $Q_v(s)$ is the steam rate.

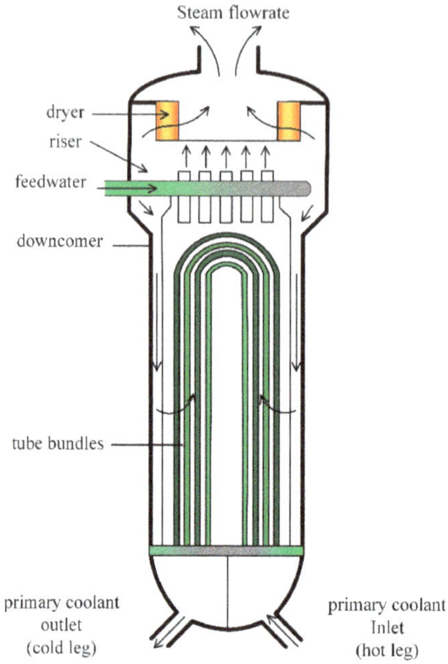

Figure 10. Structure of the U-type steam generator.

In Equation (28), the output is comprised of three terms. The first term represents the water level due to the total mass inventory in the steam generator. The second term represents the negative thermal effect of the "swell and shrinks" phenomenon. It produces an unstable zero, which accounts for the non-minimum phase behavior of the steam generator. The third term represents the mechanical oscillations caused by the feed water rate [31].

G_1, G_2, G_3, τ_1, τ_2, and T are the parameters of the steam generator model. G_1 is the magnitude of the mass capacity effects, G_2 is the magnitude of the "swell and shrinks" phenomenon, and G_3 is the magnitude of the mechanical oscillation. τ_1, τ_2 are the damping time constants. T is the period of the mechanical oscillation. All these parameters are highly related to the turbine load of the nuclear power plant. When the power level P (% Full Power, % FP) varies, the parameters should be changed. To represent the SG with the full power range, the model parameters at different power levels have been identified from experimental data by Irving, E., and the identified parameters are given in Table 1.

Table 1. Parameters of the simplified steam generator model.

P (% FP)	5	15	30	50	100
G_1	0.058	0.058	0.058	0.058	0.058
G_2	9.63	4.46	1.83	1.05	0.47
G_3	0.181	0.226	0.310	0.215	0.105
τ_1	41.9	26.3	43.4	34.8	28.6
τ_2	48.4	21.5	4.5	3.6	3.4
T	119.6	60.5	17.7	14.2	11.7
Q_v (s) (kg/s)	57.4	180.8	381.7	660	1435

A three-element control system with a cascade PID control scheme was adopted to achieve a suitable control target. The diagram of the control system is demonstrated in Figure 11.

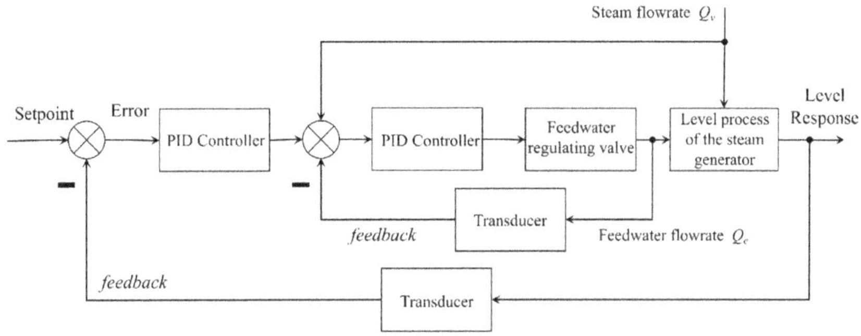

Figure 11. Cascade Proportional-Integral-Derivative (PID)-based control scheme for the SG level control.

This three-element scheme takes the feedwater flow as the internal disturbance, treats its signal as the second controlled variable, and regulates it through the vice-loop controller. Meanwhile, the scheme takes the steam flow as the external disturbance and adds it to the vice-loop controller in a feedforward way. With this architecture, the negative feedback of feedwater flow disturbance reduces the system's dynamics, while the feedforward of the steam flow weakens the phenomenon of "swell and shrinks" effects; thus, the stability and rapidity of the system operation could be improved.

However, the controller parameters of the cascade PID controller still have a significant impact on the performance of the steam generator control system. To further promote the performance of the steam generator control system, the controller parameters need to be improved.

The cascade PID scheme consists of two independent PID controllers: the principal regulator and the auxiliary regulator. Each PID controller has the control scheme as follows:

$$G(s) = k_P + \frac{k_I}{s} + k_D s \tag{29}$$

where k_P represents the proportion gain, k_I represents the integral gain, k_D represents the derivative gain.

As the process and the architecture of the control system have been determined, the system's performance will be mainly affected by the parameters of the two PID controllers. The PID parameters of the control system to be optimized can be defined as $X = [x_1, x_2, x_3, x_4, x_5, x_6]^T$. As a particular industrial process, the NPP should be kept stable during the PID parameters tuning process in the plant commission. Hence, the feasible region of the PID parameters, determined by the recommendation of the plant design specification and the engineers' experience, guarantee the system's stability. As a result, the performance optimization of the NPP control system is an optimization process on the premise of the stability of the control system. Therefore, the stability of the control system, as an implicit assumption, does not need to be further discussed in this paper. The feasible region of the PID parameters in this study is shown in Table 2.

Table 2. Feasible regions of the cascade Proportional-Integral-Derivative (PID) control system.

Variable No.	Description	Low Limits	Upper Limits
x_1	k_P of the principal regulator	0.077	0.3
x_2	k_I of the principal regulator	2.3×10^{-4}	2.3×10^{-3}
x_3	k_D of the principal regulator	−0.6	2.65
x_4	k_P of the auxiliary regulator	1	1.5
x_5	k_I of the auxiliary regulator	0.3	0.8
x_6	k_D of the auxiliary regulator	0	0.5

The above platform was built based on Simulink® v2021a. To search for the optimal PID parameters, the proposed method was formulated and verified based on the simulation platform.

A typical transient condition was adopted to verify the method at the same conditions. For each simulation experiment, the nuclear power plant is firstly operated at the initial condition for 100 s; then, the transient is actuated. After the actuation of the transient, the nuclear power plant model will run for 1100 s to ensure the plant return to a new steady state. The steam generator level response is recorded during the process, and the ITAE index is calculated after each simulation. The results can be analyzed in the next section.

5. Results and Discussion

The revised simplex search-based MFO (GK-SS-based MFO) was conducted on the above simulation platform. To verify the effectiveness of the improved method and its efficiency relative to the traditional method, two groups of tests, including the effectiveness test and the efficiency test, are designed, respectively, for the verification.

5.1. *Effectiveness Test*

Firstly, without loss of generality, a randomly initial point $x_1 = [0.077, 0.0003, 0.2, 1, 0.5, 0]^T$ was selected. The steam generator model was operated at 100%FP. The same test condition was implemented on each iteration experiment in this test. The GK-SS was conducted, and the optimization results were recorded. The optimization trajectory represented by ITAE is demonstrated in Figure 12. It can be observed that the parameters tuning with a significant performance improvement had finished in a relatively limited iteration number. With the GK-SS's optimization process, the ITAE index continued to decline. The controller parameters were perturbed dynamically during the tuning process, iteration by iteration. The trajectories of the controller parameters can be seen in Figure 13. Finally, the optimized point achieved by the GK-SS-based MFO is $x_{opt} = [0.26, 0.0023, -0.4331, 1.4883, 0.8, 0.3047]^T$. To show the performance differences between the initial guess x_1 and the optimized point x_{opt}, the level responses under the two different settings are shown in Figure 14. From the figures, it can be seen clearly that a significant improvement was obtained from x_1 to x_{opt} with the help of the GK-SS.

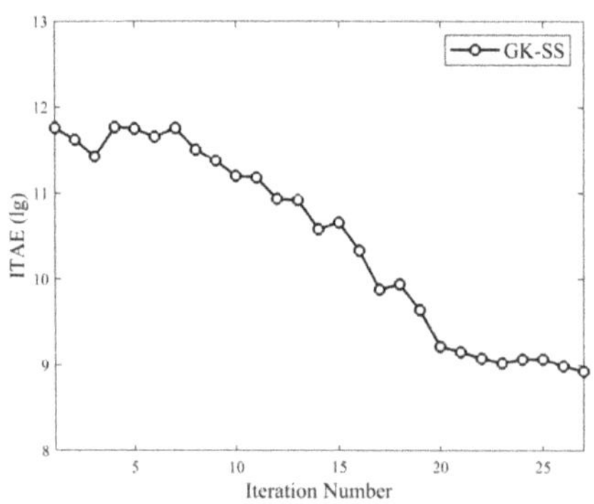

Figure 12. Iteration points trajectory of the revised simplex-search-based MFO.

The effectiveness of the iteration termination control was verified further. Another initial point was selected. The termination factor was recorded and plotted. The trajectory of the termination factor is demonstrated in Figure 15. It can be seen that the termination factor firstly went up with the optimization descending process, then the factor went down as it gradually entered the stagnation zone. The termination factor tolerance was set to 0.2. The optimization progress stopped in four successive iterations as the termination factor successively became smaller than the tolerance. A total of 24 iterations was spent on this test.

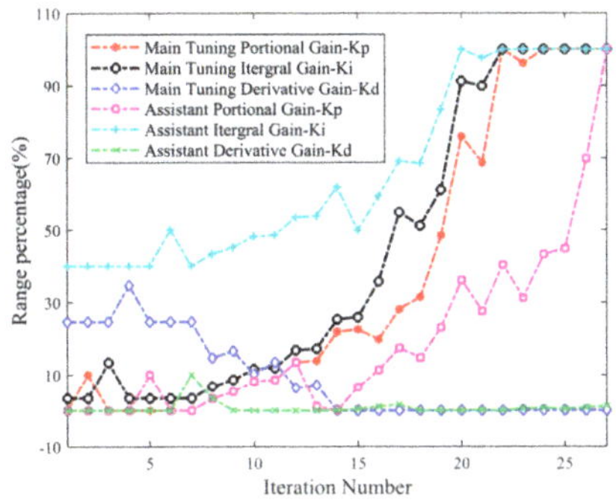

Figure 13. Trajectories of the controller parameters during the optimization.

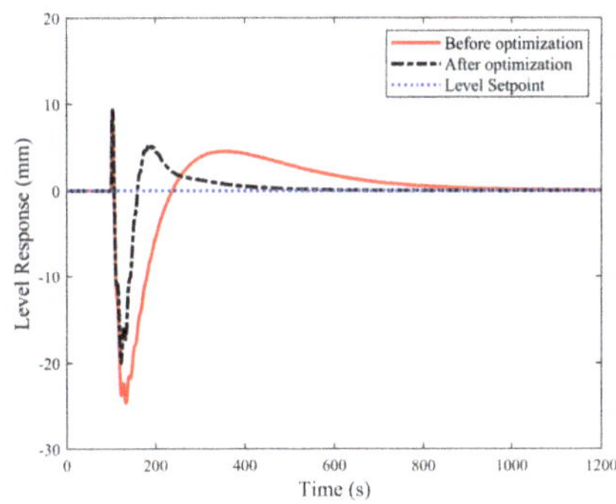

Figure 14. Level trajectories before and after the optimization.

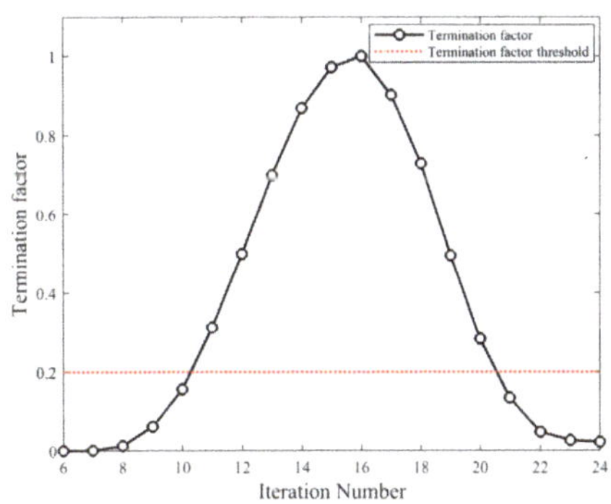

Figure 15. Trajectory of the termination factor.

To show the effectiveness from different initial points, three different initial points were chosen randomly. The optimization trajectories are demonstrated in Figure 16. All the optimization runs behaved similarly in the optimization tendency, in that the method was effective from different initial points.

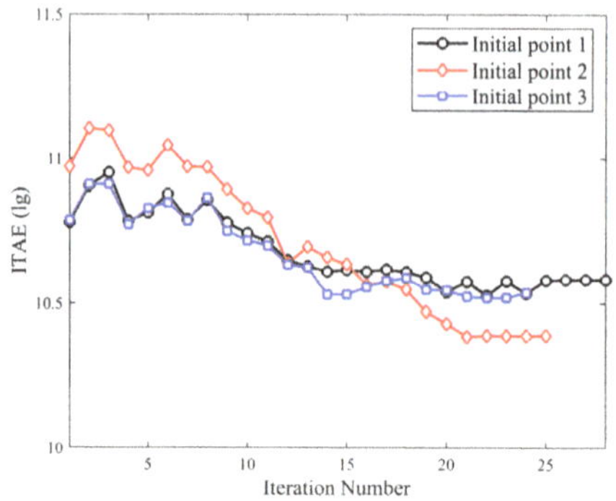

Figure 16. Optimization trajectories from different initial points.

To show the effectiveness of the GK-SS on different operation conditions, two operation scenarios under different power levels were selected—one case from the 100% FP, another one from the 50% FP. The optimization trajectories are shown in Figure 17. It can be seen clearly that the GK-SS behaved well under both scenarios.

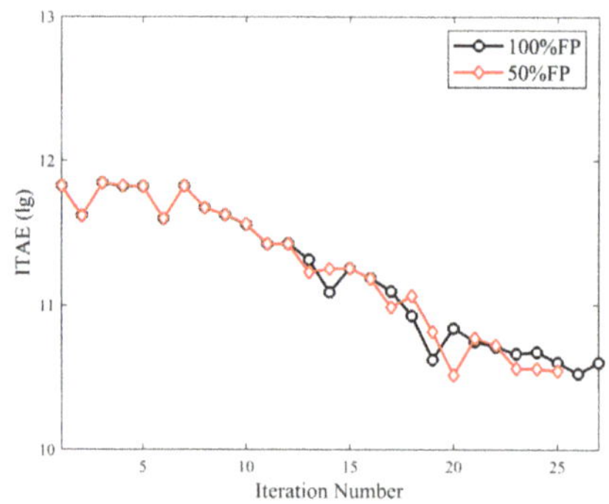

Figure 17. Level trajectories at the start and the end of the optimization.

According to the above tests, the effectiveness of the revised simplex-search-based MFO was clearly demonstrated.

5.2. *Efficiency Test*

From the above tests, the revised method, GK-SS, was showcased to be effective on the performance optimization of the SG level control system. However, the efficiency improvement of the GK-SS relative to the traditional simplex search method should be tested further. The same optimization test framework, a sequential Latin Hypercube Sampling (LHS) strategy, proposed in [17], was incorporated. In the test framework, for a single batch, the regular LHS is adopted, generating a sample of randomly distributed initial points. Then, to investigate the method's tendency with the enhancement of the sample, the same LHS operation is repeated sequentially. A repeated number of LHS tests were conducted to form a series of LHS sample tests to monitor the tendency. This kind of design of experiments (DOE) may cover enough initial points that it will provide a statistical result to show the relative efficiency of the GK-SS method.

In the test, the dimension of the process conditions was 6. Considering adopting a reasonable level for the controller parameters, each parameter was equally divided into ten levels. For a single LHS, ten independent sampling points were generated randomly. To evaluate the performance of an optimization run objectively, the iteration number of the transient process on performance optimization was selected as a critical index to show the efficiency of the method. The average iteration number was calculated to reflect the relative performance of the two methods on each batch. In this test, the batch number of the sequential design was set to 100, that is, 100 LHS batches with a total of 1000 optimization runs. This quantity of the randomly selected initial points could provide enough samples for the statistical analysis of the method's performance.

For comparison, both the GK-SS and the SS were implemented, respectively. All the experiments were carried out under identical test conditions, with only the difference in the initial points. The results of every LHS batch were recorded and observed. The dot diagram of the averaged iteration number for each LHS batch is shown in Figure 18a. It can be seen from the figure that the average iteration number per LHS batch varied considerably for both methods. However, it was observed that the iteration number of the SS was significantly greater than the GK-SS. From the perspective of statistical view, the accumulated averaged iteration number of the GK-SS is illustrated in Figure 18b. It can be seen from the figure that the accumulated average iteration number of the GK-SS was significantly decreased relative to the SS. The sequential LHS design provided an angle of view to observe the trend of the performance indices with the increase of independent initial points. Figure 18b showed the accumulated average iteration number trajectories of the GK-SS and the SS. In the initial LHS batches, due to the limited coverage of the total samples, it was shown that there were minor fluctuations in the trajectory. With the proceeding of LHS batches, the number of samples covered increases gradually. Correspondingly, the cumulative performance indices converged to a stable value gradually. The average performance in the feasible region could exhibit convergence characteristics under the cumulative effect, and the variability in the performance of the methods is thus eliminated. This could reveal the relative performance between the two methods.

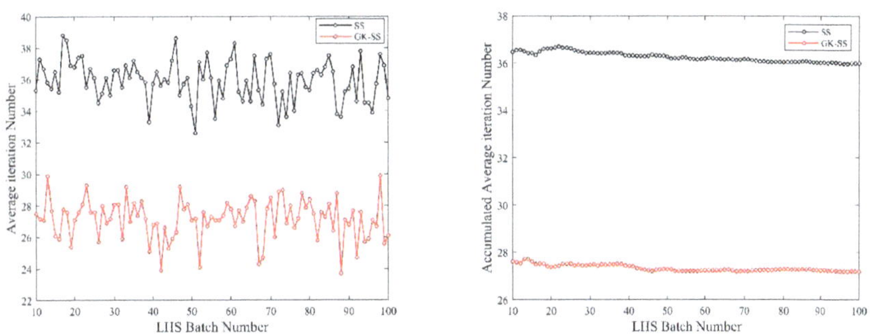

(a) (b)

Figure 18. Average iteration number distribution in the test: (**a**) dot diagram of the average iteration numbers; (**b**) Accumulated averaged iteration number.

Although the GK-SS behaved better in the optimization efficiency, the optimization results obtained by it may not be improved. Figure 19a shows the average optimal ITAE index obtained of each batch for the SS and the GK-SS, and Figure 19b shows their accumulated average optimal ITAE index. The trajectories indicated that the GK-SS behaved a little worse than the SS on the final optimal outputs. However, the deficiency was so minor that it still can be concluded that the two methods could obtain similar optimal results. The deficiency phenomenon was reasonable, because the principle of the GK-SS lies in the efficiency promotion to enhance the economy index on the iteration number, not in the promotion of the optimal results. Considering that a little sacrifice of the optimality can be exchanged for the improvement of the economy index, GK-SS is still an efficient method.

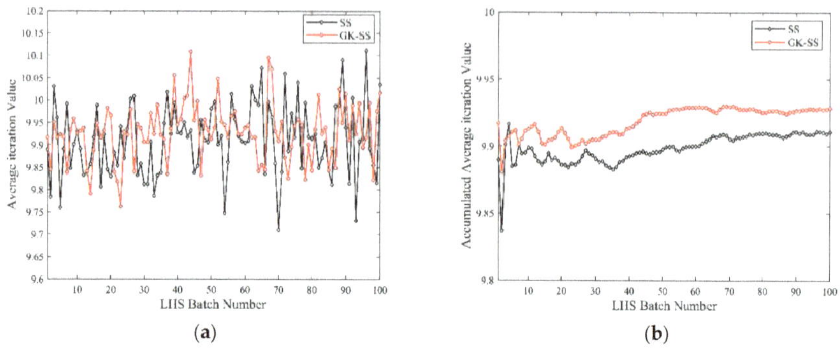

Figure 19. Average optimal ITAE index distribution in the test: (**a**) dot diagram of the average optimal ITAE index; (**b**) Accumulated averaged optimal ITAE index.

To sum up, the performance efficiency indices are demonstrated in Figure 20. Figure 20a illustrates the average iteration number of the two methods. Downward trends in iteration number statistics of the GK-SS relative to the traditional simplex search method suggested that the revised method is relatively more efficient. It can be observed that the average iteration number diminished significantly, with a reduction of 22.9%.

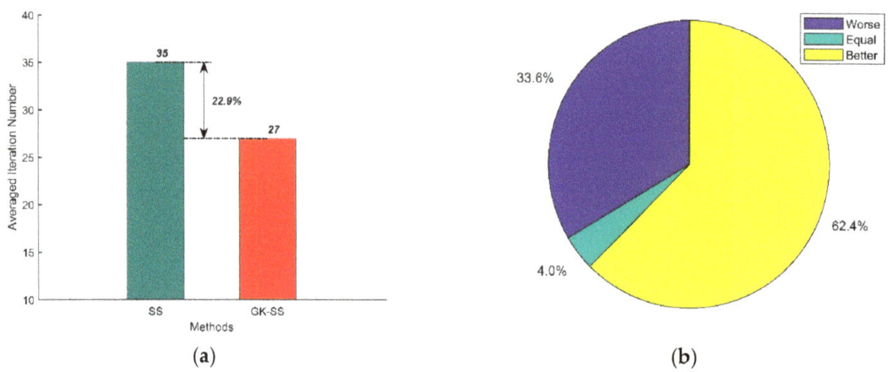

Figure 20. Performance evaluation of GK-SS vs. traditional simplex search method (SS) by statistics. (**a**) Averaged iteration number. (**b**) Superiority ratio by direct performance comparison.

Figure 20b demonstrates the performance statistics of the GK-SS relative to the traditional method. It can be seen that the GK-SS behaved better than the traditional method in almost 63% of cases. Accordingly, the results showed that the revised method had an appreciable effect on efficiency promotion. Hence, it was indicated that the GK-SS is efficient for the performance optimization of the steam generator level control.

6. Conclusions

In this study, a revised simplex search method, GK-SS, was proposed to optimize the performance of the steam generator level control system. With the data-driven optimization scheme that we proposed, the GK-SS deals with the act of data-driven optimization entirely. Hence, there are two critical modifications with the GK-SS-based data-driven method, covering all the dimensions on the efficiency promotion of the data-driven method. With the two modification mechanisms, the historical information generated in the optimization progress was fully utilized to guide the search direction and monitor the optimization progress. The simulation experiment indicated that the revised method could quickly converge to the optimal parameter settings. Further, the deliberately designed statistical experiments also showcased that the revised method significantly fortified the efficiency of the data-driven optimization method. Thus, it can be concluded that the revised simplex search method was effective and efficient in controller parameters optimization of the SG level control system. This method can be applied to other similar controller parameters optimization problems in the process control of all kinds of nuclear power plants.

Author Contributions: Conceptualization, X.K., C.S., H.L. and J.L.; data curation, X.K., P.G. and Y.F.; formal analysis, X.K.; funding acquisition, X.K., C.S., H.L. and J.L.; investigation, X.K., C.S. and P.G.; methodology, X.K. and C.S.; project administration, X.K., C.S., H.L. and J.L.; resources, X.K., C.S. and H.L.; software, P.G. and Y.F.; supervision, X.K., C.S. and H.L.; validation, X.K. and P.G.; visualization, X.K. and P.G.; writing—original draft, X.K., P.G. and Y.F.; writing—review and editing, X.K., C.S., H.L. and P.G. All authors have read and agreed to the published version of the manuscript.

Funding: This research was funded by the program of the State Key Laboratory of Nuclear Power Safety Monitoring Technology and Equipment of China (K-A2020.412), and the Natural Science Foundation of Fujian Province, grant numbers 2021J011205 and 2018J01564.

Institutional Review Board Statement: Not applicable.

Informed Consent Statement: Not applicable.

Data Availability Statement: Not applicable.

Conflicts of Interest: The authors declare no conflict of interest.

Abbreviations

CCG	compensated composite gradient for the current simplex
DOE	design of experiments
EGCS	estimated quasi-gradient for current simplex
FP	full power
GK-SS	knowledge-informed simplex search based on historical gradient approximations
ITAE	integral of time multiply by absolute error
LHS	Latin hypercube sampling
MFO	model-free optimization
NPP	nuclear power plant
OE	optimization efficiency index
PID	Proportional-Integral-Derivative
PWR	pressurized water reactor
SG	steam generator
SS	simplex search method
SPSA	simultaneous perturbation stochastic approximation

References

1. Wu, S.; Wang, P.; Wan, J.; Xinyu, W.; Zhao, F. Parameter Optimization for AP1000 Steam Generator Feedwater Control System Using Particle Swarm Optimization Algorithm. In Proceedings of the 24th International Conference on Nuclear Engineering, Charlotte, NC, USA, 26–30 June 2016. [CrossRef]
2. Irving, E.; Miossec, C.; Tassart, J. Towards Efficient Full Automatic Operation of the PWR Steam Generator with Water Level Adaptive Control. In *Boiler Dynamics and Control in Nuclear Power Stations, Bournemouth*; Thomas Telford Publishing: London, UK; pp. 309–329. [CrossRef]
3. Salehi, A.; Safarzadeh, O.; Kazemi, M.H. Fractional Order PID Control of Steam Generator Water Level for Nuclear Steam Supply Systems. *Nucl. Eng. Des.* **2019**, *342*, 45–59. [CrossRef]
4. Ablay, G. Robust Estimator-based Optimal Control Designs for U-Tube Steam Generators. *Trans. Inst. Meas. Control.* **2015**, *37*, 636–644. [CrossRef]
5. Xu, Y.; Chen, J.; Yu, H.; Wang, Z. Research on Feedforward Compensation for Steam Generator Level Control System Manual/Automatic Switch. *Hedongli Gongcheng/Nucl. Power Eng.* **2021**, *42*, 140–144. [CrossRef]
6. Hui, J.W.; Ling, J.; Dong, H.; Wang, G.X.; Yuan, J.Q. Distributed Parameter Modeling for the Steam Generator in the Nuclear Power Plant. *Ann. Nucl. Energy.* **2021**, *152*. [CrossRef]
7. Zhang, D.; Cai, W.; Xie, C.; Liu, P.; Liu, C. Research on Automatic Optimization Software for PID Parameters of Nuclear Power Digital Control System PID. *Hedongli Gongcheng/Nucl. Power Eng.* **2020**, *41*, 77–81. [CrossRef]

8. Zhuang, D.P.A. Automatic Tuning of Optimum PID Controllers. *IEE Proc. -D* **1993**, *140*, 216–224. [CrossRef]
9. Wei, Z.J.; Chai, T. A Survey of Advanced PID Parameter Tuning Methods. *Acta Autom. Sinica* **2000**, *26*, 347–355.
10. Åström, K.J.; Hägglund, T. The Fture of PID Control. *Control Eng. Pract.* **2001**, *9*, 1163–1175. [CrossRef]
11. Zheng, Y.; Zhang, Y.; You, K.; Zhao, M.; Li, Y.; Chen, G.; Yan, X. Research on Improved Auto-Tuning of a PID Controller Based on Phase Angle Margin PID. *Hedongli Gongcheng/Nucl. Power Eng.* **2020**, *41*, 108–113. [CrossRef]
12. Demerdash, N.A.; El-Hameed, M.A.; Eisawy, E.A.; El-Arini, M.M. Optimal Feed-Water Level Control for Steam Generator in Nuclear Power Plant based on Meta-Heuristic Optimization. *J. Radiat. Res. Appl. Sci.* **2020**, *13*, 468–484. [CrossRef]
13. Formentin, S.; Karimi, A. A Data-Driven Approach to Mixed-Sensitivity Control with Application to An Active Suspension System. *IEEE Trans. Ind. Inform.* **2012**, *9*, 2293–2300. [CrossRef]
14. Procházka, H.; Gevers, M.; Anderson, B.D.; Ferrera, C. Iterative Feedback Tuning for Robust Controller Design and Optimization. In Proceedings of the 44th IEEE Conference on Decision and Control, Seville, Spain, 15 December 2005. [CrossRef]
15. Sala, A.J.A. Integrating Virtual Reference Feedback Tuning into A Unified Closed-Loop Identification Framework. *Automatica.* **2007**, *43*, 178–183. [CrossRef]
16. Lei, T.; Hou, Z.; Ren, Y. Data-Driven Model Free Adaptive Perimeter Control for Multi-Region Urban Traffic Networks With Route Choice. *IEEE Trans. Intell. Transp. Syst.* **2019**, 1–12. [CrossRef]
17. Hou, Z.; Zhu, Y.I. Controller-dynamic-linearization-based Model Free Adaptive Control for Discrete-Time Nonlinear Systems. *IEEE Trans. Ind. Inform.* **2013**, *9*, 2301–2309. [CrossRef]
18. Hou, Z.-S.; Wang, Z. From Model-Based Control to Data-Driven Control: Survey, Classification and Perspective. *Inf. Sci.* **2013**, *235*, 3–35. [CrossRef]
19. Gao, Z.; Saxen, H.; Gao, C. Guest editorial: Special section on data-driven approaches for complex industrial systems. *IEEE Trans. Ind. Inform.* **2013**, *9*, 2210–2212. [CrossRef]
20. Gao, Z.; Chen, M.Z.; Zhang, D.J.P. Special Issue on "Advances in Condition Monitoring, Optimization and Control for Complex Industrial Processes". *Processes* **2021**, *9*, 664. [CrossRef]
21. Gao, Z.; Nguang, S.K.; Kong, D.-X.J.C. Advances in Modelling, Monitoring, and Control for Complex Industrial Systems. *Complexity* **2019**, *2019*, 2975083. [CrossRef]
22. Gao, Z.; Kong, D.; Gao, C. Modeling and Control of Complex Dynamic Systems: Applied Mathematical Aspects. *J. Appl. Math.* **2012**, *2012*, 869792. [CrossRef]
23. Kong, X.; Xiao, Y.; Qian, L.; Su, L.; Chen, B.; Xu, M. Performance Optimization for Steam Generator Level Control Based on A Revised Simultaneous Perturbation Stochastic Approximation Algorithm. In Proceedings of the 2018 3rd International Conference on Intelligent Green Building and Smart Grid (IGBSG), Yilan, China, 22–25 April 2018. [CrossRef]
24. Kong, X.-S.; Guo, J.-M.; Zheng, D.-B.; Yang, J.-F.; Zhang, J.; Fu, W. An Improved-SPSA Quality Control Method for Medium Voltage Insulator SPSA. *Gao Xiao Hua Xue Gong Cheng Xue Bao/J. Chem. Eng. Chin. Univ.* **2020**, *34*, 1500–1510. [CrossRef]
25. Kong, X.; Guo, J.; Zheng, D.; Zhang, J.; Fu, W. Quality Control for Medium Voltage Insulator via a Knowledge-Informed SPSA Based on Historical Gradient Approximations. *Processes* **2020**, *8*, 146. [CrossRef]
26. Kong, X.; Ma, Z. A Novel Method for Controllers Parameters Optimization of Steam Generator Level Control. In Proceedings of the 21st International Conference on Nuclear Engineering, Chengdu, China, 29 July–2 August 2013. [CrossRef]
27. Kong, X.; Zheng, D. A Knowledge-Informed Simplex Search Method Based on Historical Quasi-Gradient Estimations and Its Application on Quality Control of Medium Voltage Insulators. *Processes* **2021**, *9*, 770. [CrossRef]
28. Dai, X.; Gao, Z. From model, signal to knowledge: A Data-Driven Perspective of Fault Detection and Diagnosis. *IEEE Trans. Ind. Inform.* **2013**, *9*, 2226–2238. [CrossRef]
29. Zhang, D.; Gao, Z. Improvement of Refrigeration Efficiency by Combining Reinforcement Learning with A Coarse Model. *Processes* **2019**, *7*, 967. [CrossRef]
30. Nelder, J.A.; Mead, R. A Simplex-Method for Function Minimization. *Comput. J.* **1965**, *7*, 308–313. [CrossRef]
31. Tan, W. Water Level Control for A Nuclear Steam Generator. *Nucl. Eng. Des.* **2011**, *241*, 1873–1880. [CrossRef]

processes

Article

Remote Wind Farm Path Planning for Patrol Robot Based on the Hybrid Optimization Algorithm

Luobing Chen [1], Zhiqiang Hu [2,*], Fangfang Zhang [1,*], Zhongjin Guo [3], Kun Jiang [4], Changchun Pan [5] and Wei Ding [6]

1 School of Information and Automation Engineering, Qilu University of Technology (Shandong Academy of Sciences), Jinan 250353, China
2 College of Mechanical and Architectural Engineering, Taishan University, Taian 271000, China
3 School of Mathematics and Statistics, Taishan University, Taian 271000, China
4 College of Mechanical and Control Engineering, Guilin University of Technology, Guilin 541000, China
5 Key Laboratory of Navigation and Location Based Services, Shanghai Jiao Tong University, Shanghai 200030, China
6 Shandong Computer Science Center (National Supercomputer Center in Jinan), Qilu University of Technology (Shandong Academy of Sciences), Jinan 250353, China
* Correspondence: zhiqiangh@tsu.edu.cn (Z.H.); zhff4u@qlu.edu.cn (F.Z.)

Abstract: Globally, wind power plays a leading role in the renewable energy industry. In order to ensure the normal operation of a wind farm, the staff will regularly check the equipment of the wind farm. However, manual inspection has some disadvantages, such as heavy workload, low efficiency and easy misjudgment. In order to realize automation, intelligence and high efficiency of inspection work, inspection robots are introduced into wind farms to replace manual inspections. Path planning is the prerequisite for an intelligent inspection robot to complete inspection tasks. In order to ensure that the robot can take the shortest path in the inspection process and avoid the detected obstacles at the same time, a new path-planning algorithm is proposed. The path-planning algorithm is based on the chaotic neural network and genetic algorithm. First, the chaotic neural network is used for the first step of path planning. The planning results are encoded into chromosomes to replace the individuals with the worst fitness in the genetic algorithm population. Then, according to the principle of survival of the fittest, the population is selected, hybridized, varied and guided to cyclic evolution to obtain the new path. The shortest path obtained by the algorithm can be used for the robot inspection of the wind farms in remote areas. The results show that the proposed new algorithm can generate a shorter inspection path than other algorithms.

Keywords: wind farms; inspection; path planning; chaotic neural network; genetic algorithm

Citation: Chen, L.; Hu, Z.; Zhang, F.; Guo, Z.; Jiang, K.; Pan, C.; Ding, W. Remote Wind Farm Path Planning for Patrol Robot Based on the Hybrid Optimization Algorithm. *Processes* **2022**, *10*, 2101. https://doi.org/10.3390/pr10102101

Academic Editor: Zhiwei Gao

Received: 8 July 2022
Accepted: 13 October 2022
Published: 17 October 2022

1. Introduction

In recent years, power enterprises have been paying more attention to wind power generation, developing new energy power generation and promoting the sustainable development of the power generation industry [1,2]. Yet at the same time, the equipment used by wind farms is prone to malfunction when large-scale wind farms are affected by extreme operating environments, such as extremely high or low temperature, corrosion, and strong high pressure. Equipment failure causes a huge economic loss. Patrol inspection and maintenance of wind farms can reduce the probability of equipment failure, thereby avoiding some economic losses.

With the development of intelligent technology, inspection robots have been widely used at wind farms to complete inspection work. At present, wind farms have begun to reduce their number of inspection personnel. By using inspection robots to carry out intelligent inspections on the wind farm, inspection efficiency is improved and the safe operation of the wind farm is guaranteed. Global path planning is a prerequisite for an intelligent inspection robot to complete inspection tasks. Therefore, it is of great significance to study the global path planning of robot wind farm inspections.

Many scholars have discussed the path planning of wind farms for patrol robots and proposed a variety of algorithms. Path planning is essentially a traveling salesman problem. Generally, these algorithms can be divided into biology-based metaheuristic algorithms, physics-based metaheuristic algorithms, group-based metaheuristic algorithms and so on. In addition, there are hybrid algorithms that combine these methods [3,4], which are shown in Table 1.

Table 1. Classification of literature review on path planning.

Whether It Is a Hybrid Optimization Algorithm	Time	Main Research Work	Algorithm cATEGORY
No	2019	Liu et al. [5] used an ant colony algorithm to find the optimal path for inspection robots. Through simulation analysis, the ant colony algorithm reduces the time of shortest path searching, and increases the success rate of finding the optimal path.	Metaheuristic methods based on swarm
No	2021	Dong et al. [6] proposed a path planning method for an ultra-high voltage substation based on an ant colony optimization algorithm. Experimental results show that the algorithm could significantly improve the number of iterations.	Metaheuristic methods based on swarm
No	2022	In 2022, based on the improved biologically inspired neural network algorithm, Chen et al. [7] proposed a method of multi-mobile robot cooperative full-area coverage inspection. Simulation experiments verify the feasibility of the proposed multi-robot cooperative inspection scheme.	Metaheuristic methods based on biology
Yes	2018	He et al. [8] proposed an improved genetic simulated annealing algorithm. Compared with the optimization results of other path optimization algorithms, the proposed algorithm can obtain better travel path.	Metaheuristic methods based on biology and metaheuristic methods based on physics
Yes	2019	Yuan et al. [9] proposed a robot path-planning method based on simulated an annealing ant colony algorithm. Simulation results show that the algorithm can quickly plan the shortest and optimal inspection path.	Metaheuristic methods based on swarm and metaheuristic methods based on physics

For small-scale path-planning problems, a single algorithm [5–7] can quickly obtain the optimal solution. However, it is difficult to cope with large-scale and complicated structures. Therefore, the scholars proposed some hybrid algorithms. These algorithms can find high-quality approximate solutions to solve road-planning problems, such as hybrid algorithms based on genetic algorithms, analog annealing algorithms and other algorithms [10].

For reference [8], high quality initial solutions of genetic algorithms can improve problem-solving speed and even improve the quality of final solutions. For reference [9], the ant colony algorithm has strong robustness. The ant colony algorithm has fewer parameters and is easy to set up, so it is easy to apply to other combinatorial optimization problems. The algorithms proposed in [8,9] all involve the simulated annealing algorithm. However, the chaotic annealing mechanism of transient chaotic neural network (TCNN) is better than the traditional analog annealing algorithm in regard to global search capabilities and learning rates [11,12].

Based on the above discussions, we combine the advantages of genetic algorithms and TCNN, propose a hybrid optimization algorithm, and apply it to the path planning of a patrol robot. This algorithm can obtain a higher-quality path solution, thereby reducing the costs of wind farm operation and maintenance. The main contributions are as follows:

In order to optimize the patrol path of the wind farm in the remote area, a new hybrid optimization algorithm based on the chaotic neural network and genetic algorithm is proposed. To verify its effectiveness, the proposed algorithm is firstly applied to the traveling salesman problem, then compared with other algorithms. The simulation results show that the algorithm has better path search capabilities and can be extended to other related engineering fields, such as path planning for wind farms.

In path planning for wind farms, if there are some obstacles between any two adjacent wind farms that can be detoured around, the original distance is replaced with the detour length. The proposed algorithm can obtain a shorter patrol path than other algorithms. In this case, it is still an unconstrained problem.

In path planning for wind farms, if we consider the obstacles such as mountains and rivers that cannot be detoured around, or other constrained conditions, such as the inability of the patrol robot to move from one wind farm to the special wind farm, the model of path planning is established, and the proposed algorithm also obtains a shorter patrol path than other algorithms. In this case, it is, in essence, one constrained problem.

The content of this paper is arranged as follows: Section 2 introduces the patrol path model of wind farms. Section 3 introduces a new hybrid optimization algorithm and introduces the simulation experiment of the traveling salesman problem to verify the effectiveness of the proposed algorithm. In Section 4, we apply the proposed algorithm to the path planning for the patrol robot in remote areas. Section 5 summarizes the main work of this paper.

2. The Model of Path Planning

2.1. Path Planning

A redundant inspection path will cause the problems of heavy workload and low efficiency [13]. The location coordinates of each wind farm are known. A reasonable routing planning scheme must meet the following conditions:

Condition (1): Path length problem. The patrol robot visits each wind farm at a fixed position once, then returns to the original wind farm after the entire visit. Based on ensuring the effective completion of the inspection task, the patrol robot selects the optimal forward path to reduce the operation and maintenance cost of the wind farms.

Condition (2): If there is a barrier between the two wind farms, the robot can detour around the barrier from the current wind farm to the next wind farm.

Condition (3): If there are obstacles that cannot be detoured around between the two wind farms, the patrol robot will find other wind farms to ensure that the patrol task can be completed.

2.2. Objective Function

2.2.1. Constraints

The function corresponding to condition (1) is as follows:

$$C_1 = \frac{W_2}{2}\{\sum_{i=1}^{N}(\sum_{j=1}^{N} x_{ij} - 1)^2 + \sum_{j=1}^{N}(\sum_{i=1}^{N} x_{ij} - 1)^2\} \quad (1)$$

where $x_{i0} = x_{in}$, $x_{in+1} = x_{i1}$, W_2 is the coupling coefficient corresponding to the constrained cost function. Condition (3) discussed in this paper is embodied in the form of algorithms. For example,

$$p = \begin{bmatrix} 0 & 0 & 0 & 1 \\ 0 & 0 & 1 & 0 \\ 0 & 1 & 0 & 0 \\ 1 & 0 & 0 & 0 \end{bmatrix}$$

where p is the matrix that measures the obstacle between two points. p_{ij} stands for path condition from city i to city j, where 1 indicates that there is an obstacle, and 0 indicates that there is no obstacle. When the solution matrix dot multiplied by matrix p is 0, the scheme satisfies condition (3). The case for condition (2) is described in Section 4.1.

2.2.2. Objective Function

The essence of the wind farms inspection problem is the path-planning problem. The goal is to minimize the length of the patrol path. Assuming that the output x_{ij} is the order j of accessing the city i, the value of the output x_{ij} is initialized randomly. x_{ij} means that city i is visited on the jth order. For example,

$$x = \begin{bmatrix} 0 & 1 & 0 \\ 1 & 0 & 0 \\ 0 & 0 & 1 \end{bmatrix}$$

where 1 indicates the arrival to the city, and 0 indicates no access to the city. x_{12} means city 1 is visited at the second time.

Then, the objective function corresponding to the problem is as follows:

$$E_1 = \frac{W_1}{2}\sum_{i=1}^{N}\sum_{j=1}^{N}\sum_{k=1}^{N}(x_{k,j+1} + x_{k,j-1})x_{ij}d_{ik} + \frac{W_2}{2}\{\sum_{i=1}^{N}(\sum_{j=1}^{N}x_{ij} - 1)^2 + \sum_{j=1}^{N}(\sum_{i=1}^{N}x_{ij} - 1)^2\} \quad (2)$$

where W_1 is the coupling coefficient corresponding to the cost function of the path length, and d_{ij} is the distance between city i and city j, $x_{i0} = x_{in}$, $x_{in+1} = x_{i1}$, W_2 is the coupling coefficient corresponding to the constrained cost function.

3. The Proposed Hybrid Optimization Algorithm

3.1. Transient Chaotic Neural Network

Based on a Hopfield neural network, the self-feedback term with simulated annealing mechanism is added, which is called the transient chaotic neural network (TCNN). The dynamic equation is as follows:

$$\begin{aligned} y_i(t+1) &= ky_i(t) + \alpha[\sum_{j=1,j\neq i}^{N} w_{ij}x_j(t) + I_i] - z_i(t)g(x_i(t) - I_0) \\ x_i(t) &= S_1(y_i(t)) = 1/[1 + \exp(-y_i(t)/\varepsilon)] \\ z_i(t+1) &= (1-\beta)z_i(t) \end{aligned} \quad (3)$$

where $y_i(t)$ is the internal state of neurons, $x_i(t)$ is the output of neurons, k is the damping factor of the neural diaphragm $(0 \leq k \leq 1)$, α is a positive proportion parameter. It represents the effect of energy function on chaotic dynamics, $\alpha \in [0, \infty)$, ε is the steepness parameter of the activation function $(\varepsilon > 0)$, I_0 is the positive parameter, $z_i(t)$ is the self-feedback connection weight, β is the annealing attenuation factor of $z_i(t)$.

A positive Lyapunov exponent indicates that the model has chaotic characteristics, and the larger the Lyapunov exponent, the stronger the degree of chaos. Lyapunov exponent is defined as follows:

$$\lambda = \lim_{n\to\infty}\frac{1}{n}\sum_{i=0}^{n-1}\log\left|\frac{dy(t+1)}{dy(t)}\right| \quad (4)$$

Therefore, for the TCNN chaotic neuron model:

$$\frac{dy(t+1)}{dy(t)} = k - z(t)\frac{dx(t)}{dy(t)} = k - z(t)\frac{dS_1(y(t))}{dy(t)} \quad (5)$$

where

$$\frac{dS_1(y(t))}{dy(t)} = \frac{1}{\varepsilon}S_1(y(t))(1 - S_1(y(t))) \quad (6)$$

Set parameters: k = 1, β = 0.02, I_0 = 0.65, z (0) = 0.8, α = 0.07, ε = 0.05. The inverse bifurcation diagram and Lyapunov exponent time evolution diagram of the TCNN are shown in Figures 1 and 2, respectively.

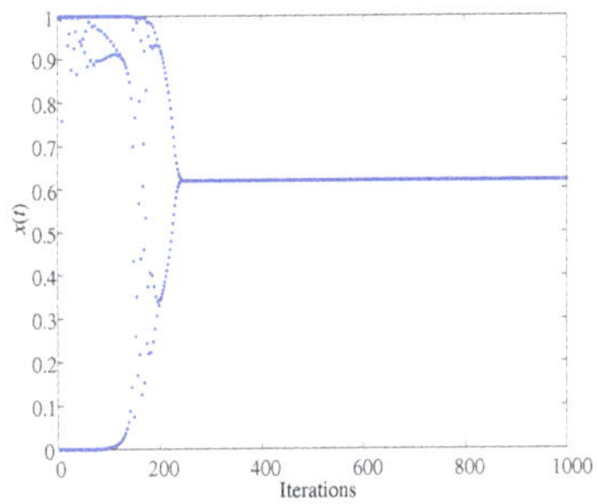

Figure 1. Inverted bifurcation diagram.

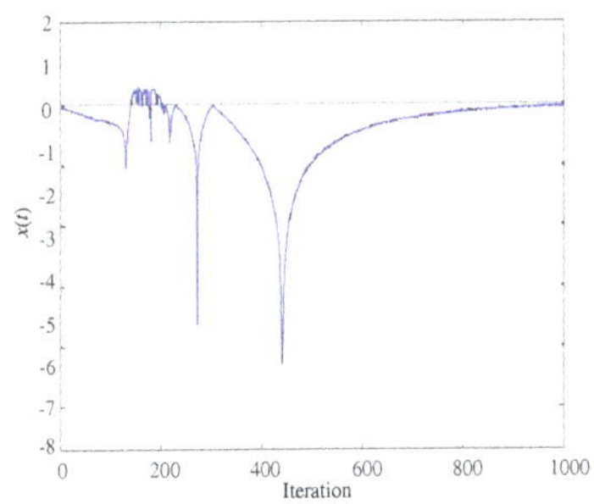

Figure 2. Lyapunov exponent diagram.

It can be seen from Figures 1 and 2 that in the early stage of the evolution of the TCNN model, z_i (t) takes a larger initial value, and the model is chaotic. Since z_i (t) decays continuously with time until it is 0, the model undergoes an inverse bifurcation transition and degenerates to a Hopfield neural network (HNN) with gradient convergent. The model finally converges to the stable point to obtain the optimal solution [14].

The mechanism of the TCNN is to map the objective function of the problem into the energy function of the network and then evolve the dynamics of the network into the process of the objective function. When the network converges to the stable point, the corresponding neuron output is the optimal/suboptimal solution to the problem. According to the following principles of TCNN:

$$\frac{dy_i}{dt} = -\frac{\partial E}{\partial x_i} \tag{7}$$

Kwok and Smith [15] proposed a modified energy function as follows:

$$E(t) = E_{Hop} + H = -\frac{1}{2}\sum_{i=1, i\neq j}^{N}\sum_{j=1, j\neq i}^{N} w_{ij}x_i(t)x_j(t) - \sum_{i=1}^{N} I_i x_i(t) + \frac{1}{\tau_i}\sum_{i=1}^{N}\int_0^{x_i(t)} f^{-1}(\xi)d\xi + H(x_i, w_{ij}, I_i) \tag{8}$$

where $i, j = 1, 2, \cdots, N$, N is the number of neurons, x_i (t) is the output of the ith neuron at time t, I_i is the threshold of the ith neuron, and W_{ij} is the connection weight between neuron i and neuron j, τ_i is the time constant of the ith neuron, $f^{-1}(\bullet)$ is the inverse function of the activation function, and H is the additional energy term. H represents the energy value of the self-feedback term, and its selection form determines the variation characteristics of chaotic dynamics. The variation characteristics of different chaotic dynamics will obtain different CNN models.

3.2. Genetic Algorithm

The genetic algorithm (GA) is a population intelligent meta-heuristic algorithm, which is similar to evolution theory. The process of the genetic algorithm is as follows: firstly, a population is initialized to determine the individual size of the population and the chromosome of each individual in the population; secondly, the learning model is established according to the learning object, and the iteration times of the algorithm are determined; thirdly, the chromosome information of all individuals in the population is changed through selection, crossover, variation and changes in the information of all individual chromosomes in the population. Among them, we use the roulette method to select individuals to enter the next generation; Finally, when the number of iterations of the algorithm reaches designated number of iterations, the solution with the highest fitness function value is the global solution [16,17].

3.3. New Hybrid Optimization Algorithm

Genetic algorithms have strong path-planning ability, and they are extensively used in practical engineering. However, they have some shortcomings, such as partial search capability and poor convergence performance. These shortcomings hinder the promotion and application of genetic algorithms.

The transient chaotic neural network has ergodicity and pseudo-randomness of chaos. It can improve the problem of HNN falling into local minimum value. Therefore, the optimization result obtained by transient chaotic neural network is better than that obtained by HNN [18]. However, no matter how slow the annealing speed of TCNN may be, it may not converge to the global optimum [19].

With the continuous increase of complexity, a single algorithm has exhibited limitations in its convergence and solving speed. Therefore, the mixing learning strategy used in combination with multiple algorithms has gradually become a new research hotspot [20,21]. The algorithm proposed in this paper is a hybrid optimization algorithm for combining transient chaotic neural networks with a genetic algorithm. The algorithm flow is as follows:

Step 1: The appropriate TCNN parameters and the maximum number of iterations are selected. Set parameters of TCNN: memory constant $k = 1$, positive proportion parameter $\alpha = 0.07$, positive parameter $I_0 = 0.65$, initial annealing value $z\ (0) = 0.8$, annealing attenuation factor $\beta = 0.008$, Sigmoid steepness parameter $\varepsilon = 0.05$. The initial state of a network is randomly selected.

Step 2: Run the TCNN dynamics Equation (1).

Step 3: Judge whether the discrete output of the network is an effective solution. The judging rules are as follows.

Every row and every column of a valid solution has only one 1. The valid solution is a binary square matrix. This rule indicates that each city traversed is visited once.

If the discrete output of the network is not an effective solution, turn to Step 2. If the discrete output of the network is an effective solution, calculate the value of the energy function corresponding to the discrete output. If the energy function value does not change for a number of iterations, turn to Step 4, otherwise, turn to Step 2.

Step 4: The fitness of the effective solution obtained by Step 3 is compared to the fitness of the worst individual in the genetic algorithm. If the fitness of the effective solution is better, encode and express the obtained effective solution as a chromosome, and replace the least adaptable individuals in the genetic algorithm's initial population.

Step 5: Evaluate the fitness of the individual corresponding to each chromosome.

Step 6: According to the principle that the higher the fitness is, the greater the selection probability is, select two individuals from the population as the parents, apply a crossover operation on the chromosomes of the parents to produce the offspring. Repeat Step 5 and Step 6 until the optimal solution is obtained.

The flow chart of the optimization algorithm is shown in Figure 3.

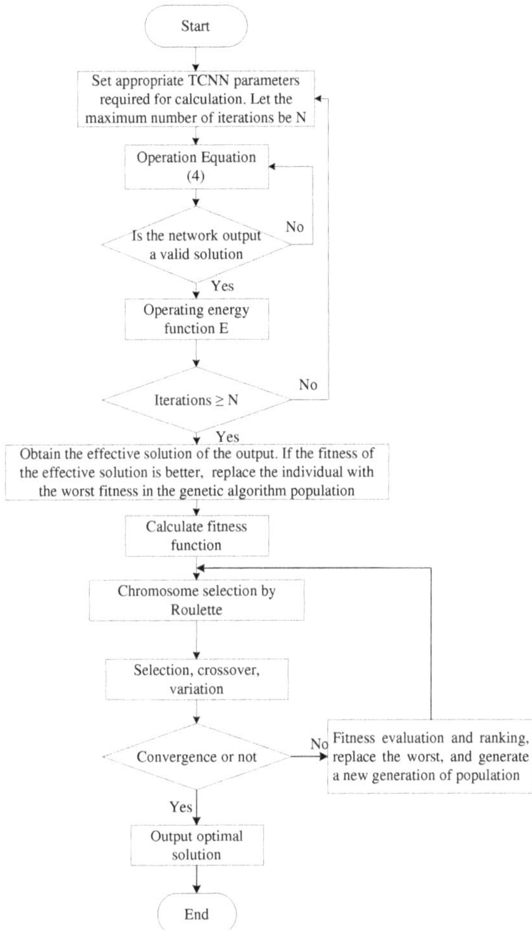

Figure 3. Algorithm flow chart.

3.4. The Simulation of Travelling Salesman Problem (TSP)

In this section, the simulation experiment of the travelling salesman problem is used to verify whether the proposed hybrid algorithm has better route-planning ability. The travelling salesman problem (TSP) is a typical NP problem, and it is often used to test the optimization performance of an algorithm. If the algorithm has a better search capability, it can be extended to the actual industrial scene.

It is assumed that there are N cities with known locations and mutual distance. A closed path is sought. Each city in the closed path is visited once. After accessing all cities, the traveler goes back to the starting city; thus, a closed path is formed. Among all closed paths, we adopt the proposed hybrid optimization algorithm to search for the closed path with the shortest distance.

3.4.1. TSP of 75 Cities

Select 75 cities to normalize the coordinates. See Appendix A for normalized coordinates. In many references, the ideal shortest path for this instance is 5.434474 [22].

It is assumed that the output x_{ij} of neurons is the order j of visiting city i. The value of the neuron output x_{ij} is randomly initialized. The objective function of TSP is Equation (1). From Equation (3), the internal state dynamics equation of TCNN neurons describing the model is as follows:

$$\begin{aligned} y_{ij}(t+1) = ky_{ij}(t) - z(t)(x_{ij}(t) - I_0) + \alpha\{-W_1[\sum_{k\neq i}^{N} d_{ik}(x_{k,j+1}(t) + x_{k,j-1}(t))] \\ -W_2[\sum_{l\neq j}^{N} x_{il}(t) + \sum_{k\neq i}^{N} x_{kj}(t)] + W_2\} \end{aligned} \quad (9)$$

The selection parameters are as follows: $k = 1$, $\alpha = 0.07$, $\beta = 0.08$, $I_0 = 0.65$, $z(0) = 0.8$, $\varepsilon = 0.05$, $W_1 = 1$, $W_2 = 1$, $P_c = 0.8$, $P_m = 0.07$. The sum of distances in this problem is used as the fitness function to measure whether the solution is optimized. The optimized path is shown in Figure 4.

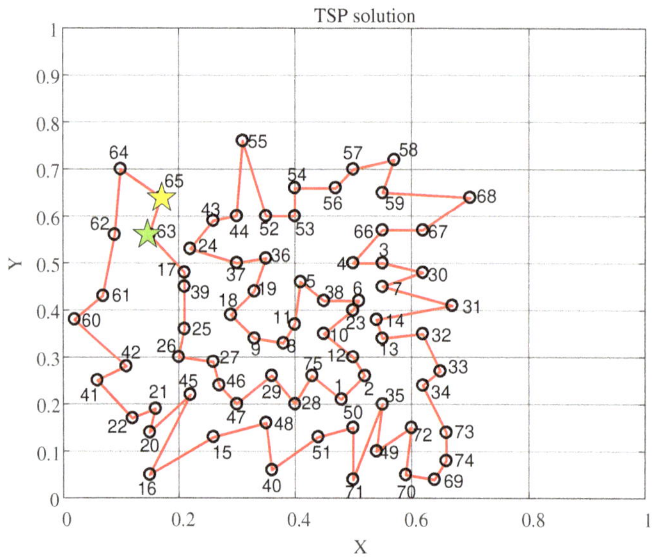

Figure 4. Optimal path of 75 cities (Note: yellow five-pointed star 65 indicates the starting city, green five-pointed star 63 indicates the second city).

The parameters in Equation (9) determine the number of legal paths of the algorithm. The more legal paths the algorithm obtains, the more efficient it is to obtain the optimal path. The number of legal paths varies with parameters as shown in Tables 2–8. The parameter corresponding to the maximum number of valid paths is selected.

Table 2. The number of legal paths varies with α.

The Value of α	Number of Valid Paths n
$0 < \alpha \leq 0.03$	$0 < n \leq 25$
$0.03 < \alpha \leq 0.05$	$30 < n \leq 60$
$0.06 < \alpha \leq 0.08$	$80 < n < 100$
$0.09 < \alpha \leq 0.1$	$30 < n \leq 50$
$0.11 < \alpha \leq 0.13$	$0 < n \leq 10$
$\alpha > 0.13$	$n = 0$

Table 3. The number of legal paths varies with β.

The Value of β	Number of Valid Paths n
$0.001 < \beta \leq 0.004$	$0 < n \leq 15$
$0.005 < \beta \leq 0.007$	$30 < n \leq 70$
$0.008 < \beta \leq 0.009$	$80 < n < 100$
$0.01 < \beta \leq 0.02$	$30 < n \leq 50$
$0.025 < \beta \leq 0.04$	$0 < n \leq 10$
$0.4 < \beta < 1$	$n = 0$

Table 4. The number of legal paths varies with I_0.

The Value of I_0	Number of Valid Paths n
$0 < I_0 \leq 0.1$	$0 < n \leq 15$
$0.1 < I_0 \leq 0.45$	$30 < n \leq 75$
$0.45 < I_0 \leq 0.65$	$80 < n < 100$
$0.65 < I_0 \leq 0.75$	$30 < n \leq 40$
$0.75 < I_0 \leq 0.8$	$0 < n \leq 10$
$0.8 < I_0 < 1$	$n = 0$

Table 5. The number of legal paths varies with z (0).

The Value of z (0)	Number of Valid Paths n
$0 < z\ (0) \leq 0.3$	$0 < n \leq 20$
$0.3 < z\ (0) \leq 0.6$	$30 < n \leq 50$
$0.6 < z\ (0) \leq 0.8$	$80 < n < 100$
$0.8 < z\ (0) \leq 0.9$	$45 < n \leq 60$
$0.9 < z\ (0) \leq 1$	$10 < n \leq 30$
$z\ (0) > 1$	$n = 0$

Table 6. The number of legal paths varies with ε.

The Value of ε	Number of Valid Paths n
$0 < \varepsilon \leq 0.02$	$0 < n \leq 30$
$0.02 < \varepsilon \leq 0.04$	$40 < n \leq 60$
$0.04 < \varepsilon \leq 0.1$	$80 < n < 100$
$0.1 < \varepsilon \leq 0.3$	$20 < n \leq 30$
$0.3 < \varepsilon \leq 0.5$	$10 < n \leq 30$
$\varepsilon > 0.5$	$n = 0$

Table 7. The number of legal paths varies with P_c.

The Value of P_c	Algorithm Optimization Performance
$P_c < 0.4$	The speed of producing new individuals is slow
$0.4 \leq P_c \leq 0.99$	Obtain the optimal solution
$P_c > 0.99$	The excellent pattern of population is easily destroyed

Tables 7 and 8 give the influence of parameters in genetic algorithm on the optimization effect [23]. In different simulation experiments, different values of P_c and P_m have different effects on the optimization performance of the algorithm. Therefore, the parameters corresponding to the best optimal result of the algorithm are selected.

Table 8. The number of legal paths varies with P_m.

The Value of P_m	Algorithm Optimization Performance
$P_m < 0.01$	The ability to generate new individuals and inhibit premature phenomenon will be poor, which will affect the optimization performance of the algorithm.
$0.01 \le P_m \le 0.1$	Obtain the optimal solution
$P_m > 0.1$	More new individuals can be generated, and many good patterns may be destroyed. The performance of the genetic algorithm is similar to that of a random search algorithm.

In order to reflect the fairness of experimental simulation, we added the element of optimization rate. The lower the optimization rate, the closer the path-planning scheme calculated by the algorithm is to the ideal scheme. The formula of optimization rate is [22]

$$Optimization\ rate = \frac{real\ solution - optimal\ solution}{optimal\ solution}$$

In the simulation experiment, 100 independent experiments are set and each independent experiment is iterated 3000 times. In our proposed method, the first 3000 iterations are the path-planning process of the TCNN, and the last 300 iterations are the path-planning process of the genetic algorithm. The simulation results are shown in Table 9. As can be seen from Table 9, there is a 0.000026 difference between the TSP shortest path result obtained by our proposed algorithm and the known optimal TSP path result. Under the condition that each city is visited once, the average operation time of our proposed algorithm is relatively shorter and the path optimization result is better.

Table 9. Comparison of TSP City Travel Business Problem (75 cities).

Algorithm \ Operation Results	Best Results	Worst Result	Average Time	Average Optimization Rate
Grey Wolf Optimizer (GWO)	9.5523	15.0516	56.619 s	75.77%
Firefly Algorithm (FA)	7.1349	7.0772	159.6186 s	31.29%
Ant Colony Algorithm (ACA)	5.6281	6.5804	107.784 s	3.56%
Genetic algorithm (GA)	6.0926	6.9553	77.56 s	12.11%
TCNN	6.124	8.9322	26.545 s	12.69%
proposed algorithm	5.4345	6.2969	33.94 s	≈ 0

Note: $k = 1$, $\alpha = 0.07$, $\beta = 0.08$, $I_0 = 0.65$, $z(0) = 0.8$, $\varepsilon = 0.05$, $W_1 = 1$, $W_2 = 1$, $P_c = 0.8$, $P_m = 0.07$.

3.4.2. TSP of Other Cities

In order to further test the optimization ability of the model for medium–large scale problems, TSP of 50,100 and 1000 cities are selected.

Select 50 cities to normalize the coordinates. See Appendix B for normalized coordinates. The ideal shortest path for these 50 cities' TSP is 5.4604.

In the simulation experiment, 150 independent experiments are set and each independent experiment is iterated 2500 times. In our proposed method, the first 2500 iterations are the path-planning process of the TCNN, and the last 500 iterations are the path-planning process of the genetic algorithm. The simulation results are shown in Table 10.

As can be seen from Table 10, there is a 0.1196 difference between the TSP shortest path result obtained by our proposed algorithm and the known optimal TSP path result. Under the condition that each city is visited once, the average operation time of our proposed algorithm is relatively shorter and the path optimization result is better.

Select 100 cities to normalize the coordinates. See Appendix C for normalized coordinates. The ideal shortest path for these 100 cities' TSP is 7.9782.

Table 10. Comparison of TSP City Travel Business Problem (50 cities).

Algorithm \ Operation Results	Best Results	Worst Result	Average Time	Average Optimization Rate
Grey Wolf Optimizer (GWO)	7.4258	8.4002	34.87 s	35.99%
Firefly Algorithm (FA)	7.30	11.38	120.1204 s	33.69%
Ant Colony Algorithm (ACA)	5.58	7.204	90.435 s	2.19%
Genetic algorithm (GA)	5.7495	7.6771	64.608 s	5.29%
TCNN	6.3398	7.153	21.7 s	16.11%
proposed algorithm	5.5551	6.6146	28.94 s	1.73%

Note: $k = 1$, $\alpha = 0.06$, $\beta = 0.08$, $I_0 = 0.5$, $z\ (0) = 0.8$, $\varepsilon = 0.04$, $W_1 = 1$, $W_2 = 1$, $P_c = 0.65$, $P_m = 0.06$.

In the simulation experiment, 150 independent experiments are set and each independent experiment is iterated 3000 times. In our proposed method, the first 3000 iterations are the path-planning process of the TCNN, and the last 300 iterations are the path-planning process of the genetic algorithm. The simulation results are shown in Table 11.

Table 11. Comparison of TSP City Travel Business Problem (100 cities).

Algorithm \ Operation Results	Best Results	Worst Result	Average Time	Average Optimization Rate
Grey Wolf Optimizer (GWO)	17.6394	20.1142	76.809 s	94.45%
Firefly Algorithm (FA)	14.65	27.03	2905.2109 s	58.56%
Ant Colony Algorithm (ACA)	8.1728	16.0399	115.314 s	2.44%
Genetic algorithm (GA)	8.6926	13.4465	99.79 s	8.96%
TCNN	16.7864	19.0772	58.545 s	72.8%
proposed algorithm	8.0345	10.2969	66.94 s	0.71%

Note: $k = 1$, $\alpha = 0.08$, $\beta = 0.09$, $I_0 = 0.65$, $z\ (0) = 0.8$, $\varepsilon = 0.08$, $W_1 = 1$, $W_2 = 1$, $P_c = 0.85$, $P_m = 0.09$.

As can be seen from Table 11, there is a 0.0563 difference between the TSP shortest path result obtained by our proposed algorithm and the known optimal TSP path result. Under the condition that each city is visited once, the average operation time of our proposed algorithm is relatively shorter and the path optimization result is better.

Select 1000 cities to normalize the coordinates. See Appendix D for normalized coordinates.

In the simulation experiment, 150 independent experiments are set and each independent experiment is iterated 3500 times. In our proposed method, the first 3500 iterations are the path-planning process of the TCNN, and the last 500 iterations are the path-planning process of the genetic algorithm. The simulation results are shown in Table 12.

Table 12. Comparison of TSP City Travel Business Problem (1000 cities).

Algorithm \ Operation Results	Best Results	Worst Result	Average Time
Grey Wolf Optimizer (GWO)	421.7637	482.7657	690.728 s
Firefly Algorithm (FA)	328.65	437.03	11,322.9213 s
Ant Colony Algorithm (ACA)	265.5747	328.1901	915.314 s
Genetic algorithm (GA)	287.3142	326.639	444.598 s
TCNN	331.8651	349.1945	271.265 s
proposed algorithm	251.0795	279.5731	300.807 s

Note: $k = 1$, $\alpha = 0.08$, $\beta = 0.09$, $I_0 = 0.65$, $z\ (0) = 0.8$, $\varepsilon = 0.1$, $W_1 = 1$, $W_2 = 1$, $P_c = 0.95$, $P_m = 0.1$.

Under the condition that each city is visited once, the average operation time of our proposed algorithm is relatively shorter and the path optimization result is better.

It can be seen from Tables 9–12 that the algorithm in this paper has the best path-planning effect and a short operation time. Although the operation time of the TCNN is shorter than that of our proposed algorithm, its patrol path is much longer than that of our proposed algorithm. The actual operation and maintenance cost of the TCNN will be higher than that of our proposed algorithm. This will reduce the efficiency of wind farm inspection planning. Compared with the TCNN and the genetic algorithm, the experimental results of our algorithm are significantly improved. This shows that the hybrid optimization algorithm improves the ability to jump out of local minima. The initial population quality generated by the hybrid algorithm is significantly better than that generated by the completely random population. From Tables 9–12, it can be seen that the path scheme obtained by our proposed algorithm is closer to the best solution. The proposed algorithm is feasible and effective.

By setting the number of iterations, we can obtain the scheme that is closest to the optimized path, as shown in Tables 9–12. In addition, in the simulation experiments of traveling salesman problem in different cities, we added two groups of instances respectively. To further make the algorithms comparable, each algorithm is run for the same amount of time to compare the metric "% Improvement". The experimental results are shown in Tables 13–15. The Improvement can reflect the effect of the optimization algorithm. The higher the value of improvement is, the better the optimization performance is. The formula of Improvement is defined as:

$$Improvement = \frac{Initial\ path\ length - Optimize\ path\ length}{Initial\ path\ length}$$

Table 13. "Improvement" of different instances for 75 cities (The running time is 110s).

% Improvement / Algorithm	Instance 1	Instance 2	Instance 3
Grey Wolf Optimizer (GWO)	52.696%	54.47%	57.64%
Firefly Algorithm (FA)	17.42%	18.74%	18.97%
Ant Colony Algorithm (ACA)	18.41%	18.92%	19.62%
Genetic algorithm (GA)	76.79%	79.67%	79.51%
TCNN	45.24%	47.31%	48.54%
proposed algorithm	79.65%	80.99%	82.35%

Table 14. "Improvement" of different instances for 50 cities (The running time is 100s).

% Improvement / Algorithm	Instance 1	Instance 2	Instance 3
Grey Wolf Optimizer (GWO)	55.40%	57.89%	47.71%
Firefly Algorithm (FA)	20.20%	21.38%	19.42%
Ant Colony Algorithm (ACA)	16.74%	17.66%	12.91%
Genetic algorithm (GA)	70.26%	77.34%	69.22%
TCNN	49.10%	53.47%	47.81%
proposed algorithm	77.09%	79.03%	76.45%

Table 15. "Improvement" of different instances for 100 cities (The running time is 150 s).

% Improvement / Algorithm	Instance 1	Instance 2	Instance 3
Grey Wolf Optimizer (GWO)	54.63%	52.31%	53.36%
Firefly Algorithm (FA)	19.58%	16.52%	17.34%
Ant Colony Algorithm (ACA)	20.16%	17.99%	18.67%
Genetic algorithm (GA)	81.48%	78.75%	79.26%
TCNN	53.45%	50.78%	51.48%
proposed algorithm	84.52%	81.96%	82.49%

It can be seen from Tables 13–15 that compared to other algorithms, our proposed algorithm has a higher improvement value. It shows that our algorithm has the best optimization performance.

3.4.3. Friedman Test

The Friedman test was introduced to further verify the best results of the algorithms.

D_1, D_2, D_3 and D_4 are used to represent the datasets corresponding to the traveling salesman problem for 50 cities, 75 cities, 100 cities and 1000 cities, respectively. According to the previous experiments, the test results of each algorithm on each dataset are ranked from best to worst. If the test performance of the algorithm is the same, the sequence value is bisected. The obtained algorithm comparison sequence values are shown in Table 16.

Table 16. The obtained algorithm comparison sequence values.

Data Set	Proposed Algorithm	TCNN	GA	ACA	FA	GWO
D_1	1	4	3	2	5	6
D_2	1	4	3	2	5	6
D_3	1	5	3	2	4	6
D_4	1	5	3	2	4	6
The average sequence value	1	4.5	3	2	4.5	6

The Friedman test is used to determine whether these algorithms all perform equally. Let r_i denote the average sequence value of the ith algorithm. The variable τ_{χ^2} is as follows:

$$\tau_{\chi^2} = \frac{k-1}{k} \cdot \frac{12N}{k^2-1}\sum_{i=1}^{k}\left(r_i - \frac{k+1}{2}\right)^2 = \frac{12N}{k(k+1)}\left(\sum_{i=1}^{k} r_i^2 - \frac{k(k+1)^2}{4}\right) \tag{10}$$

The variable τ_F is as follows:

$$\tau_F = \frac{(N-1)\tau_{\chi^2}}{N(k-1) - \tau_{\chi^2}} \tag{11}$$

where τ_F follows the F distribution with degrees of freedom $(k-1)$ and $(k-1)(N-1)$. Table 17 shows some commonly used critical values of τ_F.

Table 17. Commonly used critical values for F-test.

$\alpha = 0.1$									
Number of Data Sets N	Number of Algorithms k								
	2	3	4	5	6	7	8	9	10
4	5538	3.463	2.813	2.480	2.273	2.130	2.023	1.940	1.874
5	4.545	3.113	2.606	2.333	2.158	2.035	1.943	1.870	1.811
8	3.589	2.726	2.365	2.157	2.019	1.919	1.843	1.782	1.733
10	3.360	2.624	2.299	2.108	1.980	1.886	1.814	1.757	1.710
15	3.102	2.503	2.219	2.048	1.931	1.845	1.779	1.726	1.682
20	2.990	2.448	2.182	2.020	1.909	1.826	1.762	1.711	1.668

If the performance of the algorithms is significantly different, a "follow-up test" should be carried out to further distinguish the performance of each algorithm. The Nemenyi test is commonly used.

The Nemenyi test calculates the critical range of the difference in mean serial values. The critical range is defined as follows.

$$CD = q_\alpha \sqrt{\frac{k(k+1)}{6N}} \tag{12}$$

Table 18 shows the q_α values commonly used for $\alpha = 0.1$. If the difference between the average sequence values of the two algorithms exceeds the critical range CD, the assumption that the performance of the two algorithms is the same is denied with corresponding confidence.

Table 18. q_{α} values commonly used in Nemenyi test.

α	Number of Algorithms k								
	2	3	4	5	6	7	8	9	10
0.1	1.645	2.052	2.291	2.459	2.589	2.693	2.780	2.855	2.920

According to Equations (11) and (12), we calculated $\tau_F = 102.26$. According to Table 17, when τ_F is greater than $\alpha = 0.1$, the critical value of F-test is 2.273. Therefore, the assumption that all algorithms perform equally is rejected. Using the Nemenyi test, we found in Table 18 that $q_{0.05} = 2.589$ when k = 6, and the critical range $CD = 3.42096$ was calculated according to Equation (13).

The above test comparison can be visually shown with the Friedman test figure. According to the sequence value results in Table 17, Figure 5 can be drawn. In the figure, the vertical axis shows each algorithm, and the horizontal axis is the average sequence. For each algorithm, a dot displays its average sequence value. The horizontal line with the dot as the center indicates the size of the critical range. According to Figure 5, if the horizontal lines of the two algorithms overlap, it means that there is no significant difference between the two algorithms; otherwise, there is a significant difference between the two algorithms.

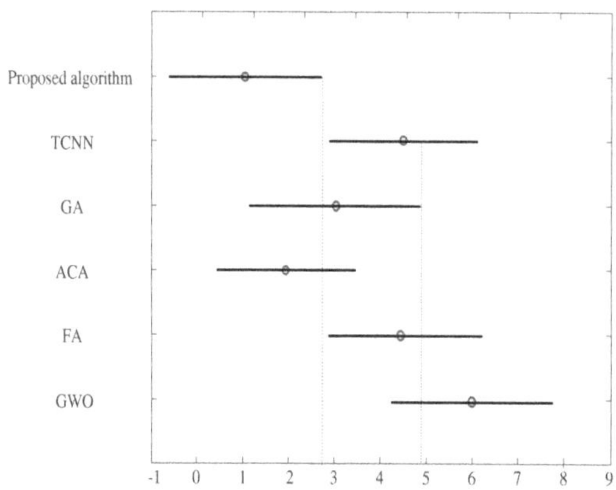

Figure 5. Friedman test figure.

From Figure 5, it can be easily seen that the proposed algorithm significantly outperforms the TCNN algorithm, FA algorithm and GWO algorithm, because their horizontal line segments have no overlapping regions.

4. The Application of Path Planning for Patrol Robot

Select the actual locations of the 30 wind farms in remote areas to normalize the coordinates, and the values are: (0.41,0.94), (0.37, 0.84), (0.54, 0.67), (0.25, 0.62), (0.07, 0.64), (0.02, 0.99), (0.68, 0.58), (0.71, 0.44), (0.54, 0.62), (0.83, 0.69), (0.64, 0.60), (0.18, 0.54), (0.22, 0.60), (0.83, 0.46), (0.91, 0.38), (0.25, 0.38), (0.24, 0.42), (0.58, 0.69), (0.71, 0.71), (0.74, 0.78), (0.87, 0.76), (0.18, 0.40), (0.13, 0.40), (0.82, 0.07), (0.62, 0.32), (0.58, 0.35), (0.45, 0.21), (0.41, 0.26), (0.44, 0.35), (0.04, 0.50). In this section, simulation experiments are conducted for conditions (2) and (3) to verify the performance of the hybrid optimization algorithm proposed.

The path-planning diagram is shown in Figure 6.

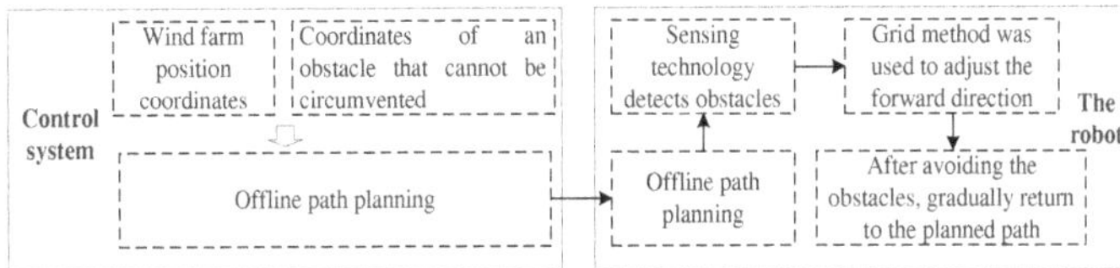

Figure 6. The path-planning diagram.

The experimental computer is configured with Inter Intel (R) Core (TM) I5-8300H CPU @ 2.30GHz and NVIDIA GTX 1050Ti 4G video memory (Beijing, China). The simulation calculation is carried out on MATLAB 2017 platform. The robot obtains the planned path offline from the wind farm inspection auxiliary intelligent system. The system uses the algorithm proposed in this paper to formulate the optimal path scheme for wind farm inspection and eliminate the scheme with non-detour obstacles. From the starting point of the planned path, the robot moves along the planned offline path for inspection.

The wind farm belongs to a time-varying environment to some extent, and there may be detour obstacles on the planned path in a certain period of time. Even for the optimal offline path planning, it is difficult to fully consider all obstacles in the robot path plan. Therefore, sensor technology is also needed to track the obstacles in the path of the robot in real time.

Sensing technology is applied to path planning, which can effectively make up for the unavoidable obstacles in offline path planning. Through position sensors and speed sensors, the robot can obtain its own trajectory and information about surrounding obstacles, and calculate the ideal path to avoid encountering obstacles. The robot only needs to offset the shortest distance without touching the obstacle. When the robot bypasses an obstacle, it returns to the planned path offline and resumes the original inspection path.

When there are devious obstacles, the system adopts the grid method to adjust the forward direction. The concrete principle of the grid method is as follows.

Since the environment is known, the number of obstacles and the location of obstacles are known. The workspace of the robot is set as a two-dimensional plane, denoted as *SG*. The upper left corner of *SG* is the two-dimensional plane coordinate origin. The horizontal direction of the origin to the right is the positive direction of the *X*-axis of the coordinate plane, and the vertical direction of the origin is the positive direction of the *Y*-axis of the coordinate plane. The maximum range that the mobile robot can move in the horizontal direction *X* and vertical direction *Y* are denoted as X_{Max} and Y_{Max}, respectively. The environment model obtained through MATLAB programming is shown in Figure 7, where obstacle grids are represented in black and free grids are represented in white.

In this paper, the grid is represented by serial number method. In the constructed raster map, each raster is numbered separately from left to right and from top to bottom, where "0" represents the start point of the robot path and "63" represents the end point of the path. The serial number of the raster model was coded by MATLAB, as shown in Figure 8.

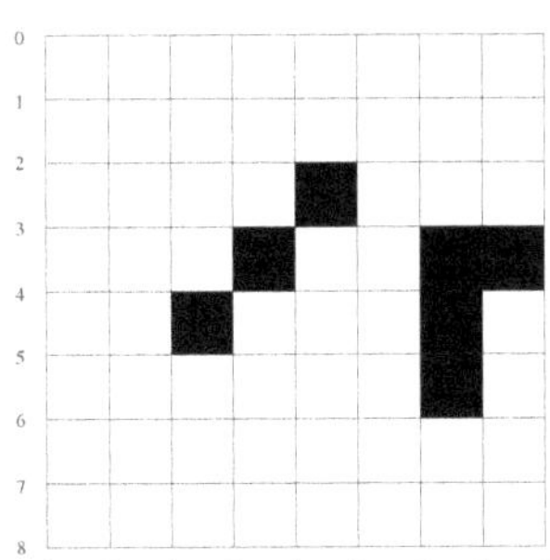

Figure 7. Environmental model represented by raster method.

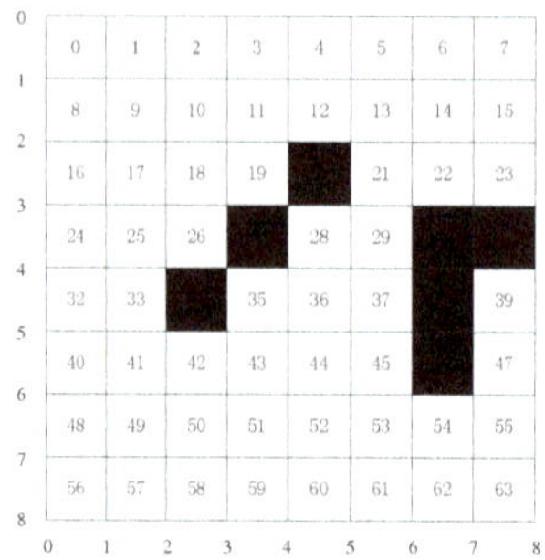

Figure 8. The correspondence between raster coordinates and serial numbers.

In Figure 7, coordinates corresponding to grids and serial numbers corresponding to grids are mapped one by one. The cartesian coordinates of serial number *P(i,j)* can be expressed as follows:

$$\begin{cases} X(i,j) = \text{mod}(P(i,j), 8) + 0.5 \\ Y(i,j) = \text{floor}(\frac{P(i,j)}{8}) + 0.5 \end{cases} \tag{13}$$

where mod denotes the remainder operation and floor denotes the rounding down operation. When calculating the fitness of each path individual, the serial number is transformed into cartesian coordinates.

As shown in Figure 9, a feasible path of the mobile robot is taken as an individual and represented by cartesian coordinates. The robot inspects from point 0 to point 63. These coordinates are {(0.5,0.5), (1.5,1.5), (1.5,2.5), (1.5,3.5), (1.5,4.5), (2.5,5.5), (3.5,5.5), (4.5,5.5), (5.5,5.5), (6.5,6.5), (7.5,7.5)}. If the raster number method is used, the path can be expressed as [0 9 10 11 12 21 29 37 45 54 63].

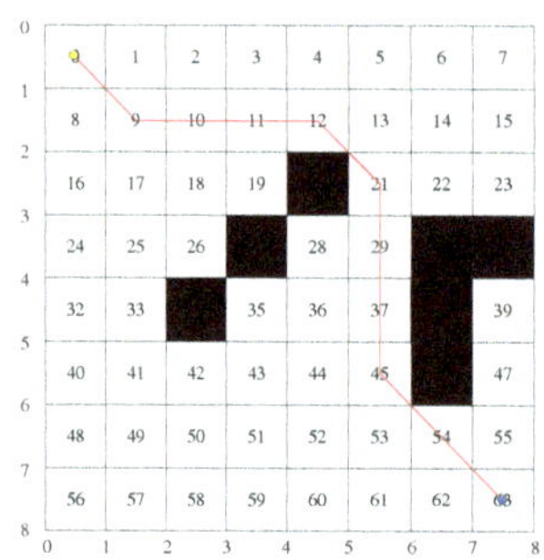

Figure 9. Path diagram.

4.1. Condition (2): Obstacles That Can Be Detoured

In order to improve the complexity of the actual environment and make the algorithm more practical, the simulation test environment is modified based on the existing grid graph. Here, we consider condition (2) in Section 2.1.

When there are some detour obstacles between two wind farms, the straight-line distance can be replaced by the detour distance during robot inspection. The constrained problem is transformed into an unconstrained problem.

The path-planning simulation environment is an 8 × 8 grid environment. Figure 10 shows the path of robot inspection when there are obstacles between wind farms 8 and 9. The yellow point is the starting point, the blue point is the target point, and the path is the red line.

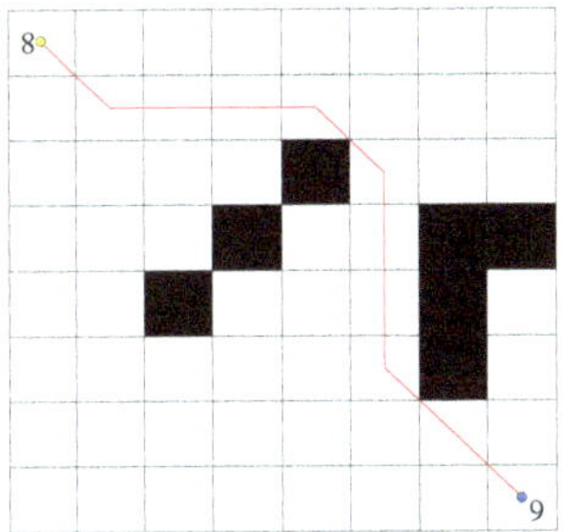

Figure 10. The path image generated by the proposed algorithm.

If we set up one barrier between wind farm 8 and wind farm 9, one barrier between wind farm 25 and wind farm 26, one barrier between wind farm 1 and wind farm 6, and one barrier between wind farm 3 and wind farm 9, the comparison of the hybrid optimization algorithm, heuristic algorithm, and TCNN algorithm in the routing results of wind farms in 30 remote areas is shown in Table 19.

Table 19. Comparison of wind farm path inspection results with detour obstacles.

Operation Results / Algorithm	Best Results	Worst Result	Average Time
Grey Wolf Optimizer (GWO)	7.1136	9.8808	18.219 s
Firefly Algorithm (FA)	5.2821	7.6176	86.3244 s
Ant Colony Algorithm (ACA)	4.7551	6.7835	30.939 s
Genetic algorithm (GA)	4.8965	5.0656	46.053 s
TCNN	4.9696	5.8965	23.354 s
proposed algorithm	4.5366	4.7446	22.09 s

Note: $k = 1$, $\alpha = 0.07$, $\beta = 0.08$, $I_0 = 0.6$, $z\ (0) = 0.8$, $\varepsilon = 0.06$, $W_1 = 1$, $W_2 = 1$, $P_c = 0.85$, $P_m = 0.07$.

In the simulation experiment, 200 independent experiments are set and each independent experiment is iterated 2000 times. In our proposed method, the first 2000 iterations are the path-planning process of the TCNN, and the last 200 iterations are the path-planning process of the genetic algorithm. The simulation results are shown in Table 19. From Table 19, the optimization algorithm in this paper is better than meta-heuristic algorithms such as the genetic algorithm and the TCNN. Therefore, it has better global search ability.

4.2. Condition (3). Obstacles That Cannot Be Detoured

Considering the actual route of remote wind farms, for example, the two wind farms are separated by the Yangtze River, the patrol path cannot only consider the straight-line distance. It is assumed that there are some mountains between the 8th wind farm and the 9th wind farm, there are some rivers between the 25th wind farm and the 26th wind farm, there are some rivers between the 1st wind farm and the 6th wind farm, and there are some mountains between the 3rd wind farm and the 9th wind farm. This means that inspection robots cannot directly move from one wind farm to the next special wind farm. The positions of wind farms are shown in Figure 11.

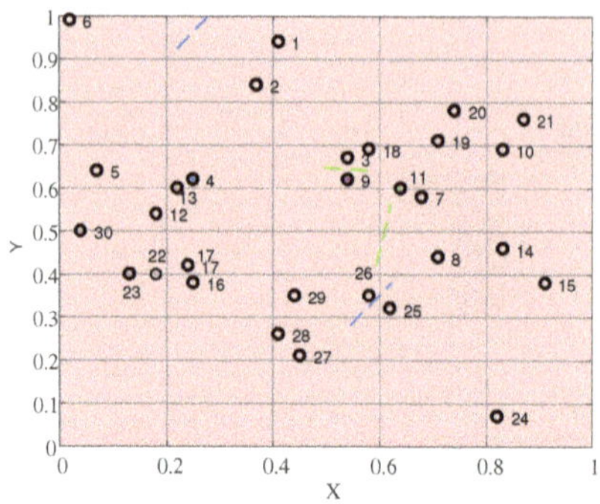

Figure 11. Location of remote wind farms. The green line represents mountains and the blue line represents rivers.

The energy function of the problem is Equation (3).

From Equation (3), the internal state dynamics equation of TCNN neurons describing the model is Equation (9).

(1) The first step of feasible path is obtained by Equation (14), and the selected parameters are as follows: $k = 1$, $\alpha = 0.07$, $\beta = 0.008$, $I_0 = 0.65$, $z(0) = 0.8$, $\varepsilon = 0.05$, $W_1 = 1$, $W_2 = 1$.
(2) In a genetic algorithm, the crossover probability is 0.5, the variation probability is 0.05, and the fitness function takes the reciprocal of the path length.

200 independent experiments were conducted to optimize the patrol path of wind farms in remote areas by using the new algorithm and each independent experiment is iterated 2200 times. The path length of the TCNN, the genetic algorithm and our proposed method changes with the number of iterations, as shown in Figure 12.

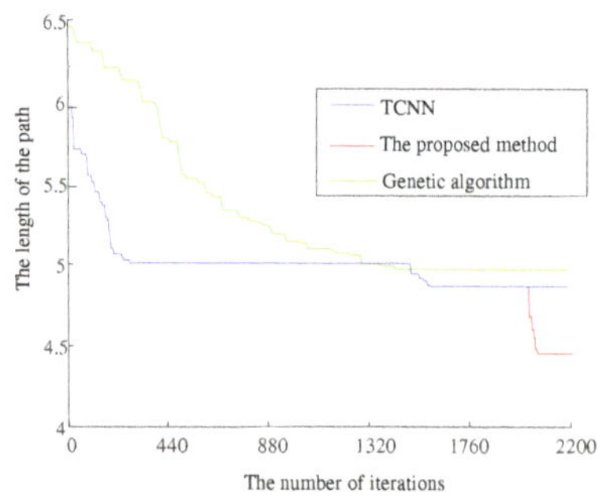

Figure 12. Solution of path planning.

In our proposed method, the first 2000 iterations are the path-planning process of the TCNN, and the last 200 iterations are the path-planning process of the genetic algorithm. As can be seen from Figure 12, our hybrid optimization algorithm has better optimization ability than a single algorithm.

The results of the path planning are shown in Figure 13. The yellow five-pointed star in Figure 8 represents the starting point.The shortest path length of the proposed hybrid optimization algorithm for patrol inspection is 4.5755.

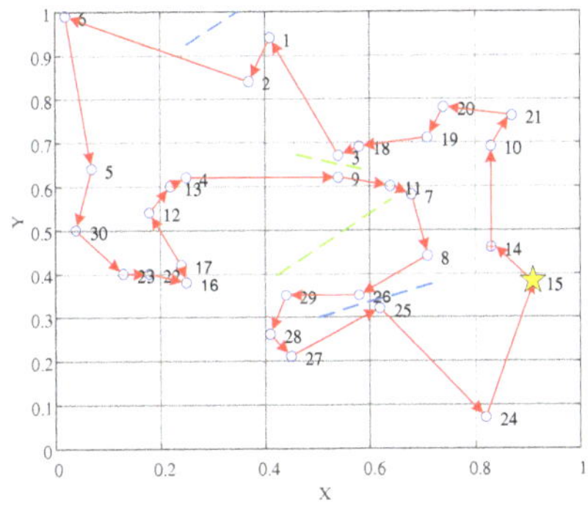

Figure 13. The results of path planning (Note: The yellow star represents the starting point).

Since the patrol path from 15 is the shortest, this can be set as their starting point.

The comparison of the hybrid optimization algorithm, genetic algorithm and TCNN algorithm in the path planning of 30 wind farms in remote areas is shown in Table 20. From Table 20, our proposed algorithm is better than other algorithms.

Table 20. Comparison of wind farm path inspection results with non-detour obstacles.

Operation Results / Algorithm	Best Results	Worst Result	Average Time
Grey Wolf Optimizer (GWO)	7.1525	8.9197	10.42 s
Firefly Algorithm (FA)	5.0085	7.6565	65.2261 s
Ant Colony Algorithm (ACA)	4.793	6.8224	22.939 s
Genetic algorithm (GA)	4.9354	5.1454	39.77 s
TCNN	5.321	6.4954	17.616 s
proposed algorithm	4.5755	4.7835	21.768 s

Note: $k = 1$, $\alpha = 0.07$, $\beta = 0.08$, $I_0 = 0.55$, $z(0) = 0.8$, $\varepsilon = 0.06$, $W_1 = 1$, $W_2 = 1$, $P_c = 0.8$, $P_m = 0.07$.

It can be seen from Tables 19 and 20 that the algorithm in this paper has the best path-panning effect and a short operation time. Although the operation time of the TCNN is shorter than that of our proposed algorithm, its patrol path is much longer than that of our proposed algorithm. The actual operation and maintenance cost of the TCNN will be higher than that of our proposed algorithm. Compared with the TCNN and the genetic algorithm, the experimental results of our algorithm are significantly improved. This shows that the initial population quality generated by the hybrid algorithm is significantly better than that generated by the completely random population. Therefore, the path scheme obtained by our proposed algorithm is closer to the best solution. The proposed algorithm is feasible and effective.

5. Conclusions

Global path planning is a prerequisite for robot inspection of wind farms. The design and implementation of global path planning is beneficial to the development, application and commercial promotion of wind farm inspection robots. In order to reduce the inspection cost of wind farms in remote areas, a hybrid optimization algorithm is proposed based on a chaotic neural network algorithm and a traditional genetic algorithm. The algorithm is applied to the path planning of wind farms for patrol robots. This algorithm has better search performance and shorter running time than a heuristic algorithm based on TSP and TCNN, and thus, has higher application value.

However, there are still some shortcomings that can be further studied.

(1) The hybrid algorithms proposed in this paper are based on theoretical analysis and simulation. We could consider more possible influencing factors for modeling and apply our hybrid algorithm to the actual project.

(2) This paper refers to the patrol path planning of a single robot. We could deeply study the collaborative scheduling of multiple robots to improve the efficiency of wind farm inspection.

Author Contributions: Conceptualization, L.C., F.Z. and Z.H.; methodology, L.C. and F.Z.; software, L.C. and W.D.; validation, L.C., F.Z. and Z.H.; formal analysis, L.C., F.Z. and Z.G.; investigation, L.C., F.Z. and Z.H.; resources, L.C., C.P. and W.D.; data curation, L.C., K.J. and W.D.; writing—original draft preparation, L.C. and F.Z.; writing—review and editing, L.C. and F.Z.; visualization, L.C., K.J. and Z.G.; supervision, F.Z. and Z.H.; project administration, F.Z., Z.H., K.J., Z.G. and C.P.; funding acquisition, Z.H., K.J., Z.G. and C.P. All authors have read and agreed to the published version of the manuscript.

Funding: This research was funded by National key research and development program under grant number 2019YFB1705800, International Collaborative Research Project of Qilu University of Technology under grant number QLUTGJHZ2018020, the Shandong Provincial Natural Science Foundation of China under Grant ZR2021MF022, ZR2021QF065, and ZR2019MA017, and Major scientific and technological innovation projects of Shandong Province under grant number 2019JZZY010731 and 2020CXGC010901.

Institutional Review Board Statement: Not applicable.

Informed Consent Statement: Not applicable.

Data Availability Statement: Not applicable.

Conflicts of Interest: The authors declare no conflict of interest.

Appendix A

Select the actual locations of the 75 cities in remote areas to normalize the coordinates, and the values are: (0.48,0.21), (0.52,0.26), (0.55,0.50), (0.50,0.50), (0.41,0.46), (0.51,0.42), (0.55,0.45), (0.38,0.33), (0.33,0.34), (0.45,0.35), (0.40,0.37), (0.50,0.30), (0.55,0.34), (0.54,0.38), (0.26,0.13), (0.15,0.05), (0.21,0.48), (0.29,0.39), (0.33,0.44), (0.15,0.14), (0.16,0.19), (0.12,0.17), (0.50,0.40), (0.22,0.53), (0.21,0.36), (0.20,0.30), (0.26,0.29), (0.40,0.20), (0.36,0.26), (0.62,0.48), (0.67,0.41), (0.62,0.35), (0.65,0.27), (0.62,0.24), (0.55,0.20), (0.35,0.51), (0.30,0.50), (0.45,0.42), (0.21,0.45), (0.36,0.06), (0.06,0.25), (0.11,0.28), (0.26,0.59), (0.30,0.60), (0.22,0.22), (0.27,0.24), (0.30,0.20), (0.35,0.16), (0.54,0.10), (0.50,0.15), (0.44,0.13), (0.35,0.60), (0.40,0.60), (0.40,0.66), (0.31,0.76), (0.47,0.66), (0.50,0.70), (0.57,0.72), (0.55,0.65), (0.02,0.38), (0.07,0.43), (0.09,0.56), (0.15,0.56), (0.10,0.70), (0.17,0.64), (0.55,0.57), (0.62,0.57), (0.70,0.64), (0.64,0.04), (0.59,0.05), (0.50,0.04), (0.60,0.15), (0.66,0.14), (0.66,0.08), (0.43,0.26).

Two additional sets of normalized coordinates for 75 cities will be uploaded as attachments for spatial reasons.

Appendix B

Select the actual locations of the 50 cities in remote areas to normalize the coordinates, and the values are: (0.41,0.94), (0.37,0.84), (0.54,0.67), (0.25,0.62), (0.07,0.64), (0.02,0.99), (0.68,0.58), (0.71,0.44), (0.54,0.62), (0.83,0.69), (0.64,0.60), (0.18,0.54), (0.22,0.60), (0.83,0.46), (0.91,0.38), (0.25,0.38), (0.24,0.42), (0.58,0.69), (0.71,0.71), (0.74,0.78), (0.87,0.76), (0.18,0.40), (0.13,0.40), (0.82,0.07), (0.62,0.32) (0.58,0.35), (0.45,0.21), (0.41,0.26), (0.44,0.35), (0.04,0.50), (0.48,0.21), (0.52,0.26), (0.55,0.50), (0.50,0.50), (0.41,0.46), (0.51,0.42), (0.55,0.45), (0.38,0.33), (0.33,0.34), (0.45,0.35),(0.40,0.37), (0.50,0.30), (0.55,0.34), (0.54,0.38), (0.26,0.13), (0.15,0.05), (0.21,0.48), (0.29,0.39), (0.33,0.44), (0.15,0.14).

Two additional sets of normalized coordinates for 50 cities will be uploaded as attachments for spatial reasons.

Appendix C

Select the actual locations of the 100 cities in remote areas to normalize the coordinates, and the values are: (0.48, 0.21), (0.52, 0.26), (0.55, 0.50), (0.50, 0.50), (0.41, 0.46), (0.51, 0.42), (0.55, 0.45), (0.38, 0.33), (0.33, 0.34), (0.45, 0.35), (0.40, 0.37), (0.50, 0.30), (0.55, 0.34),

(0.54, 0.38), (0.26, 0.13), (0.15, 0.05), (0.21, 0.48). (0.29, 0.39), (0.33, 0.44), (0.15, 0.14), (0.16, 0.19), (0.12, 0.17), (0.50, 0.40), (0.22, 0.53), (0.21, 0.36), (0.20, 0.30). (0.26, 0.29), (0.40, 0.20), (0.36, 0.26), (0.62, 0.48), (0.67, 0.41), (0.62, 0.35), (0.65, 0.27), (0.62, 0.24), (0.55, 0.20). (0.35, 0.51), (0.30, 0.50), (0.45, 0.42), (0.21, 0.45), (0.36, 0.06), (0.06, 0.25), (0.11, 0.28), (0.26, 0.59), (0.30, 0.60). (0.22, 0.22), (0.27, 0.24), (0.30, 0.20), (0.35, 0.16), (0.54, 0.10), (0.50, 0.15), (0.44, 0.13), (0.35, 0.60), (0.40, 0.60). (0.40, 0.66), (0.31, 0.76), (0.47, 0.66), (0.50, 0.70), (0.57, 0.72), (0.55, 0.65), (0.02, 0.38), (0.07, 0.43), (0.09, 0.56), (0.15, 0.56), (0.10, 0.70), (0.17, 0.64), (0.55, 0.57), (0.62, 0.57), (0.70, 0.64), (0.64, 0.04), (0.59, 0.05), (0.50, 0.04). (0.60, 0.15), (0.66, 0.14), (0.66, 0.08), (0.43, 0.26), (0.10,0.10), (0.90,0.50), (0.90,0.10), (0.45,0.90), (0.90,0.80), (0.70,0.90), (0.10,0.45), (0.45,0.10), (0.40,0.44), (0.24,0.15), (0.17,0.23), (0.23,0.71), (0.51,0.94), (0.87,0.65), (0.68,0.52), (0.84,0.36), (0.66,0.25), (0.61,0.26), (0.91,0.45), (0.83,0.72), (0.16,0.82), (0.66,0.10), (0.79,0.79), (0.82,0.70), (0.22,0.98).

Two additional sets of normalized coordinates for 100 cities will be uploaded as attachments for spatial reasons.

Appendix D

Select the actual locations of the 1000 cities in remote areas to normalize the coordinates. Due to space reasons, the coordinates will be uploaded as an attachment.

References

1. Gao, Z.W.; Liu, X.X. An Overview on Fault Diagnosis, Prognosis and Resilient Control for Wind Turbine Systems. *Processes* **2021**, *9*, 300. [CrossRef]
2. Gao, Z.W.; Sheng, S. Real-time monitoring, prognosis, and resilient control for wind turbine systems. *Renew. Energy* **2018**, *116*, 1–4. [CrossRef]
3. Alatas, B.; Bingol, H. A Physics Based Novel Approach for Travelling Tournament Problem: Optics Inspired Optimization. *Inf. Technol. Control* **2019**, *48*, 373–388. [CrossRef]
4. Akyol, S.; Alatas, B. Plant intelligence based metaheuristic optimization algorithms. *Artif. Intell. Rev.* **2016**, *47*, 417–462. [CrossRef]
5. Liu, X.K.; Liu, H.B.; Wang, J. Research on path optimization of substation inspection robot based on ant colony algorithm. *Elect. Apps.* **2019**, *38*, 113–117.
6. Dong, X.Y.; Ji, K.; Zhu, J.; Yang, B. A retrofitted ant colony algorithm for inspection robot path planning in UHV substations. *Power Syst. Prot. Control.* **2021**, *49*, 154–160.
7. Chen, N.K.; Wang, Y.N.; Jia, L. Multi-mobile robot cooperative inspection operation based on improved biological excitation neural network algorithm in substation. *Control Decis.* **2022**, *37*, 1453–1459.
8. He, Q.; Wu, Y.L.; Xu, T.W. Application of improved genetic simulated annealing algorithm in TSP optimization. *Control Decis.* **2018**, *33*, 219–225.
9. Yuan, J.Q.; Li, S.; Wu, Y.F.; Guo, J. Robot Path Planning Method Based on Simulated Annealing Ant Colony Algorithm. *Comput. Integr. Manuf. Syst.* **2019**, *36*, 329–333.
10. Liu, J. Applied Research of Hybrid Genetic Algorithm and Simulated Annealing Algorithm in Traveling Salesman Problem. Master's Thesis, South China University of Technology, Guangzhou, China, 2014.
11. Hu, Z.Q.; Li, W.J.; Qiao, J.F. Frequency-conversion sinusoidal chaotic neural network with disturbance feature. *CAAI Trans. Intell. Syst.* **2018**, *13*, 493–499.
12. Wang, L.; Li, S.; Tian, F.; Fu, X.J. A noisy chaotic neural network for solving combinatorial optimization problems: Stochastic chaotic simulated annealing. *IEEE Trans. Syst. Man Cybern.* **2004**, *34*, 2119–2125. [CrossRef]
13. Wang, L.; Zhang, Z.J. Automatic Detection of Wind Turbine Blade Surface Cracks Based on UAV-Taken Images. *IEEE Trans. Ind. Electron.* **2017**, *64*, 7293–7303. [CrossRef]
14. Hu, Z.Q.; Li, W.J.; Qiao, J.F. Variable frequency sinusoidal chaotic neural network and its application. *Acta Phys. Sin.* **2017**, *66*, 12–22.
15. Kwok, T.; Smith, K.A. A unified framework for chaotic neural-network approaches to combinatorial optimization. *IEEE Trans. Neural Netw.* **2004**, *10*, 978–981. [CrossRef] [PubMed]
16. Liu, Q.H.; Li, X.Y.; Gao, L.; Li, Y.L. A Modified Genetic Algorithm With New Encoding and Decoding Methods for Integrated Process Planning and Scheduling Problem. *IEEE Trans. Cybern.* **2021**, *51*, 4429–4438. [CrossRef] [PubMed]
17. Ling, Z.X.; Tao, X.Y.; Zhang, Y.; Chen, X. Solving Optimization Problems Through Fully Convolutional Networks: An Application to the Traveling Salesman Problem. *IEEE Trans. Syst. Man Cybern.* **2021**, *51*, 7475–7485. [CrossRef]
18. Hu, Z.Q.; Li, W.J.; Qiao, J.F. Variable frequency sinusoidal chaotic neural network based on adaptive simulated annealing. *Acta Electron. Sin.* **2019**, *47*, 613–622.
19. Cong, S.; Wang, Z.N. Improved Algorithm of Transient Chaotic Neural Network and Its Application in TSP. *Sci. Technol. Rev.* **2009**, *27*, 60–63.

20. Jia, S.J.; Du, B. Hybrid optimized algorithms based on the Rosenbrock search method and dynamic inertia weight PSO. *Control Decis.* **2011**, *26*, 1060–1064.
21. Zhao, G.Q.; Peng, X.Y.; Sun, N. A Modified Differential Evolution Algorithm with Hybrid Optimization Strategy. *Acta Electron. Sin.* **2006**, *34*, 2402–2405.
22. Yu, Y.Y.; Chen, Y.; Li, Y.Y. Improved genetic algorithm for solving TSP. *Control Decis.* **2014**, *29*, 1483–1488.
23. Jin, H.D.; Leung, K.S.; Wong, M.L.; Xu, Z.B. An efficient self-organizing map designed by genetic algorithms for the traveling salesman problem. *IEEE Trans. Syst. Man Cybern. Part B* **2003**, *33*, 877–888.

processes

MDPI

Article

Material Removal Optimization Strategy of 3D Block Cutting Based on Geometric Computation Method

Hui Shao [1,2,*], Qimeng Liu [1,2] and Zhiwei Gao [3]

[1] School of Information Science and Engineering, Huaqiao University, Xiamen 361021, China; lqimeng2021@163.com
[2] Fujian Engineering Research Center of Motor Control and System Optimal Schedule, Xiamen 361021, China
[3] Faculty of Engineering and Environment, University of Northumbria at Newcastle, Newcastle upon Tyne NE1 8ST, UK; zhiwei.gao@northumbria.ac.uk
* Correspondence: shaohuihu11@163.com

Abstract: During the material removal stage in stone rough processing, milling type has been widely explored, which, however, may cause time and material consumption, as well as substantial stress for the environment. To improve the material removal rate and waste reuse rate in the rough processing stage for three-dimensional stone products with a special shape, in this paper, circular saw disc cutting is explored to cut a convex polyhedron out of a blank box, which approaches a target product. Unlike milling optimization, this problem cannot be well solved by mathematical methods, which have to be solved by geometrical methods instead. An automatic block cutting strategy is proposed intuitively by considering a series of geometrical optimization approaches for the first time. To obtain a big removal block, constructing cutting planes based on convex vertices is uniquely proposed. Specifically, the removal vertices (the maximum thickness of material removal) are searched based on the octree algorithm, and the cutting plane is constructed based on this thickness to guarantee a relatively big removal block. Moreover, to minimize the cutting time, the geometrical characteristics of the intersecting convex polygon of the cutting plane with the convex polyhedron are analyzed, accompanied by the constraints of the guillotine cutting mode. The optimization algorithm determining the cutting path is presented with a feed direction accompanied by the shortest cutting stroke, which confirms the shortest cutting time. From the big removal block and shortest cutting time, the suboptimal solution of the average material removal rate (the ratio of material removal volume to cutting time) is generated. Finally, the simulation is carried out on a blank box to approach a bounding sphere both on MATLAB and the Vericut platform. In this case study, for the removal of 85% of material with 19 cuts, the proposed cutting strategy achieves five times higher the average material removal rate than that of one higher milling capacity case.

Keywords: block cutting; data reconstruction; convex polyhedron (CPH); convex polygon (CPC); path optimization; average material removal rate (AMRR)

Citation: Shao, H.; Liu, Q.; Gao, Z. Material Removal Optimization Strategy of 3D Block Cutting Based on Geometric Computation Method. *Processes* **2022**, *10*, 695. https://doi.org/10.3390/pr10040695

Academic Editor: Gade Pandu Rangaiah

Received: 3 March 2022
Accepted: 30 March 2022
Published: 2 April 2022

1. Introduction

The stone processing industry has adverse effects on the environment, economy and sustainability. Stone processing causes heavy pollution from dust and CO_2 emissions, and high water and energy consumption, which brings tremendous pressure and threats to the ecological environment, especially in natural stone mining areas [1]. On the other hand, with the development of modern civilized society, the stone industry is indispensable and has become increasingly important. Moreover, the demands and varieties of stone products are increasing day by day. Therefore, the demand for the stone industry puts forward higher requirements for stone processing, especially in the rough processing stage. Building a green manufacturing system and process scheme, developing energy-saving and emission reduction optimization technologies and improving the processing efficiency of the stone industry have attracted more and more attention recently [1,2].

In the machining process of special-shaped stone products, the traditional rough processing adopts diamond wire sawing or circular sawing for simple tasks [2,3], or conventional milling for simple tasks [4], or automatic milling for complex tasks [5] with a diamond grinding wheel, etc., but such processing modes cannot realize automation completely, or are accompanied by low waste utilization, a large processing time, high consumption and serious tool wear even for straight cuts. Although the milling process has reflected automation to some extent, a large amount of dust will be generated in the machining process. The above mentioned processes can often not meet the needs of green manufacturing and high-efficiency machining in the rough machining stage.

For a material removal optimization strategy, the analysis methods of different machining tools are completely different. Owning to the distinct advantages of having a big cutting depth and high cutting linear velocity, the diamond circular saw blade is widely used in the stone cutting process, which provides a possible way to achieve efficient green processing [1,6]. It is noted that circular saws are known to likely be the cheapest and fastest motorized saw available [7,8]. With the development of the multi axis linkage technology of machining machinery, the applications of 5–6 axis NC machine tools and robots, some interesting results have been reported in recent years on the material removal method during the rough machining process using circular saw blades. In [1], an energy consumption prediction model of the stone sawing process of a circular saw was proposed. By predicting the power and energy consumption in the whole sawing process, the optimal scheme considering the variable material removal rate (MRR) could be discussed for stone processing to achieve energy saving and emission reduction. In [8], a technique to cut freeform curves with a flexible circular saw was addressed by setting the width of the cutting edge larger than the width of the saw body to ensure there was no friction between the machined surface and the saw body while cutting. Moreover, cutting any polygon down to an inner complex nonconvex shape was achieved by a sequence of straight cuts with linear-time algorithms in [9], where the cuttability of a small saw and large saw was analyzed attentively. Ref. [10] studied the algorithm for cutting polyhedral shapes with a hot wire cutter, utilizing computational geometry techniques to solve the problems of lines and segments in the cutting process. Particularly used in recent years when fabricating freeform geometries, in order to find collision-free tangential cutting directions, a conservative algorithm for line cutting with a wire cutter was presented by [11], which provided advanced techniques to remove large amounts of material. Exploring the material cutting of 2D or 3D geometric shapes, in [12], an approximation algorithm for cutting out convex polygons was presented, which can cut convex polygons from the plane at a minimum cost by designing an optimal cutting sequence. Their algorithm can achieve a constant approximation ratio of the paper diameter to the polygon diameter. On the basis of [12], in [13], an approximate algorithm for cutting out a convex polyhedron from a sphere was surveyed, in which several approximate algorithms were discussed to find the plane sequence with the minimum cutting cost. Ref. [14] proposed a method of 3D curved block cutting analysis by utilizing the advantages of topology and computational geometry in geological solid modeling, where 3D curved blocks were formed with less calculation and memory. These studies not only provide feasibility for a stone cutting plan with a diamond circular saw, but also provide a geometrical analysis method for solving the geometrical characteristics of block cutting optimization strategies to some extent.

Despite some of the new algorithms and analysis methods being investigated in [11,14], to the best of our knowledge, the theoretical research available on the optimization strategies of block sawing with a circular saw blade for special-shaped stone in the rough machining stage is currently sparse. This is due to the need to touch upon the convex polyhedron (CPH) reconstruction techniques of computational geometry and computer graphics, and the constraints of cutter head feeding along a straight line in the cutting process. The geometric challenges have also been stated for multi axis machining including the material properties, tools shapes, accessibility, collision detection, etc. [11]. Although line cutting [11] was an available new and flexible style for automatic block cutting, the cutting technique was completely different from saw disc cutting. Consequently, these studies have motivated us to find out more about block cutting optimization. According to the characteristics of the stone rough machining stage, making full use of the advantages of the diamond circular saw and taking into account the average material removal rate (AMRR) as the cost, we propose a cutting optimization strategy for special-shaped stone blocks with a large material removal amount and a short cutting time, so as to automatically realize removing materials with high efficiency while ensuring energy saving and emission reduction.

The main contributions in this study can be summarized as follows: (i) In order to automatically obtain a set of relatively optimal cutting planes to ensure a big removal block for each cutting, the method of constructing a cutting plane (CP), in turn, is proposed according to the geometrical characteristic of the CPH in space and with the data reconstruction of the CPH, where the octree algorithm is used to search for the removal vertices of the CPH for each cutting to reduce the amount of calculation. (ii) Except for the cutting of a big block, another key point is to reduce the cutting time, whose optimization model is thus established. The convex polygon (CPG) generated by the intersection of the CP with the parent CPH is analyzed, where the optimization objective and optimization algorithm for determining the feed direction and starting point of the cutting path are addressed to produce the shortest cutting time. From (i) and (ii), the suboptimal solution of the AMRR for block cutting is obtained.

Cutting strategies with a circular saw will play a particularly important role not only in 3D stone processing but also in wood, metal and harder diamond 3D processing. This strategy may be not very satisfactory, nevertheless, which leads us to explore more feasible geometric techniques for efficient 3D cutting, whether for convex or nonconvex polyhedrons.

The rest of this paper is organized as follows: In Section 2, preliminary information regarding the block cutting mechanism is described. Section 3 addresses the reconstruction description of the CPH with vertex–face information. Section 4 investigates the scheme to design cuttable big blocks by constructing the CP. In Section 5, the cutting time optimization method is presented by considering the geometrical analysis of the cutting path. Before the conclusion in Section 7, validation studies are addressed in Section 6.

2. Preliminaries: Block Cutting Mechanism

The definitions of the symbols and units used are shown in Table 1.

Table 1. Nomenclature.

Symbols	Quantity	Unit	Symbols	Quantity	Unit
Q_0	Blank box (Blank workpiece)	/	P_{csi}	Vertex set of intersection CPG	/
Q_{i-1}	Parent CPH	/	S_i	The ith intersection CPG	/
Q_i	Child CPH	/	T	Homogeneous transformation matrix	/
Q_M	Final CPH	/	R	Radius of sawblade	mm
Q_{ri}	Removal block by the ith cutting	/	v_f	Feeding speed	mm/min
π	Cutting plane	/	V_{ri}	The ith material removal volume	mm^3
π_i	The ith CP	/	T_i	The ith cutting time	min
P_{i-1}	Vertex set of Q_{i-1}	/	W	Width of CPG	mm
P_{ri}	Removal vertex set of the ith cutting	/	L_i	Cutting path	mm/min

In the rough machining process of removing materials, due to the significant differences in shape between the original blank and the final product, rapidly removing most of the extra materials on the original blank to form a rough blank has become one of the most important processes. For any 3D special-shaped product, in the rough machining stage of removing materials, it can be representatively wrapped as a compact bounding sphere (BS) or bounding ellipsoid (BE) to perform further cutting exploration. A BS or BE is selected depending on the 3D shape of the special-shaped product. If the product is a thin, long shape we would choose the corresponding blank box and a BE as a target to perform the rough cutting, conversely, for a short, round shape we would choose a BS. We would hope not to have any unnecessary material consumption. For instance, in Figure 1, the 3D symmetrical penguin and its BS are shown simultaneously.

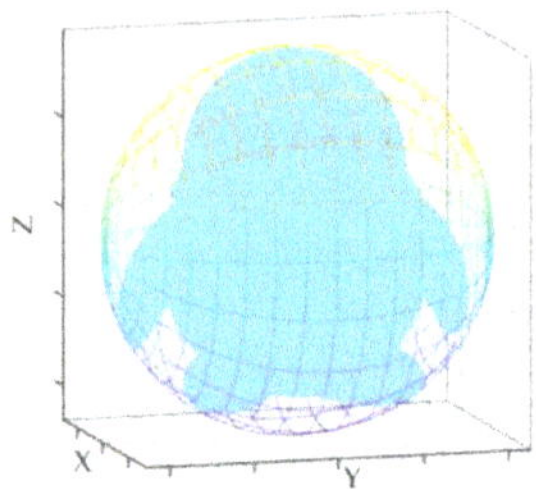

Figure 1. 3D penguin and its BS.

Hence, when confirming the machining allowance, this paper considers the compact BS to be a typical target shape for the rough cutting of a blank CPH (a blank box), Q_0, which reflects the oriented bounding box (OBB) of the BS This guarantees the efficient removal of the extra materials without overcutting and excessive material consumption. In the blank CPH cutting, we need to design the optimal cutting strategy to cut out the final CPH, Q_M, from Q_0 to approximate the target BS, as shown in the cutting schematic diagram in Figure 2, thus satisfying the characteristics of the circular saw processing. To solve this problem, the cutting process should be analyzed and monitored, not only combining the theory from computational geometry and computer graphics, but also taking into account the technique of the processing plan and design, which may bring complexity and challenges to the design and implementation of the cutting strategy. If π_i denotes the ith CP, the cutting optimization process reflects a series of optimization processes of π_i intersections with the CPH and the optimization process of each cutting path. The problem of the material removal following cutting can be described as follows:

- Reconstruction description of the CPH: During a continuous block cutting process, it is inevitable to be confronted with the problem of the CPH reconstruction and intersection calculation between a plane and the CPH combining with computational geometry. Therefore, it is important but challenging to update the CPH information including removal vertices, intersection vertices, intersection faces and updated faces using an appropriate, available data structure. We hope these can be efficiently implemented to carry out dynamic data storage so that any queries about the workpiece during the cutting process can be answered with the need for less data and calculation. Meanwhile, the visual image of the live cutting can be displayed to demonstrate the reconstruction of the CPH after each cutting.
- Design cuttable big blocks: For block cutting, aiming for the highest efficiency for the removal of materials, it is better when removal blocks are designed larger, and as far as possible, with the minimum number of cutting times. The difficult question is how to construct the CP so as to determine the angle of the circular saw, which is related to the vertices of the blank CPH and the BS surface.
- Optimize cutting time: The cutting time under the condition with a constant feed speed can be confirmed by investigating the cutting path. Therefore, in order to obtain the shortest cutting time, it is necessary to analyze the geometrical characteristics of the CPG generated by the intersection of the CP with the CPH, so as to calculate the cutting feed direction and cutting point of the shortest path under constraints.

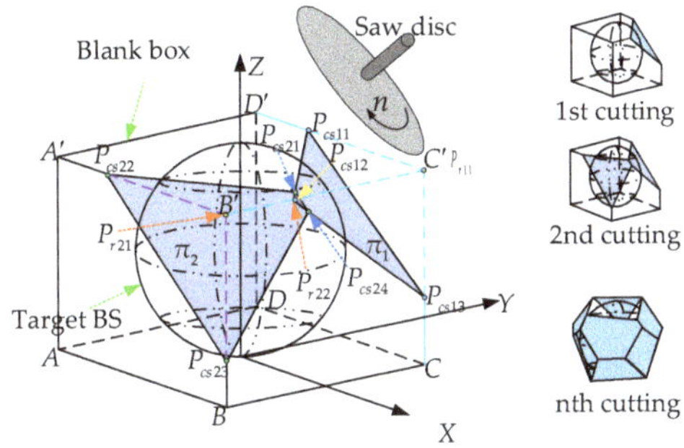

Figure 2. Schematic diagram of block cutting.

In the real machining process, it is necessary to comprehensively consider the machining efficiency and machining allowance alongside the capacity of the machine tools or robots. Therefore, the optimization strategy of 3D block cutting discussed in the paper makes the following assumptions: (1) The BS of the special-shaped product is regarded as the shape of the target object; (2) the convex polyhedral block is cuttable by a circular saw when its angle satisfies the geometrical constraints of the mechanism of the machine tools or robots and it is set up properly. (3) In the cutting process, there is no interference between the cutter and the stone. Moreover there are no obstacles around.

3. Reconstruction Description of the CPH

The essence of the removal of material when block cutting is found in the process of the intersection of the CP with the parent CPH to produce the child CPH. The vertices, edges and faces of the CPH will be updated dynamically in the cutting process, which is a complex process with tremendous and heavy computation and data storage [14,15]. In order to realize continuous cutting and the dynamic visual effect automatically, the updating of information with a relatively simple data structure is introduced to reconstruct the CPH in the updating process.

3.1. Data Structure of CPH

In the cutting process, updating the polyhedron experiences tedious and algorithmically complex updates of the data structure, which is used to describe the significant geometrical features of the convex polyhedron and bounding sphere [14]. As the cutting is completed, the cut edge causes changes in the number of faces, edges and vertices on the child polyhedron. In the literature, some data structures for a polyhedron in 3D space have been proposed, such as single level, 2 or 3 levels or half-edge data structures [15,16], which are for a face list, vertex–face list, vertex–edge–face list or doubly connected edge list. In order to realize the dynamic storage, querying and management of polyhedron data efficiently, a double level data structure for the vertex–face list is established, as shown in Figure 3, to describe the geometrical characteristics of the polyhedron, which retains the vertex–face information to guarantee that the volume of the CPH can be calculated and the cutting calculation can be implemented dynamically with a relatively small calculation and amount of storage. Here, every vertex and face of a polyhedron are indexed separately and the array data of each face lists the allocated vertices of each face so that they meet in a counter clockwise (CCW) order. Namely, the vertices are listed in a CCW order and the array of face lists is filled with the index of the vertex list.

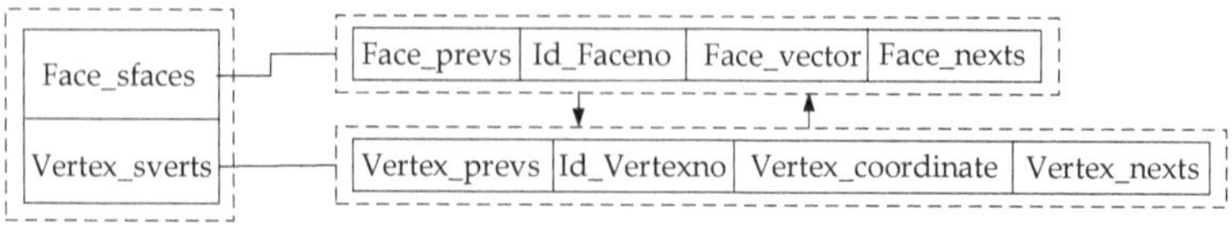

Figure 3. Vertex–face data structure.

3.2. CPH Model Reconstruction

Based on the vertex–face data structure, the reconstruction process of the CPH is as follows: Let $i = 1, 2, \ldots, M$ denote the number of cuts. Given $(i-1)$th vertex set $P_{i-1} = \{P_{i-1,1}, P_{i-1,2}, \ldots P_{i-1,k} \ldots\}$ of the parent polyhedron Q_{i-1} and the ith CP π_i in 3D space, where k denotes k-th vertex. It is then possible to compute renewed vertex and face lists of the cutting process when the removal vertices and intersection face of Q_{i-1} are confirmed; in this way, we complete the dynamic data management of the continuous cutting process and the quantitative evaluation of the cutting algorithm. The updating process follows three steps:

Step1: Judge removal vertices

In order to yield the update vertices in the cutting process, the first thing is to judge the vertices to cut off. Intuitively, from the schematic diagram of block cutting as shown in Figure 2, π_i for each cutting divides the whole space into two half spaces [15,17]. According to binary space partition algorithm (BSP), the positive and negative half space can be defined by the normal vector of π_i as the boundary. The half space pointed by the normal vector is the positive half space, which is the half space to be removed, and vice versa, the negative half space is the reserved CPH part, which is the child CPH. Aiming at vertex set $P_{i-1} = \{P_{i-1,1}, P_{i-1,2}, \ldots P_{i-1,k} \ldots\}$ of the parent CPH, the vertex partition equation is established as follows

$$\begin{cases} \pi_i = K_{Ai} \cdot x + K_{Bi} \cdot y + K_{Ci} \cdot z + K_{Di} \\ \pi_{ik} > 0, P_{i-1,k} \in P_{ri} \\ \pi_{ik} \leq 0, P_{i-1,k} \notin P_{ri} \end{cases} \tag{1}$$

where $K_{Ai}, K_{Bi}, K_{Ci}, K_{Di}$ represent the equation coefficients of π_i; π_{ik} are the solutions of the plane equation for each vertex of Q_{i-1}; hence, the removal vertex set $P_{ri} = \{P_{ri1}, P_{ri2}, \cdots, P_{rin}\}$ by the ith cutting can be obtained, and n is the number of vertices to be cut off.

Step2: Calculate intersection vertices (CPG)

If one knows the vertices that should be removed, the edges that should be cut can be deduced. This means that the CP can be calculated based on some of that information. For each cutting, one CP intersects with one CPH, then an intersection CPG is generated. Calculating CPG vertices need to judge the edges of the CP intersecting with the parent CPH and find its intersection. As can be seen from Section 3.1, the data structure stores the vertex–face list, ignoring the storage of the edge list. However, the face list is composed of vertices allocated by the right-hand rule. Therefore, we can connect two adjacent vertices in the face list to determine the edges. To judge whether the edge of the parent CPH intersects with π_i, Equation (1) can be used. If the two vertices of the edge are located on the positive and negative half space of the CP separately, the edge is intersected by the CP. Otherwise, there is no intersection. In addition, by vector parallel condition, the equation of the intersection edges of the CPH can be written as follows

$$\frac{(x - x_{iq})}{m_{iq}} = \frac{(y - y_{iq})}{n_{iq}} = \frac{(z - z_{iq})}{p_{iq}} \tag{2}$$

where (x_{iq}, y_{iq}, z_{iq}) is any vertex on the intersection edge E_{iq} of the parent CPH, $\overrightarrow{s_{iq}} = (m_{iq}, n_{iq}, p_{iq})$ is the direction vector of E_{iq} by calculating two adjacent vertices, and $q = 1, 2, \ldots, h$ represents the number of the intersection edges of ith cutting.

For the intersection edges of the parent CPH and the CP, the intersection point can be solved by synthesizing Equations (1) and (2). The vertex set of the intersection CPG is expressed as $P_{csi} = \{P_{csi1}, P_{csi2}, \cdots, P_{csih}\}$. It is worth noting that since each edge of the CPH is shared by two faces, they intersect with the CP to obtain the intersection point. After yielding the intersection point, we first store and query the intersection calculation in the temporary list before obtaining all intersection points in one cutting, then update the vertex list to avoid redundant and incorrect calculations.

Step3: Update faces of child CPH

Based on steps 1 and 2, the faces of the child CPH can be constructed. As can be seen from Figure 2, the vertices in the negative half space of π_i and the intersection points in the parent CPH constitute the vertex list of the child CPH. Hence, one can see that the faces in the child CPH can also be divided into two categories: one is the original face (remained face, e.g., $A'D'DA$ in Figure 2) or a part of the original face (renewed face, e.g., $A'ABP_{cs23}P_{cs22}$ in Figure 2) of the parent CPH, and the other is the new face of the intersection CPG, e.g., $P_{cs21}P_{cs22}P_{cs23}P_{cs24}$ in Figure 2. To construct the first type face, we need to delete the vertices of the parent CPH on the basis of steps 1 and 2, and reserve or add the vertices that do not belong to the data set P_{ri} to the child CPH. If the adjacent vertices lie on both sides of the CP, the indices of the intersection point $P_{csi} = \{P_{csi1}, P_{csi2}, \cdots, P_{csih}\}$ are added to the face list of the child CPH replacing data information of all P_{ri} and replenishing renewed vertex indices in the corresponding face list.

To construct the second type face, since a disordered vertices set of the intersection CPG has been obtained in step 2, and the data structure follows CCW order rule, it is necessary to reorder the CPG vertices obtained in step 2. It is difficult to sort a random point with a feature in 3D space even if they are in one plane. Therefore, utilizing a uniform linear coordinate transformation method, we hope that the 3D data description of the CPG vertices can be converted into a 2D data description in a plane to analyze where the centroid (x_c, y_c, z_c) of the transformed CPG coincides with the coordinate origin. If the normal vector N_i of the CPG, as shown in Figure 4, which is calculated from the vertices of CPG, is not perpendicular to any plane of XOY, YOZ and XOZ in frame O of CPH, it is possible to express the coordinate transformation between the frames O and O'. Let the centroid O' be the origin of the frame of the CPG, and the normal vector N_i be a coordinate axis. Referring to Figure 4, frame O' is obtained from the frame O by translating it along X, Y and Z axes by x_c, y_c and z_c, respectively, followed by two times rotation of β about Z and γ about X with respect to the current frames. Therefore, the 4×4 homogeneous transformation matrix can be written as $T = T_{P_c} \cdot R_\beta \cdot R_\gamma$ where

$$R_\beta = \begin{bmatrix} \cos\beta & \sin\beta & 0 & 0 \\ -\sin\beta & \cos\beta & 0 & 0 \\ 0 & 0 & 1 & 0 \\ 0 & 0 & 0 & 1 \end{bmatrix}, R_\gamma = \begin{bmatrix} 1 & 0 & 0 & 0 \\ 0 & \cos\gamma & \sin\gamma & 0 \\ 0 & -\sin\gamma & \cos\gamma & 0 \\ 0 & 0 & 0 & 1 \end{bmatrix}, T_{P_c} = \begin{bmatrix} 1 & 0 & 0 & x_c \\ 0 & 1 & 0 & y_c \\ 0 & 0 & 1 & z_c \\ 0 & 0 & 0 & 1 \end{bmatrix}$$

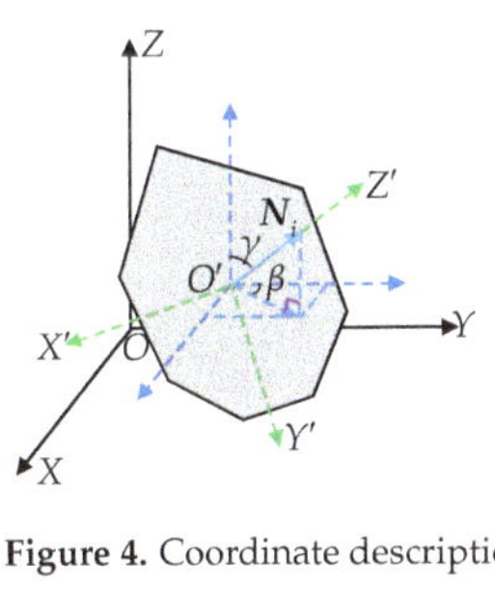

Figure 4. Coordinate description.

R_β, R_y and T_{P_c} denote the rotation and translation transformation matrices separately. Completing the transformation to reach frame O' from frame O, the vertices of CPG lie in plane $X'O'Y'$, $Y'O'Z'$ or $X'O'Z'$ of the coordinate system with the centroid origin, so that its vertices can be sorted simply according to the angle between the vector of the origin to the corresponding vertex and the Z', X' or Y' axis. Otherwise, the vertices of the CPG can be sorted directly based on the axes of the CPH without coordinate transformation. At this point, the vertex list of the second type face has been updated completely.

The dynamic updating algorithm of the CPH, i.e., the CPH list updating, can be summarized as follows (Algorithm 1):

Algorithm 1 CPH List Updating.

Input: Vertex-Face List of Q_{i-1}, π_i
Output: Vertex-Face List of Q_i
function ConstructCPHList(Vertex-Face List of Q_{i-1}, π_i)
  // Construct the Vertex-Face List of the first type face of Q_i
  for all F in Face List of Q_{i-1} **do**
    for all $vertex_{i-1,j}$ in F_m **do**
      $vertex_{i-1,j}$ ←Vertex of $Q_{i-1}[F_m[j]]$ //Index the corresponding vertex
      $vertex_{i-1,j+1}$ ←Vertex of $Q_{i-1}[F_m[j+1]]$
      if $\pi_i(vertex_{i-1,j}) \leq 0 \wedge \pi_i(vertex_{i-1,j+1}) \leq 0$ **then**//Two vertices are in negative half-space
        Vertex List of Q_i ←$vertex_{i-1,j}$, $vertex_{i-1,j+1}$
        Face List of Q_i ←indexes of $vertex_{i-1,j}$ and $vertex_{i-1,j+1}$
      else if $\pi_i(vertex_{i-1,j}) \leq 0 \wedge \pi_i(vertex_{i-1,j+1}) > 0$ **then**//Two vertices lie on either side of π_i
        $E_{i-1,j}$ ←*line*($vertex_{i-1,j}$, $vertex_{i-1,j+1}$)
        P_{csij} ←$E_{i-1,j} \cap \pi_i$ //Find the intersection of $E_{i-1,j}$ and π_i
        if $P_{csij} \notin$ *T empList* **then**
          T empList←P_{csij} //*T empList* is used to store the P_{csij}
          Vertex List of Q_i ←P_{csij}
        end if
        Face List of Q_i ←index of P_{csij}
      end if
    end for
  end for
  // Construct the Vertex-Face List of the second type face of Q_i
  [N_i]←ComputeNormalVector(P_{csi})
  [P_{csi}]←3DCoordinateTransformation(N_i, P_{csi}) //3D Coordinate transformation for vertex sorting
  Face List of Q_i ←index of P_{csi}
  return Vertex-Face List of Q_i
end function

4. Design Cuttable Big Blocks

In order to realize a high efficiency of cutting, we propose a strategy including a set of reasonable cutting schemes to ensure a large amount of cutting materials with a shorter cutting time. The cost of one cut is the MRR (i.e., the ratio of removal block volume to cutting time) originated by the saw disc intersecting with Q_{i-1}. Our objective is to find a series of cuts whose total cost–AMRR is relatively large. First, in order to obtain a large material removal amount, according to the geometrical characteristics of the symmetric convex bounding box and the BS, the removal vertices of the CPH are searched whose distance from the surface of the BS is at maximum. The direction is regarded as the normal vector of the CP for each cut. After that, the CP is constructed based on the extracted normal vector at the corresponding tangent point on the surface of the BS. It is noted that, in this process, to ensure a small machining allowance and no overcut at the same time, the data amount of the BS saved in the triangular mesh format with a certain accuracy is not optimistic. In particular, due to the increase in the convex vertices after being cut again and again, the search process causes a problem by requiring a large amount of calculation, thus increasing the burden of searching. The octree decomposition of the 3D model to reduce the computational complexity has been considered in many applications [18–20], such as image processing, collision detection algorithms, mesh generation procedures, and so on, which allows the search time to be easily reduced. For this reason, for each cutting, the algorithm of the octree partition is introduced to divide the OBB and BS into subdivisions at the same time, in order to search for the vertices of the CPH that are to be removed, which will reduce the load of the calculation and increase the speed and the search efficiency.

4.1. Octree Space Partition

To build an octree for the given set of 3D mesh points in the geometric space, firstly, it is necessary to decide the root node, which can be defined as an OBB. Then, we can subdivide it into multi-level equally sized cubes, called octants as shown in Figure 5, where each cube region in the space is a node of the octree [18,21]. Different from the binary tree, in which each node has two branches, each node of the octree has eight branches. The octree representation of 3D objects recursively subdivides the root cubic data into eight sub-cube arrays.

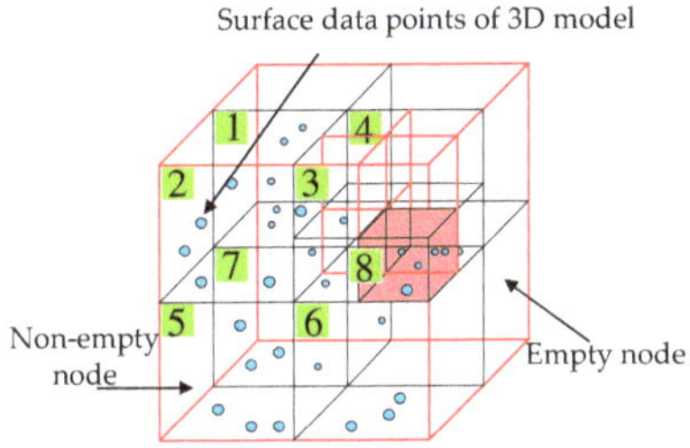

Figure 5. Schematic diagram of space partition.

Based on the above description, we need to build a compact OBB wrapping the BS with a reasonable machining allowance. Generally, the size of the blank workpiece is selected as the size of the OBB. From this point, a compact CPH Q_M, bounding the target BS, can be cut out from the blank workpiece (box). Therefore, we obtain the root node of the octree, which contains a data set P_Σ consisting of all points of the blank workpiece and spherical shell of the BS. Then, we calculate the geometrical centroid (x_{bc}, y_{bc}, z_{bc}) of the blank workpiece, which overlaps the center of the BS. Based on this, P_Σ can be generally decomposed. The subdivision nodes of the octree are recursively divided along the three coordinate axes according to the octree depth, which are determined according to the labeled sub-cube, while any point $P(x_s, y_s, z_s)$ of the data set are subdivided into a different sub-cube according to the coordinate location. When the number of the octree level increases, the cost of the octree storage occupation and node query time consumption will also increase. Comprehensively considering the factors of the search efficiencyand node query time in this paper, an octree with a depth of two levels and eight subspaces is established, which subdivides all the data in the data set into eight groups, assigning the node cubes $G_o(o = 1, 2, \ldots, 8)$ of the subdivision depending on the coordinate range, respectively, as in Equation (3). Since each octree node has eight branches, it is convenient to number a child node using an appropriate index ranging from one to eight. This index denotes the subregion covered by each child. An example of the space partition of a BB and BS is shown in Figure 6, which is used to reduce the calculation amount for searching in the subdivision region. The data points in each sub-cube are shown in a different color. For complex cases of octrees with many levels, we can divide subdivision by discussing solid angles distributed in a corresponding space region.

$$\begin{array}{l}
G_1 : \{P \in P_\Sigma : x_s < x_{bc} \wedge y_s < y_{bc} \wedge z_s > z_{bc}\} \\
G_2 : \{P \in P_\Sigma : x_s < x_{bc} \wedge y_s > y_{bc} \wedge z_s > z_{bc}\} \\
G_3 : \{P \in P_\Sigma : x_s > x_{bc} \wedge y_s > y_{bc} \wedge z_s > z_{bc}\} \\
G_4 : \{P \in P_\Sigma : x_s > x_{bc} \wedge y_s < y_{bc} \wedge z_s > z_{bc}\} \\
G_5 : \{P \in P_\Sigma : x_s < x_{bc} \wedge y_s > y_{bc} \wedge z_s < z_{bc}\} \\
G_6 : \{P \in P_\Sigma : x_s > x_{bc} \wedge y_s > y_{bc} \wedge z_s < z_{bc}\} \\
G_7 : \{P \in P_\Sigma : x_s < x_{bc} \wedge y_s < y_{bc} \wedge z_s < z_{bc}\} \\
G_8 : \{P \in P_\Sigma : x_s > x_{bc} \wedge y_s < y_{bc} \wedge z_s < z_{bc}\}
\end{array} \quad (3)$$

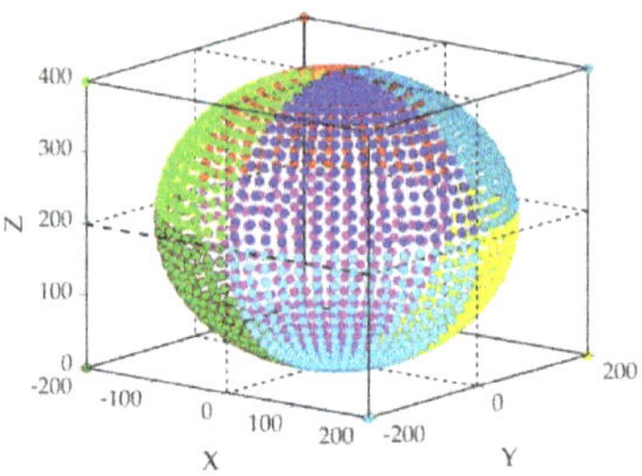

Figure 6. Space partition of the BB and BS.

4.2. Construct Cutting Plane

To achieve a short total cutting time, under the principle of "cutting big blocks as few times as possible" in the cutting process, we construct a series of CPs to ensure that the material amount is removed as much as possible. The idea is to intuitively remove the most convex vertex of the convex blocks from a practical point of view. Firstly, the maximum distance $d_{i\max}$ from the vertex of the parent CPH Q_{i-1} to the surface of the BS can be found. On the basis of Section 4.1, we can subdivide the vertices of Q_{i-1} into the sub-cube region G_o by the octree algorithm. Then, each d_o in the sub-cube G_o can be further calculated, where d_o is defined as the distance of a normal vector on the BS to the vertex of Q_{i-1} in the G_o. As shown in Figure 7, from the schematic diagram of cutting, we can see a data point on the surface of the BS is represented as P_{ti}, and a corresponding vertex of Q_{i-1} is represented as P_{vi}, where the normal vector of the tangent plane at P_{ti} is $\overrightarrow{P_{ti}P_{vi}}$ and the distance $|P_{ti}P_{vi}|$ is d_o in G_o. Choosing the maximum value of $\{d_o\}$ as the cutting thickness $d_{i\max}$ of the removal block, in that way the tangent plane of the BS at the corresponding P_{ti} is selected as the CP π_i. $\overrightarrow{P_{ti}P_{vi}}$ indicates $N_i(A_i, B_i, C_i)$ also.

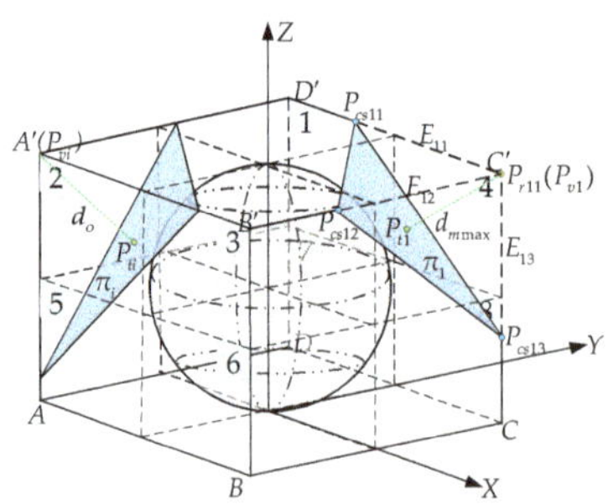

Figure 7. Diagram of constructing CP.

The coefficient of π_i in Equation (1) can be calculated based on the point P_{vi} or P_{ti} and normal vector $N_i(A_i, B_i, C_i)$. For instance, a CP π_i is shown in Figure 7 constructed in the sub-cube region G_4. The cutting along π_i will produce a big block removal effect by adjusting the orientation of the disc saw. Accordingly, using the CPH-updating algorithm of Section 3, the child CPH Q_i can be obtained.

4.3. Calculate Material Removal Volume (MRV)

Some methods have been proposed to calculate the volume of the CPH [16,22,23]. Combining these, we introduce a volume calculation method using vertex coordinates and face information. As above mentioned, the vertices of the removal block are $P_{ri} = \{P_{ri1}, P_{ri2}, \cdots, P_{rin}\}$ and $P_{csi} = \{P_{csi1}, P_{csi2}, \cdots, P_{csih}\}$; nevertheless, for the convenience of describing the calculation of the volume of the block to be removed, Q_{ri}, we assume it has n_l polygonal faces $S_{r1}, S_{r2}, \cdots, S_{rl}, \cdots$ each with a different amount of n_q vertices $P_{r1}, P_{r2}, \cdots, P_{rq}, \cdots$. The simplest contour of Q_{ri} is a pyramid feature. If Q_{ri} is not a pyramid, we can decompose the polyhedron Q_{ri} into multiple pyramids Q_{ril} with a common tip O_p in a geometrical centroid of Q_{ri} as shown in Figure 8. After that, to obtain the volume of the pyramid, if S_{rl} is not a triangle, we can take the polygon S_{rl} and decompose it into triangles from any vertex as a common vertex P_{r1}. The vertices $P_{r1}, P_{r,q-1}, P_{r,q}$ of each triangle are kept in a CCW order to the triangles with respect to their outward direction. Moreover, the triangle arrays are stored in the vertex order of CPG. Eventually, a pyramid Q_{ril} is decomposed into multiple tetrahedrons $O_p - P_{r1}P_{r,q-1}P_{rq}$ with a common pyramid tip. When O_p is assigned at the origin, one-sixth of the mixed product of $\overrightarrow{O_pP_{r1}}, \overrightarrow{O_pP_{r,q-1}}, \overrightarrow{O_pP_{r,q}}$, namely, one-sixth of the dot product between the normal vector of each triangle $\Delta(P_{r1}, P_{r,q-1}, P_{r,q})$ and any vector of $\overrightarrow{O_pP_{r1}}, \overrightarrow{O_pP_{r,q-1}}, \overrightarrow{O_pP_{r,q}}$, can be denoted as the determinant

$$\frac{1}{6}D(P_{r1}P_{r,q-1}P_{r,q}) = \frac{1}{6}\begin{vmatrix} x_{r1} & x_{r,q-1} & x_{r,q} \\ y_{r1} & y_{r,q-1} & y_{r,q} \\ z_{r1} & z_{r,q-1} & z_{r,q} \end{vmatrix}$$

which is the volume of $O_p - P_{r1}P_{r,q-1}P_{rq}$. Thus, by the vector method the following volume equation of the pyramid Q_{ril} for the face S_{rl} can be derived

$$V_{S_{rl}} = \sum_{q=2}^{n_q} V(O_pP_{r1}P_{r,q-1}P_{rq}) = \sum_{q=2}^{n_q} \frac{1}{6}D(P_{r1}P_{r,q-1}P_{r,q}) \tag{4}$$

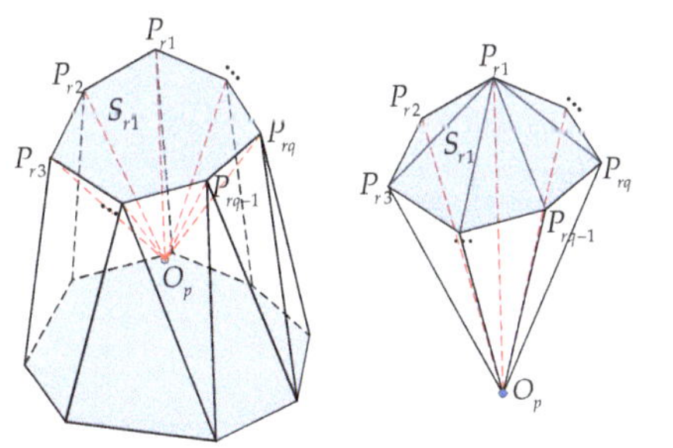

Figure 8. CPH decomposition and face decomposition.

Obviously, the volume equation of the polyhedron Q_{ri} is

$$V_{ci} = \sum_{l=1}^{n_l} V_{S_{rl}} = \sum_{l=1}^{n_l}\sum_{q=2}^{n_q} \frac{1}{6}D(P_{r1}P_{r,q-1}P_{r,q}) \tag{5}$$

From the above, we can see that it is easier to calculate the CPH volume from the face index and all vertex information on the faces with less calculation. Of course, one can also finally calculate the total removal volume directly from the remaining block. Here, in order to observe each cutting, we choose to analyze the removal block each time.

5. Optimize Cutting Time

When the material removal volume is constant, the shorter the cutting time is and the greater the MMR is. In order to minimize the cutting time, we summarize the cutting process into the second problem finding the shortest path. By analyzing the geometrical characteristics of the CPG, a path optimization algorithm to determine the feed direction and cutting point is proposed.

5.1. Time Optimization Model

By cutting with a guillotine style accompanying the aforementioned strategy, the CPH is completely separated into two convex blocks: one is the removal block Q_{ri} and the other is the child CPH Q_{i-1}. To sum each V_{ci} of Q_{ri}, we can obtain the total MRV. Due to the cutting block being relatively big, we establish the objective function with minimum cutting time as follows

$$\begin{aligned}
&\min \sum_{i=1}^{M} (T_i),\ T_i = \frac{L_i}{v_f} \\
&s.t.\ \pi_i = K_{Ai} \cdot x_i + K_{Bi} \cdot y_i + K_{Ci} \cdot z_i + K_{Di} \\
&Q_0 = \bigcap_{m=1}^{z} F_{0m} \\
&\{S_i, Q_i\} = Q_{i-1} \cap \pi_i \\
&L_i = W(S_i) \\
&L_i < R
\end{aligned} \tag{6}$$

where M is the aforementioned number of cuts, generated automatically according to the required material removal percentage (MRP). T_i is the time consumed at ith cutting. L_i is the feed stroke along each cutting path. v_f is the feed speed, which is normally set as a constant. R is the radius of the circular saw blade. F_{0m} represents a face of the Q_0, and m is the face number of the initial blank box, selected as six here. S_i is the intersection CPG generated by the ith cut. The width W of S_i needs to be calculated by the optimization algorithm.

Assuming the time consumption of the motion in space is ignored, the cutting time is only related to the feed stroke along the intersection with the CPG, which is decided by the geometrical characteristics of the CPG satisfying the shape constraint of the circular saw at the same time.

5.2. Intersection CPG Analysis

It can be seen from Section 5.1 that in order to obtain the shortest cutting time, it is necessary to seek out the shortest cutting path on the CPG. Through analysis, it can be found that here the path is generally characterized by the span of the CPG along a certain direction. Therefore, confirming the shortest cutting path can be worked out using the minimum span (i.e., width) of the CPG [24]. In recent years, the width calculation of the CPG has been widely used in collision detection and other calculations [19], but few researches and applications have explored its use in the field of machining. For the intersection CPG S_i, its width is defined as the minimum distance between the supporting parallel lines of S_i, decided by the vertex–edge (V–E) pairs here, in which the relative edge is formed by the intersection of two faces. The vertices set of the CPG is $P_{csi} = \{P_{csij}\}, j = 1 \cdots h$. The width calculation is as follows. Utilizing the supporting parallel lines, the V–E pairs can be scanned to obtain the maximum distance $D_j = \max\{D_{j,j+1}\}$ between any vertices of P_{csi} and one edge or extended edge $E_{j,j+1}$ in $O(h)$ time for each edge, which is the span of one edge. Once the scan is completed, we can compare each of these pairs $\{D_j\}$ and note the smallest span distances, $\min\{D_j\}$. As a result, that distance is the width $W(S_i)$, whose direction decides an initial feed direction. Here, the corresponding vertex is represented by P_{dj}, and the corresponded edge is E_w. For instance, in Figure 9, for the intersection CPG $P_{cs1}P_{cs2}P_{cs3}P_{cs4}$, the width is $D_{4,3}(W)$.

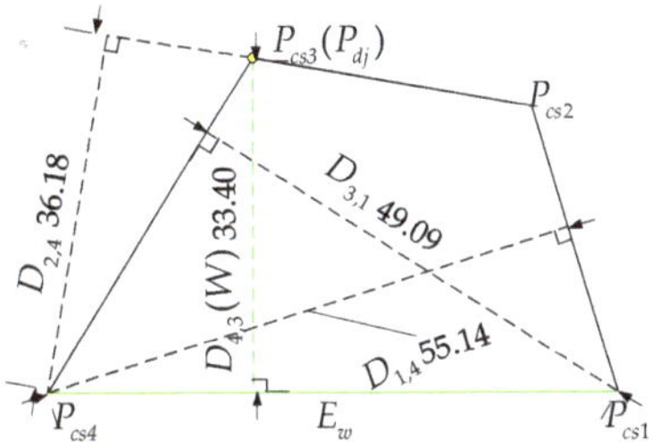

Figure 9. The width of intersection CPG.

5.3. Cutting Path Optimization

According to the analysis of the CPG in Section 5.2, if the width direction of the CPG is used as the cutting feed direction, and the relative vertex and the perpendicular point on the edge are chosen as the starting point and ending point of the cutting, respectively, then sometimes, the cutting requirements are unable to be met. In other words, the cutting range of the circular saw is not able to cover the whole CPG area without a guillotine cut; an example as shown in Figure 10a. Therefore, a bounding rectangle method of the CPG is proposed to determine the feed direction and starting point for the cut, as shown in Figure 10b, where the black dash line is the bounding rectangle of the CPG, and the red dash line demonstrates the feed direction. From the length E_w and width W of the bounding rectangle, we can calculate its centroid O_r. Along O_r, the vertical line to E_w can be drawn, intersecting E_w with point P_s, and intersecting the opposite edge with point P_e. P_s and P_e are chosen as the starting point and the ending point of the cut respectively. The vector direction from P_s to P_e is the optimal feeding direction here. If the cutting with P_e fails to cut through the block, a certain cut depth compensation should be considered to recalculate P_e, which can be found by calculating the intersection chord length between the edge P_eP_{dj} and the circular saw, achieved by using the bottom vertices of the polygon.

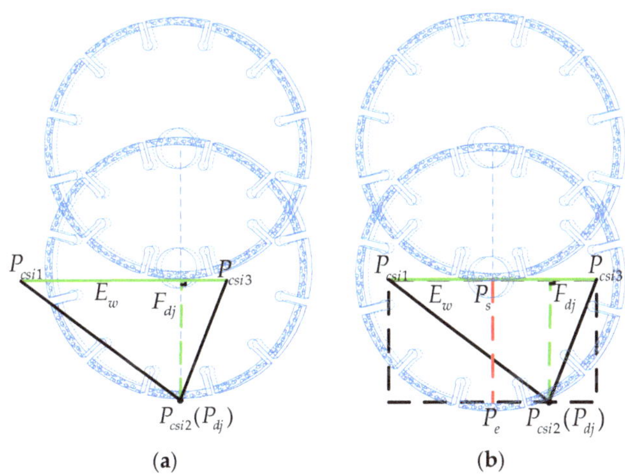

Figure 10. Feed direction of cutting. (**a**) Feed direction along W; (**b**) feed direction along $\overrightarrow{P_sP_e}$.

6. Simulation Results and Analysis

6.1. Simulation Verification

A minimum BS of a three-dimensional penguin is used as the target for the block cutting simulation. The triangular mesh model of the BS with a radius 200 mm is generated by using CAD/CAM software, which is stored in STL format. All of the data is loaded on to the MATLAB platform, which is used to verify the proposed strategy, thus benefiting from the powerful computing and drawing ability of MATLAB. In order to avoid any impact on the finished machined product due to the brittle fracturing of materials in the machining process, and ensuring a certain machining allowance, we select a $400 \times 400 \times 400$ mm^3 blank box, considering that polyhedron is able to be cut by a large enough circular saw. To verify the proposed strategy for the general cutting condition, the sawing parameters are given as follows: a rotation speed of the saw blade is 1400 r/min; the radius of the circular saw blade is 400 mm; the feed speed is set to 180 mm/min. When the MRP reaches 85% of the total materials that should be removed, the cutting search stops. Importing the data set of the BS, the data of the spatial region is subdivided according to the octree algorithm in Section 4.1. For contrast, the simulation experiments are carried out with and without the octree partition. The search time after data partition is reduced by 31.64% compared with that without data partition. Some effect graphs of the workpiece's dynamic reconstruction during cutting are shown in Figure 11, resulting in a total of 19 cuts. It can be seen from Figure 11c that the contour of the CPH after cutting is closer to the target BS. Through the visualization analysis, it can be seen that there is no overcutting phenomenon in the sawing process, and the proposed methods are feasible. Figure 12 shows the cutting time after optimization, which is less than or equal to that before optimization. Moreover, we can find that the total cutting time after path optimization is about 21.3 min, which is about 11.33% less than that without optimization. In Figure 13, the MRV based on the maximum removal thickness and MRR are shown, where, in order to show these clearly, the blue solid line and red dash line have been employed to illustrate them with different labels on the left and right longitudinal axes, respectively. One can see that after nine cuts, more than 70% of MRP has been reached. After this the increase in MRP becomes slower with the increase in cutting time, i.e., the removal volume of each cut becomes smaller. This is an inevitable result of block cutting, in which each cut leads to the rough blank moving closer and closer to the target BS. By balancing the cutting times and the removal volume, in this study, we targeted 85% of the total MRP according to some engineering experience in the milling process [5] and the characteristics of the saw disc. After 19 cuts, the MRV reaches 2.6×10^7 mm^3, accounting for about 85.34% of the total blank materials that should be removed. Accordingly, there is no need to set a higher desired MRP for rough cutting. Through the further analysis of Figure 13 (the 12th–14th cuts bring an increase in MRV), it can be seen that the cutting algorithm based on the maximum thickness of the removal block cannot completely ensure the complete removal of the material for each cutting, but it is nearly at maximum, which is why we also call this process the relatively big block cutting method. The reason that this phenomenon arises is because the contour of the removal block becomes more and more irregular with the increase in the number of cuts. Therefore, the maximum thickness of the removed block materials only reflects a relatively big block not the maximum removal amount.

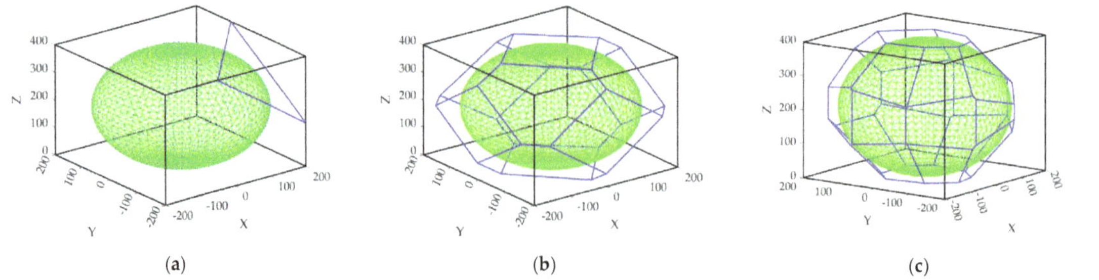

Figure 11. Dynamic reconstruction effect of CPH. (**a**) 1st cutting; (**b**) 8th cutting; (**c**) 19th cutting.

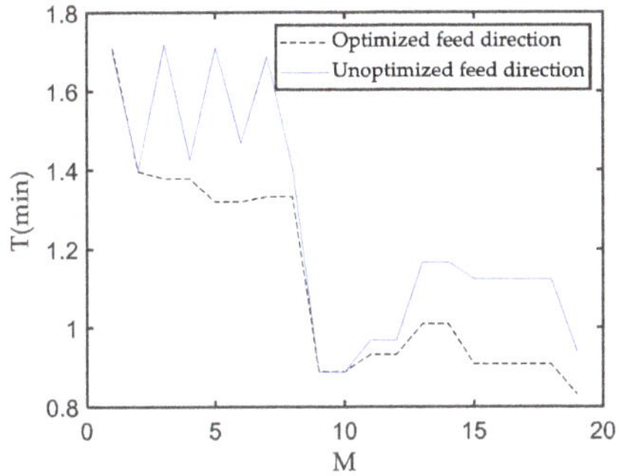

Figure 12. Cutting time.

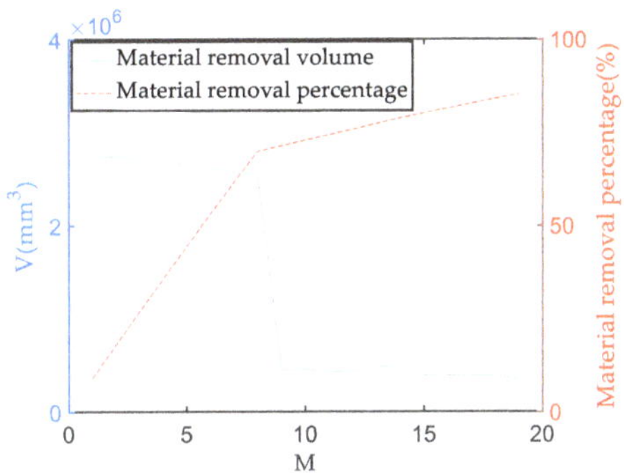

Figure 13. MRV and MRP.

Additionally, an accurate simulation model of the NC (numerical control) machine tool is established with a 5-axis cutting characteristic and circular saw blade as a simulation platform in a Vericut environment. Given the same cutting parameters as above mentioned, the cutting simulations are performed by a generated G code on the NC cutting machine model to verify the proposed cutting strategy. Some cutting results are shown in Figure 14. The cutting time displayed in the Vericut environment is about 23.4 min. Ignoring the travel time in space, it is almost the same as the cutting time in the MATLAB platform. Moreover, compared with the reconstructed CPH of each cut in the MATLAB platform, the features of each CPH processed by the 5-axis NC machine tool in the Vericut platform are roughly the same in shape and size, which further verifies the effectiveness and feasibility of the cutting strategy for removing blocks proposed in the paper.

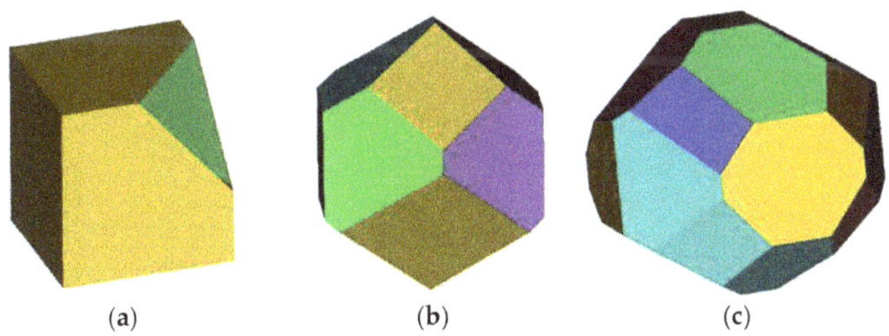

Figure 14. Cutting effect in Vericut platform. (**a**) 1st cutting; (**b**) 8th cutting; (**c**) 19th cutting.

6.2. Comparison of Sawing and Milling

For the removal of materials during stone rough machining, a milling mode is often employed in the newest automatic level. In order to verify the effectiveness of the block cutting strategy proposed in the paper, the milling mode with a high capacity during stone rough machining is chosen as a comparison. The milling parameters are given as follows [5]: a rotation speed of 6000 r/min; a milling depth of $a_p = 2$ mm; a milling width of $a_w = 30$ mm, which is the tool diameter of a diamond grinding wheel; a feed speed of $v_f = 4000$ mm/min. For the same MRV 2.6×10^7 mm^3, the milling time is 108.4 min and the AMRR is 2.4×10^5 mm^3/min. In this case, compared with the cutting results, it can be seen that the AMRR of the cutting strategy is more than five times that of the milling. The main reason of the difference is that milling is limited by its processing mode and milling depth. For the case of removing the materials of a big block, layered milling must be adopted, which consumes a lot of time and produces dust. However, the block sawing method with a saw blade can directly carry out the cutting operation with a large feed stroke and a large block thickness; thereby, the efficiency can be significantly improved. For special shapes, if a higher cutting performance saw blade [25] is employed, AMRR will be improved further. More tool cases with different machining capacities in the simulation will be performed in the future.

7. Conclusions and Future Scope

To address block cutting with a saw disc in the 3D space usually needs with a lot of hard labor and time. This article has addressed a series of works concerned with analyzing the geometrical characteristics of convex polyhedrons and convex polygons in order to complete block cutting automatically and rapidly. This has made efficient and continuous block cutting available. The optimization strategy for cutting a convex polyhedron out of a blank box has been presented by combining computational geometric theory and computer graphics knowledge. Dynamic cutting data management has been implemented with the vertex–face polyhedron data structure, which means the convex polyhedron model reconstruction is completed in the updating process. A range of cutting planes with the maximum thickness of the removal block have been constructed, and the space partition with the octree algorithm has been used in the process of searching for the vertices to be removed, which can reduce the search time. The geometrical characteristics of the convex polygon cutting plane generated by the cutting intersection have also been analyzed, and the method for an optimized cutting time has been presented. Finally, the suboptimal solution of the average material removal rate at the rough machining stage has been quantitatively analyzed. Simulation and comparison results in MATLAB and the Vericut platform have been provided to demonstrate the effectiveness of the proposed strategy. In particular, the Vericut platform is able to reflect the real processing environment. We have investigated a block cutting strategy with a practicable automatic strategy for the first time. Realistically, it is necessary to utilize multi axis machine tools or robots with good capacity.

In this work, our discussion concentrated on the problem of cutting a symmetric convex polyhedron with a bounding sphere target. However, there are some directions that can be extended further: (i) While we constructed cutting planes, we selected the suboptimal solutions with relatively big blocks and high efficiency. This leads us to think deeply about the optimal methods for constructing a cuttable maximum block for each cutting; (ii) as well as typical bounding sphere targets, we can explore the changing factors that affect the AMRR when cutting the symmetric convex polyhedron with bounding ellipsoid targets; (iii) as only one high capacity milling comparison has been completed, some different saw disc parameters and milling modes can be analyzed to obtain a more general conclusion of the higher MRR with block cutting under a similar mechanical level; (iv) instead of the bounding sphere and the bounding ellipsoid, we can explore cutting strategies according to the target polyhedrons of the compact bounding convex or nonconvex contours of special-shaped products.

Finally, we hope this work can stimulate research and applications in this field.

Author Contributions: Conceptualization, H.S. and Q.L.; writing—original draft preparation, H.S., Q.L. and Z.G.; writing—revision, H.S., Q.L. and Z.G.; supervision, H.S.; project administration, H.S.; All authors have read and agreed to the published version of the manuscript.

Funding: This research was funded by the Natural Science Foundation of Fujian Province, China, grant number 2021J01291; and the Science and Technology Planning Project of Quanzhou City, Fujian Province, China, grant number 2017T001.

Institutional Review Board Statement: Not applicable.

Informed Consent Statement: Not applicable.

Data Availability Statement: Not applicable.

Acknowledgments: The authors would like to acknowledge the research support from the Natural Science Foundation of Fujian Province, China, grant number 2021J01291; the Science and Technology Planning Project of Quanzhou City, Fujian Province, China, grant number 2017T001; the research support from Northumbria University, UK.

Conflicts of Interest: The authors declare no conflict of interest.

References

1. Bai, S.W.; Qiu, H.; Rabe, L.; Jia, S.; Elwert, T.; Wang, Q.Y.; Wang, Q.; Sun, Y. A study on energy-saving optimization strategy for the stone processing industry—An improved method for modeling cutting power and energy consumption: A case study of block sawing process. *J. Clean. Prod.* **2021**, *300*, 126922. [CrossRef]
2. Şirin, E.; Bonduà, S.; Elkarmoty, M. Environmental and economic optimization for block cutting of dimension stones in a limestone quarry. *Resour. Policy* **2021**, *74*, 102396. [CrossRef]
3. Kim, D.; Kim, H.; Lee, S. Characterization of diamond wire-cutting performance for lifetime estimation and process optimization. *J. Mech. Sci. Technol.* **2016**, *30*, 847–852. [CrossRef]
4. Kumar, V.; Saraswat, P.; Kumar, A. Experimental investigation of rotary ultrasonic face milling on red granite: A comparison with conventional grinding. *Mater. Today Proc.* **2022**. [CrossRef]
5. Yin, F.C.; Ji, Q.Z.; Wang, C.Z. Research on machining error prediction and compensation technology for a stone-carving robotic manipulator. *Int. J. Adv. Manuf. Technol.* **2021**, *115*, 1683–1700. [CrossRef]
6. Buyuksagis, I.S.; Rostami, J.; Yagiz, S. Development of models for estimating specific energy and specific wear rate of circular diamond saw blades based on properties of carbonate rocks. *Int. J. Rock Mech. Min. Sci.* **2020**, *135*, 104497. [CrossRef]
7. Gao, Z.; Chen, M.Z.Q.; Zhang, D. Special Issue on "Advances in Condition Monitoring, Optimization and Control for Complex Industrial Processes". *Processes* **2021**, *9*, 664. [CrossRef]
8. Yamada, Y.; Sasahara, H. Free-form curves cutting using flexible circular saw. *Precis. Eng.* **2014**, *38*, 611–616. [CrossRef]
9. Demaine, E.D.; Demaine, M.L.; Kaplan, C.S. Polygons cuttable by a circular saw. *Comput. Geom.* **2001**, *20*, 69–84. [CrossRef]
10. Jaromczyk, J.W.; Kowaluk, M. Sets of lines and cutting out polyhedral objects. *Comput. Geom.* **2003**, *25*, 67–95. [CrossRef]
11. Van Sosin, B.; Bartoň, M.; Elber, G. Accessibility for line-cutting in freeform surfaces. *Comput.-Aided Des.* **2019**, *114*, 202–214. Available online: https://www.researchgate.net/publication/332670481 (accessed on 13 January 2022). [CrossRef]
12. Dumitrescu, A. An approximation algorithm for cutting out convex polygons. *Comput. Geom.* **2004**, *29*, 223–231. [CrossRef]
13. Tan, X.; Wu, G. Approximation algorithms for cutting a convex polyhedron out of a sphere. *Theor. Comput. Sci.* **2013**, *508*, 66–73. [CrossRef]
14. Zhang, Q.H.; Su, H.D.; Lin, S.Z.; Shi, G.H. Algorithm for three-dimensional curved block cutting analysis in solid modeling. *Comput. Methods Appl. Mech. Eng.* **2020**, *360*, 112721. Available online: https://www.researchgate.net/publication/337304188 (accessed on 13 January 2022). [CrossRef]
15. Boon, C.W.; Houlsby, G.T.; Utili, S. A new rock slicing method based on linear programming. *Comput. Geotech.* **2015**, *65*, 12–29. [CrossRef]
16. Chiodi, R.; Desjardins, O. General, robust, and efficient polyhedron intersection in the Interface Reconstruction Library. *J. Comput. Phys.* **2022**, *449*, 110787. [CrossRef]
17. Chen, Z.; Tagliasacchi, A.; Zhang, H. Learning mesh representations via binary space partitioning tree networks. *IEEE Trans. Pattern Anal. Mach. Intell.* **2020**, *1*, 1. [CrossRef]
18. Kim, Y.H.; Ko, S.L. Improvement of cutting simulation using the octree method. *Int. J. Adv. Manuf. Technol.* **2006**, *28*, 1152–1160. [CrossRef]
19. Zhou, K.; Gong, M.; Huang, X. Data-parallel octrees for surface reconstruction. *IEEE Trans. Vis. Comput. Graph.* **2010**, *17*, 669–681. [CrossRef]
20. Xu, J.; Liu, Z.; Yang, C. A pseudo-distance algorithm for collision detection of manipulators using convex-plane-polygons-based representation. *Robot. Comput. -Integr. Manuf.* **2020**, *66*, 101993. [CrossRef]
21. Trinidad, M.C.; Vinacua, A.; Carruesco, A. Sweep Encoding: Serializing Space Subdivision Schemes for Optimal Slicing. *Comput.-Aided Des.* **2022**, *146*, 103189. [CrossRef]

22. Goldman, R. IV.1 Area of Planar Polygons and Volume of Polyhedra. In *Graphics GEMS II*, 2nd ed.; Arvo, J., Ed.; Academic Press/Morgan Kaufmann: San Francisco, CA, USA, 1991; pp. 169–171. [CrossRef]
23. Berg, M.; Cheong, O. *Computational Geometry: Algorithms and Applications*, 3rd ed.; Springer: Berlin/Heidelberg, Germany, 2008; pp. 1–17.
24. Lassak, M. Spherical Geometry—A Survey on Width and Thickness of Convex Bodies. In *Surveys in Geometry*; Papadopoulos, A., Ed.; Springer: Cham, Germany, 2022; pp. 7–47. [CrossRef]
25. Sánchez Egea, A.; Martynenko, V.; Martínez Krahmer, D.; López de Lacalle, L.; Benítez, A.; Genovese, G. On the Cutting Performance of Segmented Diamond Blades When Dry-Cutting Concrete. *Materials* **2018**, *11*, 264. [CrossRef] [PubMed]

processes

Article

Successful Pass Schedule Design in Open-Die Forging Using Double Deep Q-Learning

Niklas Reinisch *, Fridtjof Rudolph, Stefan Günther, David Bailly and Gerhard Hirt

Institute of Metal Forming, RWTH Aachen University, 52072 Aachen, Germany;
fridtjof.rudolph@ibf.rwth-aachen.de (F.R.); stefan.guenther@ibf.rwth-aachen.de (S.G.);
david.bailly@ibf.rwth-aachen.de (D.B.); gerhard.hirt@ibf.rwth-aachen.de (G.H.)
* Correspondence: niklas.reinisch@ibf.rwth-aachen.de; Tel.: +49-(0)241-8097624

Abstract: In order to not only produce an open-die forged part with the desired final geometry but to also maintain economic production, precise process planning is necessary. However, due to the incremental forming of the billet, often with several hundred strokes, the process design is arbitrarily complicated and, even today, often only based on experience or simple mathematical models describing the geometry development. Hence, in this paper, fast process models were merged with a double deep Q-learning algorithm to enable a pass schedule design including multi-objective optimization. The presented implementation of a double deep Q-learning algorithm was successfully trained on an industrial-scale forging process and converged stably against high reward values. The generated pass schedules reliably produced the desired final ingot geometry, utilized the available press force well without exceeding plant limits, and, at the same time, minimized the number of passes. Finally, a forging experiment was performed at the institute of metal forming to validate the generated results. Overall, a proof of concept for the pass schedule design in open-die forging via double deep Q-learning was achieved which opens various starting points for future work.

Keywords: open-die forging; reinforcement learning; double deep Q-learning; process design; process optimization

Citation: Reinisch, N.; Rudolph, F.; Günther, S.; Bailly, D.; Hirt, G. Successful Pass Schedule Design in Open-Die Forging Using Double Deep Q-Learning. *Processes* **2021**, *9*, 1084. https://doi.org/10.3390/pr9071084

Academic Editor: Zhiwei Gao

Received: 25 May 2021
Accepted: 18 June 2021
Published: 22 June 2021

1. Introduction

1.1. Process Design in Open-Die Forging

Open-die forging is an incremental forming process to produce predominantly longitudinally oriented components such as turbine shafts. Despite its long history, the open-die forging process has not lost any of its importance to this day. Open-die forging is characterized by the fact that not only a defined final geometry is generated, but excellent mechanical properties are also produced. Due to the incremental forming of the workpiece by commonly hundreds of strokes, the high flexibility and, at the same time, the high complexity of the process design in open-die forging arises.

In order to ensure consistent quality in industrial production, forging processes are usually designed in advance. Important process parameters, such as height reductions and bite ratios, are summed up in a so-called "pass schedule" that serves as a template for the forming process and can also be used directly to pilot the forging plant. Even today, the design of pass schedules is often based on the experience of long-standing employees or on simple mathematical models, e.g., based on volume constancy and the Tomlinson and Stringer [1] spread formulas. Although, processes designed through this approach achieve the final geometry accurately, they do not consider the expected component quality and the efficient utilization of the machine's limits.

Pass schedule calculation programs can improve on this by including not only the geometry development but also other workpiece and process parameters, such as temperature, deformation resistance, or press force, in the pass schedule design. In this way, the generated pass schedules lead to components with the desired final geometry and, at the same time, comply with restrictions such as a maximum press force or a permissible temperature range. In addition, the consideration of higher-level objectives, such as an efficient production process by minimizing the process time, is possible. Subsequently, some typical approaches for pass schedule design in open-die forging are listed.

In 1978, Mannesmann presented the pass schedule calculation program "MD-Dataforge" [2] without publishing the underlying calculation equations. The program is able to determine the workpiece dimensions depending on the material, the temperature as well as height reduction and manipulator feed. In addition, it is checked whether the forming forces exceed the maximum press force.

Bombač et al. [3] presented the pass schedule calculation program "HFS" that uses the spread equation according to Tomlinson and Stringer (cf. Equations (1)–(3)) and a temperature calculation that is not specified in detail. The comparison of an ad hoc forging of a very experienced press operator and the pass schedule calculation program HFS shows that the pass schedules made by HFS are of at least as good quality.

Kakimoto et al. [4] integrated empirically determined results for the workpiece spread into the "Forging Guidance System (FGS)", which allows pass schedule calculations from square to octagon and from octagon to round. After the development of an alternative spread equation and the implementation into the pass schedule calculation, the geometry model achieved an accuracy of 2% [5].

The program "ForgeBase®" [6] from SMS group GmbH resulted from the programs MD-Dataforge [2] and "ComForge" [7] and allows for the design of pass schedules under consideration of plant and material restrictions such as a permissible maximum press speed or a maximum height reduction. The user can select between different block cross-sections and forging strategies and defines the boundary conditions to be observed. In addition, pass schedules calculated with ForgeBase can be used directly to control the open-die forging plant during the process, whereby the press operator can select freely between either fully automatic, semi-automatic, or manual mode. Furthermore, previous versions of ForgeBase offered the option to optimize the generated pass schedules regarding, for example, a maximum throughput [8].

Using Tomlinson and Stringer's expansion equation (cf. Equations (1)–(3)), Yan et al. [9] calculated pass schedules whereby the intermediate geometries for each stroke were simplified as rectangles. The simple functionality of the program implemented in MathWorks MATLAB was expanded by the informative graphical representation of individual strokes.

1.2. Machine Learning for Process Planning and Optimization in Metal Forming

Quality requirements for dimensional accuracy and material properties of forged components are continuously increasing and, therefore, the tools for process design and monitoring need to evolve accordingly. Because methods of artificial intelligence and, in particular, artificial neural networks are able to map highly non-linear interrelations well, in recent years, they have been applied progressively in the field of process design in metal forming technology. For example, Sauer et al. [10] developed deep neural networks that are able to locally predict properties such as the true strain of sheet-bulk metal parts. This locally resolved workpiece information is then integrated into the proprietary program environment SLASSY (self-learning assistance system) that supports product developers in the creation of sheet-bulk metal parts [11]. Addona et al. trained neural networks on the basis of data sets from FE simulations to generate meta models for wear-determining parameters of tools in closed-die forging [12]. These meta models then replace the computationally intensive FE simulations in a sequential optimization of the process design of the closed-die forging process.

Methods of artificial intelligence are also applied in sheet metal forming. Guo et al. as well as Dornheim et al. coupled a deep reinforcement learning algorithm to an FE simulation of a deep drawing process and trained it to regulate the blankholder force during the process. Hereby, the algorithm was trained to control the blankholder force in such a way that the finished deep-drawn part exhibited as little internal stress as possible as well as a sufficient thickness everywhere, while the manufacturing process was as material-efficient as possible [13–15]. Furthermore, Störkle et al. used a reinforcement learning algorithm to control the springback of the material in incremental sheet metal forming processes. The algorithm does not directly calculate the tool path but creates a modified geometry from the desired geometry that is then fed into a path planning software. The path generated in this way does not create the modified geometry but rather the desired initial geometry due to the springback [16].

Kim et al. [17] trained separate functional link neural networks (FNNs) with a single layer of neurons each to predict the resistance to forming, the spread, and the forming speed in open-die hammer forging depending on the current ingot height and width, the temperature, the height reduction, and the bite length. The training data was gathered during industrial forging processes, though no information about the individual processes and the quantity of the data was given. Furthermore, besides the results of the spread prediction using an FNN, no training results or validations were presented. Finally, an algorithm for a pass schedule calculation was described that used the FNNs to design forging processes with the greatest possible height reduction while maintaining the given force limit. This algorithm has not been validated though.

In 2018, Meyes et al. [18] described the successful use of reinforcement learning algorithms for pass schedule design in hot rolling. The database used for the training, consisting of information about the workpiece and its changes within a pass, was generated by fast process models developed at the Institute of Metal Forming (IBF). The trained algorithm can design pass schedules for rolling processes that generate a workpiece with the desired final height using as few passes as possible and, at the same time, sets a predefined grain size.

1.3. Reinforcement Learning for Pass Schedule Design in Open-Die Forging

The fast process models for open-die forging presented in Section 2.1 open the possibility to predict the process and workpiece properties that result from a pass schedule quickly and, thus, to include the workpiece properties directly in the pass schedule design. In this paper, this was solved by coupling the fast models with a double deep Q-learning (DDQL) algorithm. During the training phase, the algorithm learned how to design pass schedules that, on the one hand, maintained a maximum process force and, on the other hand, led to the desired geometry as efficiently as possible. After completion of the training process, the user should be able to select any ingot geometry within the selected process window and quickly receive an optimized pass schedule. Here, the optimization goals are closely aligned to the possibilities offered by existing software for the pass schedule design mentioned above, so that comparability is ensured and, furthermore, a successfully trained DDQL algorithm can serve as a proof of concept for optimized pass schedule design in open-die forging using reinforcement learning.

For the here presented implementation of an optimized pass schedule design in open-die forging using a reinforcement learning algorithm, a rather short and simple process with approximately six passes and rectangular ingot cross-sections were selected. This enabled a broad analysis of the complex correlations between the algorithm's behavior and its effects on the forging process, while the training durations of the required multiple training processes remain acceptable.

In future work, the presented concept will be applied on more complex forging processes featuring for example, octagonal ingot cross-sections or longer processes. Furthermore, this reinforcement learning algorithm can be extended to include, for example, the microstructure evolution into the optimization. At the same time, optimized pass schedules can be calculated very quickly so that further development of the DDQL algorithm towards a process parallel control system regarding the microstructure evolution seems possible.

2. Process, Materials, and Methods

2.1. Fast Process Modeling of Geometry, Temperature, and Equivalent Strain

The ad hoc production of any workpiece geometry with good mechanical properties requires a forging press operator to have great experience gained through a high number of trial-and-error forgings over the course of his work life. For this reason, various possibilities for process design have been established to describe the forging process in advance with regard to its main target variables: geometry, strain, and temperature. A central position is taken by finite element modeling (FEM), which, on the one hand, can calculate these parameters without restrictions for each point in the component. On the other hand, FEM requires a long calculation time, which motivates the development of faster calculation models for open-die forging. These models usually derive their high speed from the fact that compared to other calculation methods simplifying assumptions have been made which are permissible for the examined process and the examined target variables.

2.1.1. Geometry

Due to the use of geometrically simple tools and free workpiece surfaces (cf. Figure 1) the material flow in open-die forging is not predetermined but depends, among other things, on the shape and dimensions of the contact area between the ingot and the tools, the decrease in height, and on the material [19]. As shown in Figure 1, this contact area is defined by the bite length, s_{B0}, which corresponds to the manipulator feed, and on the ingot width, b_0, before the forging stroke was performed. Over the course of a stroke, the initial ingot height, h_0, is reduced to h_1, while the initial ingot width, b_0, increases to b_1 due to the spread. Here, the forging press applies the force, F, that is required to perform the forging stroke.

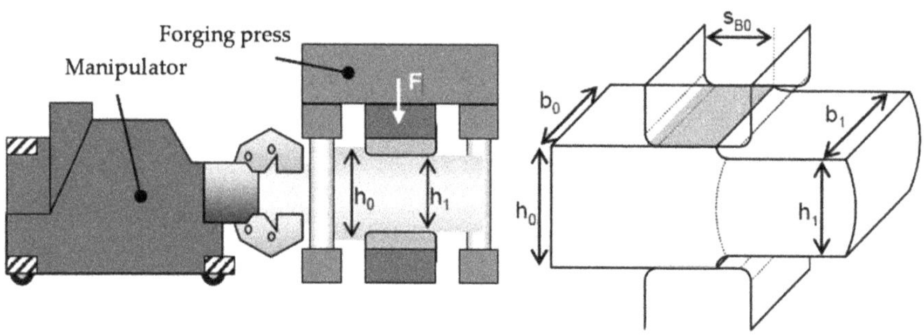

Figure 1. Outline sketch of an open-die forging process (**left**) including the most important geometric parameters (**right**) [20].

To calculate intermediate and final geometries during process design and taking the free material flow in open-die forging into account, various approaches to geometry calculation based on, for example, empirical equations or the upper bound method were established.

Tomlinson and Stringer [1,21] presented simple formulae to calculate the spread and elongation of rectangular blocks, which are nowadays among the most widely used approaches mainly because of their simplicity. Forgings of steel 1.0402 (C22) were carried out by varying temperature, pressed area (initial bite width/initial width ratio s_{B0}/b_0), specific height reduction ε_h, and initial width/start height ratio b_0/h_0. Afterwards, the measured change in length can be used to determine the mean width, b_1, of the block from Equation (1). The coefficient of expansion, s, is determined according to Equation (2). Due to the small influence of the height decrease, it can be neglected, see Equation (3).

$$\frac{b_1}{b_0} = \left(\frac{h_0}{h_1}\right)^s \tag{1}$$

$$s = 0.29 - 0.16\left(\frac{h_1}{h_0}\right) + 0.343\left(\frac{s_{B0}}{b_0}\right) - 0.048\left(\frac{s_{B0}}{b_0}\right)^2 \tag{2}$$

$$s = k_1 + k_2\left(\frac{s_{B0}}{b_0}\right) + k_3\left(\frac{s_{B0}}{b_0}\right)^2 \tag{3}$$

with $k_1 = 0.14$; $k_2 = 0.36$; $k_3 = -0.054$ for 1.0402.

The equation can also be used for alloyed steel by adjusting the parameters k_1, k_2, and k_3 [22] as shown in Table 1 for 42CrMo4.

Table 1. Tomlinson and Stringer parameters for 42CrMo4 [23].

Model	k_1	k_2	k_3	Δb_{rel}
Tomlinson Stringer, 2 parameters	0.160	0.260	0	3.9%
Tomlinson Stringer, 3 parameters	0.130	0.377	−0.075	2.0%

2.1.2. Temperature

The temperature calculation model presented by Rosenstock et al. [24] is based on the finite difference method (FDM). This approach was chosen since the 3D temperature field in open-die forging cannot be calculated with a simple analytical calculation, and it fulfills the condition of short calculation times. The FDM model considers all relevant phenomena such as convection, conduction, radiation, heat transfer to the surroundings and tools as well as dissipation during forging strokes based on the elementary theory of plasticity, see Equations (4) and (5).

$$\Delta\vartheta = 0.9 \cdot k_{res} \cdot \frac{|\varepsilon_h|}{\rho \cdot c_p} \tag{4}$$

$$k_{res} = k_f \cdot \left(1 + \frac{1}{2} \cdot \mu \cdot \frac{s_b}{h} + \frac{1}{4} \cdot \frac{h}{s_b}\right) \tag{5}$$

where k_{res} is the mean resistance to deformation of the stroke, k_f is the yield stress, ε_h is the true plastic strain in the height direction of the stroke, ρ is density, c_p is specific heat, s_b is bite length (cf. Figure 1), h is height, and μ is the friction coefficient. The temperature increase by dissipation $\Delta\vartheta$ was modeled following the elementary theory of plasticity, where a mean value for the deformed volume of each stroke was used.

Thermal conduction in the ingot was calculated by using the implicit Crank–Nicolson method, whereas thermal radiation was implemented as an explicit boundary condition of the model with the material parameters seen in Table 2. In addition, an emissivity of $\varepsilon = 0.85$, a heat transfer to the environment of $\alpha_e = 25\ \text{W}/(\text{m}^2 \times \text{K})$, and a heat transfer to the tools of $\alpha_t = 4500\ \text{W}/(\text{m}^2 \times \text{K})$ were assumed.

Table 2. Material parameters used in temperature model for 1.7725 [25].

ϑ	λ [1]	c_p	ρ
°C	W/(m × K)	J/(kg*K)	kg/m^3
800	23.44	697.1	7570.5
900	26.14	694.6	7517.4
1000	27.87	692.1	7461.9
1100	29.21	689.6	7404.2
1200	30.60	687.2	7344.4

[1] thermal conductivity.

2.1.3. Equivalent Strain

Fast models for the calculation of strain along the core fiber of a forged workpiece have been investigated on a wide range. First approaches were given by Stenzhorn [26,27] 40 years ago, which allowed for a rough estimation of the strain distribution by a linear interpolation of the values. Siemer [28] extended the model with an approximation of the evolution of the equivalent strain using a sin^2 function. Following the approach of Siemer, Franzke et al. [29] adapted a model that calculated the strain of the deformed zone between the forging dies based upon the length change during forging with two empiric parameters for two-dimensional cases. Recker et al. [30] applied this model to the calculation of strain distribution after more than one forging pass.

Since these models were developed for a two-dimensional case only, Recker [20] adapted the strain model towards three-dimensional forging processes by calculating the maximum equivalent strain of a single forging stroke based on the elementary theory, see Equation (6), with α_m as a material constant.

$$\varepsilon_{V,max} = 2 \cdot \alpha_m \cdot \ln\left(1 + \frac{\Delta l}{x_d}\right) \quad (6)$$

$$x_d = s_{B0}\left(q_{x_d} + 1\right) \quad (7)$$

He also enhanced the model to only require one empiric parameter and the bite ratio s_{B0}, cf. Equation (7). The remaining empiric parameter, q_{X_d}, shown in Table 3, which is the normalized length of the forming zone, was based on FEM studies and was used for the calculation of the forming zone length of one stroke x_d.

Table 3. Empiric parameter q_{Xd} for 1.7225 [23].

s_{B0}/h_o	0.2	0.3	0.4	0.6	0.6	0.95	1.1	1.3	1.6
q_{x_d} *for* **1.7225**	1.294	1.030	0.730	0.452	0.325	0.344	0.303	0.234	0.174

The principle of the calculation is shown in Figure 2. For each individual stroke, a $\sin^2$ distribution of equivalent strain, which is defined by $\varepsilon_{v,max}$ and x_d, was imposed on the core fiber. The maximum was placed in the center of the corresponding forming zone, while the equivalent strains for the single strokes were super-imposed for the full forging sequence to obtain the distribution for the whole process or intermediate process steps.

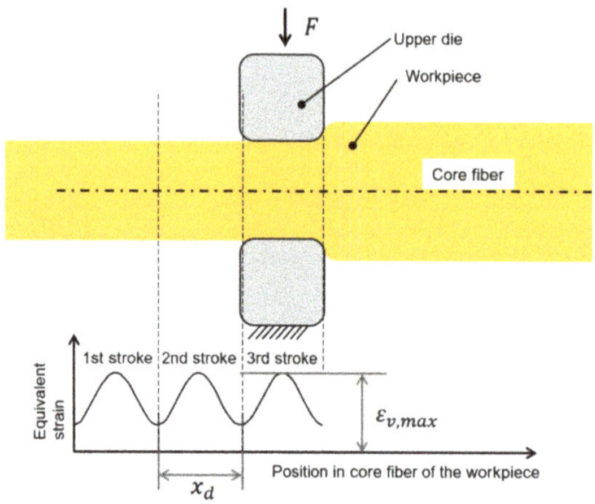

Figure 2. Calculation of the equivalent strain along the core fiber.

2.2. Reinforcement Learning

Reinforcement learning is a machine learning approach based on, among others, the "trial and error" mechanism of human as well as animal learning [31]. In this learning approach, a so-called agent interacts with its environment in a goal-oriented manner and, thus, in the long term learns to maximize the positive feedback received from the environment and to avoid negative responses.

Such goal oriented and interactive learning processes also take place in the production of open-die forged components. In order to avoid the constant repetition of already known misbehavior, a plant operator strives to adapt his behavior after negative events appear, e.g., the production of scrap. In this tangible example, the plant operator represents the agent that interacts with the forging plant via the selection of process parameters (e.g., the height reduction) in order to produce a defined component as efficiently as possible as shown in Figure 3 (variant A, red). After completion of the forging process, the plant operator receives feedback about the quality of the forged product from his environment. In the industrial everyday life, this feedback could be a visual check for surface cracks or the outcome of a final quality control that classifies whether a good part or, in the worst case, scrap was produced. In this way, the plant operator builds up a broad process knowledge over the years and learns to design the forging process reasonably.

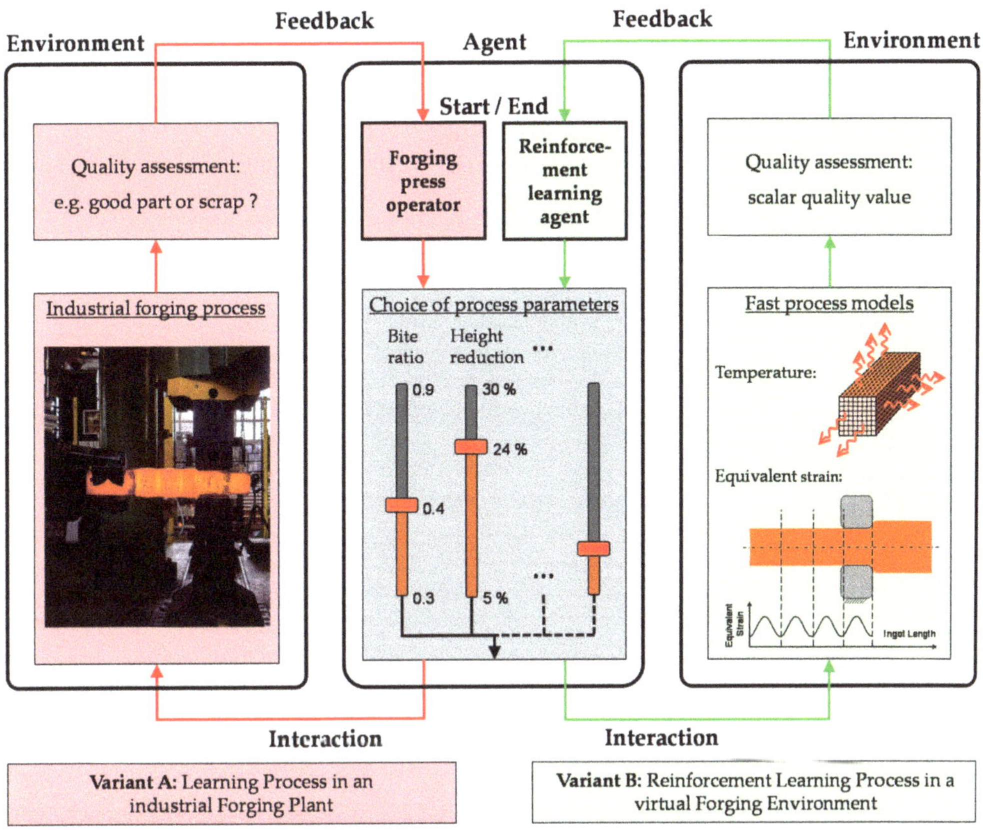

Figure 3. Schematic diagram of learning through interaction and feedback. Variant A, red: learning process in an industrial forging plant; variant B, green: reinforcement learning in a virtual forging environment.

In this publication, the described learning process is realized by an independently learning agent that interacts with a virtual forging environment by selecting various process parameters in exactly the same way as a human plant operator (cf. Figure 3; variant B, green). This virtual environment represents the open-die forging process and always has a defined state that corresponds to the current ingot state (temperature, geometry, strain, etc.). The change in the ingot state, which results directly from the behavior of the agent, is calculated promptly by the fast process models described in Section 2.1 and, thus, enables an interaction between environment and agent. Through the feedback that the agent receives from the environment after each interaction, it is able to adapt its behavior in the course of a training process and, thus, to learn, e.g., how to produce desired open-die forged parts in an optimal way.

After the basic concept of reinforcement learning has been briefly outlined with the help of a practical example, the following sections first explain the most important mathematical and theoretical principles and then present the specific implementation of the generated reinforcement learning algorithm.

2.2.1. Markov Decision Processes

The concept of Markov decision processes (MDPs) offers the possibility to describe and optimize sequential decision problems and, therefore, MDPs form the basis of reinforcement learning. Markov decision processes are characterized by the fact that the probability of a future state s' depends only on the current state, s, and the associated action, a, and not on past states and actions [32]. An MDP is defined by 5 tuples $(S,\ A, P, R, \gamma)$ that corresponds to:

- S, to the set of possible states;
- A, to the set of possible actions;
- $P = p(s'|s, a)$ to the probability of a state change from s to s' if in state s the action a is chosen;
- $r(s, a, s') \in R$ to the scalar reward for each change of state from s to s';
- γ, to the discount factor, which devalues any future reward and, thus, defines its value in the current time step.

2.2.2. Optimal Decision Making

During the optimization of an MDP, a selection rule, the so-called optimal policy π^*, is sought, which assigns exactly that action, $a \in A$, to each state, $s \in S$, which maximizes the sum of the devalued rewards, the so-called return g (cf. Equation (8)), in the long run [31].

$$\begin{gathered} g_t = f(r_t,\ r_{t+1}, r_{t+2}, \ldots,\ r_{T-1}) = \sum_{k=0}^{\mathrm{T-t-1}} \gamma^k \cdot r_{t+k} \\ with\ T = Number\ of\ last\ time\ step \end{gathered} \tag{8}$$

In Equation (9), the Bellman equation of the action values $q^\pi(s, a)$ is given, which defines a value for action choice a in a state s depending on the value of the action choice in the subsequent state $q^\pi(s', a')$. Here the value $q^\pi(s, a)$ of an action a in state s corresponds to the expected return g in the case that in state s action a is chosen and afterwards policy π is followed [31].

$$q^\pi(s, a) = E_\pi[r + \gamma \cdot q^\pi(S_{t+1}, A_{t+1}) \mid S_t = s, \qquad A_t = a] \tag{9}$$

Consequently, $q^\pi(s, a)$ directly describes how good or bad an action choice a in a state s is with respect to the long-term generated reward in the context of the underlying decision problem. Assuming the optimal action value function $q^*(s, a) = \max_\pi q^\pi(s, a)$ is known, it follows that the optimal policy π^* sought is to choose action a in each state s that maximizes $q^*(s, a)$ [31]:

$$\pi^* = argmax(q^*(s, a)) \tag{10}$$

This type of policy is called greedy policy, because the decision is only made based on values of the current time step, in this case the action value function $q^\pi(s, a)$ [31].

2.2.3. Q-Learning

Q-learning [33] is a widely used reinforcement learning algorithm that is able to approximate the optimal action value function $q^*(s, a)$ by iterative fitting of a Q-function and, thus, to solve Markov decision problems [34]. The rule for the adaption of the Q-function is based on the Bellman equation (Equation (9)) for the case of an optimal action value function $q^*(s, a) = \max_\pi q^\pi(s, a)$ and the results are as follows [31]:

$$\begin{gathered} Q_{t+1}(s, a) = (1 - \beta) \cdot Q_t(s, a) + \beta \cdot \left[r + \gamma \cdot \max_a \{Q_t(s', a')\}\right] \\ with\ \beta\ = learning\ rate\ and\ t\ =\ time\ step \end{gathered} \tag{11}$$

The Q-learning algorithm consists of an agent that interacts with an environment (cf. Figure 4) and in each discrete training time step t

- receives a state information $s \in S$ from the environment,
- then selects an action $a \in A$
- and finally receives a scalar reward $r(s, a, s') \in R$, which describes the quality of the action choice respectively the state change $s \xrightarrow{a} s'$ made [31].

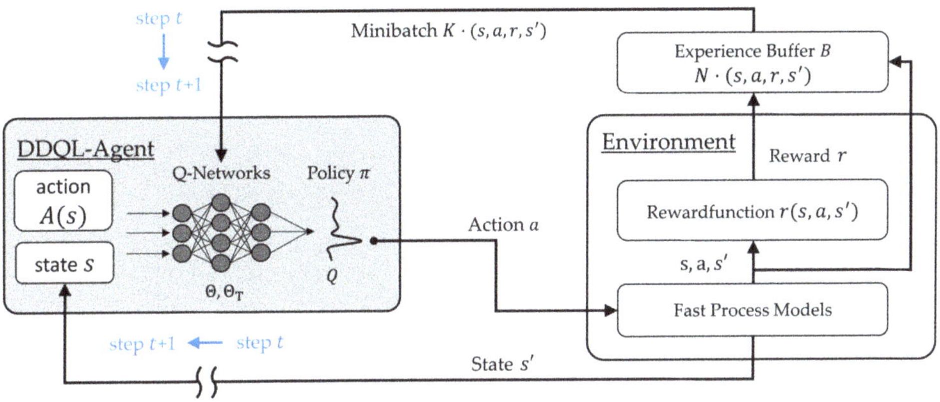

Figure 4. Schematic simplified representation of the implemented DDQL algorithm.

This procedure is repeated until after T training steps a terminal state s^+ is reached, after which the current state is transferred back to an initial state $s = s_0$ based on defined rules. Afterwards, the training process is continued starting from the new initial state $s = s_0$. The entirety of training time steps t, starting from an initial state s_0 to the next final state s^+, form a training iteration.

Due to the continuous interaction between agent and environment during the training process, the algorithm independently collects experience data tuples (s, a, r, s') that are used together with Equation (11) to continuously adjust the Q-function and, thus, find an approximation of the optimal action value function $q^*(s, a)$. Hence, a separate training data acquisition and preprocessing as needed in, for example, supervised learning algorithms is not necessary, the training data is generated during the training process.

2.2.4. Double Deep Q-Learning

Mnih et al. [35] present the deep Q-learning algorithm (DQL), which approximates the Q-function not, as usual in basic Q-learning, by a lookup table $L(S \times A)$ [34], but by a (deep) neural network (DNN) with the set of parameters Θ. Thus, the algorithm is able to map finely discretized and even continuous state spaces. However, the non-linear Q-networks in combination with model-free reinforcement learning, such as Q-learning, often tend to diverge, which is why the training process can be unstable. In order to achieve good convergence properties, nonetheless, a training data memory B was introduced, in which the last N experience data tuples (s, a, r, s') are stored. In each training step t, a minibatch of K randomly selected data tuples is compiled from the experience buffer B and is then used to adjust the parameters of the Q-network Θ by minimizing the loss function L [36]:

$$L = \frac{1}{K} \sum_{i=1}^{K} (y_i - Q(s_i, a_i|\Theta))^2 \tag{12}$$

The so-called target values y_i result from the sum of the received reward r_i and the estimated maximum Q-value of the subsequent state $Q(s_i', a_i')$, which is not directly estimated by the Q-function to be learned but by an additional target Q-function [36]:

$$y_i = r_i + \gamma \cdot \max_{a_i'} Q(s_i', a_i' | \Theta_T) \tag{13}$$

The target Q network $Q(s, a|\Theta_T)$ corresponds exactly to the Q function $Q(s, a|\Theta)$ in its structure but uses a different set of parameters Θ_T. The target-Q parameter set Θ_T is adapted to all C time steps using the parameters of the current Q function Θ. Overall, there are several advantages to the described procedure that increases the training stability of the DQL algorithm [35,36]:

- A higher data efficiency, since an experience data tuple $(s_t, a_t, r_{t+1}, s_{t+1})$ is potentially trained more than once;
- Correlations between data sets of consecutive learning steps are dissolved, since the training data sets are randomly selected from the experience data memory B;
- Preventing unintentional "feedback loops", because the current Q function no longer correlates directly with the current training data due to the use of the target Q function for the training data generation and the random compilation of minibatches.

A common problem of Q-learning algorithms is the overestimation of Q-values, which can cause a performance loss [37]. To counteract this, van Hasselt et al. [38] present the double DQL approach (DDQL), which reduces the problem of overestimation by a minor adjustment in the calculation of the target values y_i. Within the calculation of the target values, the Q-value of the subsequent state s_i' is still determined as in DQL (cf. Equation (13)) using the target Q-function $Q(s, a|\Theta_T)$, but the best action a_{max}' in the subsequent state s_i' is now calculated using the maximum of the Q-function $Q(s, a|\Theta)$:

$$\begin{aligned} y_i &= r_i + \gamma \cdot Q(s_i', a'_{max} | \Theta_T) \\ with\ a_{max} &= \underset{A'}{\operatorname{argmax}} Q(s_i', A' | \Theta) \end{aligned} \tag{14}$$

This decouples the selection of the best action a_{max}' in the subsequent time step s_i' from the estimation of the value of this selection $Q(s_i', a'_{max}|\Theta_T)$ and, thus, reduces the tendency to overestimate [37].

2.3. Implementation of the DDQL Algorithm for Pass Schedule Design in Open-Die Forging

Here, a reinforcement learning DDQL algorithm is presented that is supposed to generate optimized pass schedules following defined goals. The DDQL algorithm was chosen since deep Q- and double deep Q-learning are used successfully to solve various engineering problems [39–41] including the optimization of the production of multilayered optical thin films [42] whereby, as in the optimization of open-die forging processes, different highly non-linear physical relations need to be considered. The developed algorithm was implemented in MathWorks MATLAB and consists, as described in Section 2.2 and shown in Figure 4, of two main elements interacting with each other: The agent and the environment.

The state s of the environment represents the current state of the forging block and is characterized by five normalized values: The ingot height, h, at the start of the current state, s, the difference, Δh_f, between the target height, $h_{f,target}$, and the current height, h, the mean equivalent strain in the core fiber, $\overline{\varepsilon_v}$, the mean temperature in the core fiber, $\overline{T_{cf}}$, and the mean surface temperature, $\overline{T_s}$. These values were chosen based on their process information content. The current ingot height, h, and the height difference, Δh_f, carry needed information about the current and the target geometry and, hence, enable an estimation of the remaining process length. Since the press force is used as an optimization criterion (cf. Section 2.6), the temperature ($\overline{T_{cf}}$, $\overline{T_s}$.) and equivalent strain ($\overline{\varepsilon_v}$) information are required, allowing a prediction of the current yield stress and consequently a prediction of forces. In addition, the mean surface temperature and the mean core temperature result in a gradient that usually increases over the course of the forging process and, consequently, allows in combination with the current ingot height, h, a rough estimate of the previous process duration and the current temperature field. Based on this state information, $\left(h,\ \Delta h_f, \overline{\varepsilon_v}, \overline{T_{cf}}, \overline{T_s}\right)$, the agent selects the next action, a, which consists of the two process parameters: bite ratio, s_{B0}/h_0, and height reduction, ε_h. In contrast to the state information, s, which consists of continuous values, the action selection of the agent is limited to discrete steps. In the present algorithm, the bite ratio, s_{B0}/h_0, can be varied between 0.3 and 0.9 in steps of 0.05 and the height reduction between 3% and 30% in steps of 1%.

To ensure the accuracy of the calculations from consecutive state transitions within a pass schedule, the spatially resolved information about the temperature and the equivalent strain distribution along the core fiber was stored within the environment. In subsequent passes, this information was used as the point of departure of the state transition calculations using the fast process models. Afterwards, the above-described averaged values were determined and merged together with the current geometry information in the form of the five tuple $\left(h,\ \Delta h_f, \overline{\varepsilon_v}, \overline{T_{cf}}, \overline{T_s}\right)$ representing the resulting state s'.

In addition to the process parameters, such as the current ingot height h or the selected height reduction ε_h, which can be directly related to the open-die forging process, the algorithm also contained parameters without a direct physical background or reference to the forging process. These included, for example, the model parameters Θ and Θ_T of the Q-networks, which were first learned during the training process using experience data tuples (s, a, r, s'). Since these learned model parameters cannot be physically interpreted on their own but significantly determine the behavior of the later used algorithm and, thus, the quality of the designed pass schedules, only an indirect connection between model parameters and the forging process was present. On the other hand, the algorithm has so-called hyperparameters that are not learned during training but are predefined by the user and have no direct connection to the forging process either. Since hyperparameters significantly determine the training process of the algorithm, there is nevertheless an influence on the quality of the generated pass schedules, for example, through improved convergence properties or higher average rewards.

To approximate the Q- and target Q-values, the agent used a neural network that had two separate input branches: one for state information and one for action information. The action information was normalized and fed into the neural network by an input layer. This was followed by a fully connected layer with 24 neurons and a subsequent addition layer. This addition layer combined the pre-processed information from the two branches by an element-wise addition. In the state branch, the normalized information was also fed into the net via an input layer. Two fully connected layers followed, which were separated by a ReLu layer. The abbreviation ReLu stands for "rectified linear unit" and corresponds to the mathematical operation $f(x) = \max(0, x)$, where negative values are set to zero and non-negative values remain unchanged. The output of the second fully connected layer was connected to the addition layer, followed by another ReLu layer. Finally, a fully connected layer with a single neuron formed the output layer which had a value that corresponded to the estimated Q- or target Q-value. The used network architecture was proposed in [43] and shown in Figure 5.

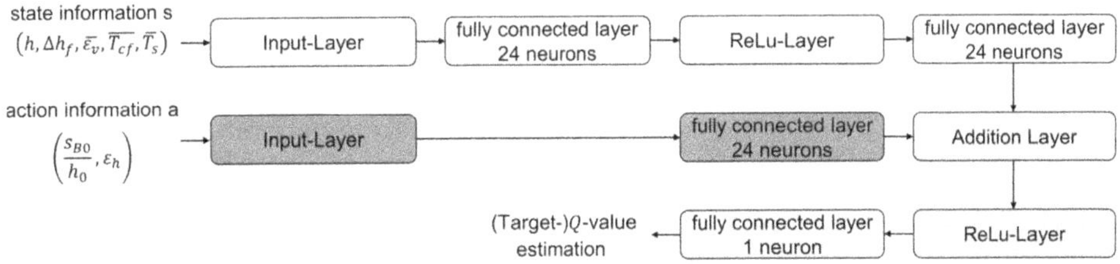

Figure 5. Graphic representation of the used (Target-)Q-Network.

To ensure the algorithm not only relies on its experience during the training process but also sufficiently explores the state space S and action space A, an ε-greedy-policy was used to maximize the return, g, obtained in the long run [31]. Here, a random action, $a(s)$, was chosen with the probability ε and, otherwise with the probability of $(1-\varepsilon)$, a greedy action with respect to the $Q(s,a)$ value according to Equation (10) was selected. For this purpose, the agent needed to calculate the $Q(s,a)$ values of each combination of the current state s and all possible actions $a \in A(s)$ separately. This was necessary because the implemented Q-network architecture (cf. Figure 5) uses not only the state information s but also action information $a \in A(s)$ to determine the $Q(s,a)$ value of a single state–action pair. The probability ε for a random action choice does not remain constant over the training process. Starting from a value of 1, ε decreases continuously in every training time step t according to the selected epsilon decay rate of 0.0001 to a minimum value $\varepsilon_{min} = 0.1$ (cf. Equation (15)). These hyperparameters are based on literature values [35] that were adapted to the conditions of the DDQL implementation in MathWorks MATLAB.

$$\varepsilon_t = \max\left(\varepsilon \cdot (1 - EpsilonDecayRate)^t,\ \varepsilon_{min}\right) \tag{15}$$

With the selected action $a = (s_{B0}/h_0,\ \varepsilon_h)$ and the present state $s = \left(h,\ \Delta h_f, \overline{\varepsilon_v}, \overline{T_{cf}}, \overline{T_s}\right)$, the environment was able to calculate the change of state of the forging block $s \xrightarrow{a} s'$ using the fast process models presented in Section 2.1. A change of state does not correspond to just one pass, but always to two passes, where the forging block is rotated by 90° around the longitudinal axis after each pass. For this purpose, the bite ratio s_{B0}/h_0 selected by the agent was used in both passes, while the selected height reduction ε_h was only used in the first pass. The height reduction in the second pass was determined in such a way that after the pass a square billet cross-section $h = b$ was created, taking the occurring width spread into account (see Equation (1)). This forging strategy with a square ingot cross-section after every second pass is called "square forging" and is widely used in industrial applications. Subsequently, fast process models were used to calculate the development of the temperature field and the equivalent strain distribution during both passes (see Section 2.1). Finally, the values of the state information s′ $\left(h',\ \Delta h'_f, \overline{\varepsilon_v}', \overline{T_{cf}}', \overline{T_s}'\right)$ were put together and transferred to the reward function and the training data memory B as shown in Figure 4.

The hyperparameters experience buffer size $N = 10{,}000$ data sets, and the size of the mini-batch $K = 64$ randomly chosen data sets was taken from the default settings of the DDQL implementation in MathWorks MATLAB [44]. As described in Section 2.2, the mini-batch was used to adjust the parameters Θ and Θ_T of the Q- and Target Q-networks. The learning rate β determines the size of the individual adaptions and had to be lowered from the default value $\beta = 0.01$ to $\beta = 0.001$ based on some test trainings that showed unstable training behavior and convergence problems.

The reward function, $r(s,a,s')$ (cf. Figure 4), is the part of the implementation in which the user defines the goals of the algorithm during the training process. For this purpose, a mathematical formulation $r = f(s,a,s')$ must be found that represents the goals in such a way that the algorithm achieves them by maximizing the received return g. Hence, the reward r corresponds to the quality of an action choice consisting of the bite ratio and height reduction within a defined ingot state s, considering the defined optimization goals.

The goals for the present application example of a pass schedule design in open-die forging were aligned to the possibilities offered by comparable pass schedule calculation software without methods of artificial intelligence (see Section 1). The first goal a pass schedule should fulfill is to reach the target geometry, whereby a permissible percentage deviation, h_{var}, is tolerated. Here, this permissible deviation is one percent and is thus within the usual range for open-die forging [45]. Two further criteria for an optimal pass schedule design are derived from the desire to make the process as fast and efficient as possible. On the one hand, the final geometry should be achieved in as few passes as possible and, on the other hand, the available machine force, F_{max}, should be used in the best possible way. For this purpose, the desired target force, F_{target}, was set at 80% of the maximum press force, F_{max}, to consider a sufficient safety margin.

Through this selection of optimization criteria with a force and a pass number component, the complicated incremental process kinematics and geometry development in open-die forgings were taken into account. If only the press force is evaluated, the algorithm could tend to apply the largest possible bite ratios while using small height reductions, resulting in long forging processes. This effect is intensified by the occurring width spread, since width spread increases with increasing bite ratios and must be forged back in the subsequent pass due to the 90° ingot rotation at the end of each pass.

On the other hand, the pure evaluation of the number of passes can lead to an algorithm selecting the largest possible height reductions combined with small bite ratios, as in this case little width spread occurs and, overall, the fewest passes are required. However, such a forging strategy can cause surface defects in the billet, such as laps, and is therefore not desirable. In addition, small bite ratios result in high numbers of strokes, which increase the process time.

Moreover, the sole evaluation of the process time is not implemented, because the process time is highly individual for each plant, as it strongly depends on the present velocities and accelerations of the manipulator and of the forging press. In addition, discrete events, such as a significant increase in time due to reheating, can hardly be weighted reasonably within the complex process optimization framework.

In Equations (16)–(18), the complete mathematical formulation of the used reward function $r(s,a,s')$ is summarized. The function is divided into two large subareas, the evaluation of the geometry and the evaluation of the occurring maximum press force:

$$r(s,a,s') = \frac{\max(r_{geo} + r_F, -10)}{10} \tag{16}$$

$$r_{geo} = \begin{cases} 0, & h > h_{f,ub} \\ -\frac{a \cdot h^2}{h^2_{f,target}} + \frac{a \cdot 2 \cdot h}{h_{f,target}} + a + 10, & h_{f,ub} \geq h \geq h_{f,lb} \\ \max\left(h - h_{f,ub}, -10\right), & h < h_{f,lb} \end{cases} \tag{17}$$

$$r_F = \begin{cases} \left.\begin{array}{ll} -3 + 2 \cdot \left(\exp - \frac{\left(\frac{F}{F_{max}} - 0.8\right)^2}{0.045}\right), & F \leq F_{target} \\ -6 + 5 \cdot \left(-\frac{\left(\frac{F}{F_{max}} - 0.8\right)^2}{0.04} + 1\right), & F_{target} < F \leq F_{max} \\ -6, & F > F_{max} \end{array}\right| & h > h_{f,ub} \\ \left.\begin{array}{ll} 0, & F \leq F_{target} \\ -5 + 5 \cdot \left(-\frac{\left(\frac{F}{F_{max}} - 0.8\right)^2}{0.04} + 1\right), & F_{target} < F \leq F_{max} \\ -5, & F > F_{max} \end{array}\right| & h \leq h_{f,ub} \end{cases} \tag{18}$$

$$with\ a = -\frac{10}{\left(\frac{h_{var}}{100\%}\right)^2};\ h_{f,lb} = h_{f,target} \cdot \left(1 - \frac{h_{var}}{100\%}\right);$$

$$h_{f,ub} = h_{f,target} \cdot \left(1 + \frac{h_{var}}{100\%}\right)$$

The reward for the generated geometry, r_{geo} (see Equation (17)), evaluates only the final generated ingot height of a pass schedule and not the intermediate geometry development during the forging process, as the continuous evaluation of the ingot height could cause contradictions with the other optimization goals. If the permissible final height range, $h_{f,target} \pm 1\%$, is met, the reward according to Equation (17) corresponds to a symmetrical parabola that reaches its maximum of 10 at $h_{f,target}$ and has the value 0 at the edges. If the algorithm generates a final height, h, that falls below the minimum permissible height, $h_{f,ub}$, the reward corresponds to the resulting difference, $h - h_{f,ub}$. The resulting overall function for the evaluation of the block geometry, r_{geo}, is shown in Figure 6 (right) showing an example target height, $h_{f,target} = 350$ mm.

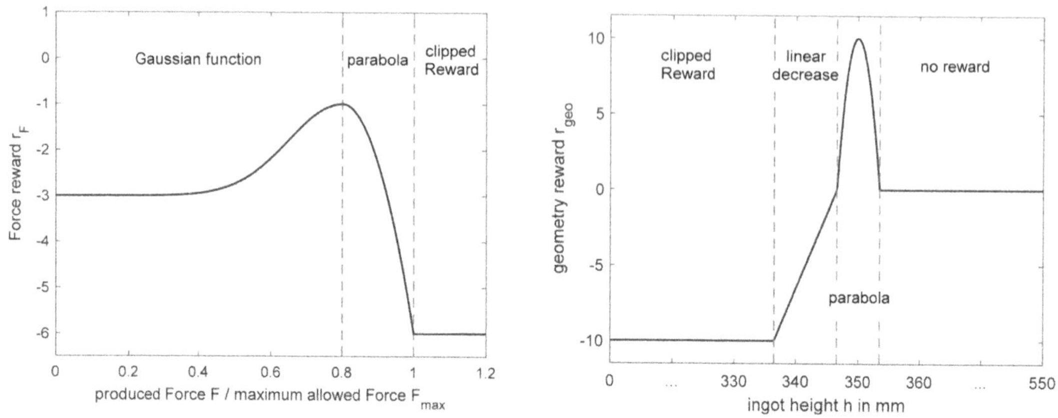

Figure 6. Left: Plot of the force reward, r_F, over the ratio of the produced force, F, and the maximum allowed force, F_{max}, in case the current height, h, is greater than the maximum final height, $h_{f,ub}$, and, thus, the related state, s, is no terminal state, s^+. **Right**: Graphic representation of the geometry reward, r_{geo}, based on one specific target final height, $h_{f,target}$. The plot ranges from zero to the maximum initial height $h_{0,max}$ = 500 mm, thus representing all possible height values. The maximum reward of 10 is located in $h_{f,target}$ = 350 mm.

Equation (17) also shows that the penalty for dropping below the final height has a lower limit of –10. This procedure is derived from the reward clipping recommended in [36] and can significantly improve the convergence properties of an algorithm, since individual rewards with high values can lead to large adjustments in the Q-function and, thus, to instability in the training process.

The evaluation of the maximum force, F, occurring in the two passes of the state transition $s \xrightarrow{a} s'$ was divided into different scenarios according to its value and according to the block height, h, in state s' (cf. Equation (18)). As long as the generated height h' is above the allowed end range $h_{f,ub}$, the reward r_F is calculated as shown in Figure 6 (left) either by a Gaussian function or a parabola, depending on the calculated force, F. If the resulting force, F, corresponds exactly to the target force, F_{target}, the reward $r_F = -1$ reaches its maximum. If the force decreases starting from the maximum $F = F_{target}$, the reward drops to -3, while for forces above F_{target}, rewards up to -6 are possible. This asymmetry is due to the fact that exceeding the target force is more critical than undercutting it. If the force, F, rises above the maximum press force, F_{max}, reward clipping comes into effect again and limits the reward, r_F, to a minimum of -6.

If a block height is generated that is in the range of the permitted final heights or below, the completed change of state describes the last two passes of a pass schedule. In these passes, the focus is primarily on setting the desired final geometry and not on the optimal use of the press force, which is why the force reward for forces below the target force, F_{target}, is omitted (cf. Equation (18) and Figure A1 (Appendix A)). However, since the maximum press force, F_{max}, must be maintained in the last two passes as well, the rest of the force evaluation ($F > F_{target}$) remains qualitatively unchanged.

The reward for the generated forces, r_F, is not arbitrarily defined negatively, but it represents the goal of using the smallest possible number of passes. This results from the fact that the algorithm tries to maximize the received return, g, in the course of the training and, consequently, shows a tendency to lower the number of passes, because at lower numbers of passes negative force rewards are received less often.

The total reward of an action choice is finally calculated according to Equation (16) from the sum of the geometry reward, r_{geo}, and the force reward, r_F. The total reward of an action choice is also limited by reward clipping to a minimum value of -10. Therefore, the clipped sum of the individual rewards is always in the range $[-10, 10]$ and can be scaled to the required order of $[-1, 1]$ using the divisor 10. The general weighting of the individual parts of the reward function arose from the prioritization of the defined goals. Therefore, the reward for reaching the final geometry was set proportionally higher than the cumulative punishment of the force utilization, respectively, the number of passes. The exact weightings were determined by many test trainings.

In summary, the general characteristic of the reward function is that the focus in each of the last passes of a pass schedule is primarily on reaching the desired final height as accurately as possible and in all preceding passes, mainly on making the best possible use of the press force. To emphasize this characteristic, the discount factor γ was reduced from the default value $\gamma = 0.99$ to 0.9. Since the geometry reward is achieved only at the end of each training pass schedule, it tends to be discounted more often by the discount factor. As a result, the influence of the geometric reward granted for hitting the target height decreases in preceding passes and using the press force optimally becomes more important. However, the discount factor must not be selected too low either, since in a state s, the algorithm can only select discrete actions, $a(s)$, and, thus, the resulting block heights, h', are also limited. In addition, the permissible deviation from the target height, $h_{var} = \pm 1\%$, is small, and the reward received rapidly decreases with increasing distance from the optimal value $h_{f,target}$, even within the permissible deviations (see Figure 6 (right)). Consequently, it may be necessary for the algorithm to adapt its actions over several passes in advance in such a way that the target height is hit as accurately as possible, thereby maximizing the total reward.

2.4. *Definition of the Process Characteristics for the Pass Schedule Design in Open-Die Forging*

According to [46], the weight of open-die forged parts can range from a few pounds to more than 150 tons. Furthermore, hydraulic open-die forging machines are usually offered with maximum press forces between five and two hundred meganewton [47–49]. Hence, the hereafter described chosen process window for the intended optimized pass schedule design using the presented DDQL algorithm with ingot weights of approximately three tons and a forging press with a maximum press force of 25 MN corresponds to a small- to medium-sized open-die forging process. Subsequently, a successfully trained DDQL algorithm can serve as a proof of concept for the pass schedule design in open-die forging via reinforcement learning.

The boundary conditions of the forging processes to be planned in an optimal manner are defined as follows:

- The assumed maximum press force, F_{max}, is 25 MN. Hence, a target force of $F_{target} = 20$ MN results;
- The initial height of the workpiece, $h_i (= b_i)$, can be selected freely by the user between $h_{i,min} = 450$ mm and $h_{i,max} = 550$ mm;
- The target height of the workpiece $h_{f,target} \left(= b_{f,target}\right)$ can also be selected freely by the user between $h_{f,min} = 250$ mm and $h_{f,max} = 350$ mm;
- The ingot start length $l_i = 1500$ mm is constant;
- The forging block consists of 42CrMo4 (1.7225), a commonly used steel for open-die forged parts. The corresponding parameters of the fast process models for temperature and equivalent strain are described in the Sections 2.1.2 and 2.1.3, and the flow curve originates in [25];
- There is no pre-deformation of the ingot and, thus, no equivalent strain at the start of the process ($\overline{\varepsilon_v} = 0$);
- The ingot temperature at process start T_0 corresponds to 1000 °C and is homogeneous in the billet;
- In every second pass a bite shift of $s_{b0}/2$ is applied. Bite shift corresponds to the offset of the contact areas between the ingot and the dies of a pass in the longitudinal direction in comparison to the contact areas of the subsequent pass;
- Since excessive cooling of the ingot during the forging process can cause damage, such as edge cracks as well as high forming forces, reheating of the forging ingot should be taken into account. This becomes necessary if the minimum temperature, T_{min},

anywhere in the workpiece, including its edges and vertices, falls below a lower limit of T_{crit} = 650 °C.

The reheating of a hypothermic workpiece is represented by reading out the minimum block temperature, T_{min}, after each change of state, respectively, for each even-numbered pass and comparing it with the minimum allowed temperature, T_{crit}. If the temperature falls below T_{crit}, the temperature field is reset to the initial state with homogeneously present temperature T_0. Reheating is not directly considered in the reward function, since the chosen optimization criteria should lead to short processes anyway and thereby indirectly prevent mandatory reheating of the ingot.

In order to generate an algorithm that calculates pass schedules that meet the above-mentioned goals while maintaining the selected process framework, the algorithm must be trained for a sufficiently long time. Therefore, the number of training iterations in the training process, which corresponds to the number of designed pass schedules, was set to 50,000 and the convergence of the algorithm was monitored (cf. Sections 2.6 and 3.1).

At the beginning of each training iteration the state, s, must be reset to the initial state, s_0. Both the mean temperature in the core fiber, $\overline{T_{cf}}$, and the mean surface temperature, $\overline{T_s}$, correspond in the initial state to the initial temperature, T_0, and there is no equivalent strain, $\overline{\varepsilon_v} = 0$. The start and target height of the forging block, h_i and $h_{f,target}$, are randomly selected within their limits for each initial state, so that the algorithm learns to generate an optimized pass schedule for all possible combinations of given input parameters.

2.5. *Application of the DDQL Algorithm for Pass Schedule Design in Open-Die Forging*

In Sections 2.3 and 2.4, the implementation of the DDQL algorithm for the design of pass schedules is presented and is now completed by a description of the application of the algorithm. Therefore, Figure 7 shows the application cycle of the DDQL algorithm, which is divides into three major steps: setup, training, and usage.

At the beginning, the user (e.g., a plant operator) defines the forging process window (process parameters), while the hyperparameters of the DDQL algorithm can often remain unchanged or can be set by a programmer or a research engineer as shown in Figure 7, "Setup". Here, together with process parameters, such as permissible geometries or material properties, the optimization goals the algorithm should pursue in the pass schedule design need to be determined, and a corresponding reward function must be formulated. The training process follows (cf. Figure 7, "Training Process"), whereby the process parameters selected by the user are incorporated into the environment and the hyperparameters into the agent as well as the framework of the training process. The environment contains not only the training goal represented by the reward function but also the fast process models (cf. Section 2.1) that calculate the change of state of the forging ingot. A training process of the implemented algorithm takes approximately two days, but once the training is successfully completed, the trained agent is able to quickly generate optimized pass schedules in the selected process window. Now, the user can select a desired initial and target geometry and receives an optimized pass schedule in a few seconds (cf. Figure 7, "Usage").

If required, the user is able to flexibly influence all sections of the algorithm and the underlying forging process to achieve his desired pass schedule design. At the same time, the user does not need to change all sections of the algorithm at once; for example, it is also possible to change only the process parameters in an existing implementation and restart the training with the same hyperparameters and reward function to obtain an optimized pass schedule design for a different process window. However, in this case, successful training is not guaranteed, which is why the transferability of the implementation described above needs to be reviewed critically. For this purpose, the algorithm was trained on an industrial scaled forging process at first (cf. Sections 3.1 and 3.2) and was transferred to a smaller experimental scale of the Institute of Metal Forming (IBF) for validation purposes later on (cf. Section 3.3).

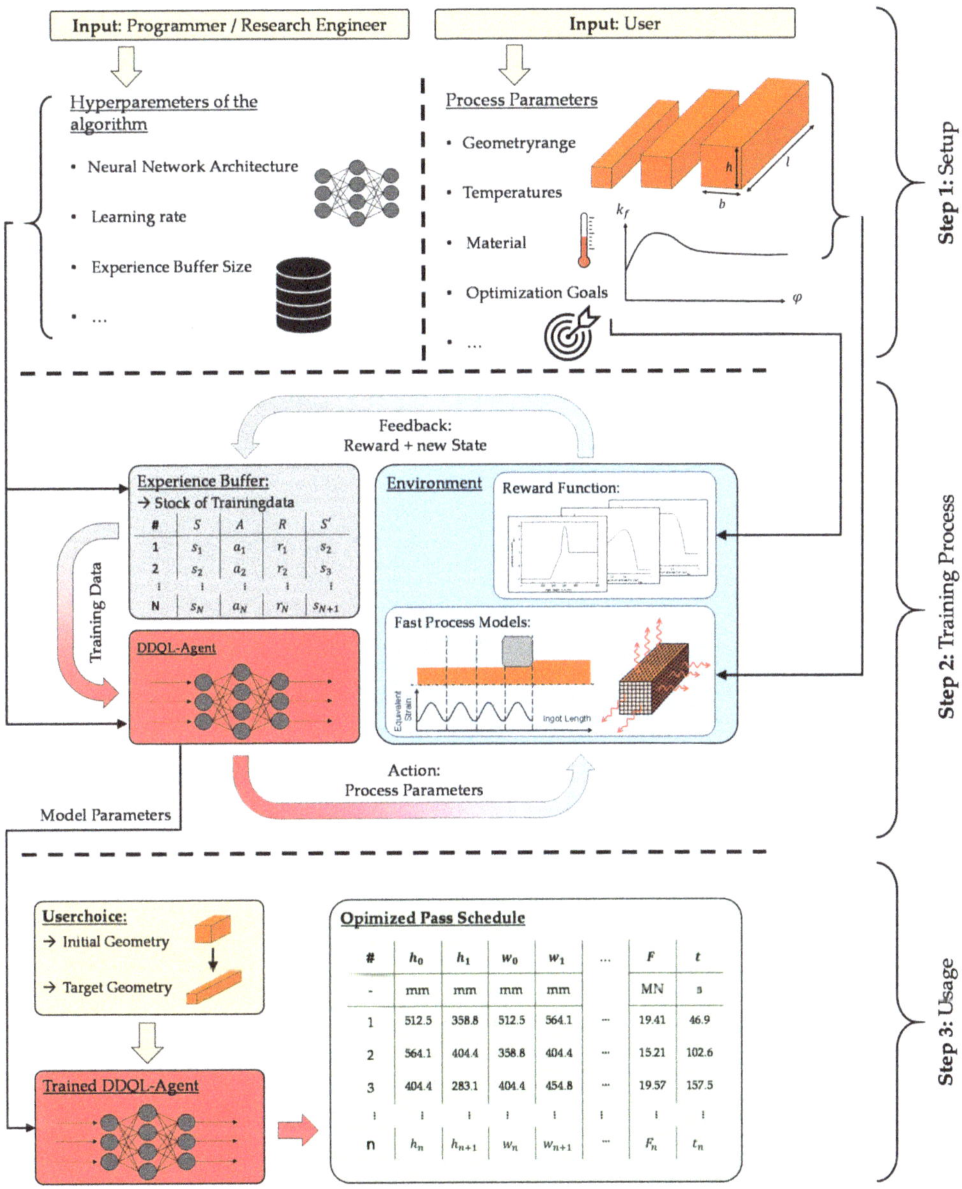

Figure 7. Application cycle of the implemented DDQL algorithm.

2.6. Evaluation of the Algorithm Performance

To evaluate the performance and the convergence of a reinforcing learning (DDQL) algorithm, commonly the generated rewards or the Q-value of the initial state, $Q(s_0)$, are plotted over the iterations of the training process [31,35–38]. If the training is successful, both values increase during the training and, in the best case, converge towards their optimum.

If an algorithm performs well measured by the generated total reward or Q-values, it is only a reliable indicator for the achievement of the defined training goals as long as the reward function is a sufficiently accurate mathematical representation of these goals. Although this condition is assumed in the present implementation, it must be checked nevertheless in the analysis of the training results. Therefore, the performance and the convergence of the algorithm are measured not only by the collected total rewards of individual iterations $r_{tot,i}$ or the $Q(s_0)$-value, but additionally by the development of the following process parameters and auxiliary variables, which directly represent the defined goals:

- Difference Δ_{h_f} between the generated final height h_f and target height $h_{f,target}$;
  - Goal: accurately achieve the desired final height $h_{f,target}$;
- Number of passes n_p;
  - Goal: minimize the number of passes;
- Maximum force occurring in the pass schedule F_{tot};
  - Goal: stay below the press force limit;
- Force utilization F_{use}
  - Goal: best possible utilization of the forging press.

The force utilization, F_{use}, corresponds to the sum of the absolute differences between the maximum generated force, F, and the target force, F_{target}, over all state changes that make up the evaluated pass schedule (see Equation (19)). Since the force of the last pass is not evaluated in the training process, it is also dismissed from the force utilization calculation. This procedure is based on the fact that the forces of the two passes, which are derived from the selection of an action a, were not selected independently of each other due to the square forging.

$$F_{use} = \frac{\sum_{k=1}^{n_p/2-1} \left| F_k - F_{target} \right|}{n_p/2 - 1} \tag{19}$$

Furthermore, due to the underlying non-continuous problem of pass schedule design in open-die forging, only short training iterations with mostly single-digit training step numbers occur in the present algorithm. At the same time, the structure of the reward function leads to the fact that the negative force reward is received less frequently with lower numbers of passes, which means not every combination of initial and target height can achieve the same maximum total reward. Since the start and target heights of a pass schedule are chosen randomly in the training process, the maximum possible reward is not constant within individual training iterations. For this reason, the convergence and performance of the algorithm is not measured by the characteristics of pass schedules of individual combinations of initial and target heights, but by parameters averaged over the entirety of input parameters.

To generate these mean values for different training states, the algorithm was evaluated every 2500 training iterations. The input parameters initial and target height were both varied in 20 steps, resulting in 400 evenly distributed parameter combinations. With the help of the current agent, a pass schedule was designed for each of these combinations and the abovementioned characteristic process and algorithm parameters were stored. Subsequently, an average value can be calculated for each characteristic variable across all 400 pass schedules of an evaluation step. The pass schedule design using a fully trained algorithm works in the same way as the pass schedule design during the training process, with the difference that instead of the ε greedy policy, a pure greedy policy according to Equation (10) is used.

Machine learning algorithms and especially neural networks represent black boxes which have a learning process and behavior that is very difficult or even impossible to comprehend. Therefore, the characteristics of the evaluated pass schedules are graphically refurbished to provide an insight into the behavior and the learning process of the algorithm. For this purpose, the respective characteristic parameter was plotted over the varied initial and target heights, and the individual areas were colored according to the value of the characteristic parameter. So-called heatmaps (cf. Section 3.2) arose, which enable the user to utilize his process knowledge of open-die forging to analyze the behavior of the algorithm more precisely. For example, the user can examine which training goal is already achieved for which combinations of initial and target heights or the user can analyze which process and algorithm parameters correlate.

Exactly the same algorithm might converge in one training delivering good results and diverge in another seemingly identical training producing no useful training results. This is related to random influences on the training process such as the initialization of the (target) Q-network and the epsilon-greedy policy. In order to ensure that the convergence of the algorithm and the generated training results are not purely by chance but resilient, the training process was carried out and analyzed 15 times with the same algorithm and process parameters but different random seeds over 50,000 training iterations each.

The training processes lasted approximately two days performed on an Intel CPU of type Xeon E-2286G. This training duration results primarily from the calculation of the environment and not from the adjustment of the network parameters. Since the DDQL algorithm has an experience data memory and consequently learns "off-policy", the generation of experience data sets (s, a, r, s') can be parallelized in the future, so that the training duration may be reduced significantly.

3. Results and Discussion

3.1. Training Outcome and Convergence

The plot of the evaluation results against the number of training iterations allows assessing the convergence of the algorithm based on the different averaged parameters. In Figure 8, this was done for all 15 trainings in form of box plots.

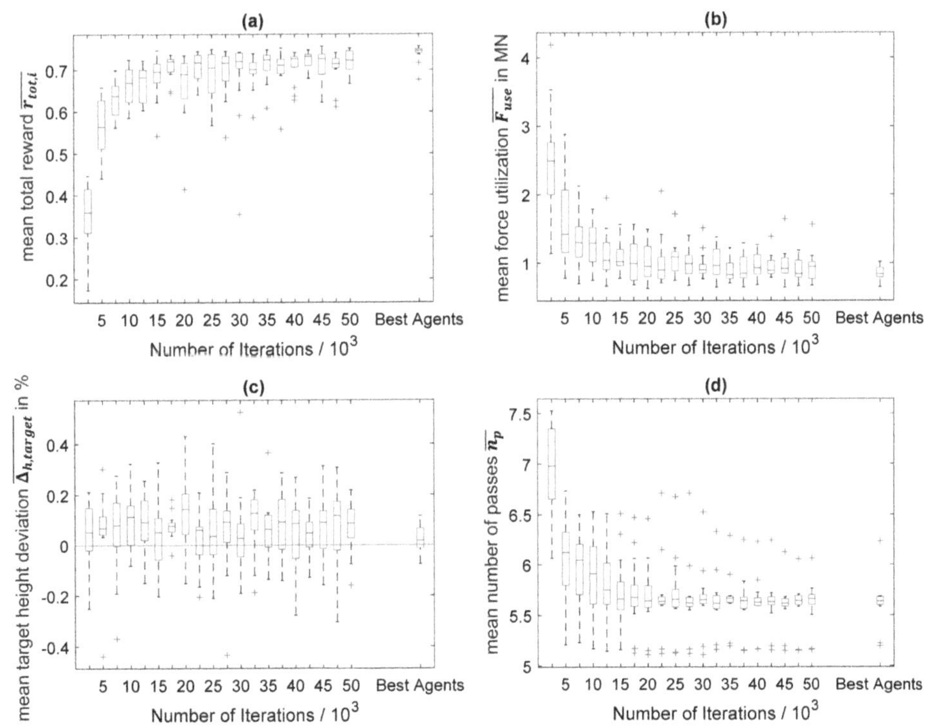

Figure 8. Boxplots of the results of all 15 Trainings over the course of the training process representing: (**a**) the mean total reward $\overline{r_{tot,i}}$, (**b**) the mean force utilization $\overline{F_{use}}$, (**c**) the mean target height deviation $\overline{\Delta_{h,target}}$, and (**d**) the mean number of passes $\overline{n_p}$.

Figure 8a shows the average total reward, $r_{tot,i}$, obtained by a pass schedule. The algorithm converges robustly against the same averaged reward in all 15 trainings. However, it was noticeable that there were single outliers, which are represented by red crosses. These can be traced back to the learning steps of the agent that were taken shortly before the evaluation in which the designed processes have changed significantly. Due to the changed behavior, the algorithm often reaches areas of the state space, S, in which it does not have a good approximation of the optimal action value function $q^*(s,a)$ yet and, therefore, performs poorly. The fact that these outliers were not transferred to the following evaluation steps shows that the algorithm quickly learns to select the right actions even in the new states and, consequently, achieves higher rewards again.

It follows that the agent of the last training iteration was not necessarily the best one. The best agent was the one that has the highest averaged reward compared to all the other evaluations of the training process [38] while always hitting the final height range and respect the maximum press force, F_{max}. Hence, hereafter, the best agent determined in this way represents the training outcome. The number of training iterations of the best agents ranged between 22,500 and 50,000 iterations and amounted to an average of ~37,500 iterations. In addition to the development of a variable, the box plot of the 15 best agents of all trainings is shown on the right-hand side of each single graphic of Figure 8. The best averaged rewards were very close together except for one value, which receives a slightly lower maximum reward. This accentuates the good convergence of the algorithm.

In Figure A2, the plot of the averaged Q-values of the initial states $\overline{Q(s_0)}$ against the number of training iterations is shown. The graph has qualitatively the same course as the graph of the generated rewards. Since the Q-function does not approximate the summed reward r_{tot}, but the received return, g, it ran below the total reward. This also emphasizes the stability of the training process and the good convergence properties of the presented algorithm.

After the general convergence of the algorithm has been demonstrated, it must be checked whether the defined goals for the optimized pass schedule design in open-die forging were achieved over the course of the trainings. Therefore, the development of the force utilization, F_{use}, the deviation from the target height, $\Delta_{h,target}$, and the number of passes, n_p, throughout all 15 trainings is shown in Figure 8 as well. Figure 8b shows that the force utilization drops to a value of approximately one meganewton during the first 20,000 iterations of the training process and remains constant in this range from then on. The 15 best agents achieved an average force utilization of less than one meganewton. Considering a target force of F_{target} = 20 MN, this corresponded to a deviation of less than 5%, showing that the goal of optimal press force utilization was reached.

The mean deviation from the target height, $\overline{\Delta_{h,target}}$, converged significantly faster compared to the mean force utilization, $\overline{F_{use}}$, and already shows deviations of less than 0.4% on average during the first evaluation after 2500 training iterations and, thus, clearly undercuts the targeted maximum permissible deviation of 1% (see Figure 8c). Due to the very early achievement of this goal, the graph does not show the typical convergence curve, but is constant at low values around 0% $\pm$ 0.5% deviation. However, a slight bias towards positive deviations, $\Delta_{h,target}$, or towards exceeding the final height, h_{target}, was evident. This bias could result from the nature of the training process, since the reward for setting the final height was symmetrical on both sides. If the algorithm exceeds the allowable final height, it only has to accept the negative force reward of an additional pass but can still receive the big positive reward for setting the final height with another pass. This is not possible if the final height is undershot, so the algorithm probably tends to exceed the target height. The box plot of the 15 best agents was very close to a 0% deviation and showed the described bias only very slightly, which confirms the set geometry target had been met.

Figure 8d shows the development of the average number of passes in the 15 training processes carried out. The mean number of passes, $\overline{n_p}$, decreased in the course of the training and finally converged after approximately 25,000 iterations in a range of approximately

5.6 passes confirming the achievement of the last goal, the minimization of the number of passes, n_p.

However, it is noticeable that the algorithms did not converge against one, but against several similarly small average numbers of passes $\overline{n_p}$ = ~6.1, $\overline{n_p}$ = ~5.6 and $\overline{n_p}$ = ~5.2. These different numbers of passes do not result in total rewards that differ to the same extend (see Figure 8a), since e.g., a higher reward from lower numbers of passes can be equaled out by lower force and geometry rewards. Consequently, there is probably no clearly defined global maximum in the current reward function.

A comparison of the convergence times of the individual process variables reveals that first the geometry target (approx. 0–2500 iterations), then the force target and finally the target of the lowest possible number of passes is attained. This sequence corresponds to the weightings that the individual reward components take up within the total reward and occurs in all 15 training sessions.

Overall, the analysis of the 15 training processes revealed that the agent trained robustly and converged reliably towards high rewards. At the same time, the development of the characteristic process variables corresponded to the formulated goals for the pass schedule design in open-die forging, which proves the suitability of the reward function.

3.2. Best Agent Results

After Section 3.1 presented the convergence properties of the algorithm and the development of the characteristic parameters in the course of the training, this chapter constitutes the overall best training result throughout all 15 accomplished trainings in more detail. This result arose in the 45,000th iteration of the fifteenth training process and achieved an average total reward of $\overline{r_{tot}} = 0.76$ in the evaluation.

As shown in Figure 9a, the deviations from the target height, $\Delta_{h,target}$, were below the limit of $\pm 1\%$ in each of the 400 pass schedules, with the maximum deviation being -0.66%. On average, the agent reached a deviation of $\overline{\Delta_{h,target}} = 0.22\%$. The force utilization shown in Figure 9b averaged at $\overline{F_{use}} = 0.75$ MN and was, at most, just over two meganewtons. As seen in Figure A3b, the maximum force of 25 MN was not exceeded in any pass schedule but always undercut by at least 3 MN. Figure 9c shows the number of passes, n_p, via the input parameter space. There were three clearly separated areas of different numbers of passes, starting from four passes at small initial heights and large target heights, over six passes to eight passes at large initial heights and small target heights. The mean number of passes, $\overline{n_p}$, was 5.20. Since more passes cause a negative force reward more often and the algorithm was able to approximate this correctly, the heat map of the Q-values in the initial state, $Q(s_0)$, shows the same three distinct ranges as the heat map of the number of passes (see Figure A3a).

On closer examination of the heat maps for the deviation from the target height, $\Delta_{h,target}$, and the force utilization, F_{use}, patterns were noticeable. In the heatmap of the force utilization (cf. Figure 9b), these slight patterns ran predominantly parallel to the ordinate and were best seen in the four-pass area on the top left. Here, it can be observed that the force utilization, from left to right, repeatedly increased at first and then abruptly dropped. In the four-pass area (top left) of the heatmap of the target height deviation (cf. Figure 9a), a diagonal pattern was present that consisted of a fixed sequence of slightly undershooting the target final height, a good agreement with the final target height and slightly overshooting the target final height. This diagonal pattern changed its orientation in the six and eight pass regions (central and bottom right), although it was not as distinct in the eight as in the six-pass area of the heatmap.

The described patterns were partially caused by direct correlations between different process parameters. For example, at pass number transitions, the press force tended to be used less efficiently and the target height was hit the worst. On the other hand, such patterns can result from the restriction of the discrete action selection. Due to the discrete action selection, $A(s)$, the set of possible geometries in the subsequent state, s', was also discrete and dependsed directly on the geometry of the current state, s, and the discretization of the actions, a.

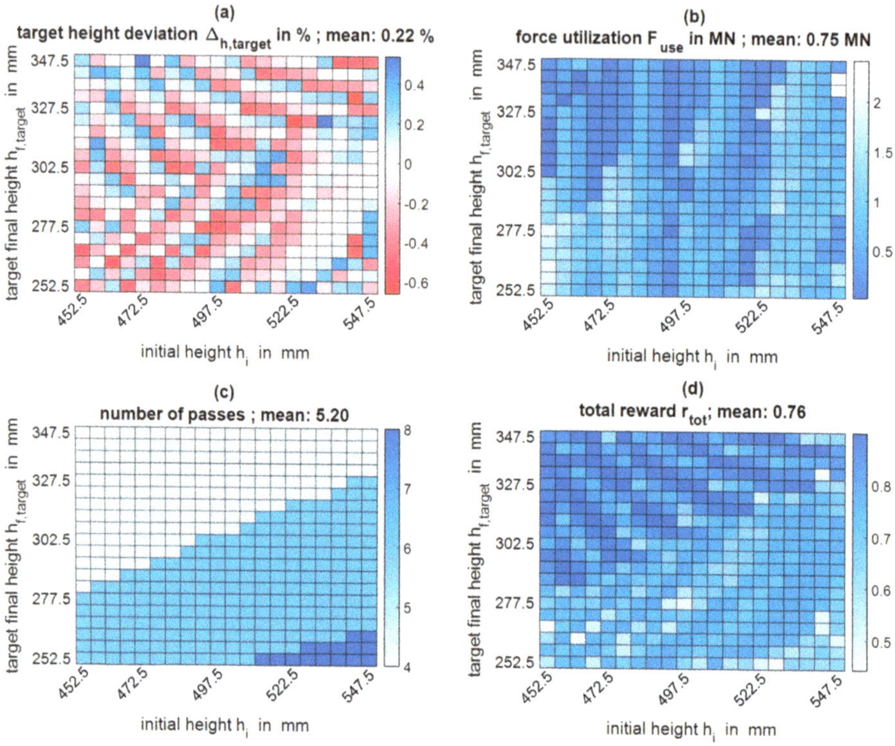

Figure 9. Heatmaps of the training results of the overall best agent: (**a**) the target height deviation, $\Delta_{h,target}$, (**b**) the mean force utilization, F_{use}, (**c**) the mean number of passes, n_p, and (**d**) the total reward, r_{tot}.

Starting from an ingot geometry in state, s, the algorithm could only realize a discrete quantity of geometries $M_{geo}(s')$ in state s' due to the limited discrete number of actions, $A(s)$. Including another action choice in state s' from the discrete set of actions, $A(s')$, the quantity of possible discrete geometries, $M_{geo}(s'')$, increases exponentially, because now the intermediate geometry in state s' can be adjusted specifically. Consequently, the quantity of possible discrete geometries $M_{geo}(s^+)$ in a final state s^+ of a pass schedule design increases exponentially with the number of action choices (i.e., passes) the algorithm can take into account before reaching the final geometry. With only four passes to reach the final geometry, however, the number of possible actions and realizable discrete geometries, M_{geo}, is significantly limited.

In addition, the reward, r_{geo}, obtained decreases rapidly over the narrow allowed final height range, so that the algorithm is usually not able to obtain the maximum geometry reward only by the action choice of the last pass. In order to be as close as possible to the target geometry and, thus, maximize its reward, the algorithm must also adjust the parameters of the previous passes in such a way that the last intermediate geometry in combination with the allowed discrete action choices can the target height be achieved as accurately as possible. While adjusting the parameters of previous passes, the algorithm accepts reduced force rewards in order to achieve higher geometry rewards, whereby the vertical patterns in the heat map of the force utilization arise (cf. Figure 9b). However, the algorithm cannot always perfectly hit the target height, even when taking previous passes into account, which results in fine patterns in the heatmap of the deviation from the final height. These patterns are significantly more distinct in the four-pass area than in the areas with six or eight passes, since the quantity of possible geometries, M_{geo}, is significantly more limited in this area as described above.

Due to the connection in the form of the reward function, $r(s,a,s')$, the described patterns can also be found in an attenuated form and partly overlaid in the heatmap of the collected total reward, r_{tot}. For example, the average reward in the four -pass area was higher than in the six- and eight-pass areas due to the negative force rewards and, also, the patterns from the deviations from the target height as well as the patterns observed in the force utilization were recognizable.

Behavior Analysis Using an Example Pass Schedule

In the two previous chapters, the results of the generated pass schedules were analyzed and evaluated on the basis of different partially averaged parameters. Therefore, in this chapter, a tangible example pass schedule is used to show how the algorithm was choosing the actions consisting of a bite ratio and height reduction in order to generate the optimized pass schedules.

The billet had an initial height and width of $h_i = b_i$ of 512.5 mm, which was reduced to a target height and width of $h_{f,target} = b_{f,target} = 267.5$ mm over the course of the forging process. The remaining process boundary conditions corresponded to those listed in Section 2.4. Since the implemented pass schedule design was based on the square forging strategy, the forging block was rotated by 90° around the longitudinal axis after each pass. The pass schedule (see Table 4) was generated using the action choices of the best agent (training process 15, iteration 45,000), and it contained all the necessary variables, such as ingot dimensions and process and plant characteristics, which are required to carry out the forging process.

Table 4. The pass schedule example generated using the overall best agent (training 15, iteration 45,000).

Pass Number	Initial Height	Final Height	Initial Width	Final Width	Initial Length	Final Length	Height Reduction	Bite Ratio	Bite Length
-	mm	mm	mm	mm	mm	mm	%	-	mm
Total	512.5	267.5	512.5	267.5	1500	5505.9	3–30	0.3–0.9	-
1	512.5	358.8	512.5	564.1	1500	1946.7	30	0.4	205.0
2	564.1	404.4	358.8	404.4	1946.7	2409.0	28.3	0.4	225.7
3	404.4	283.1	404.4	454.8	2409.0	3059.8	30	0.6	242.6
4	454.8	327.8	283.1	327.8	3059.8	3667.7	27.9	0.6	272.9
5	327.8	245.8	327.8	350.8	3667.7	4568.4	25	0.3	98.3
6	350.8	266.6	245.8	266.6	4568.4	5542.0	24	0.3	105.3
Time Stamp [1]	**Rotation Angle** [1]	**Number of Strokes**	**Forging Direction**	**Tool Speed**	**Mean Temperature** [1]	**Min. Temperature** [1]	**Reheat** [1]	**Maximum Force**	**Mean Force**
s	°	-	-	mm/s	°C	°C	-	MN	MN
470.3	-	-	feed	-	1000	1000	-	-	-
46.9	90	8	feed	100	993.7	-	-	19.41	19.11
102.6	90	9	feed	100	986.6	732.0	no	15.21	14.31
157.5	90	10	feed	100	979.2	-	-	19.57	18.52
226.4	90	12	feed	100	968.3	661.5	no	14.41	14.31
346.1	90	37	feed	100	946.2	-	-	9.16	7.05
470.3	90	44	feed	100	922.0	598.3	yes	8.32	6.10

[1] At the end of the pass.

In addition to the maximum force, a mean force was listed to provide the machine operator with a better estimation of the expected forces during the forging process. This was calculated from the individual forces of all strokes except those at the ends of the workpiece, since smaller bite widths can occur in these due to the bite shift or smaller final strokes.

An examination of the selected height reductions and bite ratios, which are representative for the training sessions performed, shows that the algorithm tended to select the largest possible height reductions, ε_h, in all but the last passes combined with rather small bite ratios, $s_{B0}h_0$. Only if the target force is not reached by this strategy, it increases the bite ratio until the desired force is attained. Since the final height is defined in the last two passes and the force must not exceed the force limit, the action selection of the algorithm directly results from this. Bite ratio and height reduction are chosen in a way that the final height is hit as exactly as possible, taking the occurring spread into account and, consequently, maximizing the received reward, r_{geo}. As already described in Section 3.2, due to the discrete action choices, it may be necessary to deviate slightly from the described selection behavior at certain times in order to obtain the maximum geometry and force reward.

Utilizing this procedure, the algorithm hits the desired final geometry, represented by the target height (=target width), excellently and uses the available press force well at the same time. In addition, the tendency towards small bite ratios results in a tendency towards less spread and, thus, a lower number of passes. Furthermore, as soon as the bite ratio is increased to generate the desired target force, the number of strokes per pass decreases, which leads to a reduction in the total process time. Overall, this procedure for the pass schedule design of the algorithm is adequate and reasonable in the context of the selected optimization goals. Nevertheless, in the shown example, the parameter selection of the last pass may be different in an experience-based pass schedule, since the algorithm selected a small bite ratio of 0.3 and, hence, did not utilize the forging press to its capacity. This choice resulted from the fact that in the last pass, the reward of the generated force was omitted and only the deviation from the final height was evaluated. In combination with the discrete action selection, the chosen combination of bite ratio and height reduction presumably maximized the received reward in the last pass. In industrial processes, a larger bite ratio may be more appropriate, because the press force is not yet exhausted, and a higher bite ratio would reduce the process time even further due to the decreasing stroke numbers in the last pass. In this case, the height reduction would also have to increase, as a larger bite ratio is associated with more width spread. Such an increase of bite ratio and height reduction is only possible and useful as long as the increased width spread does not cause higher number of passes.

3.3. Experimental Validation

In order to validate the applicability of the generated optimized pass schedule design via an DDQL algorithm, a forging experiment was carried out on the forging plant of the Institute of Metal Forming of the RWTH Aachen University (IBF). The setup consisted of a 6.3 MN forging press and a 6 axis heavy-duty robot. Since the pass schedule design presented so far was based on an industrial-scale forging process, first, the process parameters in the algorithm implementation were adapted to the conditions at the IBF and, afterwards, the adjusted algorithm was trained again. Therefore, a quadratic ingot cross-section of 150 mm × 150 mm (height × width) was chosen, since these are the biggest ingot dimensions that can be reliably gripped and manipulated. Hence, to ensure a reasonable process design and to match the ingot dimensions the maximum press force, F_{max}, was reduced from 25 MN to 3.5 MN, resulting in a target force, F_{target} = 2.8 MN. The permissible initial and target heights of the forging block were set to h_i = 135–165 mm and $h_{f,target}$ = 75–105 mm. The initial block length, l_i, corresponded to 500 mm. In addition, the press speed was halved to 50 mm/s and the maximum height reduction was reduced from 30% to 25%. The forging ingot was made of a low-cost low alloyed steel of grade C45. The material-dependent parameters for the temperature and spread calculation as well as the yield stress were changed accordingly [1,25]. Furthermore, the process time calculation was adjusted to the real process conditions at the IBF, e.g., by adding a transport time from the oven to the press of 45 s as well as break times of 15 s between individual passes. This is necessary because full automation was assumed in the so far considered industrial process window, which was not achieved at the IBF forging plant to this degree. All other process and algorithm parameters remain unchanged (cf. Sections 2.3 and 2.4) and the adapted DDQL algorithm was trained only once.

Even with the new significantly different process window, the algorithm successfully converged after 50,000 training iterations against an average reward of $\overline{r_{tot}}$ = ~0.63. Due to the changed geometry spectrum, the generated pass schedules contained more passes, which is why the overall reward was lower compared to the industrial process. The best agent occurred after 35,000 training iterations and produced a mean number of passes of $\overline{n_p}$ = 6.56, a mean force utilization of $\overline{F_{use}}$ = 0.2 MN, and a mean deviation from the target end height of $\overline{\Delta_{h,target}}$ = 0.22%. Consequently, the formulated optimization goals for the pass schedule design were also achieved for this modified process window. The associated heat maps of the characteristic process and algorithm parameters of the best agent are appended to this publication (see Figure A4).

The pass schedule used in the experiment was generated in advance using the best agent of the successfully trained adapted DDQL algorithm and is listed in Table 5.

Table 5. Pass schedule of the experiment, generated with the adapted and retrained algorithm.

Pass Number	Initial Height	Final Height	Initial Width	Final Width	Initial Length	Final Length	Height Reduction	Bite Ratio	Bite Length
-	mm	mm	mm	mm	mm	mm	%	-	mm
Total	150	100	150	100	500	1125	3–25	0.3–0.9	-
1	150	112.5	150	165.6	500	603.9	25	0.65	97.5
2	165.6	126.6	112.5	126.6	603.9	701.4	23.5	0.65	107.6
3	126.6	96.3	126.6	139.7	701.4	836.6	24	0.7	88.7
4	139.7	108.1	96.3	108.1	836.6	961.9	22.6	0.7	97.8
5	108.1	94.1	108.1	114.3	961.9	1046.3	13	0.85	91.9
6	114.3	99.9	94.1	99.9	1046.3	1126.3	12.5	0.85	97.1

Time Stamp [1]	Rotation Angle [1]	Number of Strokes	Forging Direction	Tool Speed	Mean Temperature [1]	Minimum Temperature [1]	Reheat [1]	Maximum Force	Mean Force
s	°	-	-	mm/s	°C	°C	-	MN	MN
284	-	-	feed	-	1050	1000	-	-	-
81.6	90	5	feed	50	972.7	-	-	2.79	2.61
122.4	90	6	feed	50	935.5	710.7	no	2.51	2.28
164.8	90	8	feed	50	900.4	-	-	2.72	2.43
211	90	9	feed	50	864.7	658.6	no	2.53	2.15
249.8	90	11	feed	50	832.9	-	-	2.33	2.02
284	90	11	feed	50	805.7	622.1	yes	2.31	1.96

[1] At the end of the pass.

In the forging experiment at the IBF (cf. Figure 10), the positioning of the ingot before the first stroke of each pass was done manually. Consequently, the bite length and the press force of the strokes at the ingot ends could fluctuate. For this reason, the maximum press force of each pass, which occurred at one of the block's ends due to the stronger cooling, could not be used as a reliable benchmark. Instead of this, the averaged forces of the middle strokes, which were not affected by bite length fluctuation due to the manual handling, were compared. Hence, the compared mean forces (see Figure 11) were lower than the maximum predicted forces within a pass (cf. Table 5).

During the experiment, the process time and the occurring press forces were measured and, subsequently, the abovementioned averaged measured force, $\overline{F_{i,exp}}$, as well as the averaged predicted force, $\overline{F_i}$, were evaluated for every pass (see Figure 11). In addition, Table 6 gives an overview of the strokes taken into account and the standard deviations that occurred for each pass.

Figure 10. Photo from the forging experiment at the beginning of the first stroke of pass two.

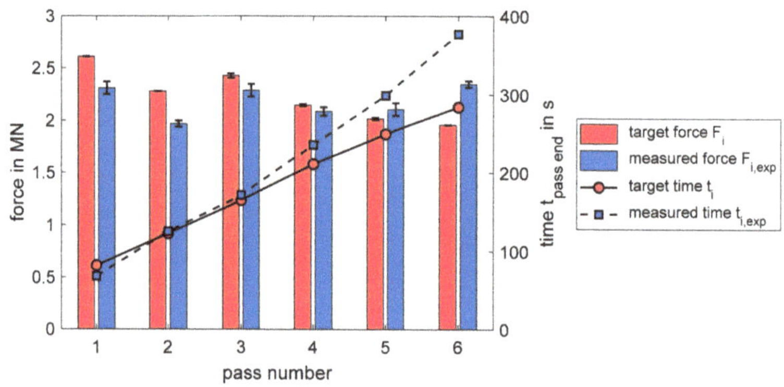

Figure 11. Plot of the expected and measured averaged forces and the measured time delay.

Table 6. Overview of the expected and measured averaged forces and their standard deviations.

Pass Number	Evaluated Strokes	Mean Target Force $\overline{F_i}$	SD $\overline{F_i}$	Mean Measured Force $\overline{F_{i,exp}}$	SD $\overline{F_{i,exp}}$	Time Difference [1]
-	-	MN	MN	MN	MN	s
1	2–4	2.61	0.00	2.31	0.06	−13.6
2	2–5	2.28	0.00	1.97	0.03	2.5
3	2–7	2.43	0.02	2.29	0.06	6.5
4	2–8	2.15	0.01	2.09	0.04	24.9
5	2–9	2.02	0.01	2.11	0.06	49.1
6	3–9	1.96	0.00	2.35	0.03	93

[1] Between the generated pass schedule and real experiment at the end of each pass.

Figure 11 shows that the created DDQL algorithm was able to predict the real press forces both qualitatively and quantitatively. The mean measured forces, $\overline{F_{i,exp}}$, deviated from the mean predicted forces, $\overline{F_i}$, by an average of 9.7% and a maximum of 19.9%, which emphasizes the estimated target force of $F_{target} = 0.8 \cdot F_{max}$ for the algorithm. Over the process, the measured force $F_{i,exp}$ increased continuously compared to the predicted force, F_i, and exceeded it from the fifth pass. The continuous increase could have resulted from the delay between the timeline of the experiment and the timeline of the designed pass schedule that increased from −13.6 s at the end of pass one to 93 s at the end of pass six (cf. Table 6 and Figure 11). This delay appeared despite the abovementioned adapted time calculation, since the break time in between different passes, needed for manual handling of the forging ingot, was slightly underestimated. The longer process duration can cause a stronger cooling of the forging ingot compared to the prediction of the algorithm, thus, to higher yield stresses and ultimately to higher press forces.

Moreover, deviating flow curves, model inaccuracies of the force and temperature calculation as well as real process conditions, such as fluctuating initial ingot temperatures or slight deviations in the chemical composition of the material, can lead to deviating press forces.

In combination with the results of the optimized pass schedule generation already presented (cf. Sections 3.1 and 3.2), the findings of the forging experiment accentuate the capability of the generated DDQL algorithm to produce optimized pass schedules for real open-die forging processes. Here, the press forces were predicted with sufficient accuracy and the generated pass schedules efficiently used the available press force. In addition, the final height was reliably reached, and the number of passes was reduced to a minimum within the given conditions.

4. Conclusions and Outlook

In this paper, a DDQL algorithm was presented that learned to design optimized pass schedules autonomously with the help of fast process models. The training process was stable and independent of the selected process window (industrial-scale forging/test forging at institute), the selected material (42CrMo4/C45), and the random seed. Moreover, the algorithm converged reliably in all training scenarios. This shows that the selected algorithm and its hyperparameters are well suited for the introduced problem of pass schedule design in open-die forging and, at the same time, emphasizes the transferability of the presented approach. The algorithm can be flexibly transmitted to other square forging processes by changing the process boundary conditions as well as the geometry, while the retraining does not require an adjustment of the hyperparameters. After the training process that takes about two days, the algorithm was able to design optimized pass schedules in two to four seconds, which makes the concept of pass schedule design in open-die forging using a DDQL algorithm altogether well suited for industrial use.

The pass schedules generated by the trained algorithms fulfilled the formulated goals. They created the desired final geometry while maintaining the permissible target height deviation of one percent. At the same time, with an average deviation of less than 5%, a very good utilization of the available press force was realized (cf. Section 3). In addition, the maximum press force was never surpassed in all 6000 pass schedules of the 15 training outcomes, and the average number of passes was reduced to a minimum $\overline{n_p} = 5.6$ within the given conditions. This also shows the developed reward function was a good mathematical formulation of the defined optimization goals, since all set goals could be achieved in the course of the trainings.

During the development of the DDQL algorithm, the important role of the reward function became clear, because the reward function first had to be extended by reward clipping (see Section 2.3) and the weightings of the individual reward parts had to be varied before a stable training process with useful results could be achieved. Consequently, the reward function not only determines the quality of the generated pass schedules but also contributes significantly to the success of the training process. In addition, the average number of passes, $\overline{n_p}$, over the training process (see Figure 8d) showed a slight scatter in the convergence limits. This could be an indication that the global maximum of the reward function was not clearly defined and an improvement in this could lead to even better convergence properties. Overall, the development of a good reward function is a crucial factor in the application of reinforcement learning, which is why a more detailed investigation and optimization of the reward function is desirable.

Despite the good results and the finely discretized evaluation with 400 sampled schedules each, the utilized neural networks represent black boxes which display behavior that cannot be predicted for each individual case. For this reason, a pass schedule designed via machine learning should not go unchecked before it is used in the production process. To assure this, for example, the calculated maximum press force and the final geometry can be output, and the observance of the strict limits can be checked beforehand either automatically or by a forging press operator.

Furthermore, the usage of heat maps to visualize evaluation results can help to gain deeper insights into the normally opaque training process of neural networks. The training results of an input parameter space can be clearly displayed and, thus, interpreted by the user. This facilitates, for example, the creation of practicable reward functions, because it is possible to comprehend which process parameters are changed in what way and at what time during the training. In addition, the use of heat maps also enables the user to find correlations between different process and algorithm parameters, such as the influence of the discrete action selection on the target height deviation and the force utilization, which caused patterns in the corresponding heat maps (cf. Section 3.2).

Due to the reliable convergence of the algorithm and the good achievements of the optimization goals, it can be concluded that the presented DDQL algorithm offers the same possibilities for optimized pass schedule design as conventional pass schedule calculation tools (see Section 1). Therefore, the presented algorithm serves as proof of concept for the optimized pass schedule design in open-die forging via double deep Q-learning and, furthermore, as a starting point for further work.

Since the presented implementation of the DDQL algorithm was based on rather simple forging processes an extension to longer or more complex forging processes with different ingot cross-sections, for example, octagonal seems useful. The required fast process models for octagonal cross-sections are available [23].

Furthermore, the presented algorithm used neural networks that can map highly non-linear relationships; hence, the chosen approach is ideally suited to include other important process and material parameters in the pass schedule design. In the future, the environment could be extended by an already existing semi-empirical fast model for microstructure calculation [50], so that, for example, the generation of a defined grain size may be included in the goals for the optimized pass schedule generation. In addition, an extension of the action selection of the agent by a variable bite shift is conceivable; hence, the agent has the possibility to create homogeneous equivalent strain distributions. With such additions, the functional range of the DDQL algorithm would exceed that of the conventional pass schedule calculation approaches and could therefore contribute to continuous quality improvement of open-die forgings.

The successful offline design of optimized pass schedules using a DDQL algorithm also opens up the opportunity for an online process optimization, since an online process control can be understood as a process design starting from an intermediate process state. Furthermore, the calculation times of the pass schedule design are sufficiently short, and the neural networks used in the reinforcement learning algorithm may be able to depict the more complex correlation in an online process control. Hence, in combination with an existing microstructure model [50] even the online control of the microstructure seems possible.

Author Contributions: Investigation, N.R.; Conceptualization; N.R., F.R., S.G., D.B., and G.H.; writing—original draft preparation, N.R. and F.R.; writing—review and editing, N.R., F.R.; S.G., D.B., and G.H.; supervision, S.G., D.B., and G.H. All authors have read and agreed to the published version of the manuscript.

Funding: This work was funded by the Deutsche Forschungsgemeinschaft (DFG, German Research Foundation)—452539735.

Data Availability Statement: The program code of the fast process models (Section 2.1) and of the described DDQL algorithm (Section 2.3) is not publicly available.

Conflicts of Interest: The authors declare no conflict of interest.

Appendix A

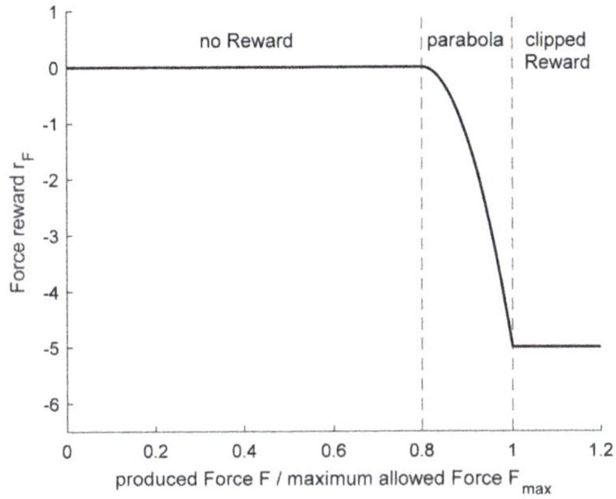

Figure A1. Plot of the Force reward, r_F, over the ratio of the produced force, F, and the maximum allowed Force, F_{max}, in case the current height, h, is equal or lower than the maximum final height $h_{f,ub}$, and thus the related state, s, is a terminal state, s^+.

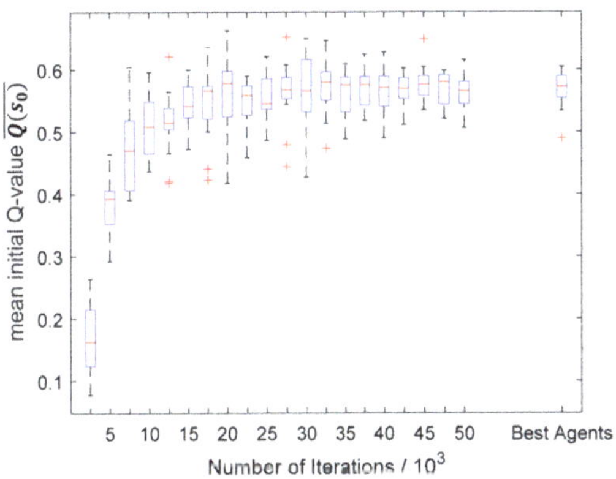

Figure A2. Boxplots of the mean Q-value of the initial state, s_0, of all 15 trainings over the course of the training process.

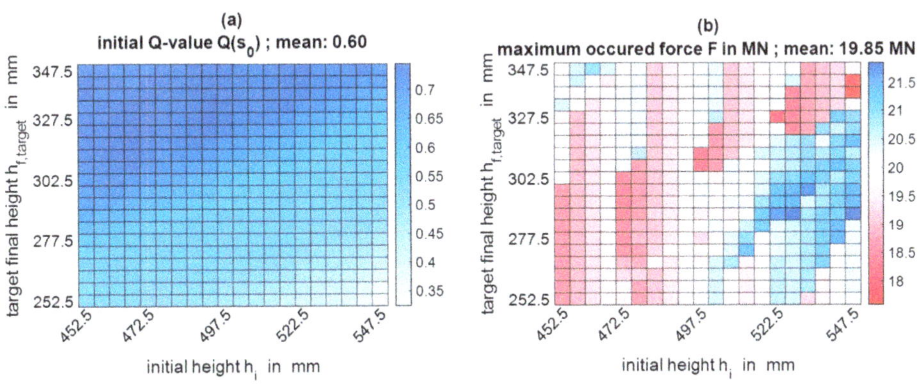

Figure A3. Heatmaps of the training results of the overall best agent (training 15, iteration 45,000): (**a**) initial Q-value $Q(s_0)$ and (**b**) the maximum occurred force in a single pass schedule, F.

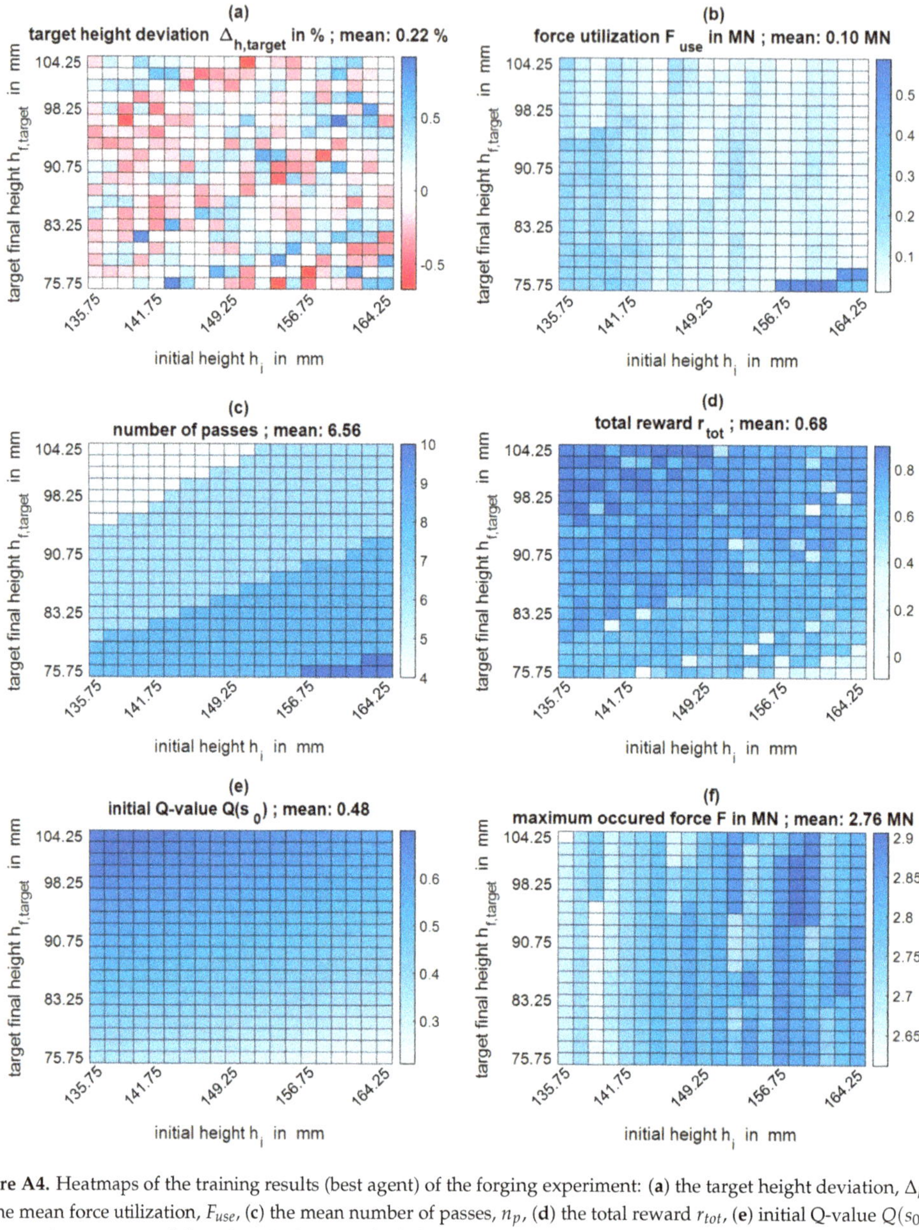

Figure A4. Heatmaps of the training results (best agent) of the forging experiment: (**a**) the target height deviation, $\Delta_{h,target}$, (**b**) the mean force utilization, F_{use}, (**c**) the mean number of passes, n_p, (**d**) the total reward r_{tot}, (**e**) initial Q-value $Q(s_0)$, and (**f**) the maximum occurred force in a single pass schedule, F.

References

1. Tomlinson, A.; Stringer, J.D. Spread and elongation in flat tool forging. *J. Iron Steel Inst.* **1959**, *193*, 157–162.
2. Mannesmann Demag Hüttentechnik. *MD-Dataforge—The Program Control System for MD-Open Die Forge Plants;* Mannesmann Demag Hüttentechnik: Mönchengladbach, Germany, 1990.
3. Bombač, D.; Fazarinc, M.; Kugler, G.; Turk, R. Tool for programmed open-die forging–case study. *RMZ-Mater. Geoenviron.* **2008**, *55*, 19–29.

4. Kakimoto, H.; Choda, T.; Takamori, H.; Takahashi, Y.; Takahara, H.; Takahashi, I. Renewal of Oil Hydraulic Press and Optimization of Forging Pass Design to Improve Productivity. In Proceedings of the 16th Internationl Forgemasters Meeting, Sheffield, UK, 15–19 October 2006; pp. 231–240.
5. Kakimoto, H.; Takashi, Y.; Takamori, H.; Tanaka, T.; Imaida, Y. Process Design of Extend Forging Process Using Numerical Simulation Development of Process Design Method for the Finish Forging Process. *Mater. Trans.* **2009**, *50*, 1998–2004. [CrossRef]
6. SMS Group GmbH. *Open-Die Forging Plants: Tailormade Solutions*; SMS Group GmbH: Mönchengladbach, Germany, 2021. Available online: https://www.sms-group.com/press-media/media/downloads/download-detail/download/17560 (accessed on 26 February 2021).
7. Nieschwitz, P.; Ecken, F.-P.; Siemer, E. COMFORGE Program for pass-schedule calculation in open-die forging. In *Verein Deutscher Eisenhüttenleute VDEh*; Stahl und Eisen, No. 12; Association of Steel and Ironworks: Düsseldorf, Germany, 1988; pp. 55–60.
8. Schmitz, W. ForgeBase—A Tool for Optimization of the Open-Die Forging Process. In Proceedings of the 15th International Forgemasters Meeting, Kobe, Japan, 26–29 October 2003.
9. Yan, C.; Gao, F. Simulation of programmed incremental open-die forging processes based on spread coefficient and geometric transformation. *Proc. Inst. Mech. Eng. Part E J. Process Mech. Eng.* **2010**, *224*, 129–135. [CrossRef]
10. Sauer, C.; Küstner, C.; Schleich, B.; Wartzack, S. Einsatz von Deep Learning zur ortsaufgelösten Beschreibung von Bauteileigenschaften. *Des. X-Beiträge Zum* **2017**, *28*, 49–60.
11. Sauer, C.; Schleich, B.; Wartzack, S. Deep learning in sheet-bulk metal forming part design. In Proceedings of the DESIGN 2018 15th International Design Conference, Dubrovnik, Croatia, 21–24 May 2018; Faculty of Mechanical Engineering and Naval Architecture, University of Zagreb: Zagreb, Croatia; The Design Society: Glasgow, UK, 2018; pp. 2999–3010.
12. D'Addona, D.M.; Antonelli, D. Neural Network Multiobjective Optimization of Hot Forging. *Procedia CIRP* **2018**, *67*, 498–503. [CrossRef]
13. Guo, P.; Yu, J. Optimal Control of Blank Holder Force Based on Deep Reinforcement Learning. In Proceedings of the 2019 IEEE International Conference on Industrial Engineering and Engineering Management (IEEM), Macao, China, 15–18 December 2019; IEEE: Piscataway, NJ, USA, 2019; pp. 1466–1470, ISBN 978-1-7281-3804-6.
14. Dornheim, J.; Link, N. Multiobjective Reinforcement Learning for Reconfigurable Adaptive Optimal Control of Manufacturing Processes. In Proceedings of the 2018 International Symposium on Electronics and Telecommunications (ISETC), Timisoara, Romania, 8–9 November 2018. Available online: https://arxiv.org/pdf/1809.06750 (accessed on 17 June 2021).
15. Dornheim, J.; Link, N.; Gumbsch, P. Model-Free Adaptive Optimal Control of Episodic Fixed-Horizon Manufacturing Processes using Reinforcement Learning. *Int. J. Control Autom. Syst.* **2019**, *18*, 1593–1604. [CrossRef]
16. Stoerkle, D.D.; Seim, P.; Thyssen, L.; Kuhlenkoetter, B. Machine Learning In Incremental Sheet Forming. In Proceedings of the ISR 2016: 47st International Symposium on Robotics, Munich, Germany, 21–22 June 2016; VDE: Berlin, Germany; IEEE: Piscataway, NJ, USA, 2016.
17. Kim, P.H.; Chun, M.S.; Yi, J.J.; Moon, Y.H. Pass schedule algorithms for hot open die forging. *J. Mater. Process. Technol.* **2002**, *130-131*, 516–523. [CrossRef]
18. Meyes, R.; Tercan, H.; Thiele, T.; Krämer, A.; Heinisch, J.; Liebenberg, M.; Hirt, G.; Hopmann, C.; Lakemeyer, G.; Meisen, T.; et al. Interdisciplinary Data Driven Production Process Analysis for the Internet of Production. *Procedia Manuf.* **2018**, *26*, 1065–1076. [CrossRef]
19. Pahnke, H.J. Grundlagen des programmierten Schmiedens. In *Verein Deutscher Eisenhüttenleute VDEh*; Stahl und Eisen; Association of Steel and Ironworks: Düsseldorf, Germany, 1983; Volume 103, pp. 547–552.
20. Recker, D. *Entwicklung von Schnellen Prozessmodellen und Optimierungsmöglichkeiten für das Freiformschmieden*; Shaker: Aachen, Germany, 2014; ISBN 9783844028478.
21. Tomlinson, A.; Stringer, J.D. The Closing of Internal Cavities in Forgings by Upsetting. *J. Iron Steel Inst.* **1958**, *196*, 209–217.
22. Nieschwitz, P. *Wege zur Qualitätssteigerung beim Freiformschmieden*; Stahleisen: Düsseldorf, Germany, 1988; ISBN 9783514004092.
23. Rosenstock, D. *Schnelle Prozessmodellierung, Online-Visualisierung und Optimierung beim Freiformschmieden*; Verlagshaus Mainz GmbH: Aachen, Germany, 2017; ISBN 9783958862043.
24. Rosenstock, D.; Recker, D.; Hirt, G. Online Analysis of Cogging Processes–Evaluation of the Equivalent Strain, Temperature and Grain Size Distribution. In Proceedings of the 19th International Forgemasters Meeting, Tokyo Bay Area, Japan, 29 September–3 October 2014; pp. 507–512.
25. Spittel, M.; Spittel, T..; Warlimont, H. (Eds.) *Steel symbol/number: 42CrMo4/1.7225, Landolt-Börnstein–Group VIII Advanced Materials and Technologies 2C2 (Metal Forming Data of Ferrous Alloys—Deformation Behaviour)*; Springer: Berlin/Heidelberg, Germany, 2009. [CrossRef]
26. Stenzhorn, F. Beitrag zur Empirsch-Theoretischen Vorausplanung des Freiformschmiedens Großer Blöcke mit Hohlräumen. Ph.D. Dissertation, RWTH Aachen, Aachen, Germany, 1982.
27. Kopp, R.; Stenzhorn, F. Zum Problem des Schließens von Hohlräumen beim Freiformschmieden-Ergebnisse aus Labor-und Betriebsversuchen. In *Verein Deutscher Eisenhüttenleute VDEh*; Stahl und Eisen; Association of Steel and Ironworks: Düsseldorf, Germany, 1983; Volume 103, pp. 377–382.
28. Siemer, E. *Qualitätsoptimierende Prozesssteuerung des Reckschmiedens*; Stahleisen: Düsseldorf, Germany, 1987; ISBN 3514003920.
29. Franzke, M.; Schneider, S.; Li, X.; Recker, D. Online Analysis of Grain Size Evolution for Open Die Forging Processes. *Int. Meet.* **2008**, 370–375.

30. Recker, D.; Franzke, M.; Hirt, G. Fast models for online-optimization during open die forging. *CIRP Ann. Manuf. Technol.* **2011**, 295–298. [CrossRef]
31. Sutton, R.S.; Barto, A.G. *Reinforcement Learning: An Introduction*, 2nd ed.; in progress; The MIT Press: Cambridge, MA, USA, 2014.
32. White, C.C.; White, D.J. Markov decision processes. *Eur. J. Oper. Res.* **1989**, *39*, 1–16. [CrossRef]
33. Watkins, C. Learning from Delayed Rewards. Ph.D. Dissertation, King's College, London, UK, 1989.
34. Watkins, C.; Dayan, P. Q-learning. *Mach. Learn.* **1992**, *8*, 279–292. [CrossRef]
35. Mnih, V.; Kavukcuoglu, K.; Silver, D.; Graves, A.; Antonoglou, I.; Wierstra, D.; Riedmiller, M. Playing Atari with Deep Reinforcement Learning. *arXiv* **2013**, arXiv:1312.5602.
36. Mnih, V.; Kavukcuoglu, K.; Silver, D.; Rusu, A.A.; Veness, J.; Bellemare, M.G.; Graves, A.; Riedmiller, M.; Fidjeland, A.K.; Ostrovski, G.; et al. Human-level Control Through Deep Reinforcement Learning. *Nature* **2015**, *518*. [CrossRef]
37. van Hasselt, H. Double Q-learning. In *NIPS'10: Proceedings of the 23rd International Conference on Neural Information Processing Systems—Volume 2*; Curran Associates Inc.: Red Hook, NY, USA, 2010; pp. 2613–2621.
38. van Hasselt, H.; Guez, A.; Silver, D. Deep Reinforcement Learning with Double Q-Learning. *arXiv* **2015**, arXiv:1509.06461.
39. Okuyama, T.; Gonsalves, T.; Upadhay, J. Autonomous Driving System based on Deep Q Learnig. In Proceedings of the 2018 International Conference on Intelligent Autonomous Systems (ICoIAS'2018), Singapore, 1–3 March 2018; Institute of Electrical and Electronics Engineering Inc.: Piscataway, NJ, USA, 2018; pp. 201–205, ISBN 978-1-5386-6329-5.
40. Han, X.; He, H.; Wu, J.; Peng, J.; Li, Y. Energy management based on reinforcement learning with double deep Q-learning for a hybrid electric tracked vehicle. *Appl. Energy* **2019**, *254*, 113708. [CrossRef]
41. Xia, K.; Sacco, C.; Kirkpatrick, M.; Saidy, C.; Nguyen, L.; Kircaliali, A.; Harik, R. A digital twin to train deep reinforcement learning agent for smart manufacturing plants: Environment, interfaces and intelligence. *J. Manuf. Syst.* **2021**, *58*, 210–230. [CrossRef]
42. Jiang, A.; Osamu, Y.; Chen, L. Multilayer optical thin film design with deep Q learning. *Sci. Rep.* **2020**, *10*, 12780. [CrossRef]
43. Mathworks. Create Policy and Value Function Representations—MATLAB & Simulink. Available online: https://www.mathworks.com/help/reinforcement-learning/ug/create-policy-and-value-function-representations.html (accessed on 9 November 2020).
44. Mathworks. Options for DQN Agent—MATLAB. Available online: https://www.mathworks.com/help/reinforcement-learning/ref/rldqnagentoptions.html (accessed on 9 November 2020).
45. Deutsches Institut für Normung. *Schmiedestücke aus Stahl: Bearbeitungszugaben und zulässige Abweichungen*; DIN 7527-6:1975-02; Beuth Verlag GmbH: Berlin, Germany, 1975. [CrossRef]
46. Forging Industry Association. *Open Die—Articles: Forging Solutions—Design Engineering Information From FIA*; Forging Industry Association: Cleveland, OH, USA, 2007. Available online: https://www.forging.org/uploaded/content/members/field_document/OpenDieArticles.pdf (accessed on 10 June 2021).
47. Kojima Iron Works Co., Ltd. Hydraulic Forging Press. Available online: https://kojimatekko.co.jp/en/product/press02.html (accessed on 10 June 2021).
48. SMS Group GmbH. Open-Die Forging Presses. Available online: https://www.sms-group.com/plants/all-plants/open-die-forging-presses (accessed on 10 June 2021).
49. Macrodyne Technologies Inc. Forging Presses. Available online: https://macrodynepress.com/hydraulic-presses/forging-presses/ (accessed on 10 June 2021).
50. Karhausen, K.; Kopp, R. Model for integrated process and microstructure simulation in hot forming. *Steel Res.* **1992**, *63*, 247–256. [CrossRef]

processes

Article

Optimization of the Sustainable Distribution Supply Chain Using the Lean Value Stream Mapping 4.0 Tool: A Case Study of the Automotive Wiring Industry

Yousra El Kihel [1,*], Ali El Kihel [2] and Soufiane Embarki [3]

[1] IUT of Bordeaux-Bastide, Department MLT, University of Bordeaux, 33072 Bordeaux, France
[2] Faculty of Sciences and Technologies, Hassan 1st University, Settat 26000, Morocco
[3] Industrial Engineering Laboratory, National School of Applied Sciences, Oujda 60000, Morocco
* Correspondence: elkihel.yosra@gmail.com; Tel.: +33-66-696-9023

Abstract: The transformation to Supply Chain (SC) 4.0 promises new opportunities for companies to gain competitiveness. The Lean Value Stream Mapping (VSM) tool allows the supervision of all the processes of the entire SC, from which we can identify the different types of waste that hinder the competitiveness of the SC. Following the existing problems detected with the help of a diagnostic, we will propose a new process design by integrating 4.0 technologies to modernize the company. For our case study, we treat the multinational SC of Automotive Wiring Equipment Morocco, where we will focus on the downstream part of the SC composed of the warehouse and the different stages of road and sea transport until the final delivery in Austria. Then, we will consider the opportunities offered by each country in terms of logistics competitiveness. In this research work, we will show how Lean VSM4.0 will contribute to sustainable development by integrating the three pillars economic, environmental, and social. With the Lean VSM 4.0 tool, all logistic processes will be visualized, from which improvements could be obtained, especially the optimization of the lead-time, the cost, the energy consumed, and the follow-up of the products during the whole SC while reducing accidents.

Keywords: VSM; durability; logistics 4.0; methodology; automobile; wiring

Citation: Kihel, Y.E.; Kihel, A.E.; Embarki, S. Optimization of the Sustainable Distribution Supply Chain Using the Lean Value Stream Mapping 4.0 Tool: A Case Study of the Automotive Wiring Industry. *Processes* **2022**, *10*, 1671. https://doi.org/10.3390/pr10091671

Academic Editor: Zhiwei Gao

Received: 20 July 2022
Accepted: 16 August 2022
Published: 23 August 2022

1. Introduction

Currently, there is increased competition between companies in the automotive industry. The automotive equipment manufacturers, cabling, and others are characterized by very high competitiveness and diversity of products and distribution on an international scale. Morocco has become the first construction hub on the African continent, ahead of South Africa and Egypt, with distribution on an international scale [1]. Therefore, it is necessary for each company to adapt to new technological changes to increase production and find solutions for improving the flexibility of its distribution system by delivering to customers at a low cost with a deadline and optimal quality [2].

Enterprise 4.0 refers to a new generation of connected and intelligent factories. It is a digital transformation characterized mainly by intelligent automation and the integration of new technologies into the enterprise value chain, giving life to an interconnected enterprise 4.0 where machines, products, and people interact [3]. The digitization of the SC, according to Industry 4.0, therefore, promises new opportunities to develop more efficient supply, production, and distribution lines, while supervising the entire supply line.

Digital twins play a crucial role in supporting and improving industrial manufacturing processes [4,5]. Traditional value stream mapping (VSM) methods are suitable for planning-based production process reengineering. Some studies introduce the methodological contribution to integrating digital twins and VSM for SMEs. This approach is based on a simulation to improve quantitative analysis in the production process by reengineering tasks and logistics [6] and provides a framework based on the digital twin to use physical product data, virtual product data, and connected data that link physical and virtual products for product design, manufacturing, and service. It is possible to design digitized views of the different actors in the chain and simulate their reactions. The digital twins involve partners, suppliers, customers, etc., and all actors in the company's supply chain.

VSM 4.0 is one of the effective tools to analyze, optimize the process, and eliminate waste [7].

The VSM 4.0 method, which is easy to implement, makes it possible to identify, with the help of a diagnostic, classic waste as well as information wastes and to design lean value flows in terms of products and, above all, of information flows [8]. Thanks to the VSM 4.0 application, it is now possible to perform a value stream analysis, to create a new value stream design based on new digital technologies [9].

Our problem involves studying, analyzing, and optimizing the efficiency and effectiveness of a logistic chain of the multinational Automotive Equipment Manufacturer EAKM located in Kenitra, which knows is in hard competition with other companies. We will first focus on the downstream part of the SC which concerns the shipping area, preparation of orders, and loading into trucks in the factory. Then, we will investigate the different stages of road and sea transport from Kenitra to Vienna according to the following route: Morocco, Spain, France, Italy until delivery to the EAVA warehouse in Austria. Following the different types of existing waste identified, we will propose modern technologies to be integrated into the logistic line as solutions for a better performance of the whole SC. The SC functions will be connected by 4.0 technologies, with a visualization of all the tasks with the VSM 4.0 tool. The results obtained are very satisfactory, particularly the collection of data, the optimization of the functions of the operators in Morocco and Vienna, automation of several workstations, tracking of products throughout the supply chain, and optimization of transport time between the two countries.

For the working methodology, we present a literature review on VSM 4.0, its usefulness in Lean Logistics, and the different modern 4.0 technologies compatible with VSM with their advantages. Next, we perform a VSM mapping of the current state and define the different tasks of the process. Finally, by analyzing the current state, we will propose improvements for establishing the future state of our VSM 4.0. The results obtained are also part of a sustainable development framework through improving the economic, social, and environmental components.

2. Literature Review

2.1. Lean Production System

In manufacturing systems, research is focused on reducing process variability, increasing productivity, and developing new processing methods to reduce costs, implement predictive models, increase quality, and adopt intelligent automation and robotics solutions by adopting lean manufacturing [10].

The Lean concept is a solution to reduce the complexity of automated production systems. One of the latest developments of the adaptive production system is Lean Automation, which allows the system to adapt to future market requirements by achieving high flexibility and short-flow information. This new theory offers the perspective of integrating the Industry 4.0 concept into a Lean production system [11].

According to Lean theory, there is a set of tools and applications through which waste is identified and eliminated or reduced. Also, a set of Industry 4.0 fundamental technologies has been identified through which losses are identified, managed, or eliminated [12].

In this research work, we are interested in the lean VSM tool and how the new technologies of Industry 4.0 can contribute to it. In addition, VSM 4.0 eliminates redundant steps in information logistics along the value chain.

In the following paragraph, we conduct a literature review on SCOPUS to further our research on this topic.

2.2. Research in VSM 4.0 Published between 2016 and 2021

We used Scopus, a scientific research platform, to determine the number of publications on VSM 4.0 between the years 2016 and 2021 (Figure 1).

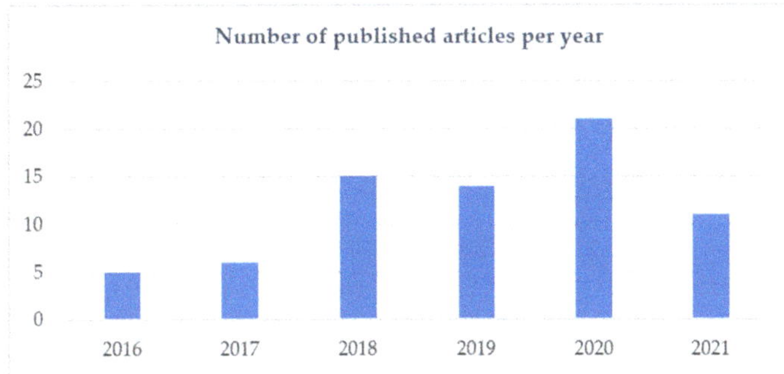

Figure 1. Number of publications from 2016 to 2021 on VSM 4.0.

According to the results of the bibliographic research, the total number of articles remains low, showing that the VSM 4.0 is a new and very interesting topic, which researchers and industrialists are developing more and more.

2.2.1. Publications of the Theme in Relation to New Technologies between 2016 and 2021

We conducted a more in-depth bibliographic study on the SCOPUS search engine only for scientific articles that were published between the years (2016–2021).

The objective of this study was to show the relationship between 4.0 technologies and the VSM tool used in the literature or in the industry. The results obtained are presented in Figure 2. We noticed that IoT technologies and Big Data are the most frequently cited. In addition, the rest of the technologies, such as cloud computing, robots, artificial intelligence, etc., remain useful for the VSM 4.0. These results can be explained by the large amount of information to manage, the physical flow of products, the number of stakeholders, and the SC.

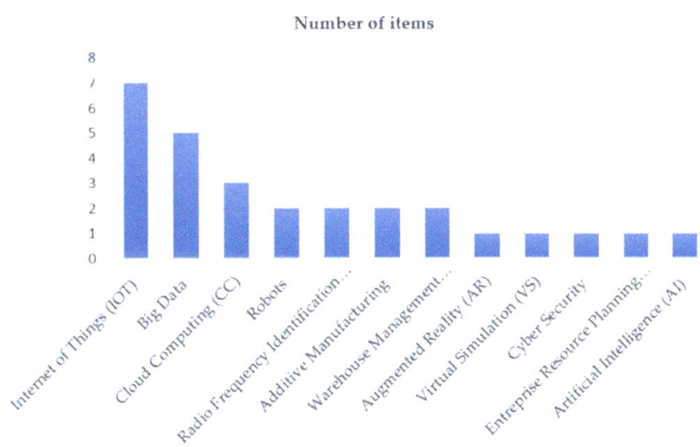

Figure 2. Number of publications of 4.0 technologies used in VSM4.0.

2.2.2. Publications sur VSM Classique et VSM 4.0

Table 1 presents the research work (17 articles) on the VSM tool with the integration of new technologies for the last few years and, particularly, the benefits it can offer to logistics activities.

Table 1. Literature review from 2015 to 2022.

N°	Year	Title	Authors	Type	Contenent
1	2022	Extending the lean value stream mapping to the context of Industry 4.0: An agent-based technology approach [12]	Ferreira, Armellini, F., Santa-Eulalia, Thomasset-Laperrière	Article	Integrating VSM with hybrid simulation (HS) and extending its scope to the Industry 4.0 context
2		A cyber-physical system architecture based on lean principles for managing industry 4.0 setups [13]	Nounou, Jaber, H., Aydin,	Article	Smart VSM 4.0 to improve material and information flow and autonomous decision making via IOT
3		An Integrated Value Stream Mapping and Simulation Approach for a Production Line: A Turkish Automotive Industry Case [14]	Aksar, O., Elgun, D., Beldek, T., Konyalıoğlu, A.K., Camgöz-Akdağ, H	Conference	To guide the automotive sectors by analyzing the current situation with the Lean concept and using the necessary technologies.
4	2020	VALUE STREAM MAPPING 4.0: A STRUCTURAL MODELING APPROACH [3]	Rania El Kammouni, Oualid Kamach, Malek Masmoudi	Article	VSM 4.0 for a collaborative work environment for management teams and production process optimization through digitization
5		Value Stream Mapping-based Logistics 4.0 Readiness for Thailand Automotive-Part Manufacturers [15]	Boonsothonsatit, G.; Tonchiangsai, K.; & Choowitsakunlert, S.	Article	The article discusses a methodology to transition to logistics 4.0 based on VSM and the adaptation of 4.0 technologies.
6		Industry 4.0-based real-time scheduling and dispatching in lean manufacturing systems [16]	Ramadan Muawia Salah Bachir	Article	An Industry 4.0 based Lean framework called DVSM dynamics to digitize Lean manufacturing with Industry 4.0 Lean tools and technologies.
7		Lean and Industry 4.0—How to Develop a Lean Digitalization Strategy with the Value Stream Method [17]	Schneider Markus	Article	Methodology that enables manufacturing companies to develop an individual digitalization strategy using Industry 4.0
8	2019	Industry 4.0: Development of a multi-agent system for dynamic value stream mapping in SMEs [18]	Huang, Z., Kim, J., Sadri, A., Dowey, S., Dargusch	Article	Application of VSM in SMEs is more difficult for complicated product processing and improvements in labor management and facility utilization.
9	2018	Sustainable value stream mapping and technologies of Industry 4.0 in manufacturing process reconfiguration [19]	Phuong, N.A., Guidat, T	Article	RFID, big data and usability improvement as solutions and the impact of their implementation on process sustainability
10		Value stream method 4.0: Holistic method to analyze and design value streams in the digital age [8]	Hartmann, L., Meudt, T., Seifermann, S., Metternich, J	Article	Promising new opportunities from digitalization and Industry 4.0 to integrate information flows.
11	2017	From value stream mapping to value stream management—how the static lean method can be further developed to a dynamic management approach using solutions of Industry 4.0 [20]	Lugert, A., Winkler, H.	Article	VSM is common in manufacturing companies and has advantages but also some shortcomings suffered by the current megatrends in the production environment.
12		Value stream mapping 4.0: Holistic examination of value stream and information logistics in production [21]	Joachim Metternich	Article	Mapping allows the analysis of process chains and helps to deduce improvement potentials

Table 1. *Cont.*

N°	Year	Title	Authors	Type	Contenent
13	2015	Reduction of Work in Process Inventory and Production Lead Time in a Bearing Industry Using Value Stream Mapping Tool [22]	Rajenthirakumar et Shankar	Article	A case study on the application of VSM in the automotive sector, resulting in a 67% reduction in cycle time by optimizing value-added activities
14		Deployment of Lean Management in a packaging workshop and change management [23]	Julia Flauder	Thèse	Lean should extend to all links in the supply chain, upstream and downstream of production, which has long been its focus.
15	2014	Reduction of Wastage Using Value Stream Mapping: Case Study [24]	Rajendra kumar	Article	The application of the VSM on the production flow shows that the non-value-added activities have been optimized such as waiting time, handling time . . .
16		Quality Value Stream Mapping [25]	B. Haefner, A. Kraemer, T. Stauss, and G. Lanza	Article	Q-VSM is a tool for the visualization, design and analysis of quality assurance measures for the manufacturing system in the electronics industry
17	2012	The transportation value stream map (TVSM) [26]	Bernardo Villarreal	Article	T-VSM modeling that provides a detailed description of storage and transportation operations

We find that the new VSM 4.0 method creates a digital and collaborative work environment for management teams and stakeholders in the Industry 4.0 era. It offers logistics production companies a huge benefit in the planning and optimization processes by digitizing the value to make improvements and more accurate decisions.

The objective of our research is to introduce digital and automatic tools to collect data, intelligent means of information processing, data storage medium, and key performance indicators (KPIs) so that the monitoring of the logistics activity and the information management system (IMS) are accurate and consistent and within the framework of sustainable development. The information collected can be visually represented in real-time to logistics managers.

In conclusion, the approach of VSM is very effective and often used in the literature for production with possible extensions to transport, quality, environment, etc. we want to use VSM as a model to support the analysis process of the downstream SC.

According to the literature review, this method is more applied to the automotive industry, especially to the procurement and production process, and very little work has begun to apply it to the downstream supply chain.

3. Lean Value Stream Mapping VSM Implementation Method

3.1. Value Stream Mapping

Value Stream Mapping was developed by engineer Taiichi Ohno at Toyota to identify and eliminate waste sources [27]. Value Stream Mapping (VSM) is a type of flowchart used to analyze, model, and propose improvements to the processes leading to the delivery of a good or service. It is one of the most frequently used formalisms in companies. Indeed, it is a key element of the Lean methodology to detect and eliminate waste. By grouping all the activities that make up the supply chain, the VSM tool aims to identify non-value-added processes that are considered wasteful. Following this teamwork, improvement solutions are developed to correct these bottlenecks to optimize the production flow [28]. The VSM tool takes into consideration physical flows, but especially information flows [29].

3.2. Advantages of the Lean VSM Tool in Logistics

According to our bibliographic study, several authors have shown that the VSM method is an essential tool in Lean Manufacturing and Supply Chains [2,28] for several reasons:

- It helps to go beyond the level of single individual processes and to visualize the entire supply chain as a whole (the VSM map starts with the arrival of the products at the warehouse and ends with the shipment of the products to the final customer);
- Detection of waste and indication of the causes;
- Providing a basis for discussing the importance of the various logistics processes;
- It constitutes a draft for discussion towards a lean approach, the outline of the plan of a future organization;
- The VSM map shows the links between the different product and information flows.

The VSM integrates three components: material flow, information flow, and scheduling. Building a VSM is often the first step in the transformation to lean. The VSM tool identifies problems and provides suggestions on how to transform the process to lean. In other words, the VSM can be used to identify where loss occurs and where value is added. It can also help a company understand what value it is actually providing to its customers [30].

3.3. VSM Map Design Process

To analyze the current process of the downstream SC, we have chosen the lean tool of VSM to map the value chain. This tool allows us to analyze all the tasks of the SC with the objective of optimizing the tasks that have a non-added value. This approach offers the possibility of federating human capital around common objectives focused on economic, social, and environmental performance. In this study, we present the development of this VSM project in 6 steps, as shown in Figure 3:

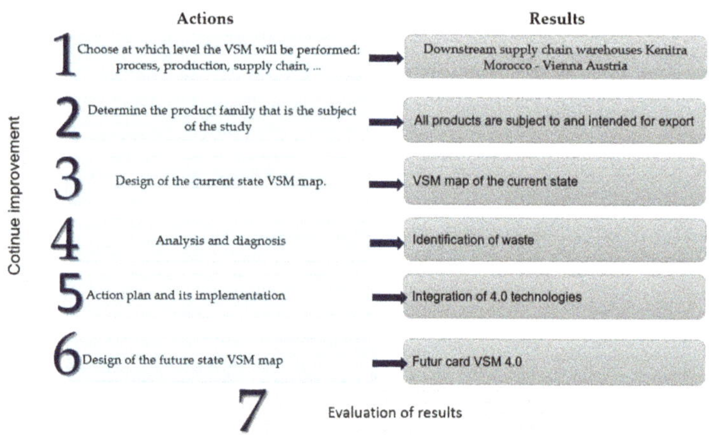

Figure 3. VSM design process.

4. Contribution of 4.0 Technologies in Logistics

Before approaching the case study of downstream logistics, it is fundamental to conduct a study that establishes the relationship between 4.0 technologies and logistics processes. This work is of great necessity and considered a decision support tool for the implementation of our approach

Table 2 presents a study of the most recent technologies and their impact on the different SC processes in different industries and modern logistics companies.

It is clear that there is a multitude of Industry 4.0 technologies and the disparity of their uses in the industrial world (Table 2). Each of the technologies mentioned can be used according to the needs of the company, the specificities of the sector, and the amount of investment that can be made by the companies. Negotiating the digital shift is imperative for companies today. Knowing the technological building blocks of Industry4.0 will allow decision-makers to ask themselves the right questions about their uses in accordance with the needs and priorities to be defined very early on.

Table 2. The different 4.0 technologies used in the industry.

	New Technologies	Contribution of the New Technologies on the Logical Processes
1	Big Data	- Comprehensive real-time data collection and evaluation - Rapid data processing - Support decision making - Analyze and separate important data - Support more effective knowledge transfer to achieve business goals [31].
2	Data storage in the cloud	- Reduces the initial investment required when implementing a WMS, since instead of acquiring the entire technical infrastructure, a monthly rent is paid for the license. - Large-scale processing - Flexible use for computing and storage [32].
3	IOT	- Related to the various sensors that collect information from the physical world and transfer it to digital databases [32]. - Improving Supply Chain Visibility - Accurate information in real time [33].
4	COBOTS	- Transport of goods - Order picking - Improve supply chain efficiency [31].
5	AM (Additive Manufacturing)	- AM refers to a set of automated processes that build products layer by layer, based on three-dimensional models designed in computer-aided design software. - Its objective is to standardize the packaging of the same size [34].
6	Drones	- Flying over the logistics sector with the aim of providing support for the performance of certain tasks, such as inventory management [28].
7	AI	- Simulate natural intelligence to interpret external data - Learn from this data for descriptive, predictive, or normative analysis in logistics. - Help generate supplies, anticipate customer orders in the warehouse [25].
8	AR (Augmented Reality)	- Works with virtual objects that overlap with the existing environment. - Virtual and real information, acquired with a camera, are digitally merged, and represented on a screen - Create an interface between employees and digital products or equipment [10].
9	VS (Virtual simulation)	VS is a computer-based modeling system that promotes real-time data to reflect the physical world in a virtual model that includes machines, products, logistics flows and humans [31]
10	RFID (Radio frequency identification)	- Identify and track without physical contact. - Read, store, and write information on electronic tags using radio waves [35]. - Composed of a chip, a tag (antenna role) and a reader. - To be read, the tags need to be present in the radio wave radius of the reader [35].
11	Smart Pack	- Connecting packages - Monitor the conditions of transport of goods - Collect time-stamped data on geolocation, shocks, temperature, and humidity - Transmit the data according to the chosen frequency [33].
12	Code-barre	- Bar code, or bar code (CAB), is the representation of a numerical or alphanumerical data in the form of a symbol made up of bars and spaces. -The thickness varies according to the symbology used and the data thus coded [36].
13	Power BI	- Power BI Warehouse Performance content was created so that warehouse and operations managers can monitor important inbound, outbound and inventory metrics. - It uses warehouse management, product, and other trans-actional data [37].

Table 2. *Cont.*

	New Technologies	Contribution of the New Technologies on the Logical Processes
14	WMS	- Guide the receipt and storage of inventory - Optimize order preparation and shipping - Advises on inventory replenishment [38].
15	Scan 3D	- A three-dimensional scanner is a device for scanning and 3D acquisition - Analyzes the objects or their close environment to collect precise information on the form and possibly on the appearance (color, texture ...) [39]. - This device helps us to facilitate the quality control of the goods, 3D modeling.
16	ERP	- Collects all data from a warehouse into a centralized database - Enable informed and intelligent decision making - By using an ERP system, inventory will be automatically integrated with all aspects of the business, from planning to production to accounting [40].
17	ATLS	- Robotic system by which pallets are automatically loaded and unloaded from trucks with little operator intervention. - The ideal solution to speed up and secure the receipt and dispatch of goods in the loading area of the warehouse [41].
18	AGV/LGV	- Self-guided vehicles, similar to forklifts, move autonomously and automatically, following a pre-determined or pre-programmed path. - To guide them, two systems are available: self-guided (AGV) or laser guidance (LGV) [25].
19	Stacker cranes	Automation of product entry and exit operations. Elimination of errors derived from manual management. Control and updating of inventory management [42].

Table 2, presents a study of the most recent technologies used in different industries and modern logistics companies.

We note that 4.0 technologies clearly contribute to the improvement of the supply chain and provide efficient solutions for forecasting the flow of supplies, sales, distribution, and location of products in the warehouse.

5. Case Study of an International Supply Chain in the Automotive Industry

5.1. Presentation of the Study

Our case study was carried out in a company of Automotive Equipment Manufacturers located in Morocco (EAKM) to deliver to its warehouse in Vienna (EAVA) and distribute products throughout Europe. Its official headquarters is in Austria. The group specializes in designing and realizing electrical cables for the automotive sector, whose production is done in Kenitra, Morocco. The production of electrical and electronic components is entirely intended for export and, more particularly, for the large car manufacturers in Europe.

The circuit through which the product passes is three stages and presented on Figure 4:

(1) Road transport from Kenitra to Tangier by truck;
(2) Sea transport from Tangier to Algeciras by a commercial ship;
(3) Road transport from Algeciras to Vienna by the same truck;
(4) It should be noted that for this transport, the goods remain in the truck from its loading in Kenitra until its unloading in Vienna, unlike the first method of transport which was the goods in containers.

Our work focuses on the downstream supply chain of the distribution process that transports the finished product to the company's warehouse in Vienna, Austria.

Figure 4. Road and sea transport route.

5.2. Analysis of the Current State

To analyze the current state of our value stream mapping, we will proceed through three steps:

- Identify the different steps in our supply chain, from order preparation to delivery to customers (see Table 3).
- Consider the logistics competitiveness of each country involved in this supply chain by proposing a new process.
- Present the map of the current state of the process. Each task is presented by a box attached with a data box that contains information such as the number of operators in each workstation with the cycle time of all tasks, Lead Time, Down Time, and Process Times.

Table 3. The tasks of the distribution process.

Activities	Definition
Product pickup	The accumulation of the stock of finished products requires several rounds of collection during the rotation of all the teams.
Grouping of packages	By grouping, we mean putting the products in racks in the shipping area.
Sorting of Finished Product (FP)/Customer	The dispatch area is divided into different zones to store one batch per customer.
Sorting PF/Reference	In the shipping area, sorting can also be done by reference
Palletizing	La mise en palette de produit finis commence dès qu'une référence atteint la quantité exigée par le client.
Transfer of packages	The palletization of finished products starts as soon as a reference reaches the quantity required by the customer. The pallets having the required standards of packaging and quantity will have to be transferred by means of the radio frequency gun from warehouse F to warehouse E
Edition of the expedition file	Preparation of the export file (Invoices …)
Packaging and shipping	The pallets transferred to the warehouse and well checked are now ready to be stretch-wrapped with the help of handling means.
Moving of finished pallets	The finished pallets must be loaded respecting the priority to the pallets that must be the first to be unloaded.
Loading of packages	The filling of the trailer begins each time there are finished pallets while respecting the loading norms in force.
Road Transport Kenitra-Tangier	Transport the goods to the port of Tangier Med.
Sea Transport Tangier-Algeciras	Checking in Tangier Med and boarding for Algeciras.
Road transport Algeria-Vienna	Transporting the goods from the port of Algeciras to the warehouse in Vienna
Reception and unloading	Unloading and storage of the finished products in the EAVA warehouse in Austria.

The different steps of the downstream process in the company are identified and projected on the VSM map. The objective is to visualize in a clearer way the positions that require the most attention, for example time management, number of operators, cycle time, and non-value-added time (see Figure 5). Following this first mapping of the current state, the company wants to improve its competitiveness. In this framework, we will propose an improvement by introducing new technologies of industry 4.0, whose objective is to optimize this downstream supply chain. Therefore, we will first propose the most used 4.0 technologies in modern companies and the most suitable technologies for each workstation; this leads us to obtain a new future state mapping called VSM 4.0 (future map), which will receive new data and information from the whole SC, rationalizing both the management of information flows and product flows.

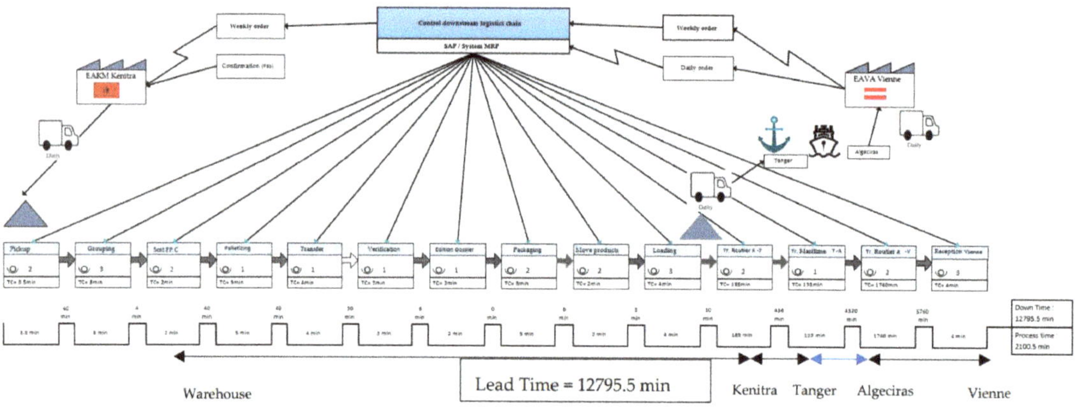

Figure 5. Mapping of the current state.

5.3. Implementation of the Methodology

In order to implement the research work, a methodology of implementation in 4.0 was implemented. It was divided into three phases:

First, according to the previous approach Figure 3, the first phase is to study and analyze the existing process in order to obtain the objectives, identify the process, and understand the industrial context of the company, then determine the sources of waste in the process and finally propose solutions.

The second phase consists in implementing 4.0 technologies in the supply chain. Study the compatibility of these technologies with each process and establish a VSM 4.0 mapping, where each process will be connected to this mapping (Future Map).

Finally, phase III corresponds to the collection and processing of data by the WMS, ERP, and intermediate communication tools and their connectivity. We present in Figure 6 below the results of the methodology implementation.

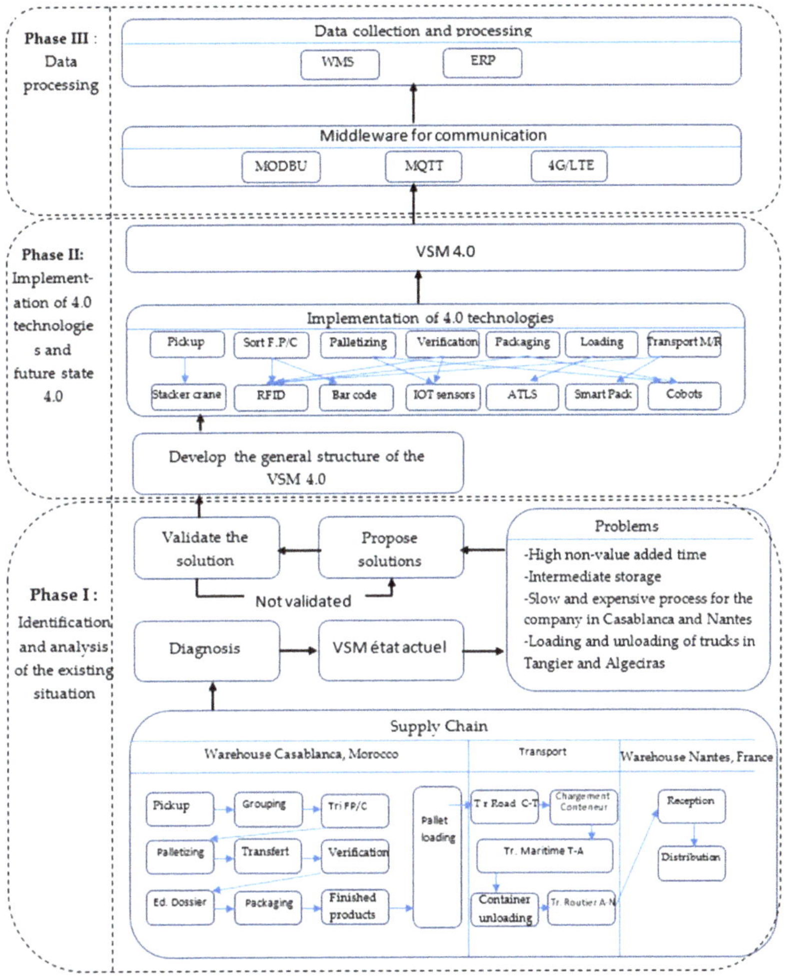

Figure 6. VSM 4.0 implementation methodology.

5.4. Implementation of Technologies

We have established the relationship between the 4.0 technologies and the different tasks of the downstream logistics process (see Table 4). The company has selected the technologies according to the following criteria:

- The most efficient way to optimize time and gain more cost.
- The investment cost is not too high for the implementation of these technologies.
- The training and mastery of these technologies by its employees.
- The transition time between the current state and the future state is not long.

Table 4. Assignment and adaptability of technologies to process functions.

Technologies \ Functions	Big Data	Cloud Computing	IoT	Cobots	AM	Drones	AI	AR	VS	RFID	Smart Pack	Bar Code	Power BI	WMS	Scan-3D	ERP	ATLS	AGV/LGV	Transtockeur
Pick-up & grouping of products				X				X		X				X		X			X
Product sorting	X	X		X	X		X	X		X		X	X		X				
Palletization & transfer of packages			X	X						X								X	X
Editing the file	X	X					X	X		X			X	X		X			
Packaging & Shipping				X			X	X		X						X			
Moving finished pallets	X	X								X								X	
Loading and unloading				X	X									X			X		X
Road, sea and warehouse delivery	X	X								X	X								

The 4.0 technologies chosen by the company are implemented in the different functions of the logistic process. A training of the personnel took place for a mastery of these technologies to accompany this digital transformation. After six months of work the results were very satisfactory (see Table 5).

Table 5. Contribution of technologies for each process.

Tasks	Contribution of New Technologies
Pickup & consolidation	**Stacker cranes 1, and Augmented Reality** - Product location and identification in the warehouse - Products are scanned and registered in real time, - Transport of products from production according to traceability to the grouping area - Optimize the role of the operator and optimize space - Merging of picking and grouping - The storage capacity is doubled - The circuit of movements is optimized - Reduced number of accidents - Reduced energy consumption - Reduced number of damaged packages **WMS** - Receipt of goods from imported files, the characteristics of goods received and purchase orders (order number, reference, quantities, etc.) are recorded in a file - Direct connection between the warehouse and the company's ERP system, which has led to better order management and a significant reduction in the delivery time to the customer - Possibility to follow each product inside the warehouse by controlling the inventory.
Sorting products	**Bar codes** - Automatic sorting where you already know where each product is and to which category it belongs - Reduce the verification time for the operator
Palletizing & transfer	**AGV and augmented reality** - Optimize the role of the operators in this stage by automatic palletizing system - Reduce largely the time of palletization in this stage - Reduced number of accidents - Eeduced energy consumption

Table 5. *Cont.*

Tasks	Contribution of New Technologies
Verification and editing of the file	**RFID & Scan** - Instantly detect which product is which and fill in the file data automatically - The shipping warehouse manager here will have the role of confirming the exit of the product without the need for manual entry. - It will save time for the final verification of the products. - Follow-up of the product until the loading in the trucks.
Packing and shipping	**RFID** - In this step, RFID technology allows us to track each finished product after wrapping and labeling, the WMS software allows us to control the goods before shipment and their allocation to the truck, we can add a third operator to speed up the operation.
Moving of finished pallets	**AGV** - For fast and organized movement without the need for operator intervention - Reduction of movements inside the warehouse, and facilitates the work of the operators - The movement circuit is optimized - Reduced number of accidents - Reduced energy consumption
Charging	**ATLS** - Reduction of the loading time in the trucks - Rare intervention of the operator whose role will be to supervise - All the boxes are of standardized size which will allow to gain in storage space and to optimize the distances between boxes - Optimize the company's logistic resources …
Road and sea transport and warehouse delivery	**RFID & Smart pack** - Traceability and tracking of products - Optimize the time of search and transport of boxes - Locate where each product is at any time
Distribution	**AGV and stacker crane** - Gain in reactivity and delivery time - The circuit of movements is optimized - Reduced number of accidents - Reduced energy consumption - Reduced number of damaged packages

5.5. Future State Mapping VSM 4.0

Figure 7 shows the future state 4.0 mapping of the downstream supply chain. Our value stream mapping is optimized with fewer tasks. The total number of workstations within the company is reduced to 9. In summary, after implementing this new mapping along the supply chain we obtained the following improvement results by implementing VSM 4.0 within the international supply chain:

- Automation and fusion of several processes: handling, transport, data collection, etc.
- The use of 4.0 technologies offers a fast product flow, and where data can be transmitted in real time.
- The VSM will continuously receive new data and information from the entire supply chain, allowing both the management of flows and the identification of waste that hinders the competitiveness of the company.
- Track the product throughout the process to identify and locate it quickly and above all to eliminate the possibility of loss. Simultaneous updating facilitates the consolidation of data, making it easier to make decisions.
- Change and improvement in the transportation of goods.

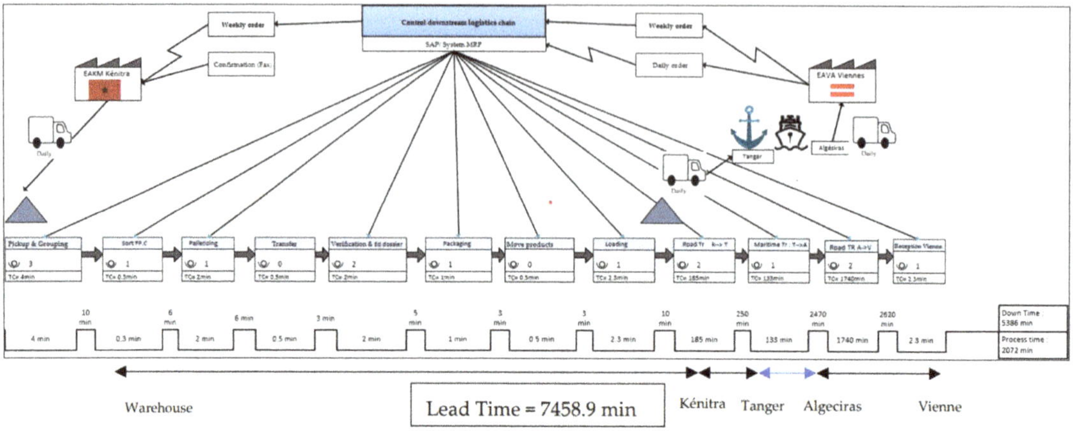

Figure 7. Future State 4.0 Mapping.

6. Analysis of the Results

6.1. *Implementation of the Technologies (Future State) in the SC*

After implementing the previously proposed solutions, Table 6 presents the various improvements on the downstream part of the logistic chain from Kenitra-Morocco to Vienna-Austria. We merged the collection and grouping using stacker cranes and the verification with the edition of the file. This action allowed us to optimize the process cycle time, the information flow in the company, as well as the role of the operators. By using new 4.0 technologies, automation, and connectivity, we have managed to optimize these processes reducing time, risks, and energy consumption.

Table 6. Comparison table between current state and future state 4.0 in warehouses.

Process	Current State			Improvement through New Technologies	Future State 4.0 (min)		
	Operators	Cycle Time (min)	Non-Value Added Time (min)		Operators	Temps de Cycle (min)	Temps de Non-Valeur Ajoutée (min)
Pickup	2	3,5	40	Stacker cranes, RA	3	4	10
Grouping	3	8	4				
Sort FP/C	2	2	40	Bar codes	1	0,3	6
Palletizing	1	5	40	AGV	1	2	6
Transfer	1	4	30	AGV	0	0,5	3
Verification	1	3	6	RFID & Scan	2	2	5
Edition Dossier	1	2	0				
Packaging	2	5	6	AGV	1	1	3
Move products	2	2	3	AGV	0	0,5	3
Loading EAKM	3	4	10	Stacker cranes, ATLS, AGV, RA	1	2,3	10
EAVA unloading and distribution	3	4			1	2,3	
Total	21	42,5	179		10	14,9	46

6.2. *SC Performance after Integration of New Technologies*

Currently, an SC's design must consider the interests of all integrated actors. Logistics is part of a "sustainable" movement, and Sustainable Development (SD) is becoming increasingly important within companies. According to this vision, companies are obliged to think about the extended economic, environmental, and social/societal performance. In this context, we present the improvement of the company's performance after the integration of new technologies.

6.2.1. Economic Performance

We have seen a strong benefit in time saving such as:

- Time and cost savings in Tangier and Algeciras: loading and unloading avoided;
- Reduction of loading and unloading time by using ATLS tools.

Table 7, we compare some indicators after the implementation of this new card, and we note the improvements and efficiency introduced.

Table 7. Gain between current state and future state 4.0.

Comparison Table			
Indicator	**Current State**	**Future State 4.0**	**% Gain**
Number of positions (tasks)	11	9	20%
Lead Time	8 days 21 h	5 days 4 h	41%
Number of operators	21	10	52%

These results are obtained thanks to the technologies introduced in this supply chain Table 6. For transportation, the value-added time and the non-value-added time is presented in Table 8.

Table 8. Transportation Comparison Table between Current State and Future State 4.0.

Process	**Process Time (min)**	**Non-Value-Added Time (Min)**	
		Current State	**Future State 4.0**
Road Transport Kenitra-Tanger	185	436	250
Sea Transport Tangier-Algeciras	133	4320	2470
Road Transport Algeciras-Vienna	1740	5760	2620
Total Time	2058	10,516	5340

Finally, the Lead Time and the time between the order and final delivery of the product to the customer are compared.

This section may be divided by subheadings. It should provide a concise and precise description of the experimental results, their interpretation, and the experimental conclusions that can be drawn.

In order to analyze the current state of our value stream mapping, we will proceed by:

- Identifying the different steps in our supply chain, from order preparation to delivery to customers.
- Taking into account the logistics competitiveness of each country involved in this supply chain by proposing a new process.
- Presenting a map of the current state of the process. Each task is presented by a box attached with a data box that contains information such as the number of operators in each workstation with the cycle time of all tasks, Lead Time, Down Time, and Process Times. (see Figure 8).

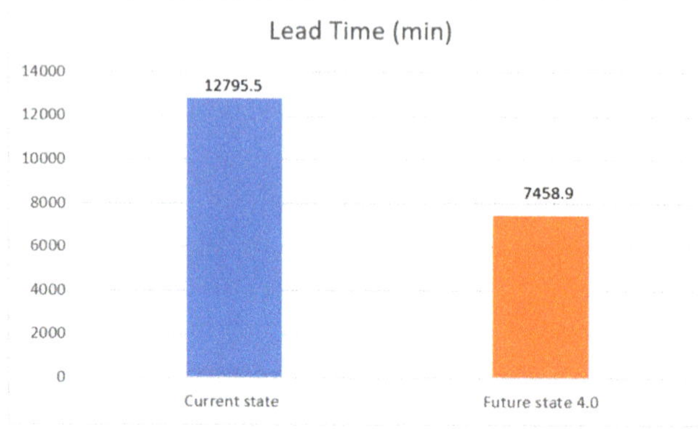

Figure 8. Comparison between Time to Value Added current and future state.

With the integration of these technologies, we witness a performance in time, distance travelled, truck fill rate, control of the distance travelled, and stop management.

The reactivity and flexibility led to improved logistics services, customer satisfaction, and increased turnover.

6.2.2. Social Performance

With the introduction of the Lean VSM method combined with 4.0 technologies, automated handling, and the motivation and development of the staff through the number of trainings, the number of conferences/seminars on sustainability, and the level of effort and communication to understand all stakeholders of the company and the concept of sustainability, it has been noted for a duration of the year 2021:

A space-saving in the warehouse of 20% is due to the use of modern handling equipment such as stacker cranes. The new organization and the space-saving have significantly reduced the travel circuit in the warehouse. The use of augmented reality has facilitated the location and inventory of products. The movement of products is reduced by the AGVs and the number of staff too. All these technologies had a very positive impact on the logistics staff. We have identified three indicators to show this impact on the social environment of the company as presented in Table 9.

Table 9. Social indicators.

Indicators	The Year 2020	The Year 2021
The number of accidents	- 25 accidents - 23 minors less than 10 days sick leave - 2 major more than two months of sick leave	5 minors
Absenteeism rate	63	9
the number of hours of maintenance intervention	242	27

We noticed that the number of accidents and the rate of absenteeism was very low after one year. With the introduction of a preventive maintenance of the material, the number of hours of intervention on the machines was much reduced.

As a result of these changes, we noticed a climate of trust, team spirit, motivation, employee satisfaction, and a collective intelligence that developed.

6.2.3. Environmental Performance

We noticed a positive impact on the management of resources in particular:

- Reduction in the cost of energy consumption that can be explained by the switch to an electric automated handling that replaced the traditional fuel-based handling means;
- Elimination of noise (noise pollution) and CO_2 emission;
- Reduction of waste (hydrocarbon waste, sludge, grease, used oil);
- The optimization of the movement circuit at the warehouse level has eliminated the risk of damage to packages during movement and consequently reduced the number of claims by customers for obsolete products.

7. Discussion

There was a gap of 5336.6 min, equivalent to three days and some hours in Lead Time. This was the result of optimization of the supply chain by modifications to the transport process and implementation of new 4.0 technologies in the warehouse of Kenitra and Vienna.

First of all, in EAKM Kenitra-Morocco, by using 4.0 technologies such as automatic handling equipment: stacker cranes, AGVs, and ATLS, we facilitated the role of the operators, and, as a result, these tasks will no longer be carried out manually but by machines and more quickly than before. This means that the number of operators in the process can be optimized, especially in the receiving area of the EAVA warehouse in Austria.

Secondly, through connectivity technologies, such as augmented reality, RFID, and Smart pack, connected to the WMS of the warehouse and the ERP of the company to follow the quantity, location, and state of the product at every moment in the warehouses of Kenitra and Vienna and the transport are tracked.

For the existing means of transport, the container was transported by a truck to Tangier, unloaded at the port, and loaded on a commercial ship. Then, in Algeciras, again, it is loaded in a truck which transports the goods to Vienna.

The proposed solution is to transport the goods and store them in a truck that makes the whole trip from EAVA to Vienna. This will avoid the costs of loading and unloading operations in both ports, which are very expensive.

Finally, at the transport level, we add a second transport staff to reduce the non-value-added time at this level for the journey to Europe.

8. Conclusions

The objective of our study was to provide a methodology based on Value Stream Mapping 4.0 for the continuous improvement of the downstream SC distribution process.

We established a VSM map to analyze the current state, then the design of the future state with optimization of some tasks considered competitive for the country of origin and the integration of 4.0 technologies accepted by the company from the point of view of implementation time and budget. Through this work, we were able to improve economic, social, and environmental performance. With the Lean Value Stream Mapping 4.0 tool, all products and information flow in a value chain from suppliers to customers are analyzed and optimized.

The integration of 4.0 technologies makes the VSM tool capable of monitoring value streams in real time to resolve potential waste quickly. This new intelligent organization provides a fully integrated logistics environment where data is transmitted in real time between warehouses in the countries concerned and the customers. The results obtained are encouraging and have an impact on economic and social performance. The introduction of new technologies in a company requires a significant investment cost but is quickly paid back by the improvements offered.

Author Contributions: Conceptualization; methodology, software, formal analysis, investigation, data curation, writing—original draft preparation, Y.E.K.; writing—review and editing, A.E.K. and S.E. All authors have read and agreed to the published version of the manuscript.

Funding: This research received no external funding.

Acknowledgments: We would like to thank the EAKM distribution team for allowing us to collaborate with our research team by providing us with the downstream logistics process data. We would like to thank the engineers and managers of the logistics distribution chain for the time they devoted to us, and for agreeing to implement the methodology developed within their international distribution chain.

Conflicts of Interest: The authors declare no conflict of interest.

References

1. Industrie Automobile au Maroc: Le Chiffre D'affaires a L'export Depasse 83.7 Milliards de Dirhams 2021, UN Record. Available online: https://fr.le360.ma/economie/industrie-automobile-au-maroc-le-chiffre-daffaires-a-lexport-depasse-837-milliards-de-dirhams-en-254131 (accessed on 17 April 2022).
2. El Kihel, Y.; Amrani, A.; Ducq, Y.; Amegouz, D. Implementation of Lean through VSM modeling on the distribution chain: Automotive case. In Proceedings of the 2019 International Colloquium on Logistics and Supply Chain Management (LOGISTIQUA), Paris, Franc, 12–14 June 2019; pp. 1–7. [CrossRef]
3. El Kammouni, R.; Kamach, O.; Masmoudi, M. Value Stream Mapping 4.0: A Structural Modeling Approach. In Proceedings of the 13ème Conference Internationale De Modelisation, Optimisation Et Simulation (MOSIM2020), Agadir, Maroc, 12–14 November 2020.
4. Gao, Z.; Castiglione, A.; Nappi, M. Biometrics in Industry 4.0: Open Challenges and Future Perspectives. *IEEE Trans. Ind. Inform.* **2022**, 1–4. [CrossRef]
5. Gao, Z.; Paul, A.; Wang, X. Éditorial invité: Jumelage numérique: Intégration de l'IA-ML et de l'analyse big data pour la représentation virtuelle. *IEEE Trans. Ind. Inform.* **2022**, *18*, 1355–1358. [CrossRef]
6. Tao, F.; Sui, F.; Liu, A.; Qi, Q.; Zhang, M.; Song, B.; Guo, Z.; Lu, S.C.-Y.; Néant, A.Y.C. 2018b Digital twin-driven product design framework. *Int. J. Prod. Res.* **2019**, *57*, 3935–3953. [CrossRef]
7. Demir, S.; Yilmaz, I.; Paksoy, T.S. Augmented Reality in Supply Chain Management. In *Logistics 4.0*; CRC Press: Boca Raton, FL, USA, 2020; pp. 136–145. [CrossRef]
8. Hartmann, L.; Meudt, T.; Seifermann, S.; Metternich, J. Value stream method 4.0: Holistic method to analyse and design value streams in the digital age. *Procedia CIRP* **2018**, *78*, 249–254. [CrossRef]
9. Sherwood, J.A.; Dignam, L.J.J.; Dobbins, T.; Boeman, R.G. US Composites Manufacturing Industry Technical Roadmap. Facilitating Industry by Engineering, Roadmapping and Science (FIBERS) to Advance US Manufacturing of Composites. 2017, pp. 1–53. Available online: https://www.uml.edu/docs/FIBERS%20Composites%20Manufacturing%20Roadmapping%20Report_Published-31Aug2017_tcm18-286787.pdf (accessed on 18 July 2022).
10. Kolberg, D.; Zühlke, D. Lean automation enabled by industry 4.0 technologies. *IFAC-Pap.* **2015**, *48*, 1870–1875. [CrossRef]
11. Satoglu, S.; Ustundag, A.; Cevikcan, E.; Durmusoglu, M.B. Lean production systems for industry 4.0. In *Industry 4.0: Managing the Digital Transformation*; Springer: Cham, Switzerland, 2018; pp. 43–59.
12. de Paula Ferreira, W.; Armellini, F.; de Santa-Eulalia, L.A.; Thomasset-Laperrière, V. Extending the lean value stream mapping to the context of Industry 4.0: An agent-based technology approach. *J. Manuf. Syst.* **2022**, *63*, 1–14. [CrossRef]
13. Nounou, A.; Jaber, H.; Aydin, R. A cyber-physical system architecture based on lean principles for managing industry 4.0 setups. *Int. J. Comput. Integr. Manuf.* **2022**, 1–19. [CrossRef]
14. Aksar, O.; Elgun, D.; Beldek, T.; Konyalıoğlu, A.K.; Camgöz-Akdağ, H. An Integrated Value Stream Mapping and Simulation Approach for a Production Line: A Turkish Automotive Industry Case. In *Digitizing Production Systems*; Springer: Cham, Switzerland, 2022; pp. 357–371. [CrossRef]
15. Boonsothonsatit, G.; Tonchiangsai, K.; Choowitsakunlert, S. Value stream mapping-based logistics 4.0 readiness for Thailand automotive-part manufacturers. In Proceedings of the 2020 2nd International Conference on Management Science and Industrial Engineering, Osaka, Japan, 7–9 April 2020; pp. 309–313. [CrossRef]
16. Ramadan, M.; Salah, B.; Othman, M.; Ayubali, A. Industry 4.0-based real-time scheduling and dispatching in lean manufacturing systems. *Sustainability* **2020**, *12*, 2272. [CrossRef]
17. Schneider, M.; Michalicki, M.; Rittberger, S. Lean and Industry 4.0—How to Develop a Lean Digitalization Strategy with the Value Stream Method. In *European Lean Educator Conference*; Springer: Cham, Switzerland, 2019; pp. 127–135. [CrossRef]
18. Huang, Z.; Kim, J.; Sadri, A.; Dowey, S.; Dargusch, M.S. Industry 4.0: Development of a multi-agent system for dynamic value stream mapping in SMEs. *J. Manuf. Syst.* **2019**, *52*, 1–12. [CrossRef]
19. Phuong, N.A.; Guidat, T. Sustainable value stream mapping and technologies of Industry 4.0 in manufacturing process reconfiguration: A case study in an apparel company. In Proceedings of the 2018 IEEE International Conference on Service Operations and Logistics, and Informatics (SOLI), Singapore, 18 July 2018; pp. 85–90.

20. Lugert, A.; Völker, K.; Winkler, H. Dynamization of Value Stream Management by technical and managerial approach. *Procedia CIRP* **2018**, *72*, 701–706. [CrossRef]
21. Meudt, T.; Metternich, J.; Abele, E. Value stream mapping 4.0: Holistic examination of value stream and information logistics in production. *Cirp. Ann.* **2017**, *66*, 413–416. [CrossRef]
22. Saraswat, P.; Kumar, D.; Sain, M.K. Reduction of work in process inventory and production lead time in a bearing industry using value stream mapping tool. *Int. J. Manag. Value Supply Chain. (IJMVSC)* **2015**, *6*, 27–35. [CrossRef]
23. Flauder, J. Déploiement du Lean Management dans un atelier de conditionnement et conduite du changement. *Sci. Pharm.* **2015**, Dumas-01202623. Available online: https://dumas.ccsd.cnrs.fr/dumas-01202623 (accessed on 28 April 2022).
24. Gupta, R.K.; Singh, M.P.; Sharma, L.K. Reduction of Wastage Using Value Stream Mapping: Case Study. *Int. J. Res. Mech. Eng. Technol.* **2014**, *4*, 52–55.
25. Haefner, B.; Kraemer, A.; Stauss, T.; Lanza, G. Quality value stream mapping. *Procedia Cirp.* **2014**, *17*, 254–259. [CrossRef]
26. Villarreal, B. The transportation value stream map (TVSM). *Eur. J. Ind. Eng.* **2012**, *6*, 216–233. [CrossRef]
27. Dushyanth Kumar, K.R.; Shivashankar, G.S.; Rajeshwar, S.K. Application of value stream mapping in pump assembly process: A case study. *Ind. Eng. Manag.* **2015**, *4*, 3.
28. Hirsch-Kreinsen, H. Industrie 4.0. In *Handbuch Innovationsforschung*; Springer: Wiesbaden, Germany, 2021; pp. 811–826.
29. Mosser, J. Cartographie 4.0 Pour la Transformation Numérique Des Processus. Ph.D. Thesis, Ecole Polytechnique, Montreal, QC, Canada, 2020.
30. Manning, J.; et Sörlin, F. *Value Stream Mapping as a Basis for Process Improvement in the Pharmaceutical Industry*; Lund University Faculty of Engineering Departement of Industrial Management and Logistics: Lund, Sweden, 2017.
31. Deniaud, I.; Marmier, F.; Michalak, J.L. Méthodologie et outil de diagnostic 4.0: Définir sa stratégie de transition 4.0 pour le management de la chaîne logistique. *Logistique Manag.* **2020**, *28*, 4–17. [CrossRef]
32. Transformation-Digital. Available online: https://www.mecalux.fr/blog/transformation-digitale (accessed on 2 June 2022).
33. Bai, C.; Dallasega, P.; Orzes, G.; Sarkis, J. Industry 4.0 technologies assessment: A sustainability perspective. *Int. J. Prod. Econ.* **2020**, *229*, 107776. [CrossRef]
34. Ahuett-Garza, H.; Kurfess, T. A brief discussion on the trends of habilitating technologies for Industry 4.0 and Smart manufacturing. *Manuf. Lett.* **2018**, *15*, 60–63. [CrossRef]
35. Barz, C.; Jalba, C.K.; Erdei, Z.; Hahn, S.M.L. Approaches for the planning and implementation of Industry 4.0. Period. *Eng. Nat. Sci. PEN* **2019**, *7*, 375–380. [CrossRef]
36. Singh, D. Use of RFID System and Improvement in Library Services. *IJR* **2022**, *8*, 283. [CrossRef]
37. Valamede, L.S.; Akkari, A.C.S. Lean 4.0: A new holistic approach for the integration of lean manufacturing tools and digital technologies. *Int. J. Math. Eng. Manag. Sci.* **2020**, *5*, 851. [CrossRef]
38. Crestan, L. Implementation of a Dashboard for Measuring Warehouse Performance-Proposal Based on a Business Intelligence Software. ING—École de Génie Industriel et de L'information. Available online: https://www.politesi.polimi.it/handle/10589/187128 (accessed on 28 April 2022).
39. La Fonction WMS Gestion d'entrepôt Préparation de Commande. Available online: https://www.e-prelude.com/ (accessed on 25 May 2021).
40. Samuel, K.E.; Ruel, S. Systèmes d'information et résilience des chaînes logistiques globales. *Syst. D'information Manag.* **2013**, *18*, 57–85. [CrossRef]
41. Automatic-Truck-Loading. Available online: https://technicaintl.com/solutions/automatic-truck-loading/ (accessed on 13 April 2021).
42. Transstockeur-Pour-Palettes. Available online: https://www.mecalux.fr/stockage-automatise/stockage-automatise-palettes/transstockeur-pour-palettes (accessed on 20 June 2021).

processes

Article

A Real-Time Optimization Strategy for Small-Scale Facilities and Implementation in a Gas Processing Unit

Pedro A. Delou [1,*], Leonardo D. Ribeiro [2], Carlos R. Paiva [1], Jacques Niederberger [2], Marcos Vinícius C. Gomes [2] and Argimiro R. Secchi [1,*]

[1] Chemical Engineering Program, LADES—Software Development Laboratory, COPPE, Universidade Federal do Rio de Janeiro, Cidade Universitária, Rio de Janeiro 21941-914, RJ, Brazil; carlosrodriguespaiva@gmail.com
[2] PETROBRAS/CENPES—Av. Horácio de Macedo, 950, Cidade Universitária, Rio de Janeiro 21941-598, RJ, Brazil; dorigo@petrobras.com.br (L.D.R.); jack@petrobras.com.br (J.N.); mvcg@petrobras.com.br (M.V.C.G.)
* Correspondence: pdelou@peq.coppe.ufrj.br (P.A.D.); arge@peq.coppe.ufrj.br (A.R.S.)

Abstract: The rise of new digital technologies and their applications in several areas pushes the process industry to update its methodologies with more intensive use of mathematical models—commonly denoted as digital twins—and artificial intelligence (AI) approaches to continuously enhance operational efficiency. In this context, Real-time Optimization (RTO) is a strategy that is able to maximize an economic function while respecting the existing constraints, which enables keeping the operation at its optimum point even though the plant is subjected to nonlinear behavior and frequent disturbances. However, the investment related to the project of commercial RTOs may make its application infeasible for small-scale facilities. In this work, an in-house, small-scale RTO is presented and its successful application in a real industrial case—a Natural Gas Processing Unit—is shown. Besides that, a new method for enhancing the efficiency of using sequential-modular simulator inside an optimization framework and a new method to account for the economic return of optimization-based tools are proposed and described. The application of RTO in the industrial case showed an enhancement in the stability of the main variables and an increase in profit of 0.64% when compared to the operation of the regulatory control layer alone.

Keywords: RTO; data reconciliation; natural gas processing; economic evaluation; sequential-modular approach

Citation: Delou, P.A.; Ribeiro, L.D.; Paiva, C.R.; Niederberger, J.; Gomes, M.V.C.; Secchi, A.R. A Real-Time Optimization Strategy for Small-Scale Facilities and Implementation in a Gas Processing Unit. *Processes* **2021**, *9*, 1179. https://doi.org/10.3390/pr9071179

Academic Editor: Zhiwei Gao

Received: 14 June 2021
Accepted: 2 July 2021
Published: 7 July 2021

1. Introduction

Real-time optimization is a model-based adaptive optimization technique that attempts to find the optimal operating condition accordingly to an economic index of a plant subjected to a process model and a set of constraints that might be, for instance, physical limits, environmental restriction, product quality, or safety criteria [1]. The so-called "two-step" approach proposed by Jang et al. [2] has become the most widespread RTO strategies in industry [3–5]. In this approach, a parameter estimation step is performed followed by an economic optimization step, so that the available static model of the plant can be adjusted considering the most recent set of plant information and the optimization may be carried out considering a rigorous model with minimum plant-model mismatch. It is true that this approach may fail when there is model structural uncertainty [6,7]. However, when the sources of uncertainty are mainly parametric, the approach has great potential to increase the economic performance of process operations [8] and to provide considerable economic benefit that overly surpass the cost of investment on the RTO design. Therefore, the development of a rigorous process model is the backbone of the two-step RTO approach and, in fact, it is possible to show that whenever a model satisfies the set of model adequacy criteria, the model-based optimization problem is able to drive the plant to its true optimum [9–11].

In the classical scheme of the two-step RTO, a parameter estimation subjected to the same static model used in the optimization step is carried out once the input data is guaranteed to be in a stationary state. To cope with this requirement, a steady-state detection (SSD) method is implemented. There are several SSD methods available, most of them are statistical-based methods, such as the F-test [12] and the Student's *t*-test [13,14]. Recently, two research groups have proposed removing the SSD requirement by proceeding a dynamic estimation, such as the Extended Kalman Filter, in a framework called Hybrid RTO (HRTO) [15,16]. Although this method presents a high potential to speed up RTO cycles and to reduce the suboptimal operation time, it lacks industrial validation.

Recently, the academic literature of RTO has been focused on solving the problem of finding the true-plant optimum in cases related to serious structural plant-model mismatch. Since the proposition of the Integrated System Optimization and Parameter Estimation (ISOPE) algorithm by Roberts [17], which led to the Constraint Adaption (CA) [18] and the Modifier Adaption (MA) [19] algorithms, several derivations have been proposed so far to cope with different methodological paths, a detailed review on MA approaches and applications can be found in Marchetti et al. [7]. These approaches are based on the introduction of zeroth-and first-order modifiers to adapt the model-based optimization problem. The modifiers are calculated from plant measurements and estimation of the plant gradients in relation to the decision variables. In the CA algorithm, the modifiers are used to adapt the constraints of the problem, so that the model-based constraints meet the plant constraints upon convergence. In the MA algorithm, both the constraints and objective function are adapted, so that the model-based optimization problem matches the optimality condition of the plant upon convergence. However, the requirement of estimating the plant gradients in the absence of a reliable model is one of the main reasons that justify the low number of industrial applications of MA approaches. Recently, an industrial application of CA in a solid-oxide fuel-cell system has been disclosed [20,21], but the problem was formulated in a manner such that the first-order modifiers were not necessary because the plant optimum was known to be located at the intersection of active constraints, so the gradient estimation was not required. In spite of that, the work was a proof of concept and may indicate the increase of initiatives of MA in the industry, especially with the rise of new methodologies that merge concepts of Machine Learning and RTO to produce better estimates of the first-order modifiers, such as the MA with Quadratic Approximation (MAWQA) [22,23], the use of neural networks [24], Gaussian process [25,26], and Bayesian optimization [27].

In the process industry, RTO is typically implemented as the two-step approach and in a hierarchical control pyramid manner [28]. In a simplified and summarized description, the regulatory control layer is implemented to deal with high-frequency disturbances of the process, typically in a multiloop PID controllers approach. This layer is able to interact with the plant in the scale of seconds, rejecting fast disturbances and keeping the stability of the operation. In the upper layer, the supervisory control layer, a model-based approach is used to determine the optimal trajectories of the process by the use of simple dynamic models, typically linear ones, with the objective to track setpoints and send control actions as setpoints to the regulatory layer. This layer works in the time scale of minutes and the most-used technique is model predictive control (MPC). The optimization layer comes above the previous control layers in order to determine the economic optimal operating point of the process. This layer typically runs in the time scale of hour due to the complexity of the models that are used.

In the context of modeling strategies, there are mainly two types of simulators available for developing the process flow-sheets, they are the so-called Equation-Oriented (EO) simulators and Sequential-Modular (SM) simulators. Despite the undoubted advantages of using EO simulators [29] and the vast quantities academic works using and developing methodologies based on them, the availability of SM models in industry is still high due to their ability to solve a problem with low initialization effort and an easier flow-sheet design [30]. However, the computational cost to handle nested recycles can hinder their use in the context of real-time optimization. Early works proposed the use of a method called the Modular Continuous (MC) approach [31], or *infeasible path* [32,33], in which the convergence loops are removed from the SM simulator and convergence of the model is assured by a higher-level optimization layer. Recent works have proposed the use of surrogate models to enhance reliability and reduce the computational effort of the optimization [34,35]. The present work proposes an extension of the concept of the MC approach, by adding an acceleration step of successive substitution in the model flow-sheet.

Despite significant advances in the field of controller performance assessment since the 1960s, these advances have been focused on univariable control structures and on obtaining a technical metric [36]. In fact, there has been very little discussion regarding the economic assessment of advanced process control and optimization applications [37]. There are many works that disclose the economic return of RTO's application [38], but it is not clear whether these numbers are trustworthy or not since the authors do not discuss their methods of assessing this economic benefit. Beyond the standard discount cash flow analysis [39], which can be used both to support decision-making at the investment stage and to perform economic assessment in an existing application, three methodologies are worth mentioning: the performance assessment of MPC proposed by Xu et al. [40]; the framework proposed by Bauer and Craig [37]; and the systematic method proposed by Udugama et al. [36]. However, the three methods have the similarity of not considering an optimization layer in their assumptions. They can account for the benefit of reducing the process variability and the distance from the desired setpoints, however, they do not account for true optimal operation. Therefore, reliable methods to assess the economic benefit of advanced control structures and RTO frameworks are still an open issue.

Petrobras started investing in RTO technology in 2004. Their early developments were based on commercial tools, such as Aspen Plus (AspenTech) and Romeo (Aveva), but using self-developed models. Since then, the company has achieved great expertise in the technology and successful applications have been reported in Gas Processing Units [41], Fluidized Catalytic Cracker Units [42], and Crude Distillation Units [29]. However, applications with small-scale scopes may not benefit from the application of these tools since the economic return may be smaller than the cost of investment in the project stage, annual licensing of the software, and often the need for external consultancy during the operation phase. Therefore, Petrobras and LADES, the Software Development Laboratory (LADES) of COPPE/UFRJ, collaborated to develop an in-house RTO software, which was called SoraIA, an acronym in Portuguese for "System of Revenue Optimization and Artificial Intelligence". The software presented some advantages due to the fact of being totally based on open-source or acquired tools, having ease of maintenance and also adaptability for applications in new processing units, allied with high flexibility for different problem formulations and small investment requirements compared with the great economic return.

Besides the description of the developed software and its application in a real industrial facility, this paper also discloses two contributions to the RTO literature in the sense of process modeling with the proposed "Modular Continuous with Successive Substitution" and a new method to assess RTO's economic return after its implementation, as well as its potential further economic benefit that would be possible with system improvement.

The software was tested in an industrial Gas Processing facility owned by Petrobras. RTO was implemented in a closed loop with the control system of the unit, and the results and discussion are provided in this paper. Further, after three months of operation, the economic return of the system was evaluated accordingly with the new proposed method and the result overcame the initial expectations of the project, even though the unit is highly instrumented and very well operated in the open loop—that is, considering only the regulatory control under the supervision of the operation team, without the action of any advanced control strategy.

The present paper is organized as follows: Section 2 presents the methodology proposed in this paper in a generalized manner; Section 3 presents how the methodology was applied to the industrial case that is object of the present application; Section 4 presents the results and discussion of the present application regarding the reconciliation problem, the optimization problem, the computational cost, and the economic benefit of RTO; finally, Section 5 presents the conclusion of the present work.

2. Methodology

2.1. Modular Continuous with Successive Substitution

The quality of the model used in the two-step RTO is crucial for the success of the application—that is, the model should be simple enough so its computational cost can be suitable for optimization purposes and detailed enough to mitigate any serious structural uncertainty, since the approach is only able to deal with parametric uncertainty. Although it is difficult to measure, Forbes and Marlin [43] developed a criteria to determine whether a model is adequate to be used in the two-step approach based on its capability to match the Karush–Kuhn–Tucker (KKT) optimality conditions [44] of the plant at the optimal point.

The EO and SM strategies have their benefits and drawbacks for the application in an RTO framework. For example, SM has dedicated numerical methods to converge each process unit, which makes this approach more robust for poor initial estimates. However, this type of model can be very costly to run, demanding several loops of convergence, which frequently make them unsuitable for optimization purposes. On the other hand, EO models are fast and efficient to be evaluated, they also present the advantage of providing accurate gradient estimates from automatic differentiation strategies. However, this type of modeling is highly dependent on the initial estimates of the dependent variables, since its convergence algorithms provide the solution of the full system of equations simultaneously. Therefore, a common choice is to use both approaches together, running the SM model first to provide better initial estimates for the EO model that is used inside the optimization framework. However, this common choice comes with the cost of having to develop two compatible models, frequently in different platforms.

In this work, a modeling approach was developed to handle an SM model inside the optimization framework in an efficient manner. The idea is derived from the early proposition by Berna et al. [31] and Biegler and Hughes [45] in the Modular Continuous (MC) approach, where the main idea consists on opening mass and energy loops and transforming them into new decision variables and constraints to be met by the optimization problem. As the model would not be converged along the optimization steps, this approach was also known as *infeasible path*, due to the fact that the model convergence is only assured upon optimization convergence [32,33]. Here, the proposed approach, "Modular Continuous with Successive Substitution" (MCSS), follows the same philosophy proposed by the early works of the MC approach.

Figure 1 illustrates the algorithmic procedures that are performed for any module i inside loops in order to carry out the optimization based on the SM, MC, and the proposed MCSS approaches, in which $\boldsymbol{u} \in \mathbb{R}^{n_u} = \mathbb{U}$ is the set of all input variables, $\boldsymbol{y} \in \mathbb{R}^{n_y} = \mathbb{Y}$ represents the set of all output variables of the model, $\boldsymbol{z} \in \boldsymbol{y}$ is a subset of the output variables that represents the material and energy loop streams of the model, and $\boldsymbol{\rho} \in \mathbb{R}^{n_\rho} = \mathbb{P}$ is the set of independent variables introduced by the opened loops.

Following the nomenclature introduced in Figure 1, the i loops represented by variable z_i in Figure 1a are torn into two variables ρ_i and z_i^{ref} in Figure 1c. In the MCSS approach, all modules inside the opened loops are replicated into the reference modules: variables ρ_i are the degrees of freedom of the optimizer and z_i^{ref} are the loop output variables of the reference module. z_i^{ref} is returned as input for the second module. In this way, a first step of successive substitution is forced within the process model. The convergence is guaranteed by the addition of constraints that enforce that all loops are closed upon convergence. In the MC approach, this set of constraints are

$$\rho - z = 0, \tag{1}$$

while, in the proposed MCSS, the set of constraints are

$$z - z^{ref} = 0. \tag{2}$$

The main goal of the MC approach is to reduce the number of convergence loops in the simulation layer by transferring the convergence conditions of the model to the higher-level optimization layer. The purpose of the additional successive substitution step in the MCSS approach is to accelerate the convergence by approximating the feasible path characteristic of the SM approach, but still keeping the faster convergence provided by the infeasible path of the MC approach.

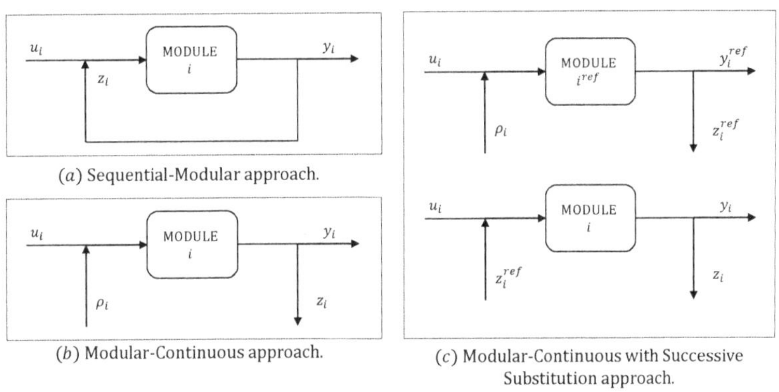

Figure 1. Schemes of the modeling approaches: (**a**) Sequential Modular; (**b**) Modular Continuous; the proposed (**c**) Modular Continuous with Successive Substitution.

PETROX, the simulator used in this work, is a reliable SM simulator developed for giving support to process design at Petrobras [46]. The success of this software can be addressed for the vast library of process units and the reliability of its internal numerical methods. Nowadays, the software is also applied to develop reliable digital twins of the different units of the company. This work opened a new perspective for its application in optimization by following the philosophy of the Modular Continuous with Successive Substitution. Some rules of thumb of the modeling approach used in the present work are summarized in the following:

1. Parameterization of the feed composition—with the composition of the product streams as specification, the feed composition can be constructed by setting some criteria, such as meeting dew or boiling temperatures in flash units and meeting components ratio with the product.
2. Simplification of process units that are not essential for the optimization scope—distillation columns before the actual scope of the optimization can be replaced by a simple approach based on flashes with parameterized fractioning ratio between cut components;
3. Avoid loops for mass and energy integration—opening loops and forcing a first successive substitution;
4. No additional convergence loops—remove any controller loops to meet specification criteria, this task should be passed to the parameter-estimation layer.

In this work, the model developed in PETROX is denoted as a stationary map of the inputs and parameters into the outputs:

$$y = \mathcal{F}(\rho, \theta, u), \tag{3}$$

where $\theta \in \mathbb{R}^{n_\theta} = \mathbb{T}$ is the set of adjustable parameters. For the sake of simplifying notation, and without loss of generality, the static mapping will be frequently denoted only by $y(\rho, \theta, u)$. Additionally, all objective functions and nonlinear constraints are also provided by the SM simulator.

2.2. RTO Architecture

An in-house software was developed with the purpose of performing all RTO stages. Apart for some specific details, the presented structure is very similar to what has been done since the 1980s in the two-step RTO, the main contribution of this work is the use of the MCSS approach, as it is described in the formulation of the optimization problems in Sections 2.2.4 and 2.2.5. Figure 2 illustrates the architecture of the SoraIA software.

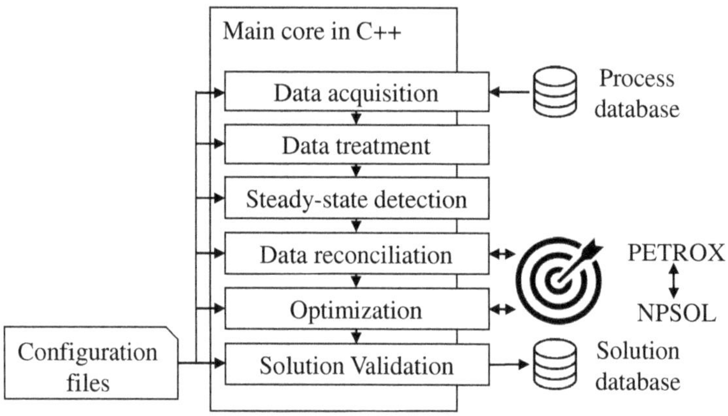

Figure 2. Scheme of the SoraIA architecture.

The main core was written in C++ and its purpose is to manage the information flow in order to perform the desired tasks that are required by RTO methodology. Six main routines are executed in sequence: data acquisition, data treatment, steady-state detection, data reconciliation, optimization, and solution validation. The software is able to communicate with the process database and the solution database, and perform the interface between the process simulator (PETROX) and the nonlinear optimization solver (NPSOL). In addition, the user can configure some features of the run by a web interface. This interface compiles the set of user configurations into some standard text files that can be read by the C++ core in order to shape each routine accordingly.

Every run cycle of RTO starts with the data acquisition routine and follows the routine's sequence until the optimal solution is obtained, except in the case where some major problem is found during any stage of the run. In this case, the software is aborted and a new cycle is started. In the following, each of the main RTO routines are further described.

2.2.1. Data Acquisition

In the data acquisition stage, the user can specify whether the software is going to run offline, so a specific operational point can run for tests independently of the actual state of the plant, or whether the software is going to run online, reading the values directly from the process database. In this stage, the user can define the size of the acquired data window and which instrument tags are related to which input variables from the process model simulator.

2.2.2. Data Treatment

In the data treatment stage, each input variable is treated by a linear transformation specified by the user in order to scale or to perform any required unit change. In addition, there are some verifications in the read information in order to exclude any unwanted null values, strings, or unexpected out-of-limit values. If any abnormal value is found in this verification, the software is restarted, so its cycle is repeated until there is no error in the input values.

2.2.3. Steady-State Detection

It is very important to assure that the data used in the data reconciliation stage is in stationary state in order to reliably estimate the model parameters. Therefore, the steady-state detection is mandatory and whenever it is detected that the read data window presents some dynamics, its cycle is restarted until the steady-state is detected in the input data window. There are several steady-state detection (SSD) techniques, Cao and Rhinehart [47] briefly reviewed some of the existing methods and proposed an efficient new statistical method. Later, Bhat and Saraf [12] also outlined aspects of the available SSD techniques and extended the proposition of Cao and Rhinehart [47].

In this work, two statistical tests were applied. The first test is based on cutting the input window in three periods and comparing these windows to each other in order to check whether their means are equivalent by a Student's *t*-test. The second is an analysis where the variance of the variables are verified within some desirable limits. These limits are estimated using past historical data by selecting periods where the variables were in the steady-state condition. Both tests are well-established in the literature [12]. However, the combination of these methods are a contribution of the present work, to the best of the authors' knowledge. This proposition is made in order to overcome the limitation of the *t*-test when too-high variance values are observed, in which false detections are observed. Therefore, the stationary condition must be detected in both tests so the data window is allowed to go to the next stage; otherwise, a new RTO cycle starts.

2.2.4. Data Reconciliation and Parameter Estimation

The data reconciliation problem was formulated in a manner that it can perform a simultaneous parameter estimation, as proposed by Rod and HanČl [48]. Differently from the usual approach in the literature, not only the adjustable parameters of the process unit models are estimated, but also the parameters that were added to ensure convergence of the model, following the one-pass modeling philosophy described in Section 2.1. The data reconciliation problem is formulated as a nonlinear programming problem:

$$\bar{\boldsymbol{\rho}}_k, \bar{\boldsymbol{\theta}}_k, \bar{\boldsymbol{u}}_k^{rdf} = \arg \min_{\boldsymbol{\rho},\boldsymbol{\theta},\boldsymbol{u}^{rdf}} \mathcal{J}_{rec}(\boldsymbol{y}^{rec}, \boldsymbol{y}_k^m, \boldsymbol{u}^{rec}, \boldsymbol{u}_k^m) \tag{4a}$$

$$\text{s.t. } \boldsymbol{y} = \mathcal{F}(\boldsymbol{\rho}, \boldsymbol{\theta}, \boldsymbol{u}) \tag{4b}$$

$$\mathcal{G}(\boldsymbol{\rho}, \boldsymbol{\theta}, \boldsymbol{u}) \leq \mathbf{0} \tag{4c}$$

$$z - z^{ref} = 0 \tag{4d}$$

$$\boldsymbol{\theta}_{lb} \leq \boldsymbol{\theta} \leq \boldsymbol{\theta}_{ub} \tag{4e}$$

$$\boldsymbol{u}_{lb}^{rdf} \leq \boldsymbol{u}^{rdf} \leq \boldsymbol{u}_{ub}^{rdf}, \tag{4f}$$

where $\boldsymbol{\rho} \in \mathbb{P}$ is the set of additional decision variables added by the MCSS approach; $\boldsymbol{\theta} \in \mathbb{T}$ is the set of all adjustable parameters to be estimated; the superscripts $(\cdot)^{rdf}$ mean "reconciliation degree of freedom", so $\boldsymbol{u}^{rdf} \subset \mathbb{U}$ is the set of n_u^{rdf} uncertain inputs that are considered degrees of freedom in the reconciliation problem; $\boldsymbol{u}^{rec} \subset \boldsymbol{u}^{rdf}$ is the set of n_{u_r} reconciled inputs, which is a subset of the uncertain inputs that have available measurements; $u \in \mathbb{R}^{n_u}$ is the set of all input variables; and $y^{rec} \subset y$ is the set of n_{y_r} reconciled output variables, which is a subset of the output variables that have available measurements. The vectors $u^m \in \mathbb{R}^{n_u^m}$ and $y^m \in \mathbb{R}^{n_y^m}$ are the measurements of the input and output variables, respectively. The function $\mathcal{G} : \mathbb{R}^{n_\theta} \times \mathbb{R}^{n_u} \rightarrow \mathbb{R}^{n_g}$ is the set of static mappings of the parameters and inputs into the nonlinear constraints, and the objective function $\mathcal{J}_{rec} : \mathbb{R}^{n_y^m} \times \mathbb{R}^{n_y^m} \times \mathbb{R}^{n_u^m} \times \mathbb{R}^{n_u^m} \rightarrow \mathbb{R}$ is the weighted sum of squared errors between predicted and measured variables. z and z^{ref} are the output of the actual and the reference modules of the opened loops added for the MCSS approach. Finally, the subscript $(\cdot)_k$ denotes the kth run of RTO, the bar emphasis $(\bar{\cdot})$ denotes the optimal values obtained in the run, and the subscripts $(\cdot)_{lb}$ and $(\cdot)_{ub}$ denote lower and upper bounds, respectively. It is noteworthy that the kth values inserted in the problem for the sets of input and output measurements, u_k^m and y_k^m, are the arithmetic mean of the time window acquired in the data acquisition stage.

After running the reconciliation problem, it is possible to define the offsets of the output measurements, which are defined by the deviation between the measured and the predicted variable:

$$\varepsilon_k^y = y_k^m - y(\bar{\rho}_k, \bar{\theta}_k, u(\bar{u}_k)) \tag{5}$$

and the offsets of the input variables, which are defined by the deviation between the measured input and the optimal input:

$$\varepsilon_k^u = u_k^m - \bar{u}_k^m. \tag{6}$$

2.2.5. Optimization Problem

In the optimization problem, as in the reconciliation problem, the objective function and constraints are evaluated by the process simulator and the main core of the software is responsible for managing the information flow between the simulator and the optimizer. The formulation of the optimization problem is presented below:

$$\bar{\rho}_k, \bar{u}_k^{odf} = \arg \min_{\rho, u^{odf}} \mathcal{J}_{opt}(u, y) \tag{7a}$$

$$\text{s.t. } y = \mathcal{F}(\rho, \bar{\theta}_k, u) \tag{7b}$$

$$\mathcal{G}(\rho, \bar{\theta}_k, u) \leq 0 \tag{7c}$$

$$z - z^{ref} = 0 \tag{7d}$$

$$y_{lb} - \varepsilon_k^y \leq y \leq y_{ub} - \varepsilon_k^y \tag{7e}$$

$$u_{lb}^{odf} - H\varepsilon_k^u \leq u^{odf} \leq u_{ub}^{odf} - H\varepsilon_k^u, \tag{7f}$$

where the function $\mathcal{J}_{opt} : \mathbb{U} \times \mathbb{Y} \rightarrow \mathbb{R}$ is the economic objective function of the optimization problem; this function is usually an economic balance between the incomes with products and the costs with raw materials and energy consumption. The superscript $(\cdot)^{odf}$ means "optimization degree of freedom", so $u^{odf} \subset \mathbb{U}$ is the set of n_u^{odf} inputs for which the objective function is sensible to variations within their feasible space, corresponding to the set of degrees of freedom of the optimization problem. It is important to highlight that all adjustable parameters and uncertain inputs—which are not degrees of freedom of the optimization—that were estimated in the data reconciliation problem are kept fixed in the optimization problem. In addition, the estimated offsets, ε_k^y and ε_k^u, are used to shift the feasible space of the output variables, as in Equation (7e), and of the degrees of freedom for which there are available measurements, as in Equation (7f). Matrix $H \in \{0;1\}^{n_u^{odf}} \times \{0;1\}^{n_u^m}$ is a rectangular matrix of zeros and ones only to select which offset is related to which degree of freedom, considering that some degrees of freedom might not be related to any offset and the opposite can also be true.

2.2.6. Solution Validation

In the solution validation stage, there is an optimization flag check to verify whether the solution of the optimization problem was found within the constraints or whether it reached an infeasible solution, which is rarely expected to occur. If the solver returns an infeasible solution, the software will disregard this point and the RTO cycle will restart. On the contrary, the solution is shifted by the estimated offsets:

$$y^{irv} = y(\bar{\rho}_k, \bar{\theta}_k, u(\bar{u}_k^{odf})) + \varepsilon_k^y, \tag{8}$$

$$u^{irv} = \bar{u}_k^{odf} + H\varepsilon_k^u, \tag{9}$$

where the superscript $(\cdot)^{irv}$ means "ideal value at rest"; these values are written in the solution database. Once the new solution is written in the solution database, it is instantaneously available to be read by other instances of the control hierarchy.

2.2.7. Interface between PETROX and NPSOL Solver

The optimization solver used in the RTO is NPSOL version 5.0 [49]. This is a nonlinear programming solver written in Fortran 66 Version 2.1 that performs a Sequential Quadratic Programming (SQP) algorithm. The model was developed in PETROX 3.8 and its interface with the main core of RTO in C++ is carried out by PetroxTR 3.4R0. In order to improve the numerical robustness of the optimization, the decision variables, constraints, and objective function are normalized between the interval $(-1, 1)$ according to a normalization space defined by the user. In addition, the version of the simulator PETROX used has no resource to provide information about the gradients of the objective function and the constraints with respect to the decision variables, since the simulator was not developed for optimization purposes. Therefore, the method used to estimate the gradient is based on finite difference approximations. For each decision variable, if the lower bound is active, a first-order forward finite difference is performed; if the upper bound is active, a backward first-order finite difference is performed; otherwise, a second-order central finite difference is performed.

2.3. Method for Accounting Economic Return

The task to account for the economic return of the implementation of any optimization strategy applied to an industrial facility is not simple, considering the several disturbances to which the plant and market are subjected. Here, a novel method is proposed in order to evaluate the economic benefit of control and optimization schemes that were implemented in the real plant. It is considered that a significant amount of data are available for periods of open-loop operation, in which the solution of the RTO is calculated but not applied to the plant, and in closed loop, in which the solution of the RTO is implemented in the plant. Each period of operation may be subjected to different disturbances, feed flow rates, feed composition, and even different feasible regions for each decision variable, so it would not be accurate to compare the periods directly based on the measurement of the economic objective function. Therefore, the idea of the developed method is to obtain a performance index that is a measure of how far the operation is from the maximum profit return value on each operation period, or the ideal economic performance for each operation period, which can be obtained by dividing the average of the profit function calculated in the solution of the reconciliation problem by the average of the same function calculated in the solution of the optimization problem:

$$\Omega_i = \frac{\sum_{k=n_{i0}}^{n_{i1}} \mathcal{J}_{profit}(\bar{\theta}_k, u(\bar{u}_k^{rdf}))}{\sum_{k=n_{i0}}^{n_{i1}} \mathcal{J}_{profit}(\bar{\theta}_k, u(\bar{u}_k^{odf}))}, \tag{10}$$

where Ω_i denotes the performance index calculated based on a data window of $N_i = n_{i1} - n_{i0} + 1$ past points, given that i is the operation mode depending on the level of automation of the unit in study. $\mathcal{J}_{profit} : \mathbb{R}^{n_\theta} \times \mathbb{R}^{n_u}$ is the profit function, which can be the objective function of the optimization problem when oriented for the economic analysis, or it can be an economic monitoring function when the optimization has other goals. This difference in nomenclature is introduced in order to avoid losing generality, since the objective function of the optimization step might not be directly suitable for the economic return analysis in some applications; for example, when it represents a measure of the operational efficiency or when it accounts for additional environmental criteria. In general terms, we consider three operation modes:

1. Regulatory control: The plant operates in manual mode, where the operators directly decide the setpoints of the PID controllers. RTO runs in open loop.
2. Supervisory control: An advanced control, frequently an MPC layer, acts on the PID setpoints and the operators decide directly the setpoints and targets of the MPC. RTO runs in open loop.
3. Optimization: RTO runs in closed loop with the supervisory control layer.

The performance index Ω_i is a measure of the distance between the actual economic performance of mode i and ideal economic performance. As the averaged reconciled economic balance will always be inferior compared with the averaged optimal economic balance, considering that the optimization constraints are respected even in Regulatory and Supervisory operational modes; then, the performance index is always inferior to 1. Hence, $(1 - \Omega_i)$ can be interpreted as a potential benefit margin that can be reduced by improving operational aspects, such as the reduction of process variability.

A hypothesis of the proposed method is that the performance index of the operational mode i obtained in a significantly large data window of N_i points can be generalized for any data window with size greater than N_i and, therefore, the performance indexes can be compared with each other. In other words, the length of each data window must be chosen to be long enough in order to properly characterize the operational mode in terms of economic performance of the operation. It is expected that these indexes can be compared as

$$\Omega_1 < \Omega_2 < \Omega_3. \quad (11)$$

Even though comparing the indexes values is enough to verify whether there are benefits in implementing a specific supervisory control layer or an optimization layer, this comparison alone does not provide a measure of this benefit in monetary terms. Therefore, an average ideal optimal economic balance value is taken as reference:

$$\mathcal{J}^{ref}_{profit,i} = \frac{1}{N_i} \sum_{k=n_{i0}}^{n_{i1}} \mathcal{J}_{profit}(\bar{\boldsymbol{\theta}}_k, \boldsymbol{u}(\bar{\boldsymbol{u}}^{odf}_k)) \quad (12)$$

and a quantitative return can be estimated by multiplying the performance index of each operational mode by this reference value, which is equivalent to evaluate the average of the actual profit:

$$\bar{\mathcal{J}}_{profit,i} = \Omega_i \, \mathcal{J}^{ref}_{profit,i}. \quad (13)$$

3. Industrial Case

3.1. Process Description

The RTO system developed in the present study was applied to an industrial Natural Gas Processing Unit (NGPU) owned by Petrobras in Brazil. In general terms, the unit is responsible for processing the NG from different sources to produce Residue Gas (RG), Fuel Gas (FG), Liquefied Petroleum Gas (LPG), and a stream containing components heavier than pentane, here denoted as Naphtha. Several unit operations are carried out in sequence—to name a few: NG dryier, Demethanizer, RG compressor, Deethanizer, and Debutanizer. These operations are supported by a Propane Refrigeration System and a Thermal Oil Heating System.

Initially, the collected NG with controlled admission pressure is cooled. Then, the gas follows to molecular sieves units where it is dried. The dry gas is then routed to the fractioning section, which corresponds to the feed of the simplified scheme illustrated in Figure 3.

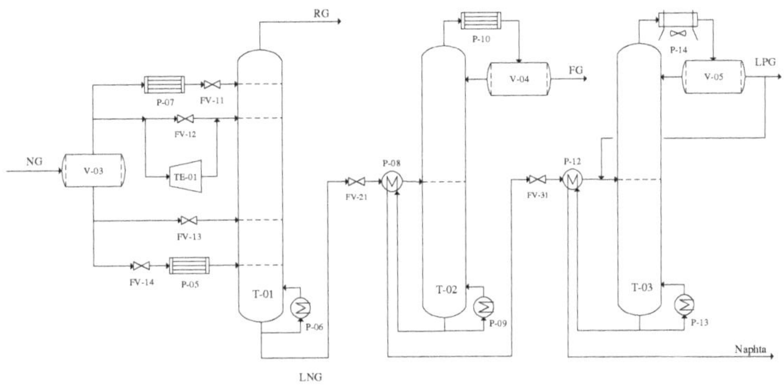

Figure 3. Simplified process flow diagram of the NGPU. V-03—feed accumulator vessel of the Demethanizer tower (T-01); FV-11/12/13/14—feed flow control valves of T-01; TE-01—turboexpander of T-01; P-05/07—heat exchanger of the cold-box of T-01; P-10—condenser of the Deethanizer tower (T-02); V-04—condenser accumulator vessel of T-02; P-14—condenser air-cooler of the Debutanizer tower (T-03); V-05—condenser accumulator vessel of T-03; P-06/09/13—reboiler heat exchangers of T-01/02/03; FV-21/31—feed flow control valves of T-02/03; P-08/12—feed heat exchanger of T-02/03.

This section is mainly composed by three separation units: T-01, in which RG is produced; T-02, which produces FG; and T-03, in which LPG and Naphtha are produced. In the first section, the dry gas is flashed into two vapor streams and two liquid streams. The first vapor stream is totally condensed and injected at the top of the Demethanizer column, acting as a reflux stream. The second vapor stream is divided into two streams, the main one is expanded in the turboexpander in order to reach even lower temperatures and the other goes to the Joule–Thomson valve, which is normally closed. These streams are then mixed and injected into the top section of the column. The fist liquid stream is injected directly as a feed-side stream at the bottom section of the column, while the other is first used in an energy integration scheme, being partially vaporized, and is then also injected into the bottom section of the column. It is noteworthy that all heat exchangers illustrated in Figure 3, upstream of T-01, are part of a single cold-box, heat integration scheme that is mischaracterized in the flowchart.

The top product of the T-01 is a methane-rich gas that is heated at the cold-box in the heat integration scheme and then sent to the compression stage associated with the turboexpander, while the bottom product, rich in components heavier than ethane, will then be fed to the Deethanizer tower—firstly passing though a flow control valve and a heat exchanger that energetically integrates the bottom product with the feed streams of the column. This unit is designed to operate in two modes depending on the current specification of the ethane in the LPG stream. With the low content of ethane in LPG, the condenser of T-02 works with the Propane Refrigeration System in order to produce the FG stream. On the contrary, the refrigeration is kept shut-down and the column loses its separation function, working only as an accumulation tank in the process. During the execution of the present RTO project and its implementation test period, the Propane Refrigeration System was off and the tower T-02 worked only as an accumulation tank.

The feed of the Debutanizer tower comes from the bottom product of tower T-02, again, after passing through a flow control valve and a heat exchanger that energetically integrates the bottom product with the feed streams of the column. The top product of the column is condensed in an air-cooler condenser and directed to the reflux drum, where, besides the split between the reflux and product streams, there is also a third stream that is mixed with the feed of the tower—the recirculation stream. This stream was added to the project of the unit due to the need to specify a large range of ethane content in the LPG stream. However, currently, the tower operates with a lower feed flow as that designed, so the recirculation also contributes to enhance the tower's hydraulics. The system presents relatively slow dynamics and is subjected to frequent disturbances, mostly in feed composition and ambient temperature.

3.1.1. Economic Interests of the Operation

The Natural Gas Processing Unit of the present application receives gas from different sources, mainly including offshore oil production. The continuity of gas processing is vital for the offshore plant to continue operating with a proper destination of the produced gas. As previously mentioned, column T-02 was acting only as an accumulator tank during the design and tests of the present work; therefore, the NGPU produced only RG, LPG, and Naphtha. According to the most frequent economic configuration, the product with greater market value is LPG, followed by Naphta, and than RG, which is commonly used as fuel in the Thermal Oil Heating System.

Hence, the economic objective is to determine the optimal operating point that is able to produce LPG with maximum efficiency, acting as follows:

1. Minimizing the loss of propane at the top of the Demethanizer column;
2. Maximizing the content of ethane in LPG with respect to the established upper bound;
3. Maximizing the content of pentanes in LPG with respect to the established upper bound;
4. Minimizing the consumption of energy demanded by the process.

3.2. Process Model in PETROX

The scope of the present RTO application was defined to be the optimization of the Debutanizer column T-03; therefore, tower T-01 and tower T-02 were described in a simplified way. This choice was carried out following the philosophy of small-scale RTO in order to reduce computational cost and focus on optimizing the efficiency of the fractioning to produce LPG and Naphta. The model was developed in PETROX 3.8 following the "Modular Continuous with Successive Substitution" approach discussed in Section 2.1 and its flow sheet can be visualized in Figure 4.

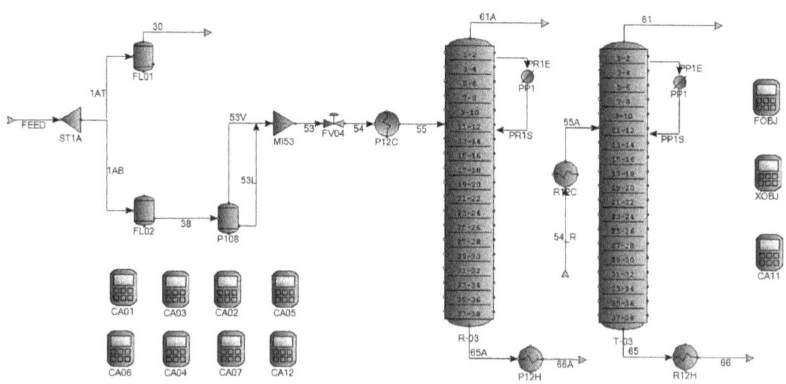

Figure 4. Flow sheet of the process model in PETROX.

The input of data to the simulation is carried out by the several unconnected streams added to the simulation. The function of the calculators at the left-hand side of the flow sheet is to manage the information from the input streams to the main core of the simulation;

they are also able to manage the information from a previous unit to a posterior unit. In addition, the function of the calculators at the right-hand side of the flow sheet is to manage the results of the simulation and to externalize results of interest, such as objective functions, nonlinear constraints, and output variables, for instance. Further description of the simulation calculators are provided in Appendix A.

The model has 13 components: nitrogen (N_2), carbon dioxide (CO_2), methane (C1), ethane(C2), propane (C3), isobutane ($iC4$), n-butane ($nC4$), isopentane ($iC5$), n-pentane ($nC5$), n-hexane (C6), n-heptane (C7), n-octane (C8), n-nonane (C9). The simulation starts with the restoration of the feed stream; since the online chromatography is not complete, it does not discriminate compounds heavier than C6 and its measurements are not synchronized with the LPG analyzer, since there is an uncertain delay between these two analyzers. Therefore, the feed composition is parameterized to be estimated in the parameter estimation step. This is achieved by considering the components C3, $nC4$, and $iC4$ in such a way that the proportion between C3 and the sum of $nC4$ and $iC4$ can be controlled and the distribution between $nC4$ and $iC4$ would be the same as in the LPG composition:

$$x_{C3}^{feed} = (x_{C3,m}^{feed} + x_{nC4,m}^{feed} + x_{iC4,m}^{feed})\ \theta_1 \tag{14a}$$

$$x_{nC4}^{feed} = (x_{C3,m}^{feed} + x_{nC4,m}^{feed} + x_{iC4,m}^{feed})\ (1-\theta_1)\ \frac{x_{nC4,m}^{LPG}}{x_{nC4,m}^{LPG} + x_{iC4,m}^{LPG}} \tag{14b}$$

$$x_{iC4}^{feed} = (x_{C3,m}^{feed} + x_{nC4,m}^{feed} + x_{iC4,m}^{feed})\ (1-\theta_1)\ \frac{x_{iC4,m}^{LPG}}{x_{nC4,m}^{LPG} + x_{iC4,m}^{LPG}}, \tag{14c}$$

where x_i^j and $x_{i,m}^j$ represent the molar composition of component i in the stream j in the simulation and measured, respectively; $\theta_1 \in \{0,1\}$ is the estimated parameter to control the proportion between propane and butanes.

A similar strategy is applied to the heavier components of the feed stream that are not directly measured but lumped in the $x_{C6+,m}^{feed}$:

$$x_{C6}^{feed} = x_{C6+,m}^{feed}\ (1-\theta_2) \tag{15a}$$

$$x_{C7}^{feed} = x_{C6+,m}^{feed}\ \theta_2\ p_{C7} \tag{15b}$$

$$x_{C8}^{feed} = x_{C6+,m}^{feed}\ \theta_2\ p_{C8} \tag{15c}$$

$$x_{C9}^{feed} = x_{C6+,m}^{feed}\ \theta_2\ p_{C9}, \tag{15d}$$

where $\theta_2 \in \{0,1\}$ is the estimated parameter to control the proportion between C6 and the heaviest components, and p_i is a typical proportion between component i and the heaviest components obtained from lab reports. The feed measurements are directly used for the other components.

The description of the Demethanizer model was carried out by a simplified approach similar to the proposition of Ito et al. [50]. In order to reduce complexity and computational cost, the tower is represented by a splitter and two flash drums to produce the two top and bottom product streams using, respectively, a dew point and boiling point flash units. A parameter is added to control the fractioning of T-01 considering the cut between ethane and propane. However, since the reflux of the column is injected directly at the top tray, part of the reflux stream is vaporized in the moment that it enters the tower, so the top product is not in fact in the dew point due to the mixture between the saturated vapor that comes from the first tray and the vaporized portion of the reflux stream. Therefore, to account for this modeling characteristic, the top temperature cannot be used as a reconciliation variable and more importance is given to the measurement of the C3 content in the RG stream and the content of C2 and C3 in the LPG stream. With a simple mass balance, it is possible to fully specify the split flow rates considering the loss of C3 in the RG stream and the flow rates of C2 and C3 in the LPG stream:

$$r_{C2/3} = \theta_3 \frac{x_{C3,m}^{LPG}}{x_{C2,m}^{LPG} + x_{C3,m}^{LPG}} \tag{16a}$$

$$F_{C3}^{RG} = x_{C3,m}^{RG} \, (F_{N_2}^{feed} + F_{CO_2}^{feed} + F_{C1}^{feed} + F_{C2}^{feed} - r_{C2/3} \, F_{C3}^{feed}) \tag{16b}$$

$$F_{C2}^{LNG} = r_{C2/3} \, (F_{C3}^{feed} - F_{C3}^{RG}), \tag{16c}$$

where F_i^j corresponds to the molar flow rate of component i in stream j, $r_{C2/3}$ is the ratio of the fraction of C3 in the LPG stream over the sum of the fractions of C2 and C3 in the LPG stream and $\theta_3 \in \{0, 1\}$ is the parameter responsible for adjusting the fraction of C3 in RG and C2 in LPG. The superscript LGN refers to the bottom product of tower T-01 that feeds tower T-03. This strategy is able to match the loss of C3 in the RG with the plant measurements. Additionally, as during the estimation of θ_3, the value of Equation (16c) might be negative, a constraint is added to prevent this value from being negative.

The LNG stream passes through an expansion valve upstream of the feed preheat exchanger (P-12) of the T-03. This heat exchanger is simulated in three steps in order to avoid any loop that would include excessive iterations in the simulation. In the first step, the simulation receives the input of the heat duty of the cold side (ρ_1), which is a decision variable added by the MCSS approach, and then goes to a reference tower to simulate an approximation of the T-03 because the energy balance in the feed heat exchanger is not satisfied. Then, the hot side of the feed heat exchanger is simulated by an HOCI (hot-out cold-in) temperature difference approach based on the bottom product of the reference tower. This HOCI approach is added in order to enhance the robustness of the simulation, avoiding physical inconsistencies in the temperature differences. The difference between the outlet and the inlet temperatures are calculated from measurement, and this difference is summed to the inlet temperature calculated by the simulation to produce the outlet specification of the hot side of the heat exchanger. After the heat duty calculation, the cold side is simulated again specifying the heat duty of the hot side and then the T-03 tower is properly simulated. It is important to highlight that each heat exchanger must be analyzed individually and, here, the HOCI approach is appropriate because the flow rate of the hot fluid will always be inferior to the flow rate of the cold fluid.

The recirculation stream is also simulated by a pump-around in order to avoid unwanted loops. A parameter is added to vary the Murphree efficiency of the inner trays of the tower, so that the internal profile of the tower may approximate the measured internal profile. The reboiler and condenser efficiencies are kept fixed and equal to 1. The flow rates of the reflux, recirculation, and bottom product were chosen as specifications of the tower to improve robustness, since choosing temperatures might reduce the chances of converging the simulation considering the whole operational range.

Finally, the pressures were defined following a backpressure propagation strategy. The final pressure points were specified accordingly to the measurements for the top pressure of T-01 and the condenser pressure of T-03 and these pressures were back-propagated by typical pressure drops of the system. This approach showed to be very efficient for this case study due to the larger values of pressures compared with the values of pressure drops.

It is important to highlight that all modeling choices were considered in order to provide a suitable model developed in a sequential-modular simulator for optimization purposes. Therefore, all excessive iterations of numerical methods are avoided and the main responsibility of converging the model is given to the optimization layer, or through the addition of constraints, either through a successive substitution generated naturally by the several calls of the model function by the optimization algorithm.

3.3. Formulation of the Data Reconciliation Problem

The formulation of the objective function for the data reconciliation simultaneously with parameter estimation was carried out in a weighted least squares estimator with normalized variables:

$$\mathcal{J}_{rec} = \sum_{i=1}^{n_{yr}} w_i^y \left(\frac{y_{k,i}^m - y_i^{rec}}{y_{k,i}^m} \right)^2 + \sum_{i=1}^{n_{ur}} w_i^u \left(\frac{u_{k,i}^m - u_i^{rec}}{u_{k,i}^m} \right)^2, \quad (17)$$

where $w^y \in \mathbb{R}^{n_{yr}}$ and $w^u \in \mathbb{R}^{n_{ur}}$ are weight vectors of the output and input variables, respectively.

Table 1 presents the measured output variables and their respective weights considered in the objective function.

Table 1. Measured output variables.

Variable Description	Weight (w^y)
Volumetric fraction of $nC5$ and $iC5$ in LPG	10^3
Top temperature of tower T-03	10^4
Bottom temperature of tower T-03	10^5
Condenser temperature of tower T-03	10^5
Fraction $C3/(nC4 + iC4)$ in LPG	10^4
Temperature of the control tray of tower T-03	10^4

As previously commented, the developed model specifies the C3 loss in the RG stream given the chromatographic measurement, so this value cannot be reconciled. The top temperature and the volumetric fractions of $nC5$ and $iC5$ are considered reconciled variables. Both of these measurements compete to specify the top of tower T-03, since they represent the same information in essence. In Table 1, more importance is given to the top temperature rather than the analyzer, because the analyzer has lower measurement frequency and is potentially more noisy. The bottom temperature of tower T-03 is an estimated output that is influenced by the content of heavier components of the feed stream, which is controlled by adjusting parameter θ_2. Similarly, the condenser temperature is an estimated output that is influenced by the content of C2 in the LPG stream; this composition is controlled by adjusting parameter θ_3, which represents the cut between C2 and C3 in tower T-01. The fraction $C3/(nC4 + iC4)$ in LPG is also an estimated output; this variable is influenced by the ratio between propane and butanes in the feed stream, which is controlled by the parameter θ_1. Finally, the temperature of the control tray is an estimated output controlled by the Murphree efficiency of the column.

Table 2 presents the input variables and parameters considering degrees of freedom in the reconciliation simultaneously with the parameter estimation problem, their bounds, and their weights in the objective function, in the case where measurements are available. It is important to highlight that when the abbreviation DCS is assigned for a bound value, this value is defined by the operator in the Digital Control System (DCS) of the plant. The specific values of the limits defined in the DCS are not disclosed because these limits varied considerably during the analyzed time period. The flow rates of the column (bottom, reflux, and recirculation) are considered degrees of freedom to specify the column variables. For the reflux and recirculation streams, the available volumetric measurements are uncertain, so these variables are also reconciled but with smaller weights than the reconciled outputs, so the optimization may have more flexibility to deviate from the measurements.

Table 2. Input variables and parameters for the data reconciliation simultaneous with the parameter estimation problem.

Degree of Freedom	Weight (w^u)	Lower Bound	Upper Bound
Bottom flow rate of tower T-03 (kmol/h)	–	0	100
Reflux flow rate of tower T-03	10^2	DCS	DCS
Recirculation flow rate of tower T-03	10^2	DCS	DCS
Murphree Efficiency of tower T-03	–	0.8	1
Heat duty of P-12 ρ_1 (Mcal/h)	–	10	500
Parameter θ_1	–	0	1
Parameter θ_2	–	0	1
Parameter θ_3	–	0	1

Table 3 presents the constraints of the data reconciliation problem and their bounds.

Table 3. Constraints of the Data Reconciliation Problem.

Constraint Description	Lower Bound	Upper Bound
Molar fraction of C4− in Naphta	0	0.015
Pressure loss on the feed valve of T-03 (kgf/cm^2)	1	∞
Ratio of C2/C5+ in LPG	0	16
Vapor flow rate after flash P108	0	0
Molar flow rate of C2 in LNG	0	∞
Heat duty difference of cold and hot side of P-12	0	0

The added constraints are the minimum requirements to guarantee that the simulation converges in an expected way. This is achieved by forcing the pressure loss on the valve upstream from the T-03 to be greater than 1 kgf/cm^2, guaranteeing that the molar flow of C2 in the LNG stream is greater than 0, as already discussed in Section 3.2, and making sure that the liquid product stream of the flash P-08 is saturated; this flash unit is downstream of tower T-01. In addition, the constraints in the maximum molar content of components lighter than C4 (C4−) to be 0.015 and in the ratio of the composition of ethane and components heavier than pentanes (C2/C5+) to be less then 16 are redundancies to guarantee product quality requirements. Finally, the last constraint is added due to the MCSS approach in order to guarantee the energy balance upon convergence of the optimization.

3.4. Formulation of the Optimization Problem

In order to meet the economic interests of the operations described in Section 3.1.1—except for objective 1, which would require a more detailed model for the system of column T-01—the following objective function was designed:

$$\mathcal{J}_{opt} = \frac{1}{W^{LNG}_{\text{-C2}}} \left(p^{LPG} W^{LPG} + p^N W^N - p^{RG} \frac{Q_r}{\eta \, \text{LHV}^{RG}} - p^E Q_c \right), \tag{18}$$

where p^{LPG}, p^N, p^{RG}, and p^E represent the price per unit of mass of LPG, Naphta, RG, and electric energy, respectively; $W^{LNG}_{\text{-C2}}$ is the mass flow rate of LNG free of ethane that feeds tower T-03; W^{LPG} and W^N are, respectively, the mass flow rates of LPG and Naphta streams; Q_r and Q_c are the heat flow rates of the reboiler and condenser of the tower T-03, respectively; η is the efficiency of the thermal oil furnace; and LHV^{RG} is the lower heat value of the residue gas used as fuel.

The idea of the developed objective function is that the optimization may be able to maximize the efficiency of the fractioning process of the column for a given feed flow rate. That is the reason for dividing the expression by the mass flow rate of the LNG stream free of ethane. This value should be free from C2 because the content of ethane in the LNG stream is a decision variable of the problem, so it could take the wrong path minimizing this variable.

The decision variables, or degrees of freedom, of the optimization problem are presented in Table 4, as well as the respective bounds.

Besides the constraints used in the data reconciliation problem, which are presented in Table 3, the optimization problem has additional constraints, as presented in Table 5.

Table 4. Decision variables of the Optimization Problem.

Decision Variables	Lower Bound	Upper Bound
Bottom flow rate of tower T-03 (kmol/h)	0	100
Reflux flow rate of tower T-03	DCS	DCS
Recirculation flow rate of tower T-03	DCS	DCS
Condenser Pressure	DCS	DCS
Heat duty of P-12 ρ_1 (Mcal/h)	10	500
Parameter θ_3	0	1

The volumetric composition of ethane and pentanes have different specification values defined by the operation. These specifications are provided by the technical team of the operation in the DCS and the optimization must respect them. A constraint is added to the value of the condenser temperature; as it is an air-cooler, in which the cold fluid is air at ambient temperature, it would not be reasonable to allow the temperature of the hot fluid at the outlet to be inferior to 30 °C. Even though this temperature is also constrained by the composition of C2 in the LPG stream, this redundancy is added in order to prevent the solution from finding an unreachable temperature.

Table 5. Additional constraints of the Optimization Problem.

Constraint Description	Lower Bound	Upper Bound
Volumetric composition of C2 in LPG	DCS	DCS
Volumetric composition of *n*C5 and *i*C5 in LPG	DCS	DCS
Condenser Temperature (°C)	30	∞
Temperature of the control tray of tower T-03	DCS	DCS
Bottom temperature of tower T-01	DCS	DCS

In addition to those constraints, two more are added with respect to safety operational limits defined by the operator in the DCS—one for the temperature of the control tray of T-03 and another at the bottom temperature of T-01. It is worth mentioning that this last constraint is another redundancy with the content of C2 in the LPG stream, as the bottom stream of T-01 is saturated and its temperature depends mostly on the content of the lighter component—C2 in this case.

3.5. Integration between RTO and the DCS

The purpose of this paper is not to deeply describe the control system of the unit. However, some points are interesting to be noted in order to clarify how the integration between the RTO and the DCS is carried out.

The regulatory control system is composed by a multiloop single-input single-output PID controller mainly designed with cascades feedback loops. The operation team of the plant can decide whether to manually define the setpoints of these controllers or to turn on the supervisory control layer. This supervisory control layer is a model predictive controller that solves an optimization problem aiming to minimize the quadratic deviation between the measured outputs and inputs from reference trajectories and target trajectories, respectively, subjected to input–output models and constraints. The algorithm uses the step response model and is based on the DMC algorithm proposed by Cutler and Ramaker [51]. However, the algorithm was modified to account for an adaptive strategy that is able to adjust the internal model of the MPC based on the current operational condition of the plant.

Above the MPC layer, there is a simple optimization layer that aims to maximize an economic objective function subjected to the same adaptive input–output model of the MPC layer, resulting in quadratic programming (QP). When the loop of the RTO is closed to the supervisory control layer, which is also a decision of the plant operator, that simple optimization layer becomes an intermediate QP problem between MPC and RTO, as proposed by Rotava and Zanin [52]. Therefore, when the RTO is in closed loop, the intermediate optimization problem aims to minimize the distance between the solution of the RTO and the achievable values based on the input–output model of the MPC. This is a way to translate the solution of the economic optimization subjected to a detailed nonlinear model to the space of linear models, preventing infeasible targets to be sent to the control layer.

The chosen variables to perform the integration were the content of ethane and pentanes in LPG, the reflux and recirculation of T-03, and the bottom temperature of T-01.

3.6. Estimation of the Economic Return

The economic return was evaluated following the novel methodology proposed in Section 2.3. The profit function was defined as the sum of the incomes with products minus the costs with Natural Gas:

$$\mathcal{J}_{profit} = p^{LPG}W^{LPG} + p^{N}W^{N} + p^{RG}W^{RG} - p^{NG}W^{NG}, \tag{19}$$

where p^{NG} and W^{NG} are the price per mass unit and mass flow rate of the Natural Gas stream, respectively.

Three data windows were collected for the analysis of each operation mode, as defined in Section 2.3, and a data treatment was performed to remove any gross errors, outliers, and regions where the unit was not operating, resulting in the following:

1. Regulatory control—an interval of 97 days of operation resulting in 60,350 points with a sampling time of 2 min;
2. Supervisory control—an interval of 138 days of operation resulting in 20,583 points with a sampling time of 2 min;
3. Optimization—an interval of 96 days of operation resulting in 49,043 points with a sampling time of 2 min.

In the three data windows, the variables frequently violate the allowable bounds due to measurement noise. Although this violation was expected, it is alarming, especially for the upper bounds of the content of ethane and pentanes in the LPG stream, since values above these upper bounds would result in values of the profit function greater than the optimum solution. This effect would enable the performance index to be greater than 1, which is unacceptable. Hence, all points above these upper bounds were excluded from the analysis for the supervisory and optimization operating modes, resulting in the removal of 28.5% of the points in the supervisory data window and 24.8% of points in the optimization data window. However, in the regulatory mode, the bounds of the manipulated and output variables were not well specified in the system, since in this mode, the operation team is mainly focused in the PID's setpoints. So, these bounds were arbitrarily chosen in order to remove some percentage of points from the analysis. As it is expected that the variability of data would be higher in the Regulatory control, this percentage was conservatively chosen as equal to the RTO data window.

Another issue that hinders the effort to calculate the performance index is that in the regulatory and supervisory control modes, the RTO was not operating in open loop. Therefore, no results of the data reconciliation and the optimization problem were available. Therefore, in order to enable the analysis, a mass balance spreadsheet was developed in order to calculate the profit function for all points in each operation mode, in which the reconciled solutions were obtained by the use of the available chromatography measurements and the optimal solutions were approximated using the upper bounds as desirable specifications. This way, the computation of the performance index was made possible regarding the specificity of each operating condition.

4. Results and Discussion

In the following, the results presented in Sections 4.1 and 4.2 are related to the first month of the RTO's operation. It is noteworthy that no fixed time frequency was set for the RTO run time. Instead, RTO runs in its own variable frequency, depending on the computational cost to run each iteration. This practice is usually avoided by commercial applications aiming not to excessively disturb the supervisory controller with frequent changes of reference values and targets. In the present study, this fact is not an issue due to the fact that the MPC controller is imbued with constraints of minimal movement, which prevents sudden transients. Additionally, due to the nondisclosure agreement with the industrial partner, the results are presented in a normalized form.

4.1. Data Reconciliation

This section presents the results related to the simultaneous data reconciliation and parameter estimation problem. As the RTO ran $N = 920$ times within the 30-day window herein analyzed, only a fragment of the data window is provided to compare the plant data with the reconciled value in the figures of this section. Therefore, in order to have a sensibility of the whole picture, a relative average error in percentage was calculated for each variable of interest as follows:

$$\epsilon_i^y(\%) = \frac{100}{N}\sum_{k=1}^{N}\frac{|y_{k,i}^m - y_i(\bar{\theta}_k, u(\bar{u}_k^{rdf}))|}{|y_{k,i}^m|}, \tag{20}$$

where ϵ_i^y is the relative average error of output i. The error of the reconciled inputs is calculated analogously.

Figure 5 illustrates the results related to the content of ethane in the LPG stream—that is, the condenser temperature of T-03, the bottom temperature of T-01, and the volumetric fraction of ethane in LPG itself. The dark-gray regions around the plant data points represent a deviation of ±5% of the plant data at instant k, while the light-gray represents a deviation of ±10%. For this set of variables, it is interesting to note that only the temperature of the condenser of tower T-03 is considered in the objective function of the problem and the other two are adjusted as a consequence of it. The relative error of these variables are 0.63%, 1.34%, and 3.81%, respectively. The higher error on the volumetric fraction was expected since this measurement has a low sampling frequency and considerably high noise.

Figure 6 shows the result of the bottom temperature of T-03, which presented a low relative error of 0.60%. This result is a consequence of the estimation of the content of the heavier components, which are not measured. Figure 6 also illustrates the variables that are related to the content of pentanes in LPG, which are the top temperature of column T-03 and the volumetric fraction of pentanes in LPG; both variables were reconciled.

The relative error of the top temperature was 4.01%, while the error of the volumetric fraction of pentanes was 7.84%. Moreover, a bias is observed in the reconciled top temperature value, consistently remaining below the plant data, which can be explained by the position of the sensor near the second tray. The sensor may be presenting interference due to the higher temperature of the tray below; therefore, the use of another temperature sensor will be considered in the future.

Figure 7 shows the results of the temperature of the control tray of tower T-03, the ratio of propane over butanes in LPG, and the reflux flow rate.

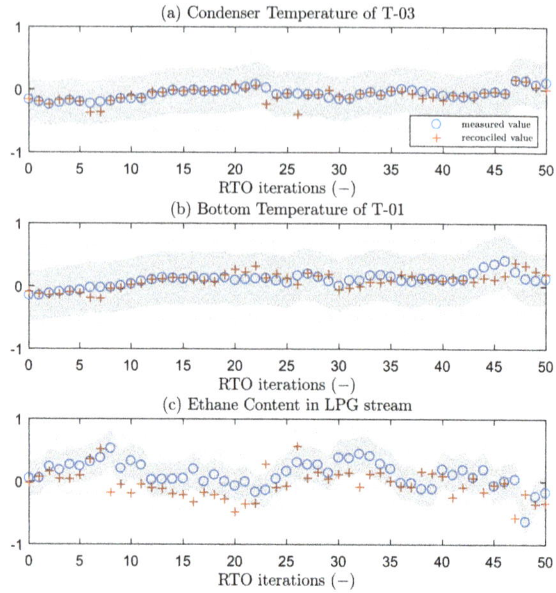

Figure 5. Data reconciliation results: (**a**) condenser temperature of T-03; (**b**) bottom temperature of T-01; (**c**) ethane content in LPG stream.

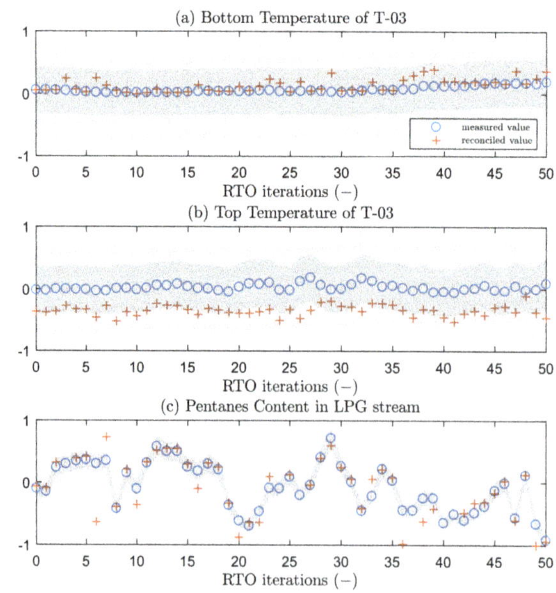

Figure 6. Data reconciliation results: (**a**) bottom temperature of T-03; (**b**) top temperature of T-03; (**c**) pentanes content in LPG stream.

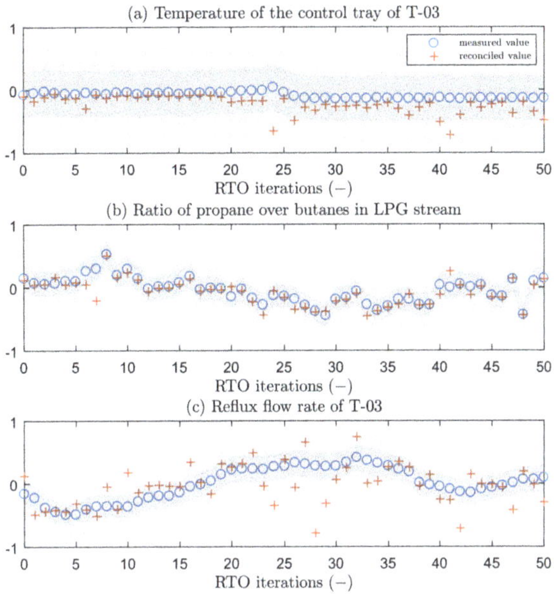

Figure 7. Data reconciliation results: (**a**) temperature of the control tray of T-03; (**b**) ratio of the content of propane over butanes in LPG; (**c**) Reflux flow rate of T-03.

The temperature of the control tray of T-03 is an estimated output variable that is adjustable by the estimation of the Murphree efficiency of the column, which explains the low relative error of 1.97% of this variable. In addition, the ratio of propane over butanes in the LPG is also an estimated output, which is adjusted by the manipulation of the ratio of propane over butanes in the feed stream of the process; this variable presented a relative error of 2.12%. Lastly, the reflux flow rate of tower T-03 is an estimated input, but a relatively low weight is assigned to this variable due to a high uncertainty in the measurement of the liquid streams. However, even with this uncertainty, an error of 5.02% is considered low.

The results herein presented attest to the quality of the developed model and its ability to represent the variables of the system. Therefore, the model is adherent to plant in study.

4.2. Optimization

During the first month of RTO operation, there were two moments that are delimited in the results of this section. First, the RTO was set to run in open loop with the DCS, so its behavior could be observed and any required adjustment could be made. Then, after approximately 17 days of operation, the RTO loop was closed with the DCS—that is, the control system started reading the solution of the RTO and tracking the optimal operation.

Figure 8 shows the result of the objective function variation in percentage, that is, the optimal value of the objective function minus the value calculated in the reconciliation problem over this last value. The vertical dashed line indicates the moment when the RTO loop was closed.

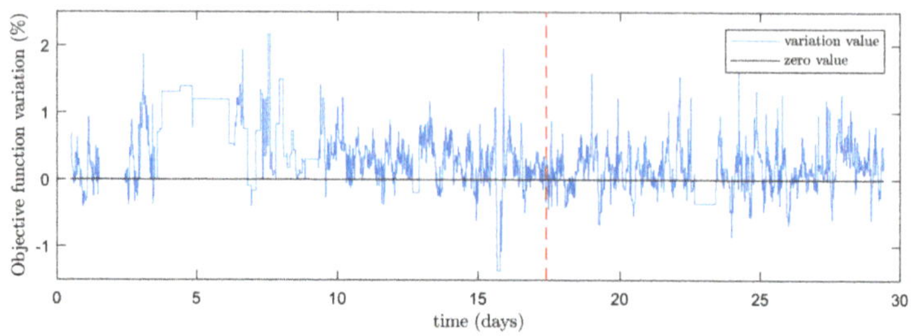

Figure 8. Optimization results: variation of the optimal objective function from the reconciliation solution.

The zero line is marked to denote the boundary between actual economic gain or loss provided by the optimal solution. When the value is positive, there are economic gains, otherwise there are economic losses. It is possible to see that there is a considerable number of times that the optimization crossed the zero line to the economic loss region. However, this can be explained by the violation of the constraints where the plant may operate, so the optimization may decide to reduce the economic gain in order to bring the plant back to a feasible region. Despite this fact, the overall economic gain is achieved and is verified by a positive value on the numerical integration of the curve.

Figure 9 shows the trend of the volumetric fraction of ethane in LPG displaced by the optimal value, which is the most sensitive variable in the economic function. The horizontal dashed lines illustrate the resultant standard deviation upward and downward in both time windows. The results show that the content of ethane, which is sent to the DCS as a reference value, was more concentrated around the optimal value after closing the loop of the RTO, with a reduction of 31.9% on the standard deviation value.

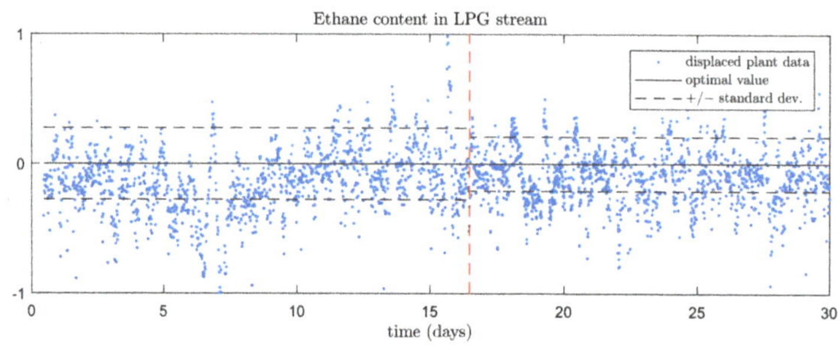

Figure 9. Optimization results: ethane content in LPG deviated from the optimal value before and after the RTO loop closing.

Figure 10 illustrates the result related to the content of pentanes in the LPG stream, which is sent to the DCS as a reference value. This result shows a high concentration of the data points around the optimal value, with a reduction of 217% in the standard deviation value after closing the loop—the right-hand section after the vertical dashed line. This result reinforces the operational benefits resulted from the implementation of the RTO.

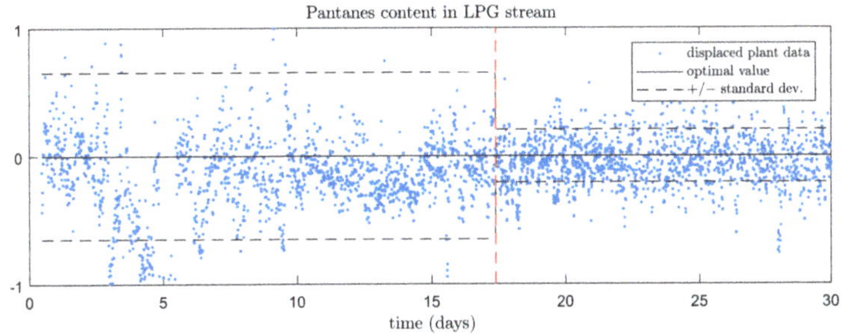

Figure 10. Optimization results: pentanes content in LPG deviated from the optimal value before and after the RTO loop closing.

A similar result is also observed in the bottom temperature of tower T-01, as showed in Figure 11, and in the reflux flow rate, in Figure 12, with decreases in the standard deviation around the optimal values of 59.9% and 19.7%, respectively. The bottom temperature of tower T-01 and the reflux flow rate are sent to the DCS as a reference value and a target, respectively.

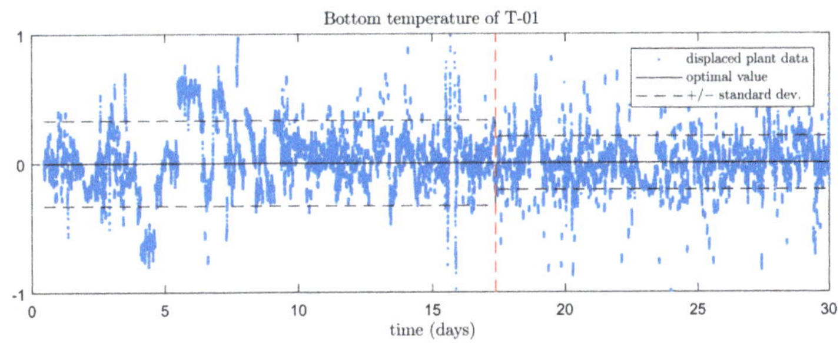

Figure 11. Optimization results: bottom temperature of T-01 deviated from the optimal value before and after the RTO loop closing.

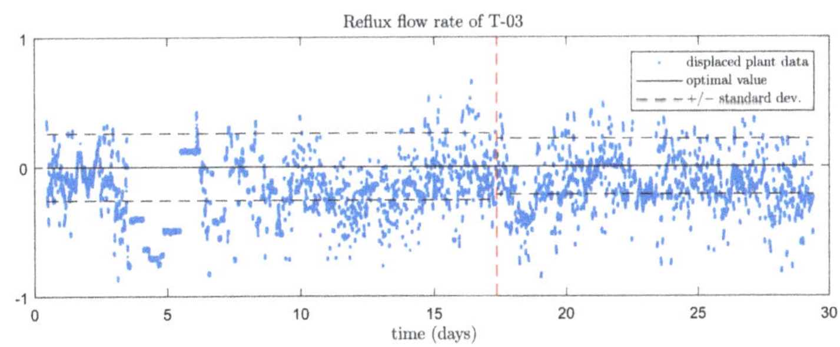

Figure 12. Optimization results: reflux flow rate of T-03 deviated from the optimal value before and after the RTO loop closing.

Regarding the recirculation flow rate, which is sent as a target to the DCS, it possible to see in Figure 13 that, even though there is a decrease in the standard deviation around the optimal value of 19.7%, the data points kept consistently below the optimal value. This suggests that the control layer may be neglecting the optimal target of this variable and adjustments must be carried out in the future.

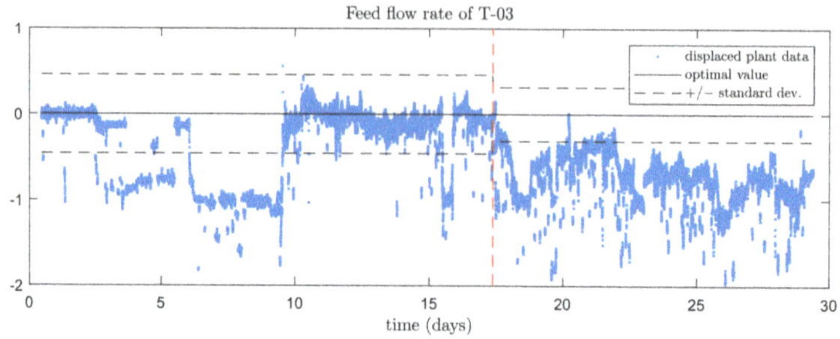

Figure 13. Optimization results: recirculation flow rate of T-03 deviated from the optimal value before and after the RTO loop closing.

The results shown support the fact that the RTO provides an economic benefit by the positive variation of the objective function and also provides an operational benefit by the reduction of the variability of the data points. In general, the communication between RTO and the control layer was successfully achieved.

4.3. Computational Cost and MCSS Performance

As previously mentioned, the cycle of the RTO is not fixed. Therefore, there is no specific frequency for each RTO run; in fact, each run will have its own time length depending on the computational cost needed for the reconciliation problem and the optimization problem, which are the most costly stages of the cycle. This computational cost is dependent on the set of initial points given for each problem, if this set is far from the optimal solution, the problem may take more time to converge. As the initial guess is constituted by the actual plant data, the computational cost is highly dependent on the quality of the input data. Table 6 shows the average, the minimum, and the maximum time spent in each stage in the first month of operation.

Table 6. Computational cost of each stage of the RTO cycle.

Stage	Average Time (min)	Minimum Time (min)	Maximum Time (min)
Reconciliation problem	13.5	1.9	46.3
Optimization problem	1.9	0.1	16.0
Total cycle	15.4	3.1	47.1

The presented small-scale RTO presents a fast cycle compared to full-scale commercial solutions. This result is not only due to the fact that the scope is reduced, but also due to the proposed "Modular Continuous with Successive Substitution" approach adopted in the simulation. With the philosophy to avoid convergence loops, the cost to run the model was around 2 s, if that was considered otherwise, the cost to converge the model could vary between 30 s to 60 s, which could significantly delay the optimization algorithm—especially in the gradient estimation stage, where the model has to be called twice for each decision variable at each optimization iteration.

Therefore, the present RTO has potential to be spread not only to other small-scale applications, but also to applications where there is already a slow commercial RTO implemented; in this case, the small-scale RTO can provide intermediate optimal solutions during the long cycle of the full-scale optimization. This could be in order to improve the robustness to frequent disturbances and fast dynamics in the whole optimization approach.

The proposed MCSS approach was compared to the classical MC approach in terms of computational cost. The data reconciliation problem and the optimization problem were run offline for a single point and for different values of the optimizer feasibility tolerance. Tables 7 and 8 show the obtained results.

Table 7. Comparison between MCSS and MC: Data Reconciliation problem.

Approach	Feasibility Tolerance	Loop Constraint	Optimizer Iterations	Simulator Calls	Total Time (min)
MCSS	10^{-1}	2.0×10^{-2}	53	908	13.0
	10^{-2}	5.5×10^{-3}	102	1747	26.4
	10^{-4}	1.4×10^{-4}	158	2689	37.2
MC	10^{-1}	4.4×10^{-2}	147	2513	28.3
	10^{-2}	8.8×10^{-3}	152	2587	30.1
	10^{-4}	6.0×10^{-4}	196	3346	37.7

Table 8. Comparison between MCSS and MC: Optimization problem.

Approach	Feasibility Tolerance	Loop Constraint	Optimizer Iterations	Simulator Calls	Total Time (min)
MCSS	10^{-1}	2.2×10^{-2}	9	129	2.1
	10^{-2}	1.0×10^{-2}	11	165	2.5
	10^{-4}	1.5×10^{-4}	20	283	4.3
MC	10^{-1}	2.6×10^{-2}	20	272	3.5
	10^{-2}	1.0×10^{-2}	36	480	5.7
	10^{-4}	3.4×10^{-4}	36	492	6.2

As can be seen, the proposed MCSS resulted in less numbers of optimizer iterations and simulator calls, which implied a significant reduction in the total spent time when compared to the classic MC approach for all values of the feasibility tolerance tested. An acceleration of the convergence is also observed, as the number of optimizer iterations was significantly reduced. The MCSS spent, on average, 0.24 min per optimizer iteration; the MC spent, on average, 0.18 min per iteration, due to the fact that the model used in the MC approach is computationally cheaper since it does not perform the extra calculations required for the successive substitution method. Therefore, it might have a trade-off regarding the feasibility tolerance and the additional computational time of the reference module (see Figure 1) for the acceleration to be advantageous.

In the present application, the MCSS approach presented a real benefit in improving the computational cost of RTO, enabling to fasten RTO cycles and decrease the chances of operating in suboptimal conditions.

4.4. Economic Return

The evaluation of the economic return of RTO was performed following the method proposed and described in Section 2.3 and applied to the industrial case in the study as presented in Section 3.6. The economic return is measured by the performance index for each operational mode. These indexes are calculated by the fraction of the profit function calculated in the data reconciliation problem over the value calculated in the optimization problem. This can be interpreted as the distance between the actual economic performance of the operational mode from the utopian economic performance resulted from the optimization. Table 9 presents the performance index calculated for each operational mode.

Table 9. Performance index for each operational mode.

Operational Mode	Performance Index
Regulatory control	0.9888
Supervisory control	0.9913
Optimization	0.9951

Just by analyzing the performance index, it is possible to note that the economic benefit follows the expected tendency—that is, the supervisory control presents higher return compared with the regulatory control, and the optimization presents higher return compared with the supervisory control. As previously commented, it is possible to have a better sensibility of the economic return by fixing a reference scenario of optimal profit return—not shown due to the nondisclosure agreement. In spite of that, it is possible to evaluate the relative return by directly analyzing the performance indexes, as shown in Table 10.

Table 10. Relative return comparing the operational modes.

Comparison	Relative Return
Supervisory control—regulatory control	0.253%
Optimization—supervisory control	0.383%
Optimization—regulatory control	0.636%
Potential gain of the optimization	0.492%

The potential gain of the optimization, mentioned in Table 10, is measured by the distance between the performance index of the optimization mode and a utopian operational mode with performance index equal to 1. Although this utopian operation is not achievable, the distance between the actual operation and this utopian operation can be decreased by improving the synergism between the layers, the tuning of the control layers, and possibly the quality of the models used in the supervisory control layer.

Another way of visualizing this result is to construct a "utopian operational path chart" is presented in Figure 14.

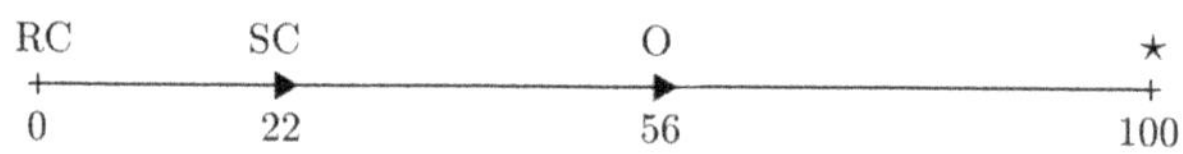

Figure 14. "Utopian operational path" chart. RC—regulatory control; SC—supervisory control; O—optimization; ⋆—utopian value.

This graphic is constructed by normalizing the performance index between 0 and 100, with 0 being the performance of the Regulatory Control and 100 being the utopian performance. It is possible to see that the Supervisory control was able to move 22 points in the utopian path and the Optimization was able to move 54 points. This is a good way of visualizing the benefit of the implementation of RTO and the potential for improvement of the control and optimization strategies.

5. Conclusions

The present paper describes the development of an RTO strategy for small-scale applications and its implementation to a Debutanizer section of a Natural Gas Processing Unit. The whole methodology is disclosed, including a proposition of a new modeling method for enhancing the effectiveness of the use of sequential-modular models inside an optimization framework and a new method to account for the economic benefit of applications based on optimization frameworks after a period of operation. The developed model proved to be adherent to measured data, which emphasizes its adequacy to be applied for optimization purposes. The implementation of RTO in a closed loop presented not only an economic benefit, but also operational benefits observed from the reduction of variability of the key variables of the process. Moreover, the low computational cost and the great economic benefits indicate the success of the proposed Modular Continuous with Successive Substitution in enhancing the efficiency of sequential-modular simulators in optimization schemes. In addition, it also indicates the potential to spread the application of small-scale RTO to other applications and even to applications where a slow RTO has already been implemented. Finally, RTO showed an increase in operational aspects, as observed by the reduction in variability of the main variables of the system, and in economic aspects, as shown by an increase of 0.64% in profit return when compared with the operation of the regulatory control layer alone.

Author Contributions: Conceptualization, P.A.D., L.D.R., C.R.P. and A.R.S.; methodology, P.A.D., L.D.R., C.R.P. and A.R.S.; software, P.A.D. and L.D.R.; validation, C.R.P., J.N. and M.V.C.G.; formal analysis, P.A.D., L.D.R., C.R.P. and A.R.S.; investigation, P.A.D., L.D.R. and C.R.P.; resources, J.N., M.V.C.G. and A.R.S.; data curation, P.A.D.; writing—original draft preparation, P.A.D.; writing—review and editing, P.A.D., L.D.R., C.R.P., J.N. and A.R.S.; visualization, P.A.D.; supervision, L.D.R., J.N., M.V.C.G. and A.R.S.; project administration, L.D.R., J.N. and A.R.S.; funding acquisition, J.N. and M.V.C.G. All authors have read and agreed to the published version of the manuscript.

Funding: This research was funded by Cenpes/Petrobras collaboration terms PT-112.01.12343 and PT-112.01.13164.

Institutional Review Board Statement: Not applicable.

Informed Consent Statement: Not applicable.

Data Availability Statement: Restrictions apply to the availability of the data. The data is not publicly available due to a Confidentiality agreement between COPPE/UFRJ and CENPES/Petrobras.

Conflicts of Interest: The authors declare no conflict of interest.

Appendix A. Description of the Simulation Calculators

As presented in Section 3.2, the management of information throughout the model flow-sheet is done via calculators modules inserted into the simulation. Here, the purpose of each calculator is detailed:

- CA01—receives the measurement of the top pressure of tower T-01 (DFLP) and substitutes the pressure of modules FL01 and FL02, such that

$$\begin{aligned} \text{FL01(Pressure)} &\longleftarrow \text{DFLP(Pressure)} \\ \text{FL02(Pressure)} &\longleftarrow \text{DFLP(Pressure)} + 0.3\ \text{kgf/cm}^2. \end{aligned}$$

- CA02—receives the measurement of the feed composition, the composition of propane and butanes in LPG, the values of parameters θ_1 and θ_2, and substitutes the values of the feed stream of the simulation accordingly to the procedure discussed in Section 3.2 in Equations (14) and (15);
- CA03—implement the fractioning cut in module ST1A accordingly to the procedure described in Section 3.2, Equation (16);
- CA05—receives the measured temperature and pressure and substitutes them into module P108;
- CA06—receives the measured top pressure of tower T-03 and substitutes this value into the top pressure of this tower and the condenser pressures of modules R-03 and T-03. Further, this value, incremented by 2.5 kgf/cm^2, is inserted into the valve module FV04;
- CA04—receives the simulated temperate of the stream downstream of module P12C, the measured temperatures of the hot-out and cold-in streams and the value of the decision variable ρ_1, and implements the HOCI approach strategy described in Section 3.2;
- CA07—receives the measurement of the reflux and recirculation flow rates and the input of the molar flow rate of the bottom of tower T-03 and substitutes these values into modules R-03 and T-03;
- CA12—receives the simulated duty of module P12H and inserts it into module R12C;
- FOBJ—calculates the objective function of the optimization problem;
- XOBJ—calculates all constraints of the optimization and data reconciliation problems;
- CA11—calculates the objective function of the data reconciliation problem.

References

1. Trierweiler, J.O. Real-Time Optimization of Industrial Processes. In *Encyclopedia of Systems and Control*; Springer: Berlin, Germany, 2014; pp. 1–12. [CrossRef]
2. Jang, S.; Joseph, B.; Mukai, H. On-line optimization of constrained multivariable chemical processes. *AIChE J.* **1987**, *33*, 26–35. [CrossRef]
3. Câmara, M.M.; Quelhas, A.D.; Pinto, J.C. Performance evaluation of real industrial RTO systems. *Processes* **2016**, *4*, 44. [CrossRef]
4. Müller, D.; Dercks, B.; Nabati, E.; Blazek, M.; Eifert, T.; Schallenberg, J.; Piechottka, U.; Dadhe, K. Real-Time Optimization in the Chemical Processing Industry. *Chem. Ing. Tech.* **2017**, *89*, 1464–1470. [CrossRef]
5. Quelhas, A.D.; de Jesus, N.J.C.; Pinto, J.C. Common vulnerabilities of RTO implementations in real chemical processes. *Can. J. Chem. Eng.* **2013**, *91*, 652–668. [CrossRef]
6. Marchetti, A.G.; de Avila Ferreira, T.; Costello, S.; Bonvin, D. Modifier Adaptation as a Feedback Control Scheme. *Ind. Eng. Chem. Res.* **2020**, *59*, 2261–2274. [CrossRef]
7. Marchetti, A.G.; François, G.; Faulwasser, T.; Bonvin, D. Modifier adaptation for real-time optimization—Methods and applications. *Processes* **2016**, *4*, 55. [CrossRef]
8. Marlin, T.; Hrymak, A. Real-Time Operations Optimization of Continuous Processes. In *AIChE Symp Series CPC V*; American Institute of Chemical Engineers: New York, NY, USA, 1997; Volume 93.
9. Biegler, L.; Grossmann, I.; Westerberg, A. A note on approximation techniques used for process optimization. *Comput. Chem. Eng.* **1985**, *9*, 201–206. [CrossRef]
10. Forbes, J.; Marlin, T.; MacGregor, J. Model adequacy requirements for optimizing plant operations. *Comput. Chem. Eng.* **1994**, *18*, 497–510. [CrossRef]
11. Forbes, J.F.; Marlin, T.E. Model Accuracy for Economic Optimizing Controllers: The Bias Update Case. *Ind. Eng. Chem. Res.* **1994**, *33*, 1919–1929. [CrossRef]
12. Bhat, S.A.; Saraf, D.N. Steady-State Identification, Gross Error Detection, and Data Reconciliation for Industrial Process Units. *Ind. Eng. Chem. Res.* **2004**, *43*, 4323–4336. [CrossRef]
13. Rhinehart, R.R. Automated steady and transient state identification in noisy processes. In Proceedings of the 2013 American Control Conference, Washington, DC, USA, 17–19 June 2013; pp. 4477–4493. [CrossRef]
14. Øyvind Øksnes Dalheim.; Steen, S. A computationally efficient method for identification of steady state in time series data from ship monitoring. *J. Ocean. Eng. Sci.* **2020**, *5*, 333–345. [CrossRef]
15. Krishnamoorthy, D.; Foss, B.; Skogestad, S. Steady-state real-time optimization using transient measurements. *Comput. Chem. Eng.* **2018**, *115*, 34–45. [CrossRef]
16. Matias, J.O.; Le Roux, G.A. Real-time Optimization with persistent parameter adaptation using online parameter estimation. *J. Process. Control.* **2018**, *68*, 195–204. [CrossRef]
17. Roberts, P.D. An algorithm for steady-state system optimization and parameter estimation. *Int. J. Syst. Sci.* **1979**, *10*, 719–734. [CrossRef]
18. Chachuat, B.; Marchetti, A.; Bonvin, D. Process optimization via constraints adaptation. *J. Process. Control.* **2008**, *18*, 244–257. [CrossRef]
19. Marchetti, A.; Chachuat, B.; Bonvin, D. Modifier-adaptation methodology for real-time optimization. *Ind. Eng. Chem. Res.* **2009**, *48*, 6022–6033. [CrossRef]
20. de Avila Ferreira, T.; Wuillemin, Z.; Marchetti, A.; Salzmann, C.; Herle, J.V.; Bonvin, D. Real-time optimization of an experimental solid-oxide fuel-cell system. *J. Power Sources* **2019**, *429*, 168–179. [CrossRef]
21. de Avila Ferreira, T.; Wuillemin, Z.; Faulwasser, T.; Salzmann, C.; herle, J.V.; Bonvin, D. Enforcing optimal operation in solid-oxide fuel-cell systems. *Energy* **2019**, *181*, 281–293. [CrossRef]
22. Gao, W.; Wenzel, S.; Engell, S. A reliable modifier-adaptation strategy for real-time optimization. *Comput. Chem. Eng.* **2016**, *91*, 318–328. [CrossRef]
23. Gao, W.; Hernández, R.; Engell, S. A Study of Explorative Moves during Modifier Adaptation with Quadratic Approximation. *Processes* **2016**, *4*, 45. [CrossRef]
24. Matias, J.; Jäschke, J. Using a neural network for estimating plant gradients in real-time optimization with modifier adaptation. *IFAC-PapersOnLine* **2019**, *52*, 808–813. [CrossRef]
25. de Avila Ferreira, T.; Shukla, H.A.; Faulwasser, T.; Jones, C.N.; Bonvin, D. Real-Time optimization of Uncertain Process Systems via Modifier Adaptation and Gaussian Processes. In Proceedings of the 2018 European Control Conference (ECC), Limassol, Cyprus, 12–15 June 2018; pp. 465–470.
26. del Rio Chanona, E.; Graciano, J.A.; Bradford, E.; Chachuat, B. Modifier-Adaptation Schemes Employing Gaussian Processes and Trust Regions for Real-Time Optimization. *IFAC-PapersOnLine* **2019**, *52*, 52–57. [CrossRef]
27. del Rio Chanona, E.A.; Petsagkourakis, P.; Bradford, E.; Graciano, J.E.A.; Chachuat, B. Real-time optimization meets Bayesian optimization and derivative-free optimization: A tale of modifier adaptation. *Comput. Chem. Eng.* **2021**, *147*, 107249. [CrossRef]

28. Darby, M.; Nikolaou, M.; Jones, J.; Nicholson, D. RTO: An Overview and Assessment of Current Practice. *J. Process. Control.* **2011**, *21*, 874–884. [CrossRef]
29. Liporace, F.S.; Gomes, M.V.; Katata, A.C.; Zanin, A.C.; Moro, L.F.; Porfírio, C.R. PETROBRAS Experience Implementing Real Time Optimization. In *International Symposium on Process Systems Engineering: Part A*; de Brito Alves, R.M., do Nascimento, C.A.O., Biscaia, E.C., Eds.; Computer Aided Chemical Engineering; Elsevier: Hoboken, NJ, USA, 2009; Volume 27, pp. 1245–1250. [CrossRef]
30. Straus, J.; Skogestad, S. Minimizing the complexity of surrogate models for optimization. In *26th European Symposium on Computer Aided Process Engineering*; Kravanja, Z., Bogataj, M., Eds.; Computer Aided Chemical Engineering; Elsevier: Amsterdam, The Netherlands, 2016; Volume 38, pp. 289–294. [CrossRef]
31. Berna, T.J.; Locke, M.H.; Westerberg, A.W. A new approach to optimization of chemical processes. *AIChE J.* **1980**, *26*, 37–43. [CrossRef]
32. Biegler, L. Improved infeasible path optimization for sequential modular simulators—I: The interface. *Comput. Chem. Eng.* **1985**, *9*, 245–256. [CrossRef]
33. Biegler, L.; Cuthrell, J. Improved infeasible path optimization for sequential modular simulators—II: The optimization algorithm. *Comput. Chem. Eng.* **1985**, *9*, 257–267. [CrossRef]
34. Gomes, M.V.; Bogle, I.D.L.; Biscaia, E.C.; Odloak, D. Using kriging models for real-time process optimisation. In *European Symposium on Computer Aided Process Engineering*; Braunschweig, B., Joulia, X., Eds.; Computer Aided Chemical Engineering; Elsevier: Hoboken, NJ, USA, 2008; Volume 25, pp. 361–366. [CrossRef]
35. Carpio, R.R.; Giordano, R.C.; Secchi, A.R. Enhanced surrogate assisted framework for constrained global optimization of expensive black-box functions. *Comput. Chem. Eng.* **2018**, *118*, 91–102. [CrossRef]
36. Udugama, I.A.; Taube, M.A.; Mansouri, S.S.; Kirkpatrick, R.; Gernaey, K.V.; Yu, W.; Young, B.R. A Systematic Methodology for Comprehensive Economic Assessment of Process Control Structures. *Ind. Eng. Chem. Res.* **2018**, *57*, 13116–13130. [CrossRef]
37. Bauer, M.; Craig, I.K. Economic assessment of advanced process control—A survey and framework. *J. Process. Control.* **2008**, *18*, 2–18. [CrossRef]
38. Alkaya, D.; Vasantharajan, S.; Biegler, L.T. Successive quadratic programming: applications in the process industry Successive Quadratic Programming: Applications in the Process Industry. In *Encyclopedia of Optimization*; Floudas, C.A., Pardalos, P.M., Eds.; Springer: Boston, MA, USA, 2009; pp. 3853–3866.
39. Mochizuki, S.; Saputelli, L.; Kabir, S.; Cramer, R.; Lochmann, M.; Reese, R.; Harms, L.; Sisk, C.; Hite, J.; Escorcia, A. Real Time Optimization: Classification and Assessment. *Spe Prod. Oper.* **2006**, *21*, 455–466. [CrossRef]
40. Xu, F.; Huang, B.; Akande, S. Performance Assessment of Model Pedictive Control for Variability and Constraint Tuning. *Ind. Eng. Chem. Res.* **2007**, *46*, 1208–1219. [CrossRef]
41. Campos, M.; Gomes, M.; Souza, A.; Barros, A. Optimisation of natural gas plant—Gains in profitability, stability and energy efficiency. *Int. Gas Union World Gas Conf. Pap.* **2012**, *3*, 2089–2120.
42. Camolesi, V.J.; Moro, L.L.F.; Zanin, A.C. Implantação de um otimizador em tempo real (RTO) no conversor de uma unidade de craqueamento catalítico. *Sba Controle Automação Soc. Bras. Autom.* **2008**, *19*, 128–137. [CrossRef]
43. Forbes, J.; Marlin, T. Design cost: a systematic approach to technology selection for model-based real-time optimization systems. *Comput. Chem. Eng.* **1996**, *20*, 717–734. [CrossRef]
44. Biegler, L.T. *Nonlinear Programming*; Society for Industrial and Applied Mathematics: Philadelphia, PA, USA, 2010. [CrossRef]
45. Biegler, L.T.; Hughes, R.R. Infeasible path optimization with sequential modular simulators. *AIChE J.* **1982**, *28*, 994–1002. [CrossRef]
46. Niederberger, J.; Gama, M.S.; dos Santos, L.C.; da Silva, J.A.; Vargas, C.E.; Ahón, V.R.R.; Silva, E.P.; de Santana Souza, D.F.; Aquino, C.; Aires, J.S.S. PETROX—PETROBRAS Technology in Process Simulation. In *10th International Symposium on Process Systems Engineering: Part A*; de Brito Alves, R.M., do Nascimento, C.A.O., Biscaia, E.C., Eds.; Computer Aided Chemical Engineering; Elsevier: Hoboken, NJ, USA, 2009; Volume 27, pp. 675–680. [CrossRef]
47. Cao, S.; Rhinehart, R. An efficient method for on-line identification of steady state. *J. Process. Control.* **1995**, *5*, 363–374. [CrossRef]
48. Rod, V.; HanČl, V. Iterative estimation of model parameters when measurements of all variables are subject to error. *Comput. Chem. Eng.* **1980**, *4*, 33–38. [CrossRef]
49. Gill, P.E.; Murray, W.; Saunders, M.A.; Wright, M.H. *User's Guide for NPSOL 5.0: A Fortran Package for Nonlinear Programming*; Technical Report SOL 86-1; University of California: Auckland, CA, USA, 1998; pp. 1–44.
50. Ito, E.H.; Secchi, A.R.; Gomes, M.V.; Paiva, C.R. Development of a gas composition soft sensor for distillation columns: A simplified model based and robust approach. In *International Symposium on Process Systems Engineering (PSE 2018)*; Eden, M.R., Ierapetritou, M.G., Towler, G.P., Eds.; Talylor Francls; Computer Aided Chemical Engineering; Elsevier: Hoboken, NJ, USA, 2018; Volume 44, pp. 661–666. [CrossRef]
51. Cutler, C.R.; Ramaker, B.L. Dynamic matrix control—A computer control algorithm. *Jt. Autom. Control. Conf.* **1979**, *17*, 72. [CrossRef]
52. Rotava, O.; Zanin, A. Multivariable control and real-time optimization—An industrial practical view. *Hydrocarb. Process.* **2005**, *84*, 61–71.

processes

Article

The Effect of Changes in Settings from Multiple Filling Points to a Single Filling Point of an Industry 4.0-Based Yogurt Filling Machine

Jinping Chen [1], Razaullah Khan [2,*], Yanmei Cui [3,*], Bashir Salah [4], Yuanpeng Liu [1] and Waqas Saleem [5]

1 School of Aerospace Engineering, Zhengzhou University of Aeronautics, Zhengzhou 450046, China
2 Engineering Management Program, Institute of Manufacturing, University of Engineering and Applied Sciences, Swat 19060, Pakistan
3 School of Mechanical Engineering, Shanghai Dianji University, Shanghai 201306, China
4 Industrial Engineering Department, College of Engineering, King Saud University, P.O. Box 800, Riyadh 11421, Saudi Arabia
5 Department of Mechanical and Manufacturing Engineering, Institute of Technology, F91 YW50 Sligo, Ireland
* Correspondence: razaullah@ueas.edu.pk (R.K.); hongyong83@126.com (Y.C.)

Abstract: In process optimization, a process is adjusted so as to optimize a set of parameters while meeting constraints, with the objective to either minimize the total processing time or maximize the throughput. This article focused on the process optimization of a fully automated yogurt and flavor-filling machine developed based on the industrial revolution 4.0 concept. Mathematical models were developed for minimizing the total processing time or maximizing the throughput of an Industry 4.0-based yogurt filling system with two different machine settings called Case-I and Case-II. In Case-I, the yogurt and flavors are filled at two distinct points while Case-II considers the filling of yogurt and flavors at a single point. The models were tested with real data and the results revealed that Case-II is faster than Case-I in processing a set of customer orders. The results were used as inputs for the single-dimension rules to check which one results in more intended outputs. Additionally, different performance measures were considered and the one with most importance to the management was selected.

Keywords: mathematical modeling; production scheduling; process optimization; advanced optimization methodologies; modeling of industrial processes

Citation: Chen, J.; Khan, R.; Cui, Y.; Salah, B.; Liu, Y.; Saleem, W. The Effect of Changes in Settings from Multiple Filling Points to a Single Filling Point of an Industry 4.0-Based Yogurt Filling Machine. *Processes* **2022**, *10*, 1642. https://doi.org/10.3390/pr10081642

Academic Editor: Zhiwei Gao

Received: 25 July 2022
Accepted: 16 August 2022
Published: 18 August 2022

1. Introduction

The Industry 4.0 concept has sparked worldwide attention from various production sectors. In order to conceptualize the idea in the manufacturing sectors, policy makers of different countries initiated their funded mega-research projects to uplift and transform existing technologies according to the modern needs of industry to achieve dynamic customer demands [1–8]. The idea of Industry 4.0 was proposed by Germany as one of the key initiatives towards its high-tech strategy to uplift the manufacturing sector and achieve its desired goals [9]. This policy shift posed a threat to European policies and compelled the leading nations to initiate their own strategies towards developing their own dominant technological manufacturing sectors. For instance, the United States has initiated similar initiatives, known as Smart Manufacturing, to compete in the new technological revolution [10]. In 2014, China proposed the "Made in China 2025" vision to break China's reliance on the foreign technology and transform China into a world-leading manufacturing power [11,12]. In 2016, Japan introduced an advanced variant to Industry 4.0, called Society 5.0, which went far beyond Industry 4.0, whose aim was the digitization of all sectors of the Japanese society [13]. In parallel, Saudi Arabia announced its strategic framework known as "Vision 2030" to update the existing industries (refinery, petrochemical, fast-moving consumer goods, etc.) within the kingdom according to modern technological revolutions [14].

The basic purpose of the Vision 2030 framework is to renovate, standardize, and digitally transform the Kingdom of Saudi Arabia. A step towards this strategy was taken by the laboratory of Computer Integrated Manufacturing (CIM) at the Department of Industrial Engineering, King Saud University, Saudi Arabia by designing an automatic yogurt-filling system for automatic filling of yogurt and addition of different flavors as per customers' demands [15]. This approach has produced consistent outcomes and studies have been published in peer-reviewed journals. For example, in Virtual Reality (VR)-based engineering education, the precision of robotic arms and conveyor belts was shown to improve production sustainability in Industry 4.0 [16]. A thorough description of the yogurt-filling method was also offered in Industry 4.0-based real-time scheduling and dispatching of lean production environments [17].

Industry 4.0 systems involve a variety of agents, from factories to end customers, in which the high-connectivity and cooperation can optimize operational processes, products, and services [18]. B. Salah [19] performed a research study of the yogurt filling system capable of mixing the yogurt with three different flavors as demanded by the customers. The number of containers used in the machine were four. In one container, the base yogurt was stored, while in the remaining three constrainers, three different flavors were stored. The empty cups were automatically moved to a specific location called the bottle feeding point and were then entered into the machine through the entry point over the conveyor belt. Upon reaching the single filling point, the bottles were filled with the required volumes of yogurt and flavors. The feed rate of the nozzles was controlled through solenoid valves. Four diaphragm pumps were used to deliver the yogurt and flavors from the tanks to the unified head of nozzles. The valves with a specific feed rate value were opened for a certain time to fill the bottles with the required volume of yogurt and flavors. The completely filled bottles were then moved to the exit point over the belt where a robotic arm was used to remove the bottles from the system. The whole process was made automatic with increased throughput and less human involvement. Additionally, Node-RED was used to run the system, which has the capability of machine-to-machine communication as it can connect multiple devices. There were several challenges faced during the project and few of them were minimizing the overall cost of the system, programming expertise while managing the Raspberry Pi, and the slow speed of the internet. The design was divided into two phases called the pre- and post-production phases. In the study, only the pre-production phase was considered while the post production phase in which the completely filled bottles are to be stored in the refrigerator storage system will be studied in future research. M. Ramadan et al. [20] created a novel real-time manufacturing cost-monitoring system (RT-MCT) that combines lean manufacturing and RFID ideas. The RT-MCT was designed to connect lean operational characteristics with financial expenses in real time. F. A. German et al. [21] conducted a survey in several manufacturing businesses to investigate the deployment of Industry 4.0 technologies and developed an Industry 4.0 technology layer structure. They demonstrated the levels of adoption of these technologies as well as their implications for manufacturing firms. F. Longo et al. [22] provided a human-centered strategy to improve operators' skills and competencies in the context of the new smart factory. The proposed research activity focused on the design and implementation of a practical solution called Sophos-MS, which is capable of integrating augmented-reality contents and intelligent teaching systems with cutting-edge fruition technologies to assist operators in complicated man–machine interactions. M. E. Leusin et al. [23] suggested a system with self-configuring characteristics to deal with production line disruptions. The performance of the suggested framework was evaluated in a simulation study based on a real-world industrial example. Their findings supported benefits in flexibility, scalability, and efficiency achieved by data sharing between manufacturing levels. D. A. Rossit et al. [24] proposed Smart Scheduling, a novel decision-making schema designed to provide flexible and efficient production plans on the fly while taking advantage of the peculiarities of these new contexts. P. Chen et al. [25] sought to investigate how environmental collaboration across organizational borders influences green innovation from the standpoint of social capital. The study provided an important contribution to the literature by thoroughly exploring how environmental collaboration in emerging nations affects green innovation from the standpoint of social capital. P. Zawadzki et al. [26] developed a broad notion of smart design and production control as key determinants for efficient and reliable operation of a smart factory. Several strategies were offered to help in the design process of personalized goods and the organization of their manufacturing in the context of realizing the mass customization strategy, allowing for a shorter period of creation for a new product. A. Grassi et al. [27] developed a revolutionary architecture for production planning and control with a semi-hierarchical structure in which various management levels were distinguished by their physical identity as well as their functional scope. The suggested architecture represented an improved application of the Industry 4.0 decentralized decision idea, allowing for a better understanding and management of system performance to yield higher profit and response time predictability in highly customized scenarios. P. Spenhoff et al. [28] established an "every product every cycle" (EPEC) 4.0 production control approach, which intended to schedule the production system as effectively as possible while providing the required flexibility and minimum schedule disturbance. L. E. Quezada et al. [29] provided operational excellence as a means of achieving sustainable development goals via Industry 4.0. C. Santos [30] examined some of the important European Union (EU) industrial standards, roadmaps, and scientific publications that led to the portrayal of the phrase Industry 4.0, as well as how key technologies and concepts have evolved over time. B. Salah et al. [31] presented the second phase design of control architecture for the yogurt filling machine based on Industry 4.0 principles, which included a near field communication platform to improve consumer service. Capaci et al. [32] presented an innovative performance monitoring system, specifically devoted to control loops, based on cloud technology by focusing on three different aspects: describing the entire cloud architecture and its implementation issues, illustrating basic techniques and features installed in the updated analytics tool, and presenting significant case studies. The system illustrated a successful example of a cloud-based platform for performance monitoring and assessment of process plants specifically oriented to proportional-integral-derivative control loops. Capaci et al. [33] designed and tested an automated system for modeling and controlling color quality of dyed lathers. The proposed software was fully integrated with the machineries of the finishing line and an automated tintometer system. A set of company data was used to validate the identified colorimetric models and the proposed color correction strategy. The paradigm shifts in manufacturing that Industry 4.0 brings forth with new advanced technologies and the rapid growth of sensing and controlling technologies enable further visualization and optimization that can contribute to achieving improved decision-making in manufacturing. A significant new capability is the ability to construct a Digital Twin that connects the physical and virtual space [34]. Z. Han et al. [35,36] explored the multi-queue limited buffers scheduling problems in a flexible flow shop with setup times in a bus manufacturer and designed several local scheduling rules to control the moving process of the work pieces.

In a previous published study, a mathematical model was developed for the yogurt filling system with different processing layout and process parameters [37]. The main objective of model was to maximize the speed of the conveyor belt within the allowable limit. This speed was linked with the feed rate of nozzles of the yogurt and flavor valves. Increasing the feed rates of the nozzles increased the speed of the conveyor belt and vice versa. Two different points over the conveyor belt were considered for filling the yogurt and flavors into cups of different volumes demanded by customers. The total length of the conveyor belt was divided into three equal segments. The three segments included the distance from the entry point to the yogurt filling point, the distance from the yogurt filling point to the flavor filling point, and the distance from the flavor filling point to the exit point. In the mathematical model, the decision variables considered were the feed rates of yogurt and flavor valves linked with the speed of the conveyor belt.

In the present article, the setting of the yogurt filling machine was changed from multi filling points (Case-I) to a single filling point (Case-II) to check which setting results in more intended outputs. A mathematical model was developed for the new machine setting with an objective to maximize the throughput or minimize the processing time and the results were used as inputs in the single-dimension rules for the purpose to select a better setting in Case-I and Case-II. Additionally, the outcomes of the performance measures for Case-I and Case-II were compared using single-dimension rules (EDD, SPT, or FCFS) to find a rule which results in a better performance measure preferred by the management.

The article is divided into different sections as follows. In Section 1, an introduction including a literature survey of the research work is provided while the problem description is written in Section 2, where operational and technological constraints as well as the different assumptions for both Case-I and Case-II are presented. In Section 3, the methodology adopted for the process optimization problem, the possible combinations of yogurt mixing with different flavors, and the equations for different processing times are explained in detail. Section 4 illustrates the solution procedure of the problem and a customer order problem is solved while Section 5 provides details of the sequencing of customer order processing based on the single-dimension rules. The results are discussed in Section 6 while conclusion and future research directions appear in Section 7.

2. Problem Description

The research work presented in this article focused on the process optimization of a yogurt and flavor filling machine based on Industry 4.0 concept. The speed of the belt is linked with the feed rates of the nozzles of the valves. The speeds of the conveyor belts carrying either the empty, or only yogurt-filled, or yogurt mixed with any flavor-filled cups and feed rates of the yogurt and flavor nozzle valves were determined to minimize the filling process time. The cups were filled with required volumes of yogurt and different flavors according to the customer demands in the minimum possible time. In this study, the following assumptions and constraints were considered:

2.1. Operational Constraints (Management and Customers' Specifications)

1. Consideration of the minimum cup volume to optimize the sequence of processing the cups.
2. Satisfaction of customer specifications with minimum and maximum yogurt volumes with three different flavors.
3. Filling of the same volumes of yogurt mixed with required flavors in batches.
4. Filling of yogurt and flavors at a single point results in mixed yogurt and flavors, while filling yogurt and flavors at different points results in flavors above the yogurt.

2.2. Technological Constraints (Machinery Characteristics)

1. Using minimum and maximum capacity cups.
2. Limited number of yogurt and flavor nozzles.
3. Minimum and maximum feed rates of the valves of yogurt and different flavors.
4. Limited number of conveyor belts at the machines.
5. Minimum and maximum speed of the conveyor belt carrying the cups.
6. Limited number of flavors and yogurt types.
7. Limited number of holders for cups over the conveyor belt.

There may be some standard tolerances in use in the industry of yogurt and flavor filling, which must be considered while fulfilling an order. The customers are obliged to accept deviations of the quantity ordered in specific ranges and in cases of over-production, the marketing department negotiates its acceptance by the consumer. During the planning phase, under-production is never considered due to the losses inherent in production.

Two cases were considered. In Case-I, the yogurt and flavor filling points are at different locations while in Case-II, the yogurt and flavors are filled into the cup from a single location. In Case-I, an empty cup is entered into the system by placing it on the entry point. It moves over the belt to the yogurt filling nozzle and is filled with required volume of yogurt. After that, it covers some distance to reach the flavor filling point. Once the desired flavor of required volume is filled, the completely filled cup then moves towards the leaving point over the belt and leaves the filling system. In Case-II, an empty cup enters the system through the placement point, it then moves towards the yogurt and flavor filling point located at a single point over the belt in the machine, where the cup is filled with required volumes of yogurt and flavors simultaneously, and the cup filled with the desired volumes then moves towards the exit point to leave the system. A few assumptions were considered in both cases and are given below.

2.3. Case-I Assumptions

1. The cups move over the conveyor belt and the belt is divided into three equal segments.
2. The yogurt and flavors are filled at two different points.
3. Any of the three flavors can be filled in the cups at a single point after the yogurt is filled.
4. There are five distinct times in the processing of cups: the traveling time of cup from the entry point to the yogurt filling point, the yogurt filling time, the time in which the only yogurt-filled cup moves towards the flavor filling point, the flavor filling time, and the time in which a completely filled cup moves to the exit point.
5. There are two conveyor belts moving in parallel and carrying cups between the entry and exit points. When one belt stops for the filling process, the other one remains in motion and moves the cups.
6. A cup is filled with yogurt and only one type of flavor.

2.4. Case-II Assumptions

1. The conveyor belt is divided equally into two segments.
2. The filling process of yogurt and all three different flavors is performed at a single point.
3. There are three equal times: the traveling time of cups from placement to the filling point, the time in which a cup is filled simultaneously with the required volumes of yogurt and flavors, and the time in which the completely filled cup travels from the filling point to the exit point.
4. There are two conveyor belts which move in parallel and when one stops for the filling process, the other one carries the cups between the entry and exit points.
5. A cup can be filled with yogurt and multiple flavors.

There were a few other assumptions considered in both cases. The common assumptions include that the processing is uninterrupted, no cancellation and arrival of orders occur once the filling process has started, and the processing times are deterministic.

3. Production Line Architecture

Although a mathematical model for the process optimization of the Industry 4.0-based yogurt filling machine was developed in the current article, the details of the architecture of production line also needs to be discussed. The machine setting was changed from multiple filling points to a single filling point and a straight conveyor belt was considered. In the previous setting, the arrangement of the conveyor belt was either L- or U-shaped.

In the previous setting, the conveyor belt was semi-automatic and human intervention was often required for the production process completion. The semi-automatic conveyor belt is now shifted to complete the automatic mode, which increased the throughput and reduced the process time by using the latest technology IR 4.0 enablers.

For flavor identification in the previous machine setting, each cup had a color code. Additionally, for quantity grouping, optical sensors were used. In the new setting, the Near-Field Communication (NFC) technology was used for quantity categorization and color identification. NFC tags comprising of all details replaced the different colored bottles with uniform bottles from the production line.

The Fanuc LR Mate 200ic robotic arm was used for loading and unloading of NFC-tagged empty cups and yogurt-filled cups, respectively. The robotic arm was programmed in a way such that it could gently lift the empty cups and place them on the entry point of the machine. It was also used for the removal of the completely filled cups from the exit point of the machine and placing them outside the system.

In the new setting, all filling nozzles were unified into a single head which was designed in solidworks software. The drilling and milling operations were performed in the laboratory to make the unified head from a stainless steel metal block. The block was used to connect the filling nozzles with the yogurt and flavor containers through tubes.

A PN532 NFC/RFID controller breakout board manufactured by Adafruit was used for reading and writing data. The Future Technology Devices International (FTDI) chip was used for the interfacing of devices and to power the PN532 board. The information pasted on the empty cups transforms with the motion of the cup to the conveyer belt. The tag information pasted on the empty cups is decoded by the PN532 NFC tag reader, transmitted to the Raspberry Pi, and commands are given to the WAGO Programmable Field Controller (PFC). The Raspberry Pi was programmed in such a way that it could provide digital control signals for defining the customer order.

To improve the functionality of the machine, a few new devices and components were added to it. In order to maintain a steady flow rate of the yogurt and flavors, diaphragm pumps were installed between the unified head of nozzles and the yogurt and flavors tanks. Electrically controlled solenoid valves were used to release the yogurt and flavors for a certain time depending on the feed rates of the valves. To avoid any spillage of the yogurt and flavors, a pneumatic manipulator arm was used for the alignment of the filling nozzle position in front of the cups. Additionally, a push button was used to start and stop the conveyor belt when needed.

The details about the major parts used in the automatic yogurt filling machine are provided in Table 1 below.

Table 1. Major parts used in the automatic yogurt filling machine.

Part	Reason	Quantity
Piston	One for the NFC station, one for the filling station, and two for the start and end stations	4
Photoelectric Proximity Sensor	One for the NFC station, one for the filling station, and two for the start and end stations	4
Proximity Sensor Normally Closed	For the notification of process finishing	1
Solenoid Valve	Three for the flavors and one for the yogurt	4
Diaphragm Pump	One for each solenoid valve	4
Solid State Relay	For switching of the load that comes from the WAGO PFC	4
WAGO PFC	Managing the system	1
NFC Module	Read tags	1
Raspberry Pi	To control the signal from NFC	1
5.5-inch OLED Touch Screen	To display the function of the Raspberry Pi	1
FTDI Chipped Board	To supply the power to the NFC Module	1
Switched Mode Power Supply (787–1602)	Switch the power source of the diaphragm WAGO controller (only 0.5 A)	1
Switched Mode Power Supply (787–1717)	Switch the power source of the diaphragm pump (only 2.0 A)	1
Stack Light	To provide the visual status for the system	1
Control Panel	For protection of the electrical and electronics parts	1
Air Filter Regulator	To filter the air coming from the air laboratory source	1
Tank	Three for the flavors and one for the yogurt	4

4. Mathematical Modeling

The development of the mathematical model considers the process optimization of the overall system. Two processes were performed once a cup had been placed at the entry point and until it left the system. These two are the filling process and the movement of cups over the conveyor belt between any points. The speed of the belt is directly linked with the feed rates of the yogurt and flavor valves. The speed of the conveyor belt could be maximized within allowable limits, which resulted in an increased throughput. For every valve, there was a maximum feed rate value beyond which the feed rate could not be exceeded. In Figure 1 below, the previous setting in which the filling processes were performed at two distinct locations and the conveyor belt was divided into three equal segments (Case-I) can be seen.

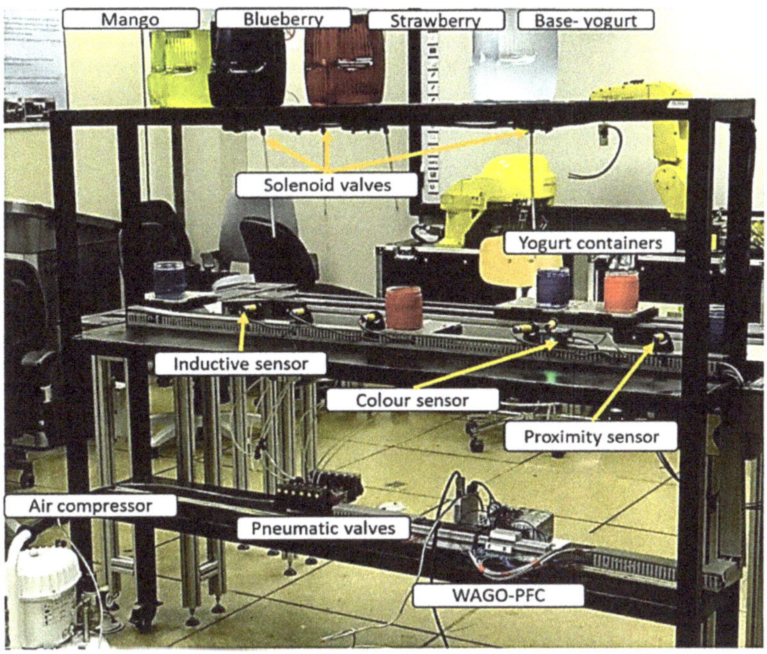

Figure 1. The yogurt filling machine with filling points at two distinct locations (Case-I). **Figure source:** B. Salah et al. [15].

A linear programming mathematical model was developed for the yogurt filling machine with the objective to maximize the speed of the conveyor belt and hence throughput, while also filling the cups with the required volumes of yogurt and different flavors to meet the customer demand in a given time. A few constraints were considered during the filling process. Some indices, parameters, and decision variables were considered in the mathematical modeling for process optimization and are given below.

Indices

i	yogurt percentage	$i = 1, 2, \ldots, I$
y	yogurt type	$y = 1, 2, \ldots, Y$
j	flavor percentage	$j = 1, 2, \ldots, J$
f	flavor type	$f = 1, 2, \ldots, F$
k	total volume	$k = 1, 2, \ldots, K$

Process Parameters

S_b conveyor belt speed
L_t conveyor belt total length
V_{iyjfk} yogurt volume in a total volume
v_{iyjfk} flavor volume in a total volume
D_{iyjfk} demand from customers
W_{iyjfk} pickup time in minutes
V_{max} upper limit of total volume of yogurt and flavors
V_{min} lower limit of total volume of yogurt and flavors

Decision variables

β_{iyjfk} yogurt valve feed rate
γ_{iyjfk} flavor valve feed rate

Mathematically, the objective function for maximizing the speed of conveyor belt in terms of length between any two points on the conveyor belt, feed rate of the yogurt valve, and the required yogurt volume can be stated as follows:

$$\textbf{Maximize: } Z = l\sum_{i=1}^{I}\sum_{y=1}^{Y}\sum_{j=1}^{J}\sum_{f=1}^{F}\sum_{k=1}^{K}\left(\frac{\beta_{iyjfk}}{V_{iyjfk}}\right) \quad (1)$$

As the ratio of the feed rate of the yogurt solenoid valve to the yogurt volume and the ratio of the feed rate of the flavor solenoid valve to the flavor volume are equal, the objective function for the speed of the conveyor belt can also be written as follows:

$$\textbf{Maximize: } Z = l\sum_{i=1}^{I}\sum_{y=1}^{Y}\sum_{j=1}^{J}\sum_{f=1}^{F}\sum_{k=1}^{K}\left(\frac{\gamma_{iyjfk}}{v_{iyjfk}}\right) \quad (2)$$

To manage the movement of cups over the belt, five times were considered equal in Case-I. These times are the movement time of an empty cup from the placement to the yogurt filling point, the yogurt filling time, the movement time of the yogurt-filled cup from the yogurt to flavor filling points, the flavor filling time, and the movement time of the cup filled with yogurt and flavors from the flavor filling to exit points. Due to these equal times, the ratio of the yogurt feed rate to required volume of yogurt remained equal to the ratio of flavor feed rate to the required volume of flavor.

A few constraints and equations were considered in the model relevant to the speed of the conveyor belt, the feed rates of the valves, customer waiting time, and volumes of yogurt and flavors. These constraints and equations are given as follows:

$$\frac{\beta_{iyjfk}}{V_{iyjfk}}l \le Maximum\ S_b \quad I = 1,2,\ldots,I \quad y = 1,2,\ldots,Y \quad j = 1,2,\ldots,J \quad f = 1,2,\ldots,F \quad k = 1,2,\ldots,K \quad (3)$$

$$\frac{\gamma_{iyjfk}}{v_{iyjfk}}l \le Maximum\ S_b \quad i = 1,2,\ldots,I \quad y = 1,2,\ldots,Y \quad j = 1,2,\ldots,J \quad f = 1,2,\ldots,F \quad k = 1,2,\ldots,K \quad (4)$$

$$\beta_{iyjfk} \le Maximum\ \beta_{iyjfk} \quad i = 1,2,\ldots,I \quad y = 1,2,\ldots,Y \quad j = 1,2,\ldots,J \quad f = 1,2,\ldots,F \quad k = 1,2,\ldots,K \quad (5)$$

$$\gamma_{iyjfk} \le Maximum\ \gamma_{iyjfk} \quad i = 1,2,\ldots,I \quad y = 1,2,\ldots,Y \quad j = 1,2,\ldots,J \quad f = 1,2,\ldots,F \quad k = 1,2,\ldots,K \quad (6)$$

$$\frac{60W_{iyjfk}\beta_{iyjfk}}{V_{iyjfk}} - 4 \ge D_{iyjfk} \quad i = 1,2,\ldots,I \quad y = 1,2,\ldots,Y \quad j = 1,2,\ldots,J \quad f = 1,2,\ldots,F \quad k = 1,2,\ldots,K \quad (7)$$

$$\frac{V_{iyjfk}}{\beta_{iyjfk}} = \frac{v_{iyjfk}}{\gamma_{iyjfk}} \quad i = 1,2,\ldots,I \quad y = 1,2,\ldots,Y \quad j = 1,2,\ldots,J \quad f = 1,2,\ldots,F \quad k = 1,2,\ldots,K \quad (8)$$

The objective functions (1) and (2) of the model resulted in equal values and maximized the conveyor belt speed. The constraints (3) and (4) were for the conveyor belt speed while the inequalities (5) and (6) were used for the feed rates of yogurt and flavor valves, respectively. Constraint (7) is written to make sure that the processing time is less than or equal to the customer waiting time, whereas Equation (8) satisfies that the yogurt filling time should be equal to the flavor filling time.

The possibility of processing time minimization was noted in the yogurt filling machine setting (Case-I). It was decided to change the setting and the filling operation of yogurt and flavors to be performed at a single location over the conveyor belt (Case-II). The modified system with yogurt and flavor filling points at a single location (Case-II) can be seen in Figure 2.

Figure 2. Modified setting of the yogurt filling system (Case-II).

All other constraints for Case-II remained the same as those for Case-I, except constraint (7), which was related to the customer waiting time for filling the required number of cups with the desired volumes of yogurt and flavors. Constraint (7) can be changed to constraint (9) for the new setting (Case-II) and can be written as follows:

$$\frac{60W_{iyjfk}\beta_{iyjfk}}{V_{iyjfk}} - 2 \geq D_{iyjfk} \quad i = 1,2,\ldots,I \quad y = 1,2,\ldots,Y \quad j = 1,2,\ldots,J \quad f = 1,2,\ldots,F \quad k = 1,2,\ldots,K \tag{9}$$

As already discussed in Section 2, there is a constraint of maximum and minimum total volume of a cup and orders for greater and lower volumes than the available sizes of cups must not be accepted. To optimize the use of the available machine, the constraints (10) are written in one line for both minimum and maximum limits of the total volume of yogurt and flavors in a cup.

$$V_{min} \leq (V_{iyjfk} + v_{iyjfk}) \leq V_{max} \quad i = 1,2,\ldots,I \quad y = 1,2,\ldots,Y \quad j = 1,2,\ldots,J \quad f = 1,2,\ldots,F \quad k = 1,2,\ldots,K \tag{10}$$

The v_{iyjfk} is used for the volume of flavors, but it can be either zero or non-zero. The zero value of the v_{iyjfk} means customer demand for pure yogurt and non-zero value of the v_{iyjfk} means flavor(s) mixed with yogurt.

Depending on the customer demand, either a single flavor or a combination two or more than two flavors can be mixed with the yogurt. All customer orders for pure yogurt or a combination of flavors mixed with yogurt can be found by the formula given in Equation (11).

$$\text{Number of orders for yogurt or combination of flavors mixed with yogurt} = 2^n \quad (11)$$

where n is the number of available flavors.

All customer orders for yogurt and three available flavors are summarized in Table 2. In all customer orders, yogurt is demanded and hence in all combinations its value is 1. A value of 1 for a flavor in a combination means the presence of that flavor. The customer may demand pure yogurt, which is the combination in which yogurt takes the value of 1 and all three flavors are 0 while yogurt mixed with all three flavors is a combination in which the yogurt as well as all three flavors have the value of 1. The percentages of yogurt and flavors in a total volume also depend on the customer demand and can take any value (in percentages) provided that the minimum total volume is greater than or equal to the lower limit and less than or equal to the upper limit.

Table 2. All combinations of yogurt mixing with three flavors.

Yogurt	Flavor 1	Flavor 2	Flavor 3	V_{min} (mL)	V_{max} (mL)
1	0	0	0	250	1000
	1	0	0		
	0	1	0		
	1	1	0		
	0	0	1		
	1	0	1		
	0	1	1		
	1	1	1		

While placing the very first cup on the entry point of the machine, there is no waiting time for the cup to enter into the system while the waiting time of the second cup for placement is equal to the time in which the first cup reaches from the entry point to the yogurt or flavor filling point. Similarly, the waiting time of the third cup to enter into the system in the sum of the filling time of the very first cup and the movement time of the second cup from the entry point to the filling point. A general relation for the waiting time of the nth empty to enter into the system is given in Equation (12) in terms of required volume and feed rate of the yogurt valve.

$$E_n = (n-1)\frac{V_{iyjfk}}{\beta_{iyjfk}} \quad i = 1,2,\ldots,I \;\; y = 1,2,\ldots,Y \;\; j = 1,2,\ldots,J \;\; f = 1,2,\ldots,F \;\; k = 1,2,\ldots,K \quad (12)$$

The two processes performed in the machine to fulfill customer orders are the movement of the cups over the conveyor belt and the filling of cups. In Case-I, the filling processes and movements of cups are performed in equal times. There are two filling processes and three movements of a cup in the machine; hence, the time taken by a cup to move from the entry to the exit point and complete the filling process is given in Equation (13).

$$P = 5\frac{V_{iyjfk}}{\beta_{iyjfk}} \quad i = 1,2,\ldots,I \;\; y = 1,2,\ldots,Y \;\; j = 1,2,\ldots,J \;\; f = 1,2,\ldots,F \;\; k = 1,2,\ldots,K \quad (13)$$

The very first cup, when moving from the entry to the exit point, takes time in its movement and filling processes. As there are three movements and two filling processes and all are performed in equal times, the first cup reaches the exit point in the sum of five equal times. The waiting time for the second cup until it reaches the exit point is the sum of the five equal times and the time in which the second cup reaches from the filling point to the exit point. A general relation to wait for the nth completely filled cup at the exit point is given in Equation (14) in terms of the required volume and feed rate of the yogurt valve.

$$F_n = (n+4)\frac{V_{iyjfk}}{\beta_{iyjfk}} \quad i = 1,2, \ldots, I \;\; y = 1,2, \ldots, Y \;\; j = 1,2, \ldots, J \;\; f = 1,2, \ldots, F \;\; k = 1,2, \ldots, K \tag{14}$$

It can be seen in the above equations that the ratio of the required yogurt volume and the yogurt feed rate was used. Instead of this, the ratio of the required flavor volume and flavor feed rate can also be used, as both the ratios are considered equal in the model, as can be seen in Equation (8).

Similarly, for Case-II, Equation (12) was used for the waiting of the nth empty cup to enter the machine for the movement and filling processes of cups, while Equation (15) shows the total time taken by a cup from the entry to the exit point in the machine. Equation (16) is used to find the waiting time for a cup at the exit point of the machine.

$$P = 3\frac{V_{iyjfk}}{\beta_{iyjfk}} \quad i = 1,2, \ldots, I \;\; y = 1,2, \ldots, Y \;\; j = 1,2, \ldots, J \;\; f = 1,2, \ldots, F \;\; k = 1,2, \ldots, K \tag{15}$$

$$F_n = (n+2)\frac{V_{iyjfk}}{\beta_{iyjfk}} \quad i = 1,2, \ldots, I \;\; y = 1,2, \ldots, Y \;\; j = 1,2, \ldots, J \;\; f = 1,2, \ldots, F \;\; k = 1,2, \ldots, K \tag{16}$$

It can be noted that Equations (12) was used in both Case-I and Case-II, as the waiting time of the cups to enter the system can be the same. The difference can be seen in the processing time and waiting time at the exit point of the machine during the filling process of the cups.

The benefits of the proposed model are as follows.

- Less computational burden while solving through software packages
- It is used for the flavor valve's size selection by finding the maximum feed rate of the flavor valve
- It is used for finding the optimal solution and satisfying all constraints
- It is used for finding the maximum values of the speed of the conveyor belt and feed rates of the yogurt and flavor nozzles
- It is used to find the processing rate of cups demanded by customers

5. Solution Procedure

The solution procedure adopted for the problem is divided into several stages. First, the upper limits of the conveyor belt speed and the feed rates of the yogurt and flavor valves were evaluated. The total length of the conveyor belt and the length between any two points (entry, yogurt and flavor filling, and exit point) were calculated. As orders from customers are received in the form of required volumes of yogurt and flavors in the total volume of a cup and the total number of cups, all values were put in the model and the objective function and constraints were set to determine the feasibility of the orders. The customer order must satisfy the upper- and lower-volume limits. Once the model was solved through the Simplex algorithm and the optimal solution is found, the values of all measures (feed rates of yogurt of flavor valves, speed on the conveyor belt, etc.) were calculated and results were documented. The movement and filling of cups with the required volumes of yogurt and flavors was then started to meet the customer demand.

For illustration purposes, a real life problem previously solved using the Case-I model was solved again using the Case-II model. A problem comprising six customer orders was solved for both cases. The set of orders is described in Table 3. The customer demand was for yogurt of a single type and three different types of flavors. In the mathematical model, the yogurt is represented by 1 and the flavors, i.e., strawberry, blueberry, and mango, are represented by 1, 2, and 3, respectively. It is to be noted that orders 1 and 5, orders 2 and 3, and orders 4 and 6 are customer orders for yogurt mixed with strawberry, blueberry, and mango flavors, respectively. The set of orders must satisfy the upper and lower limits of 1000 mL and 250 mL, respectively, on the total volume of yogurt and flavors.

Table 3. Orders from customers.

Order No.	Volume (mL)	Yogurt (%)	Flavor 1 (%)	Flavor 2 (%)	Flavor 3 (%)	D_{iyjfk} (Units)	W_{iyjfk} (Minutes)
1	300	93	7	0	0	100	10
2	300	90	0	10	0	80	9
3	600	90	0	10	0	25	7
4	500	95	0	0	5	35	20
5	900	95	5	0	0	20	15
6	900	93	0	0	7	30	25

To solve the problem using Case-I and Case-II, the maximum speed, feed rate of the yogurt valve, and feed rate of the flavor valve considered were 10 cm/s, 50 mL/s, and 25 mL/s, respectively. In Case-I, the total length of the conveyor belt (90 cm) was divided into three equal parts and each part was equal to 30 cm. In Case-II, the total length of the conveyor belt (90 cm) was divided into two equal parts and each part was equal to 45 cm.

The objective functions and constraints were written for both cases and the models were solved simultaneously to determine the optimal values of all decision variables satisfying all constraints and maximizing the throughput. The solutions of the models which resulted in optimal values of the decision variables are presented in Table 4.

Table 4. Optimal solutions for Case-I and Case-II.

Case No.	Order No.	β_{iyjfk} (mL/s)	γ_{iyjfk} (mL/s)	S_b (cm/s)	E_n (s)	P (s)	F_n (s)
I	1	50	3.76	5.38	552.4	27.9	580.3
	2	50	5.56	5.56	426.6	27.0	453.6
	3	50	5.56	2.78	259.2	54.0	313.2
	4	50	2.63	3.16	323.0	47.5	370.5
	5	50	2.63	1.75	324.9	85.5	410.4
	6	50	3.76	1.79	485.5	83.7	569.2
II	1	50	3.76	5.38	552.4	16.74	569.16
	2	50	5.56	5.56	426.6	16.2	442.8
	3	50	5.56	2.78	259.2	32.4	291.6
	4	50	2.63	3.16	323.0	28.5	351.5
	5	50	2.63	1.75	324.9	51.3	376.2
	6	50	3.76	1.79	485.5	50.22	535.68

The output of the model showed that the feed rates of both yogurt and flavor valves were equal to or less than the maximum allowable values of the feed rates of the solenoid valves. Normally, the yogurt volume in a cup is demanded more than the flavor volume, and this higher feed rate of the yogurt valve is needed more than the flavor valve's feed rate. The feed rates vary according to the percentage of yogurt or flavors in a cup. It is to be noted that both the yogurt and flavors were filled into the cups in equal times.

The results also showed that the speed of the conveyor belt was less than the maximum allowable limit while fulfilling all the customer orders. The speed is linked directly with the feed rate of the solenoid valves. The speed of the belt could not be increased further once the maximum allowable value of the feed rate of the solenoid valves was reached.

All constraints of the model were satisfied. The ratio of the required volume to feed rate of yogurt is equal to the ratio of the required volume to feed rate of flavors in filling all customer orders. Additionally, it can be noted that the customer waiting in all orders was more than the processing time of an order.

6. Sequencing the Processing of Orders

Sequencing is the order in which cups of required volumes are processed to meet customer demand. The processing can be performed easily in case of simple customer orders, but for large orders, the situation becomes complicated. The processing order is considered crucial, as it affects the machine idle time, waiting time of cups in queues, and customer waiting time. The single-dimension rules comprise one of the categories in priority sequencing which determines priority based on a single aspect such as arrival time, due date, or processing time.

The results of the problem solved for Case-I and Case-II in Section 4 were considered for single-dimension priority rules, as can be seen in Table 5. The same set of customer orders for both cases was considered with the time since an order is received. The processing time (in minutes) for each order was calculated and it can be noted that the processing time on each order was lower than the pickup time for each order.

Table 5. The processing and pickup times of the set of orders for Case-I and Case-II.

Case No.	Order No.	Minutes Since Order Arrival	Processing Time (Minutes)	Pickup Time (Minutes)
I	1	0	9.67	10
	2	1	7.56	9
	3	1	5.22	7
	4	0	6.18	20
	5	3	6.84	15
	6	2	9.49	25
II	1	0	9.49	10
	2	1	7.38	9
	3	1	4.86	7
	4	0	5.86	20
	5	3	6.27	15
	6	2	8.93	25

As can be seen in Tables 6–8, the sequence of the set orders was set according to the single dimension rules. The time since an order was received was noted and the first order to be processed started at time zero. The finish time is the sum of the starting time and processing time while flow time is the sum of finish time and time since an order arrived. The scheduled pickup time was provided by customers at the time when an order had been placed while actual pickup time is the maximum time in finish and scheduled pickup times. The minutes early or minutes past due is the difference between the scheduled pickup time and the finish time. It is to be noted that only positive values of the difference of minutes early and minutes past due were considered.

The order in which the set of customer orders can be processed in the EDD rule in both cases (Case-I and Case-II) can be seen in Table 6. According to the EDD rule for Case-I, order 3 was the one with the shortest pick up time (7 min). Hence, it was processed before all other orders. The processing of order 3 started at time zero and finished at 5.22 min. The flow time (6.22 min) is the summation of the finish time (5.22 min) and the time since the order had arrived (1 min). As the scheduled pickup time was 7 min and the finish time was 5.22 min, the order was prepared 1.78 min earlier than the scheduled pickup time. The last order to be processed in the EDD rule was order 6, with the longest pickup time (25 min). The processing of order 6 started at time 35.47 min with a processing time of 9.49 min and finished at 44.96 min. The flow time (46.96 min) is the summation of the finish time (44.96 min) and the time since the order had arrived (2 min). As the scheduled pickup time was 25 min and the finish time was 44.96 min, the order was prepared 19.96 min later than the scheduled pickup time. Similarly, for Case-II, order 3 was prepared 2.14 min earlier and order 6 was prepared 17.79 min later than the scheduled pickup times.

Table 6. Sequence of orders based on the EDD rule.

Case No.	Order No.	Minutes Since Order Arrival	Starting Time	Processing Time	Finish Time	Flow Time	Scheduled Pickup Time	Actual Pickup Time	Minutes Early	Minutes Past Due
I	3	1	0.00	5.22	5.22	6.22	7	7.00	1.78	-
	2	1	5.22	7.56	12.78	13.78	9	12.78	-	3.78
	1	0	12.78	9.67	22.45	22.45	10	22.45	-	12.45
	5	3	22.45	6.84	29.29	32.29	15	29.29	-	14.29
	4	0	29.29	6.18	35.47	35.47	20	35.47	-	15.47
	6	2	35.47	9.49	44.96	46.96	25	44.96	-	19.96
II	3	1	0.00	4.86	4.86	5.86	7	7.00	2.14	-
	2	1	4.86	7.38	12.24	13.24	9	12.24	-	3.24
	1	0	12.24	9.49	21.73	21.73	10	21.73	-	11.73
	5	3	21.73	6.27	28.00	31.00	15	28.00	-	13.00
	4	0	28.00	5.86	33.86	33.86	20	33.86	-	13.86
	6	2	33.86	8.93	42.79	44.79	25	42.79	-	17.79

Where required, the unit of time is "minutes".

Table 7. Sequence of the orders based on the SPT rule.

Case No.	Order No.	Minutes Since Order Arrival	Starting Time	Processing Time	Finish Time	Flow Time	Scheduled Pickup Time	Actual Pickup Time	Minutes Early	Minutes Past Due
I	3	1	0.00	5.22	5.22	6.22	7	7.00	1.78	-
	4	0	5.22	6.18	11.40	11.40	20	20.00	8.60	-
	5	3	11.40	6.84	18.24	21.24	15	18.24	-	3.24
	2	1	18.24	7.56	25.80	26.80	9	25.80	-	16.80
	6	2	25.80	9.49	35.29	37.29	25	35.29	-	10.29
	1	1	35.29	9.67	44.96	45.96	10	44.96	-	34.96
II	3	1	0.00	4.86	4.86	5.86	7	7.00	2.14	-
	4	0	4.86	5.86	10.72	10.72	20	20.00	9.28	-
	5	3	10.72	6.27	16.99	19.99	15	16.99	-	1.99
	2	1	16.99	7.38	24.37	25.37	9	24.37	-	15.37
	6	2	24.37	8.93	33.30	35.30	25	33.30	-	8.30
	1	1	33.30	9.49	42.79	43.79	10	42.79	-	32.79

Where required, the unit of time is "minutes".

In the SPT rule, the order in which the set of customer orders can be processed in both cases (Case-I and Case-II) can be seen in Table 7. According to the SPT rule for Case-I, order 3 was the order with the shortest processing time (5.22 min). Hence, it was processed before all other orders. The processing of order 3 started at time zero and finished at 5.22 min. The flow time (6.22 min) is the summation of finish time (5.22 min) and time since the order had arrived (1 min). As the scheduled pickup time was 7 min and the finish time was 5.22 min, order 3 was prepared 1.78 min earlier than the scheduled pickup time. The last order to be processed in the SPT rule was order 1, with the longest processing time (9.67 min). The processing of order 1 started at 35.29 min with a processing time of 9.67 min and finished at 44.96 min. The flow time (45.96 min) is the summation of the finish time (44.96 min) and the time since the order had arrived (1 min). As the scheduled pickup time was 10 min and the finish time was 44.96 min, the order was prepared 34.96 min later than the scheduled pickup time. Similarly, for Case-II, order 3 was prepared 2.14 min earlier and order 1 was prepared 32.79 min later than the scheduled pickup times.

Table 8. Sequence of orders based on the FCFS rule.

Case No.	Order No.	Minutes Since Order Arrival	Starting Time	Processing Time	Finish Time	Flow Time	Scheduled Pickup Time	Actual Pickup Time	Minutes Early	Minutes Past Due
I	5	3	0.00	6.84	6.84	9.84	15	15.00	8.16	-
	6	2	6.84	9.49	16.33	18.33	25	25.00	8.67	-
	3	1	16.33	5.22	21.55	22.55	7	21.55	-	14.55
	2	1	21.55	7.56	29.11	30.11	9	29.11	-	20.11
	4	0	29.11	6.18	35.29	35.29	20	35.29	-	15.29
	1	0	35.29	9.67	44.96	44.96	10	44.96	-	34.96
II	5	3	0.00	6.27	6.27	9.27	15	15.00	8.73	-
	6	2	6.27	8.93	15.20	17.20	25	25.00	9.80	-
	3	1	15.20	4.86	20.06	21.06	7	20.06	-	13.06
	2	1	20.06	7.38	27.44	28.44	9	27.44	-	18.44
	4	0	27.44	5.86	33.30	33.30	20	33.30	-	13.30
	1	0	33.30	9.49	42.79	42.79	10	42.79	-	32.79

Where required, the unit of time is "minutes".

Likewise, the order in which the set of customer orders can be processed in the FCFS rule in both cases (Case-I and Case-II) can be seen in Table 8. According to the FCFS rule for Case-I, order 5 was the one which arrived earlier than all the other orders (3 min since order arrival). Hence, it was processed before all other orders. The processing of order 5 started at time zero and finishes at 6.84 min. The flow time (9.84 min) is the summation of the finish time (6.84 min) and the time since the order arrived (3 min). As the scheduled pickup time was 15 min and the finish time was 6.84 min, the order was prepared 8.16 min earlier than the scheduled pickup time. The last order to be processed in the FCFS rule was order 1. The processing of order 1 started at 35.29 min with a processing time of 9.67 min and finished at 44.96 min. The flow time (44.96 min) was equal to finish time (44.96 min) as the time since the order had arrived was zero. As the scheduled pickup time was 10 min and the finish time was 44.96 min, the order was prepared 34.96 min later than the scheduled pickup time. Similarly, for Case-II, order 5 was prepared 8.73 min earlier and order 1 was prepared 32.79 min later than the scheduled pickup times.

In the three abovementioned single-dimension rules, the setup time was assumed to be independent of the processing sequence. Normally, this is not used and orders with similar setups are sequenced back-to-back in order to reduce the setup time. Some other assumptions considered during the problem solution using Case-I and Case-II were the deterministic filling time, uninterrupted processing, and no cancellation and arrival of new orders once the machine started to fill the cups with yogurt and flavors.

7. Results and Discussion

All information and data used in the solution of the problem using Case-I and Case-II were provided by the laboratory of the Computer Integrated Manufacturing (CIM), Department of Industrial Engineering, King Saud University, Saudi Arabia.

In Case-I and Case-II, the upper limit of the feed rate of the yogurt was considered as 50 mL/s and, hence, in both cases, the models resulted in the maximum allowable value (50 mL/s). The conveyor belt speed was controlled by the yogurt feed rate, which constitutes a directly proportional relationship. In all orders, the flavor volume was lower than the yogurt volume in the total volume of an order and, hence, a lower feed rate of the flavor valve is needed to fill the cup with the required volume of flavor. In both cases, the speed of the conveyor belt and waiting time for a cup to enter the system remained the same. While comparing the two cases, it was found that the processing time of a cup and the total processing time of a set of customer orders were different.

In Figure 3, it can be seen that for each order, the total processing time in Case-I was greater than Case-II. This is because in Case-I, the yogurt and all flavors were filled at two distinct locations, while in Case-II, the filling of yogurt and flavors was at a single location. In both cases, the total processing time initially decreased from order 1 to order 3 and then increased from order 3 to order 6. In the case of order 1 and 2, the total volumes were equal, but the processing time in the case of order 1 was higher than that of order 2. This is due to the difference in the number of units demanded by the customer. In order 1, the number of units demanded (100 units) was greater than in order 2 (80 units). Similarly, in the case of order 5 and 6, the total processing time of order 6 was higher than that of order 5 for the same total volume (900 mL). This is because in order 6, the number of units demanded (30 units) was greater than the number of units demanded (20 units) in order 5. Hence, the total processing time of an order depends on the total volume and the number of units demanded. The greater the total volume and the number of units demanded, the greater the total processing time of an order. In the cases of order 1 and 6, the total processing times were higher than those of other orders. In order 1, the total number of cups needed (100 units) was higher than in the other orders, although the total volume was small (300 mL); in order 6, the total volume of a cup (900 mL) was higher than in the other orders, although the total number of cups (30 units) was smaller.

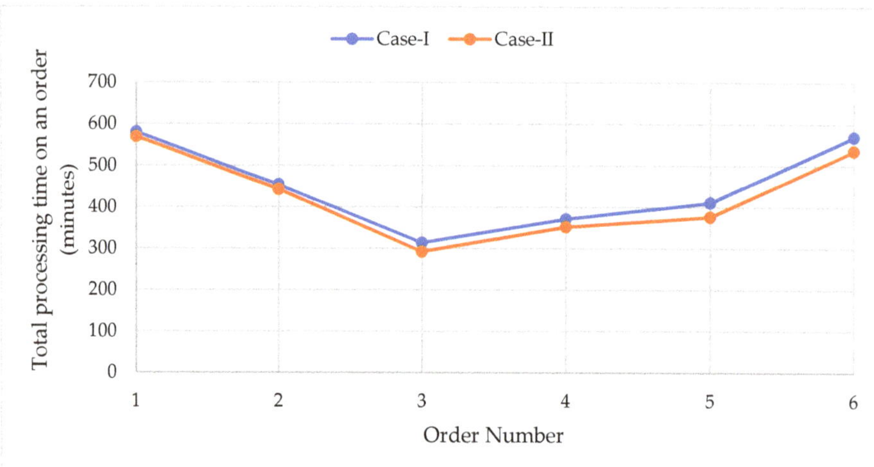

Figure 3. The total processing time on an order in Case-I and Case-II.

As shown in Table 3 in Section 5, F_n is the total processing time of a set of orders. In Case-I, the processing times for orders 1, 2, 3, 4, 5, and 6 were 580.3 s, 453.6 s, 313.2 s, 370.5 s, 410.4 s, and 569.2 s, respectively. In Case-II, the processing times for orders 1, 2, 3, 4, 5, and 6 were 569.16 s, 442.8 s, 291.6 s, 351.5 s, 376.2 s, and 535.68 s, respectively.

By converting the times from seconds to minutes, the processing times in Case-I for orders 1, 2, 3, 4, 5, and 6 were 9.67 min, 7.56 min, 5.22 min, 6.18 min, 6.84 min, and 9.49 min, respectively. In Case-II, the processing times for orders 1, 2, 3, 4, 5, and 6 were 9.49 min, 7.38 min, 4.86 min, 5.86 min, 6.27 min, and 8.93 min, respectively.

A = Sum of the processing times in Case-II = 42.78 min
B = Sum of the processing times in Case-I = 44.95 min

$$\frac{B}{A} = 1.05 \text{ or } B = 1.05\text{ A} \quad (17)$$

Based on Table 9 and Equation (17), it can be stated that processing time of the machine in Case-II was 1.05 times faster than that of Case-I.

Table 9. Comparison of the processing times of Case-I and Case-II.

Order No.	Processing Time in Case-II	Multiplicative Factor	Processing Time in Case-I
1	9.49	1.02	9.67
2	7.38	1.02	7.56
3	4.86	1.07	5.22
4	5.86	1.05	6.18
5	6.27	1.09	6.84
6	8.93	1.06	9.49
Average of Multiplicative Factor		1.05	

Where required, the unit of time is "minutes".

The set of customer orders was processed through Case-I and Case-II. It can be seen in Figures 4–6 that during the processing of each order, Case-II resulted in a lesser total processing time than in Case-I. Applying any single-dimension rules, the total time taken to process the set of orders in Case-I was 44.96 min, while in Case-II, it took 42.79 min. Hence, the results showed that Case-II was 1.05071 times faster than Case-I in processing the set of customer orders.

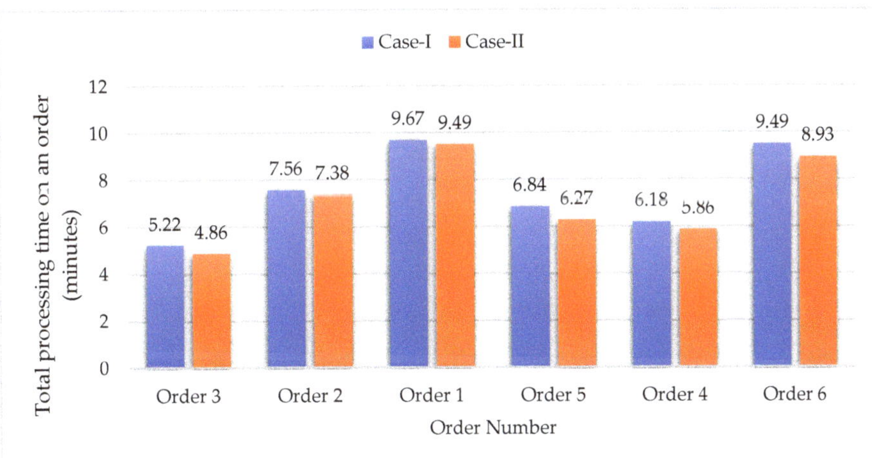

Figure 4. Comparison of the processing times for Case-I and Case-II using the EDD rule.

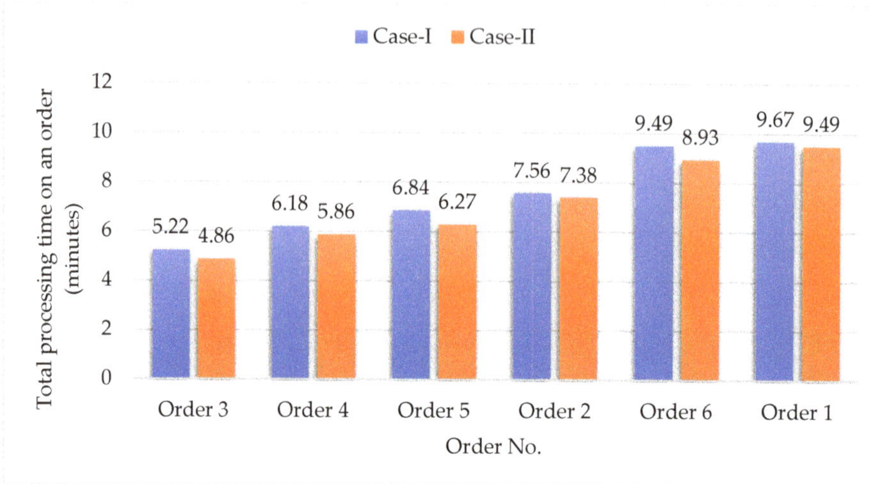

Figure 5. Comparison of the processing times for Case-I and Case-II using the SPT rule.

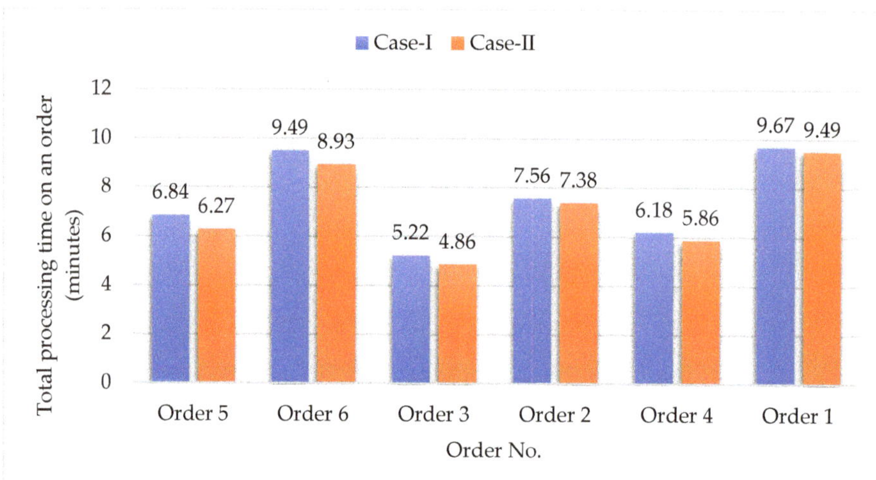

Figure 6. Comparison of the processing times for Case-I and Case-II using the FCFS rule.

The performance measures considered for the abovementioned priority sequencing rules for Case-I and Case-II were average flow time, average minutes early, and average minutes past due. The management tries to set the performance measures to the desired values and hence the minimum values of average flow time and average minutes past due, while the maximum value of average minutes early is preferred in the single-dimension rules when filling the cups with the required volumes of yogurt and flavors in Case-I and Case-II.

All the single-dimension rules resulted in slightly smaller values in Case-II than Case-I for average flow time and average minutes past due, while they resulted in slightly higher values in Case-II than in Case-I for average minutes early, as can be seen in Figures 7–9. In all single-dimension rules, the SPT rule resulted in minimum values of average flow time and average minutes past due performance measures, while the FCFS rule resulted in the highest values of average minutes early. It is to be noted that in all single-dimension rules, Case-II resulted in better values than Case-I. Hence, the filling process was completed using Case-II.

Keeping promises to customers, the SPT rule provided better results than the EDD and FCFS rules with respect to average minutes past due and average flow time while the FCFS rule produced better results than the EDD and SPT rules with respect to average minutes early. As the management prefers improving the average minutes past due over the other performance measures, the SPT sequencing rule was used.

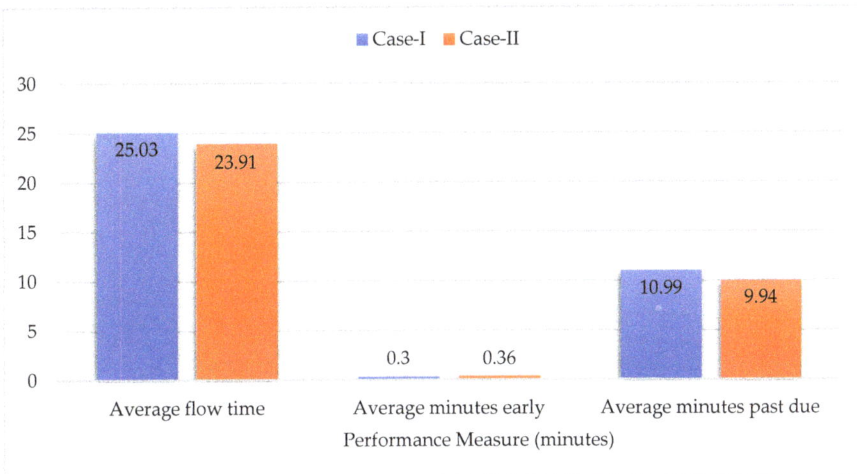

Figure 7. Comparison of the performance measures for Case-I and Case-II using the EDD rule.

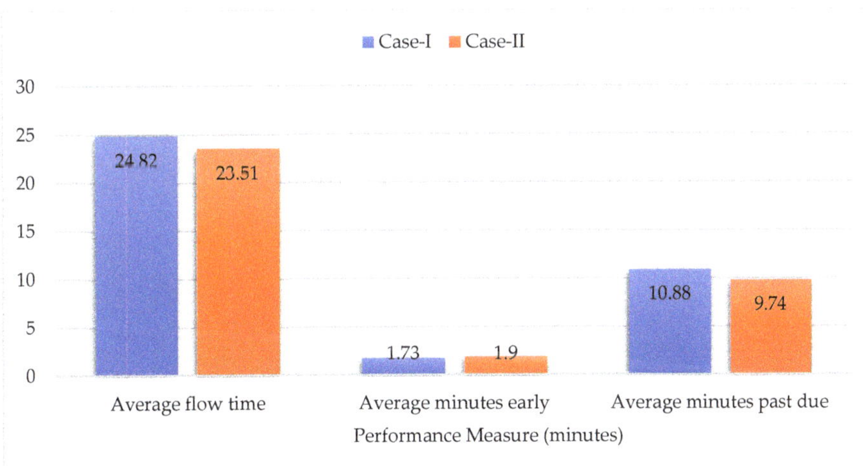

Figure 8. Comparison of the performance measures for Case-I and Case-II using the SPT rule.

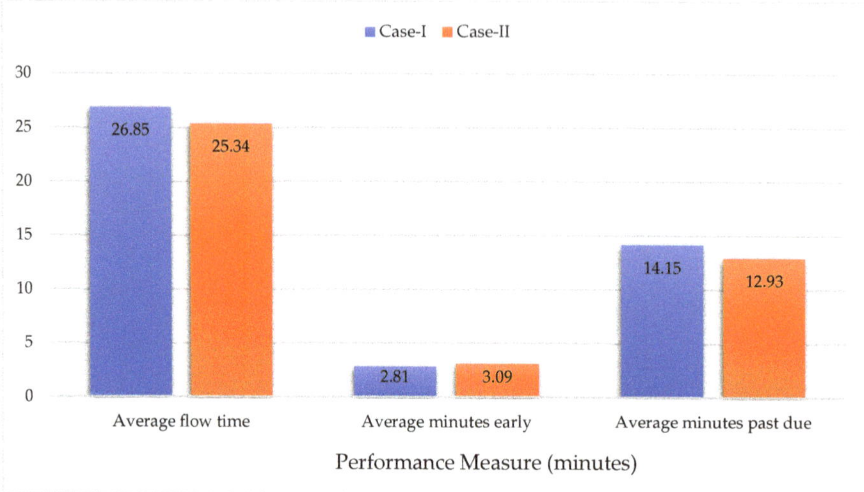

Figure 9. Comparison of the performance measures for Case-I and Case-II the using FCFS rule.

8. Conclusions

In this article, mathematical models for two settings (Case-I and Case-II) of an Industry 4.0-based yogurt filling system were developed and compared. The objectives of the models were the conveyor belt speed maximization or minimizing the processing time on an order or maximizing the throughput. In Case-I, the filling processes were performed at two different locations, while in Case-II, the filling processing was performed at a single location. The decision variables were the feed rates of the yogurt and flavor valves, and these variables controlled the speed of the conveyor belt.

The results showed that the processing time of an order in Case-II was lower than in Case-I due to the change in machine settings. Normally, the minimum customer waiting time is needed, and hence the machine setting in Case-II is preferred over the machine setting in Case-I. The results were also checked and it was found that Case-II resulted in the minimum total processing time using any single-dimension rule. Additionally, the outcomes of the performance measures for Case-I and Case-II were compared using single-dimension rules and it was found that SPT rule is used for the sequencing of orders as the management prefers improving the average minutes past due performance measure.

In the future, three cases can be compared to find which one provides the desired results. In Case-I, the yogurt- and flavor-filling operations are performed at two distinct locations. In Case-II, both filling operations are performed in a single location, while in Case-III, there will be a dedicated set of conveyor belts for the filling of each flavor and a base yogurt.

Author Contributions: Conceptualization, R.K. and B.S.; methodology, R.K. and B.S.; validation, J.C., Y.C. and Y.L.; formal analysis, W.S.; investigation, B.S.; resources, B.S.; writing—original draft preparation, R.K.; writing—review and editing, J.C., Y.C. and Y.L.; supervision, W.S.; project administration, R.K. All authors have read and agreed to the published version of the manuscript.

Funding: This study received funding from King Saud University, Saudi Arabia, through researchers supporting project number (RSP-2021/145). Additionally, the APCs were funded by King Saud University, Saudi Arabia, through researchers supporting project number (RSP-2021/145). This study also received funding from Teaching Research and Reform Fund from SDJU Grant No. 20216 and Shanghai Multidirectional Forging Engineering Technology Research Center Grant No. 20DZ2253200.

Institutional Review Board Statement: Not applicable.

Informed Consent Statement: Not applicable.

Data Availability Statement: The data presented in this study are available on request from the corresponding authors.

Acknowledgments: The authors extend their appreciation to King Saud University, Saudi Arabia for funding this work through researchers supporting project number (RSP-2021/145).

Conflicts of Interest: The authors declare no conflict of interest.

References

1. Thoben, K.D.; Wiesner, S.; Wuest, T. "Industrie 4.0" and smart manufacturing-a review of research issues and application examples. *Int. J. Autom. Technol.* **2017**, *11*, 4–16. [CrossRef]
2. Oztemel, E.; Gursev, S. Literature review of Industry 4.0 and related technologies. *J. Intell. Manuf.* **2020**, *31*, 127–182. [CrossRef]
3. Zhong, R.Y.; Xu, X.; Klotz, E.; Newman, S.T. Intelligent Manufacturing in the Context of Industry 4.0: A Review. *Engineering* **2017**, *3*, 616–630. [CrossRef]
4. Molins, X.; de Mesquita, M.A. Scheduling in the industry 4.0: A systematic literature review. *Int. J. Prod. Res.* **2019**, 1–25. Available online: https://upcommons.upc.edu/handle/2117/180742 (accessed on 24 July 2022).
5. Bueno, A.; Filho, M.G.; Frank, A.G. Smart production planning and control in the Industry 4.0 context: A systematic literature review. *Comput. Ind. Eng.* **2020**, *149*, 106774. [CrossRef]
6. Waschneck, B.; Altenmüller, T.; Bauernhansl, T.; Kyek, A. Production scheduling in complex job shops from an industrie 4.0 perspective: A review and challenges in the semiconductor industry. In Proceedings of the CEUR Workshop at i-KNOW' 16, Graz, Austria, 18–19 October 2016.
7. Parente, M.; Figueira, G.; Amorim, P.; Marques, A. Production scheduling in the context of Industry 4.0: Review and trends. *Int. J. Prod. Res.* **2020**, *58*, 5401–5431. [CrossRef]
8. Zhang, J.; Ding, G.; Zou, Y.; Qin, S.; Fu, J. Review of job shop scheduling research and its new perspectives under Industry 4.0. *J. Intell. Manuf.* **2019**, *30*, 1809–1830. [CrossRef]
9. Andulkar, M.; Le, D.T.; Berger, U. A multi-case study on Industry 4.0 for SME's in Brandenburg, Germany. In Proceedings of the 51st Hawaii International Conference on System Sciences, Hilton Waikoloa Village, HI, USA, 3–6 January 2018.
10. Posada, J.; Toro, C.; Barandiaran, I.; Oyarzun, D.; Stricker, D.; De Amicis, R.; Vallarino, I. Visual computing as a key enabling technology for industrie 4.0 and industrial internet. *IEEE Comput. Graph. Appl.* **2015**, *35*, 26–40. [CrossRef]
11. Li, L. China's manufacturing locus in 2025: With a comparison of "Made-in-China 2025" and "Industry 4.0". *Technol. Forecast. Soc. Change* **2018**, *135*, 66–74. [CrossRef]
12. Xu, L.D.; Xu, E.L.; Li, L. Industry 4.0: State of the art and future trends. *Int. J. Prod. Res.* **2018**, *56*, 2941–2962. [CrossRef]
13. From Industry 4.0 to Society 5.0: The Big Societal Transformation Plan of Japan. 2020. Available online: https://www.i-scoop.eu/industry-4-0-society-5-0 (accessed on 1 January 2020).
14. Alzahrani, A. Markets and language policy in Saudi Arabia: How the English language can contribute to the success of the Saudi vision 2030. *Int. J. Engl. Lang. Linguist. Res.* **2017**, *5*, 1–12.
15. Salah, B.; Khan, S.; Ramadan, M.; Gjeldum, N. Integrating the concept of industry 4.0 by teaching methodology in industrial engineering curriculum. *Processes* **2020**, *8*, 1007. [CrossRef]
16. Salah, B.; Abidi, M.H.; Mian, S.H.; Krid, M.; Alkhalefah, H.; Abdo, A. Virtual reality-based engineering education to enhance manufacturing sustainability in industry 4.0. *Sustainability* **2019**, *11*, 1477. [CrossRef]
17. Ramadan, M.; Salah, B.; Othman, M.; Ayubali, A.A. Industry 4.0-based real-time scheduling and dispatching in lean manufacturing systems. *Sustainability* **2020**, *12*, 2272. [CrossRef]
18. Gao, Z.; Castiglione, A.; Nappi, M. Biometrics in Industry 4.0: Open Challenges and Future Perspectives. *IEEE Trans. Ind. Inform.* **2022**, 1–4. Available online: https://ieeexplore.ieee.org/abstract/document/9852663 (accessed on 24 July 2022).
19. Salah, B. Real-Time Implementation of a Fully Automated Industrial System Based on IR 4.0 Concept. *Actuators* **2021**, *10*, 318. [CrossRef]
20. Ramadan, M.; Al-Maimani, H.; Noche, B. RFID-enabled smart real-time manufacturing cost tracking system. *Int. J. Adv. Manuf. Technol.* **2017**, *89*, 969–985. [CrossRef]
21. Frank, A.G.; Dalenogare, L.S.; Ayala, N.F. Industry 4.0 technologies: Implementation patterns in manufacturing companies. *Int. J. Prod. Econ.* **2019**, *210*, 15–26. [CrossRef]
22. Longo, F.; Nicoletti, L.; Padovano, A. Smart operators in industry 4.0: A human-centered approach to enhance operators' capabilities and competencies within the new smart factory context. *Comput. Ind. Eng.* **2017**, *113*, 144–159. [CrossRef]
23. Leusin, M.; Frazzon, E.; Maldonado, M.U.; Kück, M.; Freitag, M. Solving the Job-Shop Scheduling Problem in the Industry 4.0 Era. *Technologies* **2018**, *6*, 107. [CrossRef]
24. Rossit, D.A.; Tohmé, F.; Frutos, M. Industry 4.0: Smart Scheduling. *Int. J. Prod. Res.* **2019**, *57*, 3802–3813. [CrossRef]
25. Tortorella, G.L.; Giglio, R.; van Dun, D.H. Industry 4.0 adoption as a moderator of the impact of lean production practices on operational performance improvement. *Int. J. Oper. Prod. Manag.* **2019**, *39*, 860–886. [CrossRef]
26. Zawadzki, P.; Zywicki, K. Smart product design and production control for effective mass customization in the industry 4.0 concept. *Manag. Prod. Eng. Rev.* **2016**, *7*, 105–112. [CrossRef]

27. Grassi, A.; Guizzi, G.; Santillo, L.C.; Vespoli, S. A semi-heterarchical production control architecture for industry 4.0-based manufacturing systems. *Manuf. Lett.* **2020**, *24*, 43–46. [CrossRef]
28. Spenhoff, P.; Wortmann, J.C.; Semini, M. EPEC 4.0: An Industry 4.0-supported lean production control concept for the semi-process industry. *Prod. Plan. Control* **2020**, 1–18. [CrossRef]
29. Quezada, L.E.; Chiu, A.S.F.; da Costa, S.E.G.; Tan, K.H. Operational Excellence towards Sustainable Development Goals through Industry 4.0. *Int. J. Prod. Econ.* **2017**, *190*, 1–2. [CrossRef]
30. Santos, C.; Mehrsai, A.; Barros, A.C.; Araújo, M.; Ares, E. Towards Industry 4.0: An overview of European strategic roadmaps. *Procedia Manuf.* **2017**, *13*, 972–979. [CrossRef]
31. Salah, B.; Alsamhan, A.M.; Khan, S.; Ruzayqat, M. Designing and Developing a Smart Yogurt Filling Machine in the Industry 4.0 Era. *Machines* **2021**, *9*, 300. [CrossRef]
32. Bacci di Capaci, R.; Scali, C. A cloud-based monitoring system for performance assessment of industrial plants. *Ind. Eng. Chem. Res.* **2020**, *59*, 2341–2352. [CrossRef]
33. Di Capaci, R.B.; Pannocchia, G.; Kenny, C.; Pacchi, G.; Lupi, R.; Scali, C. An Automatic System for Modeling and Controlling Color Quality of Dyed Leathers in Tanneries. *IFAC-PapersOnLine* **2021**, *54*, 164–169. [CrossRef]
34. Gao, Z.; Paul, A.; Wang, X. Guest editorial: Digital twinning: Integrating AI-ML and big data analytics for virtual representation. *IEEE Trans. Ind. Inform.* **2022**, *18*, 1355–1358. [CrossRef]
35. Han, Z.; Zhang, Q.; Shi, H.; Zhang, J. An improved compact genetic algorithm for scheduling problems in a flexible flow shop with a multi-queue buffer. *Processes* **2019**, *7*, 302. [CrossRef]
36. Han, Z.; Han, C.; Lin, S.; Dong, X.; Shi, H. Flexible flow shop scheduling method with public buffer. *Processes* **2019**, *7*, 681. [CrossRef]
37. Salah, B.; Khan, R.; Ramadan, M.; Ahmad, R.; Saleem, W. Lab Scale Implementation of Industry 4.0 for an Automatic Yogurt Filling Production System—Experimentation, Modeling, and Process Optimization. *Appl. Sci.* **2021**, *11*, 9821. [CrossRef]

processes

Article

Dynamic Cooperation of the O2O Supply Chain Based on Time Delays and Bidirectional Free-Riding

Jing Zheng [1,2,*] and **Qi Xu** [1]

[1] Glorious Sun School of Business and Management, Donghua University, Shanghai 200051, China
[2] Logistics and E-commerce College, Zhejiang Wanli University, Ningbo 315100, China
* Correspondence: zhengjing003@zwu.edu.cn

Abstract: Advertising and service investment can enhance brand goodwill to increase the sales of branded goods. However, the impact of advertising and services on brand goodwill is not immediate but delayed. At the same time, due to the different service characteristics provided by various channels, the phenomenon of bidirectional free-riding occurs. Therefore, this paper studies the dynamic cooperation between service and advertising in the O2O (online to offline) supply chain dominated by brand owners and explores the impacts of advertising, service delay and service free-riding among channels on the dynamic cooperation decisions of the O2O supply chain. A differential game model between brands and retailers is constructed by incorporating the delay effect and the bidirectional free-riding phenomenon. The optimal advertising and service strategies and performance problems of O2O supply chain enterprises under a centralized decision, brand cost-sharing decision and bilateral cost-sharing decision are compared and analyzed. The influence of delay time, showrooming and webrooming effects on the profit of each firm is investigated by example. The results show that the service strategy, advertising strategy and brand goodwill of the O2O supply chain members are optimal under a centralized decision. Still, the supply chain profit is not necessarily optimal under the delay time, showrooming and webrooming effect coefficients. Bilateral cost-sharing contracts can achieve Pareto improvement of supply chain performance. Appropriate setting of a bilateral cost-sharing ratio can adjust the adverse effects of delay and bidirectional free-riding. The long-term strategies to deal with the delay and bidirectional free-riding phenomena are as follows: the bilateral cost-sharing contract can improve corporate profits. Setting the wholesale price, online direct-selling price and service-sharing ratio by brand owners can effectively promote retailers' investment in service, achieving a win-win situation. Retailers maintain high pricing and service levels to enhance the brand premium ability of physical stores and achieve long-term development.

Keywords: time delay; bidirectional free riding; differential game; bilateral cost-sharing decisions

Citation: Zheng, J.; Xu, Q. Dynamic Cooperation of the O2O Supply Chain Based on Time Delays and Bidirectional Free-Riding. *Processes* **2022**, *10*, 2424. https://doi.org/10.3390/pr10112424

Academic Editor: Zhiwei Gao

Received: 9 October 2022
Accepted: 10 November 2022
Published: 16 November 2022

1. Introduction

The rapid development of mobile technology and the Internet has enriched the shopping channels of consumers [1]. Channel services have their characteristics, the convenience of online channels and the experience of offline channels, so channel integration has become a development trend [2,3]. It is known as the showrooming phenomenon when consumers experience the product and then select the online channel to make a purchase [4]. When consumers search for product information (model, price, reviews, etc.) on an e-commerce platform and then choose an offline channel to make a purchase, it is called the webrooming phenomenon [5]. The combination of online and offline shopping has become common practice in omnichannel consumer behavior [6]. Van Baal and Dach [7] showed that 24.6% of consumers visited offline brick-and-mortar stores before completing transactions online, while 20.4% of consumers collected product information on the Internet before making offline purchases. Thus, this shows that the different service characteristics of the online and offline channels have made the bidirectional free-riding phenomenon (i.e., showrooming and webrooming) inevitable. To alleviate service competition between channels, a new business strategy, the O2O (online to offline) model has been introduced. The concept of O2O was first proposed by Alex Rampell, the founder of TrialPay in the United States, in August 2010. The earliest form of O2O is the "online purchase, offline consumption" model represented by online travel, catering, leisure and entertainment group purchase websites. In addition to the channel cooperation model in the service industry, the O2O model has also gradually attached importance in the retail industry. For example, retail brands such as Uniqlo and New Look have begun implementing this business strategy.

Meanwhile, studies have shown that advertising investment can alleviate channel conflict to a certain extent while increasing market demand [8]. The brand owners conduct national advertising through social media or mobile advertising to enhance the brand goodwill and promote the market demands of the brands [9]. At the same time, the retailers will improve the regional market demands of the brands by means of traditional promotional advertising. This vertical advertising cooperation among supply chain members is considered effective in enhancing the market demands [10–14]. The cost-sharing mechanism of cooperative advertising, in which manufacturers share part of the advertising costs of retailers, is a vital tool to mitigate competition and expand demand [15,16]. For example, in 2015, the manufacturer's retailer provided USD 36 billion in cooperative advertising, accounting for 12% of its total advertising costs [9]. Without cooperative advertising, the retailer's advertising efforts are often lower than the manufacturer's expectations [17]. Therefore, the optimal design of advertising cooperation in the O2O supply chain has attracted much attention from scholars [18–20]. As a result, service and advertising cooperation become the focal point of channel integration in the O2O model. This paper studies the O2O supply chain, which consists of the brands with online channels and the offline retailers and discusses the issue of advertising and service cooperation between channels.

In the O2O retail market, there are company-owned stores and franchised stores. For example, most of Zara's offline stores are owned by the company. Some small and medium-sized brands are more inclined to join the form, such as the Chinese clothing brand Metersbonwe, whose offline stores are mainly franchised. In addition to selling authorized brand products, franchise stores sell products as independent retailers and compete with the online channels of brand vendors [21]. Therefore, in the O2O model, when channels belong to different subjects, there are channel conflicts among channels.

In practice, traditional supply chain service and advertising cooperation are generally achieved by the upstream brand owners of the supply chain by sharing the cost and incentivizing the downstream retailers to invest in service and advertising levels, thus realizing vertical supply chain cooperation [9,15,21,22]. Chinese clothing brand INMAN, for example, shares some of the costs of servicing its offline franchises. However, in the O2O supply chain, in addition to the above vertical collaboration, there is horizontal competition among the supply chain members. The brand owners will share part of the service and advertising costs of the retailers, which, while improving the offline sales profits, will be horizontal competition to the online channel of the brand owners. Therefore, the question is raised: how do we coordinate the conflicts between the O2O supply chain channels through service and advertising cooperation? How do we design the service and advertising cost-sharing mechanism to achieve coordination to increase profits? There is little literature on both service and advertising cooperation, and the design of a service and cost-sharing mechanism to affect the decision-making and profit of the O2O supply chain has not been fully discussed.

The concept of SCM (supply chain management) first emerged in the 1980s; SCM has now become a hot issue in business management and practical application. Fang et al. [23] found that sustainable supply chains have become a hot research direction in supply chain management since 2010, but particularly in the last three years. Salas-Navarro et al. [24] found that developing an inventory model with environmental protection variables and parameters helps companies create a complex and solid inventory structure for sustainable development. Acevedo-Chedid et al. [25] identified the benefits of collaborative planning for production systems that can be used to make decisions for profit maximization for all participants in the supply chain. Related studies mainly focus on two areas: the bidirectional free-riding phenomenon of service and the dynamic cooperation of supply chain advertising.

The free-rider phenomenon was first proposed by Telser [26], who pointed out that it was not conducive to retailers' sales and would prevent them from providing pre-sale information services. Singley and Williams [27] proposed that free-riding reduces the profits of retailers providing comprehensive services. However, Wu et al. [28], Guan et al. [29] and Shin et al. [30] put forward different views, believing that retailers providing information services may benefit even if there is a free-riding phenomenon. The above research mainly focuses on free-riding between offline physical stores. E-commerce has emerged as a new sales method, and the emergence of homogeneous products in different or dual channels leads to free-riding among consumers [7]. In the dual-channel supply chain, most studies mainly focus on channel conflicts among supply chain members [31–34]. In contrast, Aubrey et al. [34] argued that with the increasing popularity of online shopping, online and offline channels can complement each other rather than compete. Zhang et al. [35] also studied how to manage the conflict between online and offline channels, pointing out that retailers can achieve the best returns by coordinating the two channels and adopting appropriate pricing strategies and channel combinations. To address the channel conflict, retailers should design coordination mechanisms to weaken the hitchhiking problem to alleviate channel conflicts. Common coordination mechanisms are cost-sharing contracts [35,36], the repurchase contract and sales rebate contract [37] and inventory subsidy method [38].

The above studies mainly focus on the showrooming phenomenon, while research on the webrooming phenomenon has also increased in recent years. For example, Wang et al. [39], Yuan et al. [40], and Ma et al. [41] studied the impact of the webrooming phenomenon on the supply chain sales model, pricing decision-making and the relationship with quality expectations. The above literature, whether in respect of the showrooming phenomenon or the webrooming phenomenon, studied the one-way free-riding phenomenon between channels, and research on the directional free-riding phenomenon is rare. Liu et al. [42] looked at the impact of free-riding behavior on dual-channel supply chain pricing. Luo et al. [43] studied the impact of bidirectional free-riding behavior based on information services on manufacturers' introduction of direct sales network channels. Li et al. [44] and Gong et al. [45] studied the bidirectional free-riding behavior in the dual-channel supply chain. The above research on the phenomenon of bidirectional free-riding takes the static supply chain system as the research object. It rarely considers the impact of bidirectional free-riding on the supply chain operation from the long-term dynamic perspective. These studies suggest that the integration of online and offline channels is preferred to competition between channels.

Unlike channel competition in the dual-channel supply chain, the O2O supply chain emphasizes the cooperation between online and offline channels, which can take advantage of the advantages of both channels [46]. The bidirectional free-riding phenomenon in this study, including showrooming and webrooming phenomena, is consistent with cross-channel shopping [7,47–49] research. However, the current research on showrooming and webrooming focuses more on the shopping experience and driving factors from consumers' perspectives. Sahu et al. [1] found that showroom or webrooming factors can be divided into consumer-led, company-led and situational. The company's dominant factors include price, quality, customer service, channel integration and media richness. The above studies all studied the service cooperation of the O2O supply chain from a static perspective. In contrast, the service cooperation among members of the O2O supply chain is long-term, complex and dynamic [21]. Therefore, this paper studies the impact of the bidirectional free-riding phenomenon on the optimal decision of service cooperation and supply chain profits in the O2O supply chain from a long-term dynamic perspective.

Another related field of this paper is the research on the dynamic cooperation of supply chain advertising. Nerlove and Arrow [50] developed the classic Nerlove–Arrow (N–A) advertising model, in which the goodwill is a dynamic equation that evolves. Jørgensen and Zaccour [16] adapted the N–A model and introduced the concept of advertising cooperation into supply chain decision-making. Zhang and Zhang [51] analyzed the cooperative advertising problem of the supply chain members by using the differential game to build a dynamic model. The above research is based on the assumption of the immediacy of the advertising effects. However, there is a time difference between consumers' contact with advertising and their actual demands, which is called the time delay. Gao et al. [52] present the current status of research on the application of complex dynamical systems in the modeling and control of systems with time delays. Zheng et al. [53] constructed a three-stage supply chain model using a system of differential equations to reveal the interplay among producers, distributors and end customers. Gao et al. [54] outlined emerging research and application directions in condition measurement, optimization and advanced control of complex industrial processes. Berkowitz et al. [55] and Baack et al. [56] proposed the dynamic advertising model, and Chen et al. [57] applied the delayed dynamic advertising model to the advertising cooperation among the supply chain members. Yu et al. [58] studied the effects of both delay and memory effects on the advertising decisions of supply chain members. In terms of the dual-channel supply chain, Chen et al. [59] and Cao et al. [60] considered the advertising cooperation strategy when both manufacturer and retailer advertising have delay effects. The above research mainly focuses on the advertising dynamic cooperation model of the dual-channel supply chain, but there are few studies on the O2O supply chain advertising dynamic cooperation. For the literature related to O2O supply chain advertising cooperation in this study, please refer to Table 1. The above literature discussed the advertising cooperation strategy of the O2O supply chain but did not consider the impact of advertising delay time on cooperation. Therefore, this study evaluates the effect of advertising delay on dynamic collaboration in the O2O supply chain. In addition, we consider the simultaneous service cooperation situation.

Table 1. Literature related to advertising cooperation in the O2O supply chain.

Sources	Broad Theme	Research Questions
Li et al. [11]	Bilateral participation in advertising cooperation	• How cooperative advertising is applied in the O2O business model • Who can improve O2O supply chain performance more with unilateral or bilateral advertising cooperation • Practical application of the O2O supply chain bilateral cooperation advertising strategy in the industry
Li et al. [22]	BOPS (buy-online-and-pick-up-in-store)	• The effect of BOPS on advertising cooperation between manufacturers and retailers • The effect of BOPS on the optimal strategy of the O2O supply chain • The effectiveness of advertising cooperation in the O2O supply chain
Wang and Shu [61]	Fairness concerns	• The online retail platforms' Media Resource Replacement Plans (MRRPs) impact manufacturer and platform advertising partnerships • The optimal decisions of advertising efforts and participation rates between the supply chain members • Conditions for manufacturers to participate in the MRRP of online retailer platforms • The impact of manufacturers' participation in MRRP on the profitability of O2O supply chain members

In contrast to the above studies, the main contributions of this paper are as follows: (1) introducing time delay and bidirectional free-riding into the O2O supply chain dynamic model at the same time; (2) considering the time delay of brand goodwill caused by national advertising of the brand owners and offline experience service of the retailers; (3) designing the centralized decision, the brand cost-sharing decision and the bilateral cost-sharing decision to respond to three types of cooperative relationships among the O2O supply chain members: full cooperation, one-way participation cooperation and two-way participation cooperation. We consider the impact of the time delay phenomenon and bidirectional free-riding phenomenon on the cooperative decision choices of the O2O supply chain members.

The following is the structure of this paper. Section 2 describes the research questions and related hypotheses in detail. Optimization models are developed and explored in Section 3. Section 4 provides a comparative analysis of the various models. Finally, Section 5 summarizes the results and their practical implications and proposes directions for future research.

2. Model Summary and Relevant Assumptions

This paper considers the O2O supply chain system composed of a single brand owner M and a single retailer R. The brand owner sells the products through direct online sales and offline retailers. The brand owner and the retailer cooperate in advertising and service. The brand owner invests in national advertising and provides the online information service, and the retailer supports regional promotional advertising and provides the offline experience service. Among them, the national advertising of the brand owner and the offline experience service of the retailer has an impact on the brand goodwill, which is the time delay phenomenon. The game process of O2O supply chain is shown in Figure 1.

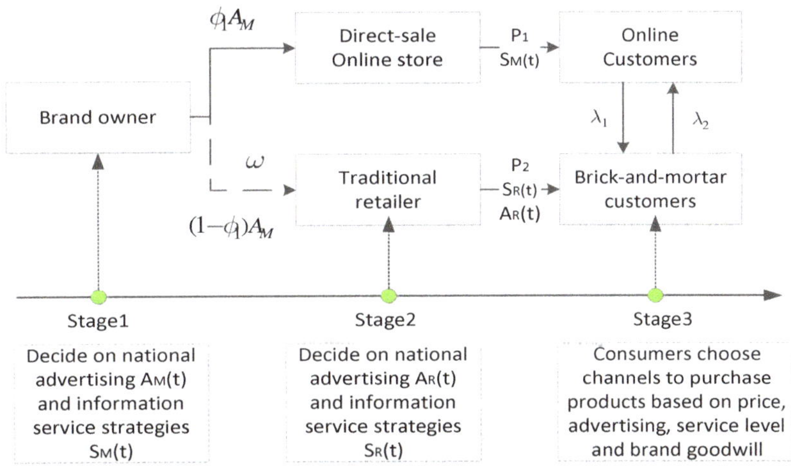

Figure 1. Game process of O2O supply chain members.

Descriptions of the model symbols in the paper are as follows:

$G(t), G(0)$: the brand goodwill at the moment t and the original goodwill level, respectively, and $G(0) = G_0$;

$A_M(t), A_R(t)$: the national advertising level of the brand owner and the regional advertising level of the retailer at the moment t, respectively;

$S_M(t), S_R(t)$: the information service level of the brand owner and the experience service level of the retailer at the moment t, respectively;

$d_1(t), d_2(t)$: the advertising delay time of the brand owner and the service delay time of the retailer, respectively;

$D_E(t)$,$D_R(t)$: the demand functions of the online channel and offline channel at the moment t, respectively;

$C(A_M(t)), C(A_R(t))$: the advertising cost functions of the brand owner and the retailer at the moment t, respectively;

$C(S_M(t)), C(S_R(t))$: the service cost functions of the brand owner and the retailer at the moment t, respectively;

$\omega(t), P_1(t), P_2(t)$: the wholesale price, online retail price, and offline retail price of the product at the moment t, respectively, and $\omega(t) < P_1(t) < P_2(t)$;

δ, ρ: the decay rate and discount factor of the brand goodwill, respectively, and $\delta > 0, \rho > 0$;

γ_1, γ_2: the influence coefficients of the brand owner's advertising level and retailer's service level to the brand goodwill, respectively, and $\gamma_1 > 0, \gamma_2 > 0$;

α, θ: the market capacity and consumers' brand preference, respectively, and $\alpha > 0, \theta > 0$;

β_1, β_2: the price demand flexibility coefficients of the online and offline channels, respectively, and $0 < \beta_1 < 1, 0 < \beta_2 < 1$;

ϕ_1, ϕ_2: the conversion rate of the brand owner's advertising to the online channel demand and the conversion rate of the retailer's advertising to the offline channel demand, respectively, and $0 < \phi_1 < 1, 0 < \phi_2 < 1$;

λ_1, λ_2: webrooming effect coefficient and showrooming effect coefficient, respectively, and $0 < \lambda_1 < 1, 0 < \lambda_2 < 1$;

k_i, η_i $(i = 1,2)$: the correlation coefficient between the advertising level and advertising cost and the correlation coefficient between the service level and service cost, respectively, and $k_i > 0, \eta_i > 0$.

Assumption 1. *By referring to the literature* [59,60], *to further consider the impact of the offline experience service of the retailer on the promotion of the brand goodwill based on the N–A advertising model and to describe the dynamic change of the brand goodwill at the moment t by using the delay differential as follows:*

$$\dot{G}(t) = \gamma_1 A_M(t - d_1(t)) + \gamma_2 S_R(t - d_2(t)) - \delta G(t), G(0) = G_0 \tag{1}$$

Assumption 2. *The purchase behavior of the consumers is comprehensively affected by the price, brand goodwill, advertising, service and other factors. Therefore, drawing on the literature* [44,45], *the demand functions of the online channel and offline channel are given as follows:*

$$D_E(t) = \alpha - \beta_1 P_1(t) + \theta G(t) + \phi_1 A_M(t) + (1 - \lambda_1) S_M(t) + \lambda_2 S_R(t) \tag{2}$$

$$D_R(t) = \alpha - \beta_2 P_2(t) + \theta G(t) + (1 - \phi_1) A_M(t) + \phi_2 A_R(t) + (1 - \lambda_2) S_R(t) + \lambda_1 S_M(t) \tag{3}$$

Assumption 3. *The cost function of the advertising and service shows a quadratic curve relationship, and relevant literature on this assumption is more common.*

$$C(A_M(t)) = \frac{1}{2} k_1 A_M^2(t), C(A_R(t)) = \frac{1}{2} k_2 A_R^2(t) \tag{4}$$

$$C(S_M(t)) = \frac{1}{2} \eta_1 S_M^2(t), C(S_R(t)) = \frac{1}{2} \eta_2 S_R^2(t) \tag{5}$$

The brand owner and the retailer have the same profit discount factor ρ at any time, and the objective is to maximize the net discounted profit within an infinite time horizon. To simplify the solution process, assume that the delay time is constant, i.e., $d_1(t) = d_1$, $d_2(t) = d_2$. Considering that the product price information is relatively transparent, assume that the wholesale price of the product and the retail prices of online and offline channels are also constant, i.e., $\omega(t) = \omega, P_1(t) = P_1, P_2(t) = P_2$.

3. Model Solution and Analysis

3.1. Centralized Decision

Under this decision, the brand owner and the retailer are in a fully cooperative relationship. Therefore, the supply chain members make optimal advertising and service decisions based on maximizing the supply chain profits. This decision is represented by superscript "C", and the objective function of the supply chain profits is as follows:

$$J_{MR}^{C}=\int_{0}^{\infty}e^{-\rho t}[P_1(t)D_E(t)+P_2(t)D_R(t)-\frac{k_1}{2}A_M{}^2(t)-\frac{\eta_1}{2}S_M{}^2(t)-\frac{k_2}{2}A_R{}^2(t)-\frac{\eta_2}{2}S_R{}^2(t)]dt$$

Proposition 1. *Under centralized decision-making, the optimal advertising and service strategies of the brand owner and the retailer are as follows:*

$$A_M^{C*}(t)=\frac{\theta\gamma_1 e^{\delta d_1}(P_1+P_2)+(\delta+\rho)[P_2(1-\phi_1)+\phi_1 P_1]}{k_1(\delta+\rho)},\ A_R^{C*}(t)=\frac{\phi_1 P_2}{k_2},$$

$$S_M^{C*}(t)=\frac{P_1+\lambda_1(P_2-P_1)}{\eta_1},\ S_R^{C*}(t)=\frac{\theta\gamma_2 e^{\delta d_2}(P_1+P_2)+(\delta+\rho)[\lambda_2(P_1-P_2)+P_2]}{\eta_2(\delta+\rho)}.$$

Brand goodwill level is as follows:

$$\begin{aligned}G^{C*}(t)=\ & e^{-\delta t}G_0+\frac{\theta\gamma_1{}^2e^{\delta d_1}(P_1+P_2)+\gamma_1(\delta+\rho)[P_2(1-\phi_1)+\phi_1P_1]}{k_1\delta(\delta+\rho)}(1-e^{-\delta t})\\ &+\frac{\theta\gamma_2{}^2e^{\delta d_2}(P_1+P_2)+\gamma_2(\delta+\rho)[\lambda_2(P_1-P_2)+P_2]}{\eta_2\delta(\delta+\rho)}(1-e^{-\delta t})\end{aligned}$$

The supply chain profits are as follows:

$$\begin{aligned}J_{MR}^{C}=\ & \tfrac{1}{\rho}(P_1[\alpha-\beta_1P_1+\phi_1A_M^{C*}+(1-\lambda_1)S_M^{C*}+\lambda_2S_R^{C*}]+P_2[\alpha-\beta_2P_2+(1-\phi_1)A_M^{C*})+\phi_2A_R^{C*}+(1-\lambda_2)S_R^{C*}+\lambda_1S_M^{C*}]\\ &-\tfrac{k_1}{2}A_M^{C*2}-\tfrac{\eta_1}{2}S_M^{C*2}-\tfrac{k_2}{2}A_R^{C*2}-\tfrac{\eta_2}{2}S_R^{C*2})+\theta(P_1+P_2)[\tfrac{G_0}{\rho+\delta}+\tfrac{\gamma_1A_M^{C*}}{\rho(\rho+\delta)}+\tfrac{\gamma_2S_R^{C*}}{\rho(\rho+\delta)}]\end{aligned}$$

Proving: the optimal decision problem is characterized as the optimal control problem.

$$\max_{A_M>0,S_R>0}J_{MR}^{C}$$

$$\text{s.t. }\dot{G}(t)=\gamma_1A_M(t-d_1(t))+\gamma_2S_R(t-d_2(t))-\delta G(t)$$

Construct the Hamiltonian function such that

$$\begin{aligned}H_{MR}^{C}=\ & e^{-\rho t}\left\{\begin{array}{l}P_1[\alpha-\beta_1P_1+\theta G+\phi_1A_M+(1-\lambda_1)S_M+\lambda_2S_R]+P_2[\alpha-\beta_2P_2+\theta G+(1-\phi_1)A_M\\ +\phi_2A_R+(1-\lambda_2)S_R+\lambda_1S_M]-\frac{k_1}{2}A_M{}^2-\frac{\eta_1}{2}S_M{}^2-\frac{k_2}{2}A_R{}^2-\frac{\eta_2}{2}S_R{}^2\end{array}\right\}\\ &+q(t)[\gamma_1A_M(t-d_1)+\gamma_2S_R(t-d_2)-\delta G(t)]\end{aligned}\tag{6}$$

$$\frac{dH_{MR}}{dA_M(t)}=e^{-\rho t}[\phi_1P_1+P_2(1-\phi_1)-k_1A_M(t)]+q(t)\gamma_1\frac{dA_M(t-d_1)}{dA_M(t)}=0;\tag{7}$$

$$\frac{dH_{MR}}{dS_R(t)}=e^{-\rho t}[\lambda_2P_1+P_2(1-\lambda_2)-\eta_2S_R(t)]+q(t)\gamma_2\frac{dS_R(t-d_2)}{dS_R(t)}=0;\tag{8}$$

$$\frac{dq(t)}{dt}=-\frac{dH_{MR}}{dG}=\delta q(t)-\theta(P_1+P_2)e^{-\rho t}$$

so

$$q(t)=C_0e^{\delta t}+\frac{\theta(P_1+P_2)}{\rho+\delta}e^{-\rho t},C_0\in R\tag{9}$$

According to the literature [57], substitute $\frac{dA_M(t-d_1)}{dA_M(t)} = e^{\delta d_1}$, $\frac{dS_R(t-d_2)}{dS_R(t)} = e^{\delta d_2}$ and Equation (9) into Equations (7) and (8), receiving the results as follows:

$$A_M^{C*}(\mathrm{t}) = \frac{P_2(1-\phi_1)+\phi_1 P_1}{k_1} + \frac{\gamma_1}{k_1}C_0 e^{(\delta+\rho)t+\delta d_1} + \frac{\theta\gamma_1(P_1+P_2)}{k_1(\delta+\rho)}e^{\delta d_1}$$

$$S_R^{C*}(\mathrm{t}) = \frac{\lambda_2(P_1-P_2)+P_2}{\eta_2} + \frac{\gamma_2}{\eta_2}C_0 e^{(\delta+\rho)t+\delta d_1} + \frac{\theta\gamma_2(P_1+P_2)}{\eta_2(\delta+\rho)}e^{\delta d_2}$$

Considering that the national advertising of the brand owner and the experience service of the retailer cannot be unlimited, i.e.,

$\lim_{t\to\infty} A_M^{C*}(t) < \infty$, $\lim_{t\to\infty} S_R^{C*}(t) < \infty$, thus receiving the result $C_0 = 0$.

At this time, the optimal level of investment for the brand's national advertising and the retailer's experiential services are as follows:

$$A_M^{C*}(\mathrm{t}) = \frac{\theta\gamma_1 e^{\delta d_1}(P_1+P_2)+(\delta+\rho)[P_2(1-\phi_1)+\phi_1 P_1]}{k_1(\delta+\rho)}$$

$$S_R^{C*}(\mathrm{t}) = \frac{\theta\gamma_2 e^{\delta d_2}(P_1+P_2)+(\delta+\rho)[\lambda_2(P_1-P_2)+P_2]}{\eta_2(\delta+\rho)}$$

Similarly, assuming that

$$\frac{\partial H_{MR}}{\partial A_R(t)} = e^{-\rho t}[\phi_2 P_2 - k_2 A_R(t)] = 0, \frac{\partial H_{MR}}{\partial S_M(t)} = e^{-\rho t}[(1-\lambda_1)P_1 + \lambda_1 P_2 - \eta_1 S_M(t)] = 0$$

The following conclusions were drawn: $A_R^{C*}(\mathrm{t}) = \frac{\phi_1 P_2}{k_2}$, $S_M^{C*}(\mathrm{t}) = \frac{P_1+\lambda_1(P_2-P_1)}{\eta_1}$.

Substituting $A_M^{C*}(\mathrm{t})$ and $S_R^{C*}(\mathrm{t})$ into Equation (1), we receive as follows:

$$G^{C*}(t) = e^{-\delta t}G_0 + \frac{\theta{\gamma_1}^2 e^{\delta d_1}(P_1+P_2)+\gamma_1(\delta+\rho)[P_2(1-\phi_1)+\phi_1 P_1]}{k_1\delta(\delta+\rho)}(1-e^{-\delta t}) + \frac{\theta{\gamma_2}^2 e^{\delta d_2}(P_1+P_2)+\gamma_2(\delta+\rho)[\lambda_2(P_1-P_2)+P_2]}{\eta_2\delta(\delta+\rho)}(1-e^{-\delta t})$$

Substituting the optimal decision value and the brand goodwill into the equation, we can receive the optimal profit ${J_{MR}}^{C*}$ of the whole supply chain, and the evidentiary process ends.

Further, we can analyze the impact of the delay time d_1 and d_2, webrooming effect and showrooming effect coefficients λ_1 and λ_2, the conversion rates ϕ_1 and ϕ_2 of the advertising investment levels on the online channel and offline channel demand, and the online and offline retail prices P_1 and P_2 on the optimal strategies of the supply chain members.

Deduction 1. *The static analysis results of the optimal strategies of the brand owner and retailer concerning relevant parameters are as follows:*

$$\frac{\partial A_M^{C*}(\mathrm{t})}{\partial d_1} > 0, \frac{\partial A_M^{C*}(\mathrm{t})}{\partial \phi_1} < 0, \frac{\partial A_M^{C*}(\mathrm{t})}{\partial P_1} > 0, \frac{\partial A_M^{C*}(\mathrm{t})}{\partial P_2} > 0;$$

$$\frac{\partial S_M^{C*}(\mathrm{t})}{\partial \lambda_1} > 0, \frac{\partial S_M^{C*}(\mathrm{t})}{\partial P_1} > 0, \frac{\partial S_M^{C*}(\mathrm{t})}{\partial P_2} > 0;$$

$$\frac{\partial A_R^{C*}(\mathrm{t})}{\partial \phi_2} > 0, \frac{\partial A_R^{C*}(\mathrm{t})}{\partial P_2} > 0;$$

$$\frac{\partial S_R^{C*}(\mathrm{t})}{\partial d_2} > 0, \frac{\partial S_R^{C*}(\mathrm{t})}{\partial \lambda_2} < 0, \frac{\partial S_R^{C*}(\mathrm{t})}{\partial P_1} > 0, \frac{\partial S_R^{C*}(\mathrm{t})}{\partial P_2} > 0.$$

Deduction 1 indicates that under this decision: (1) the optimal advertising strategy $A_M^{C*}(t)$ of the brand owner and the optimal service strategy $S_R^{C*}(t)$ of the retailer are an increasing function of the delay time d_1 and d_2. With the growth of the delay time d_1 and d_2, the brand owner's advertising and the retailer's service investment are increasing. In contrast, the brand goodwill level cannot reach the ideal state immediately, thus affecting the market demand. At this time, if the enterprise blindly increases the investment in advertising and service, the investment cost increases while the investment return period extends, which will undoubtedly increase the business risk of the enterprise; (2) the optimal service strategy $S_M^{C*}(t)$ of the brand owner is an increasing function of the webrooming effect λ_1, and the optimal service strategy $S_R^{C*}(t)$ of the retailer is a decreasing function of the showrooming effect λ_2. Due to the fact $P_2 > P_1$, the increase in the webrooming effect λ_1 has promoted the overall profit of the supply chain on the contrary, and the brand owner shall actively improve the service level $S_M^{C*}(t)$. Similarly, when the showrooming effect λ_2 intensifies, some consumers will choose the relatively cheap online channel to complete the purchase, which will reduce the overall profit of the supply chain, and the retailer shall reduce the service level $S_R^{C*}(t)$; (3) the brand owner advertising strategy $A_M^{C*}(t)$ is a decreasing function of ϕ_1, and the retailer advertising strategy $A_R^{C*}(t)$ is an increasing function of ϕ_2. Changes in national advertising $A_M^{C*}(t)$ investment by the brand owner will impact online and offline channel demand. The lower the conversion rate ϕ_1 of online channel demand, the higher the conversion rate $1-\phi_1$ of offline channel demand, which is better for the promotion of the overall profit of the supply chain. A change in the investment level of the regional promotional advertising $A_R^{C*}(t)$ will only have an impact on the offline channel demand. The higher the conversion rate ϕ_2 of the offline channel demand, the higher the advertising investment that should be made by the retailer; (4) the advertising strategy $A_M^{C*}(t)$ and the service strategy $S_M^{C*}(t)$ of the brand owner, and the retailer's service strategy $S_R^{C*}(t)$ are the increasing function of the retail prices P_1 and P_2 of both online and offline channels. The retailer's advertising strategy ${A_R}^{C*}(t)$ of is the increasing function of P_2. Equations (2) and (3) show that channel demand is negatively related to the channel retail price. The enterprises should mitigate the impact of price on consumers' purchase behavior by improving the brand goodwill, advertising and service.

3.2. Brand Cost-Sharing Decision

Under this decision, as the supply chain leader, the brand owner participates in the cooperation unilaterally. The brand owner will bear part of the advertising and service costs to encourage the retailer to improve the advertising and service level. The supply chain members all aim to maximize their profits. First, the brand owner decides to bear the proportion ξ_1 and ξ_2 of advertising and service costs of the retailer. Then the brand owner and retailer select their advertising and service strategies independently, using the inverse induction method to solve the problem. Expressed by superscript "M", the profit functions of the brand owner and the retailer are as follows:

$$J_M^M = \int_0^\infty e^{-\rho t}[P_1(t)D_E(t) + \omega D_R(t) - \frac{k_1}{2}{A_M}^2(t) - \frac{\eta_1}{2}{S_M}^2(t) - \frac{\xi_1 k_2}{2}{A_R}^2(t) - \frac{\xi_2\eta_2}{2}{S_R}^2(t)]dt \quad (10)$$

$$J_R^M = \int_0^\infty e^{-\rho t}[(P_2(t) - \omega)D_R(t) - \frac{(1-\xi_1)k_2}{2}{A_R}^2(t) - \frac{(1-\xi_2)\eta_2}{2}{S_R}^2(t)]dt \quad (11)$$

Proposition 2. *Under this decision, the optimal strategies of the brand owner and the retailer are as follows:*

$$A_M^{M*}(\mathrm{t}) = \frac{\theta\gamma_1 e^{\delta d_1}(P_1+\omega) + (\delta+\rho)[(1-\phi_1)\omega + \phi_1 P_1]}{k_1(\delta+\rho)}, A_R^{M*}(\mathrm{t}) = \frac{\phi_2(\omega+P_2)}{2k_2}$$

$$S_M^{M*}(\mathrm{t}) = \frac{(1-\lambda_1)P_1 + \lambda_1\omega}{\eta_1}, S_R^{M*}(\mathrm{t}) = \frac{\theta\gamma_2[e^{\delta d_2}(P_2-\omega) + 2(P_1+\omega)] + (\delta+\rho)[(1-\lambda_2)(\omega+P_2) + 2\lambda_2 P_1]}{2\eta_2(\delta+\rho)} \tag{12}$$

Cost-sharing ratio of the brand owner:

$$\xi_1{}^{M*} = \begin{cases} \frac{3\omega - P_2}{\omega + P_2}, & 3\omega > P_2 \\ 0, & 3\omega \le P_2 \end{cases}$$

$$\xi_2{}^{M*} = \begin{cases} \frac{X_1\omega + Y_1P_1 - Z_1P_2}{X_2\omega + Y_1P_1 + Z_1P_2}, & X_1\omega + Y_1P_1 > Z_1P_2 \\ 0, & X_1\omega + Y_1P_1 \le Z_1P_2 \end{cases}$$

Brand goodwill level is as follows:

$$\begin{aligned} G^{M*}(t) = \quad & e^{-\delta t}G_0 + \frac{\theta\gamma_1{}^2 e^{\delta d_1}(P_1+\omega) + \gamma_1(\delta+\rho)[(1-\phi_1)\omega + \phi_1P_1]}{k_1\delta(\delta+\rho)}(1-e^{-\delta t}) \\ & + \frac{\theta\gamma_2{}^2[e^{\delta d_2}(P_2-\omega) + 2(P_1+\omega)] + \gamma_2(\delta+\rho)[(1-\lambda_2)(\omega+P_2) + 2\lambda_2P_1]}{2\eta_2\delta(\delta+\rho)}(1-e^{-\delta t}) \end{aligned}$$

In the equation:

$$X_1 = \theta\gamma_2(2+e^{\delta d_2}) + 3(\delta+\rho)(1-\lambda_2)$$
$$X_2 = \theta\gamma_2(2-e^{\delta d_2}) + (\delta+\rho)(1-\lambda_2)$$
$$Y_1 = 2\theta\gamma_2 + 2\lambda_2(\delta+\rho)$$
$$Z_1 = \theta\gamma_2 e^{\delta d_2} + (\delta+\rho)(1-\lambda_2)$$

Proving: First, confirm the cost-sharing ratios ξ_1^M and ξ_2^M of the brand owner, and determine the optimal strategies of the supply chain members. The retailer's optimal decision problem is characterized as the optimal control problem, i.e.,

$$\max_{A_M>0, S_R>0} J_R^M$$

$$\text{s.t. } \dot{G}(t) = \gamma_1 A_M(t-d_1(t)) + \gamma_2 S_R(t-d_2(t)) - \delta G(t)$$

Similar to the evidentiary process of the Proposition 1, build the Hamilton function as follows:

$$\begin{aligned} H_R^M = e^{-\rho t} & \begin{Bmatrix} (P_2-\omega)[\alpha - \beta_2P_2 + \theta G + (1-\phi_1)A_M + \phi_2 A_R + (1-\lambda_2)S_R \\ +\lambda_1 S_M] - \frac{(1-\xi_1)k_2}{2}A_R{}^2 - \frac{(1-\xi_2)\eta_2}{2}S_R{}^2] \end{Bmatrix} \\ & + q(t)[\gamma_1 A_M(t-d_1) + \gamma_2 S_R(t-d_2) - \delta G(t)] \end{aligned}$$

Assuming

$$\frac{dH_R}{dA_R(t)} = e^{-\rho t}[\phi_2(P_2-\omega) - k_2(1-\xi_1)A_M(t)] = 0$$

Receiving

$$A_R{}^{M*}(\mathrm{t}) = \frac{\phi_2(P_2-\omega)}{k_2(1-\xi_1)} \tag{13}$$

Assuming

$$\frac{dH_R}{dS_R(t)} = e^{-\rho t}[(1-\lambda_2)(P_2-\omega) - \eta_2(1-\xi_2)S_R(t)] + q(t)\gamma_2\frac{dS_R(t-d_2)}{dS_R(t)} = 0 \quad (14)$$

$$\frac{dq(t)}{dt} = -\frac{dH_R}{dG} = \delta q(t) - \theta(P_2-\omega)e^{-\rho t}$$

$$q(t) = C_1 e^{\delta t} + \frac{\theta(P_2-\omega)}{\rho+\delta}e^{-\rho t},\ C_1 \in R \quad (15)$$

Substituting $\frac{dS_R(t-d_2)}{dS_R(t)} = e^{\delta d_2}$ and Equation (15) into Equation (14).
Receiving

$$S_R^{M*}(\mathrm{t}) = \frac{[(1-\lambda_2)(\delta+\rho)+\theta\gamma_2 e^{\delta d_2}](P_2-\omega)}{\eta_2(1-\xi_2)(\delta+\rho)} + \frac{\gamma_2}{\eta_2(1-\xi_2)}C_1 e^{(\delta+\rho)t+\delta d_2}$$

As $\lim\limits_{t\to\infty} S_R^{M*}(\mathrm{t}) < \infty$, so $C_1 = 0$. Therefore:

$$S_R^{M*}(\mathrm{t}) = \frac{[(1-\lambda_2)(\delta+\rho)+\theta\gamma_2 e^{\delta d_2}](P_2-\omega)}{\eta_2(1-\xi_2)(\delta+\rho)} \quad (16)$$

The optimal decision problem of the brand owner is characterized as the optimal control problem, i.e.,

$$\max_{A_M>0, S_R>0} J_M^M$$

$$\text{s.t. } \dot{G}(t) = \gamma_1 A_M(t-d_1(t)) + \gamma_2 S_R(t-d_2(t)) - \delta G(t)$$

Hamilton function is as follows:

$$H_M^M = e^{-\rho t}\left\{ \begin{array}{l} P_1[\alpha-\beta_1 P_1+\theta G+\phi_1 A_M+(1-\lambda_1)S_M+\lambda_2 S_R]+\omega[\alpha-\beta_2 P_2+\theta G+(1-\phi_1)A_M \\ +\phi_2 A_R+(1-\lambda_2)S_R+\lambda_1 S_M] - \frac{k_1}{2}{A_M}^2 - \frac{\eta_1}{2}{S_M}^2 - \frac{\xi_1 k_2}{2}{A_R}^2 - \frac{\xi_2\eta_2}{2}{S_R}^2 \end{array} \right\}$$
$$+q(t)[\gamma_1 A_M(t-d_1)+\gamma_2 S_R(t-d_2)-\delta G(t)]$$

Similar to the evidentiary process of the optimal decision of the retailer, receiving:

$$A_M^{M*}(\mathrm{t}) = \frac{\theta\gamma_1 e^{\delta d_1}(P_1+\omega)+(\delta+\rho)[(1-\phi_1)\omega+\phi_1 P_1]}{k_1(\delta+\rho)} \quad (17)$$

$$S_M^{M*}(\mathrm{t}) = \frac{(1-\lambda_1)P_1+\lambda_1\omega}{\eta_1} \quad (18)$$

Substituting Equations (16) and (17) into Equation (1), receiving the brand goodwill as follows:

$$\begin{aligned} G^{M*}(t) = \ & e^{-\delta t}G_0 + \frac{\theta{\gamma_1}^2 e^{\delta d_1}(P_1+\omega)+\gamma_1(\delta+\rho)[(1-\phi_1)\omega+\phi_1 P_1]}{k_1\delta(\delta+\rho)}(1-e^{-\delta t}) \\ & + \frac{\gamma_2[(1-\lambda_2)(\delta+\rho)+\theta\gamma_2 e^{\delta d_2}](P_2-\omega)}{\eta_2\delta(1-\xi_2)(\delta+\rho)}(1-e^{-\delta t}) \end{aligned} \quad (19)$$

The brand owner decides the optimal cost-sharing ratio and substitutes Equations (13), (16)–(19) into the Equation (10), receiving the profit function J_M^{M*} of the brand owner. Find the maximums of ${\xi_1}^M$ and ${\xi_2}^M$ by J_M^{M*}, i.e.,

Assume $\frac{\partial J_M^{M*}}{\partial \xi_1} = 0$ and $\frac{\partial J_M^{M*}}{\partial \xi_2} = 0$, the conclusion of ξ_1^{M*} and ξ_2^{M*} will obtained. The evidentiary process ends.

Further, we can analyze the impact of the delay time d_1 and d_2, webrooming effect and showrooming effect coefficients λ_1 and λ_2, the conversion rates ϕ_1 and ϕ_2 of the advertising investment levels on the online channel and offline channel demand, and the impacts of the online and offline retail prices P_1 and P_2, and the wholesale price ω on the optimal strategies of the brand owner and retailer.

Deduction 2. *The static analysis results of the optimal strategies of the brand owner and retailer concerning relevant parameters are as follows:*

$$\frac{\partial A_M^{M*}(t)}{\partial d_1} > 0, \frac{\partial A_M^{M*}(t)}{\partial \phi_1} > 0, \frac{\partial A_M^{M*}(t)}{\partial P_1} > 0, \frac{\partial A_M^{M*}(t)}{\partial \omega} > 0;$$

$$\frac{\partial S_M^{M*}(t)}{\partial \lambda_1} < 0, \frac{\partial S_M^{M*}(t)}{\partial P_1} > 0, \frac{\partial S_M^{M*}(t)}{\partial \omega} > 0;$$

$$\frac{\partial A_R^{M*}(t)}{\partial \phi_2} > 0, \frac{\partial A_R^{M*}(t)}{\partial P_2} > 0, \frac{\partial A_R^{M*}(t)}{\partial \omega} > 0;$$

$$\frac{\partial S_R^{M*}(t)}{\partial d_2} > 0, \frac{\partial S_R^{M*}(t)}{\partial P_1} > 0, \frac{\partial S_R^{M*}(t)}{\partial P_2} > 0,$$

$$\begin{cases} \frac{\partial S_R^{M*}(t)}{\partial \lambda_2} > 0, & P_1 < (\omega + P_2)/2 \\ \frac{\partial S_R^{M*}(t)}{\partial \lambda_2} < 0, & P_1 > (\omega + P_2)/2 \end{cases}, \begin{cases} \frac{\partial S_R^{M*}(t)}{\partial \omega} > 0, & d_2 < L_2 \\ \frac{\partial S_R^{M*}(t)}{\partial \omega} < 0, & d_2 > L_2 \end{cases};$$

$$\frac{\partial \xi_1^{M*}}{\partial P_2} < 0, \frac{\partial \xi_1^{M*}}{\partial \omega} > 0; \frac{\partial \xi_2^{M*}}{\partial d_2} < 0, \frac{\partial \xi_2^{M*}}{\partial \lambda_2} > 0, \frac{\partial \xi_2^{M*}}{\partial P_1} > 0, \frac{\partial \xi_2^{M*}}{\partial P_2} < 0, \frac{\partial \xi_2^{M*}}{\partial \omega} > 0;$$

Among which $L_2 = \frac{1}{\delta} \ln \frac{(\delta+\rho)(1-\lambda_2)+2\theta\gamma_2}{\theta\gamma_2}$.

Deduction 1 indicates that under this decision:

(1) The optimal advertising strategy $A_M^{M*}(t)$ of the brand owner and the optimal service strategy $S_R^{M*}(t)$ of the retailer are an increasing function of the delay time d_1 and d_2, which is consistent with the centralized decision. The cost-sharing ratio ξ_2^{M*} of the brand owner is a decreasing function of the delay time d_2; as the delay time d_2 increases, the brand owner is willing to share a smaller share of the retailer's service cost;

(2) The brand owner's optimal service strategy $S_M^{M*}(t)$ is a decreasing function of the webrooming effect λ_1, while the brand owner's cost-sharing ratio ξ_2^{M*} is an increasing function of the showrooming effect λ_2 The impact of the showrooming effect λ_2 on the retailer's optimal service strategy $S_R^{M*}(t)$ is influenced by the brand owner's cost-sharing ratio ξ_2^{M*} and the retail price P_1 of the online channel. Under the condition of $P_1 < (\omega + P_2)/2$, the price gap between the channels decreases and the showrooming effect λ_2 is relieved, while the service $S_R^{M*}(t)$ of the retailer is promoted. This shows that if the online channel price of the brand owner is properly decided, ξ_2^{M*} can relieve the impact of the showrooming effect on the retailer's service strategy;

(3) The brand owner's advertising strategy $A_M^{M*}(t)$ is an increasing function of ϕ_1 and the lower the online channel demand conversion rate ϕ_1, the lower the brand owner's advertising level should be. A change in the level of the regional promotional advertising $A_R^{M*}(t)$ by the retailer will only affect the offline channel demand and will not be affected by the decision-making; thus, the retailer shall improve the advertising level to promote the conversion rate ϕ_2 of the offline channel demand;

(4) The wholesale price ω is an important factor influencing the optimal service and advertising strategies of the supply chain members. The higher the wholesale price ω, the higher the revenue the brand owner receives from the offline channel, while the retailer's product cost will be increased. To motivate the retailer to actively promote the advertising and service levels, the brand owner shall increase the cost-sharing ratios ξ_1^{M*} and ξ_2^{M*} with the retailer. The retailer shall promote the advertising level

A_R^{M*}(t), but the delay time d_2 also affects whether to increase the level of service. Under the condition of $d_2 > L_2$, ξ_2^{M*} cannot motivate the retailer to promote its service. Thus, under the condition of $d_2 < L_2$, that the brand owner can promote the optimal advertising and service strategies of the supply chain members if it sets a comparatively high wholesale price ω.

3.3. Bilateral Cost-Sharing Decision

Under this decision, by referring to the literature [51], the brand owner and the retailer are participating in cooperation together and share the cost in a bilateral way, namely, the retailer shares the national advertising cost ratio ξ_1^D of the brand owner, and the brand owner shares the experience service cost ratio ξ_2^D of the retailer. Presented by the superscript "D", the objective profit functions of the brand owner and retailer are as follows:

$$J_M^D = \int_0^\infty e^{-\rho t}[P_1(t)D_E(t) + \omega D_R(t) - \frac{(1-\xi_1)k_1}{2}A_M{}^2 - \frac{\eta_1}{2}S_M{}^2 - \frac{\xi_2\eta_2}{2}S_R]dt \quad (20)$$

$$J_M^D = \int_0^\infty e^{-\rho t}[P_1(t)D_E(t) + \omega D_R(t) - \frac{(1-\xi_1)k_1}{2}A_M{}^2 - \frac{\eta_1}{2}S_M{}^2 - \frac{\xi_2\eta_2}{2}S_R]dt \quad (21)$$

Proposition 3. *Under this decision, the optimal strategies of the brand owner and the retailer are as follows:*

$$A_M^{D*}(t) = \frac{\theta\gamma_1[e^{\delta d_1}(P_1+\omega) + 2(P_2-\omega)] + (\delta+\rho)[(1-\phi_1)(P_2-\omega) + \phi_1 P_1]}{2k_1(\delta+\rho)}, A_R^{D*}(t) = \frac{\phi_2(P_2-\omega)}{k_2},$$

$$S_M^{D*}(t) = \frac{(1-\lambda_1)P_1 + \lambda_1\omega}{\eta_1}, S_R^{D*}(t) = \frac{\theta\gamma_2[e^{\delta d_2}(P_2-\omega) + 2(P_1+\omega)] + (\delta+\rho)[(1-\lambda_2)(\omega+P_2) + 2\lambda_2 P_1]}{2\eta_2(\delta+\rho)}(t).$$

$$\xi_1^{D*} = \begin{cases} \frac{Z_2P_2 - Y_2P_1 - X_3\omega}{Z_2P_2 + Y_2P_1 - X_4\omega}, & X_3\omega + Y_2P_1 < Z_2P_2 \\ 0, & X_3\omega + Y_2P_1 \geq Z_2P_2 \end{cases}, \xi_2^{D*} = \begin{cases} \frac{X_1\omega + Y_1P_1 - Z_1P_2}{X_2\omega + Y_1P_1 + Z_1P_2}, & X_1\omega + Y_1P_1 > Z_1P_2 \\ 0, & X_1\omega + Y_1P_1 \leq Z_1P_2 \end{cases}.$$

Bilateral cost-sharing ratios are as follows:
In the equation:

$$X_3 = \theta\gamma_1(2 + e^{\delta d_1}) + 3(\delta+\rho)(1-\phi_1)$$
$$X_4 = \theta\gamma_1(2 - e^{\delta d_1}) + (\delta+\rho)(1-\phi_1)$$
$$Y_2 = \theta\gamma_1 e^{\delta d_1} + \phi_1(\delta+\rho)$$
$$Z_2 = 2\theta\gamma_1 + 2(\delta+\rho)(1-\phi_1)$$

The brand goodwill:

$$G^{D*}(t) = e^{-\delta t}G_0 + \frac{\theta\gamma_1{}^2[e^{\delta d_1}(P_1+\omega) + 2(P_2-\omega)] + \gamma_1(\delta+\rho)[(1-\phi_1)(P_2-\omega) + \phi_1 P_1]}{2k_1\delta(\delta+\rho)}(1-e^{-\delta t}) + \frac{\theta\gamma_2{}^2[e^{\delta d_2}(P_2-\omega) + 2(P_1+\omega)] + \gamma_2(\delta+\rho)[(1-\lambda_2)(\omega+P_2) + 2\lambda_2 P_1]}{2\eta_2\delta(\delta+\rho)}(1-e^{-\delta t})$$

Proving: under this decision, after the brand owner decides its service cost-sharing ratio ξ_2^D for the retailer and the retailer decides its advertising cost ratio ξ_1^D for the brand owner, the brand owner and retailer decide their optimal service and advertising strategies, respectively. The evidentiary process is similar to Proposition 2 and will not be repeated here.

$S_M^{M*}(t) = S_M^{D*}(t)$, $S_R^{M*}(t) = S_R^{D*}(t)$ follows from Propositions 2 and 3. The impacts of the webrooming effect and showrooming effect coefficients λ_1 and λ_2 on the supply chain members' service strategies are the same as Proposition 2. The impacts of the delay time d_1 and d_2, the wholesale price ω, and the brand owner's advertising conversion rate ϕ_1 on the supply chain members' optimal strategies are further analyzed.

Deduction 3. *The static analysis results of the optimal strategies of the brand owner and retailer concerning relevant parameters are as follows:*

$$\frac{\partial A_M^{D*}(\mathrm{t})}{\partial d_1} > 0, \frac{\partial A_M^{D*}(\mathrm{t})}{\partial \phi_1} < 0, \frac{\partial A_M^{D*}(\mathrm{t})}{\partial P_1} > 0, \frac{\partial A_M^{D*}(\mathrm{t})}{\partial P_2} > 0,$$

$$\begin{cases} \frac{\partial A_M^{D*}(\mathrm{t})}{\partial \omega} > 0, & \phi_1 > L_1 \\ \frac{\partial A_M^{D*}(\mathrm{t})}{\partial \omega} < 0, & \phi_1 < L_1 \end{cases};$$

$$\frac{\partial S_M^{D*}(\mathrm{t})}{\partial \lambda_1} < 0, \frac{\partial S_M^{D*}(\mathrm{t})}{\partial P_1} > 0, \frac{\partial S_M^{D*}(\mathrm{t})}{\partial \omega} > 0;$$

$$\frac{\partial A_R^{D*}(\mathrm{t})}{\partial \phi_2} > 0, \frac{\partial A_R^{D*}(\mathrm{t})}{\partial P_2} > 0, \frac{\partial A_R^{D*}(\mathrm{t})}{\partial \omega} < 0;$$

$$\frac{\partial S_R^{D*}(\mathrm{t})}{\partial d_2} > 0, \frac{\partial S_R^{D*}(\mathrm{t})}{\partial P_1} > 0, \frac{\partial S_R^{D*}(\mathrm{t})}{\partial P_2} > 0,$$

$$\begin{cases} \frac{\partial S_R^{D*}(\mathrm{t})}{\partial \lambda_2} > 0, & \omega - 2P_1 + P_2 < 0 \\ \frac{\partial S_R^{D*}(\mathrm{t})}{\partial \lambda_2} < 0, & \omega - 2P_1 + P_2 > 0 \end{cases}, \begin{cases} \frac{\partial S_R^{D*}(\mathrm{t})}{\partial \omega} > 0, & d_2 < \mathrm{L}_2 \\ \frac{\partial S_R^{D*}(\mathrm{t})}{\partial \omega} < 0, & d_2 > \mathrm{L}_2 \end{cases};$$

$$\frac{\partial \xi_1^{D*}}{\partial d_1} < 0, \frac{\partial \xi_1^{D*}}{\partial \phi_1} < 0, \frac{\partial \xi_1^{D*}}{\partial P_1} < 0, \frac{\partial \xi_1^{D*}}{\partial P_2} > 0, \frac{\partial \xi_1^{D*}}{\partial \omega} < 0;$$

$$\frac{\partial \xi_2^{D*}}{\partial d_2} < 0, \frac{\partial \xi_2^{D*}}{\partial \lambda_2} > 0, \frac{\partial \xi_2^{D*}}{\partial P_1} > 0, \frac{\partial \xi_2^{D*}}{\partial P_2} < 0, \frac{\partial \xi_2^{D*}}{\partial \omega} > 0.$$

Among which, $\mathrm{L}_1 = \frac{1}{\delta} \ln \frac{(\delta+\rho)(1-\phi_1)+2\theta\gamma_1}{\theta\gamma_1}$.

Deduction 3 indicates that under this decision: (1) ξ_1^{D*} and ξ_2^{D*} are a decreasing function of d_1 and d_2, respectively. The longer the delay in the brand advertising and the retailer service to the brand goodwill, the lower the rate of cost-sharing between the brand and the retailer participation in a bidirectional sense; (2) the brand owner's advertising strategy is a decreasing function of the conversion rate ϕ_1. As the conversion rate ϕ_1 decreases and the conversion rate $1 - \phi_1$ of the offline channel demand increases, the retailer increases its cost-sharing ratio ξ_1^{D*} to encourage the brand owner to invest in the advertising. Therefore, although ϕ_1 decreases, the brand owner shall also increase its advertising level; (3) with the wholesale price ω increase, the brand owner promotes its service level $S_M^{D*}(\mathrm{t})$ and cost-sharing ratio ξ_2^{D*}, and the retailer decreases its advertising level $A_R^{D*}(\mathrm{t})$ and cost-sharing ratio ξ_1^{D*}. However, the delay time influences whether the cost-sharing ratios ξ_1^{D*} and ξ_2^{D*} can affect the brand owner's advertising strategy $A_M^{D*}(\mathrm{t})$ and the retailer's service strategy $S_R^{D*}(\mathrm{t})$. When $d_1 < L_1$, $d_2 < L_2$ controlled by ξ_1^{D*} and ξ_2^{D*}, the brand owner should reduce its advertising level and the retailer should improve its service level. Under the condition of $d_1 > L_1$, $d_2 > L_2$, with the rise in the wholesale price ω, the brand owner should improve its advertising level and the retailer should lower its service level.

4. Comparative Analysis

4.1. Comparison of the Optimal Strategies of the Supply Chain Members under Three Decisions

Deduction 4. *As compared under the three decisions, it is constant as follows:*

$$A_M^{C*}(t) > A_M^{D*}(t) > A_M^{M*}(t);\ A_R^{C*}(t) > A_R^{M*}(t) > A_R^{D*}(t); \tag{22}$$

$$\begin{gathered} S_M^{C*}(t) > S_M^{M*}(t) = S_M^{D*}(t); S_R^{C*}(t) > S_R^{M*}(t) = S_R^{D*}(t); \\ G^{C*}(t) > G^{D*}(t) > G^{M*}(t);\ \xi_1^{M*} = \xi_2^{D*}. \end{gathered} \tag{23}$$

Deduction 4 indicates that: compared to the other two decisions, under the Centralized Decision "C", they are the optimal ones regardless of the service or advertising strategy of the supply chain members, or the goodwill of the brand owner. The brand owner's advertising strategy $A_M(t)$ is better under the Bilateral Cost-Sharing Decision "D", which indicates that the bilateral cost-sharing decision can motivate the brand owner to improve its advertising level. The retailer's advertising strategy $A_R(t)$ is better under the brand cost-sharing decision "M", which can encourage the retailer to improve its advertising level. No matter whether it is the brand cost-sharing decision or the bilateral cost-sharing decision, the brand owner's cost-sharing ratios on the retailer's service are consistent, i.e., $\xi_1^{M*} = \xi_2^{D*}$. Meanwhile, the service investment levels of the brand owner and the retailer are also compatible, i.e., $S_M^{M*}(t) = S_M^{D*}(t), S_R^{M*}(t) = S_R^{D*}(t)$.

4.2. Profit Comparison under Three Decisions

The impact of time delay and bidirectional free-riding phenomenon on the profit of the supply chain and the supply chain members is observed through analysis of the arithmetic examples. It is assumed that the parameters included in the above model are as follows: $\delta = 0.01, \rho = 0.05, \theta = 0.4, \gamma_1 = 0.5, \gamma_2 = 0.6, \phi_1 = 0.5, \phi_2 = 0.6, \lambda_1 = 0.3, \lambda_2 = 0.4,$ $P_1 = 3, P_2 = 5.5, \omega = 2, G_0 = 0, \alpha = 1, \beta_1 = 0.6, \beta_2 = 0.4, k_1 = k_2 = 1, \eta_1 = \eta_2 = 1.$

4.2.1. Impact of Time Delay on the Profit of the Supply Chain and Supply Chain Members

Under the above parameter settings, the value range of time delay takes values in the range $d_1, d_2 \in (0, 30)$. This paper analyzes the impact of time delay on the supply chain, the brand owner and the retailer profit under different decisions.

Figure 2 indicates the impact of the change of delay time d_1, d_2 on the profit of the supply chain under different decisions. With the increase in the delay time d_1, d_2, the supply chain profit under the centralized decision shows a downward trend, while it shows an upward trend under the other two decisions. Therefore, when the delay time d_1, d_2 is comparatively short, the supply chain profit under the centralized decision is the highest. When the delay time d_1, d_2 is comparatively long, the supply chain profit under the bilateral cost-sharing decision becomes the optimal value. In the delay time d_1, d_2 range, the supply chain profit under a bilateral cost-sharing decision is always higher than that under the brand cost-sharing decision.

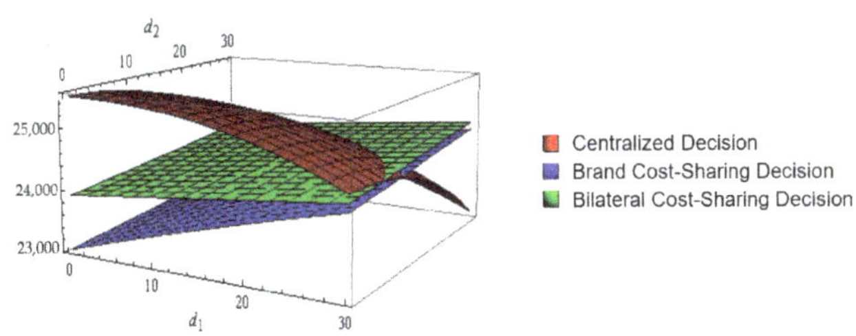

Figure 2. Impact of Time Delay on the Supply Chain Profit.

As can be seen in Figure 3, the increase in delay time d_1, d_2 will have different impacts on the profit of the brand owner. With the growth of the delay time d_1 for national advertising, the brand owners' advertising investment level is increasing. In contrast, the impact on brand goodwill and market demand is not immediately apparent, resulting in a decreasing trend in the brand owners' profits.

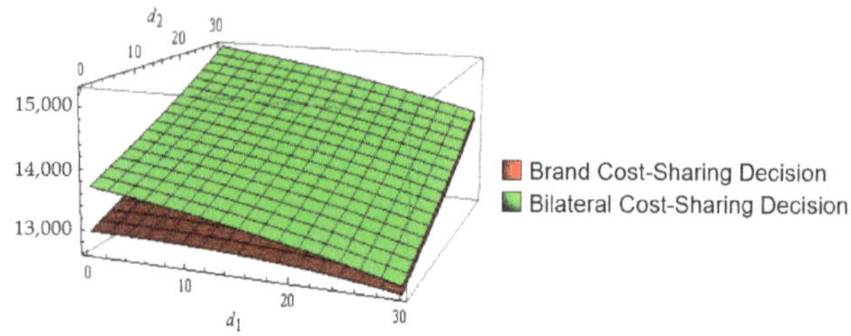

Figure 3. Impact of Time Delay on the Brand Owner's Profit.

As the service delay time d_2 of the retailer increases, the brand owner benefits from the improvement of the retailer's service level. According to Deductions 2 and 3, the brands' share of the retailer service cost ratio ξ_2 decreases at this moment, so the profit of the brand owner increases continuously.

As can be seen in Figure 4, similar to the case of the brand owner, the increase in the delay time d_1, d_2 will have different impacts on the retailer's profit. What is different is that with the rise in the national advertising delay time d_1, the retailer benefits from the improvement of the advertising level of the brand owner. Meanwhile, according to Deduction 3, the brand owner's advertising cost ratio ξ_1 decreases at this moment, increasing the retailer's profit. As the retailer's service delay time d_2 increases, the retailer's cost of providing the service increases while the brand owner's share of the service cost ratio ξ_2 decreases, decreasing the retailer's profit.

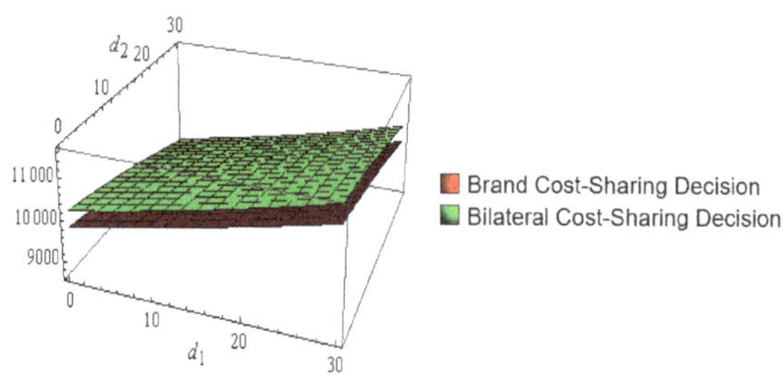

Figure 4. Impact of Time Delay on the Retailer's Profit.

Comparing Figure 3 with Figure 4, the bilateral cost-sharing decision can promote the profits of the brand owner and retailer at the same time and realize the Pareto improvement. Therefore, when there is a time delay, the supply chain members should prefer to implement the bilateral cost-sharing decision.

4.2.2. Impact of Bidirectional Free-Riding on the Profit of Supply Chain and Supply Chain Members

The webrooming effect and showrooming effect coefficients λ_1, λ_2 will impact the service decisions of the brand owner and the retailer, respectively, affecting the supply chain's profit. From Deduction 2, under the condition of $P_1 > (\omega + P_2)/2$, that the showrooming effect λ_2 is weakening, the brand owner reduces its cost-sharing ratio ξ_2 and the retailer reduces its service level. Therefore, on the premise that other parameters are unchanged, we have chosen the two different conditions of $P_1 = 3 < (\omega + P_2)/2$ and

$P_1 = 4 > (\omega + P_2)/2$ to analyze the impact of the changes of the webrooming effect and showrooming effect coefficients λ_1, λ_2 on the profits of the supply chain, the brand owner and the retailer. Set $d_1 = d_2 = 20$ and $P^* = (\omega + P_2)/2$.

From Figures 5 and 6, the changes in the webrooming effect and showrooming effect coefficients λ_1, λ_2 will have different impacts on the supply chain profit. Regardless of $P_1 < P^*$ or $P_1 > P^*$, the supply chain profit is about the increasing function of the webrooming effect coefficient λ_1 and the decreasing function of the showrooming effect coefficient λ_2; the change in the showrooming effect coefficient λ_2 will have a more significant impact. With the decrease in λ_1 and the increase in λ_2, the supply chain profit under the centralized decision tends to decrease gradually than under the other two decisions. Under the condition of $P_1 < P^*$, the supply chain profit under the bilateral cost-sharing decision is higher than that under the brand cost-sharing decision; and under the condition of $P_1 > P^*$, the supply chain profit under the brand cost-sharing decision is higher than that under the bilateral cost-sharing decision.

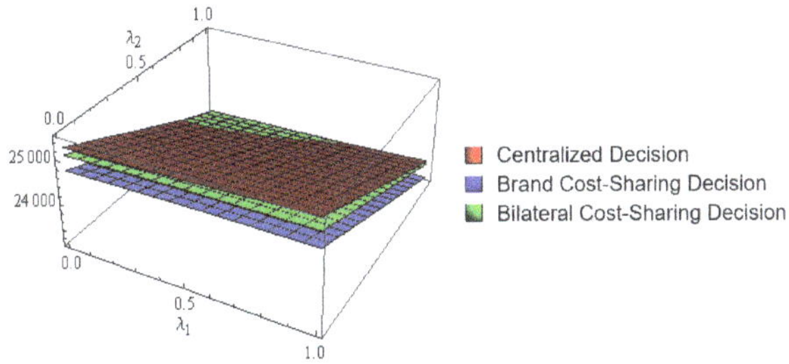

Figure 5. Impact of λ_i on the Supply Chain Profit When $P_1 < P^*$.

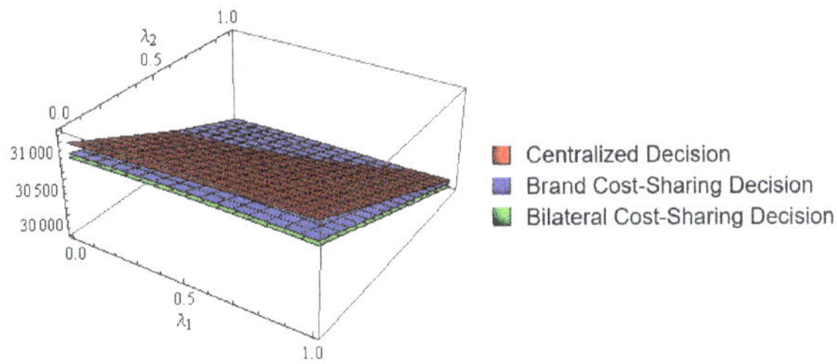

Figure 6. Impact of λ_i on the Supply Chain Profit When $P_1 > P^*$.

Figures 7 and 8 have reflected the impact of the change in the webrooming effect and showrooming effect coefficients λ_1, λ_2 on the brand owner's profit under different value ranges of P_1. Under the condition of $P_1 < P^*$, the difference in the coefficient λ_1, λ_2 on the brand owner's profit is the same when the profit of the brand owner is a decreasing function in λ_1, λ_2 and the brand owner's profit will be higher under the bilateral cost-sharing decision.

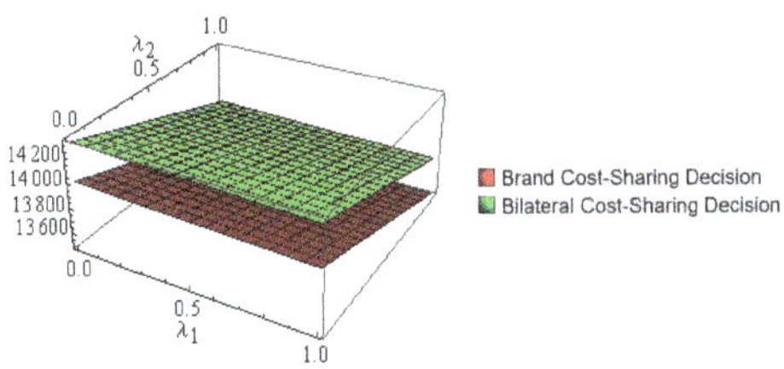

Figure 7. Impact of λ_i on the Brand Owner's Profit When $P_1 < P^*$.

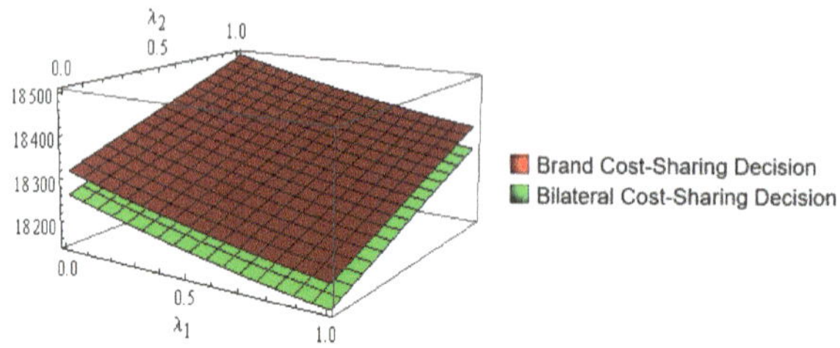

Figure 8. Impact of λ_i on the Brand Owner's Profit When $P_1 > P^*$.

Under the condition of $P_1 > P^*$, changes in λ_1, λ_2 will have different impacts on the profit of the brand owner. Under the condition that λ_1 remains unchanged, the profit of the brand owner will increase with the increase in λ_2; and under the condition that λ_2 remains unchanged, the profit of the brand owner will decrease with the increase in λ_1, and the profit of the brand owner will be higher under the brand cost-sharing decision.

Figures 9 and 10 illustrate the impacts of the changes of the webrooming effect and showrooming effect coefficients λ_1, λ_2 on the retailer's profit. It is clear from the figures that no matter whether the condition of $P_1 < P^*$ or $P_1 > P^*$, the result from the changes in the webrooming effect and showrooming effect coefficients λ_1, λ_2 on the retailer's profit are different. With the growth in λ_1, the retailer's profit shows a slight increase; while with the growth in λ_2, the retailer's profit shows a rapid decrease. Under both conditions, the retailer's profit under the bilateral cost-sharing decision is the higher one.

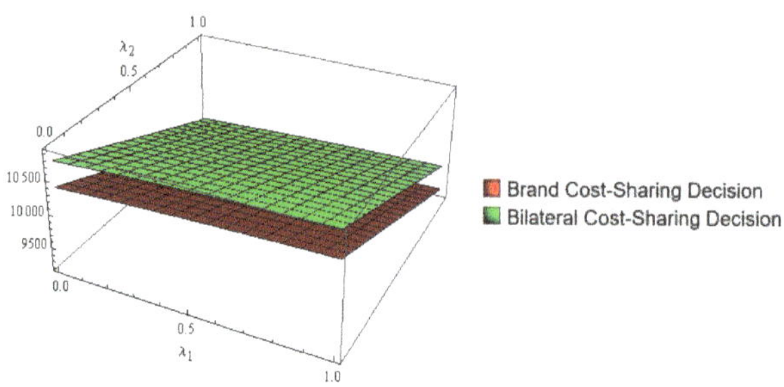

Figure 9. Impact of λ_i on the Retailer's Profit When $P_1 < P^*$.

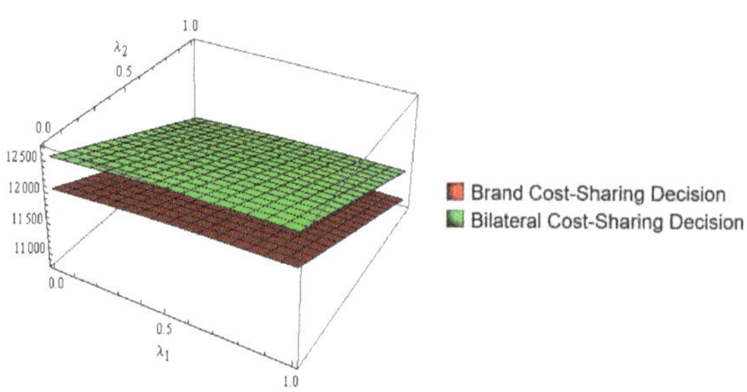

Figure 10. Impact of λ_i on the Retailer's Profit When $P_1 > P^*$.

5. Conclusions

5.1. Research Conclusions

Channel convergence is the focal issue of the O2O supply chain, and inter-channel advertising and service cooperation is the leading research area of channel convergence. This paper considered the impact of time delay and bidirectional free-riding on inter-channel service and advertising cooperation strategies. We adopted differential game theory to solve the optimal advertising and service decisions of the brand owners and retailers under a centralized decision, the brand owners' cost-sharing decision, and bilateral cost-sharing decision. We also explored the effects of delay time, the bidirectional free-riding coefficient, price and other factors on cooperative advertising and service decisions. The analysis leads to the following conclusions:

(1) Among these three decisions, the supply chain members' advertising strategy, service strategy and brand goodwill are optimal under a centralized decision. However, the supply chain profit is not necessarily optimal due to the time delay and bidirectional free-riding phenomenon. The supply chain profit depends on the delay time length and the bidirectional free-riding coefficient size.
(2) Under the two kinds of cost-sharing decisions, the supply chain members and the brand owner's service decision and the brand owner's cost-sharing of the retailer's service are the same. The brand cost-sharing decision can promote the advertising level of the retailer, while the bilateral cost-sharing decision can promote the brand owner's national advertising and the brand goodwill. From the perspective of the supply chain members' profit, the bilateral cost-sharing decision can promote the profits of the brand owner and the retailer at the same time; regardless of the length of delay time. Irrespective of the bidirectional free-riding coefficient size, the bilateral cost-sharing decision can still promote the profit of the retailer; however, whether this decision can promote the profit of the brand owner is affected by the retail price of the online channel.
(3) As the delay time grows, the brand owners' national advertising level and the retailer's experience service level continue to improve, which causes the brand owner's and retailer's investment costs to continue to increase, while it is difficult to achieve the expected effect in the short term, raising the enterprise's operational risks. The bilateral cost-sharing ratio is a decreasing function of the delay time, which can coordinate the impact of time delay on the brand owner's advertising and retailer's service decisions to a certain extent.
(4) Under a centralized decision, brands are willing to cooperate fully and improve their service levels even if the webrooming phenomenon is intensified to increase the profitability of the whole supply chain. Under the two cost-sharing decisions, the brand owner is willing to cooperate partially. The brand owner can moderate the impact of the showrooming phenomenon on retailers' service strategies by setting retail prices and sharing retail service costs proportionally in the online channel.
(5) The supply chain members' optimal service and advertising strategy is an increasing function of the retail prices of the online and offline channels. In contrast, the impact of the wholesale price on the optimal strategy of the supply chain members is affected by the decision mode. Under the brand cost-sharing decision, by setting a comparatively high wholesale price, the brand owner can adjust the impact of the time delay on the service strategy of the retailer and promote the service and advertising levels of the supply chain members. Under the bilateral cost-sharing decision, a higher wholesale price enhances the service level of supply chain members; however, it reduces the advertising level of supply chain members and exacerbates the impact of the delay phenomenon on their national advertising strategies.

This study can further explore the following aspects of the following elements. First, this study only considers the self-run online channels of brand owners and does not consider the cooperative relationship with e-commerce platforms. At present, the cooperation between brands and e-commerce platforms mainly includes two modes: resale and commission, which will become the direction for the improvement of cooperation subjects in the future of this study. Secondly, this paper only considers the pre-sale advertising and service cooperation of the O2O supply chain and can further evaluate the collaboration of after-sale delivery and return.

5.2. Practical Implications

The above research results provide some management suggestions for the service and advertising cooperation of O2O supply chain members.

(1) The research in this paper shows that the bilateral cost-sharing contract is a compelling long-term strategy to solve the adverse effects of delay and two-way free-riding. The bilateral cost-sharing contract can increase the profits of each node enterprise in the O2O supply chain, thus improving the performance of the whole Pareto.
(2) Brands should reasonably set wholesale prices, online direct-selling prices and cost-sharing ratio, which can effectively promote retailers' investment in offline experience services, to achieve a win-win situation and improve the entire supply chain performance.
(3) From the perspective of long-term development, the key for retailers to deal with showrooming behavior is to improve their brand premium ability, highlight the experience services provided by physical stores to consumers, and create the uniqueness of physical sales.

Author Contributions: Conceptualization, J.Z. and Q.X.; methodology, J.Z.; software, J.Z.; validation, J.Z. and Q.X.; formal analysis, J.Z. and Q.X.; investigation, J.Z.; resources, J.Z.; data curation, J.Z.; writing—original draft preparation, J.Z. and Q.X.; writing—review and editing, J.Z. and Q.X.; visualization, Q.X.; supervision, J.Z.; project administration, Q.X.; funding acquisition, Q.X. All authors have read and agreed to the published version of the manuscript.

Funding: This research was supported by the National Social Science Foundation of China (grant number: 21BGL014).

Data Availability Statement: The data used to support the findings of this study are available from the corresponding authors upon request.

Acknowledgments: Thanks to Xu Q. for her participation in this paper.

Conflicts of Interest: The authors declare that there are no conflicts of interest regarding the publication of this paper.

References

1. Sahu, K.C.; Khan, M.N.; Das Gupta, K. Determinants of Webrooming and Showrooming Behavior: A Systematic Literature Review. *J. Internet Commer.* **2021**, *20*, 137–166. [CrossRef]
2. Brynjolfsson, E.; Hu, Y.J.; Rahman, M.S. Competing in the age of omnichannel retailing. *MIT Sloan Manag. Rev.* **2013**, *54*, 23–29.
3. Bell, D.R.; Gallino, S.; Moreno, A. How to win in an omnichannel world. *MIT Sloan Manag. Rev.* **2014**, *56*, 45–53.
4. Neslin, S.A.; Jerath, K.; Bodapati, A.; Bradlow, E.T.; Deighton, J.; Gensler, S.; Lee, L.; Montaguti, E.; Telang, R.; Venkatesan, R.; et al. The interrelationships between brand and channel choice. *Mark. Lett.* **2014**, *25*, 319–330. [CrossRef]
5. Flavián, C.; Gurrea, R.; Orús, C. Choice confidence in the webrooming purchase process: The impact of online positive reviews and the motivation to touch. *J. Consum. Behav.* **2016**, *15*, 459–476. [CrossRef]
6. Flavián, C.; Gurrea, R.; Orús, C. Combining channels to make smart purchases: The role of webrooming and showrooming. *J. Retail. Consum. Serv.* **2020**, *52*, 101923. [CrossRef]
7. Van Baal, S.; Dach, C. Free riding and customer retention across retailers' channels. *J. Interact. Mark.* **2005**, *19*, 75–85. [CrossRef]
8. Jørgensen, S.; Sigué, S.P.; Zaccour, G. Dynamic cooperative advertising in a channel. *J. Retail.* **2000**, *76*, 71–92. [CrossRef]
9. Karray, S.; Martín-Herrán, G.; Sigué, S.P. Cooperative advertising for competing manufacturers: The impact of long-term promotional effects. *Int. J. Prod. Econ.* **2017**, *184*, 21–32. [CrossRef]
10. Aust, G.; Buscher, U. Vertical cooperative advertising and pricing decisions in a manufacturer–retailer supply chain: A game-theoretic approach. *Eur. J. Oper. Res.* **2012**, *223*, 473–482. [CrossRef]
11. Li, X.; Li, Y.; Cao, W. Cooperative advertising models in O2O supply chains. *Int. J. Prod. Econ.* **2019**, *215*, 144–152. [CrossRef]
12. Li, D.; Yu, H.; Tee, K.P.; Wu, Y.; Ge, S.S.; Lee, T.H. On time-synchronized stability and control. *IEEE Trans. Syst. Man Cybern. Syst.* **2021**, *52*, 2450–2463. [CrossRef]
13. Thiruchelvam, S.; Ismail, M.F.; Ghazali, A.; Mustapha, K.N.; Norkhair, F.F.; Yahya, N.; Isa, A.A.M.; Muda, Z.C. Development of humanitraian supply chain performance conceptual framework in creating resilient logistics network. *Malays. J. Geosci.* **2018**, *2*, 30–33. [CrossRef]
14. Yan, L.; Yin-He, S.; Qian, Y.; Zhi-Yu, S.; Chun-Zi, W.; Zi-Yun, L. Method of reaching consensus on probability of food safety based on the integration of finite credible data on block chain. *IEEE Access* **2021**, *9*, 123764–123776. [CrossRef]

15. Aust, G.; Buscher, U. Cooperative advertising models in supply chain management: A review. *Eur. J. Oper. Res.* **2014**, *234*, 31–63. [CrossRef]
16. Jørgensen, S.; Zaccour, G. A survey of game-theoretic models of cooperative advertising. *Eur. J. Oper. Res.* **2014**, *237*, 1–14. [CrossRef]
17. Zhao, L.; Zhang, J.; Xie, J. Impact of demand price elasticity on advantages of cooperative advertising in a two-tier supply chain. *Int. J. Prod. Res.* **2015**, *54*, 2541–2551. [CrossRef]
18. Martín-Herrán, G.; Sigué, S.P. An integrative framework of cooperative advertising: Should manufacturers continuously support retailer advertising? *J. Bus. Res.* **2017**, *70*, 67–73. [CrossRef]
19. Chutani, A.; Sethi, S.P. Dynamic cooperative advertising under manufacturer and retailer level competition. *Eur. J. Oper. Res.* **2018**, *268*, 635–652. [CrossRef]
20. Chatterjee, P.; Karray, S.; Sigué, S.P. Cooperative advertising programs: Are accrual constraints necessary? *Int. Trans. Oper. Res.* **2017**, *26*, 2230–2247. [CrossRef]
21. Xu, Q.; Fu, G.; Fan, D. Service sharing, profit mode and coordination mechanism in the Online-to-Offline retail market. *Econ. Model.* **2020**, *91*, 659–669. [CrossRef]
22. Li, M.; Zhang, X.; Dan, B. Cooperative advertising and pricing in an O2O supply chain with buy-online-and-pick-up-in-store. *Int. Trans. Oper. Res.* **2020**, *28*, 2033–2054. [CrossRef]
23. Fang, H.; Fang, F.; Hu, Q.; Wan, Y. Supply Chain Management: A Review and Bibliometric Analysis. *Processes* **2022**, *10*, 1681. [CrossRef]
24. Salas-Navarro, K.; Serrano-Pájaro, P.; Ospina-Mateus, H.; Zamora-Musa, R. Inventory Models in a Sustainable Supply Chain: A Bibliometric Analysis. *Sustainability* **2022**, *14*, 6003. [CrossRef]
25. Acevedo-Chedid, J.; Salas-Navarro, K.; Ospina-Mateus, H.; Villalobo, A.; Sana, S.S. Production System in a Collaborative Supply Chain Considering Deterioration. *Int. J. Appl. Comput. Math.* **2021**, *7*, 69. [CrossRef]
26. Telser, L.G. Why should manufacturers want fair trade? *J. Law Econ.* **1960**, *3*, 86–105. [CrossRef]
27. Singley, R.B.; Williams, M.R. Free Riding in Retail Stores: An Investigation of Its Perceived Prevalence and Costs. *J. Mark. Theory Pract.* **1995**, *3*, 64–74. [CrossRef]
28. Wu, D.; Ray, G.; Geng, X.; Whinston, A. Implications of reduced search cost and free riding in e-commerce. *Mark. Sci.* **2004**, *23*, 255–262. [CrossRef]
29. Guan, X.; Liu, B.; Chen, Y.; Wang, H. Inducing Supply Chain Transparency through Supplier Encroachment. *Prod. Oper. Manag.* **2019**, *29*, 725–749. [CrossRef]
30. Shin, J. How Does Free Riding on Customer Service Affect Competition? *Mark. Sci.* **2007**, *26*, 488–503. [CrossRef]
31. Carlton, D.W.; Chevalier, J.A. Free Riding and Sales Strategies for the Internet. *J. Ind. Econ.* **2001**, *49*, 441–461. [CrossRef]
32. Mittelstaedt, R.A. Sasquatch, the Abominable Snowman, Free Riders and other Elusive Beings. *J. Macromarketing* **1986**, *6*, 25–35. [CrossRef]
33. Umit Kucuk, S.; Maddux, R.C. The role of the Internet on free-riding: An exploratory study of the wallpaper industry. *J. Retail. Consum. Serv.* **2010**, *17*, 313–320. [CrossRef]
34. Aubrey, C.; Judge, D. Re-imagine retail: Why store innovation is key to a brand's growth in the 'new normal', digitally-connected and transparent world. *J. Brand Strategy* **2012**, *1*, 31–39.
35. Zhang, L.; Wang, J. Coordination of the traditional and the online channels for a short-life-cycle product. *Eur. J. Oper. Res.* **2017**, *258*, 639–651. [CrossRef]
36. Basak, S.; Basu, P.; Avittathur, B.; Sikdar, S. A game theoretic analysis of multichannel retail in the context of "showrooming". *Decis. Support Syst.* **2017**, *103*, 34–45. [CrossRef]
37. Mehra, A.; Kumar, S.; Raju, J.S. Competitive strategies for brick-and-mortar stores to counter "showrooming". *Manag. Sci.* **2018**, *64*, 3076–3090. [CrossRef]
38. Wang, Y.; Gerchak, Y. Supply chain coordination when demand is shelf-space dependent. *Manuf. Serv. Oper. Manag.* **2001**, *3*, 82–87. [CrossRef]
39. Wang, Z.Q.; Yang, D.F.; Ran, L. Relationship between webrooming and consumer quality expectations. *J. Manag. Sci. China* **2021**, *24*, 71–88.
40. Yuan, W.W. Pricing decision of dual-channel competition in presence of webrooming effect. *Sci.-Technol. Manag.* **2021**, *23*, 70–80.
41. Ma, D.Q.; Wang, X.Q.; Hu J, S. Sales format selection for platform-based supply chain considering consumers' webrooming phenomenon in multi-channel retailing. *Chin. J. Manag. Sci.* **2022**, 1–12. [CrossRef]
42. Liu, C.; Dan, Y.; Dan, B.; Xu, G. Cooperative strategy for a dual-channel supply chain with the influence of free-riding customers. *Electron. Commer. Res. Appl.* **2020**, *43*, 101001. [CrossRef]
43. Luo, M.L.; Li, G.; Zhang, W.J. The bidirectional free-riding in a dual-channel supply chain. *J. Syst. Manag.* **2014**, *23*, 314–323+338.
44. Li, J.B.; Zhu, M.P.; Dai, B. Optimal pricing and sales effort decisions in a dual-channel supply chain in case of bidirectional free riding. *Syst. Eng.-Theory Pract.* **2016**, *36*, 3046–3058.
45. Gong, Y.H.; He, Y.; Zhang, M. Decision-making strategy for service effort of dual-channel retailers under condition of bidirectional free riding. *Stat. Decis.* **2020**, *36*, 162–166.
46. Herhausen, D.; Binder, J.; Schoegel, M.; Herrmann, A. Integrating Bricks with Clicks: Retailer-Level and Channel-Level Outcomes of Online–Offline Channel Integration. *J. Retail.* **2015**, *91*, 309–325. [CrossRef]

47. Heitz-Spahn, S. Cross-channel free-riding consumer behavior in a multichannel environment: An investigation of shopping motives, sociodemographics and product categories. *J. Retail. Consum. Serv.* **2013**, *20*, 570–578. [CrossRef]
48. Huré, E.; Picot-Coupey, K.; Ackermann, C.-L. Understanding omni-channel shopping value: A mixed-method study. *J. Retail. Consum. Serv.* **2017**, *39*, 314–330. [CrossRef]
49. Zheng, J.; Xu, Q. Service Sharing Decisions between Channels considering Bidirectional Free Riding Based on a Dual-Equilibrium Linkage Algorithm. *Comput. Intell. Neurosci.* **2022**, *2022*, 1540820. [CrossRef]
50. Nerlove, M.; Arrow, K.J. Optimal advertising policy under dynamic conditions. *Economica* **1962**, *29*, 129. [CrossRef]
51. Zhang, S.P.; Zhang, S.Y. Dynamic cooperative advertising strategies based on differential games in a supply chain. *Control. Decis.* **2006**, *21*, 153–157+162.
52. Gao, Z.; Kong, D.; Gao, C. Modeling and control of complex dynamic systems: Applied mathematical aspects. *J. Appl. Math.* **2012**, *2012*, 869792. [CrossRef]
53. Zheng, J.; Zhang, Q.; Xu, Q.; Xu, F.; Shi, V. Synchronization of a supply chain model with four chaotic attractors. *Discret. Dyn. Nat. Soc.* **2022**, *2022*, 6390456. [CrossRef]
54. Gao, Z.; Chen, M.Z.Q.; Zhang, D. Special issue on "Advances in Condition Monitoring, Optimization and Control for Complex Industrial Processes". *Processes* **2021**, *9*, 664. [CrossRef]
55. Berkowitz, D.; Allaway, A.; D'Souza, G. Estimating differential lag effects for multiple media across multiple stores. *J. Advert.* **2001**, *30*, 59–65. [CrossRef]
56. Baack, D.W.; Wilson, R.T.; Till, B.D. Creativity and memory effects: Recall, recognition, and an exploration of nontraditional media. *J. Advert.* **2008**, *37*, 85–94. [CrossRef]
57. Chen, D.Y.; Yu, H.; Hou, L. Dynamic cooperative advertising strategies in a supply chain with lagged effect. *J. Manag. Sci. China* **2017**, *20*, 25–35.
58. Yu, H.; Chen, D.Y.; Huang, C.L. Advertising strategies for supply chain with lagged and memory effects. *Control Decis.* **2018**, *33*, 1871–1878.
59. Chen, G.P.; Zhang, X.M.; Xiao, J. Coordination model for cooperative advertising between manufacturers and retailers in dual-channel supply chain. *J. Syst. Manag.* **2017**, *26*, 1168–1175.
60. Cao, D.Y.; Xiao, J.; Zhang, X.M. Cooperative advertising strategy in dual-channel supply chain with lagged effect. *Comput. Integr. Manuf. Syst.* **2020**, *26*, 2253–2265.
61. Wang, Y.B.; Shu, L.Y. Cooperative advertising models in O2O supply chains with fairness concerns. In Proceedings of the IEEE 2019 16th International Conference on Service Systems and Service Management (ICSSSM), Shenzhen, China, 13–15 July 2019; pp. 1–6. [CrossRef]

processes

Article

Music Generation System for Adversarial Training Based on Deep Learning

Jun Min [1], Zhaoqi Liu [1], Lei Wang [1,*], Dongyang Li [1], Maoqing Zhang [1] and Yantai Huang [2]

1 College of Electronics and Information Engineering, TongJi University, Shanghai 201804, China
2 College of Automation and Electrical Engineering, Zhejiang University of Science and Technology, Hangzhou 310023, China
* Correspondence: wanglei@tongji.edu.cn

Abstract: With the rapid development of artificial intelligence, the application of this new technology to music generation has attracted more attention and achieved gratifying results. This study proposes a method for combining the transformer deep-learning model with generative adversarial networks (GANs) to explore a more competitive music generation algorithm. The idea of text generation in natural language processing (NLP) was used for reference, and a unique loss function was designed for the model. The training process solves the problem of a nondifferentiable gradient in generating music. Compared with the problem that LSTM cannot deal with long sequence music, the model based on transformer and GANs can extract the relationship in the notes of long sequence music samples and learn the rules of music composition well. At the same time, the optimized transformer and GANs model has obvious advantages in the complexity of the system and the accuracy of generating notes.

Keywords: artificial intelligence (AI); music generation; natural language processing; transformer; GANs

Citation: Min, J.; Liu, Z.; Wang, L.; Li, D.; Zhang, M.; Huang, Y. Music Generation System for Adversarial Training Based on Deep Learning. *Processes* **2022**, *10*, 2515. https://doi.org/10.3390/pr10122515

Academic Editors: Zhiwei Gao and Xiong Luo

Received: 10 October 2022
Accepted: 23 November 2022
Published: 27 November 2022

1. Introduction

Computer-based music generation can be traced as far back as the 1950s. The Russian mathematician Markov proposed the Markov model in 1906, which processed and predicted serial data. In 1955, Olson et al. [1] proposed using the Markov model to compose music for the first time, utilizing intelligent computer algorithms that calculate probabilities for the next note. In his research, Olson described the principal modules needed for analog electronic music synthesizers, providing a creative method through formal methods. Similar algorithms based on this approach are directed by a control program that must follow mathematical instructions in a fixed order. By the 1980s, various models had appeared offering grammar-based [2] and rule-based music generation [3]. Steedman et al. [2] referred to the grammatical rules of language-generation systems to build a set of music generation systems with pure grammar and rules based on music composition theory, such as harmony, polyphony, musical form, and orchestration [3,4]. These models could generate different styles of music but lacked the ability of generalization, so the rule-based definitions had to be completed manually. Compared with deep-learning models, these methods suffered from poor function and universality. However, this basic research provided ideas for optimizing a deep learning algorithm [5].

With the advancement of deep learning techniques, many researchers have recently tried to apply deep neural networks to creating music. These deep learning techniques usually have neural network architectures, which have performed well in computer vision and natural language programming (NLP). Generating music and other artistic content using deep learning is a growing area of research. Generating structured music, analyzing the quality of generated music, and building an interaction model are the extended problems that need to be solved. Some deep learning methods are autoregressive [5]. This new model tries to generate longer sequences by obtaining the information of past time steps, which is similar to music generation by humans. Convolutional neural networks (CNNs) [6] are a basic model commonly used in music generation. In the late 1990s, the CNN prototype [7] was proposed, but it was not until AlexNet [8] in 2013 that CNNs as we know them were recognized. Based on this, Google established the deep mind artificial intelligence laboratory in London to develop WaveNet [9], one of the most successful CNN music generation applications. The recurrent neural network (RNN) has become the most popular processing model for typical serialized data such as music.

Considering the similarity between music and language, some language generation models can be converted to generating music. This approach represents note embedding by vectors. Embedding [10] was a concept initially used in NLP. Its principle was to obtain the vector representation of each word, note, or chord through training a neural network with a corpus and then use the result as the input to the network for downstream tasks. In music, each major or minor chord has three notes as the tonic of the mode. In other words, the whole song surrounds the tonic and its related chords [11,12]. In addition to this obvious relationship between notes, there are also invisible relationships in music [13], corresponding to anaphora in language. This is why it is important to choose a transformer as the core model for music generation. In addition, GANs are widely used in imaging, language, and music by training the discriminator and generator in the network to optimize them, ensuring the final authenticity of the generated results [14]. GANs are now coming into use to generate music sequences [15]. A critical difficulty to consider is that the generated sequence must be fed into a discriminator because of the network architecture of the generating countermeasure network. Therefore, this study proposes a new model structure that combines a transformer [16] and GANs [17] to create music. The study also presents a unique loss function to enable the system to learn and update the parameters in the two gradient descent directions of "real music" and the target sequence [18].

The contributions of this study are as follows:

- Establishing a new music generation system that combines the transformer and GANs. Proposing a unique loss function for the proposed model to learn from the descending direction of "real music" and the target sequence and to update the parameters over time.
- Improving the input and output structures of the discriminator and the generator and solving the problems of gradient non-differentiability and mode collapse in the discriminator.
- Applying the vocabulary matching method to perfect the intricate melody generated in the time domain and generate a real and controllable long-term structure.
- Presenting a relatively objective suggestion to evaluate music based on Euler's music evaluation mechanism.

The rest of the study is organized as follows: Section 2 introduces related work on music generation. Section 3 puts forward the method proposed in this study and introduces the construction and training process of the model. Section 4 presents the results of the experiment.

2. Related Work

Long short-term memory (LSTM) developed rapidly and has been applied extensively in music processing and the music generation field. Through LSTM, automatically generated music can achieve high-quality, high-fidelity, and high-definition music effects. Ycart et al. [19] and Sheykhivand et al. [20] both used LSTM as the cornerstone neural network to realize music generation. Borodin et al. [21] proposed a multi-channel data-processing method for a chord using the many-hot encoding method. The input was optimized from single-note encoding to a multi-dimensional representation vector. The next note combination was predicted by LSTM, which enriched the production. Chen et al. [22] combined LSTM with chaos theory to optimize the tone shift in music without deformation, reduce the amount of calculation, and optimize the training efficiency. Lehner et al. [23]

combined LSTM with a restricted Boltzmann machine (RBM) [24]. These techniques and other probabilistic methods are combined with deep neural networks to help people better make music. However, LSTM methods have failed to generate long-term sequences. The authors of [25] proposed combining biaxial LSTM with GANs, which improved music quality significantly over ordinary LSTM. Many experiments have found that these models only allow the network to learn the relationship between note characteristics from actual music data; they did not learn harmony from the music as a whole or the rules composers need to follow.

Language and music have similar characteristics [26]. In natural language processing, Google put forward the transformer [27] in 2017, which used a self-attention mechanism as its main component and became a state-of-the art (SOTA) method in many projects. The transformer was a typical encoder-decoder model (i.e., a sequence-to-sequence (seq2seq) model) [28] mainly used for generating scenarios such as question-answering systems and machine translation. Given the similarity between language and music, Google applied the transformer to music generation in 2018 [29]. Because of its low memory density, the transformer could generate longer, more coherent music. However, the music transformer was imperfect and had too many redundant and sparse musical notes. Based on the transformer decoder, OpenAI proposed the second-generation Generative Pre-Training (GPT) model [30] in 2019. GPT-3 was considered a dangerous machine learning model due to its high intelligence. In April of the same year, the music generation system MuseNet appeared, based on the GPT-2 [31]. It could generate works of any genre and style, and even any composer and pianist. The generated music could be confused with the official versions. In 2020, Jin et al. [32] proposed a new scheme based on combining the transformer and GPT. Since then, artificial intelligence composition has reached a more mature stage. However, the quality of music generated by these methods has not reached an acceptable level because the neural network cannot understand the complexities of the language of music. The information in the notes needs to be transmitted to the system as part of the input, for example by marking. Many experiments have found that these models only allowed a network to learn the relationship between note characteristics from music data; they did not follow the rules of the music and composition.

3. Proposed Method

The first problem is generating note sequences and exploring the conversion relationship. In note conversion, the most commonly used method is to convert notes into one-hot encoding, but because the sample involves many notes, the digital matrix generated by one-hot encoding is sparse. Although one-hot encoding can reduce the dimensionality of the input model, the predicted label probability distribution cannot be directly transformed accurately into notes, so the sorted note sequence is digitally coded according to the occurrence order of the notes. Although this encoding method ignores the original relationship between notes (e.g., the relationship between tone C and chord C-E-G), the model structure of the transformer handles the note ID input by first going through a trainable embedding layer in the system to find a reasonable mapping and converting the note ID into a vector containing the note relationship. In addition, owing to the different lengths of music singles, ranging from dozens of notes to thousands of notes, the seq_length parameter is set to make the input sequence and output sequence into two equal continuous sequences. For example, the first ten notes of a song are converted as follows:

[6.11, E-5, E5, F#4, B3, E-5, B4, 2.6, E-5, 11.3.6].

Among them, "6.11" is a polyphonic chord divided by a period, and the rest are single notes. There are more than 1500 different notes or chords in 2786 samples of music in the GiantMIDI-Piano dataset. To reduce the difficulty of model prediction, all notes are sorted according to their frequency of occurrence, and the notes that occurred less frequently than the set threshold are marked [unk]. The process for note feature extraction is shown in Figure 1.

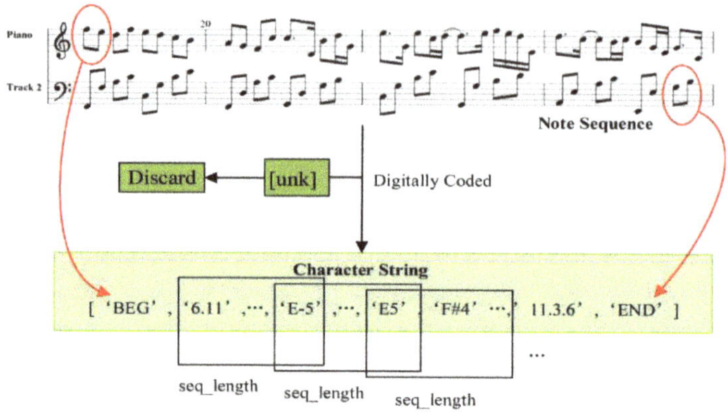

Figure 1. The process for note feature extraction.

In the recommended method, the LSTM structure in the traditional sequence-to-sequence (seq2seq) structure is replaced by relying on the attention mechanism to capture the relationship between the input and output notes. Using the attention mechanism, which was proposed by the team of Bengio in 2014 [33], overcomes the shortcomings of sequential calculation of the cyclic neural network (CNN). The system has better parallelism and improves the training efficiency.

In position encoding, different calculation methods are used for note units in odd and even positions. For note units in even positions, Equation (1) is used:

$$PE_{(pos,\ 2i)} = \sin\left(\frac{pos}{10000^{\frac{2i}{d_{model}}}}\right). \tag{1}$$

For note units in odd positions, Equation (2) is used:

$$PE_{(pos,\ 2i+1)} = \cos\left(\frac{pos}{10000^{\frac{2i}{d_{model}}}}\right) \tag{2}$$

where *pos* represents the note, and *i* is a representation dimension, with each dimension corresponding to a sine wave. The position encoding of this method can make the unit PE_{pos+k} at any position be represented by a linear function of the first unit PE_{pos}, to obtain the relative position of the whole sequence. The representation method of a trigonometric function is used to make the model infer a sequence with a longer length than that encountered in the training process. Because of the periodicity of sine and cosine, for the fixed-length note sequences, the value of *PE* in the position *pos+n* can be expressed as a linear change. In this way, it will be convenient for the model to learn the relative position relationships between each note.

The encoder is composed of *n* identical subcoding layers, and each encoding layer is composed of two sublayers. The first layer is a multiheaded self-attention mechanism composed of a multihead attention mechanism and a scaled dot-product attention unit. The second layer is a fully connected feed-forward network. The outputs of the two sublayers are connected by residuals, and each layer is normalized to avoid gradient disappearance. The output of each sublayer is $LayerNorm(x + Sublayer(x))$, where $Sublayer(x)$ is the function realized by a sublayer. In the self-attention layer of the encoder, all queries, keys, and values come from the output of the previous layer in the encoder, expressed as the three vectors Q, K, and V, respectively. The calculation formula is as follows:

$$Attention(Q, K, V) = softmax\left(\frac{QK^T}{\sqrt{d_k}}\right)V \tag{3}$$

where d_k is the dimension of the input vector. There are two common attention functions: one of them is additive attention, and the other is dot-product attention. The attention mechanism used here is dot-product attention. This attention mechanism is not only faster than additive attention but also saves more space. The weight of *Attention* is calculated according to Q and K, and d_k needs to be scaled; otherwise, when the value of dot-multiplication is too large, the gradient that is calculated by the function *softmax* will be very small, and this is not conducive to backpropagation.

Multihead attention is composed of h self-attention. The calculation formula is shown in Equation (4):

$$MultiHead(Q, K, V) = Concat(head_1, \ldots, head_h)W^O \quad (4)$$

where

$$head_i = Attention\left(QW_i^Q, KW_i^K, VW_i^V\right) \quad (5)$$

Because self-attention only learns from one perspective, it may be biased. Therefore, h different weight combinations are designed. Before calculating *Attention*, Q, K, and V are linearly transformed with the above weight combination, respectively. The weight of h angles of *Attention* is spliced, and linear transformation is performed with a new weight matrix to get the final output. Multihead attention splits the Q, K, V vector of the note unit into h word vectors with d_{model}/h dimensionality for self-attention calculation; splices the operation results; merges and adjusts them with the full connection layer; and then outputs the result. The decoder structure is roughly the same as that of the encoder and is composed of N subdecoders. The difference is that the Q and K vectors in the multihead attention input of the decoder come from the encoder. To obtain the relationship between units in the note sequence from data learning and training, the dot product of Q and key K, formed by self-attention in the encoding process, becomes the weight of V in the understanding code process. Compared with the encoder layer, a masked multihead attention unit is added to each subdecoder layer, because in generating note sequences, prediction of the next note unit needs to be performed after the prediction of the current note unit. Otherwise, the model is equivalent to directly seeing the question's answer before learning, so the training is meaningless.

A frame diagram of the whole model is shown in Figure 2.

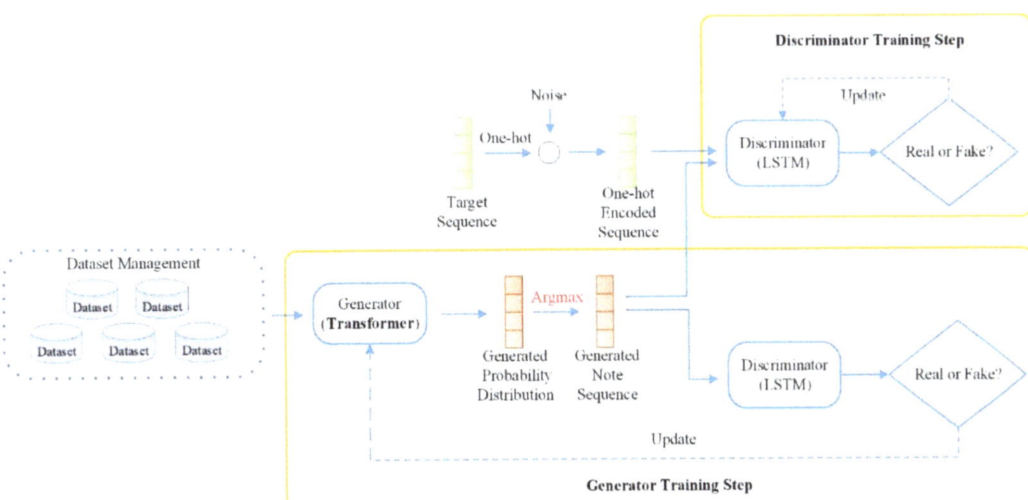

Figure 2. The structure of the music generation system.

This system performs alternating confrontation training on generator G and discriminator D in one cycle. In each cycle, the discriminator is trained first. The discrimination model comprises bidirectional long short-term memory (BLSTM), with two layers of 256 units each. The input of the discrimination model is a dataset composed of the real target sequence in the current batch and the sequence generated by the generation model. The gradient descent method is used for training when solving for the gradients of discriminator parameters.

After sequence acquisition, two continuous fixed-length note sequences can be obtained, *input_Seq* and *tar_Seq*. Next, "[BEG]" is added at the beginning of each note sequence, and "[END]" is added at the end of the note sequence; then, the label encoder encodes its ID. There is a trainable embedded layer in the input layer of the model. The mapping relationship between the input and the embedded unit is learned online during the training process.

The converted note vector is multiplied by the position code and input into the transformer encoder composed of an eight-layer subencoder. At this time, the output of the generated model is the probability distribution of dimension $[BatchSize,\ SeqLength,\ NumNotes]$, that is, the value in the last dimension of the tensor is the probability of the occurrence of the note corresponding to the digital subscript. With the updated warm-up learning rate, the Adam optimizer is used to train the generator. The calculation formula of this process is as follows:

$$LearningRate = d_{model}^{-0.5} * min\left(StepNum^{-0.5},\ StepNum * WarmupSteps^{-1.5}\right) \tag{6}$$

where $LearningRate$ is the probability distribution of notes, d_{model} is the dimension of the model input vector, and *StepNum* is the number of steps in the current workout. The reason for using the warm-up learning rate strategy is that the compensation mechanism needs to be combined at an early stage of system training. A fast update rate is used so that the model learns the parameter characteristics of notes quickly. In the middle and late stages of training, the learning rate is slowly reduced so that the model can better learn the detailed characteristics of the note distribution. Combined with GANs, the prediction sequence of notes is input to the discriminator to determine whether it is sampled from the dataset or generated. Equation (7) is the optimization function of GANs:

$$\underset{G\ \ D}{minmax}V(G,D) = \underset{G\ \ D}{minmax}\mathbb{E}_{x\sim p_{data}}[\log D(x)] + \mathbb{E}_{z\sim p_z}[\log(1-D(G(z)))] \tag{7}$$

where $x \sim p_{data}$ represents the input subject to the real note distribution, and $z \sim p_z$ represents the analog distribution. G is the generator, and D is the discriminator. As is shown in Equation (7), by term $\mathbb{E}_{x\sim p_{data}}[\log D(x)]$ and $\mathbb{E}_{z\sim p_z}[\log(1-D(G(z)))]$, the discriminator is expected to maximize the probability of sampled sequences being true and minimize the probability of generated sequences being false. Apparently, the output of the discriminator is the probability of being true music. Consequently, the only target label for $D(x)$ is 1, while for $D(G(z))$, it is 0. Therefore, Equation (7) can be seen as a variant of cross-entropy. When the generator G is fixed, the partial derivative of the objective function $V(G,D)$ yields formula (8) for discriminator D:

$$D^*(x) = \frac{p_g(x)}{p_g(x) + p_{data}(x)} \tag{8}$$

Substituting the optimal discriminator in Formula (7) into Formula (8) causes the optimization goal to optimize the Jenson–Shannon divergence (JSD) of $p_g(x)$ and $p_{data}(x)$. When $p_g(x) = p_{data}(x)$, they reach Nash equilibrium. Currently, the discrimination probability of discriminator D for actual samples or generated samples is 50%.

This process has two problems: gradient non-differentiability and mode collapse. The reason why the gradient is not differentiable is that GANs need to input the generated note sequence into the discriminator to determine authenticity, but the output of the generated model is the probability distribution with dimension $[BatchSize,\ SeqLength,\ NumNotes]$. The operation of argmax needs to be applied to each one-dimensional probability distribution to obtain the index of the note with the maximum probability, such that the generated sequence and real target have compatible and comparable shapes. However, as the function argmax returns the indices of the maximum values along an axis, the operation of argmax is nondifferentiable, which leads to a failure to solve the gradient between the judgment result of the discriminator and the trainable variables of the generator during the training step of the generator.

To make the process derivable, our solution to this problem is as follows: Firstly, the real sampling input sequence is transformed from an ID sequence into the sparse vector representation of the one-hot encoding, making the dimension of the real data and the generated probability distribution unified. Secondly, the output structure of the transformer is also optimized, and the inverse temperature parameter τ is introduced. The output optimization formula of the layer softmax is Equation (9):

$$y_i = \frac{\exp(y_i/\tau)}{\sum_{j=1}^{k} \exp(y_j/\tau)} \tag{9}$$

where y_i is the normalized output probability of the ith note, τ is the inverse temperature parameter, K is a global scalar, and $\sum_{j=1}^{k} \exp(y_j/\tau)$ is the normalizing term. As $\tau \to 0$, the probability distribution after Equation (9) approaches the one-hot encoding vector. As $\tau \to +\infty$, the output becomes a uniform probability. When τ is a finite positive value, the sample produced by Equation (9) is smooth and differentiable by the generator. To conclude, the relationship between probability distribution and extraction of the maximum value is expected to be learned by the inverse temperature parameter. During training, τ is set to a large value, which slowly decreases almost to zero.

The standard one-hot embedding represents the real sequence, and the generator's output is the probability distribution of the predicted label. There are huge differences in the expression form between the two, which the discriminator in GANs captures. Therefore, the one-hot embedding of the real sequence needs to be optimized, and noise added, as shown in Equation (10):

$$y_i = softmax\left(\frac{onehot(y_i) + g_i}{\lambda}\right) \tag{10}$$

where y_i is the sequence of the real notes, g_i is the random noise in the section $(-\varepsilon, \varepsilon)$, and λ is a constant less than 1, which is used to amplify the result of noise. The purpose of adding λ is to make the vector after softmax conversion closer to the form of one-hot encoding.

Another problem is mode collapse. The method used to evaluate the generation effect of the generator is to calculate the accuracy between the generated result and the actual result, but the system will mistake the one-hot encoding feature of the real sequence as one of the real features, resulting in the low accuracy of the generated model, and finally leading to mode collapse.

A root mean square error (RMSE) is added between the predicted and real sequence to accelerate the convergence and avoid mode collapse. When the difference between the predicted label and the real label exceeds a reasonable value, RMSE can correct the learning direction of the gradient.

Assuming the real target sequence sample $\{x_1, \ldots x_K\}$ and the real input sample $\{z_1, \ldots z_K\}$, the calculation formula of the loss function is as follows:

$$L = \frac{1}{K} \sum_{n=1}^{K} [logD(x_i) + \log(1 - D(G(z_i)))] \tag{11}$$

where G is the generator, D is the discriminator, Z is a real input sample, and n and i are constants.

Therefore, the calculation formula of the loss function L is as shown in Equation (12):

$$L = \alpha \frac{1}{K} \sum_{j=1}^{K} \| \hat{y}_j - G(z_j) \|^2 + \exp\left(\frac{1}{K} \sum_{m=1}^{K} \log(1 - D(G(z_i)))\right) \quad (12)$$

where α $(0 < \alpha < 1)$ is the preset weight coefficient, $\hat{y}_j$ represents the real sequence label of the j-th sentence, and the former is the RMSE between the predicted sequence and the real sequence.

The calculation steps of the algorithm are as follows (Algorithm 1):

Algorithm 1: The Algorithm of Transformer and GANs

Input: real input sequence $\{z_1, \ldots z_n\}$, the target output sequence $\{x_1, \ldots x_n\}$
Output: predictive output sequence $\{y_1, \ldots y_n\}$
Random initialization generator and discriminator's parameters
For i in k steps to train do
    For batch to iterate do
    Discriminator training
    • Input the real input sequence $\{z_1, \ldots z_n\}$ in G, generated $\{y_1, \ldots y_n\}$
    • Input the false sequence $\{y_1, \ldots y_n\}$ and the target output sequence $\{x_1, \ldots x_n\}$ in D
    • Used the descent gradient method to update the parameters, minimizing the loss
    Generator training
    • Input the real input sequence $\{z_1, \ldots z_n\}$ and the target output sequence $\{x_1, \ldots x_n\}$ in generator $\{y_1, \ldots y_n\}$
    • Input the generated sequence in D, judge it as real or false
    • Use the gradient descent method to update the parameters for G, minimize the combo loss
    End for
End for

4. Experimental Summary

The GiantMIDI-Piano dataset, published by Jin et al. in 2020, was used to develop the proposed method [31]. Table 1 compares several major MIDI format music datasets. The GiantMIDI-Piano dataset is dramatically improved in quantity and richness compared to the others. More than 10,000 piano pieces with a total time of more than 1200 h can be played by algorithms. It is the most extensive classical piano dataset in the world.

Table 1. Piano dataset comparison.

Dataset	Composers	Pieces	Hours	Types
Piano-midi.de	26	571	36.7	Seq.
Classical archives	133	856	46.3	Seq.
Kunstderfuge	598	-	-	Seq.
MAESTRO	62	529	84.3	Perf.
MAPS	-	270	18.6	Perf.
GiantMIDI-Piano	2786	10,854	1237	Live

In Equation (12), the former is the root mean square error, and the latter is the cross-entropy loss. The loss of the cross-entropy term results in exponential amplification, making the model converge faster during the gradient of the training process. In addition, α avoids the process of convergence of authenticity learning instability because of an excessive RMSE. The value of α is selected by the training accuracy after five epochs, and the results are shown in Table 2. Note that since α is a hyperparameter used during the training of the generator, the accuracy being compared here is the accuracy of the generator.

Table 2. Influence of different values of α on the accuracy of the model.

α	Accuracy after Five Epochs	α	Accuracy after Five Epochs
0.1	0.21	0.6	0.18
0.2	0.22	0.7	0.13
0.3	0.24	0.8	0.15
0.4	0.23	0.9	0.13
0.5	0.18	1.0	/

According to Table 2, the accuracy rate changes when different values are substituted. When α is equal to 0.3, the accuracy rate is the largest, so the value of α is 0.3 (the result rounded up to 1 decimal place).

The proposed music generation model based on transformer and GANs has two loss optimization functions, corresponding to the optimization update of discriminator D and generator G in the generation countermeasure network. For discriminator D, the output is only 0 or 1, and the accuracy rate is the ratio of the predicted number of correct tags to the number of all tags. At each time step of prediction, the generator solves a multi-classification problem with the label dimension of vocab_size. For the prediction of each unit, its output is the probability distribution of the unit label.

Figure 3a is the process of discriminator loss rate. The loss of the discriminator decreases rapidly at the beginning of training until it is finally stable. Figure 3b is the change process of discriminator accuracy. The accuracy of the discriminator also increases rapidly to about 50% after the beginning of training, reaching the optimum state. Figure 3c is the change process of the correctness of the verification set. The accuracy rate of the verification set rises during the training process and finally reaches about 90%.

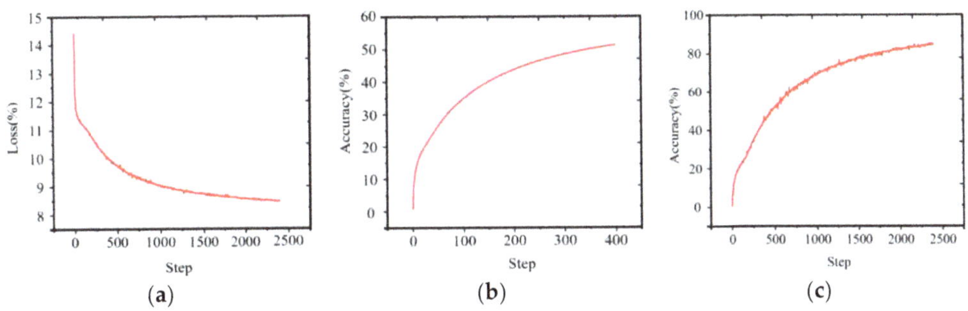

Figure 3. The result for the training process.

Figure 4a shows the input and output attention before optimization, while Figure 4b shows the input and output attention after optimization. Comparing Figure 4a,b, it can be seen that the image in Figure 4b is more complex than Figure 4a. At the same time, according to the visual experience, the color of Figure 4b is also darker than Figure 4a. Here, the image is used to reflect the relationship between the notes in generated melody and input samples. The color is darker, so the relationship between the input notes (or chords) is stronger.

In addition, the multihead attention mechanism in the transformer recognizes the relationship between input and output units. Figure 4a shows the input and output attention without using the optimization loss function, and Figure 4b shows the output attention after training with the optimization loss function.

The GAN model without the optimization function does not capture the relationship between input and output, but the optimized model learns the coupling relationship between input and output units far better. The final musical notation is shown in Figure 5.

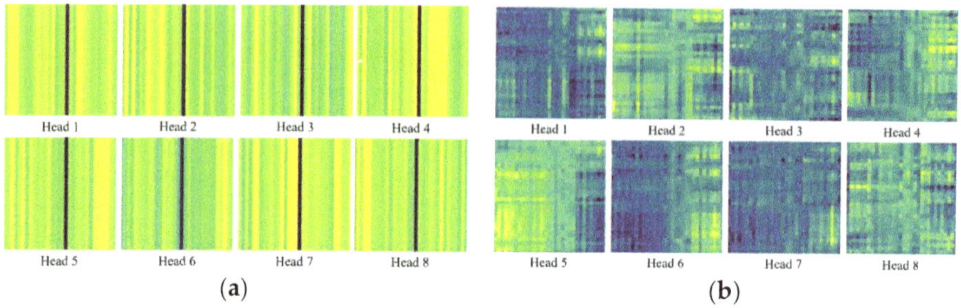

Figure 4. The input and output attention before optimization.

Figure 5. The results generated by the system based on transformer and GANs.

To make up for the defect of root mean square error in music evaluation, according to Euler's music evaluation elements, several commonly used elements in music evaluation, as shown in Table 3, were selected to quantify the advantages and disadvantages of the output music samples.

Table 3. The form for music evaluation.

Parameter	Output Mode	Explain	Weigh t
Absolute interval gradient	$w_1 * \frac{\sum_{i=0}^{n} s_g^i}{n-1}$	s_g^i: Indicates the interval difference gradient score of the second note	0.11
Extreme note number	$w_2 * min\left(4 * \frac{n_{neibor}}{n},\ 1\right)$	n_{neibor}: Indicates the average number of extreme notes in the sequence	0.29
Dissonance	$w_3 * max\left(0,\ (-4) * \frac{n_{leap}}{n-1} + 1\right)$	n_{leap}: Indicates the count value of a continuous jump	0.17
Chord monophonic ratio	$w_4 * \begin{cases} e^{10r_{chord}}, r_{chord} < 0.1 \\ 1, 0.1 \le r_{chord} < 0.5 \\ \ln(-r_{chord} + 1.5) + 1,\ r_{chord} \ge 0.5 \end{cases}$	r_{chord}: Indicates the ratio of the number of chords	0.29
Note diversity	$w_5 * \left(-7r_{div}^2 + 7.5r_{div} - 0.75\right)$	r_{div}: Indicates the ratio of the number of different notes to the length of the sequence	0.14

Note: n: sequence length. w_i: output weight of each item. The weight for each parameter is just an example; the value can be adjusted appropriately according to preference.

The calculation formula of the absolute interval gradient is shown in Equation (13):

$$s_g = \begin{cases} 1, \ x \leq 6 \\ \lg(-x+16), \ 6 < x \leq 15 \\ 0, \ x > 15 \end{cases} \tag{13}$$

where x represents the interval difference between notes.

The calculation formula for the number of extreme notes is shown in Equation (14):

$$n_{neibor} = \frac{2 * n_{min} * n_{max}}{n_{min} + n_{max}} \tag{14}$$

where n_{min} indicates the number of notes in the bass in the presence of extreme note differences; n_{max} indicates the number of notes in the range of treble; and n_{neibor} indicates the average number of extreme notes.

The calculation formula for dissonance is shown in Equation (15):

$$n_{leap} = n_{leap} + 1, \ if \ abs(p_i - p_{i-1}) \& \ abs(p_{i+1} - p_i) \tag{15}$$

where n_{leap} indicates the number of dissonances, and p_i is the tone of the i note.

The calculation formula for chord single tone ratio is shown in Equation (16):

$$r_{chord} = \frac{n_{chord}}{n_{chord} + n_{note}} \tag{16}$$

where n_{chord} indicates the number of chords, and n_{note} is the number of single notes.

The calculation formula for note diversity is shown in Equation (17):

$$r_{div} = \frac{n_{dif}}{n} \tag{17}$$

where n_{dif} indicates the number of non-repeated notes, and n is the total length of the sequence. The final score output is shown in Equation (18):

$$s = \frac{1}{\sum_{i=1}^{5} w_i} \sum_{i=1}^{5} w_i s_i \tag{18}$$

where w_i is the output's weight for each item, s_i is the output's value for each item, and S is the final score.

The same sample dataset (GiantMIDI-Piano) [31] was input into several systems, as shown in Tables 4 and 5, to compare output results (full score is 100).

Table 4. Comparison of model results.

Model	Original Sample (Average)	Output Melody (Average)	Training Accuracy
Long short-term memory (LSTM)	83.30	46.62	\
Bidirectional long short-term memory (BiLSTM)	85.75	55.83	\
Transformer and GANs (original loss)	92.60	74.51	0.034
Transformer and GANs (optimized loss)		92.33	0.562

Finally, we selected 30 volunteers with musical backgrounds from the Shanghai Conservatory of Music and 30 volunteers from the College of Electronics and Information Engineering at Tongji University for the test. Suppose that the volunteers from the Shanghai Conservatory of Music are professional and those from Tongji University are nonprofessional in the field of music. Based on Table 3, the total score is 100 points, and the scoring results are shown in Table 5. The results in Table 4 are calculated according to the elements involved in Table 3, and the calculation process refers to Equations (13)–(18). Table 5 is the result of the volunteers' manual evaluation according to the elements involved in Table 3. The original samples for Tables 4 and 5 all used melodies included in the same sample dataset (GiantMIDI-Piano).

Table 5. Result of user study.

Volunteer	Model	Original Sample (Average)	Output Melody (Average)
professional	Long short-term memory (LSTM)	94.07	42.32
	Bidirectional long short-term memory (BiLSTM)		51.44
	Transformer and GANs (original loss)		65.71
	Transformer and GANs (optimized loss)		82.58
nonprofessional	Long short-term memory (LSTM)	92.82	63.65
	Bidirectional long short-term memory (BiLSTM)		68.11
	Transformer and GANs (original loss)		75.01
	Transformer and GANs (optimized loss)		88.33

Combining Tables 4 and 5, it can be seen that, for both data evaluation based on Euler's music evaluation elements or manual evaluation, the optimized transformer and GANs model has the highest scores. At the same time, compared with other models, the optimized transformer and GANs model also has the best accuracy for the notes.

5. Conclusions

Taking aim at the challenge of music generation, this study overcame the obstacles facing the sequence generation model and GANs, proposed a music generation model based on transformer and GANs, and optimized the structure of GANs. Through experimentation, it was found that the chord processing reported in this study is not ideal, as reflected in the high proportion of chords and excessive discordant notes. A character dictionary processed by this method was constructed according to the notes and chords in real music. To reduce the size of the dictionary and improve the prediction effect, word frequency was used as the basis for filtering, and some notes and chords with low frequency, extreme notes, and complex chords were shielded. Still, the number of chord labels is much larger than the number of individual notes. Although the number of single notes in the training data is much larger than the occurrence frequency of chords, the transformer is not sensitive to the frequency in the calculation process of the model, regardless of the relative probability of single notes and chords. The dataset selected in this study is based on piano compositions that contain many different chords and notes. Using only major and minor chords is not enough to express the music in this dataset. Therefore, preprocessing of music datasets and use of algorithms to effectively summarize the rules for the occurrence of notes should be the primary goals of music creation in the future.

Author Contributions: Conceptualization, J.M. and L.W.; methodology, Z.L.; software, D.L.; validation, M.Z., J.M. and L.W.; formal analysis, Z.L. and J.M.; investigation, D.L. and M.Z.; resources, J.M. and D.L.; data curation, J.M.; writing—original draft preparation, J.M. and L.W.; writing—review and editing, D.L., Y.H. and M.Z.; visualization, Y.H.; supervision, Y.H.; project administration, J.M. and L.W.; funding acquisition, Y.H. All authors have read and agreed to the published version of the manuscript.

Funding: This work was funded by Science and Technology Winter Olympic Project (Grant number 2018YFF0300505) and Joint Fund of Zhejiang Provincial Natural Science Foundation (Grant number LHY20F030001).

Institutional Review Board Statement: The study did not require ethical approval.

Informed Consent Statement: Not applicable.

Data Availability Statement: Not applicable.

Conflicts of Interest: The authors declare no conflict of interest.

References

1. Olson, H.F.; Belar, H. Electronic music synthesizer. *J. Acoust. Soc. Am.* **1955**, *27*, 595–612. [CrossRef]
2. Steedman, M.J. A generative grammar for jazz chord sequences. *Music. Percept.* **1984**, *2*, 52–77. [CrossRef]
3. Ebcioğlu, K. An expert system for harmonizing four-part chorales. *Comput. Music. J.* **1988**, *12*, 43–51. [CrossRef]
4. Boulanger-Lewandowski, N.; Bengio, Y.; Vincent, P. Modeling temporal dependencies in high-dimensional sequences: Application to polyphonic music generation and transcription. *arXiv* **2012**, arXiv:1206.6392.
5. Gao, Z.; Chen, M.Z.; Zhang, D. Special Issue on "Advances in condition monitoring, optimization and control for complex industrial processes". *Processes* **2021**, *9*, 664. [CrossRef]
6. O'Hanlon, K.; Sandler, M.B. Fifthnet: Structured compact neural networks for automatic chord recognition. *IEEE/ACM Trans. Audio Speech Lang. Process.* **2021**, *29*, 2671–2682. [CrossRef]
7. Zou, F.; Schwarz, S.; Nossek, J.A. Cellular neural network design using a learning algorithm. In Proceedings of the IEEE International Workshop on Cellular Neural Networks and Their Applications, Budapest, Hungary, 16–19 December 1990; pp. 73–81.
8. Kingma, D.P.; Welling, M. Auto-encoding variational bayes. *arXiv* **2013**, arXiv:1312.6114.
9. Chorowski, J.; Weiss, R.J.; Bengio, S.; Van Den Oord, A. Unsupervised speech representation learning using wavenet autoencoders. *IEEE/ACM Trans. Audio Speech Lang. Process.* **2019**, *27*, 2041–2053. [CrossRef]
10. Hinton, G.E.; Salakhutdinov, R.R. Reducing the dimensionality of data with neural networks. *Science* **2006**, *313*, 504–507. [CrossRef] [PubMed]
11. Johnson, D. Composing Music with Recurrent Neural Networks. August 2015. Available online: http://www.hexahedria.com/2015/08/03/composing-musicwith-recurrent-neural-networks/ (accessed on 26 October 2015).
12. Gao, Z.; Liu, X. An overview on fault diagnosis, prognosis and resilient control for wind turbine systems. *Processes* **2021**, *9*, 300. [CrossRef]
13. Choi, K.; Fazekas, G.; Cho, K.; Sandler, M. The effects of noisy labels on deep convolutional neural networks for music tagging. *IEEE Trans. Emerg. Top. Comput. Intell.* **2018**, *2*, 139–149. [CrossRef]
14. Pelchat, N.; Gelowitz, C.M. Neural network music genre classification. *Can. J. Electr. Comput. Eng.* **2020**, *43*, 170–173. [CrossRef]
15. Lu, L.; Xu, L.; Xu, B.; Li, G.; Cai, H. Fog computing approach for music cognition system based on machine learning algorithm. *IEEE Trans. Comput. Soc. Syst.* **2018**, *5*, 1142–1151. [CrossRef]
16. Liu, C.H.; Ting, C.K. Computational intelligence in music composition: A survey. *IEEE Trans. Emerg. Top. Comput. Intell.* **2016**, *1*, 2–15. [CrossRef]
17. Sigtia, S.; Benetos, E.; Dixon, S. An end-to-end neural network for polyphonic piano music transcription. *IEEE/ACM Trans. Audio Speech Lang. Process.* **2016**, *24*, 927–939. [CrossRef]
18. Thalmann, F.; Wiggins, G.A.; Ycart, A.; Benetos, E. Sandler M. Representing Modifiable and Reusable Musical Content on the Web with Constrained Multi-Hierarchical Structures. *IEEE Trans. Multimed.* **2020**, *22*, 2645–2658. [CrossRef]
19. Ycart, A.; Benetos, E. Learning and Evaluation Methodologies for Polyphonic Music Sequence Prediction with LSTMs. *IEEE/ACM Trans. Audio Speech Lang. Process.* **2020**, *28*, 1328–1341. [CrossRef]
20. Sheykhivand, S.; Mousavi, Z.; Rezaii, T.Y.; Farzamnia, A. Recognizing emotions evoked by music using CNN-LSTM networks on EEG signals. *IEEE Access* **2020**, *8*, 139332–139345. [CrossRef]
21. Borodin, A.; Rabani, Y.; Schieber, B. Deterministic many-tomany hot potato routing. *IEEE Trans. Parallel Distrib. Syst.* **1997**, *8*, 587 596. [CrossRef]
22. Chen, J.; Pan, F.; Zhong, P.; He, T.; Qi, L.; Lu, J.; He, P.; Zheng, Y. An automatic method to develop music with music segment and long short term memory for tinnitus music therapy. *IEEE Access* **2020**, *8*, 141860–141871. [CrossRef]
23. Lehner, B.; Schlüter, J.; Widmer, G. Online, loudness-invariant vocal detection in mixed music signals. *IEEE/ACM Trans. Audio Speech Lang. Process.* **2018**, *26*, 1369–1380. [CrossRef]
24. Wang, C.; Xu, C.; Yao, X.; Tao, D. Evolutionary generative adversarial networks. *IEEE Trans. Evol. Comput.* **2019**, *23*, 921–934. [CrossRef]
25. Liang, Z.; Zhang, S. Generating and Measuring Similar Sentences Using Long Short-Term Memory and Generative Adversarial Networks. *IEEE Access* **2021**, *9*, 112637–112654. [CrossRef]
26. Arora, C.; Sabetzadeh, M.; Briand, L.; Zimmer, F. Automated checking of conformance to requirements templates using natural language processing. *IEEE Trans. Softw. Eng.* **2015**, *41*, 944–968. [CrossRef]
27. Vaswani, A.; Shazeer, N.; Parmar, N.; Uszkoreit, J.; Jones, L.; Gomez, A.N.; Kaiser, Ł.; Polosukhin, I. Attention is all you need. In Proceedings of the Advances in Neural Information Processing Systems, Long Beach, CA, USA, 4–9 December 2017; Volume 30.
28. Sutskever, I.; Vinyals, O.; Le, Q.V. Sequence to sequence learning with neural networks. In Proceedings of the Advances in Neural Information Processing Systems, Montreal, QC, Canada, 8–13 December 2014; Volume 27.
29. Huang, C.Z.A.; Vaswani, A.; Uszkoreit, J.; Shazeer, N.; Simon, I.; Hawthorne, C.; Dai, A.M.; Hoffman, M.D.; Dinculescu, M.; Eck, D. Music transformer. *arXiv* **2018**, arXiv:1809.04281.
30. Radford, A. Language Models are Unsupervised Multitask Learners. *OpenAI Blog* **2019**, *1*, 9–33.
31. Payne, C. MuseNet. *OpenAI Blog* **2019**, *3*.

32. Jin, C.; Wang, T.; Liu, S.; Tie, Y.; Li, J.; Li, X.; Lui, S. A transformer-based model for multi-track music generation. *Int. J. Multimed. Data Eng. Manag. (IJMDEM)* **2020**, *11*, 36–54. [CrossRef]
33. Bahdanau, D.; Cho, K.; Bengio, Y. Neural machine translation by jointly learning to align and translate. *arXiv* **2014**, arXiv:1409.0473.

MDPI
St. Alban-Anlage 66
4052 Basel
Switzerland
Tel. +41 61 683 77 34
Fax +41 61 302 89 18
www.mdpi.com

Processes Editorial Office
E-mail: processes@mdpi.com
www.mdpi.com/journal/processes

www.ingramcontent.com/pod-product-compliance
Lightning Source LLC
Chambersburg PA
CBHW061510220926
51850CB00004B/224
9783036563954